Digital Communications and Spread Spectrum Systems

RODGER E. ZIEMER

Department of Electrical Engineering
University of Colorado at Colorado Springs

ROGER L. PETERSON

Motorola, Inc.
Government Electronics Group
Scottsdale, Arizona

DIGITAL COMMUNICATIONS AND SPREAD SPECTRUM SYSTEMS

Macmillan Publishing Company

New York

Collier Macmillan Publishers

London

Macmillan Publishing Company
866 Third Avenue, New York, New York 10022

Collier Macmillan Canada, Inc.

Library of Congress Cataloging in Publication Data

Ziemer, Rodger E.
  Digital communications and spread spectrum systems.

  Includes bibliographical references and index.
  1. Digital communications. 2. Spread spectrum
communications. I. Peterson, Roger L. II. Title.
TK5103.7.Z54   1985   621.38'0413   84-17141
ISBN 0-02-431670-9

Printing: 1 2 3 4 5 6 7 8   Year: 5 6 7 8 9 0 1 2

ISBN 0-02-431670-9

PREFACE

The goal of this book is to carry the treatment of the implementation of digital communication systems a step beyond that given in most introductory communications systems texts. As such, it is intended for a second course in communications systems at the senior and first year graduate levels, for continuing education (short) courses, or for self study by communications engineers in industry. Previous background assumed on the part of the student is an introductory course in communications systems which has included a coverage of spectral analysis, linear system theory, basic modulation theory, and an introduction to probability and random processes. This material is included in review form in the book for the benefit of those students who have not been recently exposed to it.

A feature of the book is a treatment of digital communications techniques that includes both theoretical development and the consideration of various types of system degradations from ideal performance. These include hardware impairments introduced at the modulator and demodulator as well as channel-induced impairments.

A second feature of the book is the treatment of spread spectrum systems, including the basic spread spectrum concept, codes for spread spectrum systems, initial code synchronization and tracking, error correction coding as applied to spread spectrum systems, and performance characteristics of spread spectrum systems.

The organization of the book is as follows. The first two chapters deal with introductory and review material, the next four with topics of general concern in digital communications, and the last seven with spread spectrum communication system theory and analysis. The structure of the book allows for considerable flexibility in course arrangement, and several possible course outlines are given at the end of this preface. Each chapter includes an ample supply of problems and references for use in further study. A solutions manual is available from the publisher as an aid to the instructor.

Chapter 1 begins with an introduction to the field of digital communications, giving several reasons for the increasing popularity and use of digital communications systems. A general block diagram for a digital communications system is given and the functions of the various blocks are discussed in detail. Also introduced at this point is the concept of error-free capacity of a communication system and the power-bandwidth tradeoff which is of importance in most communication system designs. In connection with the discussion of digital information sources, the concept of a measure of average information, or entropy, is discussed, and the process whereby the theoretically maximum average source-information rate can be approached by employing variable-length coding of the source output for *redundancy reduction* is illustrated. Moving next to the transmitter, the concept of adding redundancy to the digital data stream for error correction is introduced. The use of *modulation* to suitably prepare the data for transmission through the channel is also discussed and several examples of modulation schemes are given. The final operation performed in the transmitter may involve power amplification with the possible introduction of nonlinearities and filtering.

Two approaches or levels are available for characterizing the next major component of a communication system, the channel. These two approaches are the waveform description and the transition probability description. Both are discussed and the features of both are illustrated through examples. At the waveform level of description, the most prevalent form of perturbation on the transmitted signal is thermal noise generated *internally* to the communication system. Accordingly, a brief treatment of the means for characterizing thermal noise is given in Appendix B, including the concepts of noise figure and noise temperature. Also briefly summarized in Chapter 1 are *external perturbations* on the transmitted signal.

The final important communication system component discussed in Chapter 1 is the receiver. Because the remainder of the book is concerned with the details of designing the *demodulation* and *detection* functions of the receiver, this section is relatively brief. The broad coverage of general digital communication system concepts given in Chapter 1 provides the overall perspective of the areas of digital communications systems analysis and design discussed in detail in the remainder of the book.

The next chapter of the book, ''Signals and Systems Overview,'' is intended as a review and to establish notation for the remainder of the book. However, it may also be used as a first-time coverage of the material with some expansions made by the instructor. Features of Chapter 2 are fairly comprehensive treatments of the effects of nonideal filter effects on modulated signals, nonlinear device characterization including mixers, and practical filter characteristics. Appendix A treats probability concepts, random signal characterization, and systems analysis involving random signals to provide a quick review for those requiring it.

Chapter 3 gives a treatment of basic digital data transmission concepts. The infinite bandwidth, additive white Gaussian noise (AWGN) channel is considered first. This leads to the concept of a matched filter receiver, or the alternative implementation as a correlation receiver. Several basic digital modulation methods are introduced as special cases, including biphase-shift keying (BPSK), amplitude-shift keying (ASK), frequency-shift keying (FSK), quadriphase shift keying (QPSK), and minimum-shift keying (MSK). Both the parallel and serial approaches are discussed for the latter. Bandwidths for BPSK, QPSK, and MSK are derived and compared. Following the consideration of the infinite bandwidth case, signal designs and receiver implementations for finite bandwidth channels are analyzed. This makes use of the early work by Nyquist in regard to intersymbol-interference-free transmission and the later invention by Lender of duobinary signaling. Chapter 3 closes with a consideration of several implementation questions such as nonideal filtering, carrier tracking, symbol clock tracking, and the attendant degradation introduced by nonideal realizations of these functions.

Chapter 4, which can be omitted in an introductory course, approaches the signal detection problem using the maximum *a posteriori* (MAP) criterion and vector space representation of signals. This provides a general framework for the consideration of virtually any digital signaling scheme operating in an AWGN environment. Use of the union bound in providing tight upper bounds for the probability of error for *M*-ary digital modulation schemes is introduced. Several special cases are considered including *M*-ary orthogonal signaling, *M*-ary phase-shift keying, and combined amplitude- and phase-shift keying. The subject of introducing coding to achieve efficient transmission of message sequences through AWGN channels at any rate below *channel capacity* is discussed. Contrary to intuitive notions, Shannon's *capacity theorem* shows that it is possible to simultaneously achieve bandwidth and power efficiency. The next section introduces a scheme which can simultaneously achieve good bandwidth and power efficiency; it is referred to as multi-*h* continuous phase modulation. An overview of the Viterbi algorithm as an implementation of the MAP estimator of a Markov sequence is provided in Appendix C. Its many applications include the decoding of convolutional codes and multi-*h* signals. The latter application is covered in Chapter 4.

Another important aspect of digital communication system design is that of generation of coherent references. Chapter 5 provides an overview of this area including basic phase-lock loop theory and frequency synthesizer design.

Any digital communication system requires the synchronization of clocks. These include the carrier oscillators at transmitter and receiver in a coherent communication system, the symbol timing clocks, code timing in systems employing coding, and frame timing in systems where the data is transmitted in blocks or frames. The consideration of synchronization techniques could well occupy an entire book. Accordingly, Chapter 6 can be considered only an introduction to this important area.

The remaining chapters of the book deal with spread spectrum communication systems. Chapter 7 introduces the concept of spread spectrum and the reasons for its use. The most widely used types of spread spectrum modulation are described including *direct sequence* (DS), *frequency hopped* (FH), and hybrid DS/FH spread spectrum.

The generation of pseudo-random digital sequences is important in any spread spectrum system implementation. Chapter 8 provides a comprehensive introduction to the generation of pseudo-noise (PN) sequences by means of linear feedback shift registers and the properties of PN sequences. Other types of sequences such as Gold codes, rapid acquisition codes, and nonlinear codes are considered at the end of the chapter. The latter are particularly important in spread spectrum systems where security is an issue.

An important function in any spread spectrum system is synchronization of the locally generated despreading code with the spreading code generated at the transmitter. This synchronization problem can be divided into two parts, initial synchronization and tracking. The former is the most complex to analyze mathematically. Code tracking is therefore considered in Chapter 9, with acquisition taken up in Chapter 10, even though code acquisition must chronologically precede tracking in the spread spectrum communication process. The two main code tracking methods used are referred to as the *full-time early-late tracking loop* and the *tau-dither early-late tracking loop*. With suitable manipulations and definitions of the signal and noise processes within the loop, both techniques can be reduced to conventional phase-lock loop type implementations. Once this point is reached, the treatment of code tracking loops can make use of standard phase-lock-loop analysis techniques. Also included in Chapter 9 are introductions to frequency hop tracking loops, and the double dither loop.

Initial synchronization of the spreading waveform is perhaps the most difficult spread spectrum problem. Chapter 10 treats this subject comprehensively. Beginning with the simplest technique using a swept serial search, the discussion progresses through a general analysis of stepped serial search, a discussion of multiple-dwell detection techniques, and finally to a detailed analysis of sequential detection techniques. In all cases, the student is presented with analytical techniques which enable calculation of the mean and sometimes the variance of the synchronization time. Chapter 10 finishes with a short discussion of matched filter synchronization techniques.

The analysis of the performance of spread spectrum systems in a jamming environment is the subject of Chapter 11. The chapter begins with a discussion of the system model including barrage noise, partial band noise, pulsed noise, tone, multiple tone, and repeater jamming. Following this, the most commonly used digital modulation techniques, including BPSK/BPSK,* QPSK/BPSK, FH/DPSK, and FH/MFSK, are evaluated in most types of jamming. It is concluded that error correction coding is an essential component of any spread spectrum system to provide adequate protection to jamming. Accordingly, Chapter 12 treats the performance of spread spectrum systems which employ forward error correction. Some important coding schemes, including Reed–Solomon, BCH, and convolutional, are presented. The concepts of channel capacity and computational cut-off rate as applied to spread spectrum systems are introduced to provide performance bounds for coded systems. Chapter 12 provides the reader with computational techniques which may be used to evaluate system error performance.

The discussion of spread spectrum systems is concluded in Chapter 13 with descriptions of some currently operational systems. Examples are given which apply the analytical techniques of Chapters 7–12 to actual systems.

Chapters 1–6 were written by Rodger E. Ziemer; Chapters 7–13 were written by Roger L. Peterson.

Parts of the book have been taught to engineers in industry, and portions have been used in note form as a basis for courses ranging from the senior undergraduate level to graduate level. The success of these courses has resulted from being able to select appropriate chapters from the text in order to tailor the course content to the needs and backgrounds of the students taking the particular course. Examples of chapter selections for several possible courses are given in the following table.

*Spreading Modulation/Data Modulation

Introductory Semester Course on Digital Communications for Undergraduates	Two Twenty Hour Short Courses on Spread Spectrum for Engineers in Industry	Semester Advanced Course on Digital Detection and Spread Spectrum for Graduate Students
	PART 1	
Chapter 1	Appendix A—Review	Chapter 4
Chapter 2—Last Half	Chapter 3	Chapter 6—PLL
Appendix A for Review	Chapter 5	Chapter 7
Chapter 3	Chapter 6—Review PLL	Chapter 8
Chapter 5	Chapter 7	Chapter 9
Chapter 7	Chapter 8	Chapter 10
		Chapter 11
	PART 2	Chapter 12
		Chapter 13
	Chapter 9	
	Chapter 10	
	Chapter 11	
	Chapter 12	
	Chapter 13	

The authors wish to express their thanks to the many people who have contributed to the development of this book. Thanks are due first of all to Carl Ryan, who sowed the seeds for this book while both authors worked for him in 1980–1981, and to students who took classes in which parts of the book were used in note form. These include engineers at Motorola Inc., Government Electronics Group, Scottsdale, Arizona, and students at the University of Missouri–Rolla (UMR) Electrical Engineering Department, Rolla, Missouri; the UMR Graduate Engineering Center, St. Louis, Missouri; and the Electrical Engineering Department at the University of Colorado at Colorado Springs. We also thank our colleagues at both the University of Missouri–Rolla, The University of Colorado at Colorado Springs (UCCS), and Motorola who have provided helpful suggestions. Professors J. B. Anderson, Prakash Narayan, Allan R. Hambley, Leon Couch, David L. Landis, and John N. Daigle reviewed the manuscript. Two persons in particular deserve mentioning: John Liebetreu and Mark Wickert, both of whom suffered through the book in note form at UMR and both of whom checked portions of it when a course was taught at UCCS. All errors which inevitably remain are solely the responsibility of the authors, however. The expert and fast typing of Kathy Collins at UMR is also gratefully acknowledged. Other typists who put considerable effort into various stages of the manuscript are Diane Borque and Lorrie Evans. Alice Astuto of Macmillan spent innumerable hours obtaining the permissions required for this book.

Finally, a sincere word of thanks goes to our wives Sandy and Ann for putting up with a project which to them seemed nebulous and endless at times. Without their encouragement and support, this book could not have been written.

R.E.Z.
R.L.P.

CONTENTS

1   BASIC CONCEPTS OF DIGITAL DATA TRANSMISSION   1

1-1   **Introduction**   1
1-2   **Glossary of Terms**   3
1-3   **Further Consideration of Digital Communication System Design**   5
       1-3.1   General Considerations   5
       1-3.2   Error-Free Capacity of a Communication System   7
       1-3.3   The Source in a Digital Communication System   9
       1-3.4   The Transmitter in a Digital Communication System   15
       1-3.5   The Channel   23
       1-3.6   The Receiver   36
1-4   **Prologue**   37
       References   39
       Problems   39

2   SIGNALS AND SYSTEMS: OVERVIEW   43

2-1   **Review of Signal and Linear System Theory**   43
       2-1.1   Introduction   43
       2-1.2   Classification of Signals   43

2-1.3   Fundamental Properties of Systems   45
2-1.4   Complex Exponentials as Eigenfunctions for a Fixed, Linear System; Transfer Function   47
2-1.5   Orthogonal Function Series   48
2-1.6   Complex Exponential Fourier Series   50
2-1.7   Fourier Transform   53
2-1.8   Signal Spectra   57
2-1.9   Energy Relationships   58
2-1.10 System Analysis   61
2-1.11 Other Applications of the Fourier Transform   64

2-2   Complex Envelope Representation of Signals and Systems   **67**
2-2.1   Narrowband Signals   67
2-2.2   Narrowband Signals and Narrowband Systems   69

2-3   Signal Distortion and Filtering   **72**
2-3.1   Distortionless Transmission and Ideal Filters   73
2-3.2   Group and Phase Delay   73
2-3.3   Nonlinear Systems and Nonlinear Distortion   81

2-4   Practical Filter Types and Characteristics   **86**
References   **100**
Problems   **101**

3   PERFORMANCE CHARACTERIZATION OF DIGITAL DATA TRANSMISSION SYSTEMS   **105**

3-1   Introduction   **105**
3-2   Detection of Binary Signals in White, Gaussian Noise   **106**
3-2.1   Receiver Structure and Analysis   106
3-2.2   The Matched Filter   109
3-2.3   Application of the Matched Filter to Binary Data Detection   112
3-2.4   Correlator Realization of Matched Filter Receivers   115

3-3   Quadrature-Multiplexed Signaling Schemes: QPSK, OQPSK, and MSK   **117**
3-3.1   Quadrature Multiplexing   117
3-3.2   Quadrature and Offset-Quadrature Phase-Shift Keying   118
3-3.3   Minimum-Shift Keying   120
3-3.4   Performance of Digital Quadrature Modulation Systems   120

3-4   Power Spectra for BPSK, QPSK, OQPSK, and MSK   **124**
3-5   Serial Modulation and Detection of MSK   **128**
3-5.1   Serial Approach   129
3-5.2   Terminology and Trellis Diagrams   130

3-6   Signaling Through Bandlimited Channels   **133**
3-6.1   System Model   133
3-6.2   Designing for Zero ISI: Nyquist's Pulse-Shaping Criterion   135
3-6.3   Optimum Transmitting and Receiving Filters   136

3-6.4   Quadrature Bandpass Systems and Multiple Amplitude Systems   141

3-6.5   Shaped Transmitted Signal Spectra   142

3-6.6   Duobinary Signaling   143

3-7   The Use of Eye Diagrams for System Characterization   146

3-8   Equalization in Digital Data Transmission Systems   147

3-8.1   Zero Forcing Equalizers   147

3-8.2   LMS Equalizer Application   151

3-8.3   Adaptive Weight Adjustment   155

3-8.4   Other Equalizer Structures   157

3-9   Degradations due to Realization Imperfections in Digital Modulation Systems   158

3-9.1   Phase and Amplitude Imbalance in BPSK   159

3-9.2   Phase and Amplitude Unbalance in QPSK Modulation   160

3-9.3   Power Loss due to Filtering the Modulated Signal   162

3-9.4   Imperfect Phase Reference at a Coherent Demodulator   162

3-9.5   Degradation due to a Nonideal Detection Filter   166

3-9.6   Degradation due to Predetection Filtering   169

3-9.7   Degradation due to Transmitter, or Channel Filtering; Non-Matched Detector   170

3-9.8   Degradation due to Bit Synchronizer Timing Error   171

3-10   Modulator Structures for QPSK, OQPSK, and MSK   174

3-11   Envelope Functions for BPSK, QPSK, OQPSK, and MSK   177

References   179

Problems   180

4   SIGNAL-SPACE METHODS IN DIGITAL DATA TRANSMISSION   184

4-1   Introduction   184

4-2   Optimum Receiver Principles in Terms of Vector Spaces   186

4-2.1   Maximum a Posteriori Detectors   186

4-2.2   Vector-Space Representation of Signals   188

4-2.3   MAP Detectors in Terms of Signal Spaces   192

4-2.4   Performance Calculations for MAP Receivers   195

4-3   Performance Analysis of Coherent Digital Signaling Schemes   198

4-3.1   Coherent Binary Systems   198

4-3.2   Coherent M-ary Orthogonal Signaling Schemes   200

4-3.3 M-ary Phase-Shift Keying   204

4-3.4   Multi-amplitude/Phase-Shift Keyed Systems   207

4-3.5   Bandwidth Efficiency of M-ary Digital Communication Systems   211

4-4   Signaling Schemes Not Requiring Coherent References at the Receiver   213

4-4.1   NFSK   213

4-4.2   DPSK   215

4-5	**Efficient Signaling for Message Sequences**	**222**
	4-5.1 Summary of Block-Orthogonal and *M*-ary Signaling Performance 222	
	4-5.2 Channel Coding Theorem 224	
4-6	**Multi-*h* Continuous Phase Modulation**	**228**
	4-6.1 Description of the Multi-*h* CPM Signal Format 229	
	4-6.2 Performance Bounds [12] 233	
	4-6.3 Calculation of Power Spectra for Multi-*h* CPM Signals 236	
	4-6.4 Synchronization Considerations for Multi-*h* CPM Signals 243	
	4-6.5 Application of the Viterbi Algorithm to Detection of Multi-*h* CPM Signals 246	
	References	**250**
	Problems	**251**

5 GENERATION OF COHERENT REFERENCES — 254

5-1	**Introduction**	**254**
5-2	**Description of Phase Noise and its Properties**	**255**
	5-2.1 General Considerations 255	
	5-2.2 Phase and Frequency Noise Power Spectra 255	
	5-2.3 Allan Variance 259	
	5-2.4 Effect of Frequency Multipliers and Dividers on Phase-Noise Spectra 260	
5-3	**Phase-Lock Loop Models and Characteristics of Operation**	**261**
	5-3.1 Synchronized Mode: Linear Operation 261	
	5-3.2 Effects of Noise 266	
	5-3.3 Phase-Locked-Loop Tracking of Oscillators with Phase Noise 270	
	5-3.4 Phase Jitter Plus Noise Effects 271	
	5-3.5 Transient Response 272	
	5-3.6 Phase-Locked-Loop Acquisition 275	
	5-3.7 Other Configurations 278	
	5-3.8 Effects of Transport Delay 281	
5-4	**Frequency Synthesis**	**281**
	5-4.1 Digital Synthesizers 281	
	5-4.2 Direct Synthesis 283	
	5-4.3 Phase-Locked Frequency Synthesizers 287	
	References	**289**
	Problems	**290**

6 SYNCHRONIZATION OF DIGITAL COMMUNICATION SYSTEMS — 293

6-1	**The General Problem of Synchronization**	**293**
6-2	**Application of the MAP and ML Principles to Estimation of Signal Parameters**	**296**
	6-2.1 Preliminary Definitions and Relationships 296	
	6-2.2 Expressions for Estimation of Continuous Waveform Parameters 298	

6-2.3 Generalization of the Estimator Equations to Multiple Symbol Intervals and Multiple Parameters 302

6-2.4 Data-Aided Versus Non-Data-Aided Synchronization 309

6-2.5 Joint Estimation of Parameters 309

6-2.6 Open-Loop Versus Closed-Loop Structures 311

6-2.7 Practical Timing Epoch Estimators 312

6-3 **Synchronization Methods Based on Properties of Wide-Sense Cyclostationary Random Processes**   **314**

6-3.1 Carrier Recovery Circuits 315

6-3.2 Delay and Multiply Circuits for Symbol Clock Estimation 319

References   **325**

Problems   **325**

7   INTRODUCTION TO SPREAD SPECTRUM SYSTEMS   327

7-1 **Introduction**   **327**

7-2 **Two Communications Problems**   **328**

7-2.1 Pulse-Noise Jamming 328

7-2.2 Low Probability of Detection 330

7-3 **Direct-Sequence Spread Spectrum**   **332**

7-3.1 BPSK Direct-Sequence Spread Spectrum 332

7-3.2 QPSK Direct-Sequence Spread Spectrum 340

7-3.3 MSK Direct-Sequence Spread Spectrum 344

7-4 **Frequency-Hop Spread Spectrum**   **348**

7-4.1 Coherent Slow-Frequency-Hop Spread Spectrum 348

7-4.2 Noncoherent Slow-Frequency-Hop Spread Spectrum 352

7-4.3 Noncoherent Fast-Frequency-Hop Spread Spectrum 354

7-5 **Hybrid Direct-Sequence/Frequency-Hop Spread Spectrum**   **355**

7-6 **Complex-Envelope Representation of Spread-Spectrum Systems**   **357**

References   **361**

Problems   **361**

8   BINARY SHIFT REGISTER SEQUENCES FOR SPREAD-SPECTRUM SYSTEMS   365

8-1 **Introduction**   **365**

8-2 **Definitions, Mathematical Background, and Sequence Generator Fundamentals**   **366**

8-2.1 Definitions 366

8-2.2 Finite-Field Arithmetic 368

8-2.3 Sequence Generator Fundamentals 375

8-3 **Maximal-Length Sequences**   **385**

8-3.1 Properties of m-Sequences 385

8-3.2 Power Spectrum of m-Sequences 387

8-3.3 Tables of Polynomials Yielding m-Sequences 388

	8-3.4	Partial Autocorrelation Properties of m-Sequences 392	
	8-3.5	Power Spectrum of $c(t)c(t+\epsilon)$ 396	
	8-3.6	Generation of Specific Delays of m-Sequences 396	
8-4	**Gold Codes**		**404**
8-5	**Rapid Acquisition Sequences**		**407**
8-6	**Nonlinear Code Generators**		**411**
	References		**415**
	Problems		**416**

9   CODE TRACKING LOOPS   419

9-1	**Introduction**	**419**
9-2	**Optimum Tracking of Wideband Signals**	**420**
9-3	**Baseband Full-Time Early-Late Tracking Loop**	**423**
9-4	**Full-Time Early-Late Noncoherent Tracking Loop**	**433**
9-5	**Tau-Dither Early-Late Noncoherent Tracking Loop**	**447**
9-6	**Double-Dither Early-Late Noncoherent Tracking Loop**	**456**
9-7	**Full-Time Early-Late Noncoherent Tracking Loop with Arbitrary Data and Spreading Modulation**	**459**
9-8	**Code Tracking Loops for Frequency-Hop Systems**	**467**
9-9	**Summary**	**478**
	References	**480**
	Problems	**480**

10   INITIAL SYNCHRONIZATION OF THE RECEIVER SPREADING CODE   484

10-1	**Introduction**		**484**
10-2	**Problem Definition and the Optimum Synchronizer**		**486**
10-3	**Serial Search Synchronization Techniques**		**488**
	10-3.1	Calculation of the Mean and Variance of the Synchronization Time 488	
	10-3.2	Modified Sweep Strategies 492	
	10-3.3	Continuous Linear Sweep of Uncertainty Region 494	
	10-3.4	Detection of a Signal in Additive White Gaussian Noise (Fixed Integration Time, Multiple Dwell, and Sequential Detectors) 501	
10-4	**Synchronization Using a Matched Filter**		**538**
10-5	**Synchronization by Estimating the Received Spreading Code**		**540**
10-6	**Tracking Loop Pull-In**		**543**
10-7	**Summary**		**547**
	References		**550**
	Problems		**551**

11   PERFORMANCE OF SPREAD-SPECTRUM SYSTEMS IN A JAMMING ENVIRONMENT   555

11-1	**Introduction**	**555**
11-2	**Spread-Spectrum Communication System Model**	**556**

11-3 **Performance of Spread-Spectrum Systems Without Coding** **561**
    11-3.1 Performance in AWGN or Barrage Noise
                Jamming   562
    11-3.2 Performance in Partial Band Jamming   570
    11-3.3 Performance in Pulsed Noise Jamming   582
    11-3.4 Performance in Single-Tone Jamming   586
    11-3.5 Performance in Multiple-Tone Jamming   597
    11-3.6 Conclusions   602
    References   **602**
    Problems   **604**

**12 PERFORMANCE OF SPREAD-SPECTRUM SYSTEMS
WITH FORWARD ERROR CORRECTION**   **606**

12-1 **Introduction**   **606**
12-2 **Elementary Block Coding Concepts**   **607**
    12-2.1 Optimum Decoding Rule   609
    12-2.2 Calculation of Error Probability   612
12-3 **Elementary Convolutional Coding Concepts**   **616**
    12-3.1 Decoding of Convolutional Codes   618
    12-3.2 Error Probability for Convolutional Codes   620
12-4 **Results for Specific Error Correction Codes**   **620**
    12-4.1 BCH Codes   621
    12-4.2 Reed–Solomon Codes   622
    12-4.3 Maximum Free-Distance Convolutional Codes   624
    12-4.4 Repeat Coding for the Hard Decision FH/MFSK
                Channel   624
12-5 **Interleaving**   **630**
12-6 **Random Coding Bounds**   **632**
    References   **633**
    Problems   **634**

13 EXAMPLE SPREAD-SPECTRUM SYSTEMS   **635**

13-1 **Introduction**   **635**
13-2 **Space Shuttle Spectrum Depsreader**   **636**
13-3 **TDRSS User Transponder**   **640**
13-4 **Global Positioning System**   **644**
13-5 **Joint Tactical Information Distribution System**   **647**
    References   **649**

APPENDICES

A PROBABILITY AND RANDOM VARIABLES   **650**

A-1 **Probability Theory**   **650**
A-2 **Random Variables, Probability Density Functions, and
Averages**   **654**
A-3 **Characteristic Function and Probability Generating Functions**   **658**

A-4	Transformations of Random Variables	653
A-5	Central Limit Theorem	667
A-6	Random Processes	668
A-7	Input/Output Relationships for Fixed Linear Systems with Random Inputs; Power Spectral Density	674
A-8	Examples of Random Processes	681
A-9	Narrowband Noise Representation	685
	References	687
	Problems	687

B CHARACTERIZATION OF INTERNALLY GENERATED NOISE 691

C COMMUNICATION LINK PERFORMANCE CALCULATIONS 697

D OVERVIEW OF THE VITERBI ALGORITHM 704

E GAUSSIAN PROBABILITY FUNCTION 713

F POWER SPECTRAL DENSITIES FOR SEQUENCES OF RANDOM BINARY DIGITS AND RANDOM TONES 716

G CALCULATION OF THE POWER SPECTRUM OF THE PRODUCT OF TWO M-SEQUENCES 720

H EVALUATION OF PHASE DISCRIMINATOR OUTPUT AUTOCORRELATION FUNCTIONS AND POWER SPECTRA 728

INDEX 740

CHAPTER 1

Basic Concepts of Digital Data Transmission

1-1

INTRODUCTION

This book is concerned with the transmission of information by electrical means using *digital communication techniques*. Information may be transmitted from one place to another using either digital or analog communication systems. In a digital system, the information is processed so that it can be represented by a sequence of discrete messages. Each message is one of a finite set of messages. For example, the information at the output of a sensor may be a voltage waveform whose amplitude at any given time instant may assume a continuum of values. This waveform may be processed by sampling at appropriately spaced time instants, quantizing these samples, and converting each quantized sample to a binary number (i.e., an analog-to-digital converter). Each sample value is therefore represented by a sequence of ones and zeros, and the communication system associates the message 1 with a transmitted signal $s_1(t)$ and the message 0 with a transmitted signal $s_0(t)$. During each signaling interval either the message 0 or 1 is transmitted with no other possibilities. In practice, the transmitted signals $s_0(t)$ and $s_1(t)$ may be two different phases, say $\pm\pi/2$, or two different amplitudes, say 0 and A, of a sinusoidal carrier. In an analog communication system the sensor output would be used directly to modify some characteristic of the transmitted signal, such as amplitude, phase, or frequency.

Interestingly, digital transmission of information actually preceded that of analog transmission, having been used for signaling for military purposes since antiquity through the use of signal fires, semaphores, and reflected sunlight. The invention of the telegraph, a device for digital data transmission, preceded the invention of the telephone, an analog communications instrument, by almost 40 years.*

Following the invention of the telephone, it appeared that analog transmission would become the dominant form of electrical communications. Indeed, this was true for almost a century until today, when digital transmission is replacing even traditionally analog transmission areas. For example, even though all local telephone communications are analog, it has been predicted that all short-haul and long-distance telephone transmission will be digital by the year 2000. Several reasons are given below for the increasing move toward digital transmission.

1. In the late 1940s it was recognized that *regenerative repeaters* could be used to reconstruct the digital signal *error-free* at appropriately spaced intervals.† That is, the effects of noise and channel-induced distortions in a digital communications link can be almost completely removed, whereas a repeater in an *analog* system (i.e., an amplifier) regenerates the noise and distortion together with the signal.

2. A second advantage of digital representation of information is the flexibility inherent in the handling of digital signals.** That is, a digital signal can be processed independently of whether it represents a discrete data source or a digitized analog source. This means that an essentially unlimited range of signal conditioning and processing options are available to the designer. Depending on the origination and intended destination of the information to be conveyed, these might include *source coding, encryption, pulse shaping* for spectral control, *forward error correction* (FEC) *coding*, special modulation to *spread* the signal spectrum, and *equalization* to compensate for channel distortion. These terms and others will be defined and discussed throughout the book.

3. The third major reason for the increasing popularity of digital data transmission is that it can be used to exploit the cost-effectiveness of digital integrated circuits. Special-purpose digital signal processing functions have been realized as large-scale integrated circuits for several years,†† and the time is not too distant when an entire telephone station will exist on a single chip. The development of the microcomputer and of special-purpose programmable digital signal-processing large-scale integrated (LSI) circuits means that many functions required in digital data transmission systems can now be implemented as *software* or *firmware* (see, e.g., [4]). This is advantageous in that a particular design is not ''frozen'' as hardware but can be altered or replaced with the advent of improved designs or changed requirements.

In the remainder of this chapter, some of the systems aspects of digital communications are discussed. A simplified block diagram of a digital communications system is shown in Figure 1-1. Like any communications system, it consists of a *transmitter*, a *channel* or transmission medium, and a *receiver*. The feature that distinguishes a digital communication system from an analog system is that, in the

---

*The telegraph was invented by Samuel F. B. Morse in the United States and by Sir Charles Wheatstone in Great Britain in 1837, and the first public telegram was sent in 1844. The telephone was invented by Alexander Graham Bell in 1876.

†See [1] in the References at the end of the chapter. For a more recent overview of why the accelerated movement to digital communications, see [2].

**An excellent overview of terminology, ideas, and mathematical descriptions of digital communications is provided in an article by Ristenbatt [3].

††An example is a *continuously variable-slope delta modulator* (e.g., Motorola Integrated Circuit MC3518).

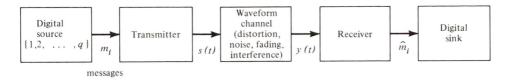

FIGURE 1-1. Simplified block diagram for a digital communication system.

digital system, one of a finite set of messages are communicated during each signaling interval. This implies that the source must emit in time sequence one of a finite number, q, of messages. These, in turn, are used to modify some attribute of a *transmitted signal* which conveys the message through the channel to the receiver. To illustrate, consider the case of *binary* messages, for which $q = 2$. The two possible messages can be represented by the set $\{0, 1\}$. If ones and zeros are emitted from the source each T seconds, a 1 might be represented by a voltage pulse of A volts T seconds in duration and a 0 by a voltage pulse of $-A$ volts T seconds in duration. The transmitted waveform appears as shown in Figure 1-2a. Assume that this waveform is distorted through filtering and noise added in the channel so that it appears at the receiver as shown in Figure 1-2c. The function of the receiver is to process the received signal so that a decision as to whether a 1 or 0 was transmitted can be made with minimum error. Because of the distortion caused by filtering and noise added in the channel, errors may be made in this decision process. Assume that the receiver processes the received signal by filtering to smooth out the noise pulses and that sampling is carried out at the end of each voltage pulse. The reconstructed data stream is shown in Figure 1-2d. The synchronization necessary to achieve proper sampling is an important detail but is not considered at this point. This and other aspects of the data communication problem will be considered further in Section 1-3 after definitions of several terms pertinent to data communications are given in Section 1-2.

1-2

GLOSSARY OF TERMS

Before embarking on a more detailed discussion of digital communication systems, it will be convenient to define several terms [5]. The reader will find it convenient to refer to this glossary as the details of a digital communication system are discussed.

Bandwidth of a signal transmission system: The difference in the frequencies at which the system response is less than that at the frequency of a reference response by a specified ratio (e.g., for the 3-dB bandwidth, this ratio is 0.707). Alternatively, the frequency difference, determined according to some distortion criterion, over which the transmission system passes signal frequencies with a distortion less than some acceptable level.

Baud: A unit of signaling speed equal to the number of discrete conditions or signal events per second.

Bit: (1) A binary digit. (2) A unit of information content equal to the information content of a message the a priori probability of which is $\frac{1}{2}$.

Channel: A single path for transmitting electric signals, usually in distinction from other parallel paths. (May signify a one-way path, providing transmission in

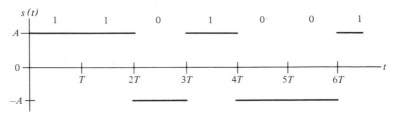

(a)   Undistorted digital signal

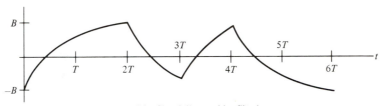

(b)   Signal distorted by filtering

(c) Distorted, noisy signal from channel

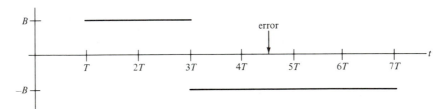

(d) Reconstructed signal at receiver (note the 1 bit delay
due to decisions at the end of each bit)

FIGURE 1-2.   Typical waveforms in a digital communications system.

one direction only, or a two-way path, providing transmission in two directions.)

Channel capacity: The maximum possible information rate through a channel subject to constraints of that channel (*see* Information content).

Code: A plan for representing each of a finite number of values or symbols by a
particular arrangement or sequence of discrete conditions or events. Examples
are: (a) source codes wherein the symbols at the output of a source (e.g.,
letters in the alphabet) may be represented by sequences of ones and zeros
termed *bits*; (b) error detection codes wherein additional bits are added in a
systematic fashion for the purpose of determining whether or not an error has
been made in transmission; (c) forward error correction (FEC) codes wherein

additional bits are added in a systematic fashion for the purpose of error correction.

Demodulation: A process wherein a wave resulting from previous *modulation* is employed to derive a wave having substantially the characteristics of the original modulating wave.

Detection: Determination of the presence of a signal.

Information content (of a message or symbol from a source): The negative of the logarithm of the probability that this particular message or symbol will be emitted from the source. If the base of the logarithm is 2, the unit of information is the *bit*; if 10, the unit of information is the *hartley*.

Message: An arbitrary amount of information whose beginning and end are defined or implied.

Modulation: The process by which some characteristic of a carrier signal is varied in accordance with a modulating wave.

Noise: Unwanted disturbances superposed on a useful signal that tend to obscure its information content.

Spectrum: The distribution of the amplitude (and sometimes phase) of the components of the wave as a function of frequency.

1-3

FURTHER CONSIDERATION OF DIGITAL COMMUNICATION SYSTEM DESIGN

The mechanization and performance considerations for digital communication systems will now be discussed in more detail. The system used throughout the text is illustrated in Figure 1-3. The functions of all the blocks of Figure 1-3 are discussed in detail in this section.

1-3.1 General Considerations

In most communication system designs, a general objective is to use as efficiently as possible the resources of bandwidth and transmitted power. In many applications, one of these resources is more scarce than the other, which results in most channels being classified as either bandwidth limited or power limited. Thus we are interested in both a modulation scheme's *bandwidth efficiency*, defined as the ratio of data rate to signal bandwidth, and its *power efficiency*, characterized by the probability of making a reception error as a function of signal-to-noise ratio. Often, secondary restrictions are imposed in choosing a modulation method, such as the requirement that the transmitted signal have a constant envelope. This constraint is often added if the channel includes nonlinear amplifiers such as a traveling-wave tube amplifier (TWTA).

In spread-spectrum communication system designs, bandwidth efficiency is not of concern. The term *spread spectrum* refers to any modulation scheme that produces a spectrum for the transmitted signal much wider than the bandwidth of the information to be transmitted *independently* of the bandwidth of the information-bearing signal. There are many such schemes for doing this, and they will be discussed in detail beginning in Chapter 7. Why would such a scheme be employed? Among the reasons for doing so are:

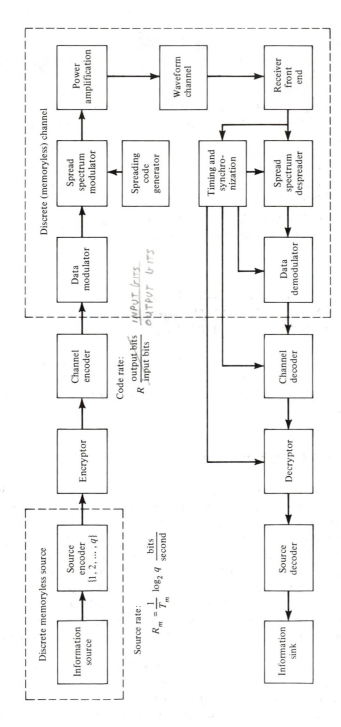

FIGURE 1-3. Block diagram of a typical digital communication system.

1. To provide some degree of resistance to interference and jamming [referred to as *jam resistance* (JR)].
2. To provide a means for masking the transmitted signal in background noise in order to lower the probability of intercept by an adversary [referred to as *low probability of intercept* (LPI)]. It is important to point out that JR and LPI are not achieved simultaneously, for the former implies that one uses the *maximum* transmitted power available, whereas the latter implies that a power level *just sufficient* to carry out the communication is used.
3. To provide resistance to signal interference from multiple transmission paths, commonly referred to as *multipath*.
4. To permit the access of a common communication channel by more than one user, referred to as *multiple access*.
5. To provide a means for measuring range, or distance between two points.

1-3.2   Error-Free Capacity of a Communication System

It is useful at this point to explore briefly the concept of the *capacity* of a digital communications link. Suppose that the communications system designer is asked to design a digital communication link which transmits no more than P watts and such that the majority of the transmitted power is contained in a bandwidth W. Assume that the only effect of the waveform channel is to add thermal noise to the transmitted signal, and that the bandwidth of this noise is very wide relative to the signal bandwidth W. The statistics of this noise are Gaussian* and the channel is called the *additive white Gaussian noise* (AWGN) channel. Given these constraints, there exists a maximum rate at which information can be transmitted over the link with high reliability. This rate is called the *error-free capacity* of a communication system. The pioneering work of Claude Shannon [6] in the late 1940s proves that signaling schemes exist such that error-free transmission can be achieved at any rate lower than capacity. The normalized error-free capacity is given by†

$$\frac{C}{W} = \log_2\left(1 + \frac{P}{N_0 W}\right) = \log_2\left[1 + \frac{E_b}{N_0}\left(\frac{C}{W}\right)\right] \qquad \text{bits} \qquad (1\text{-}1)$$

where C = channel capacity, bits/s
$\quad\; W$ = transmission bandwidth, Hz
$\quad\; P$ = received signal power, W
$\quad\; N_0$ = single-sided noise power spectral density, W/Hz
$\quad\; E_b$ = energy per bit of the received signal

More will be said about these parameters later. For now an intuitive understanding will be sufficient. For example, the capacity, C, is the maximum rate at which *information* can be put through the channel if the source is suitably matched to the channel; its full significance requires a definition of information content of a message and how the source is to be matched to the channel. The rate of information transfer may be conveniently expressed in bits/s, which is the number of binary symbols that must be transmitted per second to represent the data sequence or to represent the analog signal with a given fidelity. These units are discussed further in Section 1-3.3.

---

*See Appendix A for a review of probability and random process theory.
†This relation is developed in Chapter 4. See [6].

If the information rate, R, at the channel input is less than C, Shannon proved that it is theoretically possible through coding to achieve error-free transmission through the channel. This result, sometimes referred to as *Shannon's second theorem*, does not provide a constructive means for finding codes that will achieve error-free transmission, but it does provide a yardstick by which the performance of practical communication schemes may be measured. A graphical presentation of (1-1) is obtained by setting the rate R equal to C and plotting E_b/N_0 as a function of R/W, as shown in Figure 1-4. At points below and to the right of the curve, no amount of coding or complexity will achieve totally reliable transmission. At points above and to the left of the curve, zero error transmission is possible, although perhaps at a very high price in terms of bandwidth, complexity, or transmission delay. Note that data transmission is possible at all points in the plane of Figure 1-4, but some errors are unavoidable at rates above capacity.

This plot can be separated into a *power-limited region*, where $R/W < 1$, and a *bandwidth-limited region*, where $R/W > 1$. That is, *if* the number of bits/s/Hz is greater than unity, one has an efficient scheme in terms of utilizing bandwidth. For power-limited operation, interesting behavior is noted: As $R/W \to 0$ (i.e., infinite bandwidth) the limiting signal-to-noise ratio, E_b/N_0, approaches ln 2 or about -1.6 dB. Any signal-to-noise ratio greater than -1.6 dB results in zero probability of making a transmission error at the expense of infinite transmission bandwidth. Even more important to note, however, is that this is simply one point on the graph; *for any given rate-to-bandwidth ratio, a signal-to-noise ratio exists above which error-free transmission is possible and below which it is not.* Quite

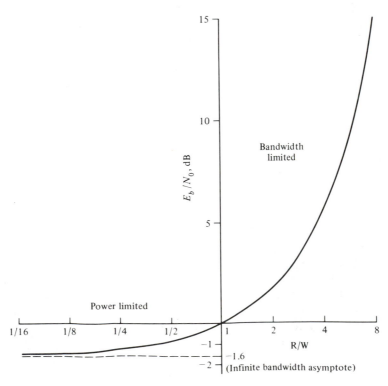

FIGURE 1-4. Power–bandwidth trade-off for error-free transmission through noisy, bandlimited channels.

often, practical communication schemes are compared with this ideal by choosing some suitable probability of error, say 10^{-6}, and finding the signal-to-noise ratio necessary to achieve it. This signal-to-noise ratio is then plotted versus R/W for the system, where W is found according to some suitable definition of bandwidth.*

With this preliminary discussion, a more meaningful consideration of the functions performed by the blocks of Figure 1-3 is possible.

1-3.3 The Source in a Digital Communication System

The source in a digital communication system emits one of a finite set $\{1, 2, \ldots, q\}$ of symbols,† each T_m seconds. If the number of messages in this set is q, they can be represented by $\log_2 q$ binary symbols, giving a source bit rate of

$$R_m = \frac{1}{T_m} \log_2 q \qquad \text{bits/s} \qquad (1\text{-}2)$$

These symbols may arise in one of two ways. First, they may originate in a naturally discrete manner, such as from English text. Second, they may be obtained by the sampling and quantization of an analog source output.

For convenience, it is assumed that successive symbols from the source are independent, or that the source is memoryless.** Although the source output sequence could be represented by a bit stream having a rate R_m, it is usually possible to represent the sequence using a lower-rate bit stream. The minimum possible bit rate is equal to the average information content, which is defined below, in the source symbol stream.

It is assumed that the ith source symbol is emitted with a priori probability $P(x_i)$. The information associated with the ith symbol is

$$I(x_i) = -\log_2 P(x_i) \qquad \text{bits} \qquad (1\text{-}3)$$

and the average information, or *entropy*, associated with the source output is

$$H(X) = -\sum_{i=1}^{q} P(x_i) \log_2 P(x_i) \qquad \text{bits/symbol} \qquad (1\text{-}4)$$

To convert these results to bits/s, all that one needs to do is to multiply by the source rate in symbols/s. That the definition of $I(x_i)$ is reasonable can be seen by considering the transmission of a sequence of symbols from the alphabet $\{1, 2, \ldots, q\}$ with widely differing probabilities of occurrence. Associating the same number of bits with each symbol results in a bit rate equal to R_m. A reduction in the average number of transmitted bits per symbol can be achieved by associating the highly likely symbols with short codewords and the least likely symbols with long codewords. Making the length of each codeword equal to $-\log_2 P(x_i)$ accomplishes this and results in an average codeword length as given by (1-4). Equation

---

*An interesting discussion of bandwidth definitions and their use in describing digital data signals is provided in [7].

†The complications of considering sources with a countably infinite set of symbols is avoided for now.

**Sources with memory are also important but are a complication that will be avoided.

(1-4) represents the minimum possible bits per symbol that can be used to represent the source output sequence.

Useful facts regarding $H(X)$ that can be proved are that:

1. It is a convex function (i.e., upside-down bowl-shaped) of the probabilities, $P(x_i)$.
2. Its maximum occurs when all the $P(x_i)$'s are equal.
3. The value of the maximum is $\log_2 q$ bits/symbol, where q is the number of possible messages.

The entropy of a source can be viewed in either one of two equivalent ways:

1. It is the average amount of information one gains in observing an output symbol from the source.
2. It is the average amount of uncertainty one has about what the source output will be before it is observed.

EXAMPLE 1-1

Consider a source that produces one of three possible symbols, x_1, x_2, or x_3, with respective probabilities $0.7, 0.2$, and 0.1 during successive signaling intervals.

(a) Find the information one gains by being told that x_i was emitted, $i = 1, 2, 3$.

(b) Find the average information, or *entropy*, of the source output.

(c) If 1000 symbols/s are emitted by the source, find the average information rate.

(d) What is the maximum possible information rate?

Solution: It is convenient for calculational purposes to use the natural logarithm, which results in a unit of information referred to as a *nat*. Nats are converted to bits using the relationship

$$\log_2 a = \frac{\ln a}{\ln 2}$$

with $\ln 2 = 0.693$. The computations for $I(x_i)$ and $H(X)$ are given in the following table:

Symbol	Probability	Information (nats)	Information (bits)
x_1	0.7	0.357	0.515
x_2	0.2	1.609	2.322
x_3	0.1	2.303	3.322
		$H(X) = 1.157$ bits/symbol	

The average information rate is therefore 1157 bits/s. The maximum possible information rate is achieved if the source output symbols are equally likely, giving

$$R_{\max} = 1000 \log_2 3 = 1585 \text{ bits/s} \qquad \blacksquare$$

The output of a source is said to possess *redundancy* if its output symbols are not equally likely or if they are not statistically independent (i.e., the source output possesses memory). Usually, the raw source output is *encoded*, which is the second block shown in Figure 1-3. This may be done for two reasons. First, it may not be convenient or possible to send the raw source output through the channel. Second, it may be desirable to remove some of the redundancy in the raw source output, thereby improving the efficiency of transmission.

An example of a code that achieves the first objective is provided in Table 1-1, which gives the *American Standard Code for Information Interchange* (ASCII code). It is the standard code used for the digital communication of individual alphabet symbols, and is used for very short range communications such as from the keyboard of a computer terminal to its processor as well as long-range computer communications and telephone modem interconnects. Note that its codewords are of a fixed length of 7 bits, thus providing $2^7 = 128$ words. This provides a word dictionary large enough to include upper- and lowercase alphabet letters, numerals, and special characters, as well as several control symbols. The systematic arrangement of the code allows one bit to be dropped if certain subsets of the word dictionary are desired. For example, setting the most significant bit (MSB), b_7, to 1 would give a dictionary consisting of upper- and lowercase letters as well as a few other symbols.

Although the ASCII code is interesting and important, its words are of fixed length and therefore do not provide a means for redundancy reduction. An example of a code that does provide redundancy reduction is the *Morse code*, in which short codewords are assigned to the most frequently occurring English letters. For example, an E is assigned the codeword DIT, while some of the more infrequently

TABLE 1-1. American Standard Code for Information Interchange

b_1	b_2	b_3	b_4	b_7=0 b_6=0 b_5=0	0 0 1	0 1 0	0 1 1	1 0 0	1 0 1	1 1 0	1 1 1
0	0	0	0	NUL	DLE	SP	0	@	P	'	p
0	0	0	1	BS	CAN	(	8	H	X	h	x
0	0	1	0	EOT	DC4	$	4	D	T	d	t
0	0	1	1	FF	FS	,	<	L	/	l	∫
0	1	0	0	STX	DC2	"	2	B	R	b	r
0	1	0	1	LF	SUB	*	:	J	Z	j	z
0	1	1	0	ACK	SYN	&	6	F	V	f	v
0	1	1	1	SO	RS	.	>	N	∧	n	~
1	0	0	0	SOH	DC1	!	1	A	Q	a	q
1	0	0	1	HT	EM	)	9	I	Y	i	y
1	0	1	0	ENQ	NAK	%	5	E	U	e	u
1	0	1	1	CR	GS	-	=	M	[	m	}
1	1	0	0	ETX	DC3	#	3	C	S	c	s
1	1	0	1	VT	ESC	+	;	K	]	k	{
1	1	1	0	BEL	ETB	'	7	G	W	g	w
1	1	1	1	SI	US	/	?	O	-	o	DEL

TABLE 1-2.   Construction of a Variable-Length Source Code

Source Symbol	Probability: Reduction 0	Codewords: Reduction 0	Probability: Reduction 1	Codewords: Reduction 1
A	0.7	0	0.7	0
B	0.2	1 0	0.3	1
C	0.1	1 1		

occurring letters, such as J, Q, and Y, are assigned combinations of three DAHs and a DIT. This is an example of a *variable-length* source code.

Interestingly, the mathematical science of information theory provides both a lower bound on the average length of a variable-length source code as well as a means for finding codes whose average length achieve or approach this lower bound. For an *r*-symbol code, the lower bound is

$$H_r(X) = -\sum_{i=1}^{q} P(x_i) \log_r P(x_i) \qquad r\text{-ary symbols/symbol} \qquad (1\text{-}5)$$

which is just the source entropy in *r*-ary units.

A method for obtaining a code with the smallest possible average codeword length is the *Huffman procedure*, named after its inventor. Although it can be used to produce an *r*-ary code for a source with *r* arbitrary, it is easiest to illustrate for *r* = 2.*

Consider a source with three output symbols, *A*, *B*, and *C*, with respective probabilities of occurrence of 0.7, 0.2, and 0.1. The entropy of this source was found to be 1.157 bits/symbol in Example 1-1.

Application of the Huffman procedure for encoding this source into a variable-length binary code begins by listing the source output symbols in descending order of their probability of occurrence, as in Table 1-2. The least probable source symbols are then combined two at a time to produce what is referred to as a *reduced source*. Source reductions are continued until only two symbols remain. At each stage of reduction the probability of the symbol resulting from the reduction is the sum of the probabilities of the symbols combined to produce it. At each reduction stage, the symbols of the reduced sources are rearranged so that they occur in order of descending probability. When the last stage is reached, the code symbols 0 and 1 are assigned to the two reduced source symbols (it makes no difference which code symbol is assigned to which reduced source symbol). These code symbols are then carried back to the preceding reduction; for example, if the code symbol 1 has been assigned to a source symbol in reduction 1 resulting from the combining of two symbols in reduction 0 (original source), it serves as a *prefix* for the codewords in reduction 0 for the symbols combined. The codewords for these symbols are then formed by appending a 0 to the prefix for symbol *B* and a 1 to the prefix for symbol *C* (both prefixes are the same), resulting in 10 being assigned to the *B* and a 11 being assigned to the *C*.

*For the general *r*-ary case, see, for example, [8].

In this fashion, the longest codewords are assigned to the least probable source symbols. If the length of the ith codeword is denoted by l_i, the average codeword length is

$$L = \sum_{i=1}^{q} l_i P_i \tag{1-6}$$

For the code obtained in Table 1-2, the average codeword length is

$$L = 1 \times 0.7 + 2 \times 0.2 + 2 \times 0.1$$

$$= 1.3 \text{ bits/source symbol} > H(X)$$

Since the average code length is greater than the entropy of the source, one might ask if there is a way to achieve this lower bound on average codeword length, or at least come closer to it. The answer is in the affirmative. Instead of encoding a single source symbol at a time, a Huffman code can be found for sequences of two or more source symbols. If the source symbols are grouped two at a time and viewed as a new source, the resultant equivalent source is called the *second extension* of the actual source. Grouping three source symbols at a time results in the *third extension*, and so on. Table 1-3 shows the second extension of the source together with the reductions necessary for obtaining the Huffman code for the extended source.

The average codeword length for the second extension of the source is

$$L_2 = 1 \times 0.49 + 2 \times 3 \times 0.14 + 2 \times 4 \times 0.07 + 4 \times 0.04$$

$$+ 5 \times 0.02 + 6 \times 0.02 + 6 \times 0.01$$

$$= 2.33 \text{ bits/extended source symbol}$$

Since each codeword for the second extension represents two source symbols, the average number of bits per symbol is

$$L' = \frac{L_2}{2} = 1.165 \text{ bits/source symbol}$$

This compares more favorably with the lower bound of $H(X) = 1.157$ bits/symbol.

The average number of bits used per source symbol could be made closer to this lower bound by encoding the third extension of the source. To get a better intuitive feeling for the reduction in the number of code symbols per unit time, consider the following example.

EXAMPLE 1-2

A communication system can transmit and receive error-free binary symbols at a rate of 1250 bits/s. It is desired to send messages over this channel that are composed of sequences of three symbols A, B, and C which occur with probabilities 0.7, 0.2, and 0.1, respectively. The source produces symbols at a rate of 1000 per second.

(a) Can a binary code with equal codeword lengths be used?
(b) Is it possible to use a variable-length code?
(c) Can a Huffman code for the raw source output be used?
(d) Can a Huffman code for the second extension be used?

TABLE 1-3.   Second Extension of the Source of Table 1-2, Showing the Huffman Encoding Procedure

Second Extension Source Symbol	Probabilities and Codewords							
	Source (Reduction 0)	Reduction 1	Reduction 2	Reduction 3	Reduction 4	Reduction 5	Reduction 6	Reduction 7
AA	0.49 1	0.49 1	0.49 1	0.49 1	0.49 1	0.49 1	0.49 1	0.51 0
AB	0.14 000	0.14 000	0.14 000	0.14 000	0.14 000	0.23 01	0.28 00	0.49 1
BA	0.14 001	0.14 001	0.14 001	0.14 001	0.14 001	0.14 000	0.23 01	
AC	0.07 0100	0.07 0100	0.07 0100	0.09 011	0.14 010	0.14 001		
CA	0.07 0101	0.07 0101	0.07 0101	0.07 0100	0.09 011			
BB	0.04 0111	0.04 0111	0.05 0110	0.07 0101				
BC	0.02 01101	0.03 01100	0.04 0111					
CB	0.02 011000	0.02 01101						
CC	0.01 011001							

TABLE 1-4.   Typical Sequence of Source Output Symbols and the Encoded Output for Three Cases

Source symbol	A	A	A	C	B	A	A	C	B	A	A	B	A	A	A	B	A	C	A	A	A
Fixed-length code	00	00	00	10	01	00	00	10	01	00	00	01	00	00	00	01	00	10	00	00	00
Huffman encoded source output	0	0	0	11	10	0	0	11	10	0	0	10	0	0	0	10	0	11	0	0	0
Huffman encoded output of the second extension	1		0100		001		000		1		000		1		001		0101		1		

Solution: The answer to part (a) is obviously no, because fixed-length codewords of length 2 are required to encode the three source output symbols. The binary digit symbol rate from the encoder output will therefore be 2000 bits/s if a fixed-length code is used.

The answer to part (b) is yes, for $H(X) = 1.157$ bits/symbol, which means that the average bit rate at the output of an ideal variable-length encoder is 1157 bits/s. Since this is less than the rate at which the communication system can accept binary symbols, it is theoretically possible to design a variable-length binary code that allows the source output to be transmitted through the channel.

The answer to part (c) is no, because it has been found previously that a Huffman code of the original source produces a code with an average codeword length of 1.3 bits/symbol. Thus the communication system would have to accept binary symbols at a rate of 1300 bits/s.

The answer to part (d) is yes, because a Huffman code for the second extension resulted in a code that required an average of 1.165 bits/source symbol. The average bit rate into the channel would therefore be 1165 bits/s, which is less than the 1250 bits/s that the communication system can handle. ∎

Note that the use of a variable-length code requires a buffer to smooth the flow of binary digits from the encoder. A typical symbol pattern and bit streams using fixed-length and Huffman codes for the raw source output of Example 1-2 and its second extension are shown in Table 1-4. The decrease in the average number of bits per symbol between the codes for the three cases is clearly evident.

1-3.4   The Transmitter in a Digital Communication System

The transmitter section of Figure 1-3 has been broken down into six blocks, which are an encryptor, a channel encoder, a data modulator, a spread-spectrum modulator, a spread-spectrum code generator, and a final power amplifier. It is not meant to be implied that all these blocks are always present, but rather, that they represent functions that might be desirable in some situations. Each block is discussed in this section.

Encryptor.   In communication systems where security of the transmitted data is important, the output of the source encoder may be encrypted. The purpose of the encryptor is to make recovery of the transmitted data as difficult as possible for an unintended receiver. A variety of encryption schemes are available. These schemes vary in complexity and in the security they provide. For obvious reasons much of the available literature on this interesting subject remains classified.*

Channel Encoder.   The purpose of the channel encoder is to insert structure into the symbol stream emitted by the source, which will be assumed to be binary for simplicity, for the purpose of detecting and correcting transmission errors at the receiver. This process is referred to as *forward error correction* (FEC) *coding*. Whether or not FEC coding is used depends on system constraints. If power is at a premium, for example, FEC coding can be used to lower the signal-to-noise ratio (SNR) required at the receiver to achieve a given bit error rate (BER). The reduction in SNR is called the *coding gain* of the code. The cost for using FEC coding is increased complexity, increased transmission delay, and usually increased band-

*For a tutorial article on the subject of cryptography, see [9].

width, all of which may be preferable to the cost of a larger final power amplifier or a more sensitive receiver.

The two most popular ways of implementing FEC coding are known as *block coding* and *convolutional coding*. Coding schemes exist for nonbinary encoder input and output alphabets. For now, however, discussion will be limited to codes using binary input and output alphabets. For both block and convolutional codes there are n binary output symbols from the encoder for k input bits and the *code rate* is defined by

$$R_{\text{code}} = \frac{k}{n} \quad \text{bits/bit} \tag{1-7}$$

with the usual case being $n > k$. The larger number of output symbols is used to structure the output sequence so that errors can be detected and corrected. In some cases, the added output symbols are appended to the input symbols in the form of *redundant* or *parity symbols*. When the input symbols are transmitted as part of the output, the code is called a *systematic code*.

EXAMPLE 1-3

A simple type of block code is repetition code, where $n - 1$ parity symbols are used which are the same as the information bit. That is, a 1 is transmitted as n ones and 0 as n zeros and the code rate is $1/n$. The received codewords are decoded by counting the ones and zeros and deciding in favor of the majority count. As a result, a total of

$$e = \tfrac{1}{2}(n - 1) \tag{1-8}$$

errors can be corrected, where n is assumed to be odd. If the channel symbol error probability is p, the probability of incorrectly decoding a codeword is

$$P_{\text{ecw}} = \sum_{i=e+1}^{n} \binom{n}{i} p^i (1 - p)^{n-i} \tag{1-9}$$

where

$$\binom{n}{i} = \frac{n!}{i! \, (n - i)!}$$

is the binomial coefficient. Each term of (1-9) is the probability of receiving a codeword that has exactly i errors. The probability of receiving a codeword that has $(n - i)$ correct symbols and i incorrect symbols in a specific order is $(1 - p)^{n-i} p^i$, and there are $\binom{n}{i}$ orderings of i incorrect symbols in a codeword of length n. Thus $\binom{n}{i}(1 - p)^{n-i} p^i$ is the probability of receiving any one of the codewords with i errors, and the sum is the probability of receiving a codeword with $e + 1$ or more errors. ∎

Many types of block codes with varying complexity and performance exist and are used in modern communication systems. In all block codes, source encoder output symbols are grouped into words k symbols long. Each distinct k-symbol input word is mapped into an n-symbol encoder output word. The mapping operation may be a simple *linear* mapping, in which case the code is said to be *linear*. Table 1-5 is an example of a simple linear block code with $k = 4$ and $n = 7$.

TABLE 1-5.   Linear Block Code

Message	Codeword
0 0 0 0	0 0 0 0 0 0 0
1 0 0 0	1 1 0 1 0 0 0
0 1 0 0	0 1 1 0 1 0 0
1 1 0 0	1 0 1 1 1 0 0
0 0 1 0	1 1 1 0 0 1 0
1 0 1 0	0 0 1 1 0 1 0
0 1 1 0	1 0 0 0 1 1 0
1 1 1 0	0 1 0 1 1 1 0
0 0 0 1	1 0 1 0 0 0 1
1 0 0 1	0 1 1 1 0 0 1
0 1 0 1	1 1 0 0 1 0 1
1 1 0 1	0 0 0 1 1 0 1
0 0 1 1	0 1 0 0 0 1 1
1 0 1 1	1 0 0 1 0 1 1
0 1 1 1	0 0 1 0 1 1 1
1 1 1 1	1 1 1 1 1 1 1

Source: Ref. 10.

This table shows the mapping of encoder input words to seven symbol output codewords. Observe that the bit-by-bit modulo-2 sum of any two codewords is another codeword, and that a minimum of three transmission errors must occur to change one codeword into another. The error correction capability of the code is due to this minimum difference between any two codewords.

Block codes can give significant improvement in BER performance. One such code, a (1023, 688) BCH (Bose–Chaudhuri–Hocquenghem) [10] code, gives about a 5-dB coding gain at a BER of 10^{-6}. Although $n = 1023$ bits requires a fairly complex encoder and decoder, the payoff in terms of coding gain is significant.

Turning next to convolutional codes, an example of a convolutional encoder is illustrated in Figure 1-5. Each output symbol is a particular modulo-2 sum of a

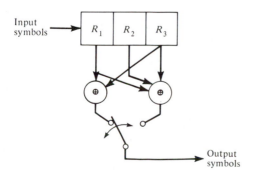

⊕ = exclusive OR

FIGURE 1-5.   Example of rate-$\frac{1}{2}$ convolutional encoder.

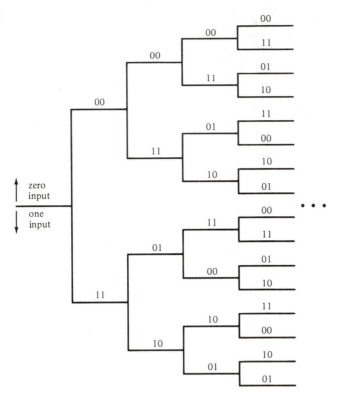

FIGURE 1-6. Code tree corresponding to the convolutional encoder of Figure 1-5.

window of k input symbols. At each shift of a new input bit into the k-stage shift register (SR), the results of the modulo-2 sums on the contents are sampled and these samples then compose the encoded bit stream. For example, if the contents of the SR are 1 1 1 and the new bit shifted in is a 0, the output bits that result by sampling the modulo-2 adder outputs are 1 0. Since each input bit results in two encoded output bits, the rate of this particular code is $\frac{1}{2}$. The error-correcting power of the code results because of the memory built into it by the SR–adder combination. A convenient tool for determining the output of a convolutional encoder is the *code tree*. The code tree for the encoder of Fig. 1-5 is shown in Fig. 1-6. The particular path taken through the tree is determined by the input bit sequence. If a 0 appears at the input, an upward branch is taken, whereas if a 1 is present at the input, a downward branch is taken. The output sequence for a given input bit is obtained by reading off the encoded symbol sequence and the particular branch one ends up on by taking either an upward route in response to an input 0, or a downward route in response to an input 1. For example, a 0 1 0 1 at the input results in the sequence 0 0 1 1 0 1 0 0 out. One way to decode a convolutional code is to do a search through the tree and find a path that differs the least in terms of ones and zeros from the received bit stream. This can be accomplished through the use of an algorithm known as the *Viterbi algorithm*, but which is really the same as forward dynamic programming. The Viterbi algorithm is described in Appendix D. An important parameter for a convolutional code that affects its performance is the constraint length of the code. It can be defined as the number of output bits that are affected by a given input bit. Figure 1-7 shows the BER performance of a selection of specific rate-$\frac{1}{2}$ convolutional codes of various constraint lengths where the decoding is by means of the Viterbi algorithm.

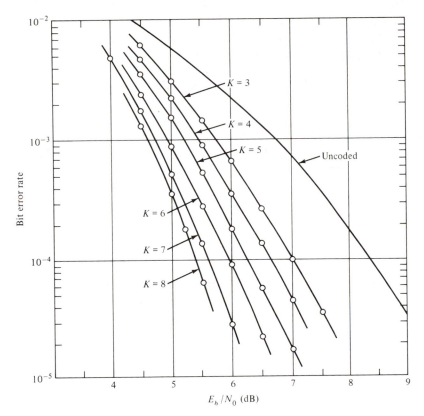

FIGURE 1-7. BER performace of rate-$\frac{1}{2}$ convolutional codes. K is the constraint length. (Reprinted with permission from Ref. 11.)

Data Modulator. The function of the modulator is to assign a particular output waveform to each possible sequence of input symbols into the modulator. In other words, a modulator performs a mapping from the input symbol stream to a set of output waveforms. If there are M possible output waveforms, where for convenience, M is assumed to be a power of 2, each possible sequence of $\log_2 M$ bits from the channel encoder is associated with a unique waveform. If R_c is the bit rate at the coder output, the duration of each waveform from the modulator output must be

$$T_s = \frac{\log_2 M}{R_c} \tag{1-10}$$

to avoid gaps or overlapping of waveforms. The example given below illustrates this discussion.

EXAMPLE 1-4

Consider an 8-ary amplitude-shift-keyed modulator with output waveforms of the form

$$s_i(t) = \begin{cases} \sqrt{\dfrac{2E_i}{T_s}} \cos \omega_0 t & 0 \le t \le T_s \\ 0 & \text{otherwise} \end{cases} \tag{1-11}$$

where $i = 1, 2, \ldots, M$ with $M = 8$. The parameter E_i is the energy of $s_i(t)$. If E is the average energy over all possible transmitted waveforms, it can be shown for equally spaced signal amplitudes which occur with equal probability that

$$E_i = \frac{6(i - 1)^2 E}{(M - 1)(2M - 1)} \qquad (1\text{-}12)$$

Since $M = 8$ is assumed to be the case here, the output of the channel encoder would be grouped $\log_2 8 = 3$ bit blocks and each block associated in some fashion with each of the eight possible $s_i(t)$'s. A possible grouping is as follows:

$$000: \; s_1(t) \qquad\qquad 100: \; s_5(t)$$

$$001: \; s_2(t) \qquad\qquad 101: \; s_6(t)$$

$$010: \; s_3(t) \qquad\qquad 110: \; s_7(t)$$

$$011: \; s_4(t) \qquad\qquad 111: \; s_8(t)$$

The association between a 3-bit sequence and a signal index i in Example 1-4 is obviously carried out by expressing $(i - 1)$ as a binary-coded number. Another modulation scheme that would be more suitable for an optical communication scheme is described in the following example, wherein another assignment scheme, referred to as a *Gray code*, is used. ∎

EXAMPLE 1-5

A fiber optic data communication system utilizes pulse-position modulation, wherein one of 16 possible messages is transmitted through an optical filter by a pulse of light in 1 of 16 possible time slots measured relative to a time reference, as shown in Figure 1-8. With proper synchronization at the receiver, it is possible to build a detector that makes a decision as to which time slot is occupied. Clearly, it takes

$$\log_2 M = \log_2 16 = 4 \text{ bits}$$

to specify which slot is occupied.

As in Example 1-4, the specification of the time slot could be made by simply grouping the bits at the channel encoder (or source) output into blocks of four and choosing the time slot number corresponding to the value of the 4-bit binary number represented by the block of 4 bits. One problem with this approach is that moving

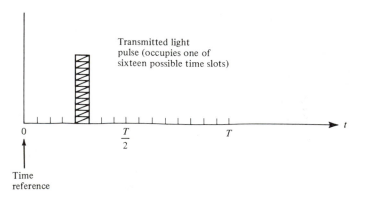

FIGURE 1-8. Modulated waveform for an optical communications link.

TABLE 1-6. Binary and Gray Codes for the Decimal Numbers 0 to 15

Digit	Binary Code				Gray Code			
	b_1	b_2	b_3	b_4	g_1	g_2	g_3	g_4
0	0	0	0	0	0	0	0	0
1	0	0	0	1	0	0	0	1
2	0	0	1	0	0	0	1	1
3	0	0	1	1	0	0	1	0
4	0	1	0	0	0	1	1	0
5	0	1	0	1	0	1	1	1
6	0	1	1	0	0	1	0	1
7	0	1	1	1	0	1	0	0
8	1	0	0	0	1	1	0	0
9	1	0	0	1	1	1	0	1
10	1	0	1	0	1	1	1	1
11	1	0	1	1	1	1	1	0
12	1	1	0	0	1	0	1	0
13	1	1	0	1	1	0	1	1
14	1	1	1	0	1	0	0	1
15	1	1	1	1	1	0	0	0

from a given time slot to an adjacent time slot corresponds to a variable number of bits in the 4-bit code changing. For example, if the time slots are numbered 0 to 15, and a pulse is transmitted in slot 7 but the receiver detects it as being in slot 8, it follows that 4 bits are received in error (the 4-bit word 0111 should have been detected, whereas the word 1000 actually was detected). The Gray code avoids the possible occurrence of such a problem because only one binary digit at a time is changed as the corresponding decimal digit is changed by one unit. The conversion from a binary-coded representation to a Gray code is easily accomplished as follows. Let $b_1 b_2 b_3 \cdots b_n$ represent a binary number representation of a number, with b_1 being the most significant bit (MSB) and b_n being the least significant bit (LSB). Let the corresponding Gray code digits be $g_1 g_2 g_3 \cdots g_n$. Then

$$g_1 = b_1$$

$$g_n = b_n \oplus b_{n-1} \qquad n \geq 2 \qquad\qquad (1\text{-}13)$$

where $\oplus$ denotes modulo-2 (exclusive OR or XOR) addition.* The binary and Gray codes for the decimal digits 0 to 15 are given in Table 1-6. From this table

*The truth table for the XOR operation is

	b_2	
b_1	0	1
0	0	1
1	1	0

it is apparent that transmission of a pulse in time slot 7 with detection in time slot 8 results in only one bit being received in error (0100 is transmitted, but 1100 is detected, resulting in an error in the MSB only).   ∎

In many cases of interest, the transmitted data are impressed on a sinusoidal carrier signal and the modulated carrier may be written

$$x_c(t) = A(t) \cos [2\pi f_0 t + \varphi(t)] \qquad (1\text{-}14)$$

where f_0 is the carrier frequency. Since a sinusoid is completely specified by its amplitude and argument, it follows that once the carrier frequency is specified, only two parameters are candidates to be varied in the modulation process: the instantaneous amplitude $A(t)$ and the instantaneous phase deviation, $\varphi(t)$. When the instantaneous amplitude $A(t)$ is linearly related to the modulating signal, the result is *amplitude modulation* (AM), also referred to as *linear modulation*. Letting $\varphi(t)$ or the time derivative of $\varphi(t)$ be linearly related to the modulating signal yields *phase modulation* (PM) or *frequency modulation* (FM), respectively. Collectively, phase and frequency modulation are referred to as *angle modulation* since the phase angle of the modulated carrier carries the information. Since a sinusoid is a non-linear function of $\varphi(t)$, phase and frequency modulation are both categorized as *nonlinear modulation* techniques. Several modulation methods suitable for transmission of digital data are considered in detail in Chapters 3 and 4.

Spread-Spectrum Modulator.   In some communications systems, the data modulator is followed by a second modulator whose purpose is to spread the data-modulated signal power over a much wider bandwidth than it would otherwise occupy. A number of reasons for employing a spread spectrum were given previously. All systems that employ a spread-spectrum modulator are called *spread-spectrum systems*. These systems are the subject of the second half of this book, where their design and application are discussed in detail.

Final Power Amplifier.   The purpose of the final power amplifier is to produce a transmitted signal of the proper power level to give the desired performance at the receiver, usually specified in terms of symbol error probability, accounting for all power losses introduced in the communications link. It may be unnecessary to employ a final power amplifier if the signal level from the modulator itself is sufficient. The two most prevalent characteristics of power amplifiers are that they bandlimit the modulated signal spectrum and they introduce nonlinear distortion into the modulated signal, even if designed to be linear. Both of these effects are defined more precisely in Chapter 2, which includes a summary of signal and systems theory, but it will be recalled that the former can best be described in the frequency domain in terms of the amplifier's frequency response or transfer function. If an amplifier is only slightly nonlinear, as a Class A amplifier would be, it can be assumed linear or described in terms of a power series transfer characteristic of the form

$$y(t) = a_0 + a_1 x(t) + a_2 x^2(t) + \cdots \qquad (1\text{-}15)$$

where $x(t)$ is its input and $y(t)$ its output. If the constants $a_2, a_3 \ldots$ in the series above are small compared with a_1, the effect of the amplifier may well be approximated by only the first two, or at most the first three, terms of the series. Such an amplifier is necessary if linear, or amplitude, modulation is employed. If angle modulation is employed, a nonlinear final power amplifier can be employed due to the data signal being conveyed by the instantaneous phase, $\varphi(t)$, of the transmitted signal. In fact, many final power amplifiers operate most efficiently if operated in a nonlinear mode. Such is the case with a traveling-wave tube amplifier (TWTA), which exhibits two distortion effects on a signal due to its nonlinear nature:

1. A nonlinear output–input power characteristic referred to as *AM/AM conversion*.
2. A nonlinear output phase–input power characteristic referred to as *AM/PM conversion*.*

For a discussion of mathematical representations of TWTA characteristics, see [13].

1-3.5   The Channel in a Digital Communication System

The *channel* is defined as a single path for transmitting electric signals either in one direction only or in both directions. The physical means by which the transmission is effected could make use of electromagnetic energy or acoustical energy, for example. If electromagnetic, the type of transmission could be further categorized as taking place in what is normally referred to as the radio spectrum (300 to 3×10^{11} Hz) or in the infrared, visible, or ultraviolet regions. Furthermore, the type of propagation can be guided or free-space. With all these possibilities, it is difficult to say anything of a generally applicable nature which applies to all modes of transmission. Indeed, the mode of transmission employed determines, to a large degree, the perturbations that the transmitted signal experiences in passing through the channel. In a digital communications system, the channel characterization is usually done at one of two levels, which will be referred to as (1) the *transition probability level* or *information-theoretic approach*, and (2) the *waveform level*. These two methods of channel characterization will be described briefly before a more in-depth discussion of properties of various types of channels is given.

Discrete Channel.   The discrete characterization of the channel includes the effects of all blocks of Figure 1-3 between the data modulator input and the data demodulator output. The effect of the waveform channel is implicitly included in the discrete model. In all digital communications systems the demodulator input is one of M possible discrete messages or symbols; thus the input is discrete. The output of the demodulator may be an estimate of the modulator input symbols or symbols from an entirely different alphabet. The reason for allowing output symbols from other alphabets is that more information can be passed to the channel decoder in this manner. When channel coding is not used, the demodulator output alphabet is always identical to the modulator input alphabet. When the input and output alphabets are the same, the demodulator is called a *hard decision demodulator;* when the alphabets are different, the demodulator is called a *soft decision demodulator*. Different output alphabets are obtained by quantizing the detector output (analog) in different ways.

It is convenient to denote the input symbols as $x_1, x_2, \ldots, x_M$ and the output symbols as $y_1, y_2, \ldots, y_N$. For example, for a binary communication system

*For an electromagnetic-field theory discussion of TWTAs, see [12].

$M = 2$ with $x_1 = 0$ and $x_2 = 1$. At the output of the channel, it may be convenient in some cases to consider three possible outputs: $y_1 = 0$, $y_2 = 1$, and $y_3 =$ undecided. If y_3 is received, a channel from output to input, or feedback channel, could be employed to request a retransmission.

With this level of description, no attention is paid to the particular waveform that represents the transmission through the channel. Rather, the only thing of interest is the probability, $p(\mathbf{y}|\mathbf{x})$, that the output symbol sequence $\mathbf{y} = (y_1, y_2, \ldots, y_k)$ is received given that the input symbol sequence is $\mathbf{x} = (x_1, x_2, \ldots, x_k)$, where each x_i is an input symbol and each y_i is an output symbol in the ith time slot.* These conditional probabilities are referred to as *channel transition probabilities*. In many cases, symbol transmissions are independent and the transition probability can be written

$$p(\mathbf{y}|\mathbf{x}) = \prod_{i=1}^{k} p(y_i|x_i)$$

where $p(y_i|x_i)$ is the probability that the output symbol y_i is received given that x_i was transmitted in the ith time slot. In this case, the channel is called a *discrete memoryless channel* (DMC). The DMC is characterized by giving the symbol transition probabilities for all possible input–output symbol pairs. This information can be provided by either a *channel transition probability matrix*, given by (1-16) below, or a *channel transition diagram*, as illustrated in Figure 1-9. The former is written as

$$[p(y|x)] = \begin{bmatrix} p(y_1|x_1) & p(y_2|x_1) & \cdots & p(y_N|x_1) \\ p(y_1|x_2) & p(y_2|x_2) & \cdots & p(y_N|x_2) \\ \cdot & & \cdot & \cdot \\ \cdot & \cdot & & \cdot \\ \cdot & \cdot & & \cdot \\ p(y_1|x_M) & p(y_2|x_M) & \cdots & p(y_N|x_M) \end{bmatrix} \qquad (1\text{-}16)$$

where each row is associated with a particular input and each column is associated with a particular output. These probabilities are obtained from a consideration of the channel properties. Because of the property of probabilities for exhaustive events, the sum over any row must be unity.† It follows that a particular output, say y_n, is obtained with probability

$$p(y_n) = \sum_{m=1}^{M} p(y_n|x_m)p(x_m) \qquad (1\text{-}17)$$

where $p(x_m)$ is the probability that input x_m is input to the channel. The *entropy* of the channel output is

$$H(Y) = -\sum_{n=1}^{N} p(y_n) \log_2 p(y_n) \qquad \text{bits/symbol} \qquad (1\text{-}18)$$

---

*Note that each $y_i \in \{0, 1, 2, \ldots, q - 1\}$, as is each x_i.
†See Appendix A for a summary of probability theory.

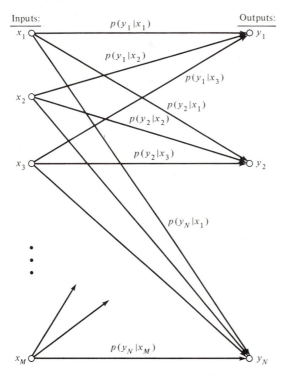

Inputs:

x_1

$p(y_1|x_1)$

$p(y_1|x_2)$

$p(y_1|x_3)$

x_2

$p(y_2|x_1)$

$p(y_2|x_2)$

$p(y_2|x_3)$

x_3

$p(y_N|x_1)$

$p(y_N|x_M)$

x_M

Outputs:

y_1

y_2

y_N

FIGURE 1-9. Transition diagram for a channel with M inputs and N outputs.

and the entropy of the channel output given that a particular input, say x_m, was present is

$$H(Y|x_m) = -\sum_{n=1}^{N} p(y_n|x_m) \log_2 p(y_n|x_m) \qquad (1\text{-}19)$$

When averaged over all possible inputs, the *conditional entropy of the output given the input*, $H(Y|X)$, is obtained, which is

$$H(Y|X) = -\sum_{m=1}^{M} \sum_{n=1}^{N} p(x_m, y_n) \log_2 p(y_n|x_m) \qquad \text{bits/symbol} \qquad (1\text{-}20)$$

where the relationship

$$p(x_m, y_n) = p(y_n|x_m)p(x_m) \qquad (1\text{-}21)$$

has been used for the probability that the joint event that the input was x_m and the output was y_n has occurred. In a similar fashion, the conditional entropy $H(X|Y)$ can be defined by replacing $p(y_n|x_m)$ by $p(x_m|y_n)$ in (1-20).

Now it will be recalled that entropy represents the average amount of information gained in observing the occurrence of a particular event, averaged over all possible events in the sample space. Thus $H(X|Y)$ represents the average information obtained by being told what the channel input was after observation of the channel output, averaged over all possible channel inputs and outputs. Since $H(X)$ represents the average information obtained by observation of the source output (i.e., channel input) with no side information (i.e., without knowing the channel output), it follows that the average amount of information conveyed through the channel is

$$I(X; Y) = H(X) - H(X|Y) \qquad \text{bits/symbol} \qquad (1\text{-}22a)$$

which is referred to as the *mutual information* between channel input and output. It can be readily shown, with the aid of Bayes' rule, that

$$I(X; Y) = H(Y) - H(Y|X) \qquad \text{bits/symbol} \qquad (1\text{-}22\text{b})$$

and that in either case, $I(X; Y)$ can be written as

$$I(X; Y) = \sum_{m=1}^{M} \sum_{n=1}^{N} p(x_m, y_n) \log_2 \frac{p(x_m, y_n)}{p(x_m)p(y_n)} \qquad \text{bits/symbol} \qquad (1\text{-}23)$$

where $p(x_m, y_n) = p(y_n|x_m)p(x_m) = p(x_m|y_n)p(y_n)$ are the joint probabilities of the event that the channel input is x_m and its output is y_n.

From (1-23) and using the theorem of total probability, it follows that the mutual information can be expressed as a function of the channel input probabilities, $p(x_m)$, and the channel transition probabilities, $p(y_n|x_m)$. For a specified channel, the transition probabilities are fixed. It is natural to ask what the maximum of $I(X; Y)$ is through proper selection of the input probabilities, $p(x_m)$. The resulting maximum for $I(X; Y)$ is called the *capacity*, C, of the channel, and represents the maximum amount of information that can be conveyed error-free through the channel if the source is matched to the channel in the sense that its output symbols occur with the proper probabilities such that the maximum mutual information is achieved and coding is used. That is,

$$C = \max_{p(x_m)} I(X; Y) \qquad (1\text{-}24)$$

Although developed for a discrete channel (finite number of inputs and outputs), $I(X; Y)$ can be appropriately generalized to channels where the inputs and outputs take on a continuum of values. Maximization of this generalized mutual information and application of appropriate time-sampling representations for continuous-time inputs and outputs then result in the capacity expression for a continuous-time waveform channel given by (1-1).

Much more could be said about the information-theoretic approach for channel descriptions. Instead, the discussion above will suffice for now, and the description of communication channels will continue with the waveform level after consideration of an example.

EXAMPLE 1-6

Consider a *binary symmetric channel* (BSC) which has a channel transition probability matrix of the form

$$P = \begin{bmatrix} q & p \\ p & q \end{bmatrix} = \begin{bmatrix} p(y_1|x_1) & p(y_2|x_1) \\ p(y_1|x_2) & p(y_2|x_2) \end{bmatrix}$$

where $q = 1 - p$, in which p is referred to as the error, or crossover, probability for the channel. Find $H(X)$, $H(Y|X)$, $H(Y)$, $I(X; Y)$, and C if $p(x_1) = \alpha$ and $p(x_2) = 1 - p(x_1) = 1 - \alpha = \beta$.

Solution: From (1-4) the source entropy is

$$H(X) = -\alpha \log_2 \alpha - (1 - \alpha) \log_2(1 - \alpha)$$

From (1-20) and the definition of conditional probability (1-21), it follows that

$$H(Y|X) = -(q\alpha \log_2 q + p\beta \log_2 p + p\alpha \log_2 p + q\beta \log_2 q)$$

$$= -(\alpha + \beta)(q \log_2 q + p \log_2 p)$$

$$= -q \log_2 q - p \log_2 p$$

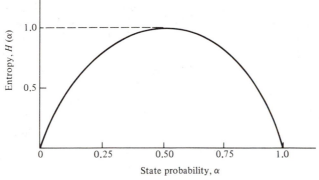

(a) Entropy for a binary source

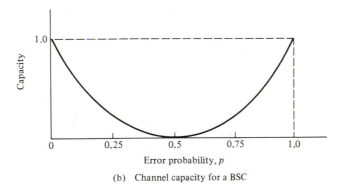

(b) Channel capacity for a BSC

FIGURE 1-10. Entropy and channel capacity functions for a binary source and channel.

From the theorem of total probability, it follows that

$$p(y_1) = p(y_1|x_1)p(x_1) + p(y_1|x_2)p(x_2)$$

$$= q\alpha + p\beta$$

$$p(y_2) = p(y_2|x_1)p(x_1) + p(y_2|x_2)p(x_2)$$

$$= p\alpha + q\beta$$

Therefore,

$$H(Y) = -(q\alpha + p\beta) \log_2(q\alpha + p\beta) - (p\alpha + q\beta) \log_2 (p\alpha + q\beta)$$

From (1-22b), the mutual information is

$$I(X; Y) = H(Y) - H(Y|X)$$

$$= \mathcal{H}_2(q\alpha + p\beta) - \mathcal{H}_2(p) \tag{1-25}$$

where $\mathcal{H}_2(u)$ is the entropy function of a binary source, defined as

$$\mathcal{H}_2(u) = -u \log_2 u - (1 - u) \log_2(1 - u) \tag{1-26}$$

which has an absolute maximum of unity for $u = \frac{1}{2}$ as shown in Figure 1-10a.*
  The channel capacity is obtained by maximizing (1-25) with respect to α with

---

*The proof of this is left for the problems.

$\beta = 1 - \alpha$. The channel is fixed, which means that $\mathcal{H}_2(p)$ has a value corresponding to the error probability, p, of the channel. Therefore, (1-25) is maximized by maximizing the first term on the right-hand side. But this term has an absolute maximum of unity which is attained for $q\alpha + p\beta = \frac{1}{2}$. For any value of p and q such that $p + q = 1$, this is indeed obtained for $\alpha = \beta = \frac{1}{2}$. Therefore, the channel capacity for the BSC is

$$C = 1 - \mathcal{H}_2(p) \qquad (1\text{-}27)$$

This function is plotted in Figure 1-10b, which shows that the channel capacity has a maximum value of 1 bit/symbol for $p = 1$ or $p = 0$, and a minimum value of 0 bits/symbol for $p = \frac{1}{2}$. In the former case, a symbol put into the channel is received without error at the output, while in the latter case, it is equally likely to be received as a 1 or a 0 (i.e., there is total uncertainty as to what the channel output will be). Therefore, the channel capacity for the BSC shown as a function of p in Figure 1-10b makes sense. Again, it is emphasized that channel capacity is the maximum amount of information per symbol that can be transmitted *error-free* using error correction coding. In the case of the BSC with $p > 0$, one bit per symbol can be transmitted; however, the bit error probability is $P_b = p > 0$, not $P_b = 0$. ∎

Waveform Description of Communication Channels. Shown in Figure 1-11 is the block diagram for a simplified description of a channel at the waveform level. Although not all possible perturbations on the input signal are shown, several representative ones are shown in order to discuss typical conditions that may prevail. The block diagram of Figure 1-11 is suggestive of a radio-wave channel, although it might be applicable to other types of channels as well.

There are several types of signal interference, which may arise in practical channels, as illustrated in Figure 1-11. These are enumerated below for convenience in later discussions.

1. Additive noise or interference at the channel input.
2. Deterministic signal processing, which may include filtering, frequency translation, and so on.
3. Multiplication by an attenuation factor, $\beta(t)$, that may be a function of time which is independent of the signal.
4. Additive noise or interference at the channel output.
5. Addition of other channel outputs which are secondary in nature.

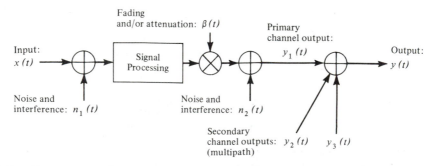

FIGURE 1-11. Channel model at the waveform level showing various purturbations.

Two specific types of channels will be considered to illustrate further possible physical originations for the perturbations listed. Before doing so, however, it should be pointed out that all the perturbations result in a *linear* channel as far as the input signal is concerned. That is, noise and interference terms are additive or, if multiplicative, are independent of the signal. Furthermore, the signal processing is assumed to be a linear operation.* Although channels that introduce nonlinear perturbations on the transmitted signal are important, the analysis of the effect of such perturbations is difficult and many practical channels can be modeled as linear. Consequently, the analysis of linear channels are focused on in this book. Two specific types of channels will now be discussed to illustrate the applications of the general model shown in Figure 1-11.

EXAMPLE 1-7

As a first example of the application of the channel model shown in Figure 1-11, consider the block diagram of Figure 1-12, which illustrates a single channel per carrier (SCPC) satellite relay link. Transmission up to the satellite is effected by a carrier of frequency f_1, and the noise $n_1(t)$ represents noise and interference added in this portion of the transmission, which is usually due primarily to the input stages of the satellite retransmission system. The function of the satellite retransmission, or relay, is to amplify the received signal and translate it in frequency to a new spectral location suitable for transmission to the destination earth stations. Noise is also added in the downlink of the transmission system as well and is represented by $n_2(t)$. Other channel perturbations could be included, such as interference due to adjacent channels, but they will not be discussed here. ∎

The additive noise and interference mentioned in Example 1-7 can fall into two possible categories: externally generated noise and noise generated internally to the communication system. Examples of the former include solar and galactic noise due to electromagnetic wave emissions from stars and other heavenly bodies, atmospheric noise which results primarily from electromagnetic waves generated by natural electrical discharges within the atmosphere, and human-made noise such as corona discharge from power lines. The modeling of this noise is, in general, difficult and imprecise due to its highly variable nature.

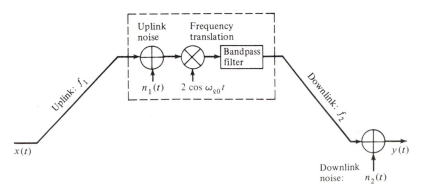

FIGURE 1-12. Model for a satellite communications link.

*The specific definition of a linear system is given in Chapter 2, which reviews important signal and system concepts.

Internally generated noise is due primarily to the random motion and random production and annihilation of charge carriers within electrical components making up a communication system. There are several treatises on the physical description and characterization of such noise (see, e.g., [14]). A short summary of its description in terms of noise figure and noise temperature is given in Appendix B.

EXAMPLE 1-8

Figure 1-13 illustrates a suitable model for certain types of terrestrial microwave communications links. In addition to additive noise, which is represented by $n(t)$, an indirect transmission path, commonly referred to as *multipath*, exists. Thus the equation relating input to output is

$$y(t) = \alpha_1 x(t - T_1) + \alpha_2 x(t - T_2) + n(t) \qquad (1\text{-}28)$$

where α_1 and α_2 are constants referred to as the attenuations of the direct and indirect transmission paths, respectively, and T_1 and T_2 are their respective delays. This channel model is an extremely simple one and yet, by virtue of the multipath term, results in two signal perturbations, known as *intersymbol interference* (ISI) and *fading*, either one of which can introduce severe degradations in system performance.

To illustrate the idea of ISI, consider Figure 1-14, which illustrates received binary data signals from the direct and indirect channels after demodulation. The differential delay $\tau = T_2 - T_1$ is assumed to be less than one bit period, although it could be several bit periods in duration. Actually, it would be impossible to observe these separate signals because they are received together at the antenna.* However, they are shown separately to illustrate that some bits destructively interfere and others reinforce each other. Unfortunately, those that destructively interfere dominate the probability of error (the average of 10^{-5}, which is a typical bit error probability for a noninterfering case, and of 10^{-2}, which might be typical for an interfering case, is very nearly 10^{-2}).

The phenomenon of fading is similar to that of ISI except that the destructive and constructive interference takes place with the high-frequency carrier. To illustrate the phenomenon of fading, consider a pure cosinusoidal input to the channel, which is representative of an unmodulated carrier. The output of the channel, ignoring the noise term, is

$$y(t) = A[\alpha_1 \cos \omega_0 t + \alpha_2 \cos \omega_0 (t - \tau)] \qquad (1\text{-}29)$$

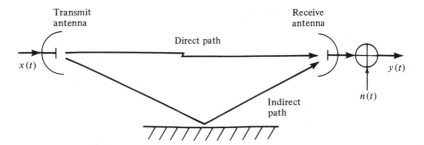

FIGURE 1-13. Model for a line-of-sight microwave relay link.

*The addition of direct and multipath signals takes place at radio frequency but is shown here at baseband for simplicity.

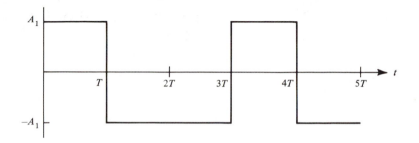

(a) Direct path signal

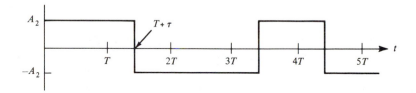

(b) Indirect path signal

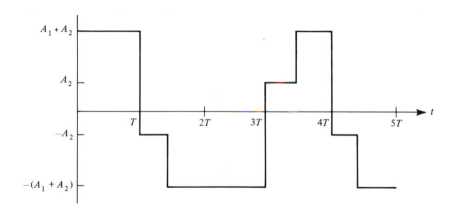

(c) Sum of direct and indirect path signals

FIGURE 1-14. Illustration of the effect of intersymbol interference.

where A is the amplitude of the input and where, for convenience, the time reference has been chosen such that the delay of the direct component is zero. Using suitable trigonometric identities, (1-29) can be put into the form

$$y(t) = AB(\tau) \cos (\omega_0 t + \theta) \tag{1-30}$$

where

$$B(\tau) = \sqrt{\alpha_1^2 + 2\alpha_1\alpha_2 \cos \omega_0 \tau + \alpha_2^2} \tag{1-31}$$

$$\theta = -\tan^{-1} \frac{\alpha_2 \sin \omega_0 \tau}{\alpha_1 + \alpha_2 \cos \omega_0 \tau} \tag{1-32}$$

Equation (1-31) shows that as the differential delay of τ changes by an integer multiple of a half carrier period, the received signal changes from a minimum amplitude of

$$AB_{\min} = |\alpha_1 - \alpha_2|A \qquad (1\text{-}33)$$

to a maximum amplitude of

$$AB_{\max} = (\alpha_1 + \alpha_2)A \qquad (1\text{-}34)$$

■

The carrier frequency in a line-of-sight (LOS) terrestrial microwave link can be of the order of 10^{10} Hz $= 10$ GHz or higher. The wavelength at 10 GHz is

$$\lambda = \frac{c}{f_0} = 3 \text{ cm}$$

where $f_0 = \omega_0/2\pi$ is the carrier frequency in hertz and $c = 3 \times 10^8$ m/s is the free-space speed of propagation for electromagnetic waves. Thus a change in differential path length of only 1.5 cm at a carrier frequency of 10 GHz means that conditions change from reinforcement to cancellation for the received carrier. This can impose severe system degradations.

A plot of $B(\tau)$ as given by (1-31) as a function of $f_0 = \omega_0/2\pi$ better illustrates the frequency-dependent nature of the channel. Figure 1-15 shows that the transmitted signal will leave "notches" placed in its spectrum each $1/\tau$ hertz of bandwidth. The bandlimited nature of the channel is therefore apparent. Any modulated

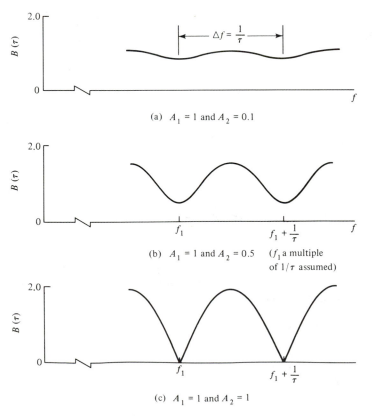

(a) $A_1 = 1$ and $A_2 = 0.1$

(b) $A_1 = 1$ and $A_2 = 0.5$ (f_1 a multiple of $1/\tau$ assumed)

(c) $A_1 = 1$ and $A_2 = 1$

FIGURE 1-15. **Received signal amplitude versus frequency for a two-path multipath channel.**

signal with bandwidth of the order of or greater than $1/\tau$ hertz will suffer degradation as it propagates through the channel because of this notching effect. The differential delay between the main and secondary paths therefore imposes an upper limit on the bandwidth of the signals that the channel can support unless compensation for the notches is accomplished somehow. Because of this frequency dependence, such channels are referred to as *frequency selective* and the degradation that they impose on the transmitted signal is referred to as *frequency-selective fading*.

A word is in order about the physical mechanisms that may give rise to multipath. The model illustrated in Figure 1-13 for the two-path case is, of course, an oversimplified view of the physical channel, as are all models. A two-path multipath channel would arise under conditions where the propagation path is highly stratified due to a temperature inversion or where stray reflections take place off of an object, such as an airplane. Such multipath is referred to as *specular* because the indirect path results in a component that is essentially a mirror-like reflection, or specular, version of the direct component. When several indirect paths combine to produce a noise-like multipath component, the resulting multipath is referred to as *diffuse*. In the specular case, alleviation of the adverse effects of the multipath can be combated by employing a filter at the receiver, referred to as an *equalizer*, which has a frequency response function that is approximately the inverse of the frequency response of the channel. In the case of diffuse multipath, the indirect-path signal power essentially acts as a signal-dependent noise.

If destructive interference of the carrier, or fading, is the predominant channel-induced perturbation, a solution is obviously to change carrier frequencies so that reinforcement results from the multipath component. Since the exact path differential is unknown or may change slowly with time, a solution is to use several frequencies simultaneously, or *frequency diversity*. Other types of diversity are also possible, such as *space diversity*, wherein several different transmission paths are used, *polarization diversity*, wherein horizontally and vertically polarized carrier waves are used, and time diversity, wherein the transmission of a symbol is spread out over time. FEC coding is a form of time diversity.

External Channel Propagation Considerations. When dealing with radio-wave propagation channels, several considerations pertaining to conditions within the propagating medium, which typically involves the earth's atmosphere for a portion or all of the propagation path, affect system design. The following radio-wave propagation factors will be discussed briefly:

1. *Attenuation* (absorption) caused by atmospheric gases.
2. *Attenuation* (scattering and absorption) by hydrometeors (rain, hail, wet snow, cloud, etc.).
3. *Depolarization* by hydrometeors, multipath, and Faraday rotation.
4. *Noise emission* due to gaseous absorption and hydrometeors.
5. *Scintillation* (rapid variations) of amplitude and phase caused by turbulence or refractive index irregularities.
6. *Antenna gain degradations* due to phase decorrelation across aperture.
7. *Bandwidth limitations* due to multipath and dispersive properties of atmosphere.

Each of these is discussed below.

1. The principal gaseous constituents of the earth's atmosphere that produce significant absorption are oxygen and water vapor. The first three absorption bands are centered at frequencies of 22.2 GHz (H_2O), 60 GHz (O_2), and 118.8 GHz (O_2)

TABLE 1-7. Average Atmospheric Attenuation Coefficient at Sea Level

Frequency (GHz)	Attenuation (dB/km)
10	0.02
15	0.04
22.2 (H_2O)	0.2
30	0.13
40	0.16
50	0.5
60 (O_2)	15

Source: Ref. 19.

[15]. The frequency dependence of the absorption has been found to depend on an empirical line-width constant which is a function of temperature, pressure, and humidity of the atmosphere. Representative values for the average attenuation coefficient for horizontal propagation at sea level are given in Table 1-7. The desirability of transmitting within certain windows between absorption lines is apparent from these data.

2. The relationship between rain rate, R (in mm/h measured at the earth's surface), and specific attenuation can be approximated by

$$\alpha = aR^b \quad \text{dB/km} \tag{1-35}$$

where a and b are frequency- and temperature-dependent coefficients. This expression was first observed from empirical observations, but has recently been given a theoretical basis (see, e.g., [16–18]). Representative values for parameters a and b are given in Table 1-8. A typical attenuation for light rain (10 mm/h) at 12 GHz is $\alpha = 0.29$ dB/km.

3. It is often desirable to use polarization of the transmitted electromagnetic wave as a means to separate two transmitted signals at the same carrier frequency. This is the case, for example, when diversity transmission is used to combat fad-

TABLE 1-8. aR^b Coefficients for the Calculation of Rain Attenuation

Frequency (GHz)	Coefficient	
	a	b
12	0.0215	1.136
15	0.0368	1.118
20	0.0719	1.097
30	0.186	1.043
40	0.362	0.972

Source: Ref. 15.

ing—for each carrier frequency or spatial path used it is possible to add another path through the use of two perpendicularly polarized waves. When the propagation is through rain or ice crystals, it is possible for a small portion of an electromagnetic wave which is, say, horizontally polarized to be converted to vertical polarization, thereby creating crosstalk between the two transmissions. This effect is generally small at frequencies below 10 to 20 GHz. For further discussion of this effect, see [15].

4. The gaseous constituents of the earth's atmosphere, and clouds or precipitation when present, all act as an absorbing medium to electromagnetic waves; therefore, they are also sources of thermal noise power radiation. The sky noise due to the atmospheric gases for an infinitely narrow beam give sky temperatures of 22, 100, and 63 K for 12, 20, and 30 GHz, respectively, at $\theta = 10$ degrees. Sky-temperature contributions caused by gaseous constituents are significant for extremely low noise receiver systems, where system noise temperatures (see Appendix B) below 300 K are required. The increase in receiver system noise figure caused by the increased sky noise from rain fades of various depths and for typical noise figures as given in the literature [15] can be of the order of 1 to 2 dB for deep fades (≥ 30 dB) and low-noise figures (≤ 4 dB). This increase in system noise adds directly to the signal loss produced by the rain attenuation in the determination of system carrier-to-noise ratio.

5. Scintillation on a radio-wave path describes the phenomenon of rapid fluctuations of the amplitude, phase, or angle of arrival of the wave passing through a medium with small-scale refractive index irregularities that cause changes in the transmission path with time. Scintillation effects, often referred to as *atmospheric multipath fading,* can be produced in both the troposphere and the ionosphere.

6. The antenna gain in a communications system is generally defined in terms of the antenna's behavior when illuminated by a uniform plane wave. Amplitude and phase fluctuations induced by the atmosphere can produce perturbations across the physical antenna aperture resulting in a reduction of total power available at the feed. The resulting effect on the antenna will look to the system like a loss of antenna gain, or a gain degradation. Gain degradation will increase as the electrical receiving aperture size increases; hence the problem will become more significant as operating frequency and/or physical aperture size increases.

7. Bandwidth limitations due to multipath and dispersive properties of the atmosphere are potentially troublesome in line-of-sight links, where multipath effects may be significant. For example, it is well known that microwave transmission over line-of-sight paths can occasionally experience periods of rapid and deep fading. This effect is ascribed to the formation of tropospheric inversion layers that permit additional paths whose vector sum makes up the received signal. It has been directly demonstrated that these separate "rays" may arrive at slightly different times, with relative delay differences typically on the order of a few nanoseconds, with the result being destructive interference, as illustrated by Figure 1-15 and the accompanying discussion.

A comprehensive theory exists in the literature for characterizing such multipath effects in terms of the time-variant transfer function of the medium and its statistical characterization in terms of correlation functions and power spectra [20]. For present purposes it is sufficient to talk about the *multipath delay* and *doppler spreads.* The former is a measure of the width of the received waveform in the time domain when a single impulse function is transmitted through the channel, and the latter is a measure of the width of the received spectrum when a single wave is transmitted. Monsen [21] points out that if the data rate is much greater than the delay spread, and the signal bandwidth is of the order of, or greater than, the doppler spread, adaptive signal processing can be used at the receiver to improve performance. The former condition ensures that the channel changes slowly enough relative to the data rate so that the receiver can "learn" the state of the channel from the received data before it changes. The latter condition means that implicit frequency diversity exists in the received signal because different parts of the frequency band fade independently.

1-3.6 The Receiver in a Digital Communication System

The last series of blocks shown in Figure 1-3 constitutes the receiver. It will be noted that the block entitled "receiver front end" has been included with the channel, for it is in this block that the major portion of the internally generated noise arises. This block typically includes preamplifiers, mixers, intermediate-frequency (IF) amplifiers, and any other subsystems necessary for preparing the received signal for extracting the data.

Although the signal-processing functions performed in the receiver are complex, this subsection will be relatively short because the various types of data detection algorithms are discussed in greater detail throughout the remainder of this book.

The purpose of the demodulator in any receiver is to remove the carrier component from the received signal and produce a baseband data-bearing signal. In some cases, a baseband signal is transmitted through the channel, and there is no need for this step. If the data-bearing signal is impressed on a carrier, one of two possible classes of modulation techniques may be chosen. A modulation scheme is termed *coherent* if it is necessary to use a locally generated reference carrier at the demodulator which is in frequency and phase coherence with the transmitted carrier (having accounted for the fixed phase shift due to transmission delay). A modulation scheme is referred to as *noncoherent* is such a locally generated reference is unnecessary. For a noncoherent scheme, it may still be necessary to perform frequency estimation on the received signal in order to center it properly within the passband of the receiver. This is especially true if the transmitter is located on a platform which is moving with respect to the receiver so that a Doppler frequency shift is introduced into the received signal.

For an analog communication system, the demodulated baseband signal is often all that is desired. For a digital communication system, however, the demodulated data-bearing signal must be processed by a *data detector*. This detection process could be as simple as passing the baseband signal through an appropriately designed filter, sampling each symbol at an appropriately chosen time instant, and comparing these samples with a properly chosen set of thresholds. In some cases, more complex signal processing operations are required. In any case, a clock signal that can be used to trigger the sampler at the proper time instants is necessary. The circuitry that generates this timing signal is referred to as a *symbol or bit synchronizer*. It may make use of an auxiliary signal to derive the timing signal, or the data-bearing signal itself may be used to produce the "bit sync signal." Additional timing signals are also sometimes required. These include code timing signals, if any type of coding was used, and synchronization with any framing used in the signal transmission process. (For additional details, see [22].)

PROLOGUE

In this chapter an attempt has been made to give an overall flavor for the general area of digital communication systems and some elementary aspects of their design. In the remainder of the book an in-depth treatment will be given of selected subject areas involved in digital communications systems design. A listing of the remaining chapters in the book together with a short description of each is given below to provide an overview of this material. Several appendices provide additional material that is not central to the treatments given in the main part of the book, but nevertheless is important in the design of digital communication systems.

Chapter	Title	Purpose
2	Signals and Systems: Overview	Provides notation and definitions regarding signals and systems; includes an overview of the sources of distortion in communication systems and analysis tools for the treatment of them.
3	Performance Characterization of Digital Data Transmission Systems	Introduces the basic concepts of digital data transmission and explores several basic digital data modulation methods. Several practical deviations from ideal models are analyzed and methods for alleviating the degradations due to them are examined.
4	Signal Space Methods in Digital Data Transmission	Provides a unified approach to analysis of digital data modulation techniques through the methodology of signal spaces. The relatively new concept of multi-h modulation, which is a special case of continuous phase modulation, is described.
5	Generation of Coherent References	Provides an overview of the important topics of phase-locked-loop modeling and analysis and frequency synthesis. Phase-locked loops and frequency synthesizers are important subsystems in digital communication systems for generation of stable references.
6	Synchronization of Digital Communication Systems	Gives the theoretical framework for the design of clock acquisition circuits for digital communication systems using the principle of maximum likelihood. Also considered is the use of the cyclostationary property of modulation and data processes in the design of such circuits.

Chapter	Title	Purpose
7	Introduction to Spread-Spectrum Systems	Gives an overview of the concept of spread-spectrum modulation, including a description of the most important spread-spectrum modulation methods.
8	Binary Shift Register Sequences for Spread-Spectrum Systems	Provides the definition, basic properties, and generators for maximal-length binary sequences. Summarizes several other types of binary sequences important to spread-spectrum system design.
9	Code Tracking Loops	Defines and analyzes systems for the tracking of the spreading codes in direct-sequence and frequency-hop systems.
10	Initial Synchronization of the Receiver Spreading Code	Defines and analyzes systems for the initial synchronization of spreading codes in direct sequence and frequency hop systems.
11	Performance of Spread-Spectrum Systems in a Jamming Environment	Considers the effects of jamming on spread-spectrum systems with no error correction coding used.
12	Performance of Spread-Spectrum Systems with Forward Error Correction	Introduces elementary coding concepts and shows how the performance of spread-spectrum systems in jamming can be improved through the use of FEC.
13	Example Spread-Spectrum Systems	Overviews several operational systems employing spread-spectrum modulation.
Appendix A	Probability and Random Variables	Provides a summary of important concepts for the analysis of systems with random inputs.
Appendix B	Characterization of Internally Generated Noise	Provides a summary of techniques for characterizing noise generated internally to systems.
Appendix C	Communication Link Performance Calculations	Illustrates the procedure for obtaining system signal-to-noise ratio under various conditions.
Appendix D	An Overview of the Viterbi Algorithm	Provides a summary of the Viterbi algorithm, which is important to the detection of Markov sequences.
Appendix E	Gaussian Probability Function	Provides tables and approximations useful for evaluating the Gaussian probability function.
Appendixes F–H		Provide details of some of the derivations omitted in the main text regarding spread-spectrum systems.

REFERENCES

[1] B. M. OLIVER, J. R. PIERCE, and C. E. SHANNON, "Philosophy of PCM," *Proc. IRE,* Vol. 36, p. 1324, November 1948.

[2] M. R. AARON, "Digital Communications—The Silent (R)evolution?" *IEEE Commun. Soc. Mag.,* Vol. 17, pp. 16–26, January 1979. [Reprinted in V. B. Lawrence, J. L. LoCicero, and L. B. Milstein, eds., *Tutorials in Modern Communications* (Rockville, Md.: Computer Science Press, 1983).]

[3] M. P. RISTENBATT, "Alternatives in Digital Communications," *Proc. IEEE,* Vol. 61, pp. 703–721, June 1973.

[4] "2920 Analog Signal Processor Design Handbook," Intel Corporation, Santa Clara, Calif., August 1980.

[5] *IEEE Standard Dictionary of Electrical and Electronic Terms* (New York: Wiley-Interscience, 1972).

[6] C. E. SHANNON, "A Mathematical Theory of Communications," *Bell Syst. Tech. J.,* Vol. 27, pp. 379–423, July 1948, and pp. 623–656, October 1948. [Reprinted in D. Slepian, ed., *Key Papers in the Development of Information Theory* (New York: IEEE Press, 1974)].

[7] F. AMOROSO, "The Bandwidth of Digital Data Signals," *IEEE Commun. Mag.,* Vol. 18, pp. 13–24, November 1980.

[8] R. J. MCELICE, *The Theory of Information and Coding,* Vol. 3 of *Encyclopedia of Mathematics and Its Applications,* Gian-Carlo Rota, ed. (Reading, Mass.: Addison-Wesley, 1977), Chap. 6.

[9] W. DIFFIE and M. E. HELLMAN, "New Directions in Cryptography," *IEEE Trans. Inf. Theory,* November 1976.

[10] S. LIN and D. COSTELLO, *Error Control Coding: Fundamentals and Applications* (Englewood Cliffs, N.J.: Prentice-Hall, 1983).

[11] J. A. HELLER and I. M. JACOBS, "Viterbi Decoding for Satellite and Space Communications," *IEEE Trans. Commun. Technol.,* pp. 835–848, October 1971.

[12] R. COLLIN, *Foundations of Microwave Engineering* (New York: McGraw-Hill, 1966).

[13] V. K. BHARGAVA, D. HACCOUN, R. MATYAS and P. NUSPL, *Digital Communications by Satellite* (New York: Wiley, 1981).

[14] A. VAN DER ZIEL, *Noise in Measurements* (New York: Wiley-Interscience, 1976).

[15] L. J. IPPOLITO, "Radio Propagation for Space Communication Systems," *Proc. IEEE,* Vol. 69, pp. 697–727, June 1981.

[16] R. L. OLSEN et al., "The aR^b Relation in the Calculation of Rain Attenuation," *IEEE Trans. Antenn. Propag.,* March 1978.

[17] R. K. CRANE, "Prediction of Attenuation by Rain," *IEEE Trans. Commun.,* Vol. COM-28, pp. 1717–1733, September 1980.

[18] D. C. HOGG and T. S. CHU, "The Role of Rain in Satellite Communications," *Proc. IEEE,* September 1975.

[19] M. I. SKOLNIK, ed., *Radar Handbook* (New York: McGraw-Hill, 1970), Chap. 24.

[20] P. A. BELLO, "A Troposcatter Channel Model," *IEEE Trans. Commun.,* Vol. COM-17, pp. 130–137, April 1969.

[21] P. MONSEN, "Fading Channel Communications," *IEEE Commun. Mag.,* Vol. 18, pp. 16–25, January 1980.

[22] R. A. SCHOLTZ, "Frame Synchronization Techniques," *IEEE Trans. Commun.,* Vol. 28, pp. 1204–1212, August 1980.

PROBLEMS

1-1. A modulation scheme is employed wherein a signal whose amplitude can assume one of 64 different levels is transmitted each 10^{-4} s through a channel.

(a) What is the baud rate?

(b) How many binary digits are necessary to specify a particular level?

(c) What is the bit rate through the channel?

1-2. Given that a certain rate-$\frac{1}{2}$ FEC code can achieve a bit error rate (BER) of 10^{-5} for $E_b/N_0 = 6$ dB. Assume that a binary signaling scheme is employed with symbols T_s seconds in duration which require a bandwidth of $W = T_s^{-1}$ hertz. Accounting for the code rate and assuming that a BER of 10^{-5} corresponds, for all practical purposes, to error-free transmission, plot the performance of this signaling scheme on the graph of Figure 1-4 and comment on its efficiency.

1-3. (a) Set $C = R$, the information rate, in (1-1), solve for E_b/N_0 in terms of R/W, and compute several points to verify the graph of Figure 1-4.

(b) A customer desires a communication system that is capable of conveying 60 kilobits/s through a channel of bandwidth 10 kHz. The channel noise is white and Gaussian with a power spectral density of 10^{-19} W/Hz. He can achieve a received signal power of $P_T = 1$ pW. Should your company bid on the job?

(c) For the noise power density and signal power of part (b), what is the theoretical maximum rate-to-bandwidth ratio below which error-free transmission is possible?

1-4. *(Useful inequalities)*

(a) Plot $\ln u$ and $u - 1$ versus u on the same graph to show that $\ln u \leq u - 1$ with equality if, and only if, $u = 1$. Equivalently, you have shown that $\ln (1/u) \geq 1 - u$ with equality if, and only if, $u = 1$.

(b) By applying the first inequality in part (a) to each term in the following sum, show that

$$\sum_{i=1}^{q} x_i \log_2 \frac{y_i}{x_i} \leq 0$$

or that

$$\sum_{i=1}^{q} x_i \log_2 \frac{1}{x_i} \leq \sum_{i=1}^{q} x_i \log_2 \frac{1}{y_i}$$

where

$$\sum_{i=1}^{q} x_i = \sum_{i=1}^{q} y_i = 1$$

with equality if, and only if, $x_i = y_i$ for all i.

1-5. Use the inequality of Problem 1-4(a) to show that $H(X) \leq \log_2 q$ with equality if, and only if, the probabilities of all source symbols are equal.

1-6. Consider a zero-memory source with q symbols, $x_1, x_2, \ldots, x_q$ with corresponding probabilities $P_1, P_2, \ldots, P_q$. The nth extension of this source is a source whose symbols are all possible n-symbol sequences of symbols from the original source, and whose symbol probabilities are the products of the corresponding symbol probabilities from the original source. For example, one possible symbol for the third extension of a four-symbol source is $x_1 x_4 x_2$ with probability $P_1 P_2 P_4$.

Denote the original source by X and its nth extension by X^n. Show that

$$H(X^n) = nH(X)$$

where the left-hand side represents the entropy of the nth extension and $H(X)$ is the entropy of the source X.

1-7. Referring to Problem 1-6, consider a three-symbol source, $X = \{x_1, x_2, x_3\}$, with symbol probabilities $\frac{1}{2}$, $\frac{1}{4}$, and $\frac{1}{4}$.

(a) List all symbols of the second extension and their corresponding probabilities.

(b) Find the entropies of the original source and of the second extension, thus verifying that $H(X^2) = 2H(X)$.

1-8. Verify the following properties of the entropy function

$$\mathcal{H}(u) = -u \log_2 u - (1 - u) \log_2 (1 - u)$$

where $0 \le u \le 1$.

(a) $\lim_{u \to 0} \mathcal{H}(u) = \lim_{u \to 1} \mathcal{H}(u) = 0$

(b) $\max_{0 \le u \le 1} \mathcal{H}(u) = 1$, which is achieved for $u = \frac{1}{2}$.

1-9. (a) Construct a Huffman code for a five-symbol source with symbol probabilities 0.4, 0.3, 0.2, 0.06, and 0.04.

(b) Find the average length of the code found in part (a) and compare with the lower bound.

1-10. (a) Obtain Huffman codes for a binary source with symbol probabilities $\frac{3}{4}$ and $\frac{1}{4}$ and its second and third extensions.

(b) Compare the average lengths of the codes found in part (a) with the lower bound.

1-11. A binary source has symbol probabilities $\frac{2}{3}$ and $\frac{1}{3}$. Symbols are emitted at a rate of 1000 per second. The channel can accept binary symbols at a rate of 950 per second.

(a) Is it possible, by suitable encoding, to transmit the source output through the channel?

(b) If the answer to part (a) is yes, find a code that will allow it.

1-12. A source output is encoded using the rate-$\frac{1}{2}$ convolutional encoder of Figure 1-5 and transmitted through a channel that causes some of the encoded symbols to be received in error. Give the most likely source output data sequences corresponding to the following channel outputs:

(a) 0 0 1 1 1 1 0 0

(b) 1 1 1 0 1 1 1 1

(c) 1 1 1 1 0 1 1 0

1-13. If $WT_s = 1$, where W is transmission bandwidth and T_s is symbol duration, plot the performance of the various constraint-length codes of Figure 1-7 on the theoretical performance curve of Figure 1-4. (Assume $P_E = 10^{-4}$ is essentially error free transmission).

1-14. Verify (1-12) under the condition of equally probable signals.

1-15. Verify the Gray code words of Table 1-6.

1-16. Using the definitions (1-22a) and (1-22b), verify that (1-23) is obtained in either case.

1-17. Using the inequality developed in Problem 1-4(b), verify that $I(X; Y)$ is nonnegative and, in fact, is zero only if $P(x_m, y_n) = P(x_m)P(y_n)$. Why is this result reasonable?

1-18. A channel has transition matrix

$$\begin{bmatrix} 0.9 & 0.1 \\ 0.3 & 0.7 \end{bmatrix}$$

Let the input symbols have probabilities P_1 and P_2, where $P_1 + P_2 = 1$.

(a) Find the probabilities, Q_1 and Q_2, of the output symbols in terms of P_1 and P_2.

(b) Solve the equations obtained in part (a) for P_1 and P_2.

(c) Obtain expressions for $P(x_m|y_n)$ for $m, n = 1, 2$.

(d) If $P_1 = P_2 = 0.5$, calculate Q_1, Q_2, and $P(x_m|y_n)$ for $m, n = 1, 2$.

1-19. It is usually a difficult task to obtain the capacity of an arbitrary channel. One case for which it is easy is for a *uniform channel*, for which every row and every column of the channel transition probability matrix is an arbitrary permutation of the probabilities in the first row. (Note that the transition matrix of a uniform channel must have the same number of rows and columns.) Show that the capacity of the uniform channel is

$$C = \log_2 M + \sum_{n=1}^{M} p(y_n|x_m) \log_2 p(y_n|x_m)$$

where M is the number of inputs or outputs.

1-20. An M-ary symmetric channel is one with transition matrix

$$
\begin{bmatrix}
p & \dfrac{1-p}{M-1} & \dfrac{1-p}{M-1} & \cdots & \dfrac{1-p}{M-1} \\[2ex]
\dfrac{1-p}{M-1} & p & \dfrac{1-p}{M-1} & \cdots & \dfrac{1-p}{M-1} \\
\cdot & \cdot & \cdot & \cdot & \cdot \\
\cdot & \cdot & \cdot & \cdot & \cdot \\
\cdot & \cdot & \cdot & \cdot & \cdot \\
\dfrac{1-p}{M-1} & \dfrac{1-p}{M-1} & \dfrac{1-p}{M-1} & \cdots & p
\end{bmatrix}
$$

Show that its capacity is

$$C = \log_2 M - (1-p)\log_2(M-1) - \mathcal{H}(p)$$

1-21. Obtain the capacity of the symmetric *binary erasure channel*, which has two inputs ± 1 and three outputs -1, 0, and 1. The channel is defined by the transition probabilities

$$p(0|+1) = p(0|-1) = p$$

$$p(+1|+1) = p(-1|-1) = q$$

with $p + q = 1$. At the output, 0 corresponds to an erasure with p being the erasure probability.

1-22. **(a)** Consider Figure 1-14c with $\tau = 0$. A modulation scheme is employed for which the BER is $P_e = \frac{1}{2}\exp(-E_b/N_0)$, where E_b is the energy per bit and N_0 is the noise power spectral density. From Figure 1-14c, $E_b = (A_1 + A_2)^2 T$. Assume that $A_2 = 0.5A_1$. Find the constant $k = A_1^2 T/N_0$ such that $P_e = 10^{-5}$.

(b) Now let $\tau = 0.25T$ and $A_2 = 0.5A_1$. Find P_e for each bit in Figure 1-14c, using the constant k obtained in part (a). Find the average P_e over the sequence of bits shown in Figure 1-14c (i.e., the sequence 1, -1, -1, 1, -1).

(c) Obtain P_e averaged over the entire 5-bit sequence as a function of τ for $0 \le \tau \le T$. (Assume $A_2 = 0.5A_1$ and $P_E = 10^{-5}$.) *Note:* The calculations above illustrate the use of a "typical data sequence" to characterize degradation due to memory effects in digital communications systems. Normally, the computation would be carried out over a much longer sequence, perhaps with the aid of a computer or programmable calculator.

1-23. Derive (1-30), where $B(\tau)$ is defined by (1-31), from (1-29). Show that $B_{\max} = \alpha_1 + \alpha_2$ and that $B_{\min} = |\alpha_1 - \alpha_2|$.

1-24. Using the data given in Tables 1-7 and 1-8 and Equation (1-35), make up a table giving attenuations at 15, 30, and 40 GHz of a 10-km path at sea level due to atmospheric attenuation and rainfall of 10, 50, and 100 mm/h.

CHAPTER 2

Signals and Systems: Overview

2-1

REVIEW OF SIGNAL AND LINEAR SYSTEM THEORY

2-1.1 Introduction

In the study of communication systems one is, of course, interested in how signals are transmitted through systems. Several concepts in this chapter should already be familiar from earlier courses on signal and system theory. The purposes of this review are to collect in one place several definitions, theorems, and formulas that will be used throughout the book as well as to establish notation that will be convenient to use later.

2-1.2 Classification of Signals

Signals are functions of time that represent any physical quantity of interest. In a communication system context signals usually represent voltages or currents but could also represent other physical quantities, such as light waves. We will be concerned primarily with *continuous-time,* or *analog, signals* in this book, or those which can be modeled as functions of a continuous-time variable.* *Discrete-time* signals, which are specified only at discrete values of the independent variable, or time, are of secondary interest.

A second way that signals may be categorized are as *deterministic* or *random*. A brief definition of this categorization is that a deterministic signal has a completely specified value for each value of time, t, whereas the value of a random signal is not precisely known for each t, but can be specified only in terms of a probability distribution. Both types of signals will be used in this book. Probabilistic concepts are reviewed in Appendix A. This categorization is exhaustive in that a signal is either deterministic or random.

Yet a third classification often used for signals is that of *finite energy* or *finite power*. The energy of a signal $x(t)$, assumed to be defined over the entire t-axis, is given by

$$E = \lim_{T \to \infty} \int_{-T}^{T} |x(t)|^2 \, dt \tag{2-1}$$

The average power of a signal is defined as

$$P = \lim_{T \to \infty} \frac{1}{2T} \int_{-T}^{T} |x(t)|^2 \, dt \tag{2-2}$$

An energy signal is one for which $0 < E < \infty$, while a power signal is one for which $0 < P < \infty$. For a power signal, $E = \infty$, and for an energy signal, $P = 0$. This categorization is not exhaustive; one can contrive examples of signals that are neither energy nor power signals.

A final classification for signals which is sometimes convenient is *periodic* and *aperiodic*. A signal $x(t)$ is periodic with *fundamental period* T_0 if

$$x(t) = x(t + T_0) \qquad \text{all } t \tag{2-3}$$

where T_0 is the smallest constant that satisfies (2-3). Often, T_0 is referred to simply as the period. All signals not satisfying (2-3) are called *aperiodic*.

Several signals which will be used often in this book are summarized in Table 2-1.

EXERCISE 1

(a) Show that $u(t)$ and sgn (t) (Table 2-1) are power signals.
(b) Show that $\Pi(t)$ is an energy signal.
(c) Show that the signal $x(t) = t^{-1/4}$, $t \geq 1$ and 0 otherwise, is neither an energy nor a power signal.

EXERCISE 2

Using the sifting integral, Table 2-1, as the definition of the unit impulse and the formal properties of integrals, show the following:

(a) $\delta(at) = 1/|a| \, \delta(t)$, where a is a constant [as a consequence, $\delta(t) = \delta(-t)$].
(b) $\int_{-\infty}^{\infty} x(\tau)\delta(t - \tau) \, d\tau = x(t)$, where $x(t)$ is continuous at $t = \tau$.
(c) $\int_{-\infty}^{\infty} x^{(n)}(t)\delta(t) \, dt = (-1)^n x^{(n)}(0)$, where the superscript (n) denotes differentiation n times and $x^{(n)}(t)$ exists and is continuous at $t = 0$.
(d) $x(t)\delta(t - t_0) = x(t_0)\delta(t - t_0)$, where $x(t)$ is continuous at $t = t_0$.
(e) $\delta(t) = du(t)/dt$.

---

*Continuous-time signals will be used to represent digital messages. Such signals are sometimes referred to as *digital signals*.

TABLE 2-1. Definitions of Several Useful Signals

Name	Definition	Comments
1. Unit step function	$u(t) = \begin{cases} 1 & t > 0 \\ 0 & t < 0 \end{cases}$	Definition $t = 0$ is finite but otherwise immaterial.
2. Unit impulse function	$\int_{-\infty}^{\infty} \delta(t)\, x(t)\, dt = x(0)$ where $x(t)$ is continuous at $t = 0$ (sifting integral)	The unit impulse, properly defined in terms of the sifting property, can be viewed as the limit of a sequence of functions each with unity area. Formally, it has the property that $\delta(t) = du(t)/dt$.
3. Signum, or sign, function	$\text{sgn}(t) = \begin{cases} 1 & t > 0 \\ -1 & t < 0 \end{cases}$	Definition at $t = 0$ is finite but otherwise immaterial. If $\text{sgn}(0) \triangleq 0$, then $\text{sgn}(t)$ is an odd function.
4. Symmetrical unit rectangular pulse	$\Pi(t) = \begin{cases} 1 & -\frac{1}{2} \le t \le \frac{1}{2} \\ 0 & \text{otherwise} \end{cases}$	$\Pi(t) = u(t + \frac{1}{2}) - u(t - \frac{1}{2})$
5. Pulse function	$P_T(t) = \begin{cases} 1 & 0 \le t \le T \\ 0 & \text{otherwise} \end{cases}$	Sometimes used in place of $\Pi(t)$; $P_T(t) = u(t) - u(t - T)$ $= \Pi\left(\dfrac{t - T/2}{T}\right)$
6. Sinc function	$\text{sinc}(x) = \dfrac{\sin \pi x}{\pi x}$	Absolute maximum of unity occurs at the origin; an oscillatory, even function whose envelope monotonically decreases as $|t| \to \infty$. Both $\text{sinc}(x)$ and $\text{sinc}^2(x)$ have unity area.

EXERCISE 3

If $x(t)$ is periodic with period T_0, show that

$$P = \frac{1}{T_0} \int_{T_0} [x(t)]^2 \, dt$$

where the integration is over any period.

2-1.3   Fundamental Properties of Systems

A *system* is mathematically represented as a transformation of one signal (or set of signals) into another signal (or set of signals). Symbolically, such a transformation is written as

$$y(t) = \mathcal{H}[x(t)] \qquad (2\text{-}4)$$

for the case of a single input $x(t)$ and a single output $y(t)$, defined over some suitable interval of time. Without further specifications or restrictions one cannot proceed further with analysis of the effect of $\mathcal{H}(\cdot)$ on $x(t)$.

One such restriction is that of linearity. A system is linear if superposition holds. That is, if $y_1(t)$ is the response of $\mathcal{H}(\cdot)$ to $x_1(t)$ and $y_2(t)$ is its response to $x_2(t)$, its response to the arbitrary, linear combination, $a_1x_1(t) + a_2x_2(t)$, of these two inputs is

$$y(t) = a_1\mathcal{H}[x_1(t)] + a_2\mathcal{H}[x_2(t)]$$

$$= a_1y_1(t) + a_2y_2(t) \tag{2-5}$$

where a_1 and a_2 are arbitrary constants. The superposition property is of fundamental importance in linear system analysis. Of course, many communication systems are not linear, but (2-5) is nevertheless a convenient starting point in many instances.

Other properties often invoked on a system are *time invariance (fixed)* and *causality*. The former is mathematically expressed as

$$\mathcal{H}[x(t - t_0)] = y(t - t_0) \tag{2-6}$$

That is, the response to $x(t)$ delayed by t_0 is the response $y(t)$ delayed by t_0 or $y(t - t_0)$. The latter property refers to a system that does not anticipate its input or, mathematically, as one where

$$x(t) = 0 \qquad \text{for } t \leq t_0$$

implies that

$$y(t) = 0 \qquad \text{for } t \leq t_0 \tag{2-7}$$

A causal system is sometimes referred to as a *realizable system* since it states that no physically realizable system can respond before its input is applied. Somewhat surprisingly, noncausal system models are often used in communication system analysis.

The mathematical representation of a linear system is conveniently accomplished in terms of *superposition integral*, which is

$$y(t) = \int_{-\infty}^{\infty} \tilde{h}(t, \lambda)x(\lambda) \, d\lambda \tag{2-8}$$

where $\tilde{h}(t, \lambda)$ is the response to a unit impulse applied at time $t = \lambda$. If the system is also time invariant, (2-8) simplifies to

$$y(t) = \int_{-\infty}^{\infty} h(t - \lambda)x(\lambda) \, d\lambda \triangleq h(t) * x(t) \tag{2-9a}$$

$$= \int_{-\infty}^{\infty} x(t - \lambda)h(\lambda) \, d\lambda \triangleq x(t) * h(t) \tag{2-9b}$$

where $h(t)$ is the response of the system to a unit impulse applied at time $t = 0$. The name *superposition* results by considering the input signal, $x(t)$, as being resolved into a sum of delayed impulses weighted by the signal values at these instants and obtaining the response of the system to this resolution of the input signal by invoking the superposition property (2-5). The resulting response is the integral (2-9a), with (2-9b) following by a simple change of variables. The operation represented by (2-9) is called *convolution*.

A second way to represent a restricted class of systems is in terms of an ordinary, integrodifferential equation relating output to input. If the system is linear and fixed, this differential equation is linear with constant coefficients. Systems representable by an ordinary differential equation are often referred to as *lumped* because the physical size of their components is small compared with the wavelength of the signals processed by them. If their physical sizes are large compared with the wavelength, such as with a transmission line, a partial differential equation is necessary to describe them.

2-1.4 Complex Exponentials as Eigenfunctions for a Fixed, Linear System; Transfer Function

The mathematical convenience of representing sinusoidal signals as the real (or imaginary) part of a complex exponential signal is well appreciated from circuit analysis courses. This idea will be extended later in regard to narrowband signals. The reason complex exponentials are convenient for linear system analysis will be considered briefly here by making use of the superposition integral (2-9b). Assume that the input to a fixed linear system is

$$x(t) = \exp{(j\omega t)} \tag{2-10}$$

[One could, of course, consider $A \exp{[j(\omega t + \varphi)]}$, which involves only an additional scale factor $A \exp{(j\varphi)}$.] When t is replaced by $t - \lambda$ and the result substituted into (2-9b), it follows that

$$y(t) = \int_{-\infty}^{\infty} h(\lambda)e^{j\omega(t-\lambda)} \, d\lambda$$

$$= \int_{-\infty}^{\infty} h(\lambda)e^{-j\omega\lambda} \, d\lambda \, e^{j\omega t}$$

$$= \tilde{H}(\omega)e^{j\omega t} \tag{2-11}$$

where

$$\tilde{H}(\omega) = \int_{-\infty}^{\infty} h(\lambda)e^{-j\omega\lambda} \, d\lambda \tag{2-12}$$

The complex function of frequency, $\tilde{H}(\omega)$, referred to as the *transfer function* of the system, can be expressed as a function of ω in rad/s or f in hertz. Recalling that $\omega = 2\pi f$, (2-12) can be written

$$\tilde{H}(\omega) = \tilde{H}(2\pi f) = H(f) = |H(f)|e^{j\underline{/H(f)}} \tag{2-13}$$

where $|H(f)|$ is called the *amplitude response* and $\underline{/H(f)}$ is referred to as the *phase response* of the systems. Usually, $\tilde{H}(\omega)$ is more convenient when analyzing circuits and $H(f)$ is more convenient when discussing filtering operations. The transfer function is the *Fourier transform* of the impulse response, which is defined by (2-12). Fourier transforms are discussed in more detail later.

Since the system output is the same form as the input when the input is a complex exponential, $\exp{(j\omega t)}$, this particular input is called an *eigenfunction* of the system. The fact that the output is of the same form as the input, together with the superposition property of a linear system, means that the *response of the system to any input which can be resolved into a summation of complex exponentials* can easily be found for a fixed linear system. Next discussed, therefore, will be a review of Fourier series and Fourier transform representations for signals. It will be approached from the standpoint of generalized vector spaces, a concept that will be developed further in regard to the detection of signals in noise.

2-1.5 Orthogonal Function Series

Consider the representation of a finite-energy signal, $x(t)$, defined on a T-second interval $(t_0, t_0 + T)$ in terms of a set of preselected time functions, $\varphi_1(t)$, $\varphi_2(t), \ldots, \varphi_N(t)$. It is convenient to choose these functions with properties analogous to the mutually perpendicular, unit vectors of three-dimensional vector space. The mutually perpendicular property, referred to as *orthogonality*, is expressed as

$$\int_{t_0}^{t_0+T} \varphi_m(t)\varphi_n^*(t)\, dt = 0 \qquad m \neq n \tag{2-14}$$

where the conjugate suggests that complex-valued $\varphi_n(t)$'s may be convenient in some cases. Further assume that the $\varphi_n(t)$'s have been chosen such that

$$\int_{t_0}^{t_0+T} |\varphi_n(t)|^2\, dt = 1 \tag{2-15}$$

In view of (2-15), the $\varphi_n(t)$'s are said to be *normalized*. This normalization will simplify future equations.

It is desired to approximate $x(t)$ as accurately as possible by a series of the form

$$y(t) = \sum_{n=1}^{N} d_n \varphi_n(t) \tag{2-16}$$

Because of the orthogonality of the $\varphi_n(t)$'s, one may think of the sum (2-16) representing a point in an N-dimensional, generalized vector space with coordinates $(d_1, d_2, \ldots, d_N)$.

The d_n's are constants to be chosen such that $y(t)$ represents $x(t)$ as closely as possible according to some criterion. It is convenient to measure the error in the integral-square sense, which is defined as*

$$\text{integral-square error} = \epsilon_N = \int_{t_0}^{t_0+T} |x(t) - y(t)|^2\, dt \tag{2-17}$$

To find $d_1, d_2, \ldots, d_N$ such that ϵ_N as expressed by (2-17) is a minimum, (2-16) is substituted into (2-17) to give

$$\epsilon_N = \int_T \left[x(t) - \sum_{n=1}^{N} d_n \varphi_n(t) \right] \left[x^*(t) - \sum_{n=1}^{N} d_n^* \varphi_n^*(t) \right] dt \tag{2-18}$$

where

$$\int_T (\cdot)\, dt \triangleq \int_{t_0}^{t_0+T} (\cdot)\, dt$$

This can be expanded to yield

$$\epsilon_N = \int_T |x(t)|^2\, dt$$

$$- \sum_{n=1}^{N} d_n^* \int_T x(t)\varphi_n^*(t)\, dt - d_n \int_T x^*(t)\varphi_n(t)\, dt$$

$$+ \sum_{n=1}^{N} |d_n|^2 \tag{2-19}$$

---

*If the d_n's are chosen to minimize the integral-square error between $x(t)$ and $y(t)$, the right-hand side of (2-16) is a generalized truncated Fourier series.

which was obtained by making use of (2-14) and (2-16) after interchanging the orders of summation and integration.

It is convenient to add and subtract the quantity

$$\sum_{n=1}^{N} \left| \int_{T} x(t)\varphi_n^*(t) \, dt \right|^2$$

to (2-19) which, after rearrangement of terms, yields

$$\epsilon_N = \int_{T} |x(t)|^2 \, dt - \sum_{n=1}^{N} \left| \int_{T} x(t)\varphi_n^*(t) \, dt \right|^2$$

$$+ \sum_{n=1}^{N} \left| d_n - \int_{T} x(t)\varphi_n^*(t) \, dt \right|^2 \qquad (2\text{-}20)$$

To show the equivalence of (2-20) and (2-19), it is easiest to work backward from (2-20).

Now the first two terms on the right-hand side of (2-20) are independent of the coefficients d_n. The last summation of terms on the right-hand side is nonnegative and is added to the first two terms. Therefore, to minimize ϵ_N through choice of the d_n's, the best that can be done is make each term of the last sum zero. That is, choose the nth coefficient, d_n, such that

$$d_n = \int_{t_0}^{t_0+T} x(t)\varphi_n^*(t) \, dt \qquad n = 1, 2, \ldots, N \qquad (2\text{-}21)$$

This choice for $d_1, d_2, \ldots, d_N$ minimizes the integral-square error, ϵ_N. The resulting coefficients are called the *generalized Fourier coefficients*.

The minimum value for ϵ_N is

$$\epsilon_{N,\min} = \int_{t_0}^{t_0+T} |x(t)|^2 \, dt - \sum_{n=1}^{N} |d_n|^2 \qquad (2\text{-}22)$$

It is natural to inquire as to the possibility of $\lim_{N\to\infty} \epsilon_{N,\min}$ being zero. For special choices of the set of functions $\varphi_1(t), \varphi_2(t), \ldots, \varphi_N(t), \ldots$ referred to as *complete sets* in the space of all integrable-square functions, it will be true that

$$\lim_{N\to\infty} \epsilon_{N,\min} = 0 \qquad (2\text{-}23)$$

for any signal that is integrable square; that is, for any signal for which

$$\int_{t_0}^{t_0+T} |x(t)|^2 \, dt < \infty$$

In the sense that the integral-square error is zero, one may then write

$$x(t) = \sum_{n=1}^{\infty} d_n \varphi_n(t) \qquad (2\text{-}24)$$

For a complete set of orthogonormal functions, $\varphi_1(t), \varphi_2(t), \ldots$, it follows from (2-22) that

$$\int_{t_0}^{t_0+T} |x(t)|^2 \, dt = \sum_{n=1}^{\infty} |d_n|^2 \qquad (2\text{-}25)$$

This formula, referred to as *Parseval's theorem*, states that the energy in $x(t)$ can be obtained by summing the energies in each $\varphi_n(t)$.

EXERCISE 4

Show that the energy of $A\varphi_n(t)$ is A^2, where A is a constant.

EXERCISE 5

Given the functions $\varphi_1(t) = -\Pi(t) + 2\Pi(2t)$ and $\varphi_2(t) = \Pi(2t + \frac{1}{2}) - \Pi(2t - \frac{1}{2})$.

(a) Sketch φ_1 and φ_2. Show mathematically that they are orthogonal and normalized.

(b) Expand $\cos 2\pi t$ in terms of φ_1 and φ_2 over the interval $|t| \leq \frac{1}{2}$. What is the integral-square error?

(c) Same questions as part (b) but consider $\sin 2\pi t$.

(d) Explain why $\cos 2\pi t$ is approximated in terms of $\varphi_1(t)$ only, and $\sin 2\pi t$ is approximated in terms of $\varphi_2(t)$ only.

2-1.6 Complex Exponential Fourier Series

Recalling that complex exponentials are eigenfunctions for fixed, linear systems, one is naturally led to a consideration of their use for the orthogonal functions in the orthogonal function series expansion of a signal. Let

$$\varphi_n(t) = e^{jn\omega_0 t} \qquad n = 0, \pm 1, \dots \tag{2-26}$$

where the interval under consideration is $(t_0, t_0 + T_0)$ and

$$\omega_0 = 2\pi f_0 = \frac{2\pi}{T_0}$$

These functions are not normalized but have the integral-square value T_0. Instead of normalizing them, a factor of $1/T_0$ will be inserted in (2-21). Thus the orthogonal function series which results from using the complex exponentials (2-26) in (2-24) is

$$x(t) = \sum_{n=-\infty}^{\infty} X_n e^{jn\omega_0 t} \qquad t_0 \leq t \leq t_0 + T_0 \tag{2-27}$$

where

$$X_n = \frac{1}{T_0} \int_{T_0} x(t) e^{-jn\omega_0 t} \, dt \tag{2-28}$$

Equation (2-27) is referred to as the *complex exponential Fourier series* of $x(t)$. Since $\exp(j\omega_0 t)$ is periodic with period, T_0, so is the sum on the right-hand side of (2-27). Thus, if $x(t)$ is not periodic, (2-19) represents it only in the interval $(t_0, t_0 + T_0)$. If $x(t)$ is periodic with period, T_0, (2-27) represents it for *all* t.

Using Euler's theorem to write

$$\exp(jn\omega_0 t) = \cos n\omega_0 t + j \sin n\omega_0 t$$

in (2-27), a trigonometric sine–cosine form of the Fourier series can be obtained, which is

$$x(t) = a_0 + \sum_{n=1}^{\infty} a_n \cos n\omega_0 t + \sum_{n=1}^{\infty} b_n \sin n\omega_0 t \qquad (2\text{-}29)$$

where it can be shown that

$$a_0 = X_0 \qquad (2\text{-}30a)$$

$$a_n = X_n + X_{-n} \qquad (2\text{-}30b)$$

$$b_n = j(X_n - X_{-n}) \qquad (2\text{-}30c)$$

Yet a third form of the Fourier series can be obtained by using the identity

$$a_n \cos n\omega_0 t + b_n \sin n\omega_0 t = A_n \cos(n\omega_0 t + \theta_n) \qquad (2\text{-}31)$$

with

$$A_n = \sqrt{a_n^2 + b_n^2} \qquad (2\text{-}32a)$$

and

$$\theta_n = -\tan^{-1} \frac{b_n}{a_n} \qquad (2\text{-}32b)$$

in (2-29). This results in the Fourier cosine series given by

$$x(t) = A_0 + \sum_{n=1}^{\infty} A_n \cos(n\omega_0 t + \theta_n) \qquad (2\text{-}33)$$

where $A_0 = a_0$.

Table 2-2 summarizes these three forms of the Fourier series together with the implications of symmetry properties of the waveform on the series coefficients, and Table 2-3 gives results for the complex exponential series of several commonly occurring signals.

Recalling the derivation of the orthogonal function series, it is seen that partial sums of exponential (and trigonometric) Fourier series minimize the integral-square error between the series and the signal under consideration.

Parseval's theorem (2-25) specializes for a complex exponential Fourier series to

$$\frac{1}{T_0} \int_{T_0} |x(t)|^2 \, dt = \sum_{n=-\infty}^{\infty} |X_n|^2 \qquad (2\text{-}34)$$

which states that the power in a periodic signal $x(t)$ is the sum of the powers in its phasor components.

EXERCISE 6

Show that if $x(t)$ is real, then $X_n = X_{-n}^*$ for its Fourier series. Therefore, if $X_n = A_n \exp(j\theta_n)$, it follows that $A_n = A_{-n}$ and $\theta_n = -\theta_{-n}$ for the Fourier coefficients of a real signal.

TABLE 2-2. Summary of Fourier Series Properties†

Series	Coefficients‡	Symmetry Properties
1. Trigonometric sine–cosine $x(t) = a_0 + \sum_{n=1}^{\infty}(a_n \cos n\omega_0 t + b_n \sin n\varphi_0 t)$ $\omega_0 = 2\pi/T_0 = 2\pi f_0$	$a_0 = \dfrac{1}{T_0}\displaystyle\int_{T_0} x(t)\, dt$ $a_n = \dfrac{2}{T_0}\displaystyle\int_{T_0} x(t)\cos n\omega_0 t\, dt$ $b_n = \dfrac{2}{T_0}\displaystyle\int_{T_0} x(t)\sin n\omega_0 t\, dt$	a_0 = average value of $x(t)$ $a_n = 0$ for $x(t)$ odd, $b_n = 0$ for $x(t)$ even $a_n, b_n = 0$, n even, for $x(t)$ odd, half-wave symmetrical
2. Trigonometric cosine $x(t) = A_0 + \sum_{n=1}^{\infty} A_n \cos(n\omega_0 t + \theta_n)$	$A_0 = a_0; \quad A_n = \sqrt{a_n^2 + b_n^2}$ $\theta_n = -\tan^{-1}\dfrac{b_n}{a_n}$	A_0 = average value of $x(t)$ $A_n = 0$, n even, for $x(t)$ odd, half-wave symmetrical
3. Complex exponential $x(t) = \sum_{n=-\infty}^{\infty} X_n e^{jn\omega_0 f}$	$X_n = \dfrac{1}{T_0}\displaystyle\int_{T_0} x(t)e^{-jn\omega_0 f}\, dt$ $X_n = \begin{cases} \frac{1}{2}(a_n - jb_n) & n > 0 \\ \frac{1}{2}(a_{-n} + jb_{-n}) & n < 0 \end{cases}$ $X_n = X_{-n}^{*}$ for $x(t)$ real	X_0 = average value of $x(t)$ X_n real for $x(t)$ even X_n imaginary for $x(t)$ odd $X_n = 0$, n even, for $x(t)$ odd, half-wave symmetrical

† $x(t)$ even means that $x(t) = x(-t)$; $x(t)$ odd means that $x(t) = -x(-t)$; $x(t)$ odd half-wave symmetrical means that $x(t) = -x(t \pm T_0/2)$.

‡ $\int_{T_0}(\cdot)\, dt$ means integration over any period T_0 of $x(t)$.

TABLE 2-3. Coefficients for the Complex Exponential Fourier Series of Several Signals

1. Half-rectified sine wave

$$X_n = \begin{cases} \dfrac{A}{\pi(1-n^2)}, & n = 0, \pm 2, \pm 4, \ldots \\ 0, & n \text{ odd and} \neq 1 \\ -\dfrac{1}{4}jnA, & n = \pm 1 \end{cases}$$

2. Full-rectified sine wave

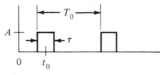

$$X_n = \begin{cases} \dfrac{2A}{\pi(1-n^2)}, & n \text{ even} \\ 0, & n \text{ odd} \end{cases}$$

3. Pulse-train signal

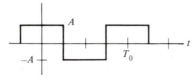

$$X_n = \begin{cases} \dfrac{A\tau}{T_0} \operatorname{sinc} nf_0\tau\, e^{-j2\pi nf_0 t_0}, & f_0 = T_0^{-1} \end{cases}$$

4. Square wave

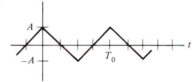

$$X_n = \begin{cases} \dfrac{2A}{|n|\pi}, & n = \pm 1, \pm 5, \ldots \\ \dfrac{-2A}{|n|\pi}, & n = \pm 3, \pm 7, \ldots \\ 0, & n \text{ even} \end{cases}$$

5. Triangular wave

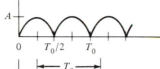

$$X_n = \begin{cases} \dfrac{4A}{\pi^2 n^2}, & n \text{ odd} \\ 0, & n \text{ even} \end{cases}$$

2-1.7 Fourier Transform

The Fourier transform of a signal, $x(t)$, is defined as

$$X(f) = \mathcal{F}[x(t)] = \int_{-\infty}^{\infty} x(t) e^{-j2\pi ft}\, dt \qquad (2\text{-}35)$$

and the inverse Fourier transform of $X(f)$ is

$$\mathcal{F}^{-1}[X(f)] = \int_{-\infty}^{\infty} X(f) e^{j2\pi tf}\, df \qquad (2\text{-}36)$$

Except for the difference in sign in the exponent, these are identical relationships. The notation $x(t) \leftrightarrow X(f)$ is often used to denote a Fourier transform pair. Since $|\exp(-j2\pi ft)| = 1$, it follows that (2-35) exists if (1) $x(t)$ is absolutely integrable, and (2) has finite discontinuities, if any. At a discontinuity of $x(t)$, the inversion integral (2-36) converges to $\frac{1}{2} x(t_0^+) + \frac{1}{2} x(t_0^-)$ where $x(t_0)$ is discontinuous; otherwise, it converges to $x(t)$.

The conditions for existence of the Fourier transform of $x(t)$, being sufficient conditions, mean that there are signals that violate either one or both conditions and yet possess a Fourier transform. An example is sinc (t), which is not absolutely integrable. We may include signals that do not have Fourier transforms in the ordinary sense by generalization to transforms in the limit. For example, to obtain the Fourier transform of a constant, we consider $x(t) = A\Pi(t/\tau)$ and let $\tau \to \infty$ after obtaining its Fourier transform.

EXAMPLE 2-1

The Fourier transform of $x(t) = A\Pi(t/\tau)$ is

$$X(f) = \int_{-\infty}^{\infty} A\Pi\left(\frac{t}{\tau}\right) e^{-j2\pi ft} \, dt$$

$$= A \int_{-\tau/2}^{\tau/2} e^{-j2\pi ft} \, dt$$

$$= A\tau \, \text{sinc} \, (f\tau)$$

after some simplification. Letting $\tau \to \infty$, it is not apparent what the result is.

∎

TABLE 2-4.   Fourier Transform Theorems

Name of Theorem	Signal	Transform
1. Superposition (a_1 and a_2 arbitrary constants)	$a_1 x_1(t) + a_2 x_2(t)$	$a_1 X_1(f) + a_2 X_2(f)$
2. Time delay	$x(t - t_0)$	$X(f)e^{-j2\pi ft_0}$
3a. Scale change	$x(at)$	$\lvert a \rvert^{-1} X\left(\dfrac{f}{a}\right)$
3b. Time reversal[a]	$x(-t)$	$X(-f) = X^*(f)$
4. Duality	$X(t)$	$x(-f)$
5a. Frequency translation	$x(t)e^{j\omega_0 t}$	$X(f - f_0)$
5b. Modulation	$x(t) \cos \omega_0 t$	$\frac{1}{2}X(f - f_0) + \frac{1}{2}X(f + f_0)$
6. Differentiation	$\dfrac{d^n x(t)}{dt^n}$	$(j2\pi f)^n X(f)$
7. Integration	$\displaystyle\int_{-\infty}^{t} x(t') \, dt'$	$(j2\pi f)^{-1} X(f) + \frac{1}{2}X(0)\delta(f)$
8. Convolution	$\displaystyle\int_{-\infty}^{\infty} x_1(t - t')x_2(t') \, dt'$ $= \displaystyle\int_{-\infty}^{\infty} x_1(t')x_2(t - t') \, dt'$	$X_1(f)X_2(f)$
9. Multiplication	$x_1(t)x_2(t)$	$\displaystyle\int_{-\infty}^{\infty} X_1(f - f')X_2(f') \, df'$ $= \displaystyle\int_{-\infty}^{\infty} X_1(f')X_2(f - f') \, df'$

[a]$\omega_0 = 2\pi f_0$; $x(t)$ is assumed to be real in 3b.

A plot reveals a function with a large central lobe which becomes narrower as $\tau \to \infty$. Since

$$\int_{-\infty}^{\infty} \tau \text{ sinc } (f\tau)\, df = \int_{-\infty}^{\infty} \frac{\sin \pi u}{\pi u}\, du = 1$$

it is deduced that $\lim_{\tau \to \infty} A\tau \text{ sinc } (f\tau) = A\delta(f)$.

The derivation of a catalog of Fourier transforms is facilitated by means of Fourier transform theorems. Several useful theorems are listed in Table 2-4. Their proofs may be found in most books on system theory. Table 2-5 summarizes several

TABLE 2-5. Fourier Transform Pairs

Pair Number	$x(t)$	$X(f)$	Comments on Derivation
1.	$\Pi\left(\dfrac{t}{\tau}\right)$	$\tau \text{ sinc } (\tau f)$	Direct evaluation
2.	$2W \text{ sinc}(2Wt)$	$\Pi\left(\dfrac{f}{2W}\right)$	Duality with pair 1
3.	$\Lambda\left(\dfrac{t}{\tau}\right)$	$\tau \text{ sinc}^2 (\tau f)$	Convolution with pair 1
4.	$\exp (-\alpha t)u(t),\ \alpha > 0$	$\dfrac{1}{\alpha + j2\pi f}$	Direct evaluation
5.	$t \exp (-\alpha t)u(t),\quad \alpha > 0$	$\dfrac{1}{(\alpha + j2\pi f)^2}$	Differentiation of pair 4 with respect to α
6.	$\exp (-\alpha\|t\|),\quad \alpha > 0$	$\dfrac{2\alpha}{\alpha^2 + (2\pi f)^2}$	Direct evaluation
7.	$\delta(t)$	1	Sifting property of $\delta(t)$
8.	1	$\delta(f)$	Duality with pair 7
9.	$\delta(t - t_0)$	$\exp (-j2\pi f t_0)$	Shift and pair 7
10.	$\exp (j2\pi f_0 t)$	$\delta(f - f_0)$	Duality with pair 9
11.	$\cos 2\pi f_0 t$	$\frac{1}{2}\delta(f - f_0) + \frac{1}{2}\delta(f + f_0)$	Exponential representation of cos and sin and pair 10
12.	$\sin 2\pi f_0 t$	$\dfrac{1}{2j}\delta(f - f_0) - \dfrac{1}{2j}\delta(f + f_0)$	
13.	$u(t)$	$(j2\pi f)^{-1} + \frac{1}{2}\delta(f)$	Integration and pair 7
14.	$\text{sgn } (t)$	$(j\pi f)^{-1}$	Pair 8 and pair 13 with superposition
15.	$\dfrac{1}{\pi t}$	$-j \text{ sgn } (f)$	Duality with pair 14
16.	$\hat{x}(t) = \dfrac{1}{\pi}\displaystyle\int_{-\infty}^{\infty} \dfrac{x(\lambda)}{t - \lambda}\, d\lambda$	$-j \text{ sgn } (f)X(f)$	Convolution and pair 15
17.	$\displaystyle\sum_{m=-\infty}^{\infty} \delta(t - mT_s)$	$f_s \displaystyle\sum_{m=-\infty}^{\infty} \delta(f - mf_s),$ $f_s = T_s^{-1}$	Example 2-2

useful Fourier transform pairs. Suggestions concerning their derivations are also given. As an example, consider the derivation of pair 17.

EXAMPLE 2-2

The Fourier transform of the signal

$$y_s(t) = \sum_{m=-\infty}^{\infty} \delta(t - mT_0)$$

is convenient from the standpoint of sampling theory, as well as in obtaining Fourier transforms of periodic signals. Since it is a periodic signal, it follows that it can be represented, in a formal sense, by a Fourier series. Thus we write

$$y_s(t) = \sum_{n=-\infty}^{\infty} Y_n e^{j2\pi n f_0 t} \qquad f_0 = \frac{1}{T_0}$$

The Fourier coefficients are given by

$$Y_n = \frac{1}{T_0} \int_{-T_0/2}^{T_0/2} \delta(t) e^{-j2\pi n f_0 t} \, dt = f_0$$

where the sifting property of the unit impulse function has been used. Therefore, $y_s(t)$ can be represented by the Fourier series

$$y_s(t) = f_0 \sum_{n=-\infty}^{\infty} e^{j2\pi n f_0 t}$$

Using the Fourier transform pair $e^{j2\pi f_0 t} \leftrightarrow \delta(f - f_0)$, the Fourier transform of this Fourier series may be taken term by term to obtain

$$Y_s(f) = f_0 \sum_{n=-\infty}^{\infty} \mathcal{F}(e^{j2\pi n f_0 t})$$

$$= f_0 \sum_{n=-\infty}^{\infty} \delta(f - nf_0)$$

Summarizing, it has been shown that

$$\sum_{m=-\infty}^{\infty} \delta(t - mT_0) \leftrightarrow f_0 \sum_{n=-\infty}^{\infty} \delta(f - nf_0) \qquad \blacksquare$$

Application of pair 17 of Table 2-5 to Fourier transformation of a periodic signal follows by observing that $\delta(t - t_0) * p(t) = p(t - t_0)$, where the asterisk denotes convolution. If $p(t)$ represents one period of a periodic signal, it follows that the signal can be written as $y_s(t) * p(t)$, which has the Fourier transform

$$y_s(t) * p(t) \leftrightarrow Y_s(f)P(f) \tag{2-37}$$

where $P(f) = \mathcal{F}[p(t)]$. That is, the Fourier transform of the periodic signal†

$$x(t) = \sum_{m=-\infty}^{\infty} p(t - mT_0) \tag{2-38}$$

†The only condition on $p(t)$ is that its Fourier transform exist.

is

$$X(f) = f_0 \sum_{n=-\infty}^{\infty} \delta(f - nf_0)P(f)$$

$$= f_0 \sum_{n=-\infty}^{\infty} P(nf_0)\delta(f - nf_0) \tag{2-39}$$

which follows since $\delta(f - f_0)P(f) = \delta(f - f_0)P(f_0)$.

2-1.8 Signal Spectra

The Fourier transform $X(f)$ of a signal $x(t)$ is, in general, complex. To represent it graphically, it is necessary to make two plots. These usually are its magnitude $|X(f)|$, referred to as the *amplitude spectrum*, and its phase $\underline{/X(f)} \triangleq \theta(f)$, referred to as the *phase spectrum*. With the result (2-39) for representation of periodic signals by the Fourier transform, we can conveniently make spectral plots for periodic or aperiodic signals. The former will consist of impulses at intervals of the fundamental frequency, f_0, while for aperiodic signals the plots will be continuous. Examples of each case are provided below.

EXAMPLE 2-3
Consider the signal

$$x_1(t) = A \exp(-\alpha t)u(t)$$

From pair 4, Table 2-5, its Fourier transform is

$$X_1(f) = \frac{1}{\alpha + j2\pi f} = \frac{1}{\sqrt{\alpha^2 + (2\pi f)^2}} \exp\left(-j \tan^{-1} \frac{2\pi f}{\alpha}\right)$$

The signal and its amplitude and phase spectra are plotted in Figure 2-1, where it is seen that $|X(f)|$ is an even function of frequency and $\underline{/X(f)}$ is odd. ∎

EXAMPLE 2-4
Consider the periodic rectangular pulse signal

$$x_2(t) = \sum_{n=-\infty}^{\infty} \Pi\left(\frac{t - nT_0}{\tau}\right)$$

shown in Figure 2-2a. From pair 1, Table 2-5, it follows that

$$\Pi\left(\frac{t}{\tau}\right) \leftrightarrow \tau \operatorname{sinc}(f\tau)$$

and from (2-39) its Fourier transform is obtained as

$$X_2(f) = f_0\tau \sum_{n=-\infty}^{\infty} \operatorname{sinc}(nf_0\tau)\delta(f - nf_0)$$

Its amplitude and phase spectra are plotted in Figure 2-2 for various ratios of τ to T_0. Note that its phase spectrum must be odd and phase shift of $\pm\pi$ rad is required whenever $\operatorname{sinc}(nf_0\tau) < 0$. ∎

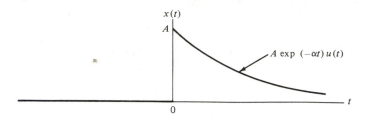

(a)   Signal

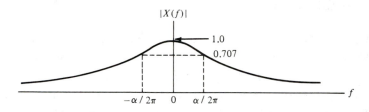

(b)   Amplitude spectrum

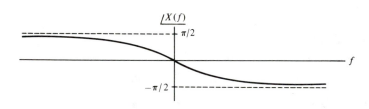

(c) Phase spectrum

FIGURE 2-1.   Signal and spectra for Example 2-3.

2-1.9   Energy Relationships

Parseval's theorem for Fourier transforms is

$$\int_{-\infty}^{\infty} x(t)y^*(t)\,dt = \int_{-\infty}^{\infty} X(f)Y^*(f)\,df \tag{2-40}$$

where $x(t) \leftrightarrow X(f)$ and $y(t) \leftrightarrow Y(f)$, and $x(t)$ and $y(t)$ are assumed to have finite energy. It can be proved by representing $y^*(t)$ in terms of its inverse Fourier transform, which is

$$y^*(t) = \left[\int_{-\infty}^{\infty} Y(f)e^{j2\pi ft}\,df\right]^*$$

$$= \int_{-\infty}^{\infty} Y^*(f)e^{-j2\pi ft}\,df$$

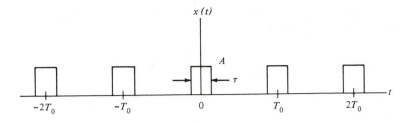

(a) Pulse-train signal

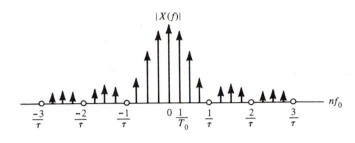

(b) Amplitude and phase spectra for $T_0 = 1$; $\tau = 0.25$; $T_0/\tau = 4$

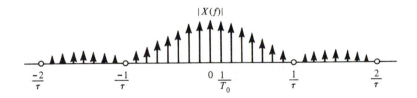

(c) Amplitude spectrum for $T_0 = 1$; $\tau = 0.125$; $T_0/\tau = 8$

FIGURE 2-2. Signal and spectra for Example 2-4.

Substitution of this on the left-hand side of (2-40) yields

$$\int_{-\infty}^{\infty} x(t)\left[\int_{-\infty}^{\infty} Y^*(f)e^{-j2\pi ft}\, df\right] dt$$

$$= \int_{-\infty}^{\infty} Y^*(f)\left[\int_{-\infty}^{\infty} x(t)e^{-j2\pi ft}\, dt\right] df$$

$$= \int_{-\infty}^{\infty} Y^*(f)X(f)\, df$$

where the second step follows by reversal of the orders of integration. A special case of (2-40) is obtained by letting $y(t) = x(t)$, which yields

$$\int_{-\infty}^{\infty} |x(t)|^2 \, dt = \int_{-\infty}^{\infty} |X(f)|^2 \, df \qquad (2\text{-}41)$$

Thus the energy in a signal may be found in the time domain or the frequency domain.

The function

$$G(f) = |X(f)|^2 \qquad (2\text{-}42)$$

is referred to as the *energy spectrum* or *energy spectral density* of $x(t)$. Equation (2-41) is convenient for analyzing the distribution of energy of a signal with frequency. For example, the fraction of total energy above a certain frequency W, referred to as out-of-band energy, contained in a lowpass signal $x(t)$ is

$$\Delta E = \frac{\displaystyle\int_{W}^{\infty} |X(f)|^2 \, df}{\displaystyle\int_{0}^{\infty} |X(f)|^2 \, df} \qquad (2\text{-}43)$$

where $|X(f)| = |X(-f)|$ has been used.

EXAMPLE 2-5

Compare the fraction of total energy above W hertz for the three signals

$$x_1(t) = \Pi\left(\frac{t}{2T}\right)$$

$$x_2(t) = \cos\left(\frac{\pi t}{2T}\right)\Pi\left(\frac{t}{2T}\right)$$

$$x_3(t) = \frac{1}{2}\left[1 + \cos\left(\frac{\pi t}{T}\right)\right]\Pi\left(\frac{t}{2T}\right)$$

Solution: The Fourier transforms of these three signals can be shown to be

$$X_1(f) = 2T \text{ sinc } (2Tf)$$

$$X_2(f) = \frac{4T}{\pi}\frac{\cos 2\pi Tf}{1 - 16T^2 f^2}$$

and

$$X_3(f) = T \text{ sinc } (2Tf) + \frac{T}{2} \text{ sinc } (2Tf - 1) + \frac{T}{2} \text{ sinc } (2Tf + 1)$$

respectively. These signals and their spectra are compared in Figure 2-3. The percent of total energy in each above W hertz is compared in Figure 2-4 as a function of WT. Note that the smoother the waveform, the faster the out-of-band energy decreases with frequency. ■

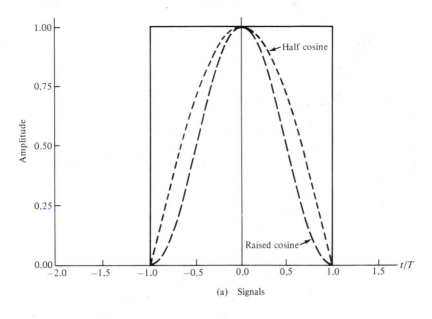

(a) Signals

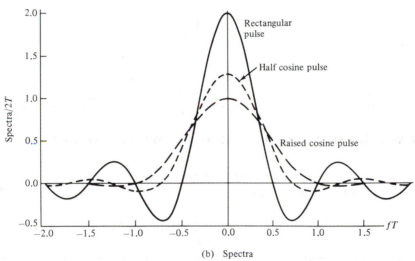

(b) Spectra

FIGURE 2-3. **Signals and spectra for Example 2-5.**

2-1.10 System Analysis

Recalling the superposition integrals (2-9), which related the output to input of a fixed, linear system through the impulse response, theorem 8 of Table 2-4 can be applied to obtain

$$Y(f) = H(f)X(f) \tag{2-44}$$

where $X(f)$, $Y(f)$, and $H(f)$ are the Fourier transforms of $x(t)$, $y(t)$, and $h(t)$, respectively. Comparison of the definition of the Fourier transform integral (2-35) with (2-12) shows that $H(f) = \mathscr{F}[h(t)]$ is the same as the system transfer function

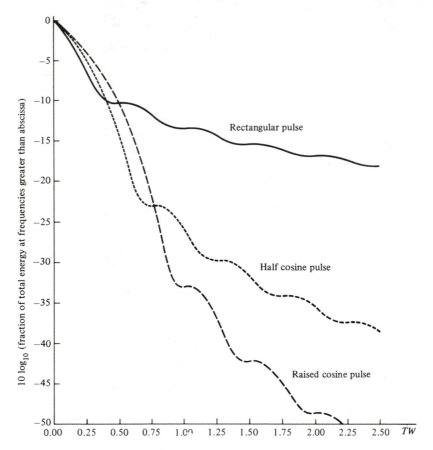

FIGURE 2-4. Out-of-band energy for rectangular, half-cosine and raised-cosine pulses.

considered earlier. Indeed, from the inverse Fourier transform integral for $y(t)$, the output signal can be written

$$y(t) = \int_{-\infty}^{\infty} H(f)X(f)e^{j2\pi ft} \, df \qquad (2\text{-}45)$$

which can be thought of as a superposition of elemental responses of the system of the form $H(f)X(f) \, df$ to the complex exponential inputs $\exp(j2\pi ft)$ as f varies from $-\infty$ to $+\infty$.

By rewriting (2-44) as

$$H(f) = \frac{Y(f)}{X(f)} \qquad (2\text{-}46)$$

yet a third way of obtaining the transfer function of a system can be identified. The three methods are:

1. $\tilde{H}(\omega) = \tilde{H}(2\pi f)$ is the multiplier of the resulting response, $\tilde{H}(\omega)e^{j\omega t}$, when the system input is $e^{j\omega t}$.
2. $H(f)$ is the Fourier transform of $h(t)$.
3. $H(f)$ is the ratio of the Fourier transforms of the output and input to the system.

In addition, if the system is lumped, ac sinusoidal steady-state analysis can be used. These methods are summarized in Figure 2-5 for a simple RC circuit.

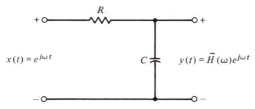

(a) Obtaining the multiplier of the response
for an input of $e^{j\omega t}$

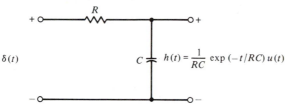

(b) Fourier transforming the impulse response

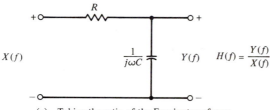

(c) Taking the ratio of the Fourier transforms
of output to input

FIGURE 2-5. Illustration of methods for obtaining the transfer function of a system.

Ordinarily, $h(t)$ is considered to be real, although it is convenient to use complex impulse responses in the analysis of narrowband systems to be considered shortly. If $h(t)$ is real, the magnitude of its Fourier transform, or the system's amplitude response, is an even function of frequency and its phase is odd. This, and several other properties of $H(f)$ are summarized in Table 2-6.

The effect of a system on any input is of interest. By expanding $h(t - \lambda)$ in (2-9a) in a Taylor series about $\lambda = 0$, the output may be expressed as the series (see [1], p. 103)

$$y(t) = m_0 x(t) - m_1 x^{(1)}(t) + \frac{m_2}{2} x^{(2)}(t)$$

$$+ \cdots \frac{(-1)^{n-1}}{(n-1)!} m_{n-1} x^{(n-1)}(t) + E_n \qquad (2\text{-}47)$$

where the m_n's are the moments of $h(t)$, which are defined as

$$m_n = \int_{-\infty}^{\infty} t^n h(t)\, dt = j^n \tilde{H}^{(n)}(0) \qquad (2\text{-}48)$$

and E_n is an error term given by

$$E_n = \frac{(-1)^n}{n!} m_n x^{(n)}(t - \tau_0) \qquad (2\text{-}49)$$

where τ_0 is some constant in the interval of integration.

TABLE 2-6. Properties of the Transfer Function of a System

Name	Mathematical Description	Comments
1. Symmetry	$\|H(f)\| = \|H(-f)\|$ $\underline{/H(f)} = -\underline{/H(-f)}$	Holds only for real $h(t)$
2. Cascaded systems	$H(f) = H_1(f)H_2(f)$ $h(t) = h_1(t) * h_2(t)$	$\xrightarrow{X(f)} \boxed{H_1(f)} \rightarrow \boxed{H_2(f)} \xrightarrow{Y(f)}$
3. Parallel systems	$H(f) = H_1(f) + H_2(f)$ $h(t) = h_1(t) + h_2(t)$	
4. Feedback	$H(f) = \dfrac{G_1(f)}{1 \mp G_1(f)G_2(f)}$	

EXAMPLE 2-6 [1]

Let $h(t) = (2a)^{-1}\Pi(t/2a)$. Then $m_0 = 1$, $m_1 = 0$, and $m_2 = a^2/3$. With $n = 2$, it follows from (2-47) and (2-49) that

$$y(t) = x(t) + \frac{a^2}{6} x^{(2)}(t - \tau_0), \qquad |\tau_0| \leq a$$

From this example and (2-47) it is seen that if $h(t)$ takes on significant values only for $|t| < \epsilon$, and $x(t)$ is sufficiently smooth, then

$$y(t) \simeq x(t) \int_{-\infty}^{\infty} h(\lambda)\, d\lambda = H(0)x(t) \qquad (2\text{-}50)$$

Such systems are normally referred to as *wideband*. ∎

2-1.11 Other Applications of the Fourier Transform

In Section 2-1.10 the convolution theorem of Fourier transforms was employed to carry out the analysis of a fixed, linear system in the frequency domain. In this subsection, applications of Fourier transform theorems to other areas of system analysis are considered.

In these considerations it will be convenient to refer to a signal, $x(t)$, as *lowpass*, *bandpass*, or *highpass* according to whether its amplitude spectrum, $|X(f)|$, has significant values for $|f| < W$, $|f \pm f_0| < W/2$, or $|f| > W$, respectively, where W is the bandwidth of $x(t)$ and f_0 is the center frequency of the bandpass spectrum.

1. *Double-sideband modulation*: The modulation theorem of Fourier transforms states that if the Fourier transform of $m(t)$ is $M(f)$, then the Fourier transform of

$$x(t) = Am(t) \cos 2\pi f_0 t \qquad (2\text{-}51)$$

is

$$X(f) = \frac{A}{2}[M(f - f_0) + M(f + f_0)] \qquad (2\text{-}52)$$

If $m(t)$, usually referred to as the *message*, is a lowpass signal of bandwidth W, then $X(f)$ is a bandpass signal of bandwidth $2W$. This is referred to as *double-sideband modulation* (DSB), which is seen to consist of multiplication of $m(t)$ by the carrier $\cos 2\pi f_0 t$. If $m(t)$ is replaced by

$$p(t) = 1 + am(t)$$

in (2-51), where a is referred to as the *modulation index*, the result is double-sideband modulation with carrier, commonly referred to as *amplitude modulation* (AM). Its spectrum is

$$X(f) = \frac{A}{2}[\delta(f - f_0) + \delta(f + f_0) + aM(f - f_0) + aM(f + f_0)] \qquad (2\text{-}53)$$

where the delta functions reflect the presence of finite power at the carrier frequency f_0. Usually, it is assumed that $|\min (m(t))| \leq 1$; in this case, if $a = 1$, the minimum of the *envelope* is zero. The *percent of modulation* is then 100.

2. *The Hilbert transform; Single-sideband modulation:* A Hilbert transform consists of a $-\pi/2$-rad phase shift (for positive frequencies only) in the frequency domain. Thus the transfer function of a Hilbert transformer is

$$H(f) = -j \operatorname{sgn}(f)$$

where $\operatorname{sgn}(f) = 2u(f) - 1$ is the signum function. The impulse response of a Hilbert transformer can be obtained as an inverse Fourier transform in the limit of a suitably chosen function that approaches $-j \operatorname{sgn}(f)$ as some parameter approaches zero.

An example is

$$H_a(f) = -j[e^{-af}u(f) - e^{af}u(-f)]$$

which has the inverse Fourier transform

$$h_a(t) = \frac{4\pi t}{a^2 + (2\pi t)^2}$$

The impulse response of a Hilbert transformer, therefore, is

$$h(t) = \lim_{a \to 0} h_a(t) = \frac{1}{\pi t} \qquad \text{(see pair 15, Table 2-5)}$$

Thus the output of a Hilbert transform filter for an arbitrary input can be written in terms of the superposition integral as

$$\hat{x}(t) = \frac{1}{\pi} \int_{-\infty}^{\infty} \frac{x(\lambda)}{t - \lambda} \, d\lambda \qquad \text{(see pair 16, Table 2-5)}$$

The applications of the Hilbert transform of most interest in this book are to modulation theory. Consider a complex signal $z(t) = x(t) + jy(t)$ with Fourier transform, $Z(f)$, which is zero for $f < 0$. It follows that $Z(f)$ can be written as

$$Z(f) = [1 - \text{sgn } (f)]X(f) \tag{2-54}$$

where $X(f)$ is the Fourier transform of $x(t)$. Using pair 16 of Table 2-5, the inverse Fourier transform of (2-54) is

$$z(t) = x(t) - j\hat{x}(t) \tag{2-55a}$$

where $\hat{x}(t)$ is the Hilbert transform of $x(t)$. If $Z(f) = 0$ for $f > 0$, the result is

$$z(t) = x(t) + j\hat{x}(t) \qquad [Z(f) = 0 \text{ for } f > 0] \tag{2-55b}$$

The signals expressed by (2-55) are referred to as *analytic signals;* that is, they are complex-valued signals whose spectra are nonzero only for $f > 0$ or $f < 0$.

Now consider the signal

$$x(t) = \text{Re} \left[\frac{A}{2} z(t)e^{j2\pi f_0 t} \right] \tag{2-56}$$

where $z(t)$ is an analytic signal of the form $z(t) = m(t) \mp j\hat{m}(t)$. It follows that

$$x(t) = \frac{A}{2} [m(t) \cos 2\pi f_0 t \pm \hat{m}(t) \sin 2\pi f_0 t] \tag{2-57}$$

where $m(t)$ represents a message signal and $\hat{m}(t)$ is the Hilbert transform of $m(t)$. Writing (2-56) as

$$x(t) = \frac{A}{4} [z(t)e^{j2\pi f_0 t} + z^*(t)e^{-j2\pi f_0 t}]$$

using the frequency translation theorem, and assuming $M(f)$ to be lowpass [i.e., $M(f) = 0, |f| > W$], it follows that

$$X(f) = \begin{cases} M(f - f_0) & f_0 - W \le f \le f_0 \\ M(f + f_0) & -f_0 \le f \le -f_0 + W \\ 0 & \text{otherwise} \end{cases} \tag{2-58a}$$

if the plus sign is chosen in (2-57) and

$$X(f) = \begin{cases} M(f - f_0) & f_0 \le f \le f_0 + W \\ M(f + f_0) & -f_0 - W \le f \le -f_0 \\ 0 & \text{otherwise} \end{cases} \tag{2-58b}$$

if the minus sign is chosen in (2-57). The former is referred to as single-sideband, lower sideband transmitted (SSBL) and the latter as single-sideband, upper sideband (SSBU) transmitted modulation.

The modulation types just considered are referred to as *linear analog modulations* because a linear operation is performed on the message to produce them. They are illustrated for typical message and spectra in Figure 2-6.

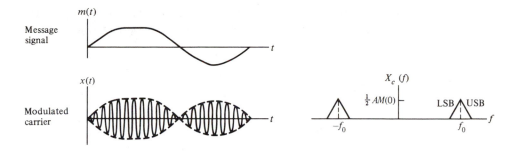

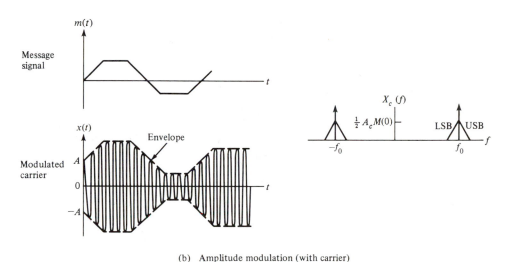

(a) Double sideband (no carrier component)

(b) Amplitude modulation (with carrier)

FIGURE 2-6. Linear analog modulation waveforms and spectra.

2-2

COMPLEX ENVELOPE REPRESENTATION OF SIGNALS AND SYSTEMS

2-2.1 Narrowband Signals

Consider a signal of the form

$$x(t) = r(t) \cos [\omega_0 t + \varphi(t)] \qquad (2\text{-}59)$$

where $r(t)$ and $\varphi(t)$ vary slowly with respect to $\cos \omega_0 t$. Such a signal is said to be narrowband, for then $X(f) \triangleq \mathscr{F}[x(t)]$ is nonzero only in the locality of $f = f_0 = \omega_0/2\pi$. Writing $x(t)$ in terms of quadrature components results in

$$x(t) = x_c(t) \cos \omega_0 t - x_s(t) \sin \omega_0 t \qquad (2\text{-}60)$$

where

$$x_c(t) = r(t) \cos \varphi(t) \qquad (2\text{-}61a)$$

$$x_s(t) = r(t) \sin \varphi(t) \qquad (2\text{-}61b)$$

It follows that these quadrature components of $x(t)$ are also slowly varying if $r(t)$ and $\varphi(t)$ are slowly varying. By defining the *complex envelope* of $x(t)$ as

$$\tilde{x}(t) = x_c(t) + jx_s(t)$$

$$= r(t)e^{j\varphi(t)} \tag{2-62}$$

$x(t)$ can be written as

$$x(t) = \text{Re}\,[\tilde{x}(t)e^{j\varphi_0 t}]$$

$$= \tfrac{1}{2}\tilde{x}(t)e^{j\omega_0 t} + \tfrac{1}{2}\tilde{x}^*(t)e^{-j\omega_0 t} \tag{2-63}$$

The spectrum of $x(t)$ is obtained by Fourier transforming (2-63). To carry out the Fourier transform, consider $\mathcal{F}[\tilde{x}^*(t)]$, which by definition is

$$\mathcal{F}[\tilde{x}^*(t)] = \int_{-\infty}^{\infty} \tilde{x}^*(t)e^{-j2\pi ft}\,dt$$

$$= \left[\int_{-\infty}^{\infty} \tilde{x}(t)e^{j2\pi ft}\,dt\right]^*$$

$$= \{\mathcal{F}[\tilde{x}(t)]\}^*_{f \to -f}$$

where $f \to -f$ signifies that f is replaced by $-f$ in the expression for $\mathcal{F}[\tilde{x}(t)]$. Letting $\tilde{X}(f) = \mathcal{F}[\tilde{x}(t)]$, it has been shown that

$$\mathcal{F}[\tilde{x}^*(t)] = \tilde{X}^*(-f) \tag{2-64}$$

Use of this result together with the frequency translation theorem given in Table 2-4 gives the Fourier transform of (2-63):

$$X(f) = \tfrac{1}{2}\tilde{X}(f - f_0) + \tfrac{1}{2}\tilde{X}^*(-f - f_0) \tag{2-65}$$

For a narrowband signal, $\mathcal{F}[\tilde{x}(t)] = \tilde{X}(f) \simeq 0$ for $|f| > W \ll f_0$ due to the slowly varying nature of $\tilde{x}(t)$.

If $\tilde{x}(t)$ is real, $\tilde{X}^*(f) = \tilde{X}(-f)$ and (2-65) becomes

$$X(f) = \tfrac{1}{2}\tilde{X}(f - f_0) + \tfrac{1}{2}\tilde{X}(f + f_0) \qquad [\tilde{x}(t)\ \text{real}] \tag{2-66}$$

which is simply the modulation theorem of Table 2-4. Thus (2-65) is a generalization of the modulation theorem for the case of narrowband signals.

EXAMPLE 2-7
Write the signal

$$x(t) = e^{-at}\cos(\omega_0 t + \Delta\omega t)u(t)$$

in terms of a complex envelope and find its spectrum.

Solution: The spectrum of $x(t)$ is found by writing it as

$$x(t) = \text{Re}\,[e^{-at}e^{j(\omega_0 + \Delta\omega)t}u(t)]$$

$$= \text{Re}\,[e^{-(a - j\Delta\omega)t}u(t)e^{j\omega_0 t}]$$

Thus the complex envelope of $x(t)$ is

$$e(t) = e^{-(a - j\Delta\omega)t}u(t)$$

$$= [e^{-at}u(t)]e^{j\Delta\omega t}$$

$\tilde{X}(f)$ is found by applying the frequency translation theorem to the transform pair

$$e^{-at}u(t) \leftrightarrow \frac{1}{a + j\omega}$$

which gives

$$\tilde{X}(f) = \frac{1}{a + j2\pi(f - \Delta f)}$$

Using this result in (2-65), the Fourier transform of $x(t)$ is

$$X(f) = \frac{\frac{1}{2}}{a + j2\pi(f - \Delta f - f_0)} + \frac{\frac{1}{2}}{a - j2\pi(-f - \Delta f - f_0)}$$

which could have been obtained by applying the modulation theorem of Table 2-4 directly to $x(t)$. ∎

2-2.2 Narrowband Signals and Narrowband Systems

Consider a fixed, linear system with impulse response $h(t)$ and transfer function $H(f)$. A system is said to be *narrowband* if its transfer function can be written in the form

$$H(f) = H_l(f - f_0) + H_l^*(-f - f_0) \qquad (2\text{-}67)$$

where $|H_l(f)| \simeq 0$ for $|f| > B << f_0$. The inverse Fourier transform of (2-67) results in the impulse response of the system:

$$h(t) = \int_{-\infty}^{\infty} [H_l(f - f_0) + H_l^*(-f - f_0)]e^{j2\pi ft}\, df$$

$$= \int_{-\infty}^{\infty} H_l(f)e^{j2\pi ft}\, df\; e^{j2\pi f_0 t}$$

$$+ \int_{-\infty}^{\infty} H_l^*(f)e^{-j2\pi ft}\, df\; e^{-j2\pi f_0 t}$$

$$= h_l(t)e^{j\omega_0 t} + h_l^*(t)e^{-j\omega_0 t}$$

$$= 2\, \mathrm{Re}\, [h_l(t)e^{j\omega_0 t}] \qquad (2\text{-}68)$$

where $h_l(t) \triangleq \mathcal{F}^{-1}[H_l(f)]$.

Thus $2h_l(t)$ is the complex envelope of the impulse response. Now consider the relationship between the input and output of a narrowband linear system using complex-envelope representation. Thus consider the narrowband linear system in Figure 2-7 with input

$$x_i(t) = \mathrm{Re}\, [\tilde{x}_i(t)e^{j\omega_0 t}]$$

$$= \tfrac{1}{2}\tilde{x}_i(t)e^{j\omega_0 t} + \text{c.c.} \qquad (2\text{-}69)$$

and output

$$x_o(t) = \mathrm{Re}\, [\tilde{x}_o(t)e^{j\omega_0 t}]$$

$$= \tfrac{1}{2}\, \tilde{x}_o(t)e^{j\omega_0 t} + \text{c.c.} \qquad (2\text{-}70)$$

where c.c. stands for the complex conjugate of the preceding term.

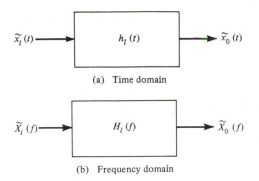

(a) Time domain

(b) Frequency domain

FIGURE 2-7. Narrowband linear system with narrowband input and output represented in terms of complex envelopes.

For a narrowband system with impulse response given by (2-68) and transfer function of the form (2-67), it can be shown that

$$\tilde{x}_o(t) = \tilde{x}_i(t) * h_l(t) \tag{2-71a}$$

and

$$\tilde{X}_o(f) = \tilde{X}_i(f)H_l(f) \tag{2-71b}$$

where

$$\tilde{X}_o(f) = \mathcal{F}[\tilde{x}_o(t)] \tag{2-72a}$$

$$\tilde{X}_i(f) = \mathcal{F}[e_i(t)] \tag{2-72b}$$

To prove (2-71a) the superposition integral (2-9b) is used with (2-68), (2-69), and (2-70) for $h(t)$, $x_i(t)$, and $x_o(t)$ substituted:

$$x_o(t) = \int_{-\infty}^{\infty} x_i(t - \tau)h(\tau)\, d\tau$$

$$= \int_{-\infty}^{\infty} \text{Re}\,[\tilde{x}_i(t - \tau)e^{j\omega_0(t-\tau)}]\, \text{Re}\,[2h_l(\tau)e^{j\omega_0\tau}]\, d\tau$$

$$= \int_{-\infty}^{\infty} [\tfrac{1}{2}\,\tilde{x}_i(t - \tau)e^{j\omega_0(t - \tau)} + \text{c.c.}][h_l(\tau)e^{j\omega_0\tau} + \text{c.c.}]\, d\tau$$

$$= \tfrac{1}{2}\, e^{j\omega_0 t} \int_{-\infty}^{\infty} \tilde{x}_i(t - \tau)h_l(\tau)\, d\tau + \text{c.c.}$$

$$+ \tfrac{1}{2}\, e^{j\omega_0 t} \int_{-\infty}^{\infty} \tilde{x}_i(t - \tau)h_l^*(\tau)e^{-j2\omega_0\tau}\, d\tau + \text{c.c.} \tag{2-73}$$

where, as before, c.c. stands for the complex conjugate of the immediately preceding term.

Since $x_i(t)$ and $h(t)$ are narrowband, their envelopes vary slowly with respect to $e^{-j2\omega_0 t}$. Consequently, the third and fourth terms of (2-73) will integrate to zero. Thus $x_0(t)$ is given by

$$x_0(t) = \text{Re}\left[\int_{-\infty}^{\infty} \tilde{x}_i(t - \tau)h_l(\tau)\, d\tau\, e^{j\omega_0 t}\right]$$

$$= \text{Re}\,[\tilde{x}_i(t) * h_l(t)e^{j\omega_0 t}] \tag{2-74}$$

from which (2-71a) follows by virtue of (2-70). Application of the convolution theorem of Fourier transforms to (2-71a) results in (2-71b).

The use of (2-71a) and (2-71b) in systems analysis will now be illustrated with an example.

EXAMPLE 2-8

Consider the system described by the differential equation

$$\frac{d^2x_o}{dt^2} + \frac{1}{Q}\frac{dx_o}{dt} + \omega_0 x_o = \frac{1}{Q}\frac{dx_i}{dt} \tag{2-75}$$

with input

$$x_i(t) = e^{-at}\cos[(\omega_0 + \Delta\omega)t]u(t) \tag{2-76}$$

The Fourier transform both sides of (2-75) allows the system transfer function to be calculated as the ratio of output to input transforms, which is

$$H(f) = \frac{1}{1 + jQ(f/f_0 - f_0/f)} \tag{2-77}$$

where $f_0 = \omega_0/2\pi$. The magnitude squared of the transfer function is

$$|H(f)|^2 = \frac{1}{1 + Q^2(f/f_0 - f_0/f)^2}$$

which is sketched in Figure 2-8. From this sketch it is apparent that the system is narrowband if $Q \gg 1$. The input signal is narrowband if (1) $|\Delta\omega| \ll \omega_0$ and (2) $a \ll \omega_0$. With these assumptions, (2-71) or (2-72) can be used to find the system output.

To proceed, note that for $Q \gg 1$, $H(f)$ can be approximated as

$$H(f) \simeq \frac{1}{1 + j2Q(f - f_0)/f_0} + \frac{1}{1 + j2Q(f + f_0)/f_0} \tag{2-78}$$

which results from the approximation

$$\frac{f}{f_0} - \frac{f_0}{f} \simeq 2\frac{f \pm f_0}{f_0} \quad \text{if } f \simeq \pm f_0$$

Comparison of (2-78) with (2-67) shows that the lowpass equivalent transfer function is

$$H_l(f) = \frac{1}{1 + j2Qf/f_0} \tag{2-79}$$

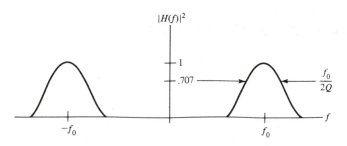

FIGURE 2-8. **Magnitude squared of the system transfer function.**

To find $x_o(t)$, it is easiest in this case to use (2-71a). Inverse Fourier transformation of (2-79) results in

$$h_f(t) = \alpha e^{-\alpha t} u(t) \tag{2-80}$$

where $\alpha = \omega_0/2Q$. Substitution of (2-80) together with

$$\tilde{x}_i(t) = e^{-(a-j\Delta\omega)t} u(t)$$

$$= e^{-\beta t} u(t) \tag{2-81}$$

where $\beta = a - j\,\Delta\omega$, into the superposition integral (2-71a) gives the output signal as

$$\tilde{x}_o(t) = \begin{cases} 0 & t < 0 \\ \alpha e^{-\alpha t} \displaystyle\int_0^t e^{(\alpha - \beta)\tau}\, d\tau & t > 0 \end{cases}$$

$$= \begin{cases} 0 & t < 0 \\ \dfrac{\alpha}{\alpha - \beta}(e^{-\beta t} - e^{-\alpha t}) & t > 0 \quad \alpha \neq \beta \\ \alpha t e^{-\alpha t} & \alpha = \beta \quad (\text{i.e., } \Delta\omega = 0 \text{ and} \\ & \qquad\qquad a = \omega_0/2Q) \end{cases} \tag{2-82}$$

For $\alpha \neq \beta$, the real output is

$$x_o(t) = \text{Re } [\tilde{x}_o(t) e^{j\omega_0 t}]$$

$$= A(\Delta\omega)\{e^{-at} \cos [(\omega_0 + \Delta\omega)t + \varphi(\Delta\omega)]$$

$$- e^{-\omega_0 t/2Q} \cos [\omega_0 t + \varphi(\Delta\omega)]\} \qquad t \geq 0 \tag{2-83}$$

where

$$A(\Delta\omega) = \frac{\omega_0/2Q}{\sqrt{(\omega_0/2Q - a)^2 + (\Delta\omega)^2}} \tag{2-84}$$

$$\varphi(\Delta\omega) = \tan^{-1} \frac{\Delta\omega}{\omega_0/2Q - a} \tag{2-85}$$

Direct application of the superposition integral to the real signals would have involved considerably more labor. It should be remembered, however, that (2-83) is accurate only if the conditions guaranteeing a narrowband situation stated previously hold. ∎

2-3

SIGNAL DISTORTION AND FILTERING

In this section the description of the effects of a system on signal transmission is considered. The concept of a distortionless system is first defined together with the artifice of ideal filters. Following this, the characterization of a filter in terms of group and phase delay is introduced and used to describe the effect of nonideal filters on signal transmission. Next, several types of practical filter designs and characteristics are summarized. The section closes with a summary of nonlinear system characteristics and techniques for their analysis.

2-3.1 Distortionless Transmission and Ideal Filters

In many signal transmission systems, the goal is to obtain an output signal that resembles the input signal or some function of it.* If the output of a system is a scaled, delayed version of its input, the system is said to be *distortionless*. Thus the input–output relationship for a fixed, linear, distortionless system in the time domain is

$$y(t) = H_0 x(t - t_0) \tag{2-86}$$

where $x(t)$ is the input, $y(t)$ is the output, and H_0 and t_0 are constants referred to as the system's gain (or attenuation) and delay, respectively. Using the delay theorem of Fourier transforms (pair 2, Table 2-4), the transfer function of a fixed, linear, distortionless system is

$$H(f) = H_0 e^{-j2\pi f t_0} \tag{2-87}$$

That is, the amplitude response of a distortionless system is a constant and its phase response is a linear function of frequency. If the input signal is bandlimited, these conditions need hold only for the frequency range over which the input signal has significant spectral content. Depending on whether a signal is lowpass, bandpass, or highpass, three types of ideal signal transmission system characteristics can be defined as shown in Figure 2-9. Such systems are referred to as *ideal filters*, a term that is in one sense a misnomer in that their impulse responses exist for $t < 0$. Thus ideal filters are noncausal, as illustrated in Figure 2-10, which shows their impulse responses.

Filters for which the amplitude response is not constant over the bandwidth of the input signal are said to introduce *amplitude distortion*; those with nonlinear phase responses introduce *phase distortion*.

2-3.2 Group and Phase Delay

Given a filter with transfer function $H(f)$ written in complex exponential form as

$$H(f) = A(f) \exp [j\theta(f)] \tag{2-88}$$

where $A(f)$ is its amplitude response and $\theta(f)$ its phase response, the *group delay* of the filter is defined as

$$t_g = -\frac{1}{2\pi}\left(\frac{d\theta}{df}\right) \tag{2-89}$$

and its *phase delay* is defined as

$$t_p = -\frac{\theta(f)}{2\pi f} \tag{2-90}$$

Figure 2-11 illustrates these concepts, where it is shown that the group delay is the magnitude of the slope of the phase response curve of the system when plotted as

---

*In some cases, such as the matched filter to be discussed later, output signal fidelity is not the primary consideration, but rather, the maximization of the output signal-to-noise ratio at time t_0.

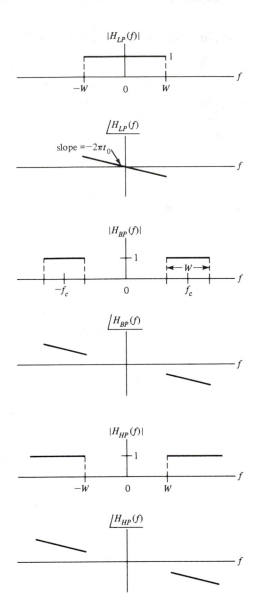

FIGURE 2-9. Ideal filter amplitude and phase responses.

a function of $2\pi f$, and the phase delay is the magnitude of the slope of a line from the origin to an arbitrary point $2\pi f$ on the same curve.

The physical significance of these two quantities is:

1. The group delay of a system describes the delay of a narrow frequency group of width Δf with center frequency f under the assumption that the system's amplitude response is constant and its phase is linear over this frequency interval, Δf.
2. The phase delay of a system at frequency f gives the delay imposed by the system on a steady-state sinusoidal input of frequency f.

From (2-87) and Figure 2-9 it is seen that signal distortion results if the group delay is not equal to the phase delay for all frequencies (i.e., a constant). Thus group delay is an important measure of *phase*, or *delay distortion* for a system.

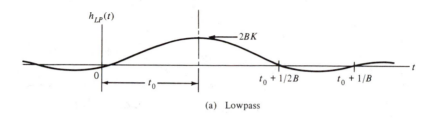

(a) Lowpass

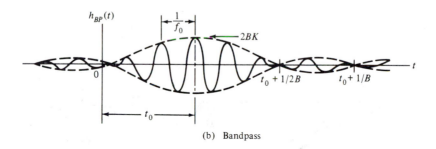

(b) Bandpass

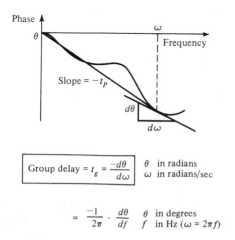

(c) Highpass

FIGURE 2-10.   Impulse responses for ideal filters.

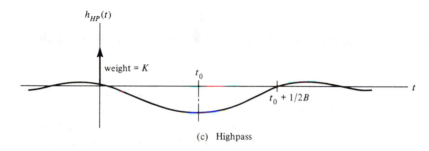

$$\text{Group delay} = t_g = \frac{-d\theta}{d\omega} \qquad \begin{array}{l} \theta \text{ in radians} \\ \omega \text{ in radians/sec} \end{array}$$

$$= \frac{-1}{2\pi} \cdot \frac{d\theta}{df} \qquad \begin{array}{l} \theta \text{ in degrees} \\ f \text{ in Hz } (\omega = 2\pi f) \end{array}$$

FIGURE 2-11.   Illustration of group and phase delay.

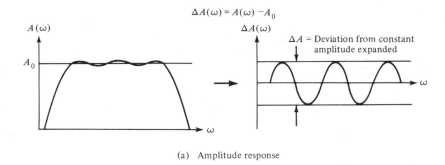

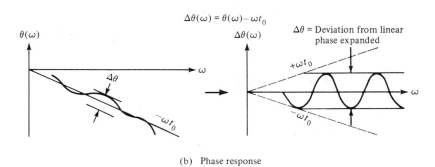

(a) Amplitude response

(b) Phase response

FIGURE 2-12. Nonideal amplitude and phase response functions for a filter.

Several example plots of group delay versus frequency are given in the following subsection, where several practical filter types are discussed.

To examine further the effect of a system with nonconstant amplitude response and group delay, consider Figure 2-12, which shows typical nonideal amplitude and phase response functions for a filter. They may be expressed as

$$A(f) = A_0 + \Delta A(f) \qquad (2\text{-}91a)$$

and

$$\theta(f) = -2\pi t_0 f + \Delta\theta(f) \qquad (2\text{-}91b)$$

where $\Delta A(f)$ and $\Delta\theta(f)$ are deviations from the ideal characteristics. The output of the filter due to an input signal $x(t)$ with Fourier transform $X(f)$ is

$$y(t) = \mathcal{F}^{-1}\{X(f)A(f)\exp[j\theta(f)]\} \qquad (2\text{-}92)$$

For $\Delta A(f)$ and $\Delta\theta(f)$ small over the passband of the signal, the filter transfer function is well approximated by

$$A(f)\exp[j\theta(f)] \simeq [A_0 + \Delta A(f) + jA_0\,\Delta\theta(f)]\exp(-j2\pi f t_0) \quad (2\text{-}93)$$

where (2-91a) and (2-91b) have been substituted, $\exp(j\,\Delta\theta)$ was expanded in a power series, and all terms above first order dropped.

Two cases can be considered for $\Delta A(f)$ and $\Delta\theta(f)$, a lowpass system and a bandpass system. Consider the case of a lowpass system first. For a lowpass system it must be true that $A(f)$ is even and $\theta(f)$ is odd. Since $\Delta A(f)$ and $\Delta\theta(f)$ are small perturbations on the ideal responses A_0 and $-2\pi t_0 f$, they may be expanded in power series as

$$\Delta A(f) = a_2 f^2 + a_4 f^4 + \cdots \qquad (2\text{-}94a)$$

and

$$\Delta\theta(f) = b_3 f^3 + b_5 f^5 + \cdots \qquad (2\text{-}94b)$$

respectively, where the even and odd symmetry conditions just mentioned have been imposed. Use of (2-94) in (2-93), and substitution of the resulting equation in (2-92), gives the expression

$$y(t) = \mathscr{F}^{-1}\{[A_0 + a_2 f^2 + jA_0 b_3 f^3 + a_4 f^4$$
$$+ jA_0 b_5 f^5 + \cdots]X(f)\exp(-j2\pi t_0 f)\} \qquad (2\text{-}95a)$$

By the time delay and differentiation theorems of Fourier transforms (Theorems 2 and 6 of Table 2-4), (2-95a) can be inverse Fourier transformed to give the series

$$y(t) = A_0 x(t - t_0) + A_2 x^{(2)}(t - t_0) + B_3 x^{(3)}(t - t_0) \qquad (2\text{-}95b)$$
$$+ A_4 x^{(4)}(t - t_0) + B_5 x^{(5)}(t - t_0)$$

where $A_2 = -a_2/(2\pi)^2$
$\qquad B_3 = -A_0 b_3/(2\pi)^3$
$\qquad A_4 = a_4/(2\pi)^4$
$\qquad B_5 = A_0 b_5/(2\pi)^5$

and $x^{(n)}(t)$ denotes the nth derivative of $x(t)$.

EXAMPLE 2-9

Consider the lowpass input signal

$$x(t) = B \exp\left(\frac{-t^2}{2\tau^2}\right)$$

where B and τ are constants, to a filter with amplitude and phase responses approximated by

$$A(f) = A_0 + a_2 f^2$$

and

$$\theta(f) = b_3 f^3$$

respectively (the delay $t_0 = 0$ for simplicity). In this case (2-95b) becomes

$$y(t) = A_0 x(t) + A_2 x^{(2)}(t) + B_3 x^{(3)}(t)$$

where A_2 and B_3 were defined previously. Taking derivatives of $x(t)$ and substituting into the expression for $y(t)$, one obtains

$$y(t) = A_0 B \exp\left(\frac{-t^2}{2\tau^2}\right) - \frac{A_2 B}{\tau^2}\,1 - t^2/\tau^2\,\exp\left(\frac{-t^2}{2\tau^2}\right)$$
$$+ \frac{B_3 B}{\tau^3}\,3t/\tau - t^3/\tau^3\,\exp\left(\frac{-t^2}{2\tau^2}\right)$$

The first term corresponds to the undistorted output signal, the second term represents amplitude distortion, and the third term represents phase distortion. Undis-

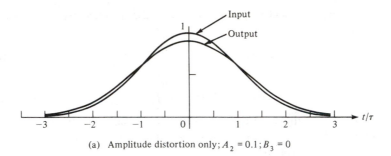

(a) Amplitude distortion only; $A_2 = 0.1$; $B_3 = 0$

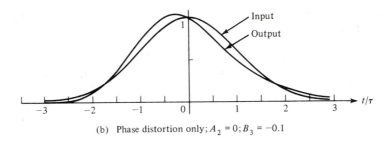

(b) Phase distortion only; $A_2 = 0$; $B_3 = -0.1$

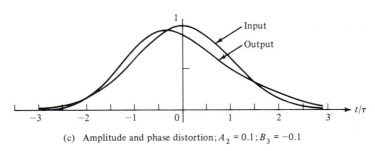

(c) Amplitude and phase distortion; $A_2 = 0.1$; $B_3 = -0.1$

FIGURE 2-13. Example showing the effect of amplitude and phase response perturbations on system response.

torted and distorted waveforms are compared in Figure 2-13 for a Gaussian pulse of amplitude $A_0 = B = 1$. ■

Consider next the bandpass case. Use of the theory developed in Section 2-2 allows (2-92) to be rewritten as

$$y_\ell(t) = \mathcal{F}^{-1}\{X_\ell(f)\mathscr{A}_\ell(f) \exp [j\theta_\ell(f)]\} \qquad (2\text{-}96)$$

where the subscript ℓ denotes complex-envelope quantities. Since $H_\ell(f) \triangleq \mathscr{A}_\ell(f)$ $\exp [j\theta_\ell(f)]$ is the Fourier transforms of a complex-envelope quantity, it is no longer true that $\mathscr{A}_\ell(f)$ must be an even function of frequency and that $\theta_\ell(f)$ must be odd. Therefore, $\Delta\mathscr{A}(f)$ and $\Delta\theta(f)$ are expanded as

$$\Delta\mathscr{A}(f) = a_1 f + a_2 f^2 + a_3 f + \cdots \qquad (2\text{-}97a)$$

and

$$\Delta\theta(f) = b_0 + b_2 f^2 + b_3 f^3 + \cdots \qquad (2\text{-}97b)$$

respectively, where $\mathcal{A}_\ell(f) = A_0 + \Delta\mathcal{A}(f)$ and $\theta_\ell(f) = -j2\pi t_0 f + \Delta\theta(f)$. By expanding (2-96) to second-order terms in $\Delta\mathcal{A}(f)$ and $\Delta\theta(f)$, and then substituting the power series (2-97), one obtains

$$y_\ell(t) = \mathcal{F}^{-1}\{[A_0(1 + jb_0 - \tfrac{1}{2} b_0^2) + a_1(1 + jb_0)f$$

$$+ (a_2 + j\mathcal{A}_0 b_2)(1 + jb_0)f^2 + \cdots] X_\ell(f) \exp(-j2\pi t_0 f)\}$$

$$= \mathcal{F}^{-1}\{[A_0(1 + jb_0 - \tfrac{1}{2} b_0^2) + \frac{a_1(1 + jb_0)}{j2\pi} (j2\pi f)$$

$$- \frac{(a_2 + jA_0 b_2)(1 + jb_0) - b_0 b_2}{(2\pi)^2} (j2\pi f)^2 + \cdots] X_\ell(f) \exp(-j2\pi t_0 f)\}$$

$$= A_0(1 + jb_0 - \tfrac{1}{2} b_0^2)x_\ell(t - t_0) + jA_1 x_\ell^{(1)}(t - t_0)$$

$$+ (A_2 + jB_2)x_\ell^{(2)}(t - t_0) + \cdots \tag{2-98}$$

where $A_1 = -a_1(1 + jb_0)/2\pi$
$A_2 = -(a_2 - A_0 b_0 b_2 b_0 b_2)/(2\pi)^2$
$B_2 = -(A_0 b_2 + a_2 b_0)/(2\pi)^2$

Thus, for an input signal of the form

$$x(t) = \text{Re } [x_\ell(t)e^{j\omega_0 t}] \tag{2-99}$$

the output

$$y(t) = \text{Re } [y_\ell(t)e^{j\omega_0 t}] \tag{2-100}$$

is obtained where $y_\ell(t)$ is given in terms of $x_\ell(t)$ by (2-98). The term $A_0 x_\ell(t - t_0)$ represents the undistorted output and all other terms are distortion.

EXAMPLE 2-10
Consider the amplitude-modulated input signal

$$x(t) = \text{Re } [(1 + a \cos \omega_m t)e^{j\omega_0 t}]$$

to a system with $b_0 = 0$; A_1, A_2, and B_2 are, for the moment, unspecified. It is also assumed that $t_0 = 0$ for simplicity. It follows that

$$x_\ell(t) = 1 + a \cos \omega_m t$$

$$x_\ell^{(1)}(t) = -\omega_m a \sin \omega_m t$$

$$x_\ell^{(2)}(t) = -\omega_m^2 a \cos \omega_m t$$

Therefore, the complex envelope of the output is

$$y_\ell(t) = A_0(1 + a \cos \omega_m t) - jA_1\omega_m a \sin \omega_m t$$

$$- (A_2 + jB_2)\omega_m^2 a \cos \omega_m t$$

The real output is obtained by multiplying $y_\ell(t)$ by $\exp(j\omega_0 t)$ and taking the real part. The result is

$$y(t) = A_0(1 + a \cos \omega_m t) \cos \omega_0 t + A_1\omega_m a \sin \omega_m t \sin \omega_0 t$$

$$- A_2\omega_m^2 a \cos \omega_m t \cos \omega_0 t + B_2\omega_m^2 a \cos \omega_m t \sin \omega_0 t$$

The first term is the undistorted output signal and the other terms represent various types of distortion. Each of these distortion terms will now be considered separately.

1. $A_1 = B_2 = 0$; $A_2 \neq 0$. In this case, the output signal becomes $y(t) = A_0[1 + a\left(1 - \dfrac{A_2\omega_m^2}{A_0}\right) \cos \omega_m t] \cos \omega_0 t$. As a result of the deviation of the amplitude response from A_0, distortion is introduced into the amplitude modulation in that the modulation index, a, is changed.

2. $A_2 = B_2 = 0$; $A_1 \neq 0$. The output signal for this case can be written

$$y(t) = A_0(1 + a \cos \omega_m t) \cos \omega_0 t + A_1 \omega_m a \sin \omega_m t \sin \omega_0 t$$

Because of the second term, which is in phase quadrature to the amplitude-modulated first term, phase modulation is introduced and a slightly distorted envelope results (*harmonic distortion,* or distortion at a frequency of $2\omega_m$, is introduced into the envelope). Such distortion is referred to as *modulation conversion* or *AM-to-PM conversion.*

3. $A_1 = A_2 = 0$; $B_2 \neq 0$. Again, as in case 2, a distortion term is introduced which is in phase quadrature to the modulated input carrier. In the present case, however, the distortion arises because of the departure of the filter's phase response characteristic from the ideal linear response $-2\pi t_0 f$. ∎

EXAMPLE 2-11

Consider the phase-modulated signal

$$x(t) = \text{Re} \{\exp [j(\omega_0 t + a \cos \omega_m t)]\}$$

with the corresponding complex envelope

$$x_\ell(t) = \exp (ja \cos \omega_m t)$$

Its derivatives are

$$x_\ell^{(1)}(t) = -j\omega_m a \sin \omega_m t \exp (ja \cos \omega_m t)$$

$$x_\ell^{(2)}(t) = -j\omega_m^2 a(\cos \omega_m t - ja \sin^2\omega_m t) \exp (ja \cos \omega_m t)$$

As in Example 2-10, assume that $A_3 = B_3 = 0$. The complex envelope of the output with $t_0 = b_0 = 0$ is

$$y_\ell(t) = [A_0 + A_1\omega_m a \sin \omega_m t + (B_2 - jA_2)\omega_m^2 a (\cos \omega_m t - ja \sin^2\omega_m t)]$$

$$\times \exp (ja \cos \omega_m t)$$

The real output signal is found by multiplying $y_\ell(t)$ by $\exp (j\omega_0 t)$ and taking the real part. The result of this operation is

$$y(t) = [A_0 + A_1\omega_m a \sin \omega_m t + \omega_m^2 a(B_2 \cos \omega_m t + A_2 a \sin^2\omega_m t)]$$

$$\cos (\omega_0 t + a \cos \omega_m t) +$$

$$\omega_m^2 a(A_2 \cos \omega_m t + B_2 a \sin^2\omega_m t) \sin (\omega_0 t + a \cos \omega_0 t)$$

As was done in Example 2-10, several special cases will be considered.

1. If $A_2 = B_2 = 0$ and A_1 is not zero, the phase-modulated input signal produces an output signal which has the same phase modulation as the input signal but

has an additional amplitude modulation. That is, *PM-to-AM conversion* has taken place. From the expression for $y(t)$, it is seen that PM-to-AM conversion results from a linear deviation of the amplitude response ($A_1 \neq 0$) from the ideal constant-amplitude response.

2. If A_2 and B_2 are not both zero, a term that is in phase quadrature to the phase-modulated input signal is produced. This results in both a nonconstant envelope as well as distortion of the phase modulation on the input signal. ∎

In this section it has been shown how nonideal amplitude and phase characteristics for linear, time-invariant systems distort signals. In the case of modulated signals, these nonideal characteristics may modify the original modulating signal by converting it from one type of modulation to another, or they may modify the modulating signal itself by introducing distortion. This distortion may involve the production of new frequencies. Another way new frequencies may be produced is through the presence of nonlinearities, either intentional or unintentional, in the system. We consider this topic next.

2-3.3 Nonlinear Systems and Nonlinear Distortion

It is indeed fortunate that linear system models have such broad applicability. However, there are many instances in communication system analysis and design that require the modeling of nonlinear systems. For purposes of modeling, a time-invariant system can be classified as having either *zero memory* or *nonzero memory*. For *zero-memory systems*, the input–output relationship is of the form

$$y(t) = G[x(t)] \qquad (2\text{-}101)$$

where $G(\cdot)$ is a single-valued function of its argument. That is, the output of the system at time t depends only on the input to the system at the same instant of time, not on past or future values. For a *nonzero memory system*, however, the output depends on past values of the input as well as the input at the present time if the system is causal, and on future values also if the system is noncausal.

A general approach for the analysis of fixed, nonlinear systems with memory is the Volterra series approach (see [2]). In this approach the input–output relationship for the system is written in the form

$$y(t) = \int_{-\infty}^{\infty} h_1(\lambda_1)x(t - \lambda_1)d\lambda_1$$

$$+ \int_{-\infty}^{\infty} \int_{-\infty}^{\infty} h_2(\lambda_1, \lambda_2)x(t - \lambda_1)x(t - \lambda_2)d\lambda_1\, d\lambda_2$$

$$+ \int_{-\infty}^{\infty} \int_{-\infty}^{\infty} \int_{-\infty}^{\infty} h_3(\lambda_1, \lambda_2, \lambda_3)x(t - \lambda_1)x(t - \lambda_2)x(t - \lambda_3)d\lambda_1,\, d\lambda_2\, d\lambda_3$$

$$+ \cdots$$

$$+ \int_{-\infty}^{\infty} \cdots \int_{-\infty}^{\infty} h_n(\lambda_1, \lambda_2, \ldots, \lambda_n)x(t - \lambda_1)x(t - \lambda_2) \cdots$$

$$x(t - \lambda_n)d\lambda_1\, d\lambda_2 \cdots d\lambda_n + \cdots \qquad (2\text{-}102)$$

The *n*th term in this series is called an *nth-order Volterra operator*. The form of the series suggests a parallel structure for the system with the first term corresponding to a linear approximation to the system, the second term corresponding to a second-order nonlinearity with memory, and so on. Another viewpoint of the Volterra representation for a system is that it is a power series expansion of the output with memory. To see this, replace $x(t)$ with $\alpha x(t)$ in (2-102). This results in a power series in α. Thus, as the input becomes small, the terms in (2-102) corresponding to the higher-order powers of α become less important. Considerable time could be spent in the study of the representation (2-102), but from now on, attention will be restricted to zero-memory nonlinear systems.

Consider a zero-memory, fixed, nonlinear system with transfer characteristic representable as

$$y(t) = a_0 + a_1 x(t) + a_2 x^2(t) + a_3 x^3(t) + \cdots$$

$$\triangleq y_0(t) + y_1(t) + y_2(t) + y_3(t) + \cdots \tag{2-103}$$

Models such as this have several uses. For example, (2-103) could represent an amplifier driven slightly nonlinear; the first term represents a dc bias, the second term the desired output, and the remaining terms the undesired nonlinearities.

Let the input in (2-103) be

$$x(t) = A_1 \cos \omega_1 t + A_2 \cos \omega_2 t \tag{2-104}$$

The output components are

$$y_0(t) = a_0 \tag{2-105a}$$

$$y_1(t) = a_1 A_1 \cos \omega_1 t + a_1 A_2 \cos \omega_2 t \tag{2-105b}$$

$$y_2(t) = a_2 \left[\frac{A_1^2}{2} + \frac{A_2^2}{2} + \frac{A_1^2}{2} \cos 2\omega_1 t \right.$$

$$+ \frac{A_2^2}{2} \cos 2\omega_2 t + A_1 A_2 \cos (\omega_1 + \omega_2)t$$

$$\left. + A_1 A_2 \cos (\omega_1 - \omega_2)t \right] \tag{2-105c}$$

$$y_3(t) = \frac{a_3 A_1^3}{4} (3 \cos \omega_1 t + \cos 3\omega_1 t)$$

$$+ \frac{3 a_3 A_1^2 A_2}{2} [\cos \omega_2 t + \tfrac{1}{2} \cos (2\omega_1 + \omega_2)t + \tfrac{1}{2} \cos (2\omega_1 - \omega_2)t]$$

$$+ \frac{3 a_3 A_2^2 A_1}{2} [\cos \omega_1 t + \tfrac{1}{2} \cos (2\omega_2 + \omega_1)t + \tfrac{1}{2} \cos (2\omega_2 - \omega_1 t)]$$

$$+ \frac{a_3 A_2^3}{4} (3 \cos \omega_2 t + \cos 3\omega_2 t) \tag{2-105d}$$

The total output is the sum of y_0, y_1, y_2, and so on. Note that the following frequencies are present in the output:

1. DC.
2. The original frequencies, ω_1 and ω_2.
3. Harmonics of ω_1 and ω_2 (i.e., frequencies at $2\omega_1$, $2\omega_2$, $3\omega_1$, $3\omega_2$, etc.).
4. Sums and differences of harmonics of ω_1 and ω_2 (i.e., frequencies at $\omega_1 + \omega_2$, $\omega_1 - \omega_2$, $2\omega_1 - \omega_2$, etc.).

If (2-103) represents an amplifier driven slightly into its nonlinear region of operation, the third set of frequencies mentioned above represents *harmonic distortion* and the fourth set represents *intermodulation distortion* components of the output.

EXAMPLE 2-12

Suppose that a nonlinear device represented by the transfer characteristic (2-103) is to be used as a *mixer*. The input is the sum of three terms, a local oscillator signal,

$$x_{\text{LO}}(t) = L \cos \omega_{\text{LO}} t$$

and two desired sinusoidal signals,

$$x_A(t) = A \cos \omega_A t$$

$$x_B(t) = B \cos \omega_B t$$

The first two terms in (2-103) give outputs at dc and feedthrough outputs,

$$a_1 x(t) = a_1(L \cos \omega_{\text{LO}} t + A \cos \omega_A t + B \cos \omega_B t)$$

The third term in (2-103), $a_2 x^2(t)$, results in the following outputs:

$$a_2 x^2{}_{\text{LO}}(t) = \frac{a_2 L^2}{2} (1 + \cos 2\omega_{\text{LO}} t)$$

$$a_2 x_A^2(t) = \frac{a_2 A^2}{2} (1 + \cos 2\omega_A t)$$

$$a_2 x_B^2(t) = \frac{a_2 B^2}{2} (1 + \cos 2\omega_B t)$$

$$2a_2 x_A(t) x_B(t) = a_2 AB[\cos (\omega_A - \omega_B)t + \cos (\omega_A + \omega_B)t]$$

$$2a_2 x_{\text{LO}}(t) x_A(t) = a_2 LA[\cos (\omega_{\text{LO}} - \omega_A)t + \cos (\omega_{\text{LO}} + \omega_A)t]$$

$$2a_2 x_{\text{LO}}(t) x_B(t) = a_2 LB[\cos (\omega_{\text{LO}} - \omega_B)t + \cos (\omega_{\text{LO}} + \omega_B)t]$$

Of these seven output components, all except the last two represent undesired output terms. In the last two terms, either the frequencies $\omega_{\text{LO}} - \omega_A$ and $\omega_{\text{LO}} - \omega_B$ or $\omega_{\text{LO}} + \omega_A$ and $\omega_{\text{LO}} + \omega_B$ would be selected at the output. The former are referred to as the *low-side frequencies* from the mixer, and the latter are referred to as *high-side frequencies*. It is important that no undesired output frequencies are close to the desired frequencies at the output. A typical spectral plot for the output is provided in Figure 2-14. If one considers the frequencies at the

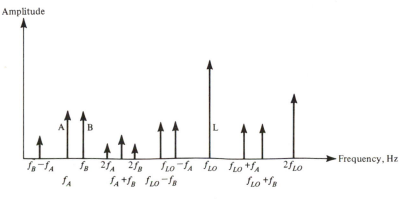

FIGURE 2-14. Typical spectral representation of a mixer output.

output resulting from the term $a_3 x^3(t)$, numerous ones are obtained. The terms of primary interest are $3x_A^2(t)x_B(t)$ and $3x_A(t)x_B^2(t)$, which can be expanded to give *third-order intermodulation products*; they contain frequencies at $2\omega_A \pm \omega_B$ and $2\omega_B \pm \omega_A$. The amplitudes of these components are proportional to the amplitude of one signal multiplied by the square of the amplitude of the other signal. Because of this, these products change amplitude three times faster than $x_A(t)$ and $x_B(t)$ if both A and B change simultaneously. Since they often fall within the passband of the system following the mixer, where they cannot be eliminated by filtering, they are of great practical importance. Balanced mixer configurations can be used to minimize these undesired output components by cancellation. Various mixer configurations are illustrated in Figure 2-15. Analysis of their output spectral characteristics is left to the problems at the end of the chapter. ∎

Several terms are used when referring to mixer performance. These are (1) conversion loss, (2) compression, (3) isolation, (4) harmonic distortion (second order), (5) intermodulation distortion (third order), (6) noise figure, and (7) voltage standing-wave ratio (VSWR). Each of these terms will now be discussed in regard to their use for specifying mixer performance.

Conversion loss: The ratio of power in the desired output signal to power in the input signal. In general, conversion loss decreases with increasing local oscillator frequency and increasing power of the local oscillator signal. Since two output sidebands are generated for a single input frequency, and only one is desired, the conversion loss of a passive mixer must be at least 3 dB. Typical values range from 5 to 8 dB.

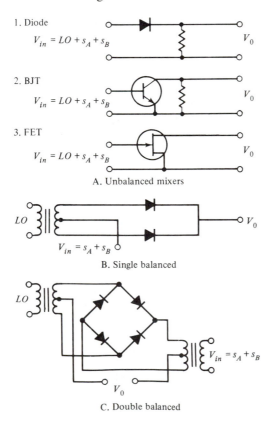

FIGURE 2-15. Various types of mixer configurations.

Compression: The 1-dB compression point of a mixer is defined as the input signal power level which causes a 1-dB increase in conversion loss. It is weakly dependent on the local oscillator power level.

Isolation: The ratio of local oscillator signal power at the output (IF) port to the local oscillator signal power at the local oscillator port is referred to as the *LO-to-IF isolation*. It typically has a minimum with frequency for double-balanced mixers. The ratio of radio-frequency (RF) signal power at the IF port to RF signal power at the RF port is referred to as *RF-to-IF isolation*. RF-to-IF isolation typically decreases with increasing frequency. Isolation values for typical double-balanced mixers can be 50 dB or more.

Harmonic distortion: The primary harmonic distortion component at a mixer output is second, and it refers to the power in second harmonic component of the IF signal due to the mixer (care must be taken to exclude second harmonic due to feedthrough of the RF or LO signal sources). Second harmonic distortion, in general, decreases with increasing LO signal power and increases with increasing input signal power.

Intermodulation distortion: These are components due to the intermodulation components at the mixer output, the primary ones being the third-order ones at the frequencies $2f_A - f_B$ and $2f_B - f_A$, where f_A and f_B are the frequencies of input signals A and B. A figure of merit used to describe the intermodulation performance of a mixer is the *third-order intercept*, which is defined as the *theoretical* intersection of the graphs (assumed linear) of powers expressed in dBm or dBW, in the desired and intermodulation output powers versus input signal power. Recall from Example 2-12 that the latter graph has a slope three times greater than that of the former as a function of input power. This concept is illustrated in Figure 2-16. The higher the intercept point, the better the

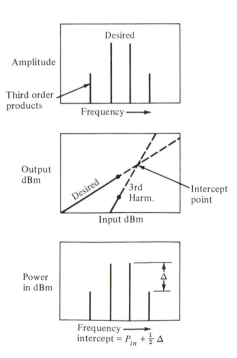

FIGURE 2-16. Plot of intermodulation distortion power and primary component output power versus input power for a mixer, which illustrates the concept of third-order intercept.

suppression of third-order products. The third-order intercept point of a physical mixer varies with input signal power level and LO power, although theoretically no dependence on these parameters is predicted.

Noise figure: This is the ratio of output port signal-to-noise ratio to input port signal-to-noise ratio. It is, in general, frequency dependent, although average values are sometimes used. Since the noise figure of an ideal attenuator at room temperature is theoretically equal to the attenuation factor of the device, it is not a bad approximation to use the conversion loss of a mixer as an estimate for its noise figure.

VSWR: This is a measure of the amount of power reflected at the input port. The VSWR is the peak amplitude to minimum amplitude of the "standing wave" on a transmission line. Since it can be shown that

$$\text{VSWR} = \frac{1 + |\Gamma|}{1 - |\Gamma|} \qquad (2\text{-}106)$$

where Γ is the reflection coefficient, VSWR can be used to determine $|\Gamma|$ whose square is proportional to reflected power. If $|\Gamma| = 0$, no power is reflected; this will be the case if the transmission line and load impedances are matched.

2-4

PRACTICAL FILTER TYPES AND CHARACTERISTICS

In this section several practical lumped-element filter types are summarized together with their frequency response characteristics.* These include Butterworth, Chebyshev, and Bessel (or linear phase). All filter frequency response characteristics considered are lowpass and, with the exception of elliptic, are assumed to have 3-dB frequencies normalized to 1 rad/s. There is no loss in generality in doing so, for frequency scaling and transformations can be used to obtain other filter types. For example, consider a filter transfer function written as a function of the Laplace transform variable s in the form

$$H(s) = \frac{N(s)}{D(s)} \qquad (2\text{-}107a)$$

where

$$N(s) = \sum_{k=0}^{m} a_k s^k = (s - z_1)(s - z_2) \cdots (s - z_m) \qquad (2\text{-}107b)$$

and

$$D(s) = \sum_{k=0}^{n} b_k s^k = (s - p_1)(s - p_2) \cdots (s - p_n) \qquad (2\text{-}107c)$$

are numerator and denominator polynomials in s of degrees m and n, respectively. The roots of $N(s)$, $z_1, z_2, \ldots, z_m$, are the *zeros* of $H(s)$ and the roots of $D(s)$,

---

*Details on realization of these filters in both lumped-element and transmission-line (strip-line) form can be found in a number of references. For lumped-element realizations, see [3]. For stripline synthesis, see [4].

$p_1, p_2, \ldots, p_n$, are the *poles* of $H(s)$. If a 3-dB cutoff frequency other than 1 rad/s is desired, s is replaced in (2-107) by

$$s' = \frac{s}{\omega_c} \tag{2-108}$$

where ω_c is the new 3-dB cutoff frequency.

If a *bandpass* filter is desired, s in (2-107) is replaced by

$$s' = \frac{s^2 + \omega_c^2}{s\omega_b} \tag{2-109}$$

The upper and lower band edge frequencies are given by

$$\omega_u = \tfrac{1}{2}\omega_b + \tfrac{1}{2}\sqrt{\omega_b^2 + 4\omega_c^2} \tag{2-110a}$$

$$\omega_l = -\tfrac{1}{2}\omega_b + \tfrac{1}{2}\sqrt{\omega_b^2 + 4\omega_c^2} \tag{2-110b}$$

The differences of these two equations results in the filter bandwidth in rad/s,

$$\omega_b = \omega_u - \omega_l \tag{2-111a}$$

and their product gives

$$\omega_c^2 = \omega_u\omega_l \tag{2-111b}$$

Thus ω_c is the geometric center frequency of the bandpass filters. If $\omega_c \gg \omega_b$, it follows from (2-110) that

$$\omega_c = \tfrac{1}{2}(\omega_u + \omega_l) \qquad \omega_c \gg \omega_b \tag{2-111c}$$

A *highpass* filter transfer function can be obtained from (2-107) by replacing s by

$$s' = \frac{1}{s} \tag{2-112}$$

and following this transformation with a frequency scaling to obtain the desired 3-dB low cutoff frequency.

If a wideband bandpass filter is desired, an alternative to using (2-109) and (2-110) is to cascade lowpass and highpass sections. This is not satisfactory for the narrowband case because of the excess attenuation thereby introduced at band center.

A band reject filter can be obtained by putting a highpass and lowpass filter in parallel as shown in Table 2-6. The combined transfer function is then

$$H_{BR}(s) = H_{LP}(s) + H_{HP}(s) \tag{2-113}$$

and it is assumed that where the magnitude of $H_{LP}(s)$ is small for $s = j\omega$, the magnitude of $H_{HP}(j\omega)$ is not, and vice versa, except in the reject band where both are small. In other words, the cutoff frequency of the lowpass arm, $H_{LP}(s)$, is below that of the highpass arm, $H_{HP}(s)$. Alternatively, a band reject transformation, given by

$$s' = \frac{s\omega_b}{s^2 + \omega_c^2} \tag{2-114}$$

can be used in (2-107) to produce a notch filter. Note that this is the inverse of (2-109).

The filter types mentioned previously will now be considered.

TABLE 2-7. Coefficients for the Denominator Polynomial of Butterworth Filters of Orders 1–5 (Unity 3-dB Frequency)

Order	b_0	b_1	b_2	b_3	b_4	b_5	b_6
1	1	1					
2	1	$\sqrt{2}$	1				
3	1	2	2	1			
4	1	2.613126	3.414214	2.613126	1		
5	1	3.236068	5.236068	5.236068	3.236068	1	

Butterworth (Maximally Flat*). The Butterworth design is an all-pole configuration with

$$N(s) = 1$$

$D(s)$ has the coefficients given in Table 2-7. The pole positions of the unity cutoff filter all lie on a unit circle and can be computed from

$$p_k = -\sin \frac{(2k - 1)\pi}{2n} + j \cos \frac{(2k - 1)\pi}{2n} \qquad k = 1, 2, \ldots, n \quad (2\text{-}115)$$

The amplitude and phase responses are found by using (2-115) in (2-107c) and (2-107a). The general form of the amplitude response for a 1-rad/s cutoff frequency is

$$A(\Omega) = -10 \log_{10}(1 + \Omega^{2n}) \qquad \text{dB} \qquad (2\text{-}116)$$

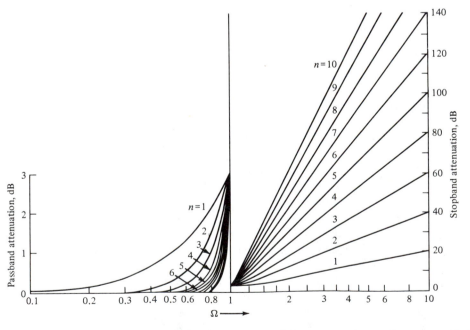

FIGURE 2-17. Attenuation versus normalized frequency for a Butterworth filter. (Reproduced from Ref. 5 with permission.)

*By *maximally flat* it is meant that the Butterworth filter transfer function has $(2n - 1)$ derivatives equal to zero at $\omega = 0$.

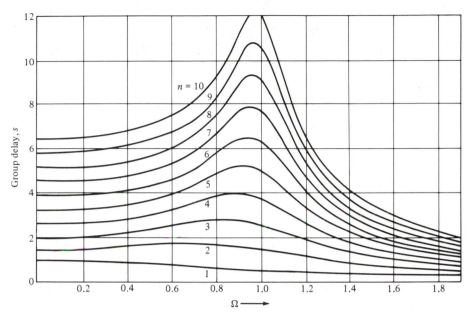

FIGURE 2-18. Group delay versus normalized frequency for Butterworth filters. (Reproduced from Ref. 5 with permission.)

where Ω denotes a normalized frequency variable for the 1-rad/s cutoff frequency. The group delay is found by obtaining the phase response and differentiating with respect to Ω. Curves showing the attenuation [the negative of (2-116)] and group delay for a 1-rad/s 3-dB cutoff frequency are shown in Figures 2-17 and 2-18, respectively. The impulse and step responses, which are useful for transient analysis, are shown in Figures 2-19 and 2-20, respectively.

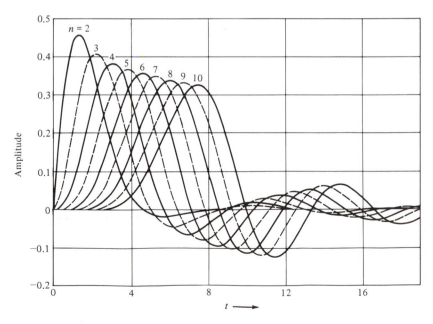

FIGURE 2-19. Impulse responses for Butterworth filters. (Reproduced from Ref. 5 with permission.)

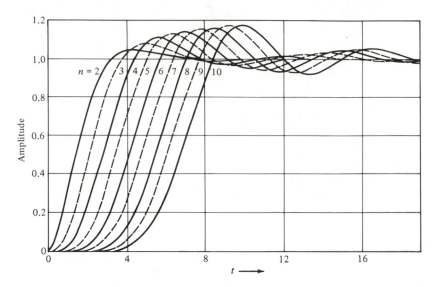

FIGURE 2-20.   Step responses for Butterworth filters. (Reproduced from Ref. 5 with permission.)

EXAMPLE 2-13

The transfer function of a third-order Butterworth filter, from Table 2-7, is

$$H(s) = \frac{1}{s^3 + 2s^2 + 2s + 1}$$

Letting $s = j\Omega$ results in

$$H(j\Omega) = \frac{1}{1 - 2\Omega^2 + j(2\Omega - \Omega^3)}$$

from which

$$A(\Omega) = -10 \log_{10}(1 + \Omega^6)$$

$$\theta(\Omega) = -\tan^{-1}\frac{2\Omega - \Omega^3}{1 - 2\Omega^2}$$

The group delay in terms of normalized frequency is

$$t_g(\Omega) = -\frac{d\theta(\Omega)}{d\Omega} = -2\left(\frac{1 + \Omega^2}{1 + \Omega^6}\right)$$

which gives the curve for $n = 3$ in Figure 2-18. Results for any cutoff frequency, say ω_c, can be obtained by letting $\Omega = \omega/\omega_c$. ∎

Chebyshev Filters (Equal Ripple).   The Chebyshev approximation to an ideal filter gives much faster rolloff in the transition region than a Butterworth filter of the same order. This is achieved by allowing ripples in the passband. The poles of a normalized Chebyshev filter are obtained by moving the poles of a normalized Butterworth response to the right by a factor

$$R_c = \tanh A \qquad\qquad (2\text{-}117a)$$

where

$$A = \frac{1}{n} \cosh^{-1}\frac{1}{\epsilon} \qquad\qquad (2\text{-}117b)$$

in which

$$\epsilon = \sqrt{10^{R/10} - 1} \qquad (2\text{-}117c)$$

and R is the passband ripple in decibels. This construction results in the poles of the Chebyshev filter lying on an ellipse with semimajor and semiminor axes

$$b, a = \tfrac{1}{2}[(\sqrt{\epsilon^{-2} + 1} + \epsilon^{-1})^{1/n} \pm (\sqrt{\epsilon^{-2} + 1} + \epsilon^{-1})^{-1/n}] \quad (2\text{-}118)$$

The amplitude response function of Chebyshev filters can be expressed as

$$A(\Omega) = -10 \log_{10}[1 + \epsilon^2 C_n^2(\Omega)] \qquad \text{dB} \qquad (2\text{-}119)$$

where $C_n(\Omega)$ is a Chebyshev polynomial of order n. Chebyshev polynomials of order 5 and less are tabulated in Table 2-8. A recursion relation for obtaining $C_n(\Omega)$ is

$$C_n(\Omega) = 2\Omega C_{n-1}(\Omega) - C_{n-2}(\Omega) \qquad n = 2, 3, \ldots \qquad (2\text{-}120)$$

where $C_o(\Omega) = 1$.

At $\Omega = 1$, Chebyshev polynomials have a value of unity, which means that the attenuation as defined by (2-119) would have a value equal to the ripple in decibels. The 3-dB cutoff is slightly greater than 1 rad/s and, in fact, is equal to $\cosh A$. In order to obtain an amplitude response function 3 dB down at $\Omega = 1$, Ω in (2-119) should be replaced by $\Omega \cosh A$. Figures 2-21 and 2-22 show attenuation and group-delay response functions, respectively, for Chebyshev filters with ripples of 0.01, 0.1, and 0.5 dB. Their impulse and step responses are shown in Figures 2-23 and 2-24, respectively.

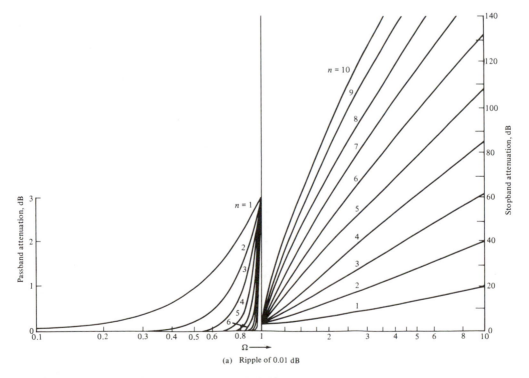

(a) Ripple of 0.01 dB

FIGURE 2-21. Attenuation as a function of normalized frequency for Chebyshev filters. (Reproduced from Ref. 5 with permission.)

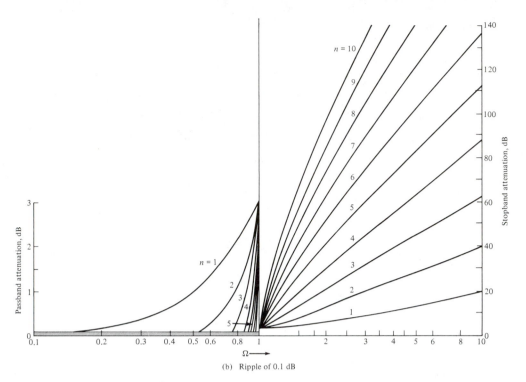

(b) Ripple of 0.1 dB

FIGURE 2-21.   continued.

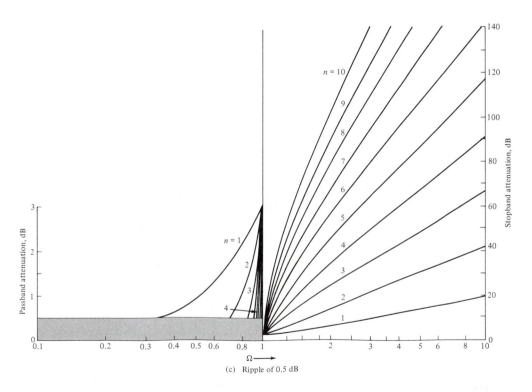

(c) Ripple of 0.5 dB

FIGURE 2-21.   continued.

TABLE 2-8.   Chebyshev Polynomials

Order	Chebyshev Polynomial
1	Ω
2	$2\Omega^2 - 1$
3	$4\Omega^3 - 3\Omega$
4	$8\Omega^4 - 8\Omega^2 + 1$
5	$16\Omega^5 - 20\Omega^3 + 5\Omega$

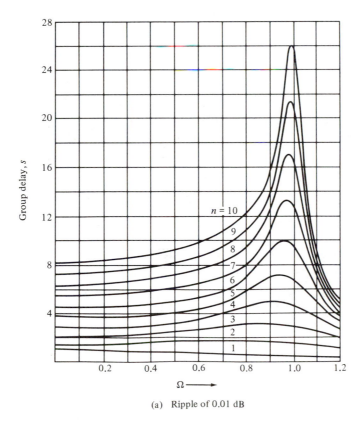

(a)   Ripple of 0.01 dB

FIGURE 2-22.   Group delay as a function of frequency for Chebyshev filters. (Reproduced from Ref. 5 with permission.)

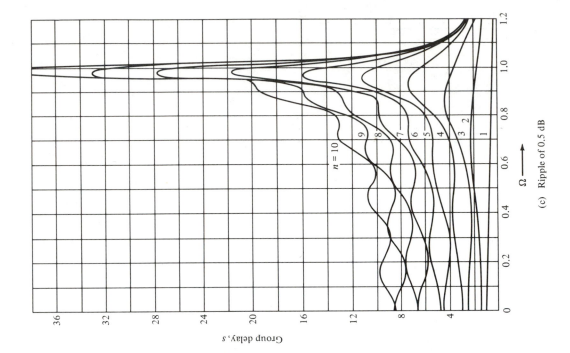

(c) Ripple of 0.5 dB

Group delay, s

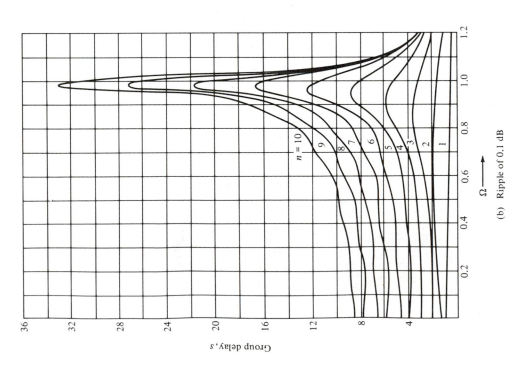

(b) Ripple of 0.1 dB

FIGURE 2-22. continued.

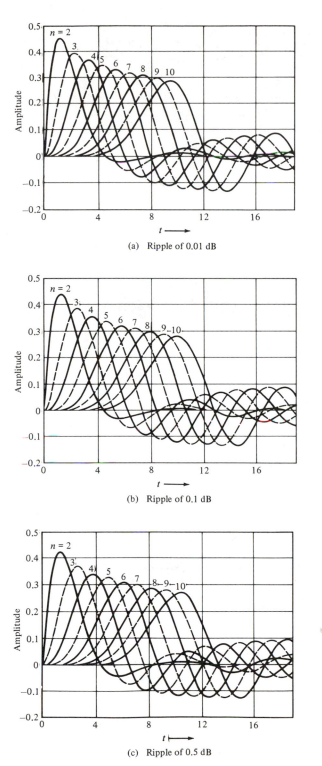

(a) Ripple of 0.01 dB

(b) Ripple of 0.1 dB

(c) Ripple of 0.5 dB

FIGURE 2-23. Impulse responses for Chebyshev filters. (Reproduced from Ref. 5 with permission.)

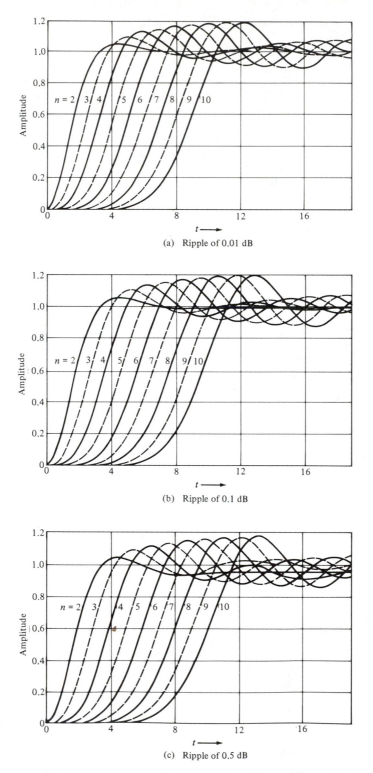

(a) Ripple of 0.01 dB

(b) Ripple of 0.1 dB

(c) Ripple of 0.5 dB

FIGURE 2-24. Step responses for Chebyshev filters. (Reproduced from Ref. 5 with permission.)

TABLE 2-9. **Factors for Calculating 3-dB Cutoff Frequencies for Cascaded Bessel Filters**

Number of Poles	Number of Filters Cascaded	Factor by Which 3-dB Frequency Exceeds 1 rad/s
2	n_f	$(-3/2 + 3\sqrt{2^{1/n}f - 3/4})^{1/2}$
3	1	1.75
3	2	1.27
3	3	1.05
4	1	2.11
4	2	1.54
4	3	1.26
5	1	2.42
5	2	1.74
5	3	1.43

Bessel (Maximally Flat Delay) Filters. The Bessel filter design is optimized to give an approximately linear phase or maximally flat group delay. The low-pass approximation to constant delay results from the general transfer function

$$H(s) = \frac{1}{\sinh s + \cosh s} \tag{2-121}$$

by expressing the hyperbolic functions as continued fraction expansions and truncating at different lengths. The resulting transfer functions are referred to as *Bessel*. As the order of a Bessel filter is increased, the region of flat delay is extended further into the stop band, but the steepness of the roll-off in attenuation in the transition region does not improve significantly. Therefore, Bessel filters are used in applications where transient properties, not rejection of undesired signals, are of main concern. The transfer function of a Bessel filter can be written in the form

$$H(s) = \frac{k_n}{B_n(s)} \tag{2-122a}$$

where $B_n(s)$ is a Bessel polynomial of order n defined recursively by

$$B_n(s) = (2n - 1)B_{n-1}(s) + s^2 B_{n-2}(s) \tag{2-122b}$$

with

$$B_o(s) = 1 \quad \text{and} \quad B_1(s) = 1 + s \tag{2-122c}$$

k_n is a constant chosen to give an amplitude response normalized to unity at $\Omega = 0$. It should be noted that application of (2-122) does not automatically result in a filter with a 3-dB cutoff frequency of 1 rad/s. The factor by which the 3-dB cutoff frequency exceeds 1 rad/s is given in Table 2-9.

The amplitude and group-delay characteristics of Bessel filters of various orders are shown in Figures 2-25 and 2-26, respectively. Their impulse and step responses are given in Figures 2-27 and 2-28, respectively.

TABLE 2-10. Normalized Pole Locations for Lowpass Butterworth, Chebyshev, and Bessel Filters (3-dB Cutoff Frequency of 1 rad/s)

Filter Type	Number of Poles, n			
	$n = 2$	$n = 3$	$n = 4$	$n = 5$
Butterworth	$p_{1,2} = -0.707(1 \pm j)$	$p_1 = 1 + j0$ $p_{2,3} = -0.5 \pm j0.866$	$p_{1,2} = -0.924 \pm j0.383$ $p_{3,4} = -0.383 \pm j0.924$	$p_1 = -1 + j0$ $p_{2,3} = -0.809 \pm j0.588$ $p_{4,5} = -0.309 \pm j0.951$
Chebyshev 0.1 dB ripple	$p_{1,2} = -0.610 \pm j0.711$	$p_1 = -0.698 + j0$ $p_{2,3} = -0.349 \pm j0.868$	$p_{1,2} = -0.526 \pm j0.383$ $p_{3,4} = -0.218 \pm j0.925$	$p_1 = -0.475 \pm j0$ $p_{2,3} = -0.384 \pm j0.588$ $p_{4,5} = -0.147 \pm j0.952$
Bessel	$p_{1,2} = -1.102 \pm j0.636$	$p_1 = -1.323 + j0$ $p_{2,3} = -1.047 \pm j0.999$	$p_{1,2} = -1.370 \pm j0.410$ $p_{3,4} = -0.995 \pm j1.257$	$p_1 = -1.502 + j0$ $p_{2,3} = -1.381 \pm j0.718$ $p_{3,4} = -0.958 \pm j1.471$

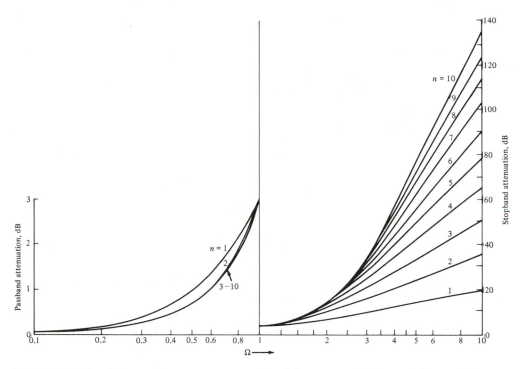

FIGURE 2-25. Attenuation versus normalized frequency for Bessel filters. (Reproduced from Ref. 5 with permission.)

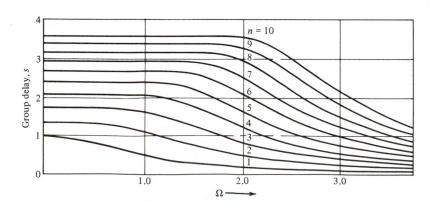

FIGURE 2-26. Group delay versus normalized frequency for Bessel filters. (Reproduced from Ref. 5 with permission.)

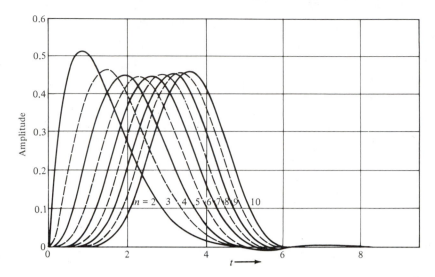

FIGURE 2-27. **Impulse responses for Bessel filters. (Reproduced from Ref. 5 with permission.)**

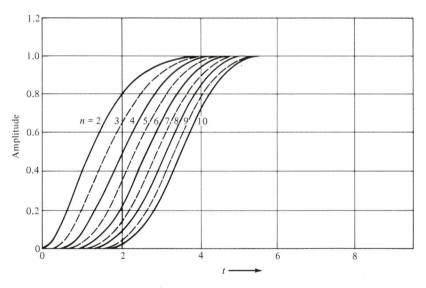

FIGURE 2-28. **Step responses for Bessel filters. (Reproduced from Ref. 5 with permission.)**

The pole locations for Butterworth, Chebyshev, and Bessel filters are given in Table 2-10 for orders 2 through 5. It should be noted that Butterworth, Chebyshev, and Bessel filters are all-pole filters. Filters that include zeros are also possible and have advantageous features not found in all-pole designs.

REFERENCES

[1] A. PAPOULIS, *Signal Analysis* (New York: McGraw-Hill, 1977).
[2] S. M. SCHETZEN, *The Volterra and Wiener Theories of Nonlinear Systems* (New York: Wiley, 1980).

[3] A. B. WILLIAMS, *Electronic Filter Design Handbook* (New York: McGraw-Hill, 1981).

[4] G. MATTHAEI, L. YOUNG, and E. JONES, *Microwave Filters, Impedance-Matching Networks, and Coupling Structures* (Dedham, Mass.: Artech House, 1980).

[5] A. I. ZVEREU, *Handbook of Filter Synthesis* (New York: Wiley, 1967).

ADDITIONAL READINGS

BRACEWELL, R. N., *The Fourier Transform and Its Applications*, 2nd ed. (New York: McGraw-Hill, 1978).

KOTZIAN, B., and A. SCHMIDT, "Understanding, Specifying, and Characterizing Mixers," Hewlett-Packard, Signal Analysis Division, 1400 Fountain Grove Parkway, Santa Rosa, CA 95404.

VIFIAN, H., "Group Delay and AM-to-PM Measurement Techniques," Hewlett-Packard, Network Measurements Division, 1400 Fountain Grove Parkway, Santa Rosa, CA 95401, March 1982.

ZIEMER, R. E., W. H. TRANTER, and D. R. FANNIN, *Signals and Systems: Continuous and Discrete* (New York: Macmillan, 1983).

PROBLEMS

2-1. Classify the following signals as to (1) finite power or finite energy; (2) periodic or aperiodic. What are the periods of those that are periodic and the energies or powers, as the case may be, for case 1?

(a) $\cos 20\pi t + \sin 26\pi t$

(b) $\exp(-10|t|)$

(c) $\Pi(t + \frac{1}{2}) - \Pi(t - \frac{1}{2})$

(d) $\dfrac{1}{\sqrt{t^2 + 1}}$

(e) $\cos 120\pi t + \cos 377t$

2-2. Classify the following systems as to (1) linear or nonlinear; (2) fixed or time varying; (3) causal or noncausal. The notation used is as follows: $x(t)$ represents the input, $y(t)$ represents the output, and $h(t)$ represents the system's impulse response.

(a) $h(t) = \dfrac{1 - \cos 20\pi t}{(20\pi t)^2}$

(b) $\dfrac{dy(t)}{dt} + 10y(t) = x^2(t)$

(c) $\dfrac{dy(t)}{dt} + ty(t) = x(t)$

(d) $\dfrac{dy(t)}{dt} + y(t) = x(t + 1)$

(e) $\dfrac{dy(t)}{dt} + y^2(t) = x(t)$

2-3. Find the response of a system with impulse response $h(t) = 10 \exp[(-10t)u(t)]$ to the following inputs:

(a) $x(t) = u(t)$

(b) $x(t) = \Pi\left(\dfrac{t}{2}\right)$

(c) $x(t) = \exp(-3t)u(t)$

(d) $x(t) = \exp(j6\pi t)u(t)$

(e) $x(t) = \cos(6\pi t)u(t)$

2-4.   (a) Show that the following set of functions is orthonormal on the interval $0 \le t \le 4$:

$$\varphi_1(t) = \Pi(t - 0.5) \qquad \varphi_2(t) = \Pi(t - 1.5)$$
$$\varphi_3(t) = \Pi(t - 2.5) \qquad \varphi_4(t) = \Pi(t - 3.5)$$

(b) Expand the signal

$$x(t) = \exp\left(\frac{-t}{2}\right)u(t)$$

as an orthogonal function series on the interval $0 \le t \le 4$ using the set of functions above.

(c) Find the integral squared error for the expansion found in part (b).

2-5.   Derive the relationships (2-30) between the exponential Fourier series coefficients and the sine–cosine Fourier series.

2-6.   Carry out the derivation of the cosine form of the Fourier series given by (2-33) starting with the exponential Fourier series. Express A_0, A_n, and θ_n in terms of X_0 and X_n.

2-7.   Derive Parseval's theorem given by (2-34).

2-8.   Derive the Fourier series given in Table 2-3.

2-9.   Use (2-39) to obtain Fourier transforms for the following periodic signals:

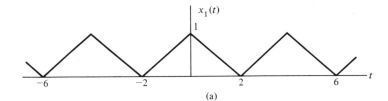

(a)

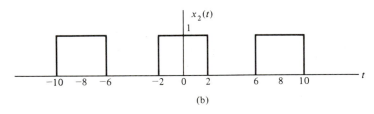

(b)

PROBLEM 2-9.

2-10.   Use pair 17 of Table 2-5 and the multiplication theorem of Fourier transforms to derive the spectrum of an impulse sampled signal.

2-11.   Use appropriate Fourier transform theorems to obtain the amplitude and phase spectra for the signal $\Pi[(t - 2)/4] \exp[j2\pi(t - 2)]$.

2-12.   Obtain the transfer function of the *RC* lowpass filter shown in Figure 2-5 using the three methods illustrated in that figure.

2-13.   Derive (2-47).

2-14.   (a) Use (2-47) to find the response of a system having impulse response

$$h(t) = \exp(-6t)u(t)$$

to an arbitrary input.

(b) Specialize the result found in part (a) to a system with input $x(t) = \exp(-5t)u(t)$ and compare the series expression derived with the exact result found by applying the superposition integral.

2-15. Given a message signal with the spectrum

$$M(f) = \Pi\left(\frac{f}{20}\right)$$

and a carrier signal

$$c(t) = \cos 200\pi t$$

sketch the spectra of the following modulated signals:
(a) Double sideband.
(b) AM with modulation index 0.5.
(c) Single sideband, upper sideband transmitted.
(d) Single sideband, lower sideband transmitted.

2-16. Derive and sketch modulated time waveforms for the cases listed in Problem 2-15(a) and (b).

2-17. **(a)** Derive the Hilbert transform of the signal $m(t) = \Pi(t/2)$.
(b) Use the result obtained in part (a) to obtain an expression for the analytic signal corresponding to $m(t) = \Pi(t/2)$.
(c) Sketch the amplitude spectrum of the analytic signal above.

2-18. **(a)** Obtain the spectrum of the narrowband signal

$$x(t) = \exp(-|t|) \cos 1000\pi t$$

(b) Obtain the time-domain response of a filter with transfer function given by (2-78) to the input signal given in part (a) for $Q = 100$ and $f_0 = 500$ Hz.

2-19. Given a filter with transfer function

$$H(f) = \left[\Pi\left(\frac{f}{10}\right) + \Pi\left(\frac{f}{20}\right)\right] \exp(-j20\pi f)$$

From a sketch of the amplitude and phase response, determine what type of distortion, if any, is imposed on the following inputs:
(a) $x_1(t) = \cos 5\pi t + \cos 15\pi t$
(b) $x_2(t) = \cos 12\pi t + \cos 18\pi t$
(c) $x_3(t) = \cos 18\pi t + \cos 30\pi t$

2-20. Analytically obtain the impulse response for the filters with transfer functions shown in Figure 2-9.

2-21. Derive the group and phase-delay functions for the following filters: (a) first-order Butterworth; (b) second-order Bessel; (c) second-order Chebyshev.

2-22. Derive the response of a filter with amplitude response

$$A(f) = (1 + 0.1f^2)\Pi\left(\frac{f}{10}\right)$$

and phase response

$$\theta(f) = -0.2\pi f + 0.05f^3$$

to the input $x(t) = \cos(2\pi t)$.

2-23. Verify the curves of Figure 2-13.

2-24. Derive (2-98).

2-25. Derive the AM-to-PM conversion term for a filter with

$$\Delta A(f) = -0.1f + 0.1f^2$$

$$\Delta\theta(f) = 0.1f^2$$

and the input

$$x(t) = (1 + 0.2 \cos 20\pi t) \cos 100\pi t$$

Let $t_0 = 0$ and $A_o = 1$. Assume that the filter passband is centered at the carrier frequency and bandwidth wide enough to pass the modulated signal.

2-26. Obtain the response of the filter of Problem 2-25 to the phase-modulated signal

$$x(t) = \cos(100\pi t + 0.1 \cos 10\pi t)$$

2-27. Consider the response of a nonlinear system as defined by (2-102) with $h_1(\lambda) = \exp(-\lambda)u(\lambda)$ and $h_2(\lambda_1, \lambda_2) = 0.1 \exp(-3\lambda_1 - 4\lambda_2)u(\lambda_1)u(\lambda_2)$ and input $x(t) = u(t) - u(t - 1)$. Assume $A_0 = 1$ and $t_0 = 0$.

CHAPTER 3

Performance Characterization of Digital Data Transmission Systems

3-1

INTRODUCTION

The basic problem of transmission of digital data through communication channels is introduced in this chapter. Initially, attention is focused on the following features of the general digital communication system:

1. Only linear modulation schemes are considered. This means that the transmitted signal can be expressed as

$$x(t) = \text{Re } [Am(t) \exp (j\omega_0 t)]$$

$$= A[m_R(t) \cos \omega_0 t - m_I(t) \sin \omega_0 t] \tag{3-1}$$

   where A is the amplitude and $f_0 = \omega_0/2\pi$ is the carrier frequency in hertz; $m(t) = m_R(t) + jm_I(t)$ is the message signal sequence, which may, in general, be complex and is described in more detail later for various modulation schemes.
2. The noise introduced by the channel is additive, white, and Gaussian (AWGN) with two-sided power spectral density $N_0/2$.
3. The phase and frequency of the received carrier are known exactly at the receiver.
4. The epoch of each symbol in the message sequence is known at the receiver.

Assuming these idealized features, two situations will be considered in regard to designing a receiver and analyzing its performance (in terms of probability of making a transmission error) for reception of a given type of modulation:

1. The channel is linear and has infinite bandwidth. This leads to the concept of a matched filter detector.
2. The channel is bandlimited with appropriate filtering included at the transmitter and receiver to compensate for the interference introduced between symbols [termed *intersymbol interference* (ISI)] by the channel filter. This analysis makes use of early work done by Nyquist in regard to ISI-free transmission.

Departures of practical digital communication systems from the idealized models above result in degradation from the ideal performance. Following are examples of some of the departures from the ideal situation, termed *impairments*:

1. Knowledge of the phase and/or frequency of the received carrier is imperfect at the receiver.
2. The timing derived at the receiver for the epochs of the digital data symbols is imperfect.
3. Nonlinearities are present in the channel either before and/or after introduction of the noise.
4. The noise may not be additive, white, or Gaussian.

The effects of some of these impairments are considered later in the chapter.

3-2

DETECTION OF BINARY SIGNALS IN WHITE, GAUSSIAN NOISE

The general binary detection problem to be considered in this section is illustrated in Figure 3-1. During a given signaling interval of T seconds' duration, one of two possible signals, denoted by $s_1(t)$ and $s_2(t)$, is present together with additive noise, $n(t)$. The energies of $s_1(t)$ and $s_2(t)$ in a T-second interval, which are assumed to be finite, are denoted as E_1 and E_2, respectively. The noise is assumed to be Gaussian and white with double-sided power spectral density $N_0/2$. A decision as to which signal is present is to be made each T-second interval.

3-2.1 Receiver Structure and Analysis

The receiver consists of linear, time-invariant filter followed by a sampler and threshold comparator. The initial conditions of the filter are set to zero just prior to the arrival of each new signal. If the sample taken at the end of a T-second interval is greater than the threshold, A, the decision is made that $s_2(t)$ was present; if the sample value is less than A, the decision that $s_1(t)$ was present is made. Denoting the sample value at time $t = T$ (without loss of generality, assume that $t_0 = 0$ and consider the first signaling interval) by V, the decision strategy just described is expressed mathematically as

$$\text{If } V > A, \text{ decide that } s_2(t) \text{ sent}$$

$$\text{If } V < A, \text{ decide that } s_1(t) \text{ sent} \qquad (3\text{-}2)$$

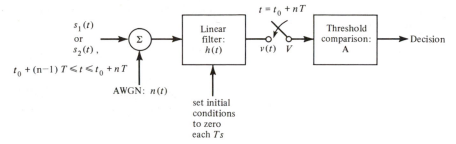

FIGURE 3-1. Receiver structure for detecting binary signals in white Gaussian noise.

Because the filter is linear and fixed, its output at time T can be expressed in terms of its impulse response, $h(t)$, as

$$V = v(T) = \int_{-\infty}^{\infty} y(\lambda)h(t - \lambda) \, d\lambda \qquad (3\text{-}3)$$

where $y(t) = s_1(t) + n(t)$ if $s_1(t)$ is present and $y(t) = s_2(t) + n(t)$ if $s_2(t)$ is present.

The question now is: How should $h(t)$ and A be selected so as to optimize receiver performance? Before answering this question, the criterion to be used to decide on optimality of performance must be chosen. The usual criterion employed is minimum probability, P_E, of making a decision error. To carry out the optimization, the probability of error is expressed in terms of $h(t)$ and A. They are then chosen so that P_E is minimized for given $s_1(t)$, $s_2(t)$, and N_0.

Given $s_1(t)$ is present at the filter input, its output at $t = T$ is

$$V = S_1 + N \qquad [s_1(t) \text{ present}] \qquad (3\text{-}4a)$$

where S_1 is the output signal component at time $t = T$ for the input $s_1(t)$, and N is the output-noise component. Similarly, given $s_2(t)$ is present at the input, the output can be expressed as

$$V = S_2 + N \qquad [s_2(t) \text{ present}] \qquad (3\text{-}4b)$$

where the notation is similar to that used for (3-4a).

Both signal and noise components could be expressed in terms of superposition integrals. For the present, however, use is made of the fact that N is a Gaussian random variable (the superposition integral is a linear transformation) and only its mean and variance are required in order to write down its probability density function.* Since $n(t)$ has zero mean (implied by its constant power spectrum), so does N. Its variance is, therefore, the same as its mean-square value, which can be expressed as

$$\sigma^2 = \int_{-\infty}^{\infty} |H(f)|^2 \, \frac{N_0}{2} \, df$$

$$= N_0 \int_0^{\infty} |H(f)|^2 \, df \qquad (3\text{-}5)$$

where $H(f)$ is the transfer function of the filter. Since (3-4a) and (3-4b) are linear transformations of Gaussian random variables, they are also Gaussian but with

*See Appendix A for a summary of probability and random process theory.

nonzero means. Given that $s_1(t)$ is present at the receiver input, the probability density function of V is

$$P(v|s_1) = \frac{e^{-(v-S_1)^2/2\sigma^2}}{\sqrt{2\pi\sigma^2}} \qquad (3\text{-}6)$$

which is a Gaussian density with mean S_1 and variance σ^2. Similarly, the conditional probability density function of V given that $s_2(t)$ is present at the receiver input is

$$P(v|s_2) = \frac{e^{-(v-S_2)^2/2\sigma^2}}{\sqrt{2\pi\sigma^2}} \qquad (3\text{-}7)$$

Recall that S_1 and S_2 are the filter outputs at time $t = T$ with $s_1(t)$ and $s_2(t)$ present, respectively, at its input. These two conditional density functions are sketched in Figure 3-2 assuming that $S_1 < A < S_2$. Note that there is no loss in generality in this assumption. If $S_2 < S_1$, the roles of $s_1(t)$ and $s_2(t)$ are simply reversed.

Figure 3-2 shows the two ways that an error can be made: in particular, if $s_1(t)$ was sent and $V > A$ or if $s_2(t)$ was sent and $V < A$. The corresponding probabilities of error are crosshatched. From Figure 3-2 it follows that the probability of error given $s_1(t)$ present is

$$P(E|s_1) = \int_A^\infty p(v|s_1)\,dv \qquad (3\text{-}8)$$

and if $s_2(t)$ is present it is

$$P(E|s_2) = \int_{-\infty}^A p(v|s_2)\,dv \qquad (3\text{-}9)$$

If the a priori probability that $s_1(t)$ was sent is p, and the a priori probability that $s_2(t)$ was sent is $q = 1 - p$, the average probability of error is

$$P_E = pP(E|S_1) + qP(E|S_2) \qquad (3\text{-}10)$$

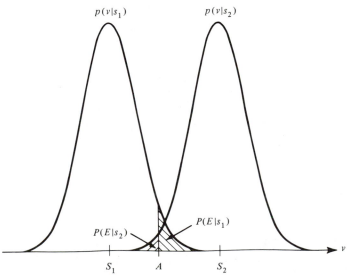

FIGURE 3-2. Conditional probability density functions of the filter output at time $t = T$.

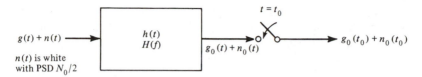

FIGURE 3-3. Block diagram of the system pertinent to the matched filter derivation.

If (3-8) and (3-9) are substituted into (3-10), the result differentiated with respect to A, and the derivative set equal to zero, the optimum choice for the threshold in terms of minimizing P_E is

$$A = A_{opt} = \frac{\sigma^2}{S_2 - S_1} \ln \frac{p}{q} + \frac{S_1 + S_2}{2} \tag{3-11a}$$

or, if $p = q$,

$$A_{opt} = \frac{S_1 + S_2}{2} \qquad (p = q) \tag{3-11b}$$

If $p = q$ and $A = A_{opt}$, the average probability of error can be expressed as

$$P_E = \frac{1}{2} \operatorname{erfc}\left(\frac{S_2 - S_1}{2\sqrt{2}\sigma}\right) = Q\left(\frac{S_2 - S_1}{2\sigma}\right) \tag{3-12}$$

where erfc $(u) = 1 - \operatorname{erf}(u)$ is the complementary error function and $Q(u)$ is the Gaussian integral, sometimes referred to as the Q-function.*

3-2.2 The Matched Filter

In order to find the filter that gives the minimum probability of error, as expressed by (3-12), what may appear to be an unrelated detour will be taken. Consider the situation depicted in Figure 3-3. A linear, time-invariant filter with a known signal $g(t)$ plus white noise at its input is followed by a sampler that samples the filter output at time $t = t_0$. At time t_0, the sample value consists of a signal-related component, $g_0(t_0)$, and a noise component, $n_0(t_0)$. Let the variance of this noise component be σ^2. The transfer function, $H_0(f)$, for the filter that provides the maximum signal-to-noise ratio at its output at time t_o is desired, where the signal-to-noise ratio is defined as

$$\zeta^2 = \frac{g_0^2(t_0)}{\sigma^2} \tag{3-13}$$

To proceed, $g_0(t_0)$ and σ^2 are expressed in terms of $H(f)$, where $H(f)$ is the transfer function of the filter before the optimization. If the two-sided power spectral density of the input noise is $N_0/2$, σ^2 may be expressed as

$$\sigma^2 = \frac{N_0}{2} \int_{-\infty}^{\infty} |H(f)|^2 \, df \tag{3-14}$$

which is the same as (3-5). Using the inverse Fourier transform, the signal component at the filter's output is

$$g_0(t) = \int_{-\infty}^{\infty} G(f)H(f)e^{j2\pi ft} \, df \tag{3-15}$$

*See Appendix E for rational approximations and tables for $Q(u)$.

where $G(f)$ is the Fourier transform of the input. Setting $t = t_0$ and taking the ratio of (3-15) squared to (3-14), an expression for ζ^2, the signal-to-noise ratio to be maximized, is obtained in terms of $H(f)$:

$$\zeta^2 = \frac{\left| \int_{\infty}^{\infty} G(f)H(f)e^{j2\pi ft_0}\, df \right|^2}{(N_0/2) \int_{\infty}^{\infty} |H(f)|^2\, df} \tag{3-16}$$

It will be verified that the maximum value for ζ^2 is achieved if, and only if,

$$H(f) = H_0(f) = kG^*(f)e^{-j2\pi ft_0} \tag{3-17}$$

where k is an arbitrary constant. For this $H_0(f)$, it can be verified by direct substitution into (3-16) that the maximum value for ζ^2 is

$$\zeta^2_{max} = \frac{2E_g}{N_0} \tag{3-18}$$

where

$$E_g = \int_{-\infty}^{\infty} |G(f)|^2\, df = \int_{-\infty}^{\infty} |g(t)|^2\, dt \tag{3-19}$$

is the total energy contained in $g(t)$.

To show that $H_0(f)$ given by (3-17) does indeed provide the maximum value for ζ^2 given by (3-18), Schwarz's inequality will be applied. It is a generalization of the inequality involving the dot product of two vectors $\mathbf{A}$ and $\mathbf{B}$, which is

$$|\mathbf{A} \cdot \mathbf{B}| = |\mathbf{A}|\, |\mathbf{B}|\, |\cos \theta| \leq |\mathbf{A}|\, |\mathbf{B}| \tag{3-20}$$

where θ is the angle between them and $|\mathbf{A}|$ denotes the magnitude of $\mathbf{A}$, and so on. That (3-20) holds is obvious, since $|\cos \theta| \leq 1$. Furthermore, since $|\cos \theta| = 1$ if, and only if, $\theta = n\pi$, where n is an integer, it follows that equality holds in (3-20) if, and only if, $\mathbf{A} = k\mathbf{B}$, where k is a constant (i.e., if $\mathbf{A}$ and $\mathbf{B}$ are collinear).

The generalization of (3-20), which is one form of Schwarz's inequality, occurs if $\mathbf{A}$ is replaced by $X(f)$, $\mathbf{B}$ is replaced by $Y(f)$, and the dot product is replaced by $\int_{-\infty}^{\infty} X(f)Y^*(f)\, df$, where the asterisk denotes a complex conjugate [both $X(f)$ and $Y(f)$ may be complex functions]. Then the inequality analogous to (3-20) is

$$\left| \int_{-\infty}^{\infty} X(f)Y^*(f)\, df \right|^2 \leq \int_{-\infty}^{\infty} |X(f)|^2\, df \int_{-\infty}^{\infty} |Y(f)|^2\, df \tag{3-21}$$

with equality if, and only if, $X(f) = kY^*(f)$, where k is a constant.† To maximize (3-16), let $G(f) = X(f)$, $H(f)e^{j2\pi ft_0} = Y^*(f)$, and replace the numerator of (3-16) by the right-hand side of (3-21) to obtain the inequality

$$\zeta^2 \leq \frac{\int_{-\infty}^{\infty} |G(f)|^2\, df \int_{-\infty}^{\infty} |H(f)|^2\, df}{(N_0/2) \int_{-\infty}^{\infty} |H(f)|^2\, df} \tag{3-22}$$

or

$$\zeta^2 \leq \frac{2}{N_0} \int_{-\infty}^{\infty} |G(f)|^2\, df = \frac{2E_g}{N_0} \tag{3-23}$$

†Schwarz's inequality is proved in Chapter 4.

Equality holds if, and only if, $X(f) = kY(f)$ or, in other words, if (3-17) holds. Thus (3-22) shows that the maximum for ζ^2 is indeed given by (3-18) and that this maximum is achieved if, and only if, (3-17) holds.

The transfer function of the optimum filter is given by (3-17), where k can be set equal to 1 since it is arbitrary. The impulse response of the optimum filter is the inverse Fourier transform of $H_0(f)$, which is

$$h_0(t) = \mathscr{F}^{-1}[H_0(f)] \tag{3-24}$$

$$= \int_{-\infty}^{\infty} G^*(f)e^{j2\pi f(t - t_0)} \, df$$

$$= \left[\int_{-\infty}^{\infty} G(f)e^{j2\pi f(t_0 - t)} \, df \right]^*$$

$$= g(t_0 - t) \tag{3-25}$$

where $g(t)$ is assumed real. In other words, the impulse response of the optimum filter is the time reverse of the input signal. For this reason, the filter is said to be *matched* to the input signal and it is referred to as a *matched filter*. A matched filter, according to (3-13), *maximizes the filter output signal squared at time t_0 divided by the output-noise variance*. This maximum value for the signal-to-noise ratio, according to (3-18), is *twice the energy of the input signal divided by the input-noise spectral density*, regardless of the input signal shape.

The output signal component from a matched filter can be found by applying the superposition integral. Using (3-25), the output is found to be

$$g_0(t) = \int_{-\infty}^{\infty} g(\lambda)h_0(t - \lambda) \, d\lambda$$

$$= \int_{-\infty}^{\infty} g(\lambda)g(t_0 - t + \lambda) \, d\lambda \tag{3-26}$$

This is essentially the autocorrelation function of the input signal. If $t = t_0$, (3-26) becomes

$$g_0(t_0) = \int_{-\infty}^{\infty} g^2(\lambda) \, d\lambda = E_g \tag{3-27}$$

which is the energy of the input signal expressed in the time domain. In other words, *the peak output signal from a matched filter*, which occurs at time t_0, *is the energy contained in the input signal*.

EXAMPLE 3-1

Consider the signal $g(t) = \Pi[(t - 1)/2]$, which is sketched in Figure 3-4a. Sketched in Figure 3-4b is the impulse response of the filter matched to this signal. The output signal is the triangular pulse with maximum value at $t = t_0$ sketched in Figure 3-4c. This maximum value is

$$g_0(t_0) = 2 \text{ (volts)}^2 - \text{seconds} \tag{3-28}$$

which is the energy in the input signal. If the input noise is white with double-sided power spectral density $N_0/2 = 1$ V^2/Hz, (3-28) is also the ratio of peak output signal squared to mean-square noise. Note that if $t_0 < 1$ s, the filter is noncausal. ∎

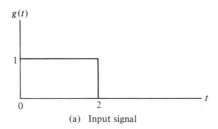

(a) Input signal

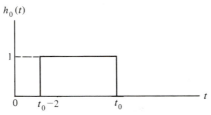

(b) Matched filter impulse response

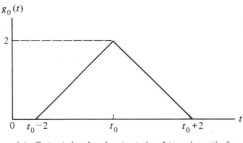

(c) Output signal and output signal-to-noise ratio for $\frac{N_0}{2} = 1$ (volt)^2/Hz.

FIGURE 3-4. Illustration of a matched filter for rectangular input signal.

3-2.3 Application of the Matched Filter to Binary Data Detection

Since the Q-function is a monotonically decreasing function of its argument, the expression for P_E for binary signaling given by (3-12) will be minimized through the choice of $H(f)$ in Figure 3-1 by maximizing $(S_2 - S_1)/2\sigma$ or equivalently by maximizing the square of this quantity. Letting $g_0(t) = S_2(t) - S_1(t)$ and $t_0 = T$ in (3-13), where $S_2(t)$ and $S_1(t)$ are the filter's response to the input $s_1(t)$ and $s_2(t)$, respectively, it is seen that the problem of minimizing P_E through choosing $H(f)$, which has just pointed out is equivalent to maximizing its argument, is identical to the matched filter problem. In particular, if $H(f)$ in Figure 3-1 is chosen to be

$$H_0(f) = [S_2^*(f) - S_1^*(f)]e^{-j2\pi fT} \tag{3-29}$$

where $S_1(f)$ and $S_2(f)$ are the Fourier transforms of $s_1(t)$ and $s_2(t)$ shown in Figure 3-1, the probability of error will be minimized.

The impulse response of the optimum filter is

$$h_0(t) = s_2(T - t) - s_1(T - t) \qquad (3\text{-}30)$$

and the corresponding signal-to-noise ratio at the matched filter output, according to (3-18), is

$$\zeta_{max}^2 = \frac{2}{N_0} \int_0^T [s_2(t) - s_1(t)]^2 \, dt \qquad (3\text{-}31)$$

where the fact that the signals are zero outside the range $(0, T)$ has been used. If the square root of (3-31) is substituted into (3-12) for $(S_2 - S_1)/\sigma$, the probability of error corresponding to the optimum receiver filter becomes

$$P_E = \tfrac{1}{2} \operatorname{erfc}(\sqrt{z}) = Q(\sqrt{2z}) \qquad (3\text{-}32)$$

where

$$z = \frac{1}{4N_0} \int_0^T [s_2(t) - s_1(t)]^2 \, dt \qquad (3\text{-}33)$$

Note that P_E depends on the dissimilarity of $s_1(t)$ and $s_2(t)$ through their difference in addition to their individual energies. Several examples will make this clear. In these examples, the signal duration, T, is denoted as T_b to indicate clearly that these are binary schemes.

EXAMPLE 3-2   ANTIPODAL BASEBAND SIGNALING

Suppose that $s_1(t) = -A$ and $s_2(t) = A$ for $0 \le t \le T_b$. Then $s_2(t) - s_1(t) = 2A$ and (3-33) becomes

$$z = \frac{1}{4N_0} \int_0^{T_b} (2A)^2 \, dt$$

$$= \frac{A^2 T_b}{N_0} = \frac{E_b}{N_0} \qquad (3\text{-}34)$$

where E_b is the energy in either $s_1(t)$ or $s_2(t)$, which is often referred to as the *bit energy*. Consequently, P_E for this case becomes

$$P_E = \frac{1}{2} \operatorname{erfc}\left(\sqrt{\frac{E_b}{N_0}}\right) = Q\left(\sqrt{\frac{2E_b}{N_0}}\right) \qquad (3\text{-}35)$$

A handy approximation to (3-35) for large values of E_b/N_0 is obtained by using the asymptotic expression

$$\operatorname{erfc}(u) \simeq \frac{e^{-u^2}}{u\sqrt{\pi}} \qquad u \gg 1 \qquad (3\text{-}36)$$

which is within 6% of the true value of $\operatorname{erfc}(u)$ for $u \ge 3$.   ∎

EXAMPLE 3-3   BIPHASE SHIFT-KEYED (BPSK) SIGNALING

Consider next the case of biphase modulation for which

$$s_1(t) = -A_c \cos 2\pi f_0 t$$

$$0 \le t \le T_b$$

$$s_2(t) = A_c \cos 2\pi f_0 t$$

The expression for z given by (3-33) then becomes

$$z = \frac{1}{4N_0} \int^{T_b} (2A_c \cos 2\pi f_0 t)^2 \, dt$$

$$= \frac{A_c^2 T_b}{2N_0}$$

$$= \frac{E_b}{N_0} \tag{3-37}$$

where the "bit energy" is now $A_c^2 T_b / 2$ because of the sinusoidal carrier. Thus P_E for biphase signaling is identical to that for baseband signaling. ∎

EXAMPLE 3-4   AMPLITUDE SHIFT-KEYED (ASK) SIGNALING

Consider the signal set

$$s_1(t) = 0$$

$$0 \le t \le T_b$$

$$s_2(t) = A_c \cos 2\pi f_0 t$$

and

$$z = \frac{1}{4N_0} \int_0^{T_b} (A_c \cos 2\pi f_0 t)^2 \, dt$$

$$= \frac{A_c^2 T_b}{8N_0} \tag{3-38}$$

At first glance, it might appear that amplitude shift keying gives the same performance as biphase modulation. However, it is important to note that a signal is being transmitted only half the time, on the average, assuming that zeros and ones are equally likely. Thus the *average* ratio of bit energy to noise spectral density is one-half of (3-38), so that

$$z = \frac{E_{b,av}}{2N_0} \tag{3-39}$$

Consequently, ASK requires twice as much average signal energy as BPSK signaling to produce the same P_E. Saying it in another way, biphase is 3 dB better than amplitude shift keying in terms of SNR required for a given P_E. ∎

EXAMPLE 3-5   COHERENT FREQUENCY SHIFT-KEYED (FSK) SIGNALING

Consider the signal set

$$s_1(t) = A_c \cos 2\pi f_0 t$$

$$0 \le t \le T_b$$

$$s_2(t) = A_c \cos 2\pi (f_0 + \Delta f)t$$

where $\Delta f = m/2T_b$ with m an integer; $s_1(t)$ and $s_2(t)$ are said to be coherently orthogonal since

$$\int_0^{T_b} s_1(t)s_2(t) \, dt = 0$$

This signaling scheme is referred to as *coherent frequency shift keying*. In this case,

$$\int_0^{T_b} [s_2(t) - s_1(t)]^2 \, dt = \int_0^{T_b} s_2^2(t) \, dt + \int_0^{T_b} s_1^2(t) \, dt$$

$$= A_c^2 T_b$$

and, therefore,

$$z = \frac{A_c^2 T_b}{4N_0}$$

$$= \frac{E_b}{2N_0} \qquad (3\text{-}40)$$

Consequently, coherent FSK signaling has the same performance as ASK signaling in terms of E_b/N_0 to provide a given probability of error. It is 3 dB worse than BPSK. Figure 3-5 shows P_E versus E_b/N_0 for these signaling schemes.

3-2.4 Correlator Realization of Matched Filter Receivers

According to (3-30) and the superposition integral, the optimum receiver performs the operations depicted in Figure 3-6a. The output signal from the matched filter

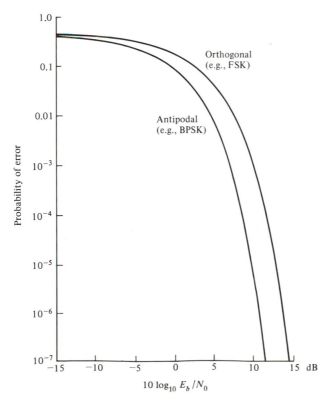

FIGURE 3-5. P_E **for BPSK (antipodal) and FSK or ASK signaling (orthogonal) data transmission.**

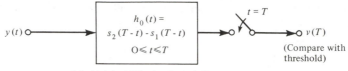

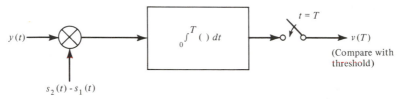

(a) Matched filter implementation

(b) Correlator implementation

FIGURE 3-6. Receiver implementations for binary data detection.

is

$$v(t) = y(t) * h_0(t) = \int_0^T [s_2(T - \lambda) - s_1(T - \lambda)]y(t - \lambda)\, d\lambda$$

which is sampled at time $t = T$ to produce

$$v(T) = \int_0^T [s_2(T - \lambda) - s_1(T - \lambda)]y(T - \lambda)\, d\lambda$$

$$= \int_0^T [s_2(\alpha) - s_1(\alpha)]y(\alpha)\, d\alpha \qquad (3\text{-}41)$$

where the change of variables $\alpha = T - \lambda$ has been made. Consequently, the matched filter receiver structure shown in Figure 3-6a is equivalent to the multiplier–integrator structure shown in Figure 3-6b. Such a realization is referred to as a *correlation receiver*.

Note that the multiplication can be carried out in two steps if carrier modulation is employed. For example, a BPSK correlation receiver can be implemented as shown in Figure 3-7, where the first stage of the receiver is a coherent demodulator followed by integration over each signaling interval and sampling the integrator output at the end of the signaling intervals. Once coherent demodulation has been performed, the only operation left is integration over the signaling interval [the constant $2A_c$ that results from taking the difference $s_2(t) - s_1(t)$ can be set to unity

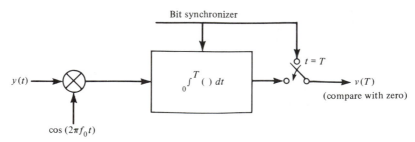

FIGURE 3-7. Receiver implementation for biphase signaling consisting of a coherent demodulator and integrate-and-dump detector.

since it multiplies both signal and noise]. After the sampling operation, the integrator initial condition is set to zero, and the process repeated for the next signaling interval. Such a series of operations is often referred to as *coherent demodulation* followed by *integrate-and-dump detection*. Note that any of the binary signaling schemes discussed in Examples 3-3 through 3-5 can be optimally detected in this fashion.

At least two levels of synchronization are necessary in such a receiver. The first is *carrier synchronization,* wherein the coherent carrier reference is established to perform coherent demodulation. The second, often referred to as *bit synchronization,* involves the establishment of a clock signal in order to perform the integrate-and-dump operation in synchronism with the incoming stream of contiguous signaling elements. Yet a third level of synchronization, referred to as *frame synchronization*, is required in many applications where the data are grouped into blocks called *frames*. This is usually accomplished by inserting a frame synchronization code between data blocks and detecting the position of this code. Synchronization techniques are discussed in Chapter 6.

3-3

QUADRATURE-MULTIPLEXED SIGNALING SCHEMES: QPSK, OQPSK, and MSK

In this section, several signaling schemes are considered which can be viewed as resulting from impressing two different data streams on two carriers in phase quadrature. The first of these is quadrature phase-shift keying (QPSK). The other two schemes to be considered are offset QPSK (OQPSK) and minimum-shift keying (MSK).

3-3.1 Quadrature Multiplexing

Consider two data-bearing signals, $m_1(t)$ and $m_2(t)$, which are to be modulated onto carriers of the same frequency. This can be accomplished so that $m_1(t)$ and $m_2(t)$ can be separated at the receiver by using carriers that are in *phase quadrature*. That is, the transmitted signal is

$$x_c(t) = A_1 m_1(t) \cos (2\pi f_0 t + \alpha) + A_2 m_2(t) \sin (2\pi f_0 t + \alpha) \quad (3\text{-}42)$$

where α is a constant phase shift to be chosen for convenience later. The block diagram of the modulator and demodulator for such a scheme is shown in Figure 3-8.

If $m_1(t)$ and $m_2(t)$ are binary digital signals with amplitudes ± 1 which may change at time intervals spaced by T_s seconds,* $x_c(t)$ can be written as

$$x_c(t) = A \cos [2\pi f_0 t + \theta(t) + \alpha] \quad (3\text{-}43a)$$

where

$$\theta(t) = \tan^{-1} \frac{-A_2 m_2(t)}{A_1 m_1(t)} \quad (3\text{-}43b)$$

$$A = \sqrt{A_1^2 + A_2^2} \quad (3\text{-}43c)$$

*T_s denotes a *symbol* interval.

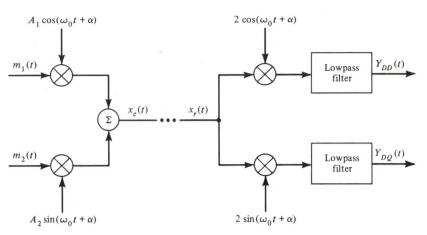

FIGURE 3-8.   Block diagram of a quadrature-multiplexed communication system.

The constellation of signal vectors corresponding to (3-43) is shown in Figure 3-9 for m_1 and m_2 binary-valued. It is seen that $x_c(t)$ is, in effect, a constant-amplitude, digitally phase-modulated signal whose phase deviation takes on the values

$$\tan^{-1} \frac{A_2}{A_1} \qquad \tan^{-1} \frac{A_2}{A_1} + \pi$$

$$-\tan^{-1} \frac{A_2}{A_1} \qquad -\tan^{-1} \frac{A_2}{A_1} + \pi \qquad\qquad (3\text{-}44)$$

3-3.2   Quadrature and Offset-Quadrature Phase-Shift Keying

If $A_1 = A_2 = A/\sqrt{2}$ and $m_1(t)$ and $m_2(t)$ have the instants when their signs can change aligned, the resulting modulated signal is called *quadrature phase-shift*

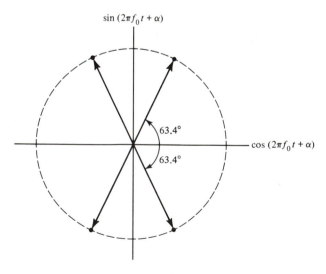

FIGURE 3-9.   Phasor diagram for a quadrature-multiplexed signal where m_1 and m_2 are constant-amplitude binary signals ($A_1 = 1$ and $A_2 = 2$ assumed for purposes of sketching).

keying (QPSK). If $A_1 = A_2$, but the time instants when $m_1(t)$ can change sign are offset by $T_s/2$ seconds from the time instants when $m_2(t)$ can change sign, the resulting phase-modulated signal is referred to as *offset QPSK* (OQPSK) or *staggered QPSK*. The former can change phase by 0, ± 90, or 180 degrees at the switching times for m_1 and m_2, whereas the phase shifts for OQPSK are limited to 0 or ± 90 degrees since only m_1 or m_2 can change sign at a given switching instant. Signal space diagrams for QPSK and OQPSK are compared in Figure 3-10, which illustrates the different phase shifts and phase transitions possible.

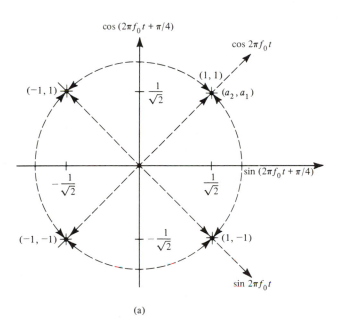

(a)

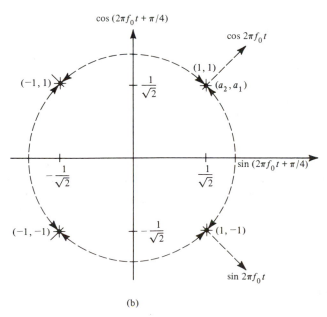

(b)

FIGURE 3-10. Signal-space (phasor) diagrams for (a) QPSK and (b) OQPSK showing possible phase transitions for each ($\alpha = \pi/4$).

The power spectra of QPSK and OQPSK are identical. However, experiments have shown that the envelope of OQPSK tends to remain more nearly constant when the waveform is filtered. Thus it is preferable over QPSK when used in applications where significant bandlimiting of the signal is performed. This point is returned to later when system impairments are discussed.

3-3.3 Minimum-Shift Keying

Another choice of $m_1(t)$ and $m_2(t)$ which allows (3-42) to be expressed in constant-envelope form is

$$m_1(t) = a_1(t) \cos 2\pi f_1 t \qquad (3\text{-}45a)$$

and

$$m_2(t) = a_2(t) \sin 2\pi f_1 t \qquad (3\text{-}45b)$$

with $A_1 = A_2$, where $a_1(t)$ and $a_2(t)$ are ± 1 binary-valued signals whose signs may change each T_s seconds with the switching times for $a_2(t)$ offset from those of $a_1(t)$ by $T_s/2$ seconds. The result for $x_c(t)$ is then of the form (3-43) with $A_1 = A_2 = A$ and

$$
\begin{aligned}
\theta(t) &= -\tan^{-1}\left[\frac{a_2(t)}{a_1(t)} \tan 2\pi f_1 t\right] \\
&= \pm 2\pi f_1 t + u_k
\end{aligned}
\qquad (3\text{-}46)
$$

where the sign on the first term in (3-46) is minus if the signs of a_1 and a_2 are the same, and plus if they are opposite. The angle $u_k = 0$ or π modulo-2π corresponding to $a_1 = 1$ or -1, respectively. Thus, for this case, (3-42) becomes

$$x_c(t) = A \cos \left[2\pi(f_0 \pm f_1)t + u_k\right] \qquad (3\text{-}47)$$

Note that the sign in front of f_1 may change at intervals of $T_b = T_s/2$ seconds since a_1 and a_2 are staggered.

With $f_1 = \frac{1}{2}T_s = \frac{1}{4}T_b$ and a_1 staggered $T_s/2 = T_b$ seconds relative to a_2, the phase of $x_c(t)$ is continuous, and the resulting waveform is referred to as a *minimum-shift-keyed* (MSK) *signal*. Its name arises from the fact that $\Delta f = 2f_1 = \frac{1}{2}T_b$ is the minimum frequency spacing for the two signals represented by (3-47) to be coherently orthogonal (i.e., the integral of the product of the two possible transmitted signals over a T_b-second interval is zero). Waveforms for MSK are illustrated in Figure 3-11.

The *peak-to-peak frequency deviation* of an angle-modulated signal can be defined as the difference between its maximum and minimum instantaneous frequencies. Thus MSK can be viewed as an FM-modulated waveform with peak-to-peak frequency deviation $\frac{1}{2}T_b$. Alternatively, it can be viewed as an OQPSK signal for which sinusoidal pulse shapes are employed for the binary message waveforms.

3-3.4 Performance of Digital Quadrature Modulation Systems

The input to the demodulator of the quadrature multiplexed system shown in Figure 3-8 consists of a quadrature-multiplexed signal with digital modulation impressed

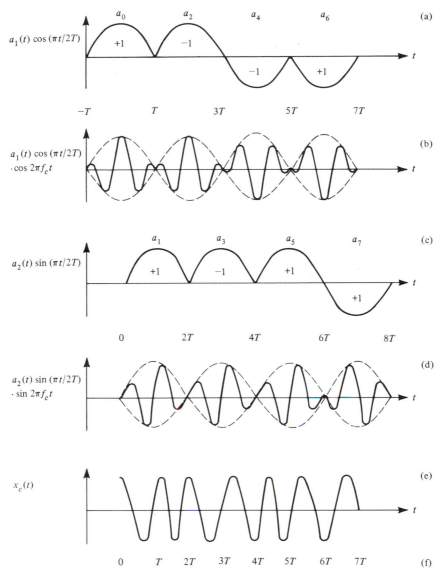

FIGURE 3-11. MSK waveforms. (From Ref. 1.)

on each component, plus white Gaussian noise, $n(t)$. Using the notation for the signal component given by (3-42) with $\alpha = 0$ for convenience, this input can be written as

$$y(t) = x_c(t) + n(t)$$

$$= A_c[m_1(t) \cos 2\pi f_0 t + m_2(t) \sin 2\pi f_0 t] + n(t) \qquad (3\text{-}48)$$

where $m_1(t)$ and $m_2(t)$ are the data-related message signals and $n(t)$ represents white Gaussian noise with spectral density $N_0/2$. Depending on the particular form chosen for $m_1(t)$ and $m_2(t)$, $x_c(t)$ could represent a QPSK, an OQPSK, or an MSK signal, as discussed previously. As shown in Figure 3-8, the quadrature components of $x_c(t)$ can be coherently demodulated independently of each other by employing two phase-quadrature sinusoidal reference signals. Following coherent demodula-

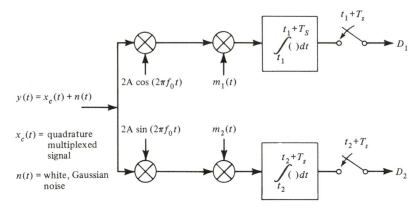

Note: $t_1 = t_2 = 0$ for QPSK
$t_1 = 0; t_2 = T_s/2$ for OQPSK and MSK

FIGURE 3-12. Block diagram of quadrature multiplexed demodulator pertinent to analyzing error probability of QPSK, OQPSK, and MSK.

tion, the next operation to be performed on each output signal from the phase-quadrature demodulators is correlation detection. Consequently, we represent the receiver as two coherent demodulators in parallel, in which quadrature references are employed, followed by correlation detectors as illustrated in Figure 3-12. From Figure 3-12 it follows that the output of the top detector at time $t = t_1 + T_s$ is*

$$D_1 = A^2 \int_{t_1}^{t_1 + T_s} m_1^2(t) \ dt + N_1 \qquad (3\text{-}49)$$

and the output of the bottom detector at time $t_2 + T_s$ is

$$D_2 = A^2 \int_{t_2}^{t_2 + T_s} m_2^2(t) \ dt + N_2 \qquad (3\text{-}50)$$

where double-frequency terms are assumed to integrate to zero and N_1 and N_2 are the output noise components to be discussed shortly. If the modulation technique employed is QPSK or OQPSK, $m_1^2(t) = m_2^2(t) = 1$, so that the first terms of (3-49) and (3-50) both evaluate to A^2T_s, which is the energy in *one quadrature component* of $x_c(t)$ during one symbol period. Because it is the energy in one quadrature component, which carries one bit per symbol, we denote it as E_b. Similarly, if $m_1(t) = a_1 \cos (\pi t/2T_b)$ and $m_2(t) = a_2 \sin (\pi t/2T_b)$, so that $x_c(t)$ represents an MSK-modulated signal, the first terms of (3-49) and (3-50) evaluate to $A^2T_b/2 = E_b$, which is again the energy in *one quadrature component* of $x_c(t)$ during a symbol period.

The second terms of (3-49) and (3-50), which represent noise components, are given by

$$N_1 = 2A \int_{t_1}^{t_1 + T_s} n(t)m_1(t) \cos 2\pi f_c t \ dt \qquad (3\text{-}51)$$

*t_1 and t_2 are chosen to correspond with the starting times of the quadrature-channel signaling intervals. Also, the scaling factor $2A$ for the quadrature carrier references in Figure 3-12 is arbitrary and chosen for convenience only.

and

$$N_2 = 2A \int_{t_2}^{t_2+T_s} n(t)m_2(t) \sin 2\pi f_c t \, dt \tag{3-52}$$

respectively. Both of these noise components are Gaussian, uncorrelated, and therefore statistically independent. Because they are statistically independent, the decisions at the upper and lower detector outputs in Figure 3-13 may be considered separately.

The variances of N_1 and N_2 can be shown to be

$$\text{var}\,(N_1) = \text{var}\,(N_2) = N_0 E_b \tag{3-53}$$

where the factor E_b may at first look strange. However, this is due to the factor of A included in the reference signals of Figure 3-12. The derivation of (3-53) is left to the problems.

The probability of error for the upper correlation detector in Figure 3-12 is

$$P_1(\epsilon) = \text{Pr}\,(-E_b + N_1 > 0) \tag{3-54}$$

where ϵ denotes the error event.

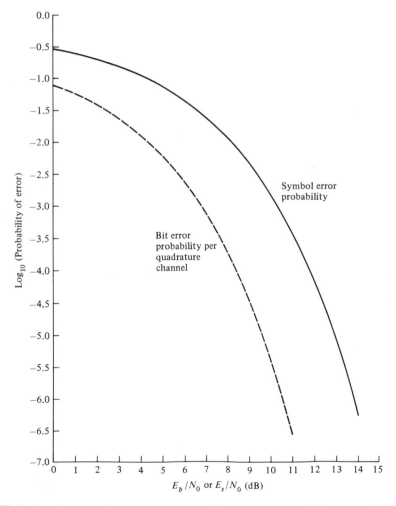

FIGURE 3-13. Comparison of symbol and bit error probabilities for QPSK.

since the signal component at its output is $-E_b$ if the upper-arm input data stream is a -1 in this interval. Using the fact that N_1 is Gaussian with zero mean and variance $N_0 E_b$, the probability of error for this correlation detector is

$$P_1(\epsilon) = Q\left(\sqrt{\frac{2E_b}{N_0}}\right) \tag{3-55}$$

The lower correlator operates with the same probability of error. Thus, when considered on a *per channel basis, quadrature modulation systems*, such as QPSK, O-QPSK, and MSK, *perform the same as binary biphase signaling.*

The symbol error probability of QPSK or OQPSK is obtained by noting that probability of *correct reception* of each phase, or symbol, is

$$P_s(c) = [1 - P_1(\epsilon)][1 - P_2(\epsilon)] \tag{3-56}$$

since both quadrature-channel signals must be received correctly in order to detect the transmitted signal phase correctly. Thus the probability of error for a symbol (i.e., the phase of the QPSK signal) is

$$P_s(\epsilon) = 1 - [1 - P_1(\epsilon)]^2 \simeq 2P_1(\epsilon) \qquad P_1(\epsilon) << 1 \tag{3-57}$$

Substituting (3-55) and letting $E_s = 2E_b$, this can be written as

$$P_s(\epsilon) \simeq 2Q\left(\sqrt{\frac{E_s}{N_0}}\right) \tag{3-58}$$

where E_s is the energy in the transmitted signal during one symbol period, T_s. When compared with BPSK, (3-58) shows that the symbol error probability of QPSK is 3 dB inferior to biphase [a factor of 2 in front of $Q(\cdot)$ has been ignored]. However, QPSK and OQPSK are modulation schemes which can be employed to double the number of bits per hertz of bandwidth over biphase. When compared on an equal data rate (bits/hertz) basis, (3-55) indicates that QPSK, OQPSK, and biphase are equivalent in terms of bit error performance. Figure 3-13 gives a comparison of the symbol and bit error probabilities for QPSK. If Gray encoding of the symbols is used, the symbol error probability and bit error probability are approximately equal for high SNRs. (Gray encoding is discussed in Chapter 4.)

3-4

POWER SPECTRA FOR BPSK, QPSK, OQPSK, AND MSK

As pointed out in Chapter 1, it is generally an objective to use the resources of bandwidth and power in a communication system as efficiently as possible. The probability of error expression just obtained for various digital modulation schemes provides a basis for power efficiency comparisons. In this section the computation of power spectra for these modulation schemes will provide a basis for choosing a particular scheme on the basis of bandwidth efficiency.

In Appendix A, the computation of the power spectrum of random sequence pulse-code-modulated signals is considered.* The general waveform considered is of the form

$$s(t) = \sum_{i=-\infty}^{\infty} A_k p(t - kT - \Delta) \tag{3-59}$$

*For the general case, see [2], pp. 12ff.

where $\{A_k\}$ is a sequence of random amplitudes, $p(t)$ is a deterministic pulse-shape function with finite energy, T is the repetition period of the pulse-train signal, and Δ is a random variable uniformly distributed in $(0, T)$.

For present purposes, this is generalized to signals having complex envelopes of the form

$$z(t) = x(t) + jy(t) \tag{3-60}$$

where

$$x(t) = \sum_{k=-\infty}^{\infty} a_k p(t - kT_s + \Delta t) \tag{3-61}$$

$$y(t) = \sum_{m=-\infty}^{\infty} b_m q(t - mT_s + \Delta t) \tag{3-62}$$

The real and imaginary parts of the complex envelope are randomly modulated pulse trains. In (3-61) and (3-62), a_n and b_m are independent, identically distributed (iid) random sequences with zero means and mean-square values

$$\overline{a_k^2} = A^2 \tag{3-63a}$$

and

$$\overline{b_m^2} = B^2 \tag{3-63b}$$

respectively. The signals $p(t)$ and $q(t)$ are basic pulse-shape functions whose Fourier transforms, $P(f)$ and $Q(f)$, exist. The time increment, Δt, which is uniformly distributed in $(0, T_s)$, is present to ensure stationarity of $z(t)$. With these assumptions, it is left to the problems to show that the power spectral density of $z(t)$ is

$$G_z(f) = \frac{A^2|P(f)|^2 + B^2|Q(f)|^2}{T_s} \tag{3-64}$$

It follows that the power spectrum of the real signal

$$s(t) = \mathrm{Re}\,[z(t)e^{j2\pi f_0 t}]$$

is

$$S(f) = \tfrac{1}{2}G_z(f - f_0) + \tfrac{1}{2}G_z(f + f_0) \tag{3-65}$$

This result will now be applied to computing the power spectra for BPSK, QPSK, OQPSK, and MSK.

For BPSK, the quadrature component $y(t)$ is identically zero and the pulse-shape function is

$$p(t) = \Pi\!\left(\frac{t}{T_b}\right)$$

where T_b is the bit period and $T_s = T_b$. Its Fourier transform is

$$\Pi\!\left(\frac{t}{T_b}\right) \leftrightarrow T_b\,\mathrm{sinc}\,(T_b f)$$

so that the baseband spectrum for BPSK, from (3-64), is

$$G_z(f) = A^2 T_b\,\mathrm{sinc}^2(T_b f) \tag{3-66}$$

For QPSK, $T_s = 2T_b$ and the basic pulse shapes are

$$p(t) = \frac{1}{\sqrt{2}} \Pi\left(\frac{t}{2T_b}\right) = q(t)$$

and for OQPSK they are

$$p(t) = q(t - T_b) = \frac{1}{\sqrt{2}} \Pi\left(\frac{t}{2T_b}\right)$$

Since the time shift of $q(t)$ by T_b seconds for OQPSK results in the factor $\exp(-j2\pi f T_b)$ in the Fourier transform of $q(t - T_b)$, which has magnitude unity, the spectrum of OQPSK is identical to that of QPSK. Using the transform pair

$$\frac{1}{\sqrt{2}} \Pi\left(\frac{t}{2T_b}\right) \leftrightarrow \sqrt{2}T_b \operatorname{sinc}(2T_b f)$$

and the fact that quadrature components for QPSK and OQPSK have equal amplitudes, it follows that the baseband power spectrum for QPSK and OQPSK is

$$G_z(f) = 2A^2 T_b \operatorname{sinc}^2(2T_b f) \qquad \text{(QPSK and OQPSK)} \qquad (3\text{-}67)$$

Finally, for MSK, the pulse-shape functions are

$$p(t) = q(t - T_b) = \cos\frac{\pi t}{2T_b} \Pi\left(\frac{t}{2T_b}\right)$$

With $A = B$ and $T_s = 2T_b$, use of these definitions together with the transform pair

$$\cos\left(\frac{\pi t}{2T_b}\right) \Pi\left(\frac{t}{2T_b}\right) \leftrightarrow \frac{4T_b}{\pi} \frac{\cos 2\pi T_b f}{1 - (4T_b f)^2}$$

and the time-delay theorem of Fourier transforms in (3-64) results in the baseband power spectrum

$$G_z(f) = \frac{16A^2 T_b}{\pi^2} \frac{\cos^2 2\pi T_b f}{[1 - (4T_b f)^2]^2} \qquad \text{(MSK)} \qquad (3\text{-}68)$$

Baseband power spectra for BPSK, QPSK, OQPSK, and MSK are compared in Figure 3-14, where it is seen that the first null is at a frequency of $\frac{1}{2}T_b$ hertz for QPSK and OQPSK, $0.75/T_b$ hertz for MSK, and $1/T_b$ hertz for BPSK. The null-to-null bandwidths for the bandpass spectra are twice these values. It will also be noted that the sidelobes for MSK fall off at a rate of 12 dB per octave, while the sidelobes for BPSK, QPSK, and OQPSK fall off at half this rate. A better comparison of bandwidth requirements for these modulation schemes is given in terms of fractional out-of-band power, which is given in terms of the power spectrum, $G_z(f)$, by

$$F = \frac{\displaystyle\int_W^\infty G_z(f)\, df}{\displaystyle\int_0^\infty G_z(f)\, df} \qquad (3\text{-}69)$$

The fractional out-of-band powers for these modulation schemes are compared in Figure 3-15.

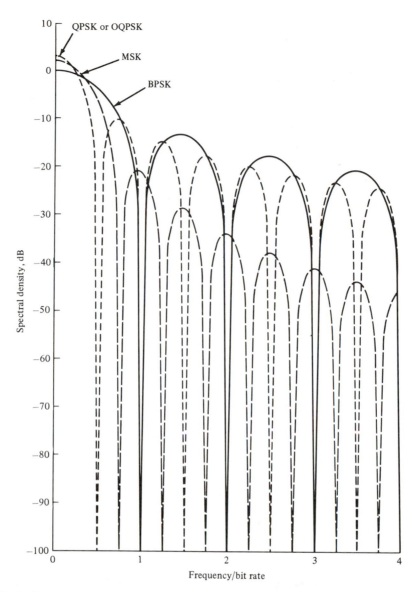

FIGURE 3-14. Baseband equivalent power spectra for **BPSK, QPSK or O-QPSK, and MSK.**

A measure of the compactness of a spectrum is the bandwidth B, which contains 99% of the total power. From graphs of out-of-band power, it follows that

$$B \simeq \begin{cases} \dfrac{1.2}{T_b} & \text{for MSK} \\[2em] \dfrac{8}{T_b} & \text{for QPSK and OQPSK} \end{cases} \tag{3-70}$$

Thus, while the main lobe of the MSK spectrum is broader than the main lobe of a QPSK or OQPSK spectrum, the sidelobes decrease much more rapidly for the former than for the latter.

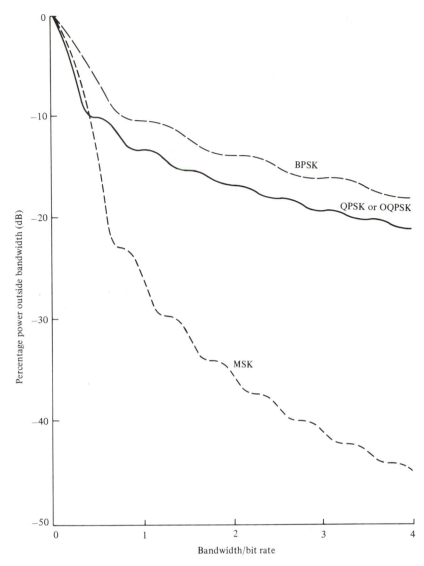

FIGURE 3-15.   Fractional out-of-band power for BPSK, QPSK, or O-QPSK, and MSK.

SERIAL MODULATION AND DETECTION OF MSK

The modulation and demodulation of MSK can be accomplished in two equivalent fashions, parallel and serial. The parallel approach was discussed in Section 3-4; the serial technique is described in this section. Its theoretical performance is identical to that of the parallel approach.

In parallel modulation of MSK, a serial data stream, $d(t)$, can be thought of as being "demultiplexed" into its even- and odd-indexed bits to produce the two bit streams, $a_1(t)$ and $a_2(t)$, which are staggered $\frac{1}{2}$ symbol and then used to biphase modulate signals $x(t)$ and $y(t)$ given by

$$x(t) = A \cos \frac{\pi t}{2T_b} \cos 2\pi f_0 t \qquad\qquad (3\text{-}71a)$$

$$y(t) = A \sin \frac{\pi t}{2T_b} \sin 2\pi f_0 t \qquad\qquad (3\text{-}71b)$$

These signals are summed to produce the MSK-modulated signal.

3-5.1 Serial Approach

The serial modulation of MSK is somewhat more subtle to grasp than the parallel method. A serial modulator structure for serial MSK is illustrated in Figure 3-16a. It is seen to consist of a BPSK modulator with carrier frequency of $f_0 - \frac{1}{4}T_b$ hertz, and a bandpass conversion filter with impulse response

$$g(t) = \begin{cases} \dfrac{1}{T_b} \sin 2\pi \left(f_0 + \dfrac{1}{4T_b} \right) t & 0 < t < T_b \\[2ex] 0 & \text{otherwise} \end{cases} \qquad (3\text{-}72)$$

which corresponds to a $(\sin x)/x$-shaped transfer function.

The serial demodulator structure for MSK is essentially the reverse of the serial MSK modulator structure and is illustrated in Figure 3-16b. It consists of a bandpass matched filter followed by a coherent demodulator and lowpass filter which eliminates double-frequency components at the mixer output. A major difference between the serial modulator and demodulator (other than the reversal of operations) is the matched filter, which has a transfer function proportional to the square root of the power spectrum of the MSK signal (with a linear phase response included). While the response of the matched filter to a single bit of duration T_b at the

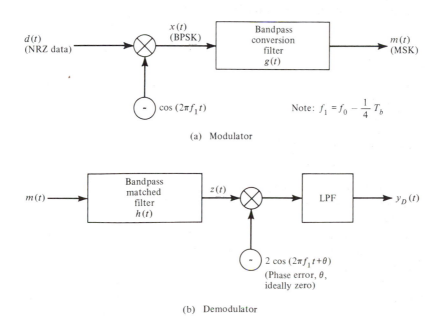

(a) Modulator

(b) Demodulator

FIGURE 3-16. Serial modulator and demodulator structures for MSK. (From Ref. 3.)

modulator input lasts for $2T_b$ seconds, the response due to a preceding symbol is ideally zero at the optimum sampling instant for the present symbol, thereby producing zero intersymbol interference (see [4]).

Serial MSK modulation and demodulation have the advantage that all operations are performed serially, and therefore offer significant implementation advantages at high data rates. The precise synchronization and balancing required for the quadrature signals of the parallel structures are no longer present. The critical system components now are the biphase modulator, the bandpass conversion and matched filters, and the coherent demodulator.*

A heuristic feel for the validity of the serial modulation technique for MSK is provided by considering the product of the power spectrum of the BPSK signal (shown in Figure 3-17a) and the magnitude squared of the transfer function of the conversion filter (Figure 3-17b), whose center frequency is offset from the BPSK carrier by one-half data rate. The resulting MSK power spectrum is shown in Figure 3-17c. When mathematically expressed in terms of single-sided spectra, the BPSK spectrum is proportional to

$$S_{\text{BPSK}}(f) = 2 \operatorname{sinc}^2[(f - f_0)T_b + 0.25] \qquad (3\text{-}73)$$

The conversion filter transfer function is

$$H(f) = \operatorname{sinc}[(f - f_0)T_b - 0.25] \exp(-j2\pi f t_0) \qquad (3\text{-}74)$$

where t_0 represents an arbitrary delay for the filter. The product of $S_{\text{BPSK}}(f)$ and $|H(f)|^2$ can be simplified to

$$S_{\text{MSK}}(f) = \frac{16}{\pi^2} \left[\frac{\cos 2\pi(f - f_0)T_b}{1 - 16T_b^2(f - f_0)^2} \right]^2 \qquad (3\text{-}75)$$

which is identical to the bandpass power spectrum for MSK obtained earlier.

3-5.2 Terminology and Trellis Diagrams

Terminology that has been used in the literature for the various frequencies used in the discussion of MSK is the following:

Apparent carrier: frequency f_0 at which the maximum of the signal spectrum occurs.
Mark frequency: the frequency $f_1 = f_0 - \frac{1}{4}T_b$, which corresponds to an all-ones or all-zeros data sequence.
Space frequency: the frequency $f_2 = f_0 + \frac{1}{4}T_b$, which corresponds to an alternating 1–0 data sequence.

That these data sequences do indeed give these frequencies will now be demonstrated with a convenient tool referred to as a *trellis diagram*. The trellis diagram of the excess phase of an angle-modulated signal is a useful way of visualizing the signal's behavior for various data inputs. The instantaneous phase of any modulated signal is defined as the argument, $\theta(t)$, of the sine or cosine representing the signal. In general, the instantaneous phase of a modulated signal consists of two terms, that due to the carrier, $2\pi f_0 t$, where f_0 is the carrier frequency in hertz, and that due to the modulation, $\varphi(t)$, called the *excess phase*. The resultant modulated signal

---

*Realization techniques for these filters are surveyed in [3].

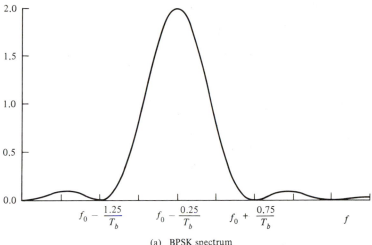

(a) BPSK spectrum

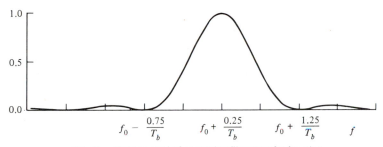

(b) Magnitude squared of conversion filter transfer function

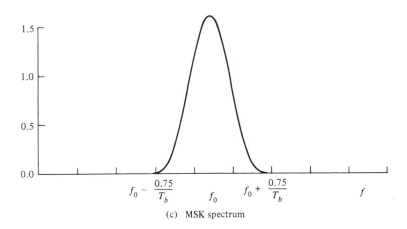

(c) MSK spectrum

FIGURE 3-17.   Spectra used in the derivation of serial MSK. (From Ref. 3.)

is

$$m(t) = A \cos [\omega_0 t + \varphi(t)] \qquad \omega_0 = 2\pi f_0 \qquad (3\text{-}76)$$

For MSK, assuming parallel modulation, the modulated signal can be written as

$$m(t) = A \left[a_1(t) \cos \frac{\pi t}{2T_b} \cos 2\pi f_0 t + a_2(t) \sin \frac{\pi t}{2T_b} \sin 2\pi f_0 t \right] \qquad (3\text{-}77)$$

where $a_1(t)$ and $a_2(t)$ are the ± 1-valued data sequences for the I and Q channels. Using trigonometric identities, (3-77) can be written in the form (3-76), where

$$\varphi(t) = -\tan^{-1}\left[\frac{a_2(t) \sin (\pi t/2T_b)}{a_1(t) \cos (\pi t/2T_b)}\right]$$

$$= -b_k(t) \frac{\pi t}{2T} + \varphi_k \tag{3-78}$$

where $b_k(t) = a_1(t)a_2(t)$ and $\varphi_k = 0$ or π corresponding to $a_1 = 1$ or -1. Therefore, if the signs of $a_1(t)$ and $a_2(t)$ are the same [i.e., $d(t)$ is an all-ones or all-zeros data sequence], the instantaneous phase of the carrier is

$$\theta(t) = 2\pi f_0 t - \frac{\pi t}{2T_b} = 2\pi\left(f_0 - \frac{1}{4T_b}\right)t \tag{3-79}$$

which corresponds to a sinusoid of frequency $f_0 - \frac{1}{4}T_b$ hertz. On the other hand, if the signs of $a_1(t)$ and $a_2(t)$ are opposite [i.e., $d(t)$ is an alternating sequence of ones and zeros], then

$$\theta(t) = 2\pi f_0 t + \frac{\pi t}{2T_b} = 2\pi(f_0 + \frac{1}{4}T_b)t \tag{3-80}$$

which corresponds to a sinusoid of frequency $f_0 + \frac{1}{4}T_b$ hertz. If the excess phase of an MSK signal is plotted versus t, as shown in Figure 3-18, it follows that it is piecewise linear and increases or decreases exactly $\pi/2$ radians each T_b seconds. Lines with positive slope represent alternating 1–0 sequences and lines with negative slope represent all-ones or all-zeros sequences. Examples of each of these situations are shown by the heavy lines in Figure 3-18.

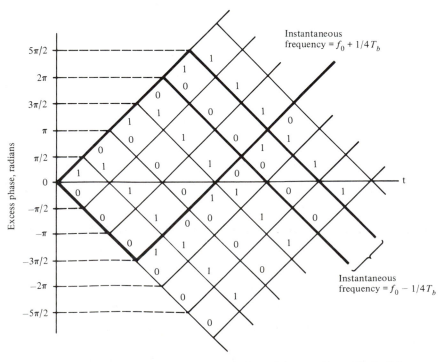

FIGURE 3-18. Trellis diagram for serial generation of MSK. (From Ref. 3.)

CHARACTERIZATION OF DIGITAL DATA TRANSMISSION SYSTEMS **132**

SIGNALING THROUGH BANDLIMITED CHANNELS

An assumption implicit in the analysis of the data transmission schemes considered so far is that there are no bandwidth restrictions imposed. Any filtering carried out in the transmitter, channel, or receiver would result in degradation of system performance. Such degradations are discussed in a later section. In this section, special signal designs and filter functions that avoid this degradation will be discussed.

3-6.1 System Model

The communications system to be considered is modeled as shown in Figure 3-19. It consists of a source that emits a binary symbol† $a_k = +1$ or $a_k = -1$ during the kth signaling interval $(k - 1)T \le t \le kT$. The signal at the source output is represented for all time as

$$x_s(t) = \sum_{k=-\infty}^{\infty} a_k \, \delta(t - kT) \qquad (3\text{-}81)$$

where $\delta(t)$ is the unit impulse function. The next subsystem is a transmitter filter with lowpass transfer function‡ $H_T(f)$ or impulse response $h_T(t) = \mathcal{F}^{-1}[H_T(f)]$. The transmitted signal is then given by

$$x_t(t) = \sum_{k=-\infty}^{\infty} a_k \, \delta(t - kT) * h_T(t)$$

$$= \sum_{k=-\infty}^{\infty} a_k h_T(t - kT) \qquad (3\text{-}82)$$

where the asterisk denotes convolution.

The channel introduces additional filtering, which is imposed by a filter with transfer function $H_c(f)$ and impulse response $h_c(t) = \mathcal{F}^{-1}[H_c(f)]$, and additive Gaussian noise represented by $n(t)$ having power spectral density $G_n(f)$. Thus the channel output, or receiver input, is given by

$$y(t) = x(t) + n(t) \qquad (3\text{-}83)$$

---

†The case of more than two values for a_k will be discussed later.

‡The bandpass case can be handled by mixing up in frequency at the transmitter (up-conversion) and down-conversion at the receiver.

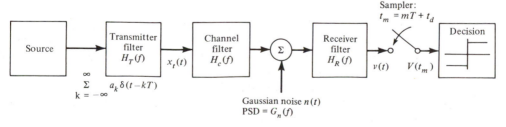

FIGURE 3-19. Idealized model for signaling through a bandlimited channel.

where

$$x(t) = x_t(t) * h_c(t) \tag{3-84}$$

is the channel filter output.

The receiver consists of a filter with transfer function $H_R(f)$ and impulse response of $h_R(t)$, a sampler, and a threshold comparator. The comparison threshold can be set at zero because of the symmetry of the data and noise amplitude distributions about zero.

The output of the receiver filter, denoted by $v(t)$, is given by

$$v(t) = y(t) * h_R(t)$$

$$= \sum_{k=-\infty}^{\infty} Aa_k p_r(t - t_d - kT) + n_0(t) \tag{3-85}$$

where A is a scale factor chosen such that $p_r(0) = 1$, $n_0(t)$ is the noise component at the receiver filter output, and $p_r(t - t_d)$ is the pulse shape at the receiver filter output, which is delayed by an amount t_d relative to $p_s(t)$ due to filtering. In terms of filter impulse responses, $n_0(t)$ and $p_r(t - t_d)$ can be written as

$$n_0(t) = n(t) * h_R(t) \tag{3-86}$$

$$Ap_r(t - t_d) = h_T(t) * h_c(t) * h_R(t) \tag{3-87}$$

The maximum of $p_r(t - t_d)$ is assumed to occur at $t = t_d$. Thus samples should be taken at times

$$t_m = mT + t_d \tag{3-88}$$

in order to sample at the peak of each received pulse. The samples at the output of the receiver filter are written as

$$V_m \triangleq v(t_m) = Aa_m p_r(0) + \sum_{\substack{k=-\infty \\ k \neq m}}^{\infty} Aa_k p_r[(m - k)T] + N_m$$

$$m = \ldots, -2, -1, 0, 1, 2, \ldots \tag{3-89}$$

where

$$N_m = n_0(t_m) = n(t) * h_R(t)\big|_{t=t_m} \tag{3-90}$$

are the noise samples at time t_m at the receiver filter output. Because $n(t)$ is Gaussian and the receiver filter is a linear system, the noise samples are Gaussian. Furthermore, it will be assumed that $G_n(f)|H_R(f)|^2$ is sufficiently wideband so that any two separate noise samples, say N_k and N_m, $k \neq m$, are uncorrelated and therefore independent.

Equation (3-89) consists of a desired signal term $Aa_m p_r(0)$, an undesired signal term which is the second term, and the noise term, N_m. The undesired signal term is referred to as *intersymbol interference* (ISI) because it originates from received signal pulses that precede and follow the desired signal pulse. Figure 3-20, which shows typical transmitted and received pulse sequences, illustrates the effect of the ISI component on detection of the desired symbol. Both the ISI and noise components cause errors in the detection process.

(a) Source output

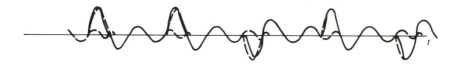

(b) Channel output

FIGURE 3-20.   Typical transmitted and received pulse trains for a bandlimited digital communication system.

3-6.2 Designing for Zero ISI: Nyquist's Pulse-Shaping Criterion

Since the channel is assumed fixed, the goal is to choose $H_T(f)$ and $H_R(f)$ to minimize the combined effects of ISI and noise on the decision process. The effects of ISI can be completely negated if it is possible to obtain a received pulse shape, $p_r(t)$, with the property

$$p_r(nT) = \begin{cases} 1 & n = 0 \\ 0 & n \neq 0 \end{cases} \qquad (3\text{-}91)$$

This condition guarantees zero ISI. The following theorem, referred to as *Nyquist's pulse-shaping criterion*, gives a condition on the Fourier transform of $p_r(t)$ which results in a pulse shape having the zero ISI property (3-91).

Theorem: *If* $P_r(f) = \mathscr{F}[p_r(t)]$ *satisfies the condition*

$$\sum_{k=-\infty}^{\infty} P_r\left(f + \frac{k}{T}\right) = T \qquad |f| \leq \frac{1}{2T} \qquad (3\text{-}92\text{a})$$

then

$$p_r(nT) = \begin{cases} 1 & n = 0 \\ 0 & n \neq 0 \end{cases} \qquad (3\text{-}92\text{b})$$

Proof: *Break the inverse Fourier transform integral for $p_r(t)$ up into contiguous intervals of length $1/T$ hertz. This results in the summation*

$$p_r(nT) = \sum_{k=-\infty}^{\infty} \int_{(2k-1)/2T}^{(2k+1)/2T} P_r(f) \exp\left(j2\pi fnT\right) df$$

By the change of variables $u = f - k/T$, *this becomes*

$$p_r(nT) = \sum_{k=-\infty}^{\infty} \int_{-1/2T}^{1/2T} P_r\left(u + \frac{k}{T}\right) \exp\left(j2\pi nTu\right) du$$

$$= \int_{-1/2T}^{1/2T} \left[\sum_{k=-\infty}^{\infty} P_r\left(u + \frac{k}{T}\right)\right] \exp\left(j2\pi nTu\right) du$$

where the order of integration and summation have been reversed to obtain the last expression. But the term in parentheses inside the integral is T by the hypothesis of the theorem. Therefore,

$$p_r(nT) = T \int_{-1/2T}^{1/2T} \exp\left(j2\pi nTu\right) du$$

$$= \text{sinc}\,(n)$$

$$= \begin{cases} 1 & n = 0 \\ 0 & n \neq 0 \end{cases}$$

This completes the proof of the theorem.

Note that the condition (3-92) does not uniquely specify $P_r(f)$. Two important considerations in selecting $P_r(f)$ are that $p_r(t)$ have a fast rate of decay with a small magnitude near the sample values for $n \neq 0$, and that shaping filters for producing the desired $P_r(f)$ be possible to realize or approximate closely. A family of spectra, $P_r(f)$, that satisfy the Nyquist pulse-shaping criterion is the *raised cosine* family, which is defined by

$$P_{\text{RC}}(f) = \begin{cases} T & |f| \leq \frac{1}{2}T - \beta \\ \frac{1}{2}T\left\{1 + \cos\left(\dfrac{\pi(|f| - \frac{1}{2}T + \beta)}{2\beta}\right)\right\} & \frac{1}{2}T - \beta < |f| \leq \frac{1}{2}T + \beta \\ 0 & |f| > \frac{1}{2}T + \beta \end{cases} \qquad (3\text{-}93)$$

The spectra given by (3-93) are illustrated in Figure 3-21 for several values of the parameter β. Note that the bandwidth of $P_r(f)$ lies between $\frac{1}{2}T$ and $1/T$ hertz depending on the value of β. It is clear from Figure 3-21 that $P_{\text{RC}}(f)$ satisfies the Nyquist pulse-shaping criterion. The inverse Fourier transform of $P_{\text{RC}}(f)$ can be shown to be

$$p_{\text{RC}}(t) = \frac{\cos 2\pi\beta t}{1 - (4\beta t)^2} \text{sinc}\left(\frac{t}{T}\right) \qquad (3\text{-}94)$$

which is graphed in Figure 3-22 for several values of β. Note that larger bandwidths for $P_{\text{RC}}(f)$ result in pulse shapes that decay faster, which is to be expected.

3-6.3 Optimum Transmitting and Receiving Filters

Taking the Fourier transform of (3-87), it follows that the spectrum of the received pulse is

$$AP_r(f) \exp\left(-j2\pi f t_d\right) = H_T(f)H_c(f)H_R(f) \qquad (3\text{-}95)$$

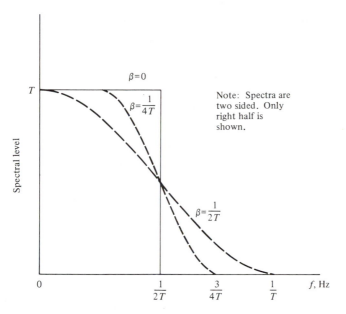

FIGURE 3-21. Raised-cosine spectra.

If the zero-ISI condition (3-91) is required, $P_r(f)$ must satisfy (3-92), which provides a constraint between the transfer functions of the transmitter and receiver filters for a fixed channel transfer function, $H_c(f)$. An additional requirement, which places a constraint on $H_T(f)$ and $H_R(f)$, is that they be chosen such that the probability of making a decision error at the receiver be minimized. If the zero-ISI constraint is satisfied, (3-89) reduces to

$$V_m = Aa_m + N_m \tag{3-96}$$

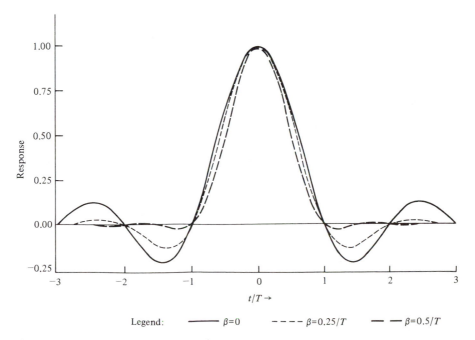

FIGURE 3-22. Pulses with raised-cosine spectra.

If $a_m = +1$ and $a_m = -1$ occur with equal probability, it follows that the probability of error is

$$P(\epsilon) = P(V_m > 0 \mid a_m = -1)$$

$$= P(N_m > A) \tag{3-97}$$

Since $n(t)$ is white with single-sided power spectral density N_0, it follows that

$$E(N_m) = 0$$

$$\text{var }(N_m) \triangleq \sigma^2 = \int_{-\infty}^{\infty} G_n(f)|H_R(f)|^2 \, df \tag{3-98}$$

Therefore, (3-97) can be written

$$P(\epsilon) = \int_A^{\infty} \frac{\exp(-u^2/2\sigma^2)}{\sqrt{2\pi\sigma^2}} \, du$$

$$= Q\left(\frac{A}{\sigma}\right) \tag{3-99}$$

Because the Q-function is a monotonically decreasing function of its argument, the probability of error is minimized if its argument is maximized. Equivalently, σ^2/A^2 can be minimized. Before doing so, however, A^2 will be expressed in terms of the transmitted energy per symbol, E_T. From (3-82), the energy of the kth transmitted pulse is

$$E_T = E(a_k^2) \int_{-\infty}^{\infty} h_T^2(t - kT) \, dt$$

$$= \int_{-\infty}^{\infty} |H_T(f)|^2 \, df \tag{3-100}$$

which results from using Parseval's theorem and the fact that $E(a_k^2) = 1$. From (3-95) it follows that this energy can be written

$$E_T = A^2 \int_{-\infty}^{\infty} \frac{|P_r(f)|^2 \, df}{|H_c(f)|^2 |H_R(f)|^2} \tag{3-101}$$

Solving for A^2 and using (3-98), σ^2/A^2 becomes

$$\frac{\sigma^2}{A^2} = \frac{1}{E_T} \int_{-\infty}^{\infty} G_n(f)|H_R(f)|^2 \, df \int_{-\infty}^{\infty} \frac{|P_R(f)|^2 \, df}{|H_c(f)|^2 |H_R(f)|^2} \tag{3-102}$$

which is to be minimized through appropriate choice of $H_R(f)$. This minimization can be accomplished through application of Schwarz's inequality, given by (3-21), with

$$|X(f)| = G_n^{1/2}(f)|H_R(f)| \tag{3-103}$$

$$|Y(f)| = \frac{|P_r(f)|}{|H_c(f)| \, |H_R(f)|} \tag{3-104}$$

Schwarz's inequality then assumes the form

$$\int_{-\infty}^{\infty} G_n(f)|H_R(f)|^2 \, df \int_{-\infty}^{\infty} \frac{|P_r(f)|^2}{|H_c(f)|^2 |H_R(f)|^2} \, df$$

$$\geq \left[\int_{-\infty}^{\infty} \frac{G_n^{1/2}(f)|P_r(f)|}{|H_c(f)|} \, df\right]^2 \tag{3-105}$$

with equality if, and only if, $|X(f)| = \alpha^2 |Y(f)|$ or if

$$|H_R(f)| = \frac{\alpha |P_r(f)|^{1/2}}{G_n^{1/4}(f) |H_c(f)|^{1/2}} \qquad (3\text{-}106)$$

where α is an arbitrary constant. That is, the minimum given by the right-hand side of (3-105) is achieved if the magnitude of the receiver transfer function is given by (3-106). From (3-95) it follows that the optimum transmitter transfer function magnitude is

$$|H_T(f)| = \frac{(A/\alpha) |P_r(f)|^{1/2} G_n^{1/4}(f)}{|H_c(f)|^{1/2}} \qquad (3\text{-}107)$$

Note that any appropriate phase response function can be used for $H_R(f)$ and $H_T(f)$.

Using the minimum value for σ/A (maximum value for A/σ) in (3-99), the probability of error minimized through appropriate choice of $H_T(f)$ and $H_R(f)$ is

$$P_{min}(\epsilon) = Q\left\{ \sqrt{E_T} \left[\int_{-\infty}^{\infty} \frac{G_n^{1/2}(f) |P_r(f)|}{|H_c(f)|} \, df \right]^{-1} \right\} \qquad (3\text{-}108)$$

A special case of interest occurs when the noise has power spectral density

$$G_n(f) = \frac{N_0}{2} |H_c(f)|^2 \qquad (3\text{-}109)$$

where $N_0/2$ is a constant. This would result if white noise were added prior to the channel filter. From (3-106), the optimum receiver and transmitter transfer-function magnitudes are

$$|H_R(f)| = \frac{\beta |P_r(f)|^{1/2}}{G_n^{1/2}(f)} = \beta' \frac{|P_r(f)|^{1/2}}{|H_c(f)|} \cdot \qquad (3\text{-}110)$$

and

$$|H_T(f)| = \beta'' |P_r(f)|^{1/2}$$

where the β's are arbitrary constants. Because $p_r(0) = 1$, it follows that $\int_{-\infty}^{\infty} P_r(f) \, df = 1$. Therefore, for cases where $P_r(f)$ is real and greater than or equal to zero for all f, the minimum probability of error for this special case becomes

$$P_{min}(\epsilon) = Q\left(\sqrt{\frac{2E_T}{N_0}} \right) \qquad (3\text{-}111)$$

which is identical to the error probability for matched filter detection of binary signals in AWGN.

Another special case of interest occurs when $|H_c(f)| = H_0$, a constant, throughout the range of frequencies where $P_r(f)$ is nonzero, and $G_n(f) = N_0/2$. In this case, the optimum receiver filter transfer function has magnitude

$$|H_R(f)| = \gamma |P_r(f)|^{1/2} \qquad (3\text{-}112)$$

where γ is an arbitrary constant, and the minimum error probability is given by (3-111).

EXAMPLE 3-6

Consider a binary communication system that transmits at a data rate of 4800 bits/s. The channel has transfer function

$$H_c(f) = \frac{1}{1 + jf/4800}$$

and the noise is white with two-sided power spectral density $N_0/2 = 10^{-12}$ W/Hz. Assume that a received pulse with raised cosine spectrum given by (3-93) with $\beta = 1/2T = 4800$ Hz is desired. Find the magnitudes of the transmit and receive filter transfer functions that give zero ISI and optimum detection. Also find the value of A^2/σ^2 and the transmitted signal energy required to give $P_{min}(\epsilon) = 10^{-6}$.

Solution: From (3-106) and (3-107) with $G_n(f) = 10^{-12}$, the magnitude of the transmit and receive filter transfer functions are

$$|H_r(f)| = \begin{cases} \left[1 + \left(\dfrac{f}{4800}\right)^2\right] \cos \dfrac{\pi f}{19,200} & 0 \le |f| \le 9600 \text{ Hz} \\ 0 & \text{otherwise} \end{cases}$$

$$|H_T(f)| = \begin{cases} \cos(\pi f/19,200) & 0 \le f \le 9600 \text{ Hz} \\ 0 & \text{otherwise} \end{cases}$$

where the arbitrary constants have been chosen in each case to make the frequency response unity at zero frequency. To provide an error probability of 10^{-6} or less requires that

$$Q\left(\frac{A}{\sigma}\right) \le 10^{-6}$$

or that $A/\sigma \ge 4.75$, which can be obtained by using a table for the Q-function. In (3-108), the factor

$$F = \int_{-\infty}^{\infty} \frac{P_r(f)}{|H_c(f)|} \, df$$

$$= \frac{1}{9600} \int_0^{9600} \left(1 + \cos \frac{\pi f}{9600}\right) \left[1 + \left(\frac{f}{4800}\right)^2\right]^{1/2} df$$

is required. By the change of variables $u = f/9600$, this integral can be written

$$F = \int_0^1 [1 + \cos(\pi u)][1 + (2u)^2]^{1/2} \, du \approx 1.21$$

which is obtained through numerical integration. Since $G_n(f) = N_0/2 = 10^{-12}$ W/Hz, the argument of the Q-function in (3-108) is

$$\frac{1}{1.21} \sqrt{\frac{2E_T}{N_0}} = 0.83 \sqrt{\frac{2E_T}{N_0}} = \frac{A}{\sigma}$$

This differs by a factor of 0.83 from the case of matched filter detection of binary antipodal signals in white noise; that is, the finite-bandwidth channel imposes a degradation of $-20 \log_{10} 0.83 = 1.62$ dB over the infinite-bandwidth case.

Using the required value of $A/\sigma = 4.75$ and the given value of $N_0/2 = 10^{-12}$ W/Hz, the required transmitted signal energy is

$$E_T = \left(1.21 \frac{A}{\sigma}\right)^2 \frac{N_0}{2}$$

$$= (1.21)^2 (4.75)^2 (10^{-12})$$

$$= 3.3 \times 10^{-11} \text{ J}$$

The average transmitter power is the transmit energy divided by the symbol duration since the symbol energy is independent of which symbol is transmitted. The resulting power is

$$P_T = \frac{E_T}{T} = 4800(3.3 \times 10^{-11})$$

$$= 15.9 \ \mu\text{W}$$

$$= -38 \text{ dBm}$$

3-6.4 Quadrature Bandpass Systems and Multiple Amplitude Systems

The use of in-phase and quadrature carriers as for QPSK and OQPSK allows the transmission of two independent data streams as shown in Figure 3-8. Or, if a single binary data stream is available, it can be demultiplexed into two binary data streams by assigning (demultiplexing) even-indexed bits to one data stream and assigning the odd-indexed bits to the other. Furthermore, it is not necessary to restrict the amplitudes used on each quadrature carrier to two levels. If more than two levels are used, properly selected multiple thresholds must be employed. The design of quadrature carrier and multiple amplitude systems, referred to as *QAM systems for quadrature amplitude modulation*, is considered in the problems.* The block diagram for a general QAM system is shown in Figure 3-23. In addition to

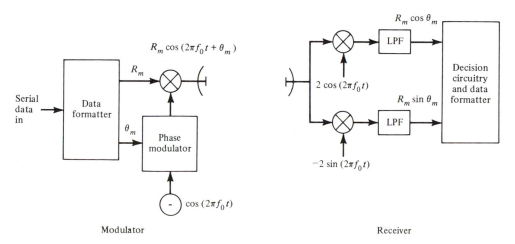

FIGURE 3-23. **Block diagram of a QAM communication system.**

*See, for example, an article by Aprille [5] for further discussion on digital data transmission through bandlimited channels.

degradation due to nonideal filtering, QAM systems are prone to *crosstalk*; that is, nonquadrature reference signals at the receiver result in a residual signal from one quadrature channel appearing in the other.

3-6.5 Shaped Transmitted Signal Spectra

If the channel frequency response has a small magnitude for certain ranges of frequencies, it may be useful to shape the transmitted signal spectrum to match the channel response; that is, where the channel response is low, the transmitted signal spectrum should be correspondingly low. This avoids the problem of having to design receiver filters with unreasonably high gains in those regions where the channel frequency response is low [see (3-106)].

The spectrum of the transmitted signal depends on both the transmitted pulse shape [transmitter filter impulse response in (3-82) and Figure 3-19] and the statistical properties of the transmitted symbols [the a_k's in (3-82) and Figure 3-19]. To investigate the effect of the latter, the power spectrum for a pulse train whose symbols are interdependent will be developed and applied to several specific cases.

Consider a transmitted signal of the form

$$x_t(t) = \sum_{n=0}^{N-1} b_n x(t - nT) \tag{3-113}$$

where $b_0, b_1, b_2, \ldots$ are constants and $x(t)$ is of the form

$$x(t) = \sum_{k=-\infty}^{\infty} a_k h(t - kT) \tag{3-114}$$

in which the a_k's are independent, identically distributed random variables. The power spectral density of $x(t)$, using methods of Section 3-6, can be shown to be

$$S_x(f) = \frac{1}{T} E(a_k^2)|H(f)|^2 \tag{3-115}$$

where $E(\cdot)$ denotes the expectation operator and $H(f) = \mathcal{F}[h(t)]$.

The power spectral density of $x_t(t)$ is desired. It can be found by expressing the autocorrelation function of $x_t(t)$ in terms of the autocorrelation function of $x(t)$ and Fourier transforming it to obtain the power spectral density. Using the definition of the autocorrelation function and (3-113) for $x_t(t)$, it follows that the autocorrelation function of $x_t(t)$ is

$$R_{x_t}(\tau) = E[x_t(t)x_t(t + \tau)]$$

$$= E\left[\sum_{n=0}^{N-1}\sum_{m=0}^{N-1} b_n b_m x(t - nT)x(t + \tau - mT)\right]$$

$$= \sum_{n=0}^{N-1}\sum_{m=0}^{N-1} b_n b_m E[x(t - nT)x(t + \tau - mT)]$$

$$= \sum_{n=0}^{N-1}\sum_{m=0}^{N-1} b_n b_m R_x[\tau - (m - n)T] \tag{3-116}$$

The Fourier transform of (3-116) gives the power spectral density of $x_t(t)$, which can be expressed as

$$S_{x_t}(f) = \sum_{n=0}^{N-1} \sum_{m=0}^{N-1} b_n b_m e^{-j2\pi(m-n)fT} S_x(f) \qquad (3\text{-}117)$$

where the frequency translation theorem of Fourier transforms has been used. The application of (3-117) will now be illustrated by an example.

EXAMPLE 3-7 DICODE PULSE TRAIN

Assume that the a_k's in (3-114) are $+1$ for a logic 1 transmitted and -1 for a logic 0 transmitted. The transmitted pulse sequence is formed by taking the difference between $x(t)$ and $x(t - T)$, which is referred to as a *dicode pulse train*. Thus $b_0 = -b_1 = 1$ and $b_2 = b_3 = \cdots = 0$. From (3-117), the power spectral density of the transmitted signal becomes

$$
\begin{aligned}
S_{x_t}(f) &= \sum_{n=0}^{1} \sum_{m=0}^{1} b_n b_m e^{-j2\pi(n-m)fT} S_x(f) \\
&= (1 - e^{j2\pi fT} - e^{-j2\pi fT} + 1) S_x(f) \\
&= (2 - 2 \cos 2\pi fT) S_x(f) \\
&= 4 \sin^2 \pi fT S_x(f) \\
&= \frac{4}{T} \sin^2 \pi fT |H(f)|^2 \qquad (3\text{-}118)
\end{aligned}
$$

The presence of $\sin^2 \pi fT$ in (3-118) guarantees that the power spectrum of the transmitted signal is zero at $f = 0$. Therefore, dicode transmission would be a useful scheme to employ for channels with poor low-frequency response. ∎

Other schemes could be employed for which the transmitted symbol stream is derived from more than two successive source bits. The dependency between bits thereby introduced amounts to the use of a *finite impulse response* (FIR) *filter*.

3-6.6 Duobinary Signaling

The baseband binary systems considered so far in this section require a transmission bandwidth of at least $\frac{1}{2}T$ hertz in order to transmit one symbol each T seconds, and a bandwidth of exactly $\frac{1}{2}T$ hertz implies the use of ideal, rectangular filters at the transmitter and receiver. In order to avoid the use of rectangular filters and yet require a transmission bandwidth of only $\frac{1}{2}T$ hertz, *partial response signaling* can be employed. Such signaling schemes, which utilize *controlled amounts* of ISI, have their origins in the early 1960s when *duobinary signaling* was invented by Adam Lender.* The concept of partial response signaling will be illustrated here with a brief introduction to duobinary signaling.

*The original work on duobinary signaling is reported in [6]. Two survey articles on correlative coding techniques are provided in [7] and [8].

Duobinary signaling uses pulse spectra which give output samples from the receiver filter of the form

$$v(t_m) = A(a_m + a_{m-1}) + N_m \qquad (3\text{-}119)$$

[see (3-89)]. A suitable received pulse spectrum is

$$P_r(f) = \begin{cases} 2T \cos \pi fT & |f| \le 1/2T \\ 0 & \text{otherwise} \end{cases} \qquad (3\text{-}120)$$

which corresponds to the time response

$$p_r(t) = \frac{4 \cos \pi t/T)}{\pi(1 - 4t^2/T^2)} \qquad (3\text{-}121)$$

The first term of (3-119) assumes the following three values depending on the values of the mth and $(m - 1)$th input bits:

$$A(a_m + a_{m-1}) = \begin{cases} 2A & \text{if } a_m = a_{m-1} = 1 \\ 0 & \text{if } a_m = -a_{m-1} \\ -2A & \text{if } a_m = a_{m-1} = -1 \end{cases} \qquad (3\text{-}122)$$

To uniquely determine the source bit in the mth signaling interval, even if an error is made on the $(m - 1)$th bit, the mth source data bit, denoted by b_m, is *precoded* according to the rule

$$d_m = b_m \oplus d_{m-1} \qquad (3\text{-}123)$$

where $\oplus$ stands for modulo-2 addition and the d_m's are the encoded bits. This rule is also referred to as *differential encoding*. Using this encoding rule, it is seen that $d_m = d_{m-1}$ if the mth input bit is 0 and $d_m = \overline{d_{m-1}}$, where the overbar denotes complement, if $b_m = 1$. The sign on the amplitude, a_m, of the mth transmitted pulse is then either $+1$ or -1 according to $b_m = 1$ or 0, respectively. It therefore follows that

$$V_m \triangleq v(t_m) = \begin{cases} +2A & \text{if } b_m = 0 \\ 0 & \text{if } b_m = 1 \end{cases} \qquad (3\text{-}124a)$$

Consequently, the source data can be detected by comparing $|v(t)|$ with a suitably chosen threshold. If it exceeds this threshold, the decision $b_m = 0$ is made, whereas if the threshold is not exceeded, the decision $b_m = 1$ is made. This is equivalent to the following sampler output values:

$$V_m = \begin{cases} \pm 2A + N_m & \text{if } b_m = 0 \\ N_m & \text{if } b_m = 1 \end{cases} \qquad (3\text{-}124b)$$

If $b_m = 0$ and $b_m = 1$ are equiprobable, the output levels $2A$ and $-2A$ occur with probabilities $\frac{1}{4}$ and the output level 0 occurs with probability $\frac{1}{2}$. If thresholds are set at $\pm A$, errors can occur in the following ways:

$$\text{If } b_m = 0, \ 2A + N_m < A \quad \text{or} \quad N_m < -A$$

$$-2A + N_m > -A \quad \text{or} \quad N_m > A$$

$$\text{If } b_m = 1, N_m > A \qquad \text{or} \quad N_m < -A$$

Therefore, the average probability of error is

$$P(\epsilon) = \tfrac{1}{4} \Pr(N_m < -A) + \tfrac{1}{2} \Pr(|N_m| > A) + \tfrac{1}{4} \Pr(N_m > A) \quad (3\text{-}125)$$

But N_m is a zero-mean Gaussian random variable with variance σ^2. Thus (3-125) can be written as

$$P(\epsilon) = \frac{1}{4} \int_{-\infty}^{-A} \frac{\exp(-u^2/2\sigma^2)}{\sqrt{2\pi\sigma^2}} \, du$$

$$+ \frac{1}{2} \left[\int_{A}^{\infty} \frac{\exp(-u^2/2\sigma^2)}{\sqrt{2\pi\sigma^2}} \, du + \int_{-\infty}^{-A} \frac{\exp(-u^2/2\sigma^2)}{\sqrt{2\pi\sigma^2}} \, du \right]$$

$$+ \frac{1}{4} \int_{A}^{\infty} \frac{\exp(-u^2/2\sigma^2)}{\sqrt{2\pi\,\sigma^2}} \, du \tag{3-126}$$

Because of symmetry of the Gaussian probability density function, this can be written in terms of the Q-function as

$$P(\epsilon) = \frac{3}{2} Q\left(\frac{A}{\sigma}\right) \tag{3-127}$$

The probability of error can again be minimized through choosing the transmit and receive filter transfer functions to maximize A^2/σ^2 subject to the received pulse response satisfying (3-121). The resulting transfer functions can be shown to be given by (3-106) and (3-107).

To compare duobinary signaling with direct baseband binary signaling, consider the ratios $(A^2/\sigma^2)_{\text{max}}$ for both cases for an ideal channel $[H_c(f) = 1]$ and AWGN. For direct binary signaling, this ratio is [see (3-111)]

$$\left(\frac{A^2}{\sigma^2}\right)_{\text{max,bin}} = \frac{2E_T}{N_0} \tag{3-128}$$

For the duobinary case, it is [see (3-108)]

$$\left(\frac{A^2}{\sigma^2}\right)_{\text{max,doubin}} = \frac{2E_T}{N_0} \left[\int_{-\infty}^{\infty} |P_r(f)| \, df \right]^{-2} \tag{3-129}$$

Therefore, the argument of the Q-function expression for the error probability of duobinary differs from that of direct binary by the factor

$$F = \left[\int_{-\infty}^{\infty} |P_r(f)| \, df \right]^{-2}$$

$$= \left[2T \int_{-1/2T}^{1/2T} \cos(\pi fT) \, df \right]^{-2}$$

$$= \left(\frac{\pi}{4}\right)^2 \tag{3-130}$$

If the factor of $\frac{3}{2}$ multiplying the Q-function for duobinary is ignored, the factor F amounts to a degradation in signal-to-noise ratio of

$$D = -20 \log_{10} \frac{\pi}{4} = 2.1 \text{ dB}$$

of duobinary over direct binary. That is, to achieve the same error probability, the transmitter power for duobinary must be 2.1 dB greater than that for direct binary, assuming ideal channel filtering and AWGN. This is the sacrifice paid for the smaller bandwidth required by duobinary.

USE OF EYE DIAGRAMS FOR SYSTEM CHARACTERIZATION

The discussion in Section 3-6 centered around ways to remove ISI from preceding and following symbols in order to improve the detectability of a given symbol. In order to characterize system performance in cases where optimum transmit and receive filters are not realized ideally, it is convenient to utilize an *eye diagram* or eye pattern. Figure 3-24a illustrates an experimental setup for measuring an eye diagram, with a typical eye diagram shown in Figure 3-24b and c for the ideal, infinite-bandwidth case and the nonideal, finite-bandwidth cases, respectively. For the nonideal case, the eye is open only $\frac{3}{4}$ of what it is for the ideal case, which implies a maximum signal-to-noise ratio degradation of $-20 \log_{10}(\frac{3}{4}) = 2.5$ dB. That is, for the bits giving the minimum eye opening $\frac{3}{4}$ the ideal, a signal power 2.5 dB greater than that used in the ideal case is required to achieve a given error probability.

This discussion shows that an eye diagram is a convenient tool for characterizing the performance of a system *experimentally*. It is also a convenient tool for characterizing a system through *computer simulation*. For example, a pseudorandom

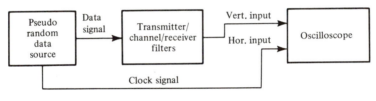

(a) Experimental setup for measuring an eye diagram

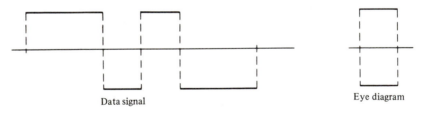

(b) Eye diagram for an infinite bandwidth system

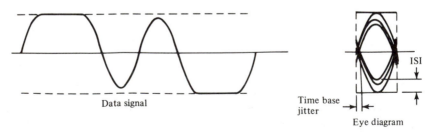

(c) Eye diagram for a finite bandwidth system

FIGURE 3-24. Illustration of eye-diagram measurement for a baseband system.

data generator can easily be programmed and used to simulate the input to a transmit/channel/receive filter chain, which is realized as a cascade of digital filters. The output of this filter chain is then sampled at multiples of the symbol duration, the ratio of the sample value to the noise standard deviation taken, and the probability of error for that symbol computed by means of (3-99). The probability of error averaged over a typical symbol pattern can then be computed by computing the average over the entire pseudorandom sequence. This method is useful and powerful for cases where no nonlinear operations are performed on the noise. If the noise is passed through a nonlinear filter (e.g., a filter-limiter-filter cascade), it is difficult to determine the probability density function of the noise at the receiver output and thereby obtain an expression corresponding to (3-99) for the probability of error on a given symbol.

3-8

EQUALIZATION IN DIGITAL DATA TRANSMISSION SYSTEMS*

In Section 3-6 it was shown how proper design of the transmitter and receiver filters of a pulse-amplitude-modulated communication system (modem) would simultaneously guarantee zero ISI and maximize the signal-to-noise ratio at the sampling time, thereby minimizing the probability of error. The filter design equations were given by (3-106) and (3-107). In some instances it may be difficult to realize filters with the frequency responses specified by these equations. In other cases, the channel may be unknown or the modem may be used to transmit data through several different channels. An example of this situation is the use of dial-up telephone channels as a communication link between a computer and a user terminal. Yet another case occurs when the channel is slowly time varying, such as a line-of-sight microwave relay link in which multipath propagation takes place due to temperature variations of the atmosphere with altitude. In all these cases, a filter with *adjustable* frequency response would be useful to employ at the receiver. Through channel measurements, its frequency response (or, equivalently, impulse response) could be adjusted to give an overall filter at the receiver which improves modem performance over that obtainable with fixed filters. This adjustable filter is referred to as an *equalization filter*, or simply equalizer. Equalizers may be either *preset* or *adaptive*. The parameters of a preset equalizer are adjusted by making measurements of the channel impulse response and solving a set of equations for the parameters using these measurements. An adaptive equalizer is automatically adjusted by sending a known signal through the channel and allowing the equalizer to adjust its own parameters in response to this known signal.

3-8.1 Zero-Forcing Equalizers

To illustrate the basic idea of the equalization process, consider the block diagram of Figure 3-25. This is a simplified version of Figure 3-19 with the channel and receiver filters combined and the noise excluded for the moment. An equalization

---

*For an easy-to-read overview of equalization, see [9].

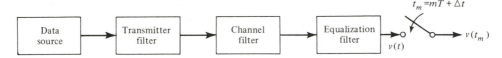

FIGURE 3-25. Simplified block diagram of a pulse-amplitude-modulated communication system with equalization filter.

filter follows the channel filter so that the overall system frequency response from transmitter filter input to equalizer output is

$$H_0(f) = H_T(f)H_c(f)H_E(f) \qquad (3\text{-}131)$$

with overall impulse response

$$h_0(t) = \mathcal{F}^{-1}[H_0(f)] \qquad (3\text{-}132)$$

Assume that the output of the equalizer is sampled each T seconds. Zero ISI results if Nyquist's first criterion is satisfied, that is, if

$$\sum_{k=-\infty}^{\infty} H_0\left(f + \frac{k}{T}\right) = \text{constant} \qquad |f| \le \tfrac{1}{2}T \qquad (3\text{-}133)$$

which follows from (3-92a) because sampling each T seconds effectively reproduces the overall frequency response about each $1/T$ hertz. The zero-ISI property (3-92b) then follows for $h_0(t)$.

From (3-133) it follows that an ideal zero-ISI equalizer is simply an *inverse filter*, which has a frequency response that is the inverse of the frequency response of the transmitter and channel cascaded and folded about the sampling frequency, $1/T$. This inverse-filter equalizer is often approximated by a *finite-impulse response* (FIR) filter or *transversal* filter as illustrated in Figure 3-26. Its impulse response is simply

$$h_E(t) = \sum_{n=-N}^{N} C_n\delta(t - nT) \qquad (3\text{-}134)$$

from which it follows that its frequency response is

$$H_E(f) = \sum_{n=-N}^{N} C_n e^{-j2\pi nTf} \qquad (3\text{-}135)$$

One problem with a transversal filter equalizer, with coefficients chosen to ap-

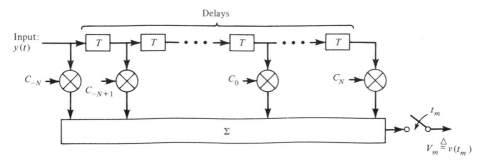

FIGURE 3-26. Transversal filter equalizer.

proximate the zero-ISI condition, is that it excessively enhances the channel noise at frequencies where the folded channel frequency response has high attenuation.

Because there are only $2N + 1$ unknown coefficients in (3-134) or (3-135), it follows that only a finite number of interfering symbols can be nulled or forced to zero. The equations for solving for the coefficients are easily found by noting that the equalizer pulse response, $p_{eq}(t)$, due to the channel output pulse response, $p_c(t)$, is

$$p_{eq}(t) = \sum_{n=-N}^{N} C_n p_c(t - nT) \qquad (3\text{-}136)$$

The zero-ISI condition (3-92b), when applied to (3-136), can hold for only $2N + 1$ sample times since only $2N + 1$ unknown constants are available for adjustment. Setting $t = mT + \Delta t$, $m = 0, \pm 1, \pm 2, \ldots, \pm N$, in (3-136), these conditions are

$$p_{eq}(mT + \Delta t) = \sum_{n=-N}^{N} C_n p_c((m - n)T + \Delta t)$$

$$= \begin{cases} 1 & m = 0 \\ 0 & m \neq 0 \end{cases} \qquad m = 0, \pm 1, \pm 2, \ldots, \pm N \quad (3\text{-}137)$$

where $t = \Delta t$ is the sampling time for which $p_{eq}(t)$ is maximum. These equations can be written in matrix form as

$$\mathbf{P}_{eq} = [P_c]\mathbf{C} \qquad (3\text{-}138)$$

where $\mathbf{P}_{eq}$ and $\mathbf{C}$ are vectors or column matrices given by

$$\mathbf{P}_{eq} = \begin{bmatrix} \left. \begin{matrix} 0 \\ 0 \\ \cdot \\ \cdot \\ \cdot \\ 0 \end{matrix} \right\} \begin{matrix} N \\ \text{zeros} \end{matrix} \\ 1 \\ 0 \\ 0 \\ \left. \begin{matrix} \cdot \\ \cdot \\ \cdot \\ 0 \end{matrix} \right\} \begin{matrix} N \\ \text{zeros} \end{matrix} \end{bmatrix} \qquad \mathbf{C} = \begin{bmatrix} C_{-N} \\ C_{-N+1} \\ \cdot \\ \cdot \\ \cdot \\ C_0 \\ C_1 \\ \cdot \\ \cdot \\ \cdot \\ C_N \end{bmatrix}$$

respectively, and $[P_c]$ is the $(2N + 1) \times (2N + 1)$ matrix of channel responses of the form

$$[P_c] = \begin{bmatrix} p_c(0) & p_c(-1) & \cdots & p_c(-2N) \\ p_c(1) & p_c(0) & \cdots & p_c(-2N + 1) \\ p_c(2) & p_c(1) & \cdots & p_c(-2N + 2) \\ \cdot & \cdot & \cdot & \cdot \\ \cdot & \cdot & \cdot & \cdot \\ \cdot & \cdot & \cdot & \cdot \\ p_c(2N) & p_c(2N - 1) & \cdots & p_c(0) \end{bmatrix}$$

where the maximum of the channel pulse response, denoted by $p_c(0)$, occurs at time $t_m = \Delta t$. Thus the $4N + 1$ sample values of the channel pulse response taken at T-second intervals can be used to determine the $2N + 1$ unknown coefficients $C_{-N}, C_{-N+1}, \ldots, C_0, \ldots, C_N$ by solving (3-138). Exactly N zeros will be forced at the sampling instants either side of the main pulse response. Because all components of the vector $\mathbf{P}_{eq}$ are zeros except for the center one, it follows that the coefficient vector $\mathbf{C}$ is the center column of $[P_c]^{-1}$.

EXAMPLE 3-8

Consider the channel pulse response shown in Figure 3-27. Determine the coefficients of a five-tap transversal filter equalizer that will force two zeros on either side of the main pulse response. Compute the sample values out to $\pm 3T$.

Solution: From Figure 3-27, the sample values for the channel response are

$$p_c(-5) = 0.01 \qquad p_c(-4) = -0.02 \qquad p_c(-3) = 0.05$$

$$p_c(-2) = -0.1 \qquad p_c(-1) = 0.2 \qquad p_c(0) \;\;\; = 1$$

$$p_c(1) \;\;\; = -0.1 \qquad p_c(2) \;\;\; = 0.1 \qquad p_c(3) \;\;\; = -0.05$$

$$p_c(4) \;\;\; = 0.02 \qquad p_c(5) \;\;\; = 0.005$$

The channel response matrix is

$$[p_c] = \begin{bmatrix} 1.0 & 0.2 & -0.1 & 0.05 & -0.02 \\ -0.1 & 1.0 & 0.2 & -0.1 & 0.05 \\ 0.1 & -0.1 & 1.0 & 0.2 & -0.1 \\ -0.05 & 0.1 & -0.1 & 1.0 & 0.2 \\ 0.02 & -0.05 & 0.1 & -0.1 & 1.0 \end{bmatrix}$$

The inverse of this matrix, by numerical methods, is found to be

$$[p_c]^{-1} = \begin{bmatrix} 0.966 & -0.170 & 0.117 & -0.083 & 0.056 \\ 0.118 & 0.945 & -0.158 & 0.112 & -0.083 \\ -0.091 & 0.133 & 0.937 & -0.158 & 0.117 \\ 0.028 & -0.095 & 0.133 & 0.945 & -0.170 \\ -0.002 & 0.028 & -0.091 & 0.118 & 0.966 \end{bmatrix}$$

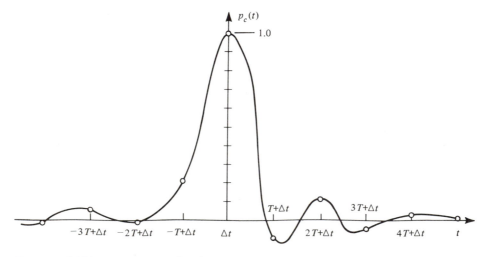

FIGURE 3-27. Received pulse from channel to be equalized.

The coefficient vector is the center column of $[P_c]^{-1}$. Therefore,

$$C_{-2} = 0.117 \qquad C_{-1} = -0.158 \qquad C_0 = 0.937$$

$$C_1 = 0.133 \qquad C_2 = -0.091$$

The sample values of the equalized pulse response, from (3-137), are

$$p_{eq}(m) = \sum_{n=-2}^{2} C_n p_c(m - n)$$

where $\Delta t = 0$. For example,

$$p_{eq}(0) = (0.117)(0.1) + (-0.158)(-0.1) + (0.937)(1)$$

$$+ (0.133)(0.2) + (-0.091)(-0.1)$$

$$= 1.0$$

which checks with the desired value of unity. Similarly, it can be verified that $p_{eq}(-2) = p_{eq}(-1) = p_{eq}(1) = p_{eq}(2) = 0$. Values of $p_{eq}(n)$ for $n < -2$ or $n > 2$ are not zero. For example,

$$p_{eq}(3) = (0.117)(0.005) + (-0.158)(0.02)$$

$$+ (0.937)(-0.05) + (0.133)(0.1) + (-0.091)(-0.1)$$

$$= -0.027$$

$$p_{eq}(-3) = (0.117)(0.2) + (-0.158)(-0.1)$$

$$+ (0.937)(0.05) + (0.133)(-0.02) + (-0.091)(0.01)$$

$$= 0.082 \qquad ∎$$

The zero-forcing equations (3-138) do not account for the effects of noise. In addition, the finite-length transversal filter equalizer can minimize worst-case ISI only if the peak distortion is less than 100% of the eye opening. Another type of equalizer which partially avoids these problems is the least-mean-square (LMS) equalizer. In an LMS equalizer, the equalizer coefficients are chosen to minimize the mean-square error, which consists of the sum of the squares of all the ISI terms plus the noise power at the equalizer output. The LMS equalizer therefore maximizes the signal-to-distortion ratio at its output within the constraints of the equalizer length and delay. The use of an LMS equalizer to compensate for multipath distortion in QPSK data transmission will be given next.

3-8.2 LMS Equalizer Application*

Because QPSK transmission basically amounts to multiplexing two binary data streams on quadrature carriers, a dispersive channel will introduce both ISI and crosstalk between channels. Accordingly, an equalizer for QPSK must correct for both of these distortions. The application given in this subsection is the design of an LMS equalizer to correct for both of these distortions in QPSK data transmission through a multipath channel. A block diagram for the receiver with equalizer is shown in Figure 3-28. It consists of parallel coherent demodulators using quadrature reference signals and transversal filter equalizers in each quadrature channel. To

*This material is adapted from [10].

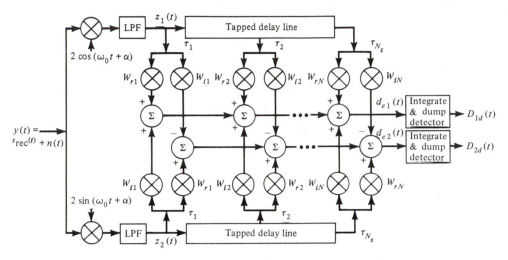

FIGURE 3-28. Transversal filter equalizer for QPSK. (From Ref. 10.)

compensate for both ISI and crosstalk, weighted addition of the delay outputs is required both within a quadrature channel as well as between quadrature channels. These weights are W_{r1}, W_{r2}, . . ., W_{rN} and W_{i1}, W_{i2}, . . ., W_{iN}, respectively.* Since the applications that motivate this equalizer implementation are to high-speed data transmission, no sampler precedes the delay lines, but rather data detection filtering is done after the channel distortion is compensated by the transversal filters.

The transmitted QPSK signal is represented as

$$s_{tr}(t) = d_1(t) \cos \omega_0 t - d_2(t) \sin \omega_0 t \qquad (3\text{-}139)$$

where $d_1(t)$ and $d_2(t)$, each with symbol duration T, are the binary data streams in each quadrature channel. The received signal plus noise is

$$
\begin{aligned}
y(t) &= s_{rec}(t) + n(t) \\
&= s_{tr}(t) + \beta s_{tr}(t - \tau_m) + n_c(t) \cos (\omega_0 t + \alpha) \\
&\quad - n_s(t) \sin (\omega_0 t + \alpha) \qquad (3\text{-}140)
\end{aligned}
$$

where β is the attenuation and τ_m the delay of the multipath channel; α represents an unknown phase offset of the received signal from the quadrature reference signals of the demodulator, and $n_c(t)$ and $n_s(t)$ are quadrature noise components which are independent and Gaussian.

From (3-139) and (3-140) it follows that the outputs of the quadrature phase detectors in Figure 3-28 are

$$
\begin{aligned}
z_1(t) &= d_1(t) \cos \alpha + d_2(t) \sin \alpha \\
&\quad + \beta d_1(t - \tau_m) \cos (\omega_0 \tau_m + \alpha) \\
&\quad + \beta d_2(t - \tau_m) \sin (\omega_0 \tau_m + \alpha) + n_c(t) \qquad (3\text{-}141)
\end{aligned}
$$

$$
\begin{aligned}
z_2(t) &= d_2(t) \cos \alpha - d_1(t) \sin \alpha \\
&\quad + \beta d_2(t - \tau_m) \cos (\omega_0 \tau_m + \alpha) \\
&\quad - \beta d_1(t - \tau_m) \sin (\omega_0 \tau_m + \alpha) + n_s(t) \qquad (3\text{-}142)
\end{aligned}
$$

*In general, four sets of weights are required. Attention is limited here to channels that induce symmetrical distortions between quadrature channels.

Equations (3-141) and (3-142) can be combined into a single complex equation as

$$z(t) = z_1(t) + jz_2(t)$$

$$= [d_1(t) + jd_2(t)]e^{-j\alpha}$$

$$+ \beta[d_1(t - \tau_m) + jd_2(t - \tau_m)]$$

$$\exp[-j(\omega_0\tau_m + \alpha)] + \tilde{n}(t) \qquad (3\text{-}143)$$

where

$$\tilde{n}(t) = n_c(t) + jn_s(t) \qquad (3\text{-}144)$$

From (3-141) and (3-142), it follows that the effect of a non-zero-phase error at the demodulator is to introduce crosstalk between channels. For zero phase error, note that the multipath component gives only ISI if $\omega_0\tau_m$ is an integer multiple of π, whereas crosstalk alone results if $\omega_0\tau_m$ is an odd integer multiple of $\pi/2$. Thus, for arbitrary multipath delays and demodulator phase errors, it is clear that both ISI and crosstalk must be equalized.

From Figure 3-28 it follows that the equalized data streams can be written as

$$d_{e1}(t) = \sum_{k=1}^{N} [W_{rk}z_1(t - k\Delta) + W_{ik}z_2(t - k\Delta)] \qquad (3\text{-}145)$$

and

$$d_{e2}(t) = \sum_{k=1}^{N} [W_{rk}z_2(t - k\Delta) - W_{ik}z_1(t - k\Delta)] \qquad (3\text{-}146)$$

where $\tau_k = k\Delta$. Letting $W_k = W_{rk} + jW_{ik}$ and $d_e(t) = d_{e1}(t) + jd_{e2}(t)$, (3-145) and (3-146) can be written as the single complex equation

$$d_e(t) = \sum_{k=1}^{N} W_k^* z(t - k\Delta) \qquad (3\text{-}147)$$

The optimum weights, according to a minimum mean-square-error criterion, can be readily found. The desired nondistorted complex modulation is

$$D(t) = (1 + \beta)[d_1(t) + jd_2(t)] \qquad (3\text{-}148)$$

where the factor $1 + \beta$ is included because the total received signal is being considered. A minimum mean-square-error equalizer is one that minimizes

$$I = E[|d_e(t) - D(t)|^2] \qquad (3\text{-}149)$$

where $E(\cdot)$ denotes expectation.

Defining the column vectors

$$\mathbf{W} = \begin{bmatrix} W_1 \\ W_2 \\ \cdot \\ \cdot \\ \cdot \\ W_{Ns} \end{bmatrix}$$

and

$$\mathbf{Z} = \begin{bmatrix} z(t) \\ z(t - \Delta) \\ \cdot \\ \cdot \\ \cdot \\ z[t - (N_s - 1)\Delta] \end{bmatrix} \qquad (3\text{-}150)$$

the integral-square error can be written as

$$I = E\{|\mathbf{W}'\mathbf{Z} - D(t)|^2\} \qquad (3\text{-}151)$$

where the prime denotes complex-conjugate transpose. The optimum W is found by setting the gradient to zero, that is, taking the derivative of (3-151) with respect to W_{rk} and W_{ik} for each k and setting the result to zero. The resulting optimum weights are given by

$$\mathbf{W}_{\text{opt}} = [E(\mathbf{Z}\mathbf{Z}')]^{-1}E[\mathbf{Z}D^*(t)] \qquad (3\text{-}152)$$

An explicit expression for $\mathbf{W}_{\text{opt}}$ in terms of the signal parameters, channel parameters, and noise power spectral density can be obtained and expressed as follows. Considering first the matrix $\mathbf{A} \triangleq E(\mathbf{Z}\mathbf{Z}')$, it can be shown that

$$\begin{aligned}
A_{kl} = {} & 2(1 + \beta^2)R_d[(l - k)\Delta] \\
& + 2\beta R_d[(l - k)\Delta + \tau_m]e^{j\omega_0\tau_m} \\
& + 2\beta R_d[(l - k)\Delta - \tau_m]e^{-j\omega_0\tau_m} \\
& + 2R_n[(l - k)\Delta)]
\end{aligned} \qquad (3\text{-}153)$$

where $R_d(\tau)$ is the autocorrelation function of $d_1(t)$ or $d_2(t)$ (assumed statistically identical) and $R_n(\tau)$ is the autocorrelation function of $n_c(t)$ [or $n_s(t)$].

The lth component of the vector $\mathbf{B} \triangleq E[\mathbf{Z}D^*(t)]$ can be shown to be

$$\begin{aligned}
B_l = {} & 2(1 + \beta)R_d(l\Delta)e^{-j\alpha} \\
& + 2\beta R_d(l\Delta + \tau_m) \exp\left[-j(\omega_0\tau_m + \alpha)\right]
\end{aligned} \qquad (3\text{-}154)$$

Finally, by inverting A, with elements (3-152) and performing the matrix multiplication $\mathbf{A}^{-1}\mathbf{B}$, $\mathbf{W}_{\text{opt}}$ for a two-component multipath channel is obtained.

Performance results for the equalizer under various channel conditions can be obtained by simulation as described in Section 3-7. Typical results for the error probability versus multipath strength are shown in Figure 3-29.

The procedure used to calculate error probability is to simulate the transmission of a typical noise-free data sequence, in particular, a 31-bit maximal-length PN sequence, through the channel and detect it by an integrate-and-dump receiver, which may be preceded by the equalizer. At the end of each bit, the error probability for that bit is calculated using the signal component obtained through simulation and the appropriate noise variance is computed from the signal-to-noise ratio desired and the equivalent noise bandwidth of the receiver. The sequence-averaged error probability is then computed.

In obtaining the simulation results, the multipath delay is normalized to the carrier period, and is written as $\tau_N + \Delta\tau$, where τ_N is the integer part of the normalized delay and $\Delta\tau$ is the fractional part. For $\Delta\tau = 0$, only ISI is present, while, for $\Delta\tau = 0.25$, crosstalk alone is present. The signal-to-noise ratio is defined as the equivalent energy per bit to single-sided noise spectral density. The improvement gained through equalization is greatest for high-crosstalk situations.

Figure 3-29 shows P_E versus β for $\tau_N/T_c = \frac{1}{4}$, $\Delta\tau = 0$, and $\alpha = 0$, 5, and 10 degrees. The equalizer is clearly capable of fully compensating for crosstalk due to demodulator phase error in addition to ISI.

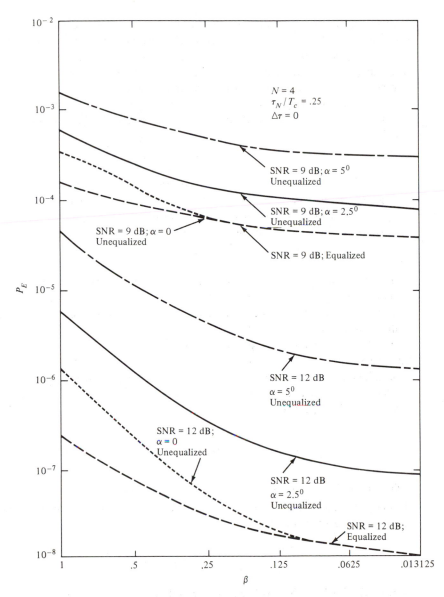

FIGURE 3-29. Error probability versus amplitude of a specular component for detection of equalized QPSK. (From Ref. 10.)

3-8.3 Adaptive Weight Adjustment

Setting the delay-line tap coefficients of the zero-forcing and LMS equalizers involves the solution of a set of simultaneous equations. In the use of the zero-forcing equalizer, adjustment of the tap coefficients involves measuring the channel filter output at $T - s$ spaced sampling times in response to a test pulse, and solving for the tap gains from (3-138). In the case of the LMS equalizer the equations to be evaluated are (3-152) through (3-154). Solution of these equations involves data-, noise-, and multipath-dependent parameters, which may be difficult to determine or may not be known at all.

Considerations such as these motivate the use of *automatic* tap coefficient adjustment algorithms. These automatic systems can utilize *preset* algorithms, wherein special sequences of pulses are used prior to data transmission or between gaps in the data transmission, and *adaptive* algorithms, which carry out the adjustment of the coefficients continuously during data transmission.

For the zero-forcing equalizer, (3-138) can be solved iteratively for the coefficient vector **C**. Let the **C**-matrix at the kth iteration be $\mathbf{C}^{(k)}$. The error in the solution to (3-138) is

$$\mathbf{E}^{(k)} = [P_c]\mathbf{C}^{(k)} - \mathbf{P}_{eq} \tag{3-155}$$

where $\mathbf{E}^{(k)}$ is a vector with $2N + 1$ components each of which represent the error in a component of $\mathbf{C}^{(k)}$. Each component of $\mathbf{C}^{(k)}$ can be adjusted in accordance with the error in it. If A is a small positive constant, an appropriate adjustment algorithm for the jth component of $\mathbf{C}^{(k)}$ is

$$\mathbf{C}_j^{(k+1)} = \mathbf{C}_j^{(k)} - A \text{ sgn } (\mathbf{E}_j^{(k)}) \tag{3-156}$$

where A is a constant determining the size of the adjustment. The iteration process is contained until $C_j^{(k+1)}$ and $C_j^{(k)}$ differ by some suitably small increment. The coefficient found by this iterative method can be shown to converge to the true values under fairly broad restrictions.

For the LMS equalizer, iterative solution for the coefficients is possible because the mean-square error (3-149) is a quadratic function of the tap gain coefficients. The adjustment of each tap gain is in a direction opposite to an estimate of the gradient of the mean-square error with respect to that tap gain. From (3-149), the gradient with respect to tap gain W_{rj} is

$$\frac{\partial I}{\partial W_{rj}} = \frac{\partial}{\partial W_{rj}} E\left[\left(\sum_k W_k^* z_k - \Delta\right)\left(\sum_k W_k z_k^* - \Delta^*\right)\right] \tag{3-157}$$

where (3-147) has been substituted and $\bar{z}_k \triangleq z(t - k\Delta) = z_1(t - k\Delta) + jz_2(t - k\Delta)$ has been used to shorten the notation. Taking the derivative first and the expectation last, it can be shown that

$$\frac{\partial I}{\partial W_{rj}} = E\left\{2 \text{ Re}\left[z_j^*\left(\sum_k W_k^* z_k - D\right)\right]\right\} \tag{3-158}$$

where $\partial W_k / \partial W_{rj} = \delta_{kj}$ has been used. Similarly, using $\partial W_k / \partial W_{ij} = -j\delta_{jk}$, it can be shown that

$$\frac{\partial I}{\partial W_{ij}} = E\left\{2j \text{ Im}\left[z_j^*\left(\sum_k W_k^* z_k - D\right)\right]\right\} \tag{3-159}$$

Equations (3-158) and (3-159) can be written more compactly using complex notation as

$$\frac{\partial I}{\partial W_j} \triangleq \frac{\partial I}{\partial W_{rj}} + j\frac{\partial I}{\partial W_{ij}} = 2E[z^*(t - j\Delta)e(t)] \tag{3-160}$$

where

$$e(t) = \sum W_k^* z(t - k\Delta) - D(t) \tag{3-161}$$

is the error between the equalizer output and the desired, undistorted output.

Equation (3-160) shows that the gradient of the mean-square error with respect to the jth tap gain is twice the *correlation* between the equalizer output and the error between actual and desired outputs. Two problems arise in applying it to adaptive weight adjustment. First, (3-160) requires the expectation or average to be taken. Since this is not available, the unbiased but noisy estimate $z^*(t - j\Delta)e(t)$ can be used. Second, the undistorted output $D(t)$ is not available unless a known data sequence is transmitted. An alternative to sending a known data sequence is to assume that the detected data is correct (which it essentially is even in a fairly bad channel giving an error probability of only 10^{-2}) and using the detected data to reconstruct an estimate of the undistorted output $D(t)$. Equalizers using this method of data estimation are called *decision directed*. A suitable decision-directed algorithm for weight adjustment of the LMS equalizer is

$$\mathbf{W}(n + 1) = \mathbf{W}(n) - A\mathbf{Z}_T^*[d_e(t_n - T) - D_d(t_n)] \qquad (3\text{-}162)$$

where $\mathbf{W}(n)$ is the complex weight vector at decision time $t_n = nT$, A is an adjustment parameter, $\mathbf{Z}_T$ is the vector (3-150) with $t = t_n - T$, $d_e(t_n - T)$ is given by (3-147), and $D_d(t_n)$ is the detected data at time t_n. In order to perform additional averaging of the estimate for the correlation in (3-160), the second term in (3-162) can be averaged over several data symbols.

Convergence behavior of the gradient-estimate update algorithm is hard to analyze. If a training period is used with known input data, the fastest convergence is obtained when the power spectrum (folded if sampling precedes equalization) of the equalizer input is flat, and the step size A is the inverse of the received signal power-step size product. The larger the variation in the power spectrum, the smaller the step size must be (see [9]).

For adaptive equalization, where a decision-directed algorithm is used, it is possible to track slow variations in the channel characteristics. The larger the step size, the faster the tracking but the larger the excess mean-square error. The excess mean-square error is that in excess of the minimum attainable mean-square error which is obtained with tap weights frozen at their optimum values and is caused by the tap gains wandering around the optimum values. It is proportional to the number of equalizer coefficients, the step size, and the channel noise power. The step size that provides the fastest convergence gives a mean-square error roughly twice that of the minimum mean-square error. A useful approach is to choose a large initial step size for fast convergence and decrease it once steady-state operation has been reached.

3-8.4 Other Equalizer Structures*

The equalizer structures discussed so far employ a transversal filter which is a linear system. Nonlinear equalizer structures may provide better performance under many circumstances.

A simple nonlinear equalizer is the *decision feedback equalizer* (DFE), which uses feedback of decisions on symbols already received to cancel the interference from symbols which have already been detected. A typical DFE structure is shown in Figure 3-30. The basic idea is that, assuming past decisions are correct, the ISI contributed by these symbols can be canceled exactly by subtracting appropriately weighted past symbol values from the equalizer output. This is the purpose of the delay line with weights $b_1, b_2, \ldots, b_M$ shown in Figure 3-30. The forward trans-

*See [9].

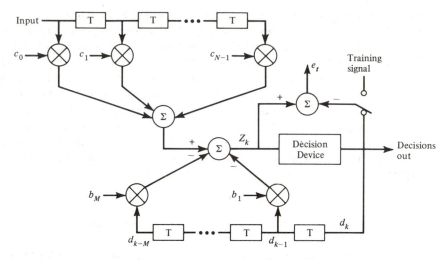

FIGURE 3-30. Decision-feedback equalizer. (From Ref. 9.)

versal filter, with weights $c_0, c_1, \ldots, c_{N-1}$, then need compensate for ISI over a smaller portion of the ISI-contaminated received signal. Both the feedback and feedforward coefficients can be adjusted simultaneously to minimize mean-square error.

Given the same number of overall coefficients, it cannot be said, in general, whether a DFE outperforms a linear, transversal filter equalizer or not. The performance of each equalizer is influenced by the particular channel characteristics, but the DFE can compensate for amplitude distortion without as much noise enhancement as a linear equalizer. When a decision error occurs in a DFE, its output reflects this error during the next few symbols because of the feedback delay line. Fortunately, the errors due to a feedback error occur in short bursts that degrade performance only slightly; the error propagation in a DFE is not catastrophic.

One problem with the DFE is startup. An approach to this problem is to open the feedback loop and let the equalizer start up as a transversal filter equalizer with adaptive weight adjustment.

Another type of device for correcting for the effects of memory in a digital communication channel is a Viterbi algorithm (VA) sequence estimator. The VA is described in Appendix D. It suffices here to mention that it is an optimum method for estimating Markov-dependent sequences.*

3-9

DEGRADATIONS DUE TO REALIZATION IMPERFECTIONS IN DIGITAL MODULATION SYSTEMS

The theoretically optimum performance, in terms of average probability of the receiver erroneously detecting the received data, of various communication systems when operating in AWGN environments has been considered in the foregoing. There are two areas of departure from this ideal situation which will result in degraded system performance from the theoretically optimum. The first is nonideal

*A comparison of VA sequence estimation and linear equalization for QPSK in a particular environment is given in [11].

realization of the various subsystems that make up the communication system. The second area is the departure of the channel from the AWGN ideal. Degradations due to non-Gaussian noise are not easy to analyze and will not be considered here. Degradations due to several commonly occurring system realization imperfections can be analyzed and are considered in this section.

The general model for considering these system impairments is shown in Figure 3-31. Consideration will be limited to coherent communication systems and to linear channels. Although this limitation is somewhat restrictive, the class of system degradations to be considered commonly occur and illustrate the type of analyses that must be made in determining a system's departure from ideal.

Returning to the block diagram of Figure 3-31, the perturbations that will be considered are the following:

1. Phase and amplitude imbalance in BPSK.
2. Phase and amplitude imbalance in QPSK modulation.
3. Power loss due to filtering modulated signals at the transmitter.
4. Imperfect phase reference at the coherent demodulator.
5. Nonideal detection filter.
6. Predetection filtering.
7. Transmitter filtering degradations—nonmatched detector.
8. Bit synchronizer timing error.

3-9.1 Phase and Amplitude Imbalance in BPSK

In some types of BPSK modulators, perfect phase switching of π radians (180 degrees) does not occur when the input data transitions occur. Let $d(t) = \pm 1$ be the input bit stream to the modulator. An imperfectly phase-switched BPSK signal can be represented as

$$y(t) = A \sin \left[2\pi f_0 t + (1 - \alpha)\frac{\pi}{2} d(t) \right] \qquad (3\text{-}163)$$

where $0 \le \alpha \le 1$ is the fractional error from perfect phase switching. If $\alpha = 0$ (3-163) represents an ideal BPSK waveform, the output of an ideal coherent demodulator in response to (3-163) is

$$y_D(t) = A \sin \left[(1 - \alpha)\pi \frac{d(t)}{2} \right]$$

$$= A \, d(t) \cos \frac{\pi\alpha}{2} \qquad (3\text{-}164)$$

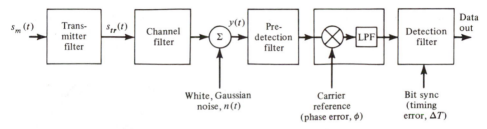

FIGURE 3-31. **Model of a coherent communication system showing various system departures from ideal model.**

The degradation due to imperfect phase switching, defined to be the relative loss in demodulation output power from ideal in decibels, is

$$D_{ps} = -20 \log_{10}\left(\cos \frac{\alpha \pi}{2}\right) \tag{3-165}$$

It is interesting to note that $\alpha = 0.5$ gives a degradation of 3.01 dB since $\cos \pi/4 = 1/\sqrt{2}$; that is, half of the demodulator output power appears as a carrier component due to imperfect phase switching.

Amplitude imbalance occurs in BPSK modulation when logic ones and zeros are represented by amplitudes for $m(t)$ that are not $+1$ or -1, respectively, but by $1 + \epsilon$ and $-1 + \epsilon$, where ϵ is the fractional amplitude error. Thus a BPSK modulated signal with amplitude imbalance can be represented as

$$y(t) = A[\epsilon + d(t)] \cos 2\pi f_0 t \tag{3-166}$$

The output of a coherent demodulator in response to this input signal is

$$y_D(t) = A[\epsilon + d(t)] \tag{3-167}$$

The ratio of power out of the demodulator with an amplitude-imbalanced signal at its input to the power for the ideal case is

$$\frac{P_0''}{P_0} = \frac{A^2\overline{[\epsilon + d(t)]^2}}{A^2\overline{d^2(t)}} = 1 + \epsilon^2 \tag{3-168}$$

where the overbar denotes an average of the quantity. The average $\overline{d^2(t)} = 1$ since $d(t)$ is either $+1$ or -1. Also, $\overline{d(t)} = 0$ under the assumption that logic ones and zeros are equally likely. The degradation in decibels due to amplitude imbalance is therefore

$$D_{ai} = 10 \log_{10}(1 + \epsilon^2) \tag{3-169}$$

The degradations due to phase switching and amplitude imbalance are additive at the demodulator output since the phase detector output is proportional to the product of the separate effects. Figure 3-32 shows a plot of degradation due to phase switching and amplitude imbalance.

3-9.2 Phase and Amplitude Unbalance in QPSK Modulation

Consider a QPSK-modulated waveform of the form

$$x_c(t) = A\left[d_1(t) \cos\left(2\pi f_0 t + \frac{\beta}{2}\right) + d_2(t) \sin\left(2\pi f_0 t - \frac{\beta}{2}\right)\right] \tag{3-170}$$

where $\beta \geq 0$ represents an unbalance in the desired quadrature relationship between the two carriers. The carrier synchronizer at the receiver is assumed to track the average phase error. With phase unbalance the outputs of the coherent demodulators (see Figure 3-12) are

$$y_{D1}'(t) = A\left[d_1(t) \cos \frac{\beta}{2} - d_2(t) \sin \frac{\beta}{2}\right]$$

$$y_{D2}'(t) = A\left[-d_1(t) \sin \frac{\beta}{2} + d_2(t) \cos \frac{\beta}{2}\right] \tag{3-171}$$

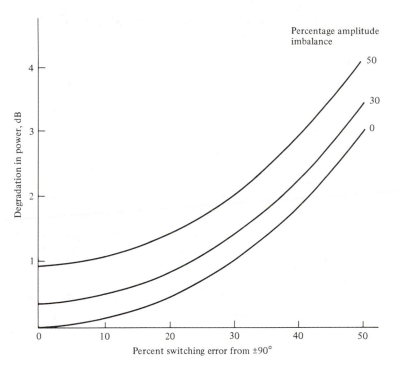

FIGURE 3-32. **Degradation due to amplitude and phase imbalance in biphase modulation.**

If no phase unbalance were present, the outputs would be

$$y_{D1}(t) = Ad_1(t)$$

$$y_{D2}(t) = Ad_2(t) \tag{3-172}$$

Thus phase unbalance introduces both amplitude degradation as well as crosstalk from the other channel. The effect of these two degradations on the receiver performance can be analyzed by noting that four combinations for the signs of $d_1(t)$ and $d_2(t)$ occur. This gives rise to the four possible outputs in each quadrature channel of the form

$$y'_{D1}(t) = \pm A \left(\cos \frac{\beta}{2} \pm \sin \frac{\beta}{2} \right) \tag{3-173}$$

with a similar set of expressions for $y'_{D2}(t)$. Equation (3-55) shows that the probability of error in one quadrature channel of a QPSK detector is dependent only on the ratio E_b/N_0. From (3-173) it follows that E_b is modified to either

$$E'_b = E_b \left(\cos \frac{\beta}{2} + \sin \frac{\beta}{2} \right)^2$$

or

$$E''_b = E_b \left(\cos \frac{\beta}{2} - \sin \frac{\beta}{2} \right)^2$$

where E_b denotes the energy per bit of an ideal demodulator, and the primes denote degraded bit energies due to phase unbalance. The average probability of error per quadrature channel for a system with phase unbalance, assuming that E_b' and E_b'' occur with equal probability, is

$$P_1'(\epsilon) = \frac{1}{2} Q\left[\sqrt{\frac{2E_b}{N_0}}\left(\cos\frac{\beta}{2} + \sin\frac{\beta}{2}\right)\right]$$

$$+ \frac{1}{2} Q\left[\sqrt{\frac{2E_b}{N_0}}\left(\cos\frac{\beta}{2} - \sin\frac{\beta}{2}\right)\right] \qquad (3\text{-}174)$$

If $\beta = 0$, (3-174) reduces to (3-55). The relative amount that E_b must be increased for a system with phase unbalance present in order to maintain the same error probability as a system with no unbalance is the degradation due to unbalance. Because (3-174) is the average of two Q-functions, there is no simple expression, such as (3-165), for this degradation; rather, a numerical solution is required for a given probability of error. A graph showing degradation due to phase unbalance in QPSK for a bit error probability of 10^{-6} is shown in Figure 3-33 as a function of the phase unbalance β.

Degradation due to amplitude unbalance occurs when the quadrature carriers for QPSK are unequal. According to (3-43b), this results in the angles between the vectors representing the four signal states being unequal. The result is degradation in the demodulator output due to amplitude attenuation similar to the degradation imposed by phase unbalance except that cross-talk is not present.

3-9.3 Power Loss due to Filtering the Modulated Signal

Filtering of the modulated signal is often done at the transmitter to limit out-of-band power. Figures 3-34 and 3-35 show the relative power loss in decibels introduced by filtering MSK and QPSK, respectively, with third-order Butterworth, Bessel, and 0.1-dB ripple Chebyshev filters as a function of filter bandwidth. Figures 3-36 and 3-37 show similar results for fifth-order filters.

3-9.4 Imperfect Phase Reference at a Coherent Demodulator

Referring to Figure 3-31, consider the effect on $P(\epsilon)$ of having the phase reference $\cos(2\pi f_0 t + \varphi)$ instead of $\cos 2\pi f_0 t$ in a BPSK system. Since the received signal

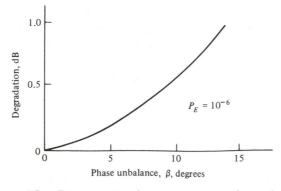

FIGURE 3-33. Degradation due to phase unbalance in a QPSK modulator.

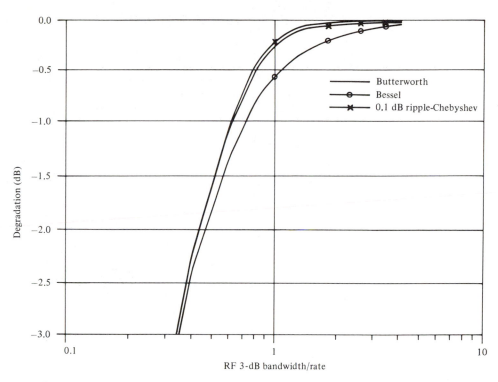

FIGURE 3-34. Power-loss degradation due to filtering MSK signals; third-order filters.

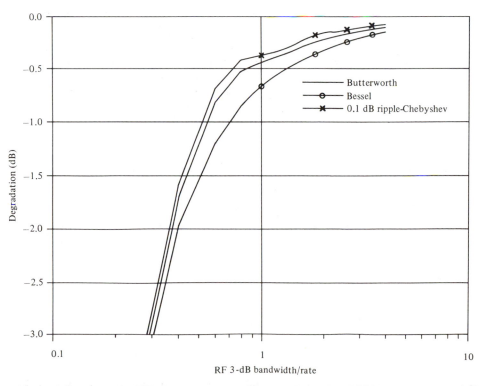

FIGURE 3-35. Power-loss degradation due to filtering QPSK signals; third-order filters.

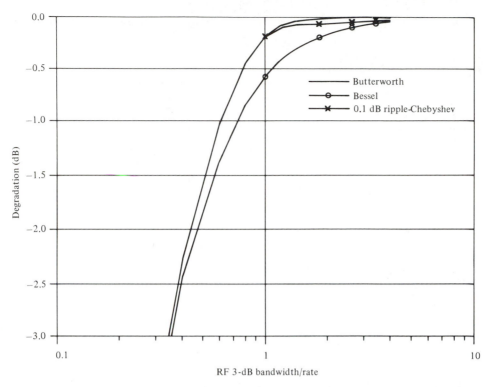

FIGURE 3-36. Power-loss degradation due to filtering MSK signals; fifth-order filters.

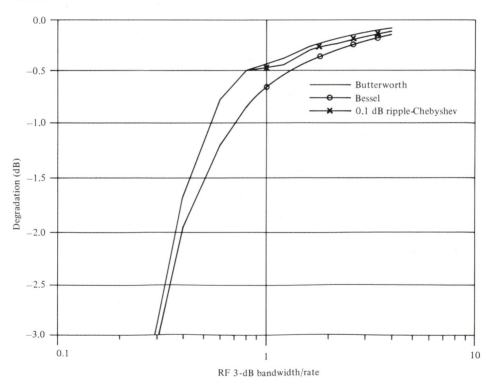

FIGURE 3-37. Power-loss degradation due to filtering QPSK signals; fifth-order filters.

(noise excluded for now) is

$$y(t) = \pm A \, d(t) \cos 2\pi f_0 t \tag{3-175}$$

the output of the coherent demodulator with imperfect phase reference is

$$y_D(t) = \pm A \, d(t) \cos \varphi \tag{3-176}$$

whereas for the ideal case, the result is $\pm A \, d(t)$. Since the amplitude is degraded by the factor $\cos \varphi$, signal-to-noise ratio is degraded by $\cos^2\varphi$ with the resulting system degradation due to demodulator phase error given by

$$D_{\text{BPSK}\varphi} = 10 \log_{10}(\cos^2\varphi)$$

$$= 20 \log_{10}(\cos \varphi) \qquad \text{dB} \tag{3-177}$$

For quadrature modulation systems, the analysis of demodulator phase error degradation is more complicated because of the crosstalk introduced between channels and is similar to that carried out for the case of phase unbalance. Degradation results due to demodulator phase error for BPSK, QPSK, and MSK are shown in Figure 3-38. Note that serial MSK and BPSK experience essentially the same degradation due to demodulator static phase error.

Computation of degradation due to random-phase perturbations requires averaging of the probability of error given the phase error with respect to the phase error probability density function. A fairly accurate model for this distribution for small phase errors is the Gaussian probability density function. (See Van Trees [12, Chap. 4] for more discussion, including a phase error density function that applies to first-order phase-locked loop carrier tracking.)

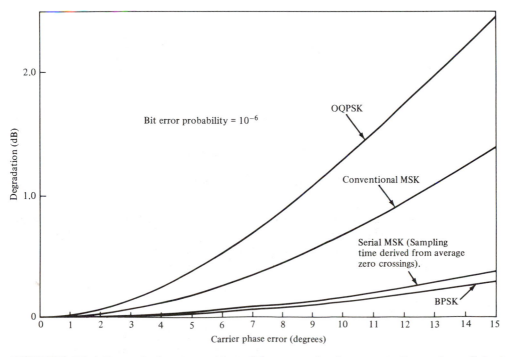

FIGURE 3-38. Degradation due to demodulator static phase error for various digital modulation schemes. (From Ref. 3.)

3-9.5 Degradation due to a Nonideal Detection Filter

A correlation detector requires precise timing of the starting and stopping times for the integration process. In view of this, it is often useful to employ as the detection filter a low-pass filter which is not the ideal correlation type. When this is done, the performance of the receiver is degraded from that of the optimum receiver for two reasons. First, the ratio of peak signal to root-mean-square (rms) noise at the output of a nonoptimum detection filter will not be the theoretically maximum value, which is $2E/N_0$ for a filter matched to the signal, E being the signal energy and N_0 the noise power spectral density. Second, a suboptimum filter will introduce intersymbol interference from one symbol period to succeeding symbol periods because of its transient response.

To see how one might analyze a system employing a suboptimum detection filter, consider the use of a first-order lowpass detection filter in a BPSK receiver. The effect of intersymbol interference between two successive bits will be analyzed. It may, in fact, be assumed that the coherent demodulation process has converted the signal to an antipodal baseband signal. Consider the response of the filter to the four possible 2-bit sequences (1) one–one, (2) one–zero, (3) zero–one, and (4) zero–zero. The step response of the first-order lowpass filter is

$$h_s(t) = (1 - e^{-t/\tau})u(t) \tag{3-178}$$

where τ is its time constant, which is related to its 3-dB frequency f_3 by

$$\tau = \frac{1}{2\pi f_3}$$

The signal component at the output of the detection filter at the end of the second bit interval for the four possible inputs listed above is

$$(1) \quad y_{D,a}(2T) = A_c(1 - e^{-2T/\tau}) \tag{3-179}$$

$$(2) \quad y_{D,b}(2T) = A_c[(1 - e^{-2T/\tau}) - 2(1 - e^{-T/\tau})]$$

$$= -A_c(1 - e^{-T/\tau})^2 \tag{3-180}$$

$$(3) \quad y_{D,c}(2T) = -y_{D,b}(2T)$$

$$(4) \quad y_{D,d}(2T) = -y_{D,a}(2T)$$

The probability of error given a particular input data sequence, say (1), is

$$P(\epsilon|a) = Q\left[\sqrt{\frac{|y_{D,a}(2T)|^2}{\sigma_N^2}}\right] \tag{3-181}$$

where σ_N^2 is the variance of the noise component at the detection filter output. Equation (3-181) can be derived by integrating the probability density function of the lowpass filter output at time $t = 2T$ from $-\infty$ to 0.

The variance of the noise at the filter output is given by

$$\sigma_N^2 = B_N N_0$$

$$= \frac{\pi}{2} f_3 N_0$$

$$= \frac{N_0}{4\tau} \tag{3-182}$$

where $B_N = (\pi/2)f_3$ is the noise equivalent bandwidth of the detection filter. Thus (3-181) can be written in terms of τ/T and $E_b/N_0 = A^2 T_s/N_0$ by substituting (3-179) and (3-182). The result is

$$P(\epsilon|a) = Q\left[2\sqrt{\frac{\tau}{T}}\,(1 - e^{-2T/\tau})\sqrt{\frac{E_b}{N_0}}\right] \qquad (3\text{-}183)$$

with an identical result for $P(E|d)$. Similarly, $P(E|b)$ and $P(E|c)$ can be written as

$$P(\epsilon|b) = P(\epsilon|c)$$

$$= Q\left[2\sqrt{\frac{\tau}{T}}\,(1 - e^{-T/\tau})^2\sqrt{\frac{E_b}{N_0}}\right] \qquad (3\text{-}184)$$

Assuming the sequences (1), (2), (3), and (4) to be equally probable, the average probability of error is

$$P(\epsilon) = \tfrac{1}{4}[P(\epsilon|a) + P(\epsilon|b) + P(\epsilon|c) + P(\epsilon|d)]$$

$$= \tfrac{1}{2}[P(\epsilon|a) + P(\epsilon|b)] \qquad (3\text{-}185)$$

To compare this result with the ideal case, curves showing probability of error versus E_b/N_0 could be plotted for both the ideal matched filter and the lowpass detection filter with τ/T employed as a parameter for the latter case. The degradation in signal-to-noise ratio from the ideal case is then the distance along the abscissa between the two sets of curves at a given probability of error. Figure 3-39 shows the degradation at $P(\epsilon) = 10^{-6}$ due to a first-order lowpass detection filter as a function of τ/T. The minimum degradation of about

$$D_{\text{LPF1,min}} \simeq 2.15 \text{ dB}$$

is achieved for $\tau/T \simeq 0.42$ or for

$$f_{3,\text{opt}}T = \frac{T}{2\pi}\,\tau_{\text{opt}}$$

$$\simeq 0.38$$

For smaller values of f_3 than this, the intersymbol interference and signal energy loss introduced by the detection filter degrades performance more than 2.15 dB,

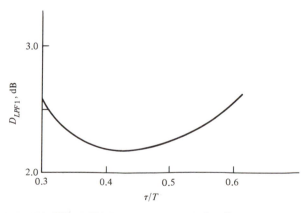

FIGURE 3-39. Degradation due to use of a first-order lowpass filter as a detection filter for antipodal signaling.

while for larger values of f_3, the noise passed by the detection filter degrades performance more than the minimum.

While a degradation of 2.15 dB may seem to be somewhat excessive, the simplicity of the lowpass filter as a detector may warrant its use in some situations. A comment is perhaps in order for those familiar with radar systems. If a first-order lowpass filter is employed in place of a matched filter detector in a pulsed radar system, the minimum degradation obtained is about 0.9 dB, or less than half the value obtained here. However, intersymbol interference is negligible in the radar application, which allows the use of a narrower bandwidth filter (the equivalent lowpass optimum bandwidth is about half that obtained here).

The question naturally arises as to what degradation is obtained if other types of detection filters are employed. Figures 3-40 and 3-41 show average probability of error versus BT product, where B is the *two-sided* or RF equivalent detection filter bandwidth, with E_b/N_0 as a parameter for single- and double-pole Butterworth filters. Note that the optimum BT, where T is the symbol duration, is weakly dependent on E_b/N_0. For the single-pole filter $(BT)_{opt} \simeq 0.75$, which compares favorably with the value obtained in the simple analysis performed above.

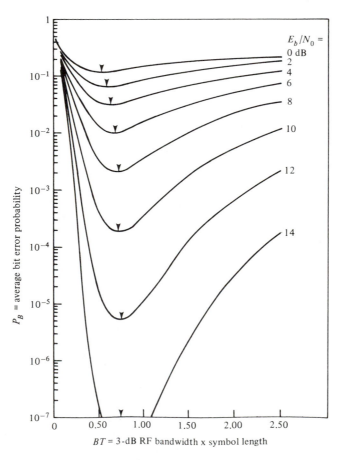

FIGURE 3-40. Single-pole *RC* data filter detection of coherent QPSK and BPSK. (From Ref. 13.)

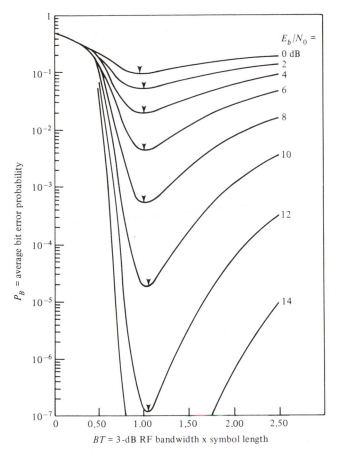

$E_b/N_0 =$
0 dB
2
4
6
8
10
12
14

P_B = average bit error probability

BT = 3-dB RF bandwidth x symbol length

FIGURE 3-41. Two-pole Butterworth data filter detection of coherent QPSK and BPSK. (From Ref. 13.)

3-9.6 Degradation due to Predetection Filtering

Since all practical receivers involve filtering prior to the detection operation, which may be due for example to an intermediate frequency amplifier, the effects of filters placed prior to an ideal matched filter receiver are now considered. The case of antipodal baseband signaling will be considered or, equivalently, BPSK signaling, which includes QPSK and OQPSK when viewed on a per quadrature channel basis. Figure 3-42 shows the effect of a five-pole, 0.1-dB ripple Chebyshev filter prior to an integrate-and-dump detector. Two curves are given showing degradation as a function of symbol rate normalized by the 3-dB RF bandwidth of the filter. The two cases depicted are those of a transmission filter and a predetection filter. In the former case, the noise enters the channel *after* the filter, whereas in the latter case, the noise enters before the filter. Thus the predetection filter rejects some of the noise added in the channel and the degradation caused by signal power loss and intersymbol interference due to filtering is *less* than for the case of a transmission filter. Note that for $BT \to \infty$ ($1/BT = 0$), the degradation is zero in both cases since this is the case of no predetection or transmission filter. The difference between the two curves for $1/BT > 0$ is due solely to the noise rejection by the predetection filter. Figure 3-42 shows that for a five-pole Chebyshev filter, the

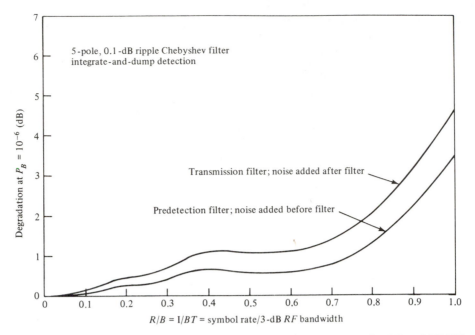

FIGURE 3-42. Comparison of transmission and predetection filtering. QPSK and BPSK symbol rates and bit energies the same. (From Ref. 13.)

degradation due to prefiltering before matched filtering becomes excessive for $1/BT \gtrsim 0.9$. Since the BPSK or QPSK spectrum main lobe occupies a bandwidth null-to-null of $2/T$, this translates to filtering out nearly half of the major lobe of the signal spectrum.

3-9.7 Degradation due to Transmitter, or Channel, Filtering; Nonmatched Detector

As explained in the preceding section, filtering at the transmitter or in the channel prior to the addition of the system noise introduces signal degradation in the form of signal power loss and intersymbol interference. None of this degradation is regained through noise power rejection by the filter. Consequently, in such cases, the use of a nonmatched filter may give better performance for certain parameter combinations than a matched filter.* Figure 3-43 compares the degradation imposed by various orders of Chebyshev transmission filters for integrate-and-dump and two-pole Butterworth filter detection of BPSK or QPSK modulated signals. Note that for $1/BT = 0$, which is the case of no transmission filter, the degradation is zero for integrate-and-dump (matched filter) detection. The loss is about 0.7 dB for the two-pole filter which corresponds to the degradation with $BT = 1$ given in Figure 3-41. From Figure 3-43, it is indeed seen that the two-pole detection filter does give less degradation than the integrate-and-dump detector for $1/BT \gtrsim 0.4$, that is, for transmission filter bandwidths which are of the order or less than the bandwidth of the main lobe of the signal spectrum. We conclude, therefore, that

---

*Another way to look at this, with ISI excluded, is that the detection filter should be matched to the *filtered* transmitted signal.

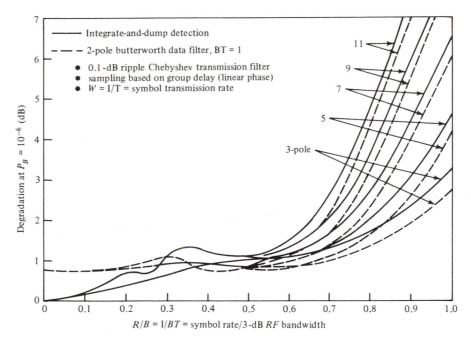

FIGURE 3-43. Bandwidth-limiting degradation of QPSK and BPSK signals. (From Ref. 13.)

in dealing with bandwidth-limited BPSK or QPSK signals, the integrate-and-dump detector is no longer the matched filter and improved performance may be achieved with simpler detection filters.

The effect of mistuning of the transmission filter is shown in Figure 3-44. Again, degradation results for both integrate-and-dump and two-pole detection filters are shown. The transmission filter considered is seven-pole Chebyshev with a bandwidth of 10 times the symbol rate. The degradation is less than 1 dB for filter center frequency displacements, normalized by the data rate, less than 4.0.

Often, cascaded transmission filters are employed in a communication system. Figure 3-45 shows degradation versus the number of cascaded five- and seven-pole Chebyshev transmission filters with $BT = 1$ and a two-pole Butterworth detection filter.

3-9.8 Degradation due to Bit Synchronizer Timing Error

The implications of bit synchronizer timing error can be inferred by considering the detection of constant-amplitude antipodal baseband signaling with an integrate-and-dump detector. As in the case of the nonideal detection filter analysis, two adjacent bits must be considered, the four possibilities being the same as in Section 3-9.5. The impulse response of the integrate-and-dump detector can be written as

$$h_{\mathrm{ID}}(t) = \begin{cases} 1 & \Delta t \le t \le \Delta t + T \\ 0 & \text{otherwise} \end{cases} \tag{3-186}$$

where Δt is the timing error. The output of the detection filter at time $\Delta t + T$ as a function of Δt is shown in Figure 3-46 for the four cases described in Section 3-9.5. Since the output noise variance from the detection filter is independent of

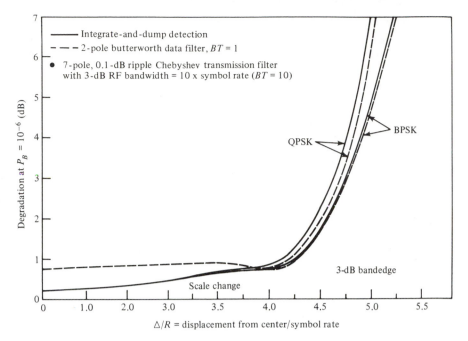

FIGURE 3-44. Mistuned broadband filtering. (From Ref. 13.)

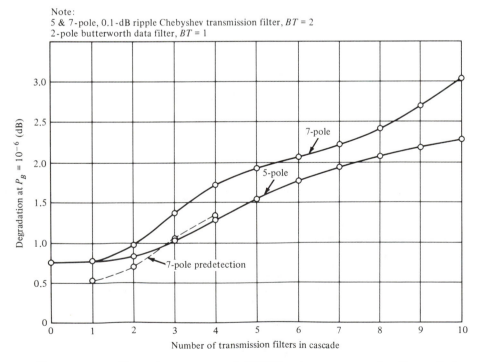

FIGURE 3-45. Cascade filter degradation of QPSK and BPSK signals. (From Ref. 13.)

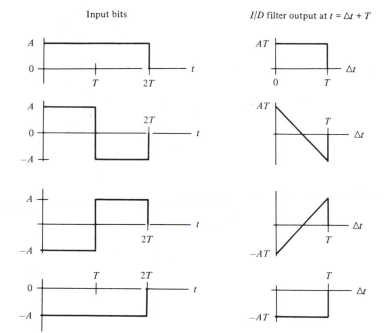

FIGURE 3-46. Waveforms for considering effects of sampling error for integrate-and-dump detection.

sampling instant, the degradation in signal-to-noise ratio as a function of Δt is proportional to the square of the matched filter output as a function of $\Delta t / T$. In particular, the relative degradation for each case is

$$D_a = D_d = 1 \qquad\qquad 0 \le \Delta t \le T$$

$$D_b = D_c = \left(1 - \frac{2\Delta t}{T}\right)^2 \quad 0 \le \Delta t \le T \qquad (3\text{-}187)$$

Some thought will lead one to the conclusion that these degradations are even functions of $\Delta t / T$. The average probability of error, assuming all sequences are equally likely, is

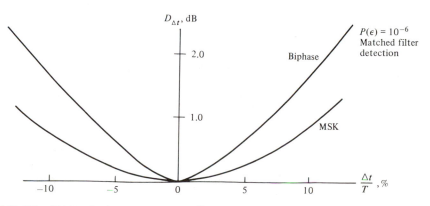

FIGURE 3-47. Degradation due to sampling error.

$$P(\epsilon) = \frac{1}{4} \operatorname{erfc}\left(\sqrt{D_a \frac{E_b}{N_0}}\right) + \frac{1}{4} \operatorname{erfc}\left(\sqrt{D_b \frac{E_b}{N_0}}\right) \qquad (3\text{-}188)$$

As in the case of the nonideal detection filter, the fractional increase in E_b/N_0 over ideal required to maintain a given $P(\epsilon)$ with timing error then defines the average degradation. The degradation at $P(\epsilon) = 10^{-6}$ is plotted in Figure 3-47 as a function of $\Delta t/T$ for ideal matched filter detection, where it is noted that a timing error of $\pm 7\%$ results in a degradation of 1 dB. Also shown in Figure 3-47 is degradation due to timing error for MSK, which was obtained through simulation.

3-10

MODULATOR STRUCTURES FOR QPSK, OQPSK, AND MSK

Two approaches can be taken to producing quadrature-multiplexed modulated signals. These are the parallel and serial approaches. In the former approach, the modulator structure is a realization of (3-42), while, for the latter approach, it is a realization of (3-43). The general form of the parallel modulator structures for QPSK, OQPSK, and MSK are shown in Figure 3-48a. Two parallel data streams, $d_1(t)$ and $d_2(t)$ with symbol duration $T_s = 2T_b$, are produced from the serial bit stream $d(t)$ each with bit duration T_b. If QPSK is to be produced, these parallel data streams, assumed to be ± 1-valued, are used to directly biphase modulate sinusoidal carriers which are in phase quadrature. (Note that the parallel data streams do not have to have the same symbol duration in the case of QPSK.) The biphase-modulated quadrature carriers are then summed to produce QPSK.

In the case of OQPSK modulation, the parallel data streams, $d_1(t)$ and $d_2(t)$, are offset one-half symbol period with respect to each other before biphase modulation of the quadrature carriers takes place. As a result of this offset, $d_1(t)$ and $d_2(t)$ must have the same symbol durations.

Parallel modulation of MSK is similar to parallel modulation of OQPSK, except that half-sinusoid weighting functions are used to multiply the symbols of the parallel data streams. Two options are available for this multiplication, resulting in what has been referred to as MSK type 1 and MSK type 2 modulation (see [14]). In MSK type 1, the weighting pulses alternate as positive and negative half-sinusoids. In MSK type 2 modulation, a positive half-sinusoid only is used. If, in addition, the serial input data are differentially encoded in an MSK type 1 modulator, the result is fast frequency-shift keying (FFSK). The differential encoding is carried out according to the rule

$$c_{n+1} = c_n \oplus d_n \qquad (3\text{-}189)$$

where c_n is the previous encoder output, d_n is the present encoder data input, and $\oplus$ denotes the logical Exclusive-OR (EXOR) function or binary add without carry. The differential encoding results in a one-to-one relationship between the modulator input data and transmitted frequency. Without differential encoding, resolution of the regenerated carrier ambiguity is not accomplished.

The serial approach to MSK modulation has already been discussed. A serial modulator for QPSK is shown in Figure 3-48b, and a serial modulator for OQPSK is shown in Figure 3-48c. The former requires one 0–90-degree phase modulator and one 0–180-degree phase modulator. The latter requires two 0–90-degree phase

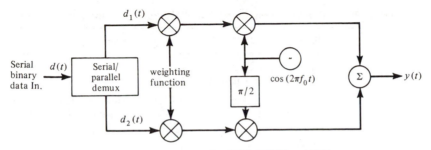

(a) Parallel modulator structure for QPSK, OQPSK, and MSK
(for QPSK and OQPSK weighting functions are unity)

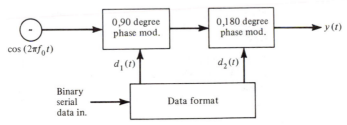

(b) Serial modulator structure for QPSK

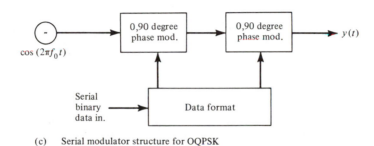

(c) Serial modulator structure for OQPSK

FIGURE 3-48. Parallel and serial modulator structures.

modulators. Table 3-1 gives a list of advantages and disadvantages for parallel and serial modulator implementations.

It is useful to visualize QPSK, OQPSK, and MSK in terms of phasor or signal space diagrams. These would be traced as Lissajous curves on an oscilloscope screen if the vertical axis has as its input the (weighted) data stream in the modulator inphase channel, and the horizontal axis input is the quadrature-channel data stream. Since these signals are ±1-valued in the case of QPSK, the Lissajous curve is ideally a square with diagonals. This is because ±90 or 180-degree excess phase transitions can take place with a QPSK signal. For OQPSK, 180-degree phase transitions are impossible, so that the diagonals of the square are absent in the Lissajous curve of an OQPSK signal. With MSK, the inphase and quadrature signals are sinusoidally weighted which, ideally, produces a circle for the Lissajous figure. Only ±90-degree phase transitions are possible, so that only the circumference of the circle is traced in response to a random data pattern. Phasor diagrams, together with allowed transitions are shown in Figure 3-49. The signal space dia-

TABLE 3-1. **Characteristics of Parallel and Serial Implementations**

		Parallel	Serial
QPSK		1. The way the equation for $x_c(t)$ is written suggests parallel	
		2. Modulator requires two 0–180° phase modulators	Modulator requires one 0–90° and 0–180° phase modulator
		3. Amplitude imbalance between channels translates to nonorthogonal signal phasors, which results in degraded performance	Amplitude imbalance does not produce nonorthogonal signal phasors (nonorthogonality is dependent on phase modulator accuracy only)
		4. Quadrature data channels can be asynchronous	Data must be serial or, if derived from two sources, they must be synchronous
		5. Quadrature phase misalignment of 90° hybrid and summer in modulator causes degradation	
		6. Two bits of ambiguity in detector due to two separate data streams	Potential of two bits of ambiguity; can be implemented so only one bit of ambiguity
OQPSK		Quadrature data channels must be synchronous. Remarks 1, 2, 3, 5, and 6 under QPSK apply here also	Suboptimum if demodulated serially
MSK		Data detection and conversion back to serial is complex. Clock can be derived as a consequence of demodulation	Simple to implement if required filters can be synthesized accurately. Can demodulate with Costas loop, or acquire mark frequency reference with squaring loop

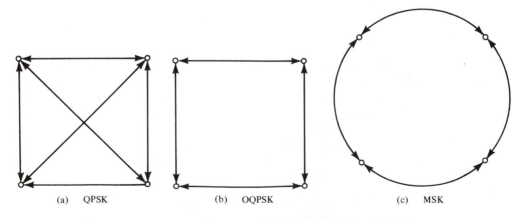

(a)   QPSK   (b)   OQPSK   (c)   MSK

FIGURE 3-49. Phase diagrams for ideal, QPSK, O-QPSK, and MSK.

grams for nonideal modulators do not consist of straight lines because the transitions from one phase state to the next are not instantaneous due to distortion. Similarly, the transitions in the signal space diagram for MSK do not follow arcs of circles in a nonideal MSK modulator because of distortion.

3-11

ENVELOPE FUNCTIONS FOR BPSK, QPSK, OQPSK, AND MSK

If constant-envelope modulation schemes such as BPSK, QPSK, OQPSK, and MSK are filtered prior to transmission, the result will be a modulated carrier with a nonconstant envelope. The reason for this can be seen by viewing the filtering as taking place on the unmodulated data streams.* For example, Figure 3-50 il-

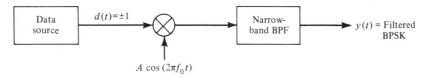

(a) BPSK modulator with bandpass filter at output

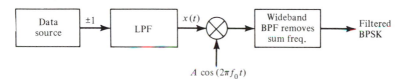

(b) BPSK modulator with filtered data; equivalent to (a)

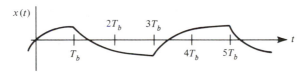

(c) Output of LPF in (b)

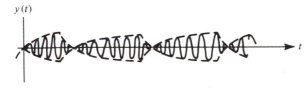

(d) Filtered output of modulator in (a) or (b)

FIGURE 3-50. Illustration of the cause for envelope deviation of a filtered BPSK modulated signal.

*Recall that in Chapter 2 it was shown how the impulse response of a narrowband, bandpass filter could be represented as $h(t) = h_R(t) \cos 2\pi f_0 t - h_I(t) \sin 2\pi f_0 t = Reh(t) \exp(j2\pi f_0 t)$, where $h_R(t)$ and $h_I(t)$ are lowpass. Thus a bandpass filter can be represented by two lowpass filters in parallel arms where the input is mixed to zero frequency and then mixed up again.

lustrates the process of BPSK. Each time the data switches sign, a transient buildup of the output takes place, which results in the modulated carrier envelope going through a null. Filtering of QPSK would result in similar nulls, followed by OQPSK, in which the dips in the envelope are not as deep, and finally followed by MSK for which the dips are less yet.

The reason that OQPSK maintains a more nearly constant envelope than QPSK when bandlimited can be intuitively seen by considering both as composed of the sum of two biphase-modulated signals on phase-quadrature carriers. If a biphase-modulated signal is filtered, the filtered output signal undergoes a transient buildup each time the phase of the input signal to the filter undergoes a 0–180-degree phase reversal. Since these phase reversals can line up in each quadrature channel of a QPSK system, the envelope of the QPSK signal goes to zero each time this happens. On the other hand, the 0–180-degree phase reversals of each quadrature channel cannot line up in an OQPSK system with the result that the envelope can, at worst, dip to 0.707 of its maximum value.

If an MSK-modulated signal is filtered, its envelope undergoes less severe fluctuations than QPSK or OQPSK due to the continuous phase of the modulated signal. Figure 3-51 shows the maximum-to-minimum deviation of the envelope of MSK when filtered by Butterworth, Bessel, and 0.1-dB ripple Chebyshev filters of various bandwidths. For a further discussion of the effect of filtering on angle-modulated signals, see [15, Chap. 11].

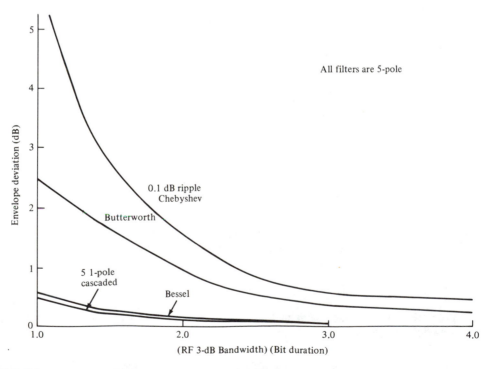

FIGURE 3-51. Envelope deviation of MSK due to filtering the modulated signal. (CRF simulation: 15-bit PN sequence used.)

REFERENCES

[1] S. PASUPATHY, "Minimum Shift Keying: A Spectrally Efficient Modulation," *IEEE Commun. Mag.*, Vol. 17, pp. 14–22, July 1979.

[2] J. K. HOLMES, *Coherent Spread Spectrum Systems* (New York: Wiley, 1982).

[3] R. E. ZIEMER and C. R. RYAN, "Minimum-Shift Keyed Modem Implementations for High Data Rates," *IEEE Commun. Mag.*, Vol. 21, pp. 28–37, October 1983.

[4] F. AMOROSO and J. A. KIVETT, "Simplified MSK Signaling Technique," *IEEE Trans. Commun.*, Vol. COM-25, pp. 433–441, April 1977.

[5] T. J. APRILLE, "Filtering and Equalization for Digital Transmission," *IEEE Commun. Mag.*, Vol. 21, pp. 17–24, March 1983.

[6] A. LENDER, "The Duobinary Technique for High Speed Data Transmission," *IEEE Trans. Commun. Electron.*, Vol. 82, pp. 214–218, May 1963.

[7] A. LENDER, "Correlative Level Encoding for Binary Data Transmission," *IEEE Spectrum*, pp. 104–105, February 1966.

[8] S. PASUPATHY, "Correlative Coding—A Bandwidth Efficient Signaling Scheme," *IEEE Commun. Soc. Mag.*, pp. 4–11, July 1977.

[9] S. QURESHI, "Adaptive Equalization," *IEEE Commun. Mag.*, pp. 9–16, March, 1982.

[10] R. E. ZIEMER and C. R. RYAN, "Equalization of QPSK Data Transmission in Specular Multipath," *IEEE Trans. Aerosp. Electron. Syst.*, Vol. AES-10, pp. 588–594, September 1974.

[11] S. H. R. RAGHAVAN and R. E. ZIEMER, "Improving QPSK Transmission in Band-limited Channels with Interchannel Interference Through Equalization," *IEEE Trans. Commun.*, Vol. COM-25, pp. 1222–1226, October 1977.

[12] H. L. VAN TREES, *Detection, Estimation, and Modulation Theory*, Vol. I (New York: Wiley, 1968).

[13] J. JAY JONES, "Filter Distortion and Intersymbol Interference Effects on PSK Signals," *IEEE Trans. Commun. Technol.*, Vol. COM-19, pp. 120–132, April 1971.

[14] V. K. BHARGAVA, D. HACCOUN, R. MATYAS, and P. NUSPL, *Digital Communications by Satellite* (New York: Wiley, 1981).

[15] J. J. SPILKER, *Digital Communications by Satellite* (Englewood Cliffs, N.J.: Prentice-Hall, 1977).

ADDITIONAL READINGS

CUCCIA, C. C., *The Handbook of Digital Communications* (Palo Alto, Calif.: EW Communications Inc. (*Microwave Syst. News*, Vol. 9, No. 11, 1979).

DE BUDA, R., "Coherent Demodulation of Frequency-Shift Keying with Low Deviation Ratio," *IEEE Trans. Commun.*, Vol. COM-20, pp. 429–435, June 1972.

FEHER, K., *Digital Communications* (Englewood Cliffs, N.J.: Prentice-Hall, 1983).

LINDSEY, W. C., and M. K. SIMON, *Telecommunications System Engineering* (Englewood Cliffs, N.J.: Prentice-Hall, 1973).

MATHWICH, H. R., J. F. BALCEWICZ, and M. HECHT, "The Effect of Tandem Band and Amplitude Limiting on the E_b/N_0 Performance of Minimum (Frequency) Shift Keying (MSK)," *IEEE Trans. Commun.*, Vol. COM-22, pp. 1525–1540, October 1974.

OSBORNE, W. P. "Coherent and Noncoherent Detection of CPFSK," *IEEE Trans. Commun.*, Vol. COM-22, pp. 1023–1036, August 1974.

PODRACZKY, E. J., ed., Special Issue on Satellite Communications, *Proc. IEEE*, Vol. 65, March 1977.

PRABHU, V. K., "Error-Rate Considerations for Digital Phase-Modulation Systems," *IEEE Trans. Commun. Technol.*, Vol. COM-17, pp. 33–42, February 1969.

RHODES, S. A., "Effect of Noisy Phase Reference on Coherent Detection of Offset-QPSK Signals," *IEEE Trans. Commun.*, Vol. COM-22, pp. 1046–1052, August 1974.

Rosen, P., ed., Special Issue on Satellite Communications, *IEEE Trans. Commun.*, Vol. COM-27, October 1979.

Ryan, C. R., A. R. Hambley, and D. E. Vogt, "760 Mbit/s Serial MSK Microwave Modem," *IEEE Trans. Commun.*, Vol. COM-28, pp. 771–777, May 1980.

Shanmugam, K. S., *Digital and Analog Communication Systems* (New York: Wiley, 1979.)

Ziemer, R. E. and W. H. Tranter, *Principles of Communications* (Boston: Houghton Mifflin, 1976), Chaps. 3 and 7.

PROBLEMS

3-1. Derive the result for the optimum threshold setting as expressed by (3-11a) of the binary receiver shown in Figure 3-1.

3-2. **(a)** Obtain the maximum output signal-to-noise ratio of a matched filter for the following input signals: A and T_c are constants. Assume that the two-sided noise power spectral density at the input is N_0.
  (1) $x_1(t) = A[u(t) - u(t - 5T_c)]$
  (2) $x_2(t) = A[u(t) - 2u(t - 3T_c) + 2u(t - 4T_c) - u(t - 5T_c)]$
(b) Find the amplitude of an input pulse of duration T_c that will give the same maximum output signal-to-noise ratio as the signals in part (a).
(c) Obtain the autocorrelation functions of the signals in parts (a) and (b). Comment on the utility of each for resolving delay or relative time differences.

3-3. Make up a table giving relative merits of binary, coherent ASK, PSK, FSK and QPSK, and MSK in regard to bandwidth efficiency (assume equal binary rates), power efficiency, and peak power requirements. Take the 90% power containment bandwidth as a measure of bandwidth.

3-4. Consider *phase-shift keying* with a carrier component which can be written as

$$x_c(t) = A[\sin 2\pi f_0 t + \cos^{-1} a d(t)]$$

Assume that $d(t)$ is a ± 1-valued binary message signal. Show that the ratio of power in the carrier and modulation components of $x_c(t)$ is

$$\frac{P_c}{P_m} = \frac{a^2}{1 - a^2}$$

3-5. **(a)** Quadrature multiplexing is to be used to transmit two data streams which differ in bit rate by a factor of 10. What must be the ratio of their amplitudes if the signal-to-noise ratios of the demodulated quadrature data streams are to be the same? (See Figure 3-8.)
(b) For the situation depicted in part (a), draw a phasor diagram like the one shown in Figure 3-9.
(c) By suitably modifying (3-64) to account for differing symbol periods in each quadrature channel, plot the power spectrum of the transmitted signal.

3-6. **(a)** Derive (3-53).
(b) Show that N_1 and N_2 are uncorrelated.

3-7. Derive (3-64) under the assumptions that $\{a_k\}$ and $\{b_m\}$ are iid sequences with $E(a_k) = E(b_m) = 0$ with Δt uniformly distributed in $(0, T_s)$.

3-8. Show that (3-75) follows by multiplying (3-73) by the magnitude squared of (3-74) and using trigonometric identities.

3-9. Sketch trellis diagrams for the excess phase of (a) QPSK; (b) OQPSK.

3-10. Consider the data sequence, read from left to right,

$$
\begin{array}{ccc}
11111 & 00110 & 10010 \\
00010 & 10111 & 01100
\end{array}
$$

Take the odd-indexed bits as $d_1(t)$ and the even-indexed bits as $d_2(t)$ in a QPSK-modulated carrier [zeros in the data sequence are represented as -1 in $d_1(t)$ and $d_2(t)$].

(a) Sketch the following waveforms below each other so that their time axes may be compared: $d_1(t)$; $d_2(t)$; the excess phase, $\theta(t)$.

(b) What are the possible changes in $\theta(t)$ that can occur between successive bits?

3-11. Same question as Problem 3-10, but sketch $d_1(t)$, $d_2(t)$, and $\theta(t)$ for an OQPSK-modulated carrier.

3-12. Same question as Problem 3-10, but sketch $d_1(t)$, $d_2(t)$, and $\theta(t)$ for an MSK-modulated carrier.

3-13. (a) Given that an all-ones or all-zeros sequence corresponds to a sinusoid of frequency $f_1 = f_0 - \tfrac{1}{2}T_b$, obtain the output of the matched filter of a serial-MSK detector for this input.

(b) Obtain the output of the matched filter for an alternating 1–0 sequence at the input to a serial-MSK detector.

3-14. (a) Can OQPSK be demodulated serially? *Hint*: Consider the output of a filter matched to the inphase and quadrature channel signals in a quadrature demodulator for OQPSK where the reference frequency is $f_0 + \tfrac{1}{4}T_b$. Note that one is proportional to $\cos(\pi t/T_b)$ and the other to $\sin(\pi t/T_b)$.

(b) Explain why serial demodulation cannot be used for QPSK.

3-15. (a) Modify the factor F found in Example 3-6 so that it applies to any roll-off factor β.

(b) Using numerical integration, compute the degradation as a function of β (i.e., $D = 20 \log_{10} F$).

3-16. Redo the derivation of Section 3-6.3 for the case where the transmitter pulses are not ideal impulses but, rather, have a shape $p_t(t)$ with $\int_{-\infty}^{\infty} |p_t(t)|^2 \, dt < \infty$. In particular, obtain equations to replace (3-106), (3-107), and (3-108).

3-17. In 16-QAM modulation, the signal set is a square grid of four by four points or loci, each pair separated by $a = 2b$, and center on the origin.

(a) Find the average energy of this multiple-amplitude signal set assuming all signal loci are equally likely to be occupied.

(b) If no noise is present, find the in-phase and quadrature signal components in the receiver, $x_m = R_m \cos \theta_m$ and $Y_m = R_m \sin \theta_m$ shown in Figure 3-23.

(c) If the decision circuitry shown in Figure 3-23 consists of an integrate-and-dump operation in each quadrature channel over the length of the signaling interval and an error occurs if either or both of the resultant in-phase- and quadrature-channel voltages are more than $b = a/2$ units away from the signal point corresponding to the transmitted signal, compute the average probability of error. Let the input noise be white with two-sided power spectral density $N_0/2$ and assume that all signal points are equally likely.

3-18. The samples of a single pulse at the output of a certain channel are 0.02, -0.01, 0.2, -0.2, 1.0, -0.1, 0.1, 0.05, and -0.01, where the spacing is the symbol duration.

(a) Find the coefficients of a three-tap transversal filter equalizer that will provide two zeros either side of the main pulse response.

(b) What are the sample values of the equalizer output at ± 2, ± 3, and ± 4 sampling times away from the maximum?

3-19. Derive (3-152) for an LMS equalizer for QPSK.

3-20. Obtain (3-153) and (3-154) for presetting of the taps of an LMS equalizer.

3-21. Show that (3-160) follows by differentiating the mean-square error (3-149).

3-22. How much additional power is required in a BPSK system with the following unbalances in order to maintain the same performance as an ideal system?
(a) 10% phase unbalance.
(b) 10% amplitude imbalance.
(c) 10% of both phase unbalance and amplitude imbalance.

3-23. Assume that in a serially modulated QPSK system the serial biphase modulators both introduce phase unbalance. Represent this unbalance as

$$x(t) = \frac{A}{\sqrt{2}} \cos \left[2\pi f_0 t + (1 - \alpha_1) \frac{\pi}{2} d_1(t) \right] + \frac{A}{\sqrt{2}} \sin \left[2\pi f_0 t + (1 - \alpha_2) \frac{\pi}{2} d_2(t) \right]$$

for the modulated quadrature carriers. Further assume that the receiver establishes the quadrature reference signals

$$\cos 2\pi f_0 t \quad \text{and} \quad \sin 2\pi f_0 t$$

(a) Obtain expressions for the quadrature demodulator output signals.
(b) Find an expression equivalent to (3-174) that can be used to compute degradation for this case.

3-24. Compute a curve giving degradation due to modulator phase unbalance in QPSK such as shown in Figure 3-23 for
(a) $P(\epsilon) = 10^{-4}$
(b) $P(\epsilon) = 10^{-5}$

3-25. **(a)** Derive an expression giving probability of error as a function of demodulator phase error for QPSK and OQPSK, or parallel demodulation of MSK.
(b) Verify the curve for QPSK shown in Figure 3-38.

3-26. Let the probability of error for a digital modulation scheme given a certain demodulator phase error be $P(E|\varphi)$, and let this phase error have a probability density function $p(\varphi)$. The probability of error averaged over φ is

$$P_E = \int_{-\infty}^{\infty} P(E|\varphi) p(\varphi) \, d\varphi$$

If BPSK is considered with

$$p(\varphi) = \frac{\exp(-\varphi^2/2\tau_\varphi^2)}{\sqrt{2\pi\tau_\varphi^2}}$$

use numerical integration to make a plot of P_E versus $\dot{E}_b/N_0$ for $\tau_\varphi = 0.03, 0.1,$ and 0.2.

3-27. **(a)** Repeat the derivation for degradation due to a nonideal detection filter for a second-order Butterworth filter.
(b) Verify the curve for $E_b/N_0 = 10$ dB shown in Figure 3-41.

3-28. Using (3-183) through (3-185), verify the curve for $E_b/N_0 = 12$ dB shown in Figure 3-40.

3-29. Using Figures 3-37 and 3-42 to estimate the degradations at $BT = 5, 2,$ and 1 for (a) signal power loss; (b) non-optimum detector including intersymbol interference.

3-30. Construct a curve for $P(\epsilon) = 10^{-4}$ giving degradation due to timing error in BPSK like the one shown in Figure 3-47.

3-31. Consider the following baseband transversal equalizer:

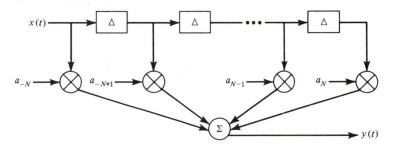

PROBLEM 3-31.

(a) Tap weights are to be chosen such that the mean-square error between the actual output $y(t) = \sum_{m=-N}^{N} a_m x(t - m\Delta)$ and a desired output, $d(t)$, is minimized; that is,

$$\epsilon \overset{\triangle}{=} E\{[y(t) - d(t)]^2\} = \text{minimum}$$

Noting that $\epsilon(a_N, \ldots, a_0, \ldots, a_N)$ is a concave function, show that the conditions for minimum ϵ reduce to

$$\sum_{n=-N}^{N} a_n R_{xx}[(n - j)\Delta] = R_{xd}(j\Delta) \qquad -N \le j \le N$$

where $R_{xx}(\lambda) = E[x(t)x(t + \lambda)]$
$R_{xd}(\lambda) = E[x(t) \, d(t + \lambda)]$

Write this set of equations in matrix form.
(b) Specialize the result found in part (a) to a baseband multipath channel for which

$$x(t) = Ad(t) + \alpha Ad(t - \tau) + n(t) \qquad (A, \, \alpha, \text{ and } \tau \text{ are constants})$$

where $d(t)$ is a binary data waveform with autocorrelation function

$$R_{dd}(\lambda) = \begin{cases} 1 - \dfrac{|\lambda|}{T_b} & |\lambda| \le T_b \\ 0 & \text{otherwise} \end{cases}$$

and $n(t)$ is noise with autocorrelation function

$$R_{nn}(\lambda) = B_N N_0 \exp(-4B_N |\lambda|)$$

B_N is the equivalent noise bandwidth.
Put your final result in terms of the parameter $E_b/N_0 = A^2 T_b/N$.
(c) Obtain the coefficients of a three-tap equalizer for $B_N = 2/T_b$, $\tau = \Delta = T_b$, $E_b/N_0 = 10$, and $\alpha = 0.1$.

CHAPTER 4

Signal-Space Methods in Digital Data Transmission

INTRODUCTION

In Chapter 3 several questions relating to digital data transmission techniques were considered. With the exception of quadrature systems, the techniques considered were binary, and the receiver was assumed to possess a reference that was phase coherent with the received signal. In this chapter the digital transmission systems considered are generalized to include nonbinary (or M-ary) as well as phase-non-coherent systems. The noise encountered in the channel is assumed to be AWGN.

The communication system model that will be employed in this chapter is illustrated in Figure 4-1. A memoryless message source emits a sequence of messages $\{\ldots, m_1, m_2, \ldots, m_k, \ldots\}$ in time at a rate R_m messages per second chosen from a finite-source alphabet $A = \{1, 2, \ldots, q\}$. For example, if $A = \{0, 1, 2\}$, a typical message sequence is $\{\ldots.2011022102\ldots\}$. By *memoryless* it is meant that a source output at a particular time, say m_j, is independent of all preceding and succeeding source outputs. The next operation in the block diagram of Figure 4-1 is to associate each possible block of J message symbols with one of M possible transmitted waveforms, $\{s_i(t) : i = 1, 2, \ldots, M\}$. For a q-symbol alphabet, it is therefore required that $q^J = M$. For example, with the ternary alphabet, $A = \{0, 1, 2\}$, all possible combinations of two-symbol messages can be associated with

184

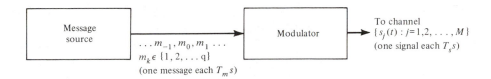

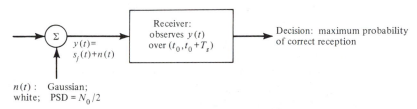

FIGURE 4-1. **Block diagram of a digital communication system.**

a nine-waveform modulator output according to the following table:

$$0 \ 0: \quad s_1(t)$$
$$0 \ 1: \quad s_2(t)$$
$$1 \ 0: \quad s_3(t)$$
$$1 \ 1 \qquad \cdot$$
$$0 \ 2 \qquad \cdot$$
$$2 \ 0 \qquad \cdot$$
$$2 \ 2$$
$$2 \ 1$$
$$1 \ 2: \quad s_9(t)$$

The signals, each assumed to be of duration T_s, are sent through the channel in accordance with the message sequence emitted from the source. To avoid storing messages or having gaps in the transmitted signal sequence, the modulator signal duration T_s must be related to the interval between messages, T_m, by

$$T_s = JT_m \qquad \text{or} \qquad R_m = JR_s \qquad (4\text{-}1)$$

where R_m and R_s are the message and symbol rate, respectively.

As mentioned previously, the channel is considered ideal in that it is infinite bandwidth and adds white Gaussian noise of two-sided power spectral $N_0/2$ W/Hz to the transmitted signal. *Memory* may be introduced in some cases into the message sequence by encoding it before the modulation step is carried out, as explained in Chapter 1. The effects of this will be considered at the end of the chapter, where a particular phase modulation technique is discussed.

In order to make a decision, the receiver observes the received signal plus noise over a T_s-second interval, which is assumed to be synchronized with the received signal sequence, and processes these data so that its probability of estimating the signal correctly, given the received signal plus noise waveform, is maximized. Such a receiver, referred to as a *maximum a posteriori* (MAP) *receiver*, also minimizes the probability of error averaged over all possible received signals.*

A general approach for finding the appropriate receiver structure is to resolve the received signal plus noise, hereafter referred to as the data, into a generalized vector space of dimension $K \leq M$. Although it may appear that this approach is unnecessary for simpler cases, the vector-space viewpoint can be made to fit virtually any signaling scheme of interest. In addition, it provides powerful insight into the relative performance of various signaling schemes. Finally, it also suggests one way to implement the optimum receiver for a given signaling scheme. In the next section, therefore, the MAP criterion is expanded on further and expressed in terms of generalized vector-space notation.

4-2

OPTIMUM RECEIVER PRINCIPLES IN TERMS OF VECTOR SPACES

4-2.1 Maximum A Posteriori Detectors

It is well known from statistical decision theory that a Bayes receiver, which is one that minimizes the average cost of making a decision, is implemented by means of the likelihood ratio test

$$\Lambda(Z) \triangleq \frac{f_Z(Z|H_2)}{f_Z(Z|H_1)} \underset{H_1}{\overset{H_2}{\gtrless}} \frac{(C_{21} - C_{11})p_1}{(C_{12} - C_{22})p_2} \triangleq \Lambda_0 \qquad (4\text{-}2)$$

where $f_Z(z|H_i)$ = probability density function of the data given that H_i was true
Z = data on which the decision is based
C_{ij} = cost of deciding hypothesis H_i was true when H_j was in actuality true
p_i = a priori probability that hypothesis H_i is true ($p_2 = 1 - p_1$)

The performance of a Bayes test is characterized by the average cost of making a decision. This in turn can be expressed in terms of two probabilities,

$$P_M = \text{Pr (say } H_1 \text{ true} \mid H_2 \text{ really true)}$$

$$P_F = \text{Pr (say } H_2 \text{ true} \mid H_1 \text{ really true)} \qquad (4\text{-}3)$$

In keeping with usage from the field of radar, these probabilities are often referred to as the probabilities of a miss and a false alarm, respectively. The average cost per decision is

$$C_{av} = p_1 C_{21} + p_2 C_{22} + p_2 (C_{12} - C_{22}) P_M - p_1 (C_{21} - C_{11})(1 - P_F) \qquad (4\text{-}4)$$

The special cost assignment

$$C_{11} = C_{22} = 0 \qquad \text{(right decisions cost zero)}$$

$$C_{12} = C_{21} \qquad \text{(either type of wrong decision is equally costly)}$$

---

*In some cases it is useful to observe the signal over a longer period than T_s.

reduces the Bayes test to[†]

$$\frac{f_Z(Z|H_2)}{f_Z(Z|H_1)} \underset{H_1}{\overset{H_2}{\gtrless}} \frac{p_1}{p_2} \triangleq \frac{P(H_1)}{P(H_2)} \tag{4-5a}$$

where the notation $P(H_i)$ is used to emphasize that $P_i = P(H_i)$ is the a priori probability of H_i. This can be manipulated, using Bayes rule, to

$$P(H_2|Z) \underset{H_1}{\overset{H_2}{\gtrless}} P(H_1|Z) \tag{4-5b}$$

where $P(H_i|Z) =$ probability of hypothesis H_i being true given the data Z. Because this modified test amounts to choosing the hypothesis corresponding to the largest a posteriori probability, $P(H_i|z)$, it is referred to as a *maximum a posteriori* (MAP) *test* or *detector*. The average cost of a decision for a MAP detector, with $C_{12} = C_{21} = 1$, is given by

$$P_E = P_M P(H_2) + P_F P(H_1) \tag{4-6}$$

which is a minimum by virtue of the MAP test being a Bayes test. In other words, a MAP detector minimizes the average probability of making an error.

The MAP hypotheses test generalizes easily to multiple observations and multiple hypothesis. If the observed data are composed of K observations $\mathbf{Z} = (Z_1, Z_2, \ldots, Z_K)$ and M hypotheses, the M a posteriori probabilities,

$$P(H_i|Z_1, Z_2, \ldots, Z_K) \qquad i = 1, 2, \ldots, M \tag{4-7}$$

are found and the hypothesis, H_j, corresponding to the largest probability is chosen as being true.

EXAMPLE 4-1
Consider the hypothesis-testing problem

$$H_1: Z = A + N$$

$$H_2: Z = -A + N$$

with $P(H_1) = p$, $P(H_2) = 1 - p$, and N Gaussian with zero mean and variance σ^2. Since $f_Z(Z) = f_Z(Z|H_1)P(H_1) + f_Z(Z|H_2)P(H_2)$ is independent of H_1 and H_2, we may express the MAP test as[†]

$$f_Z(Z|H_2)P(H_2) \underset{H_1}{\overset{H_2}{\gtrless}} f_Z(Z|H_1)P(H_1)$$

Given H_1, Z is a Gaussian random variable of mean A and variance σ^2. Given H_2, Z is a Gaussian random variable of mean $-A$ and variance σ^2. Therefore,

$$f_Z(z|H_1) = \frac{\exp\left[-(z - A)^2/2\sigma^2\right]}{\sqrt{2\pi\sigma^2}}$$

$$f_Z(z|H_2) = \frac{\exp\left[-(z + A)^2/2\sigma^2\right]}{\sqrt{2\pi\sigma^2}}$$

[†]Uppercase Z is used here because a test on the *observed data* is being formulated.

The MAP test reduces to

$$(1 - p) \exp\left[\frac{-(Z + A)^2}{2\sigma^2}\right] \underset{H_1}{\overset{H_2}{\gtrless}} p \exp\left[\frac{-(Z - A)^2}{2\sigma^2}\right]$$

or

$$Z \underset{H_1}{\overset{H_2}{\lessgtr}} \frac{\sigma^2}{2A} \ln \frac{1 - p}{p}$$

If $p = \frac{1}{2}$, the test is "decide $+A$ if $Z > 0$ and $-A$ if $Z < 0$." If $p \neq \frac{1}{2}$, the test is biased in favor of the most probable hypothesis. If $p = \frac{1}{2}$, the probability of error, by (4-6), is

$$P_E = P_M = P_F = \Pr\left[Z > 0 \mid H_2 \text{ true}\right]$$

$$= \int_0^\infty \frac{\exp\left[-(z + A)^2/2\sigma^2\right]}{\sqrt{2\pi\sigma^2}} \, dz$$

$$= \int_{A/\sigma}^\infty \frac{\exp\left(-u^2/2\right)}{\sqrt{2\pi}} \, du = Q\left(\frac{A}{\sigma}\right)$$

where

$$Q(u) = \int_u^\infty \frac{\exp\left(-t^2/2\right)}{\sqrt{2\pi}} \, dt \qquad\blacksquare$$

4-2.2   Vector-Space Representation of Signals

The MAP criterion expressed by (4-7) is based on a finite number, K, of data samples. On the other hand, the observed signal plus noise, $z(t) = s_i(t) + n(t)$, $t_0 \leq t \leq t_0 + T_s$, in a digital communication system is a function of a continuous time variable which must be reduced to N numbers in order to apply the MAP criterion. This can be accomplished in various ways, but because the noise is white, an approach will be used here which is intuitively pleasing.

A K-dimensional generalized vector space is defined by the orthonormal basis function set $\varphi_1(t), \varphi_2(t), \ldots, \varphi_K(t)$, where

$$\int_{t_0}^{t_0 + T_s} \varphi_i(t)\varphi_j^*(t) \, dt = \begin{cases} 1 & i = j \\ 0 & i \neq j \end{cases} \qquad (4\text{-}8)$$

A method for choosing this basis set will be described shortly. The received signal plus noise, when resolved into this vector space, has components

$$Z_k = S_{ik} + N_k \qquad \begin{array}{l} k = 1, 2, \ldots, K \\ i = 1, 2, \ldots, M \end{array} \qquad (4\text{-}9a)$$

where

$$S_{ik} = \int_{t_0}^{t_0 + T_s} s_i(t)\varphi_k^*(t) \, dt \qquad (4\text{-}9b)$$

$$N_k = \int_{t_0}^{t_0 + T_s} n(t)\varphi_k^*(t) \, dt \qquad (4\text{-}9c)$$

The method of choosing the set $\{\varphi_i(t)\}$ guarantees that any of the possible transmitted signals can be expressed as

$$s_i(t) = \sum_{k=1}^{K} S_{ik}\varphi_k(t) \qquad 0 \le t \le T_s; \quad K \le M \qquad (4\text{-}10)$$

The noise, however, must be viewed as composed of two components, which are

$$n(t) = n_r(t) + n_p(t) \qquad (4\text{-}11a)$$

where

$$n_r(t) = \sum_{k=1}^{K} N_k\varphi_k(t) \qquad (4\text{-}11b)$$

$$n_p(t) = n(t) - n_r(t) \qquad (4\text{-}11c)$$

Since $n(t)$ is a Gaussian process, the N_k's are Gaussian random variables.

The subscript r in $n_r(t)$ stands for "relevant," for this noise component is the only one relevant to making the decision as to which signal was sent. The subscript p in $n_p(t)$ denotes that it is statistically orthogonal, or "perpendicular" to $n_r(t)$ (this proof is left to the problems). Since $n(t)$ is Gaussian, so are $n_p(t)$ and $n_r(t)$ and, therefore, they are statistically independent random processes. Because $n_p(t)$ is independent of $n_r(t)$ it may be ignored in making the decision as to which signal was sent.

It follows that any possible transmitted signal can be expressed in the form (4-10), the problem of representing the received waveform $z(t)$ in a generalized K-dimensional vector space, with components given by (4-9a), for use in a posteriori probabilities (4-7) is solved. In some cases, the choice of an appropriate set $\{\varphi_k(t)\}$ may be obvious. For less obvious cases, the following constructive technique, referred to as the Gram–Schmidt procedure, is given.

It is convenient first to generalize the notions of the dot product of two vectors and the magnitude of a vector to signals. The former is referred to as the *scalar product*, which is denoted as (u, v) and for two signals $u(t)$ and $v(t)$ for our purposes is defined as

$$(u, v) \triangleq \int_{t_0}^{t_0+T_s} u(t)v^*(t)\, dt \qquad (4\text{-}12)$$

More generally, the scalar product is a (complex) scalar-valued function of two signals, $u(t)$ and $v(t)$, with the properties

1. $(u, v) = (v, u)^*$.
2. $(\alpha u, v) = \alpha(u, v)$ where α is a scalar (complex in general).
3. $(u + v, w) = (u, w) + (v, w)$.
4. $(u, u) \ge 0$ with equality if and only if $u \equiv 0$.

The concept of magnitude of a vector is generalized by the definition of the *norm*, $\|u\|$, of a signal, $u(t)$, which is defined in terms of the scalar product as

$$\|u\| = \sqrt{(u, u)} \qquad (4\text{-}13)$$

More generally, the norm is a nonnegative real number satisfying the properties

1. $\|u\| = 0$ if and only if $u(t) \equiv 0$.
2. $\|u + v\| \le \|u\| + \|v\|$ (triangle inequality).
3. $\|\alpha u\| = |\alpha|\, \|u\|$, where α is a scalar.

The construction of orthonormal basis function sets is considered next.

Gram–Schmidt Procedure. Given a finite set of signals $s_1(t)$, $s_2(t)$, . . ., $s_M(t)$ defined on some interval $(t_0, t_0 + T_s)$, an orthonormal basis function set may be constructed according to the following algorithm:

1. Set $v_1(t) = s_1(t)$ and define

$$\varphi_1(t) = \frac{v_1(t)}{\|v_1\|} \tag{4-14a}$$

2. Set $v_2(t) = s_2(t) - (s_2, \varphi_1)\varphi_1$ and let

$$\varphi_2(t) = \frac{v_2(t)}{\|v_2\|} \tag{4-14b}$$

3. Set $v_3(t) = s_3(t) - (s_3, \varphi_2)\varphi_2 - (s_3, \varphi_1)\varphi_1$ and let

$$\varphi_3(t) = \frac{v_3(t)}{\|v_3\|} \tag{4-14c}$$

4. Continue until all $s_i(t)$'s have been used. If one or more of the steps above yield $v_j(t)$'s for which $\|v_j(t)\| = 0$, omit these from consideration so that a set of $K \le M$ orthonormal functions is obtained.

With this procedure, an orthonormal basis set results which can be used to express *any* possible transmitted signal set in the form (4-10).†

Using the procedure just described and the definitions of scalar product and norm given by (4-12) and (4-13), one may prove several interesting and useful properties about generalized vector-space representations. Before proceeding with the detection of signals in noise, two such properties will be considered.

EXAMPLE 4-2 SCHWARZ'S INEQUALITY

If $x(t)$ and $y(t)$ are signals, then

$$|(x, y)| \le \|x\| \, \|y\| \tag{4-15}$$

with equality if and only if $x(t)$ or $y(t) \equiv 0$ or $x(t) = \alpha y(t)$, where α is a scalar (possibly complex).

Proof: Consider $\|x + \alpha y\|^2$, which is a nonnegative quantity for arbitrary α. Using the definitions of the norm and properties of the inner product, we have

$$\|x + \alpha y\|^2 = (x + \alpha y, x + \alpha y)$$
$$= \|x\|^2 + \alpha^*(x, y) + \alpha(x, y)^* + |\alpha|^2\|y\|^2$$

Since α is arbitrary, we may choose it as

$$\alpha = \frac{-(x, y)}{\|y\|^2}$$

which gives

$$\|x + \alpha y\|^2 = \|x\|^2 - \frac{|(x, y)|^2}{\|y\|^2} \ge 0$$

---

†Another way of looking at this problem is that the basis functions are to be chosen such that

$$S_1(t) = s_{11}\varphi_1(t)$$
$$S_2(t) = S_{21}\varphi_1(t) + s_{22}\varphi_2(t)$$

where the S_{ij}'s are constants. The first equation is used to find $\varphi_1(t)$, the second to find $\varphi_2(t)$, and so on.

or

$$|(x, y)| \leq \|x\| \|y\|$$

Furthermore, equality holds only if $\|x + \alpha y\| = 0$ or if $x = -\alpha y$ or if one of the signals is identically zero. ∎

EXAMPLE 4-3 PARSEVAL'S THEOREM

In terms of their generalized Fourier series components, the inner product of two signals is given by

$$(x, y) = \sum_{k=1}^{K} X_k Y_k^* \tag{4-16}$$

with the special case

$$\|x\|^2 = \sum_{k=1}^{K} |X_k|^2 \tag{4-17}$$

Proof: Represent $x(t)$ and $y(t)$ in terms of their generalized Fourier series, which are

$$x(t) = \sum_{m=1}^{K} X_m \varphi_m(t) \quad .$$

and

$$y(t) = \sum_{n=1}^{K} Y_n \varphi_n(t)$$

where $X_m = (x, \varphi_m)$ and $Y_n = (y, \varphi_n)$. Using the properties of the inner product, we have

$$(x, y) = \left(\sum_m X_m \varphi_m, \sum_n Y_n \varphi_n \right)$$

$$= \sum_m X_m \left(\varphi_m, \sum_n Y_n \varphi_n \right)$$

$$= \sum_m X_m \left(\sum_n Y_n \varphi_n, \varphi_m \right)^*$$

$$= \sum_m X_m \left[\sum_n Y_n^* (\varphi_n, \varphi_m)^* \right]$$

$$= \sum_m X_m Y_m^*$$

which follows because $(\varphi_n, \varphi_m) = \delta_{nm}$ (δ_{nm}, the Kronecker delta, is one for $n = m$ and zero for $n \neq m$). ∎

EXERCISE 4-1

Show that the two ordinary two-dimensional vectors $\mathbf{A} = \hat{\mathbf{i}} + \hat{\mathbf{j}}$ and $\mathbf{B} = 2\hat{\mathbf{i}} + 5\hat{\mathbf{j}}$ satisfy Schwarz's inequality and Parseval's theorem if the scalar product is taken as the ordinary dot product of two vectors.

EXAMPLE 4-4

Express the M-ary phase-shift-keyed signal set

$$s_i(t) = A \cos \left[\omega_0 t + \frac{(i-1)2\pi}{M} \right] \qquad i = 1, 2, \ldots, M; \quad 0 \le t \le T_s$$

in terms of a suitable orthonormal basis set. Assume that $\omega_0 T_s = (\text{integer})(2\pi)$.

Solution: Rather than use the Gram–Schmidt procedure, note, using a trigonometric identity, that

$$s_i(t) = A \cos \frac{2\pi(i-1)}{M} \cos \omega_0 t - A \sin \frac{2\pi(i-1)}{M} \sin \omega_0 t$$

Therefore, $B \cos \omega_0 t$ and $B \sin \omega_0 t$, $0 \le t \le T_s$, form an appropriate basis set. To normalize these functions, compute

$$\int_0^{T_s} B^2 \cos^2 \omega_0 t \, dt = \int_0^{T_s} B^2 \sin^2 \omega_0 t \, dt = \frac{B^2 T_s}{2} = 1$$

Therefore, $B = \sqrt{2/T_s}$ and

$$\varphi_1(t) = \sqrt{\frac{2}{T_s}} \cos \omega_0 t \qquad 0 \le t \le T_s$$

$$\varphi_2(t) = \sqrt{\frac{2}{T_s}} \sin \omega_0 t \qquad 0 \le t \le T_s$$

A plot of the signal set constellation for this signal set is provided in Figure 4-2 for various values of M, where the coordinates of the ith point are

$$\left(\sqrt{E_s} \cos \frac{2\pi(i-1)}{M} \right), \; -\left(\sqrt{E_s} \sin \frac{2\pi(i-1)}{M} \right)$$

For example, with $M = 2$, the signal coordinates are

$$\mathbf{S}_1 = (\sqrt{E_s}, 0) \qquad \text{and} \qquad \mathbf{S}_2 = (-\sqrt{E_s}, 0)$$

while for $M = 4$, they are

$$\mathbf{S}_1 = (\sqrt{E_s}, 0) \qquad \mathbf{S}_2 = (0, -\sqrt{E_s}) \qquad \mathbf{S}_3 = (-\sqrt{E_s}, 0) \qquad \mathbf{S}_4 = (0, \sqrt{E_s})$$

where $E_s = A^2 T_s/2$ is the energy of the signal. ∎

4-2.3 MAP Detectors in Terms of Signal Spaces

In Section 4-2.2 it has been shown how the received signal plus noise in a digital communication system can be resolved into a generalized vector space of dimension $K \le M$, where M is the number of possible signals transmitted. While the noise cannot, in general, be fully represented in this vector space, the nonrepresentable component is irrelevant to deciding which signal was transmitted. The computation of the signal-space coordinates of the received data, according to (4-9), is accomplished by the parallel bank of multiplier–integrators shown in Figure 4-3. Each multiplier–integrator combination is referred to as a *correlator*, and the overall receiver structure is called a *correlation receiver*. The kth correlator can also be

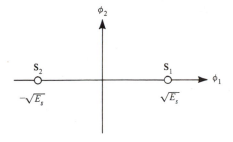

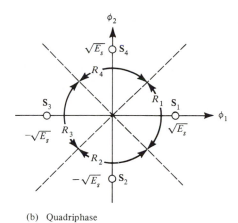

(a) Binary

(b) Quadriphase

FIGURE 4-2.   Signal constellations for *M*-ary phase-shift-keyed modulation.

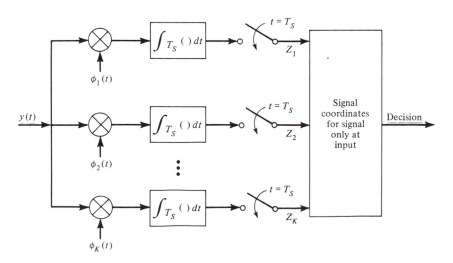

FIGURE 4-3.   Correlator—sampler bank for computing signal space coordinates of the received signal plus noise.

replaced by a linear time-invariant filter with impulse response.

$$h_k(t) = \varphi_k(T_s - t) \qquad 0 \le t \le T_s \qquad (4\text{-}18)$$

which follows since the filter output is

$$z_k(t) = h_k(t) * z(t)$$

$$= \int_0^{T_s} \varphi_k(T_s - t + \lambda)z(\lambda)\, d\lambda$$

or

$$z_k(T_s) = \int_0^{T_s} \varphi_k(\lambda)z(\lambda)\, d\lambda \qquad (4\text{-}19)$$

This is identical to the kth correlator output *at time* T_s. The receiver structure utilizing filters with impulse responses (4-18) in each arm is called a *matched filter receiver*, since the kth arm is matched to the kth term in the orthonormal series expansion of the receiver.

It remains to determine what operation should be performed on the correlator outputs in Figure 4-3 in order to implement the MAP criterion. According to (4-7), the MAP, or minimum probability of error, receiver computes the a posteriori probabilities $P(H_i|Z_1, Z_2, \ldots, Z_K)$, $i = 1, 2, \ldots, M$, and estimates the transmitted signal to have been the one corresponding to the largest probability. Using Bayes rule, each a posteriori probability may be rewritten as

$$P(H_i|Z_1, Z_2, \ldots, Z_k) = \frac{f_{\mathbf{Z}}(Z_1, \ldots, Z_k|H_i)P(H_i)}{f_{\mathbf{Z}}(Z_1, \ldots, Z_k)}$$

$$= CP(H_i)f_{\mathbf{Z}}(Z_1, \ldots, Z_k|H_i) \qquad (4\text{-}20)$$

where $f_{\mathbf{Z}}(Z_1, Z_2, \ldots, Z_k)$ is the joint probability density function (pdf) of the signal space coordinates of $z(t)$. Since this joint pdf is independent of the particular signal transmitted, its inverse is replaced by a constant C, which gives (4-20).

An expression will now be obtained for $f_{\mathbf{Z}}(Z_1, \ldots, Z_k|H_i)$, the conditional joint pdf of the received data vector given signal $s_i(t)$ was sent. According to (4-9), the kth coordinate, Z_k, is a Gaussian random variable because it is obtained as a linear operation on the Gaussian noise $n(t)$. Given $s_i(t)$ was transmitted, the mean of Z_k is just $S_{ik} = (s_i, \varphi_k)$, which is the projection of $s_i(t)$ onto the $\varphi_k(t)$ coordinate axis. The variance of Z_k, given $s_i(t)$ was transmitted, is simply the variance of the kth noise component, N_k. In fact, it is convenient to compute

$$\overline{N_k N_l} = \operatorname{cov}\{Z_k, Z_l|H_i\}$$

$$= E\left[\int_{t_0}^{t_0+T_s} n(t)\varphi_k(t)\, dt \int_{t_0}^{t_0+T_s} n(\lambda)\varphi_l(\lambda)\, d\lambda \right]$$

$$= \int_{t_0}^{t_0+T_s} \int_{t_0}^{t_0+T_s} E[n(t)n(\lambda)]\varphi_k(t)\varphi_l(\lambda)\, dt\, d\lambda$$

$$= \frac{N_0}{2} \int_{t_0}^{t_0+T_s} \varphi_k(t)\varphi_l(t)\, dt$$

$$= \frac{N_0}{2}\, \delta_{kl} \qquad (4\text{-}21)$$

which follows because $E[n(t)n(\lambda)] = (N_0/2) \delta(t - \lambda)$. This result shows that the k random variables $Z_1, Z_2, \ldots, Z_k$, which are the coordinates of the received data vector given that $s_i(t)$ was transmitted, are independent since they are Gaussian random variables with zero conditional covariances. The mean of the kth one, given that $s_i(t)$ was transmitted, is S_{ik} and its variance, from (4-21), is $N_0/2$. Thus the joint conditional pdf of the data vector $\mathbf{Z} = (Z_1, Z_2, \ldots, Z_k)$ given that $s_i(t)$ was transmitted (hypothesis H_i) is

$$f\mathbf{Z}(Z_1, Z_2, \ldots, Z_k|H_i) = \prod_{k=1}^{K} \frac{\exp\left[-(Z_k - S_{ik})^2/N_0\right]}{\sqrt{\pi N_0}}$$

$$= \frac{\exp\left[-\sum_{k=1}^{K} (Z_k - S_{ik})^2/N_0\right]}{(\pi N_0)^{K/2}}$$

$$= \frac{\exp\left(-\|\mathbf{Z} - \mathbf{S}_i\|^2/N_0\right)}{(\pi N_0)^{K/2}} \tag{4-22}$$

where

$$\mathbf{Z} \sim z(t) = \sum_{k=1}^{K} Z_k \varphi_k(t) \qquad t_0 \leq t \leq t_0 + T_s \tag{4-23a}$$

$$\mathbf{s}_i \sim s_i(t) = \sum_{k=1}^{K} S_{ik} \varphi_k(t) \qquad t_0 \leq t \leq t_0 + T_s \tag{4-23b}$$

In the last equation of (4-22), the compact representation involving the norm as expressed in (4-17) has been used. The decision rule, then, is to maximize (4-22) multiplied by $P(H_i)$. By taking the natural logarithm of $f_Z(Z_1, Z_2, \ldots, Z_k|H_i)P(H_i)$, this is equivalent to minimizing

$$\|\mathbf{Z} - \mathbf{S}_i\|^2 - N_0 \ln P(H_i)$$

If all signals are a priori equally probable, this decision rule becomes

$$\text{minimize } \|\mathbf{Z} - \mathbf{S}_i\|^2 = \sum_{k=1}^{K} (Z_k - S_{ik})^2 \tag{4-24}$$

which is the distance between the received data vector and the kth signal vector in the transmitted signal constellation. The decision rule (4-24) defines decision regions in the K-dimensional signal space such that if the received data vector $\mathbf{Z}$ falls within the ith region, R_i, the decision is made that signal $s_i(t)$ was transmitted. The decision regions for the signal set of Example 4-4 with $M = 4$ are illustrated in Figure 4-2b.

4-2.4 Performance Calculations for MAP Receivers

As discussed previously, the MAP decision strategy minimizes the average probability of error. The probability of error can be calculated as

$$P_s(\epsilon) = 1 - P_s(C) = 1 - \sum_{k=1}^{M} P(H_k)P(C|H_k) \tag{4-25}$$

where the subscript s denotes the probability of a *symbol error* (sometimes referred to as a *word error*). The probabilities $P_s(C)$ and $P(C|H_k)$ denote, respectively, the probability of correct reception averaged over all possible transmitted signals and the probability of correct reception given the hypothesis, H_k, that the kth signal was transmitted. The latter is just the probability that the data vector $\mathbf{z}$ falls within the kth decision region, R_k, given that $s_k(t)$ really was transmitted. In terms of the joint pdf (4-22), $P(C|H_k)$ is given by

$$P(C|H_k) = \int_{R_k} \frac{\exp\left(-\|\mathbf{Z} - \mathbf{S}_k\|^2/N_0\right)}{(\pi N_0)^{K/2}} \, d\mathbf{Z} \tag{4-26}$$

Several observations which are listed below are useful in the computation of $P(C|H_k)$:*

1. Rotation and translation of the coordinate system defined by the orthonormal set $\{\varphi_i(t)\}$ does not affect the computation of $P(C|H_k)$. (Average or peak signal energy for a given probability of correct reception, however, will be affected in general.) That this is the case is demonstrated by considering the data vector components $Z_k = S_{ik} + N_k$. Since the N_k's are equal-variance, uncorrelated, Gaussian random variables, the equal-probability contours surrounding each possible transmitted signal point are hyperspheres, which means that rotation or translation of the coordinate system will not affect the integral over R_i of the joint pdf $f_\mathbf{Z}(\mathbf{z}|H_k)$.

2. Peak and average energy required to achieve a given probability of error can be minimized through appropriate translation of the coordinate system. An example involving peak energy is shown in Figure 4-4. The average energy, defined by

$$\bar{E}_s = \sum_{k=1}^{M} P(H_k)E_k = \sum_{k=1}^{M} P(H_k)\|\mathbf{S}_k\|^2$$

can be minimized by subtracting a constant vector $\mathbf{a}$ from each signal point such that

$$\sum_{k=1}^{M} P(H_k)\|\mathbf{S}_k - \mathbf{a}\|^2 = \min$$

which results in

$$\mathbf{a} = \sum_{k=1}^{M} P(H_k)\mathbf{S}_k \tag{4-27}$$

The proof of this is left to the problems.

3. Noise, N_k, disturbing the kth coordinate is independent of noise in all other coordinates because $\overline{N_k N_l} = 0$, $k \neq l$. Therefore, a decision may be made on the kth coordinate *independently* of any other coordinate.

4. Given a completely symmetric signal set (i.e., one for which any relabeling of the signal points can be undone by rotation, translation, or inversion of axes), the error probability is independent of the actual source statistics. An example is provided by the signal set of Figure 4-2a. This will be demonstrated later.

5. The union bound often provides a tight upper bound for cases where the error probability is difficult or impossible to compute. It is derived by noting that an error occurs in a digital communication system if, and only if, the received data vector $\mathbf{Z}$ is closer to at least one other signal point $\mathbf{S}_l$, $l \neq k$, than it is to $\mathbf{S}_k$, which is assumed to be the correct signal point. Therefore, given H_k, the probability of

*For a fuller discussion, see [1, Chap. 4].

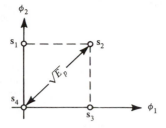

(a) E_p not minimized

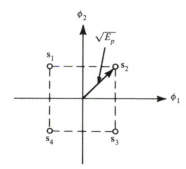

(b) E_p minimized

FIGURE 4-4. Minimizing peak energy, E_p, through appropriate translation of the signal-space coordinate system.

error can be written as

$$P_s(\epsilon|H_k) = \text{Pr}\left(\bigcup_{\substack{i=1 \\ i \neq k}}^{M} \epsilon_{ki}\right) \qquad (4\text{-}28)$$

where ϵ_{ki} is the event that a decision was made in favor of s_i when s_k was really sent and $\cup$ denotes the union of these events. But by the axiom of probability that

$$\text{Pr}\,(A \cup B) \leq P(A) + P(B)$$

$P_s(\epsilon|H_k)$ can be bounded by

$$P_s(\epsilon|H_k) \leq \sum_{\substack{i=1 \\ i \neq k}}^{M} P(\mathbf{S}_i, \mathbf{S}_k) \qquad (4\text{-}29)$$

where $P(\mathbf{s}_i, \mathbf{s}_k) = \text{Pr}\,(\epsilon_{ki})$ is the probability of error for a hypothetical communication system that uses the two signals $\mathbf{s}_i$ and $\mathbf{s}_k$ to communicate one of two equally likely messages. For an AWGN channel,

$$P(\mathbf{s}_i, \mathbf{s}_k) = Q\left[\frac{\|\mathbf{S}_i - \mathbf{S}_k\|}{\sqrt{2N_0}}\right] \qquad (4\text{-}30)$$

where

$$Q(u) = \int_u^\infty \frac{\exp\,(-t^2/2)}{\sqrt{2\pi}}\,dt = \frac{1}{2}\,\text{erfc}\left(\frac{u}{\sqrt{2}}\right) \qquad (4\text{-}31)$$

is the Q-function and

$$\text{erfc } (u) = 1 - \text{erf } (u) = \frac{2}{\sqrt{\pi}} \int_u^\infty \exp{(-t^2)} \, dt \qquad (4\text{-}32)$$

is the complementary error function.* The proof of (4-30) follows by using property 3 and choosing a coordinate direction along the vector connecting $\mathbf{S}_k$ and $\mathbf{S}_i$. Let $d = \|\mathbf{S}_i - \mathbf{S}_k\|$. Call the noise component along this coordinate direction N_1. It has mean zero and variance $N_0/2$. Therefore, the probability of error $P(\mathbf{S}_i, \mathbf{S}_k)$ is

$$P(\mathbf{S}_i, \mathbf{S}_k) = P(\epsilon|\mathbf{S}_i)$$

$$= P(\epsilon|\mathbf{S}_k)$$

$$= \text{Pr}\left(N_1 > \frac{d}{2}\right) \qquad (4\text{-}33)$$

which can be shown to be equal to (4-30).

4-3

PERFORMANCE ANALYSIS OF COHERENT DIGITAL SIGNALING SCHEMES

In this section, various types of digital modulation methods will be described and characterized in terms of symbol and bit error probabilities. The latter characterization is convenient when comparing one scheme with another. The various schemes considered can be categorized as either coherent, for which a phase-coherent carrier reference at the receiver is required, or noncoherent, for which no phase-coherent reference is required.

4-3.1 Coherent Binary Systems

The use of signal space ideas in the analysis of digital communication systems will be illustrated by first considering binary signaling schemes. Let the signaling interval under consideration be $(0, T_b)$ and assume that

$$\int_0^{T_b} |s_1(t)|^2 \, dt = \int_0^{T_b} |s_2(t)|^2 \, dt = E_b$$

where the subscript b is a reminder that this is a binary case. From (4-30), the average probability of bit error is

$$P_b(\epsilon) = Q\left(\frac{d}{\sqrt{2N_0}}\right) \qquad (4\text{-}34)$$

where the Q-function is defined by (4-31). The parameter d^2 is defined by

$$d^2 = \|\mathbf{S}_1 - \mathbf{S}_2\|^2 = \|\mathbf{S}_1\|^2 + \|\mathbf{S}_2\|^2 - 2(\mathbf{S}_1, \mathbf{S}_2)$$

$$= 2E_b(1 - \rho) \qquad (4\text{-}35)$$

---

*See [2, pp. 931ff]. A summary of useful approximations and a brief table of values is given in Appendix E.

where

$$E_b \rho = (\mathbf{S}_1, \mathbf{S}_2) = \int_0^{T_b} s_1(t)s_2(t)\, dt \qquad (4\text{-}36)$$

where ρ is the correlation coefficient between $s_1(t)$ and $s_2(t)$. To specialize (4-35) to two specific cases, consider BPSK (antipodal) and binary FSK (BFSK), which were analyzed in Chapter 3.

EXAMPLE 4-5

Binary phase-shift keying (BPSK) is defined as having the signal set

$$s_1(t) = \sqrt{\frac{2E_b}{T_b}} \cos \omega_0 t \qquad 0 \le t \le T_b$$

and

$$s_2(t) = -\sqrt{\frac{2E_b}{T_b}} \cos \omega_0 t \qquad 0 \le t \le T_b$$

where $\omega_0 T_b$ is an integer multiple of 2π (actually, $\omega_0 T_b \gg 1$ suffices). The correlation coefficient can be calculated as $\rho = -1$, so that $d^2 = 4E_b$. Therefore, the bit error probability is

$$P_b(\epsilon) = Q\left(\sqrt{\frac{2E_b}{N_0}} \right) \qquad \text{(BPSK)} \qquad (4\text{-}37)$$

EXAMPLE 4-6

Coherent binary frequency-shift keying is defined by the signal set

$$s_1(t) = \sqrt{\frac{2E_b}{T_b}} \cos \left(\omega_0 + \frac{\Delta\omega}{2} \right) t \qquad 0 \le t \le T_b$$

$$s_2(t) = \sqrt{\frac{2E_b}{T_b}} \cos \left(\omega_0 - \frac{\Delta\omega}{2} \right) t \qquad 0 \le t \le T_b$$

where $\omega_0 T_b$ is again assumed to be an integer multiple of 2π. The correlation coefficient is given by

$$\rho = \frac{1}{T_b} \int_0^{T_b} (\cos 2\omega_0 t + \cos \Delta\omega t)\, dt$$

$$= \frac{\sin \Delta\omega T_b}{\Delta\omega T_b} \qquad (4\text{-}38)$$

If $\Delta\omega T_b$ is an integer multiple of π, $\rho = 0$, so that $d^2 = 2E_b$ and the bit error probability is

$$P_b(\epsilon) = Q\left(\sqrt{\frac{E_b}{N_0}} \right) \qquad \text{(BFSK)} \qquad (4\text{-}39)$$

Because of the factor of 2 multiplying E_b in (4-37), it is seen that BFSK requires 3 dB more energy than BPSK to achieve the same bit error probability. The frequency separation $\Delta\omega$ could have been chosen to minimize ρ, in which case the performance of BFSK would have been closer to BPSK. This is left to the problems. Figure 4-5 compares $P_b(\epsilon)$ for BFSK with $\rho = 0$ and with $\rho = \rho_{min} \simeq -0.216$. ∎

4-3.2 Coherent *M*-ary Orthogonal Signaling Schemes

The signal set $\{s_i(t) : i = 1, 2, \ldots, M\}$, $0 \le t \le T_s$, with

$$\int_0^{T_s} s_i(t)s_j(t) \, dt = E_s\delta_{ij} \qquad i = 1, 2, \ldots, M \qquad (4\text{-}40)$$

is referred to an *orthogonal scheme* because the correlation coefficient of differing signals is zero. The subscript s on T_s and E_s is a reminder that, in general, we are dealing with more than two symbols per T_s-second signaling interval. Because the signals employed in this scheme are orthogonal, a suitable orthonormal basis set is defined by

$$\varphi_j(t) = \frac{s_j(t)}{\sqrt{E_s}} \qquad 0 \le t \le T_s; \quad j = 1, 2, \ldots, M \qquad (4\text{-}41)$$

The signal-space coordinates of the ith signal are therefore given by

$$S_{ij} = \sqrt{E_s}\,\delta_{ij} \qquad i, j = 1, 2, \ldots, M \qquad (4\text{-}42)$$

The decision rule (4-24) states that

$$d_i^2 = \|\mathbf{Z} - \mathbf{S}_i\|^2 = \sum_{j=1}^{M} (Z_j - S_{ij})^2 \qquad i = 1, 2, \ldots, M \qquad (4\text{-}43)$$

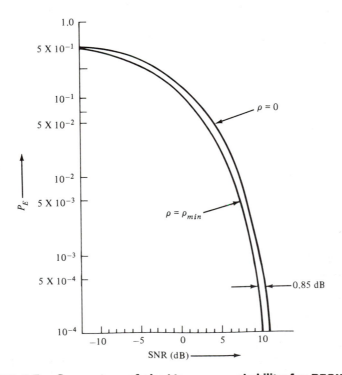

FIGURE 4-5. **Comparison of the bit error probability for BFSK with $\rho = 0$ and $\rho = \rho_{min}$**

where Z_j is the jth coordinate of the data vector, is to be minimized through appropriate choice of $s_i(t)$. The norm of the difference between $\mathbf{Z}$ and $\mathbf{S}_i$ may be expanded as

$$\|\mathbf{Z} - \mathbf{S}_i\|^2 = \|\mathbf{Z}\|^2 - 2(\mathbf{Z}, \mathbf{S}_i) + \|\mathbf{S}_i\|^2 \tag{4-44}$$

Since $\|\mathbf{Z}\|^2$ and $\|\mathbf{S}_i\|^2 = E_s$ are independent of the particular signal chosen, (4-44) may be minimized by maximizing $(\mathbf{Z}, \mathbf{S}_i)$, which can be expressed as

$$(\mathbf{Z}, \mathbf{S}_i) = \int_0^{T_s} z(t)s_i(t)\, dt \tag{4-45}$$

by virtue of the definition of the inner product (4-12) and the fact that the signals are real. Thus the decision strategy to be used on the basis of the correlator outputs of Figure 4-3 is to choose the largest as corresponding to the transmitted signal.

The probability of symbol error, $P_s(\epsilon)$, can be obtained by considering the probability of correct reception, $P_s(C|i)$, given that $s_i(t)$ was transmitted. In the light of the decision rule discussed above, this probability can be expressed as

$$P_s(C|i) = \text{Pr (all } Z_j < Z_i, j \neq i) \tag{4-46}$$

where

$$Z_j = \int_0^{T_s} z(t)\varphi_j(t)\, dt \tag{4-47}$$

Given that $s_i(t)$ was sent, $z(t) = s_i(t) + n(t)$, so that (4-46) can be expressed as

$$P_s(C|i) = \text{Pr (all } N_j < \sqrt{E_s} + N_i, j \neq i) \tag{4-48}$$

where

$$N_j = \int_0^{T_s} n(t)\varphi_j(t)\, dt \tag{4-49}$$

is a Gaussian random variable with zero mean and variance $N_0/2$. Since the N_j's are independent Gaussian random variables, (4-48) can be expressed as

$$P_s(C|i) = \overline{\prod_{\substack{i=1 \\ j \neq i}}^{M} \text{Pr } (N_j < \sqrt{E_s} + N_i)} \tag{4-50}$$

where the overbar denotes an average over N_i. Because the pdf's of all the N_j's are identical, the product in (4-50) can be written as

$$\prod_{\substack{i=1 \\ j \neq 1}}^{M} \text{Pr } (N_j < \sqrt{E_s} + N_i) = \left[\int_{-\infty}^{\sqrt{E_s}+N_i} \frac{\exp\,(-u^2/N_0)}{\sqrt{\pi N_0}}\, du \right]^{M-1} \tag{4-51}$$

Averaging with respect to N_i gives

$$P_s(C|i) = 1 - P_s(\epsilon)$$

$$= \int_{-\infty}^{\infty} \frac{\exp\,(-y^2/2)}{\sqrt{2\pi}} \left[\int_{-\infty}^{\sqrt{2E_s/N_0}+y} \frac{\exp(-x^2/2)}{\sqrt{2\pi}}\, dx \right]^{M-1} dy \tag{4-52}$$

where the substitutions $x^2 = 2u^2/N_0$ have been used and $y = \sqrt{2/N_0}\, N_i$. Since $P_s(C|i)$ is independent of i, the average probability of symbol error, $P_s(\epsilon)$, is in fact $1 - P_s(C|i)$. Unfortunately, (4-52) cannot be expressed in closed form, but must be integrated numerically.*

The union bounding technique can be applied to obtain a tight bound for $P_s(\epsilon)$ for large E_s/N_0. Choosing any two signals $\mathbf{s}_i$ and $\mathbf{s}_k$, we have

$$\|\mathbf{S}_i - \mathbf{S}_k\|^2 = \|\mathbf{S}_i\|^2 - 2(\mathbf{S}_i, \mathbf{S}_k) + \|\mathbf{S}_k\|^2$$

$$= 2E_s \qquad \text{all } i \neq k$$

so that (4-30) becomes

$$P(\mathbf{S}_i, \mathbf{S}_k) = Q\left(\sqrt{\frac{E_s}{N_0}}\right) \tag{4-53}$$

*See [3] for a numerical tabulation of values for $P_s(\epsilon)$.

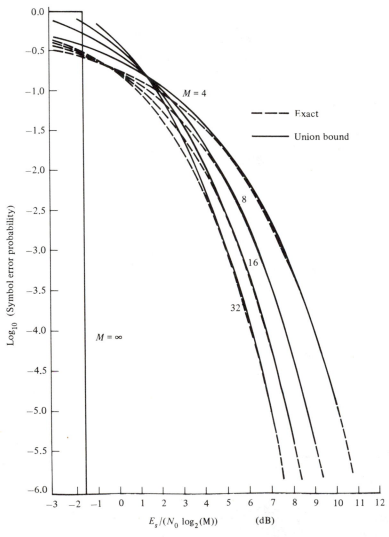

FIGURE 4-6. Symbol error probability for *M*-ary orthogonal signaling; both the exact results using numerical integration, and the upper bound, using the union bound, are shown.

independently of the two signals selected. Therefore, the union bound (4-29) for orthogonal signals becomes

$$P_s(\epsilon) \leq (M - 1)Q\left(\sqrt{\frac{E_s}{N_0}}\right) \qquad (4\text{-}54)$$

The union bound and the exact result for $P_s(\epsilon)$ are compared in Figure 4-6. Also shown is the limit of $P_s(\epsilon)$ as $M \to \infty$. Note that *error-free* transmission can be achieved in the limit as $M \to \infty$ as long as $E_s/(N_0 \log_2 M) > \ln 2 = -1.59$ dB. The signal-to-noise ratio $E_b/N_0 \triangleq E_s/N_0 \log_2 M$ is the average *energy per bit* to noise power spectral density ratio because $\log_2 M$ bits of data are transmitted per M-ary symbol.

Comparison of signaling schemes utilizing different numbers of possible transmitted signals should be done on some kind of equivalent basis. One way is on the basis of equal transmitted data rates. For example, in comparing an octal orthogonal signaling scheme with a binary scheme, the binary scheme would be required to transmit bits at three times the *symbol rate* for the octal system since each symbol in the octal system conveys one of eight possibilities (3 bits per symbol). Thus one could compare $P_s(\epsilon)$ for the octal system with the word error probability, $P_w(\epsilon) = 1 - [1 - P_b(\epsilon)]^3$, for the binary system. However, if one next wished to compare a 16-signal orthogonal system with a binary system, all error probabilities would have to be recomputed. It is therefore better to compute an equivalent bit error probability for each system under consideration and compare them on the basis of this figure of merit.

To compute this equivalent bit error probability, consider $M = 2^n$, where n is an integer. Each symbol in the M-ary system is therefore equivalent to an n-bit word for a binary system. This binary word might be chosen as the binary representation of a given symbol's index minus one, as shown below for $M = 8$:

M-ary Symbol (Signal)	Binary Representation		
0	0	0	0
1	0	0	1
2	0	1	0
3	0	1	1
4	1	0	0
5	1	0	1
6	1	1	0
7	1	1	1

Take any column, say the second, which is enclosed by a box. In this column and, in fact, any column, there are 2^{n-1} zeros and 2^{n-1} ones. If a word (symbol) is detected in error, then for any given bit position of the word there are 2^{n-1} out of a possible $2^n - 1$ ways that the chosen bit can be in error (one of the 2^n total possibilities is correct). Therefore, the probability of a given data bit being in error, given that a word (symbol) is in error, is

$$P(B|W) = \frac{2^{n-1}}{2^n - 1} \tag{4-55}$$

Since a word is in error if a bit in it is in error, the probability, $P(W|B)$, of a word error given a bit error is unity. Therefore, from Bayes' rule the equivalent bit error probability of an M-ary system can be computed from

$$P_B(\epsilon) = \frac{P(B|W)P_s(\epsilon)}{P(W|B)}$$

$$= \frac{2^{n-1}}{2^n - 1} P_s(\epsilon) \tag{4-56}$$

where $P_s(\epsilon)$ is its symbol error probability and $n = \log_2 M$ is the equivalent number of bits per symbol. Figure 4-7 gives a comparison of the equivalent bit error probability of an M-ary orthogonal system with binary orthogonal and antipodal signaling schemes.

EXAMPLE 4-7

Two examples of coherent M-ary orthogonal signaling schemes are provided in this example. The first signal set is referred to as coherent, M-ary, frequency-shift keying and consists of waveforms of the form

$$s_i(t) = \sqrt{\frac{2E_s}{T_s}} \cos 2\pi[f_0 + (i - 1)\,\Delta f]t$$

$$0 \le t \le T_s; \quad i = 1, 2, \ldots, M \tag{4-57}$$

where $f_0 T_s$ is taken as an integer for convenience and $(\Delta f)_{\min} = 1/(2T_s)$ is the minimum frequency spacing such that adjacent signals are orthogonal.

The second consists of time-displaced square pulses of the form

$$s_i(t) = \sqrt{\frac{E_s}{\tau}} \Pi\left[\frac{(t - (i - \frac{1}{2})\tau)}{\tau}\right] \qquad \tau = T_s/M; \quad i = 1, 2, \ldots, M \tag{4-58}$$

Note that the energy of each signal in both signal sets is E_s and the average power is E_s/T_s. ∎

4-3.3 *M*-ary Phase-Shift Keying

Consider a signal set of the form

$$s_i(t) = \sqrt{\frac{2E_s}{T_s}} \cos\left[\omega_0 t + \frac{2\pi(i - 1)}{M}\right]$$

$$0 \le t \le T_s; \quad i = 1, 2, \ldots, M \tag{4-59}$$

which is the same signal set considered in Example 4-4. Two orthogonal functions were shown to suffice in that example. The ith signal point with its decision region is shown in Figure 4-8a. The coordinates of the received data given $s_i(t)$ was transmitted is

$$Z_1 = \sqrt{E_s} \cos \frac{2\pi(i - 1)}{M} + N_1 \tag{4-60a}$$

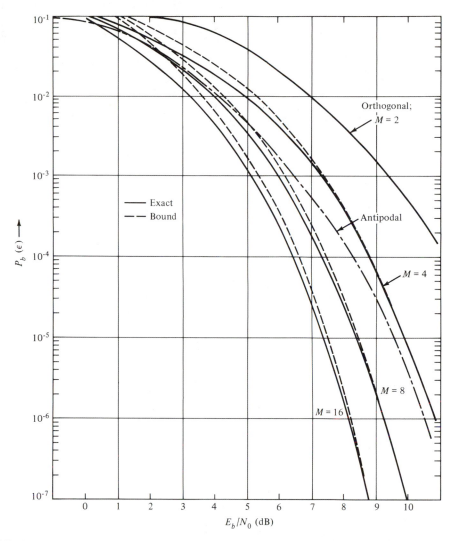

FIGURE 4-7. Comparison of equivalent bit error probability for *M*-ary orthogonal signaling compared with the bit error probability for binary antipodal and orthogonal signaling.

and

$$Z_2 = -\sqrt{E_s} \sin \frac{2\pi(i-1)}{M} + N_2 \qquad (4\text{-}60b)$$

where N_1 and N_2 are Gaussian random variables of zero mean and variance $N_0/2$. The probability of error, $P_s(\epsilon)$, is impossible to compute in closed form for this signal set except for $M = 2$ and $M = 4$. However, a lower bound for $P_s(\epsilon)$ can be computed by noting from Figure 4-8b that

$$P_s(\epsilon) \geq \Pr(\mathbf{Z} \in B_1) \qquad (4\text{-}61)$$

because the area included in B_1 is less than that included in $\overline{R}_i$, the complement of the *i*th decision region, unless $M = 2$ in which case they are equal. An upper

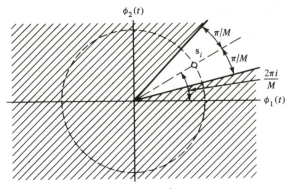

a. Decision region (unshaded) for i^{th} transmitted phase

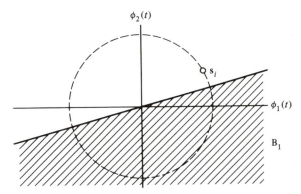

b. Integration region for computing lower bound on $P_b(\epsilon)$

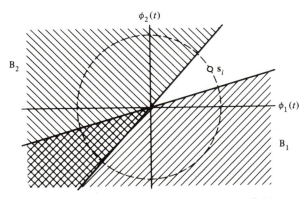

c. Integration regions for computing upper bound on $P_b(\epsilon)$

FIGURE 4-8   Signal-space diagrams for *M*-ary PSK.

bound is obtained by noting that $\overline{R}_i$ is exceeded by the sum of the areas in B_1 and B_2 as illustrated in Figure 4-8c. That is,

$$P_s(\epsilon) < \text{Pr } (\mathbf{Z} \in B_1) + \text{Pr } (\mathbf{Z} \in B_2) \qquad (4\text{-}62)$$

But both probabilities on the right-hand side of (4-62) are equal due to the rotational symmetry of the signal set and of the noise distribution. One can easily compute $\text{Pr } (\mathbf{Z} \in B_1)$, for example, by considering the noise to be resolved into components

$N_\perp$ and $N_\parallel$, which are perpendicular and parallel, respectively, to the boundary of B_1. Only $N_\perp$ is needed, since only it can cause $\mathbf{Z}$ to be in B_1. This noise component has variance $N_0/2$. Therefore, since $\sqrt{E_s}\sin(\pi/M)$ is the distance to the boundary of B_1,

$$\Pr(\mathbf{Z} \in B_1) = \Pr\left(N_\perp > \sqrt{E_s}\sin\frac{\pi}{M}\right)$$

$$= \int_{\sqrt{E_s}\sin(\pi/m)}^{\infty} \frac{\exp(-u^2/N_0)}{\sqrt{\pi N_0}}\,du$$

$$= Q\left(\sqrt{\frac{2E_s}{N_0}}\sin\frac{\pi}{M}\right) \tag{4-63}$$

and $P_s(\epsilon)$ is bounded by

$$Q\left(\sqrt{\frac{2E_s}{N_0}}\sin\frac{\pi}{M}\right) \le P_s(\epsilon) < 2Q\left(\sqrt{\frac{2E_s}{N_0}}\sin\frac{\pi}{M}\right) \tag{4-64}$$

The upper bound becomes very tight for M fixed as E_s/N_0 becomes large.

The exact expression for $P_s(\epsilon)$ is given by (see [3], pp. 228ff.)

$$P_s(\epsilon) = \frac{M-1}{M} - \frac{1}{2}\operatorname{erf}\left(\sqrt{\frac{E_s}{N_0}}\sin\frac{\pi}{M}\right)$$

$$- \frac{1}{\sqrt{\pi}}\int_0^{\sqrt{E_s/N_0}}\exp(-y^2)\operatorname{erf}\left(y\cot\frac{\pi}{M}\right)dy \tag{4-65}$$

which must be evaluated numerically except for $M = 2$ and $M = 4$. The bounds for $P_s(\epsilon)$ are compared with the exact result for several values of M in Figure 4-9. For large E_s/N_0, the bit error probability is related to $P_s(\epsilon)$ by

$$P_B(\epsilon) = \frac{P_s(\epsilon)}{\log_2 M} \tag{4-66}$$

if the M-ary PSK signals are associated with the source bits through a Gray code (see [4]). Because the dimension of the signal space is always 2 for $M > 2$, $P_B(\epsilon)$ increases for fixed E_b/N_0 as M increases. M-ary PSK is therefore a signaling scheme that is bandwidth efficient at the expense of power efficiency.

4-3.4 Multi-amplitude/Phase-Shift-Keyed Systems

Several schemes that give improved bandwidth efficiency at the expense of error probability performance are possible through combining amplitude- and phase-shift keying. One example is 16-quadrature amplitude-shift keying (16-QASK), in which the signaling waveforms are given by

$$s_i(t) = \sqrt{\frac{2}{T_s}}(A_i\cos\omega_0 t + B_i\sin\omega_0 t) \qquad 0 \le t \le T_s \tag{4-67}$$

where A_i and B_i take on the amplitudes

$$A_i = \pm a, \pm 3a$$

$$B_i = \pm a, \pm 3a \tag{4-68}$$

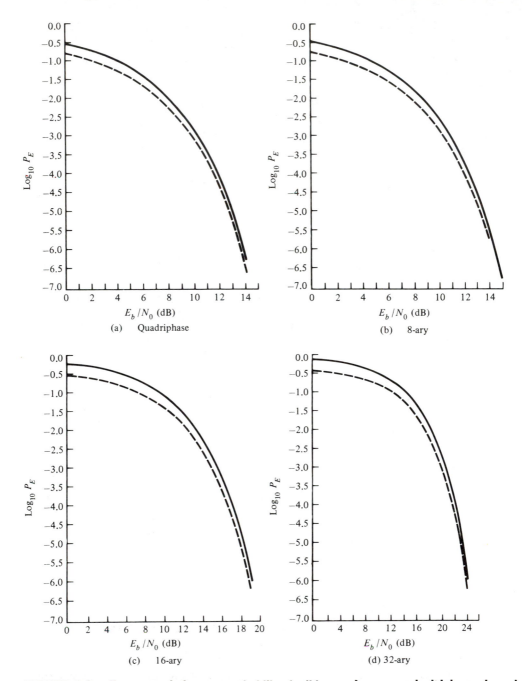

FIGURE 4-9. **Exact symbol error probability (solid curve) compared with lower bound (dashed curve) versus $E_b/N_o = E_s/(N_o \log_2 M)$ for M-ary PSK.**

where a is a parameter. By computing the average energy in (4-67), assuming all amplitudes equally likely, the parameter a can be expressed as

$$a = \sqrt{\frac{\overline{E}_s}{10}} \tag{4-69}$$

where $\overline{E}_s$ is the average energy. The constellation of signal points is shown in Figure 4-10 together with the optimum signal space partitioning to give minimum

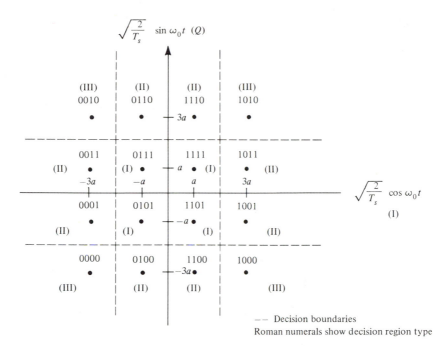

FIGURE 4-10. Signal constellation and decision regions for 16-QASK.

error probability in AWGN. A binary representation has been associated with each point by labeling the I and Q locations according to a Gray code with the first two digits denoting I and the second two Q. The demodulation and detection can be accomplished with the quadrature coherent demodulator arrangement shown in Figure 4-11. From Figure 4-10 it follows that there are three types of errors. In accordance with the Gray code labeling of the signal points, these errors can be categorized as follows:

Type I: Signal points 0111, 0101, 1101, 1111. The decision region for determining the probability of *correct reception* given these signals were transmitted is illustrated in Figure 4-12a.

Type II: Signal points 0001, 0011, 0110, 1110, 1011, 1001, 1100, 0100. The

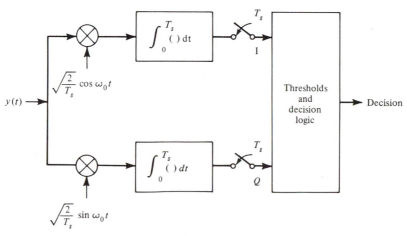

FIGURE 4-11. Detector structure for 16-QASK.

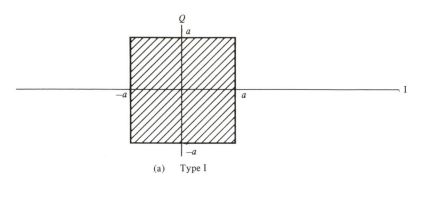

(a)   Type I

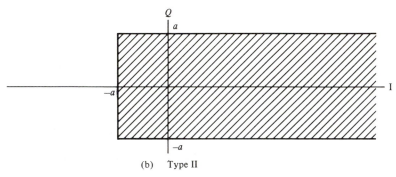

(b)   Type II

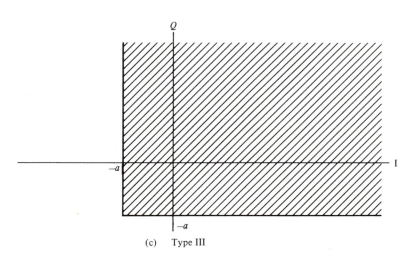

(c)   Type III

FIGURE 4-12.   Type I, II, and III decision regions for 16-QASK.

decision region for determining the correct probability of reception given these signal points were transmitted is illustrated in Figure 4-12b.

Type III: Signal points 0000, 0010, 1010, 1000. The decision region for determining the correct probability of reception given these signal points is illustrated in Figure 4-12c.

Given these conditional probabilities of correct reception, denoted by $P(C|I)$, $P(C|II)$, and $P(C|III)$, the average probability of error is

$$P(\epsilon) = 1 - [\tfrac{4}{16} P(C|I) + \tfrac{8}{16} P(C|II) + \tfrac{4}{16} P(C|III)] \qquad (4\text{-}70)$$

Using techniques which are by now standard, it can be shown that the I and Q noise components from each integrator output are zero-mean Gaussian with variances

$$\text{var } (N_I) = \text{var } (N_Q) = \frac{N_0}{2} \tag{4-71}$$

and are uncorrelated. For the type I decision regions,

$$P(C|\text{I}) = \left[\int_{-a}^{a} \frac{\exp\left(-u^2/N_0\right)}{\sqrt{\pi N_0}} \, du \right]^2$$

$$= \left[\text{erf}\left(\frac{a}{\sqrt{N_0}}\right) \right]^2 = \left[1 - 2Q\left(\sqrt{\frac{2a^2}{N_0}}\right) \right]^2 \tag{4-72}$$

where erf $(\cdot)$ is the error function. For the type II decision regions,

$$P(C|\text{II}) = \int_{-a}^{a} \frac{\exp\left(-u^2/N_0\right)}{\sqrt{\pi N_0}} \, du \int_{-a}^{\infty} \frac{\exp\left(-u^2/N_0\right)}{\sqrt{\pi N_0}} \, du$$

$$= \frac{1}{2} \text{erf}\left(\frac{a}{\sqrt{N_0}}\right) \left[1 + \text{erf}\left(\frac{a}{\sqrt{N_0}}\right) \right]$$

$$= \left[1 - 2Q\left(\sqrt{\frac{2a^2}{N_0}}\right) \right] \left[1 - Q\left(\sqrt{\frac{2a^2}{N_0}}\right) \right] \tag{4-73}$$

For the type III decision regions,

$$P(C|\text{III}) = \left[\int_{-a}^{\infty} \frac{\exp\left(-u^2/N_0\right)}{\sqrt{\pi N_0}} \, du \right]^2$$

$$= \frac{1}{4} \left[1 + \text{erf}\left(\frac{a}{\sqrt{N_0}}\right) \right]^2$$

$$= \left[1 - Q\left(\sqrt{\frac{2a^2}{N_0}}\right) \right]^2 \tag{4-74}$$

The average probability of error can be found as a function of $\bar{E}_s/N_0$ by using (4-69), (4-72), (4-73), and (4-74) in conjunction with (4-70). The results are shown in Figure 4-13, where the abscissa, given by

$$\frac{\bar{E}_b}{N_0} = \frac{1}{\log_2 M} \frac{\bar{E}_s}{N_0} = \frac{1}{4} \frac{\bar{E}_s}{N_0} = \frac{5a^2}{2N_0} \tag{4-75}$$

is the average energy per bit; that is, $a^2/N_0 = 2\bar{E}_b/5N_0$ in (4-72), (4-73), and (4-74).

4-3.5 Bandwidth Efficiency of *M*-ary Digital Communication Systems

It is often desirable to compare digital communication systems on the basis of bandwidths required to transmit equal data rates. For an *M*-ary communication system, the *bit rate*, R_b, is given in terms of the *symbol rate*, R_s, as

$$R_b = (\log_2 M)R_s \tag{4-76}$$

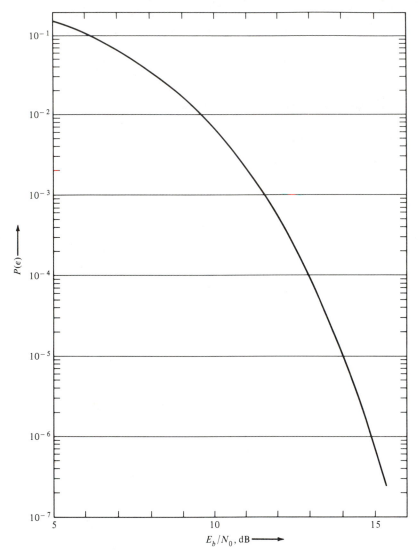

FIGURE 4-13. Symbol error probability versus $\bar{E}_b/N_o$ for 16-QASK.

Consider a BPSK system for which the null-to-null radio-frequency (RF) bandwidth is

$$B_{\text{BPSK}} = 2R_b = 2R_s \qquad \text{Hz} \tag{4-77}$$

An M-ary PSK system also has a null-to-null bandwidth of $2R_s$, but in terms of *bit rate*, its null-to-null bandwidth is

$$B_{\text{MPSK}} = \frac{1}{\log_2 M} (2R_b) \qquad \text{Hz} \tag{4-78}$$

Similarly, for 16-QASK, the null-to-null bandwidth is $2R_s$, but each symbol conveys 4 bits, giving

$$B_{\text{16QASK}} = \tfrac{1}{4} (2R_b) \qquad \text{Hz} \tag{4-79}$$

for null-to-null bandwidth in terms of bit rate. Any other bandwidth criterion in terms of out-of-band energy can be chosen by using Figure 3-15.

4.4

SIGNALING SCHEMES NOT REQUIRING COHERENT REFERENCES AT THE RECEIVER

In this section we consider the error probability performance of two modulation schemes that do not require coherent references at the receiver to demodulate the signal optimally. The first to be considered is noncoherent frequency-shift keying (NFSK) and the second is differentially coherent phase-shift keying (DPSK). It will be shown that noncoherent schemes perform worse in terms of error probability for a given signal-to-noise ratio than do their coherent counterparts. Nevertheless, there are several situations where noncoherent schemes may be employed instead of a coherent signaling method. Among these are:

1. The channel may not allow the use of a coherent signaling scheme.
2. Phase-ambiguity resolution at the receiver for establishing a locally coherent reference can be avoided with schemes such as NFSK and DPSK.
3. In certain types of spread-spectrum systems, to be discussed later, the carrier frequency is hopped during each symbol, which results in an unknown carrier phase angle being generated at each hop by the frequency synthesizer providing the carrier. Therefore, a signaling scheme is required that allows each hopped portion to be detected and used collectively to determine the data transmitted.
4. The receiver may be much simpler since the carrier reference required in a coherent system need not be generated.

4-4.1 NFSK

In an NFSK system the received data during the kth signaling interval are

$$z(t) = \sqrt{\frac{2E_s}{T_s}} \cos(\omega_i t + \alpha) + n(t) \qquad 0 \le t \le T_s; \quad i = 1, 2, \ldots, M$$

(4-80)

where α is the unknown phase which is modeled as a uniformly distributed random variable in the interval $(0, 2\pi)$ and ω_i is the radian frequency of the ith possible transmitted signal. E_s, T_s, and $n(t)$ are as defined previously. The frequency separation between adjacent signals is sufficient to guarantee that the signals are uncorrelated. This signal set can be resolved into a $2M$-dimensional signal space spanned by the basis set

$$\left. \begin{array}{l} \varphi_{xi}(t) = \sqrt{\dfrac{2}{T_s}} \cos \omega_i t \\[2em] \varphi_{yi}(t) = \sqrt{\dfrac{2}{T_s}} \sin \omega_i t \end{array} \right\} \quad 0 \le t \le T_s; \quad i = 1, 2, \ldots, M \qquad (4\text{-}81)$$

Given that $s_i(t)$ was sent, the coordinates of the received data vector denoted as $\mathbf{Z} = (X_1, Y_1, X_2, Y_2, \ldots, X_M, Y_M)$ are

$$X_j = (\varphi_{xj}, \mathbf{Z}) = \begin{cases} N_{xj} & j \neq i \\ \sqrt{E_s}\cos\alpha + N_{xi} & i = j \end{cases} \tag{4-82}$$

$$Y_j = (\varphi_{yj}, \mathbf{Z}) = \begin{cases} N_{yj} & j \neq i \\ -\sqrt{E_s}\sin\alpha + N_{yi} & i = j \end{cases} \tag{4-83}$$

where $j = 1, 2, \ldots, M$. The noise components, N_{xj} and N_{yj} are uncorrelated and have variances $N_0/2$ as for the coherent FSK case. Given that $s_i(t)$ was transmitted, a correct reception is made if

$$\sqrt{X_j^2 + Y_j^2} < \sqrt{X_i^2 + Y_i^2} \qquad \text{all } j \neq i \tag{4-84}$$

Evaluation of the probability of symbol error requires the pdf of the random variable $R \triangleq \sqrt{X_j^2 + Y_j^2}$. For $j = i$ and given α, X_j is a Gaussian random variable with mean $\sqrt{E_s}\cos\alpha$ and variance $N_0/2$. Similarly, for $j = i$, Y_j is a Gaussian random variable with mean $-\sqrt{E_s}\sin\alpha$ and variance $N_0/2$. For $j \neq i$, both have zero means. Furthermore, they are independent, so that their joint pdf is

$$f_{X_j Y_j}(x, y|\alpha) = \frac{1}{N_0}\exp\left\{-\frac{1}{N_0}[(x - \sqrt{E_s}\cos\alpha)^2 + (y + \sqrt{E_s}\sin\alpha)^2]\right\} \tag{4-85}$$

The joint pdf of X_j and Y_j unconditional on α is obtained by averaging over α, which was assumed to be uniformly distributed in $(0, 2\pi)$. A change in variables to the polar coordinates r and φ defined by

$$\left.\begin{aligned} x &= \sqrt{\frac{N_0}{2}}\, r\sin\varphi \\ y &= \sqrt{\frac{N_0}{2}}\, r\cos\varphi \end{aligned}\right\} \quad r > 0;\quad 0 < \varphi \leq 2\pi \tag{4-86}$$

results in the joint pdf

$$f_{R_j \Phi_j}(r, \varphi) = \frac{r}{2\pi}\exp\left[-\frac{1}{2}\left(r^2 + \frac{2E_s}{N_0}\right)\right] I_0\left(r\sqrt{\frac{2E_s}{N_0}}\right)$$

$$j = 1;\quad r > 0;\quad 0 < \varphi \leq 2\pi \tag{4-87}$$

The joint pdf on r alone is obtained by integration over φ. It is simply (4-87) without the 2π in the denominator and is recognized as a Rician pdf. If $j \neq i$, the pdf of R_j can be obtained by setting $E_s = 0$; the resulting pdf is Rayleigh.

In terms of the random variables R_j, $j = 1, 2, \ldots, M$, the detection criterion, given that $s_i(t)$ was transmitted, is

$$R_j < R_i \qquad \text{all } j \neq i$$

Since the R_j's are statistically independent random variables, the possibility of the compound event, given R_i, is

$$\Pr(R_j < R_i, \text{ all } j \neq i | R_i) = \prod_{\substack{j=1 \\ j \neq i}}^{M} \Pr(R_j < R_i | R_i) \tag{4-88}$$

But

$$\Pr(R_j < R_i | R_i) = \int_0^{R_i} r e^{-r^2/2} \, dr$$

$$= 1 - e^{-R_i^2/2} \qquad (4\text{-}89)$$

The probability of correct reception, given $s_i(t)$ sent, is (4-88) averaged over R_i; R_i has the Rician pdf, which is 2π times (4-87). Using (4-89) in (4-88), the result for this average probability of correct reception is

$$P_s(C|s_i \text{ sent}) = \int_0^\infty (1 - e^{-r^2/2})^{M-1} e^{-(r^2 + 2E_s/N_0)/2} I_0\left(r\sqrt{\frac{2E_s}{N_0}}\right) dr \qquad (4\text{-}90)$$

Now, by the binomial theorem,

$$(1 - e^{-r^2/2})^{M-1} = \sum_{k=0}^{M-1} \binom{M-1}{k} (-1)^k e^{-kr^2/2} \qquad (4\text{-}91)$$

This series can be substituted into (4-90), the order of summation and integration reversed, and the integral

$$\int_0^\infty x e^{-ax} I_0(bx) \, dx = \frac{1}{2a} e^{-b^2/4} \qquad (4\text{-}92)$$

used to produce the result

$$P_s(C|s_i) = e^{-E_s/N_0} \sum_{k=0}^{M-1} \binom{M-1}{k} \frac{(-1)^k}{k+1} \exp\left[\frac{E_s}{N_0(k+1)}\right] \qquad (4\text{-}93)$$

Since this result is independent of i, the probability of error is

$$P_s(\epsilon) = 1 - P_s(c|s_i) = \sum_{k=1}^{M-1} \binom{M-1}{k} \frac{(-1)^{k+1}}{k+1} \exp\left[-\frac{k}{k+1}\frac{E_s}{N_0}\right] \qquad (4\text{-}94)$$

For $M = 2$, (4-94) reduces to

$$P_b(\epsilon) = \tfrac{1}{2} e^{-E_b/2N_0}$$

which can be shown by other means to be the result for binary, noncoherent FSK (see [5], p. 331).

As in the case of coherent M-ary FSK, the equivalent bit error probability can be approximated by (4-56) if desired.

It is interesting to compare M-ary noncoherent FSK with a system that, corresponding to each M-ary signal, sends $n = \log_2 M$ bits independently through the channel by means of a binary noncoherent FSK system. The error probability for an n-bit "word" of such a system is

$$P_w(\epsilon) = 1 - [1 - P_b(\epsilon)]^n \qquad (4\text{-}95)$$

This result is compared in Figure 4-14 with (4-94) as a function of $E_b/N_0 \triangleq E_s/nN_0$.

4-4.2 DPSK

In differentially coherent detection of a PSK-modulated signal, the phase of the current symbol is compared with the phase of the preceding symbol (or, possibly,

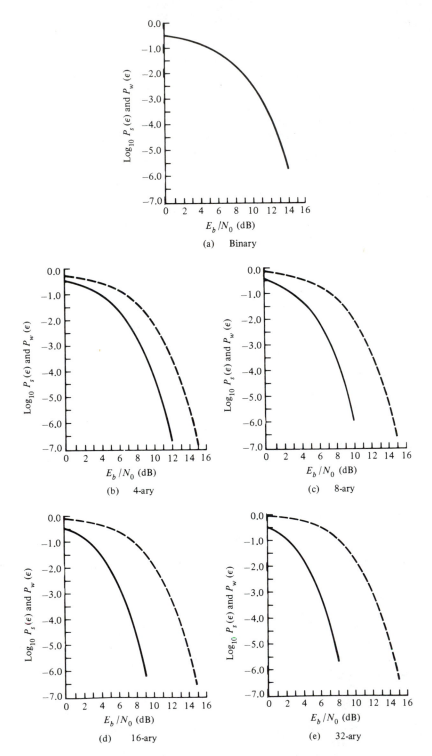

FIGURE 4-14. Comparison of symbol error probability for noncoherent *M*-ary FSK (solid curves) with the word error probability of an *n*-bit noncoherent binary system (dashed curves).

symbols) in order to make a decision. This presupposes that (1) the unknown relative phase shift of the received signal due to the channel characteristics is constant over at least two symbol periods, and (2) a known relationship exists between two successive symbol phases which depends on the input data sequence.

It is easiest to illustrate the latter condition with the binary case, where this relationship between successive symbol phases is provided by differentially encoding the data from the information source. Suppose that the nth data bit from the source is denoted by D_n, which may be either 0 or 1, and that the nth bit of differentially encoded data is denoted C_n (either a 0 or 1). The operation used to produce the differentially encoded data sequence $\{C_n\}$ is

$$C_n = D_n \oplus C_{n-1} \qquad (4\text{-}96)$$

where $\oplus$ denotes the Exclusive-OR operation (XOR), or a binary add without carry.

The *decoding* operation is performed by forming the sequence $\{\hat{D}_n\}$ with members given by

$$\hat{D}_n = C_n \oplus C_{n-1} \qquad (4\text{-}97)$$

However, when a differentially encoded data sequence is used to phase-shift key a carrier, which is referred to as DPSK, the detection operation can be implemented by comparing the present symbol with the immediately preceding symbol as shown in the block diagram of Figure 4-15.* In addition to the simplicity of differentially coherent detection of DPSK, an additional advantage accrues from the fact that the result of differentially detecting a DPSK signal *or its inversion* results in the same output data sequence. This is illustrated in the following example.

EXAMPLE 4-8

The data sequence 10111001010 can be differentially encoded by assuming an initial reference bit of $C_{-1} = 1$ and carrying out the operation (4-96) on the

*This is a suboptimum detector for DPSK. See [6].

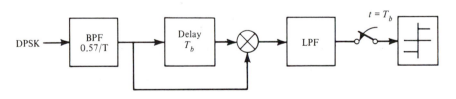

(a) IF sampling detector

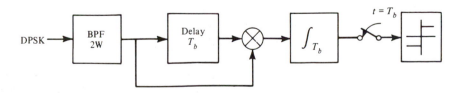

(b) IF integrate-and-dump detector

FIGURE 4-15. Suboptimum detectors for binary DPSK.

successive data bits. The results are

$$\{D_n\}: \quad 10111001010$$

$$\{C_n\}: \quad 100101110011$$

This sequence is decoded using (4-99). The resulting operations are

$$\{C_n\}: \quad 100101110011$$

$$\{C_{n-1}\}: \quad 100101110011$$

$$\{\hat{D}_n\}: \quad 10111001010$$

If the differentially encoded data sequence $\{C_n\}$ is accidentally inverted, the correct data sequence is still decoded, assuming that no errors are made, as follows:

$$\{\overline{C}_n\}: \quad 011010001100$$

$$\{\overline{C}_{n-1}\}: \quad 011010001100$$

$$\{\hat{D}_n\}: \quad 10111001010$$

Thus differential encoding and decoding can be used to solve the phase-ambiguity problem in acquisition of a coherent local reference in BPSK transmission. This problem is discussed further in Chapter 5. ∎

   It is emphasized that differential encoding/decoding of BPSK to solve the phase-ambiguity problem differs from DPSK, which makes use of the built-in phase relationship between symbols provided by differential encoding so that a detector such as illustrated in Figure 4-15 can be used. The probability of error for any modulation scheme that utilizes differential encoding and decoding of the data is given by

$$P_{\mathrm{DE}}(\epsilon) = 2P_b(\epsilon)[1 - P_b(\epsilon)] \tag{4-98}$$

where $P_b(\epsilon)$ is the raw bit error probability of the channel. This follows because an error occurs with differential encoding/decoding when the reference bit or the current bit are in error, but not both. Note that for $P_b(\epsilon)$ small, the resultant error probability is essentially twice that of the channel error probability. This amounts to an SNR degradation of about 0.3 dB for small error probabilities.

   To generalize the concept of differential PSK to the case of M-ary PSK, we suppose that the transmitted carrier phase angle at symbol time $n - 1$ is α_{n-1} and that it is desired to transmit symbol $\beta_n = \Phi$ at time n, where β_n takes on any of the values given by

$$\beta_n = \frac{(2l - 1)\pi}{M - 1} < l \leq M, \text{ modulo } 2\pi \tag{4-99}$$

The *transmitted phase* at time n is then

$$\alpha_n = \alpha_{n-1} + \Phi \tag{4-100}$$

Suppose that the phases received corresponding to α_{n-1} and α_n are $\theta_{n-1} = \alpha_{n-1} + \gamma$ and $\theta_n = \alpha_n + \gamma$, respectively, where γ is the unknown phase shift introduced by the channel. A correct decision is made at the receiver when the *received phase difference* $\theta = \theta_n - \theta_{n-1}$ is such that

$$\Phi - \frac{\pi}{M} < \theta < \Phi + \frac{\pi}{M} \qquad (4\text{-}101)$$

The receiver structure of Figure 4-16 can be used to compute the phase differences required for the decision rule (4-101). The decision rule (4-101) is identical to that used for coherent M-ary PSK except now the *phase differences* between successive symbols are used. That is, the decision procedure is to find which $2\pi/M$ wedge around each possible transmitted phase that a received phase difference falls into, as illustrated in Figure 4-17.

Assuming a uniformly distributed random phase due to the channel, the pdf of the phase difference θ in (4-101) can be shown to be [4, p. 249]

$$p(\theta) = \frac{1}{2\pi} e^{-z}[1 + \sqrt{4\pi z} \cos \theta \; e^{z \cos^2 \theta} \; Q(\sqrt{2z} \cos \theta)] \qquad -\pi \leq \theta \leq \pi \qquad (4\text{-}102)$$

where z is the signal-to-noise ratio, given by

$$z = \frac{E_s}{N_0} \qquad (4\text{-}103)$$

and $Q(u)$ is the Q-function defined by (4-31). The probability distribution function corresponding to (4-102) is

$$F(\theta) = \int_{-\pi}^{\theta} p(u) \, du \qquad -\pi \leq \theta < \pi \qquad (4\text{-}104)$$

which cannot be evaluated in closed form. The probability of symbol error given that phase Φ was transmitted, from (4-101) or Figure 4-17, can be written as

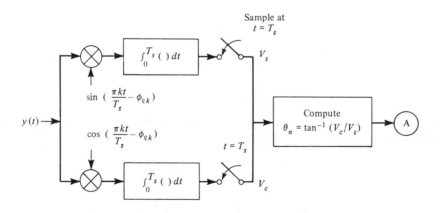

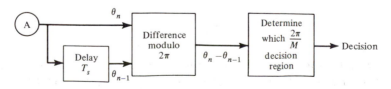

FIGURE 4-16. Block diagram of an *M*-ary differentially coherent PSK receiver.

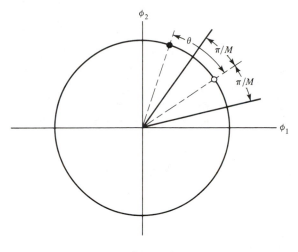

(a)   Error made

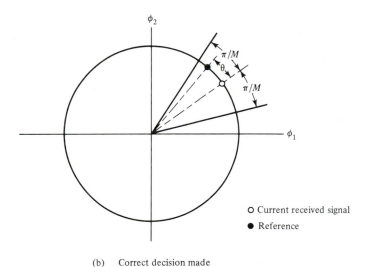

○ Current received signal

● Reference

(b)   Correct decision made

FIGURE 4-17.   Decision space for received phase differences in differentially coherent detection of *M*-ary PSK.

$$P_s(\epsilon|\Phi) = \text{Pr}\left(\theta > \Phi + \frac{\pi}{M} \text{ or } \theta < \Phi - \frac{\pi}{M}\right)$$

$$= 2\,\text{Pr}\left(\theta > \Phi + \frac{\pi}{M}\right)$$

$$= 2\left[1 - F\left(\frac{\pi}{M}\right)\right] \tag{4-105}$$

where $F(\pi/M)$ is the phase distribution function (4-104) with $\theta = \pi/M$. This can be expressed in closed form only for $M = 2$ or $M = 4$. Since (4-105) is independent of Φ, this is also the probability of error averaged over all possible transmitted phases.

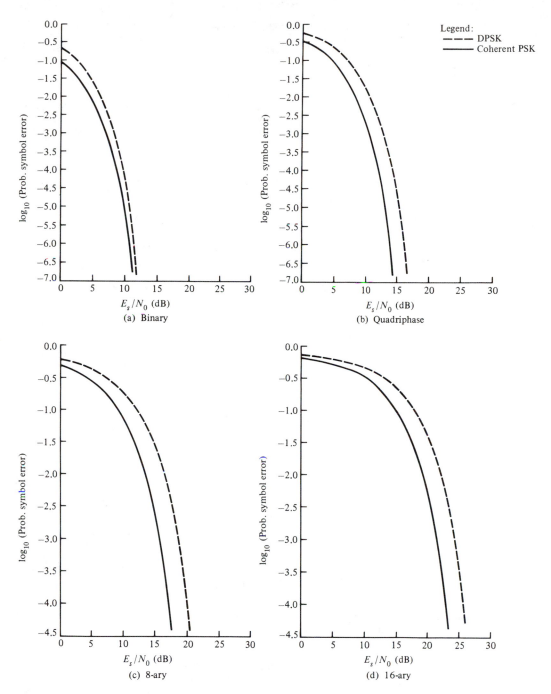

FIGURE 4-18. Probability of symbol error versus signal-to-noise ratio E_s/N_o for *M*-ary DPSK (dashed curves) compared with that for coherent *M*-ary PSK (solid curves).

It can be shown that the following bounds apply for (4-105) [7]:

$$\max(P_1, P_2) \le P_s(\epsilon|\Phi) \le P_1 + P_2 \tag{4-106a}$$

where

$$P_1 = \Pr\left[\sin\left(\theta - \Phi + \frac{\pi}{M}\right) < 0\right] \tag{4-106b}$$

$$P_2 = \Pr\left[\sin\left(\theta - \Phi + \frac{\pi}{M}\right) > 0\right] \tag{4-106c}$$

Because of the symmetry of the signal constellation and the rotational symmetry of the noise random variables it follows that $P_1 = P_2$ and

$$P_1 \le P_s(\epsilon|\Phi) = P_s(\epsilon) \le 2P_1 \qquad M > 2 \tag{4-107}$$

Prabhu [7] has shown that the probability P_1 can be expressed as

$$P_1 = \frac{1}{2}\sin\frac{\pi}{M}\int_{2z}^{\infty}\exp(-y)I_0\left(y\cos\frac{\pi}{M}\right)dy \tag{4-108}$$

where z was defined in (4-103). For $M = 2$, (4-108) reduces to

$$P_1 = P_b(\epsilon) = \tfrac{1}{2}\exp(-z) \tag{4-109}$$

which is the correct result for binary DPSK.

For large arguments, a useful asymptotic expression for P_1 is given by

$$P_1 = \frac{1}{2}\operatorname{erfc}\left(\sqrt{z}\sin\frac{\pi}{2M}\right)$$

$$= Q\left(\sqrt{2z}\sin\frac{\pi}{2M}\right) \tag{4-110}$$

Results for $P_s(\epsilon)$ versus E_s/N_0 are shown in Figure 4-18 for several values of M (see [3, pp. 250–251]).

4-5

EFFICIENT SIGNALING FOR MESSAGE SEQUENCES

4-5.1 Summary of Block-Orthogonal and *M*-ary Signaling Performance

It is useful at this point to tie together the results learned so far in regard to the performance of the *M*-ary digital communication schemes. For this purpose consider a sequential source at a rate

$$R = R_m \log_2 q \qquad \text{bits/s} \tag{4-111}$$

where R_m is the message rate mentioned in regard to Figure 4-1 and q is the number of possible transmitted messages. Suppose that the output of the source is observed for T seconds, where T is an integral multiple of $1/R$. The number of equally likely symbols in the interval T is therefore

$$M = 2^{RT} \tag{4-112}$$

Thus, if one of M symbols sent each T seconds is used to transmit the source output, R is related to M by

$$R = \frac{1}{T} \log_2 M \qquad \text{bits/s} \qquad (4\text{-}113)$$

The transmission of R bits/s by both M-ary orthogonal and bit-by-bit block orthogonal signaling has just been considered [see, e.g., (4-94) and (4-95) for the case of noncoherent signaling]. In the case of coherent signaling we can compare (4-54) with (4-95) with $P_b(\epsilon)$ replaced by the bit error probability for binary antipodal signaling. Such bit-by-bit signaling is sometimes referred to as *vertices-of-a-hypercube signaling* because the signal constellation in signal space consists of the vertices of a hypercube centered at the origin (this is the topic of a problem at the end of the chapter). In the case of vertices-of-a-hypercube signaling, it can be shown that increasing

$$K = RT \qquad (4\text{-}114)$$

forces the probability of error to unity no matter how large E_b/N_0 becomes. That is, for a block of K bits

$$\lim_{K \to \infty} \{1 - [1 - P_b(\epsilon)]^K\} = 1 \qquad (4\text{-}115)$$

no matter how small $P_b(\epsilon)$.

On the other hand, it was shown that for orthogonal signaling, $\lim_{K \to \infty} P_s(\epsilon) = 0$ as long as E_b/N_0 exceeds a threshold (ln 2 in the case of coherent orthogonal signaling).

In the case of bit-by-bit signaling, for which the signal constellation is the vertices of a K-dimensional hypercube, the distance to nearest neighbors remains fixed as K increases. This is illustrated in Figure 4-19a for $K = 3$. Therefore, $P_s(\epsilon)$ increases for fixed E_b/N_0.

In the case of orthogonal signaling, the distance to nearest neighbors in the signal constellation grows as $K^{1/2}$, as illustrated in Figure 4-19b, and $P_s(\epsilon)$ approaches zero as K increases for fixed $E_b/N_0 > \ln 2 = -1.59$ dB. Unfortunately, this behavior is obtained at the expense of increasing bandwidth as K increases.

To emphasize the dependence of bandwidth on $K = RT$ the dimensionality theorem is quoted, which is:

Theorem: *For $\{\varphi_i(t)\}$ any set of orthogonal waveforms of duration T such that each $\varphi_i(t)$ is zero outside a T-second interval and no more than $\frac{1}{12}$ of its energy is outside a bandwidth $(-W, W)$, the number, M, of different waveforms in the set is overbounded by $2.4TW$ when TW is large (see [1, p. 294]).*

For the case of a transmitter using block orthogonal signaling, use of the dimensionality theorem means that

$$M = 2^{RT} \leq 2.4TW$$

or

$$W \geq \frac{2^{RT}}{2.4T} = \frac{R2^K}{2.4K} \qquad (4\text{-}116)$$

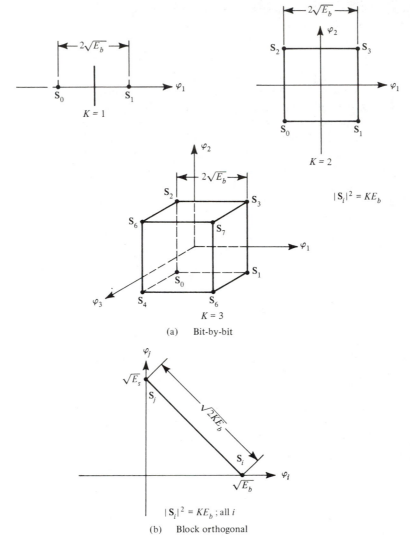

FIGURE 4-19. Signal constellations pertaining to bit-by-bit and block-orthogonal signaling.

which clearly approaches infinity with K or T. Thus error-free transmission is achieved with block orthogonal signaling at the expense of the bandwidth going to infinity exponentially.

4-5.2 Channel Coding Theorem

It is desirable to achieve error-free transmission while simultaneously imposing a constraint of finite bandwidth. That this is indeed possible is the theme of *Shannon's second theorem* or the *channel coding theorem*, which is now outlined.*

---

*See [1, pp. 297 ff.] for a more complete treatment. Strictly speaking, the development here is not a coding theorem in the sense that it is shown how to select codes. Rather, it is shown that code selections exist which allow arbitrarily low error probabilities without infinite bandwidth or power.

Assume vertices-of-a-hypercube signaling where the number of dimensions, N, is greater than $K = RT$, which is the total number of bits emitted from the source in T seconds. It is convenient to let

$$N = DT \qquad D > R \tag{4-117}$$

where D is the number of dimensions available per second. The objective is to choose a subset, $M = 2^{RT}$, of the 2^N vertices available each T seconds in such a manner that $P_s(\epsilon)$ goes to zero as T increases and yet the bandwidth remains finite. That this may be possible is indicated by the fact that the fraction of vertices that must be used is

$$\frac{2^{RT}}{2^{DT}} = 2^{-(D-R)T} \tag{4-118}$$

which approaches zero as T increases. Since the signals occupy the vertices of a hypercube, each signal has the form

$$s_i(t) = \sum_{j=1}^{N} S_{ij}\varphi_j(t) \qquad i = 1, 2, \ldots, M \tag{4-119}$$

where

$$S_{ij} = \pm\sqrt{E_N} \qquad \text{all } i \text{ and } j \tag{4-120}$$

with E_N the energy per dimension. The total symbol energy is $E_s = NE_N$. In terms of the average transmitted power, P_s, E_N can be expressed as

$$E_N = \frac{P_s}{D} = \frac{E_s}{N} \tag{4-121}$$

The problem of signal selection is one of choosing signal vector coefficients (hereafter referred to simply as a *code*):

$$\mathbf{S}_i = (S_{i1}, S_{i2}, \ldots, S_{iN}) \qquad i = 1, 2, \ldots, M \tag{4-122}$$

A good selection is difficult to find and difficult to analyze. The way around this totally unworkable problem, as done by Shannon, is to bound the probability of error averaged over all possible code selections. This bound can be shown to approach zero as T (or N) approaches infinity. Some codes will give performance appreciably worse than the average and some appreciably better. The point is that at least one code performs better than the average. The only problem with this derivation is that no constructive technique for finding a good code is provided.

To sketch the proof of the theorem further, note that with N dimensions, there are 2^N vertices from which to select M code words for a total of 2^{NM} different possible codes. Suppose that the jth code selection, $\mathbf{s}_j$, results in probability of word error $P(\epsilon|\mathbf{s}_j)$, giving an average probability of error over all codes of

$$\overline{P(\epsilon)} = 2^{-NM} \sum_{\substack{j=1 \\ \text{all codes}}}^{2^{NM}} P(\epsilon|\mathbf{S}_j) \tag{4-123}$$

where the probability of the jth code being used is 2^{-NM}. When message m_k is transmitted using a specific code $\mathbf{S}_j$, let the probability of error be $P(\epsilon|m_k, \mathbf{S}_j)$. The conditional probability of error averaged over all possible code selections given that the message m_k transmitted is

$$\overline{P(\epsilon|m_k)} = 2^{-NM} \sum_{\text{all codes}} P(\epsilon|m_k, \mathbf{S}_j) \tag{4-124}$$

where $P(\epsilon|m_k, \mathbf{s}_j)$ is the probability of error given that m_k was sent using the code selection $\mathbf{s}_j$. Application of the union bound to $P(\epsilon|m_k, \mathbf{s}_j)$ gives

$$P(\epsilon|m_k, \mathbf{S}_j) \leq \sum_{\substack{j=1 \\ j \neq k}}^{M} P(\mathbf{S}_j, \mathbf{S}_k) \qquad (4\text{-}125)$$

where $P(\mathbf{S}_j, \mathbf{S}_k)$ is the probability of error resulting when the signal vectors (code words) $\mathbf{S}_j$ and $\mathbf{S}_k$ are used to communicate one of two equally likely messages. Evaluation of the right-hand side of (4-125) requires explicit knowledge of the signal set $\{\mathbf{S}_j\}$. By considering the average probability of error, one can avoid having to have this explicit knowledge. Using (4-125) in (4-124) gives the bound

$$\overline{P(\epsilon|m_k)} \leq \sum_{\text{all codes}} 2^{-NM} \sum_{\substack{j=1 \\ (j \neq k)}}^{M} P(\mathbf{S}_j, \mathbf{S}_k)$$

$$\leq \sum_{\substack{j=1 \\ (j \neq k)}}^{M} \sum_{\text{all codes}} 2^{-NM} P(\mathbf{S}_j, \mathbf{S}_k)$$

$$= \sum_{\substack{j=1 \\ (j \neq k)}}^{M} \overline{P(\mathbf{S}_j, \mathbf{S}_k)} \qquad (4\text{-}126)$$

where the overbar denotes an average over all possible code selections. For an AWGN channel,

$$P(\mathbf{S}_j, \mathbf{S}_k) = Q\left(\frac{|\mathbf{S}_j - \mathbf{S}_k|}{\sqrt{2N_0}}\right) \qquad (4\text{-}53)$$

Now the difference between like coordinates of $\mathbf{S}_j$ and $\mathbf{S}_k$ is $2\sqrt{E_N}$ so that if they differ in p coordinates, it follows that

$$|\mathbf{S}_j - \mathbf{S}_k|^2 = \sum_{l=1}^{N} (S_{jl} - S_{kl})^2 = p(2\sqrt{E_N})^2 = 4pE_N \qquad (4\text{-}127)$$

Since the selection of each code is assumed completely random, the probability that S_{jl} equals S_{kl} is $\frac{1}{2}$, independently for all $l = 1, 2, \ldots, N$, and the probability that $\mathbf{s}_j$ and $\mathbf{s}_k$ differ in p coordinates is given by the binomial distribution: $\binom{N}{p} 2^{-N}$.

Therefore, the average of $P(\mathbf{s}_j, \mathbf{s}_k)$ over all possible codes is

$$\overline{P(\mathbf{S}_j, \mathbf{S}_k)} = \sum_{p=0}^{N} \binom{N}{p} 2^{-N} Q\left(\sqrt{\frac{2pE_N}{N_0}}\right) \qquad (4\text{-}128)$$

Since the right-hand side is independent of the indices j and k, (4-126) may be written as

$$\overline{P(\epsilon|m_k)} \leq (M - 1)\overline{P(\mathbf{S}_j, \mathbf{S}_k)} < M\overline{P_2(\epsilon)} \qquad (4\text{-}129)$$

where $P_2(\epsilon)$ is used in place of $P(\mathbf{s}_j, \mathbf{s}_k)$ to simplify notation. Finally, it follows that

$$\overline{P(\epsilon)} = \sum_{k=1}^{M} \overline{P(\epsilon|m_k)}\, P(m_k)$$

$$< M\overline{P_2(\epsilon)} \sum_{k=1}^{M} P(m_k) = M\overline{P_2(\epsilon)} \qquad (4\text{-}130)$$

To complete the bounding operation, note that [1, Prob. 2-26]

$$Q(\alpha) \le \tfrac{1}{2}\, e^{-\alpha^2/2} < e^{-\alpha^2/2} \qquad \alpha > 1$$

which, when substituted in (4-129), yields

$$\overline{P_2(\epsilon)} < \sum_{p=0}^{N} 2^{-N}\binom{N}{p} e^{-pE_N/N_0}$$

$$= 2^{-N} \sum_{p=0}^{N} \binom{N}{p} (e^{-E_N/N_0})^p$$

$$= 2^{-N}(1 + e^{-E_N/N_0})^N \qquad (4\text{-}131)$$

Using this in (4-130), the bound

$$\overline{P(\epsilon)} < M 2^{-NR_0} \qquad (4\text{-}132)$$

is obtained where

$$R_0 \triangleq \log_2 \frac{2}{1 + \exp(-E_N/N_0)} \qquad (4\text{-}133)$$

If the source rate, R, is written in terms of bits per dimension, as

$$R_N \triangleq \frac{R}{D} \qquad (4\text{-}134)$$

M can be expressed as

$$M = 2^{RT} = 2^{NR_N} \qquad (4\text{-}135)$$

so that

$$\overline{P(\epsilon)} < 2^{-N[R_0 - R_N]} \qquad (4\text{-}136)$$

where R_0, the exponential bound parameter, is plotted in Figure 4-20.

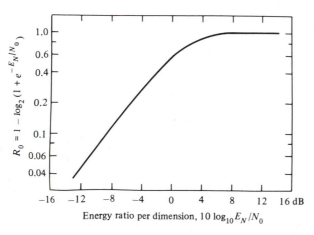

FIGURE 4-20. Exponential bound parameter versus E_N/N_0 for binary antipodal signaling.

It is clear from (4-136) that as long as $R_N < R_0$, the average probability of error can be made arbitrarily small by choosing N (or T) sufficiently large. From Figure 4-20 it is seen that $R_0 \to 1$ as E_N/N_0 becomes large. Thus the number of bits per dimension, R_N, must be less than unity to ensure an arbitrarily small $\overline{P(\epsilon)}$.

To increase the one-bit-per-dimension saturation level of R_0 found in binary (vertices of a hypercube) signaling, signal sequences using multilevel signals may be employed. The discussion so far is formalized with a statement of the capacity theorem due to Shannon [1, pp. 321ff.].

Theorem: *There exists a constant C_N given by*

$$C_N = \frac{1}{2} \log_2\left(1 + \frac{2E_N}{N_0}\right) \qquad \text{bits/dim} \qquad (4\text{-}137)$$

called the capacity of an AWGN channel with the following properties:
    (a) If $R_N > C_N$ and the number of equally likely messages, $M = 2^{NR_N}$, is large, the probability of error is close to unity for every possible set of M codewords (transmitted signals).
    (b) If $R_N < C_N$ and M is sufficiently large, there exists sets of M codewords (transmitted signals) such that $P(\epsilon)$ achieved with an optimum receiver is arbitrarily small.

The parameters C_N and R_0 are not the same, the former being referred to as the channel capacity and the latter as the exponential bound parameter. The limitation $R_N < R_0$ is not inescapable, but depends partially on the method of bounding. A proof of the capacity theorem, however, reveals that the bound $R_N < C_N$ is inescapable.*

4-6

MULTI-*h* CONTINUOUS PHASE MODULATION

In many applications, particularly in satellite communications, modulation techniques that have constant or nearly constant envelopes versus time are required due to power amplifier considerations. In addition, good communications efficiencies in terms of low error probability for a given signal-to-noise ratio, and good bandwidth efficiencies in terms of bits/s of information transmitted per hertz of bandwidth, are required due to both technological and regulatory limitations.

The constant envelope requirement is easily met by employing phase and frequency modulation formats, which is one reason for the popularity of phase-shift-keyed and frequency-shift-keyed formats. For simultaneously good bandwidth and communication efficiencies, QPSK, OQPSK, or MSK have been the options most often selected among in the past since all have identical communications efficiencies under ideal AWGN conditions, all have competitive bandwidth efficiencies, and all have roughly the same complexity in terms of implementation. In the preceding section it has been shown that it is possible to obtain arbitrarily small probability of error without the required bandwidth increasing exponentially. The capacity theorem, quoted in Section 4-5, summarizes this fact but does not provide a constructive procedure for finding signaling schemes that approach this ideal.

*See [1, pp. 323ff.] for a proof of the capacity theorem.

Historically, the means used for doing so has been to encode the data source output and use the encoder output to modulate the transmitted carrier. Improved communications efficiency was thereby achieved at the expense of increased implementation complexity due to the data encoder and decoder.

Recently, another approach to this problem, referred to as *combined modulation and encoding*, has been investigated which combines the encoding and modulation functions. Simultaneously good communications and bandwidth efficiencies can be achieved at the expense of increased complexity. It is the purpose of this section to describe briefly one approach to combined modulation and encoding, referred to as *multi-h continuous phase modulation* (CPM), and to characterize its performance.

In general, multi-*h* phase coding includes MSK as a special case. The optimum detection of multi-*h* CPM signals requires use of the Viterbi algorithm. A description of this algorithm is given in Appendix D and its application for the detection of multi-*h* CPM signals is given in Section 4-6.6.

Due to the relatively recent appearance of multi-*h* CPM in the communications literature, and due to the abbreviated treatment given here, the reader is referred to the literature if a more in-depth treatment is desired [see, e.g., [8]].

4-6.1 Description of the Multi-*h* CPM Signal Format

The general form for a multi-*h* CPM signal is

$$s(t; \boldsymbol{\alpha}) = \sqrt{\frac{2E_s}{T_s}} \cos\left[2\pi f_0 t + \varphi(t; \boldsymbol{\alpha}) + \varphi_0\right] \tag{4-138}$$

where E_s and T_s are the symbol energy and duration as before, f_0 is the carrier frequency in hertz, φ_0 is an arbitrary carrier phase, and $\varphi(t; \boldsymbol{\alpha})$ is the information-carrying phase function, which can be expressed as

$$\varphi(t; \boldsymbol{\alpha}) = 2\pi \int_{-\infty}^{t} \sum_{i=-\infty}^{\infty} h_i a_i g(\tau - iT_s) \, d\tau \qquad -\infty < t < \infty \tag{4-139}$$

This term is referred to as the *excess phase* of $s(t; \boldsymbol{\alpha})$. In (4-139), the symbols used are defined as follows:

$\boldsymbol{\alpha} = (\ldots, a_{-2}, a_{-1}, a_0, a_1, a_2, \ldots)$ represents the data sequence. Each member can, in general, assume any one of M levels. That is, $a_i = \pm 1, \pm 3, \ldots, \pm (M - 1)$ for any i.

$\{h_i : i = 1, 2, \ldots, K\}$ is a set of phase modulation indices which is cycled through periodically. $h_{i+K} = h_i$

$g(t)$ is a *frequency pulse-shape function* which is zero for $t < 0$ and $t > LT_s$, where L is an integer, and nonzero otherwise. $L = 1$ yields a *full response* signal, while $L > 1$ yields a *partial response* signal.

It is convenient to define a *phase* pulse-shaping function, $q(t)$, as

$$q(t) = \int_{-\infty}^{t} g(\tau) \, d\tau \qquad -\infty < t < \infty \tag{4-140}$$

with the normalization property that $q(LT_s) = \frac{1}{2}$.

EXAMPLE 4-9

(a) *Continuous-phase FSK* (CPFSK) is a subclass of multi-h CPM with

(1) $h_i = h$, all i

$$(2)\ g(t) = \begin{cases} 0 & t < 0 \\ \dfrac{1}{2}T_s & 0 \le t \le T_s \\ 0 & t > T_s \end{cases} \tag{4-141a}$$

(b) *Fast frequency-shift keying (FFSK)* is a special case of CPFSK with $h = \frac{1}{2}$. FFSK is identical to MSK with differential encoding of the quadrature data streams imposed. ∎

The frequency pulse shape has a considerable effect on the asymptotic behavior of the modulated-signal power spectra. In general, it can be stated that *if the nth order derivative of the phase function is the lowest-order derivative that is not everywhere continuous, the power spectral envelope decays as $f^{-2(n+1)}$*. For example, the phase function of FFSK has no continuous derivatives ($n = 1$), so that its power spectrum decays as f^{-4} or at a rate of 12 dB/octave.

Another popular frequency pulse shape is *raised cosine* (RC), for which

$$g(t) = \begin{cases} \dfrac{1}{2LT_s}\left(1 - \cos\dfrac{2\pi t}{LT_s}\right) & 0 \le t \le LT_s \\ 0 & \text{otherwise} \end{cases} \tag{4-141b}$$

This form of multi-h CPM has continuous first- and second-order derivatives so that $n = 3$ in the spectral property stated above. Thus the power spectral envelope for RC multi-h CPM decays at a rate of 24 dB/octave.

In general, a faster roll-off rate of the sidelobes for a given multi-h CPM format means a wider main lobe for the power spectrum. This will be pointed out explicitly when the calculation of spectra and out-of-band power for multi-h CPM is discussed.

For the remainder of this section, attention is focused on full-response, rectangular frequency-pulse, multi-h CPM. For this special case, we can write the modulated waveform in the $(i - 1)$st signaling interval as

$$s_i(t) = \sqrt{\frac{2E_s}{T_s}}\cos\left\{2\pi\left[f_0 t + \frac{1}{2}a_i h_i\left(\frac{t}{T_s} - (i-1)\right)\right] + \varphi_i\right\}$$

$$(i-1)T_s \le t \le iT_s \tag{4-142}$$

where φ_0 in (4-138) is zero, a rectangular frequency pulse has been assumed, and

$$\varphi_i = \pi \sum_{j=-\infty}^{i-2} a_j h_j \tag{4-143}$$

is the excess phase at a time $t = (i - 1)T_s$ due to previous information digits. Note that during the ith signaling interval, (4-142) represents a tone displaced from the nominal carrier, f_0, by $\pm h_i/2T_s$ corresponding to a binary 1 or 0 for binary data ($M = 2$); if $M > 2$, one of M tones are sent during the ith signaling interval.

Furthermore, if

$$h_i = \frac{L_i}{q} \qquad (4\text{-}144)$$

where L_i, $i = 1, 2, \ldots, K$, and q are integers, all phase values at the transition times $t = iT_s$ are multiples of $2\pi/q$ radians. Thus a set of phase states are defined for each $t = iT_s$. This phase-state information can be used to optimally (i.e., MAP) detect the data provided:

1. The detector is synchronized to the incoming signal. There are three levels of timing: (a) carrier phase, (b) signal interval (baud), and (c) superbaud (i.e., the sequence $h_1, h_2, \ldots, h_K$).
2. The set of possible phase transitions from state to state are known.

Detection of multi-h CPM signals is considered more fully in the next subsection. It is instructive at this point, however, to consider a specific sequence of h values and plot the possible phase trajectories. Such a plot is called a *phase trellis* and is an important tool in the consideration of multi-h CPM properties and receiver performance.

EXAMPLE 4-10

(a) Consider the h-sequence $\{h_1, h_2\} = \{\frac{1}{4}, \frac{2}{4}\}$. In the ith signaling interval, the excess phase changes by $\pm\pi/4$ radians if $h_1 = \frac{1}{4}$ is used and by $\pm\pi/2$ radians if $h_2 = \frac{1}{2}$ is used. Figure 4-21a shows the development of the phase trellis assuming that the data sequence starts at time $t = 0$. The phase trajectory for the data sequence 1001110 is indicated by the heavy line.

(b) Consider the h-sequence $\{\frac{3}{8}, \frac{4}{8}\}$. Now the excess phase of the transmitted signal changes by $\pm3\pi/8$ radians in intervals where h_1 is used and by $\pm\pi/2$ radians if h_2 is used. The development of the phase trellis is shown in Figure 4-21b.

(c) As a final example, consider the sequence of three h-values $\{\frac{3}{8}, \frac{4}{8}, \frac{5}{8}\}$. Now the excess phase changes $\pm3\pi/8$ radians when h_1 is used, $\pm\pi/2$ when h_2 is used, and $\pm5\pi/8$ when h_3 is used. The development of the phase trellis is shown in Figure 4-21c. ∎

The assumption that the h_i are rational [see (4-144)] means that the excess phase at baud-interval sampling times takes on values in a discrete set. The signal phase, as will be discussed later, may be coherently decoded in an optimal fashion by use of the Viterbi algorithm. The free distance, defined by

$$D_{min}^2 = \lim_{n \to \infty} \min_{i,j} \int_0^{nT_s} [s(t; \boldsymbol{\alpha}^{(i)}) - s(t; \boldsymbol{\alpha}^{(j)})]^2 \, dt \qquad (4\text{-}145)$$

where $s(t; \boldsymbol{\alpha}^{(i)})$ and $s(t; \boldsymbol{\alpha}^{(j)})$ are two signals whose phase trajectories split at $t = 0$, is a key performance measure. The larger D_{min}, the better the performance of the optimal detector [9].

The main factor influencing the magnitude of D_{min} is the minimum interval, C, over which the phase trajectories of signals corresponding to differing data sequences (with the same initial digit) remain *unmerged*. For binary data sequences it can be shown that the maximum value for this interval is $C = K$ only if (1) $q \geq 2^K$, and (2) no subset of $\{h_i\}$ have an integer sum. The parameter $C + 1$ is termed the *constraint length* of the code.

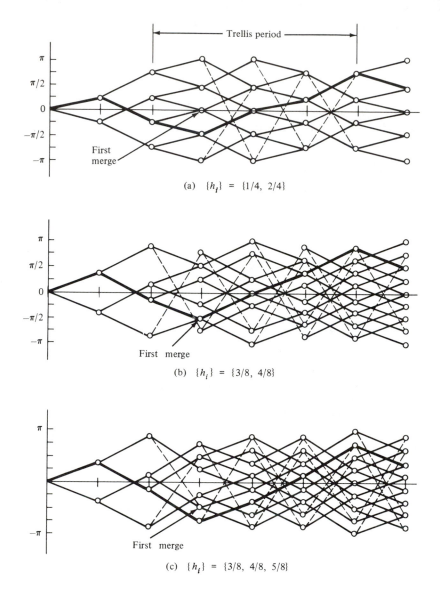

(a) $\{h_i\} = \{1/4,\ 2/4\}$

(b) $\{h_i\} = \{3/8,\ 4/8\}$

(c) $\{h_i\} = \{3/8,\ 4/8,\ 5/8\}$

FIGURE 4-21. Excess phase trellis diagrams for various multi-h CPM signals.

EXAMPLE 4-11

(a) The h-sequence $\{\frac{1}{4},\ \frac{2}{4}\}$ has $C = 2$ since $q = 4 = 2^2$ and $h_1 + h_2$ is not integer-valued.

(b) The phase code $\{\frac{3}{8},\ \frac{4}{8}\}$ has $C = 2$ since $q = 8 \geq 2^K$ and $h_1 + h_2$ is not integer-valued.

(c) The phase code $\{\frac{3}{8},\ \frac{4}{8},\ \frac{5}{8}\}$ has C less than $K = 3$ since $h_1 + h_3$ has an integer sum. From the trellis diagram of Figure 4-19c it is seen that $C = 2 < K = 3$ in this case. ∎

For the case of M-ary data q must equal or exceed M^K and weighted sums of the h_i must not be integer valued. For example, $M = 4$ and $K = 2$ requires that

$q \geq 16$ and $a_1h_1 + a_2h_2$ must not be integer valued for a_1 and a_2 in the set $\{0, 1, 2, 3\}$. The code $M = 3$, $K = 2$, and $\{h_i\} = \{\frac{3}{16}, \frac{4}{16}\}$ satisfies these constraints [10].

The second set of fundamental properties in regard to multi-h excess-phase trellises deal with their periods, or the minimum interval over which they repeat. For modulation-index sequences of the form $\{h_i\} = \{L_i/q\}$, where L_i and q are integers, the following can be shown [11]:

1. If the sum $\Gamma = \Sigma_{i=1}^{K} L_i$ is *even*, the period of the trellis is $T_p = KT_s$; if this sum is *odd*, the trellis period is $T_p = 2KT_s$.
2. The number of phase states is q if Γ is even and $2q$ if Γ is odd.

The *complexity* of the optimum detector is affected by the number of phases and the periodicity of the trellis which, as seen above, is determined by the form of the h_i's.

4-6.2 Performance Bounds [12]

In general, the exact performance characterization of an optimally detected multi-h CPM signal in terms of symbol error probability is not possible. However, it is possible to obtain performance bounds in a relatively straightforward fashion by use of union bounding techniques.

To see how this problem is approached, consider a received CPM signal plus noise of the form

$$y(t) = s(t; \boldsymbol{\alpha}) + n(t) \qquad -\infty < t < \infty$$

where $n(t)$ represents AWGN with double-sided spectral density $N_0/2$. The detector that minimizes erroneous decisions must observe $y(t)$ over all t and choose the infinitely long data sequence $\boldsymbol{\alpha}$ that minimizes the probability of error. Since this is impossible, consider a suboptimum detector that observes N symbol intervals to make a decision about a specific symbol, say a_0. That is, the receiver observes

$$y(t) = s(t; \boldsymbol{\alpha}) + n(t) \qquad 0 \leq t \leq NT_s \qquad (4\text{-}146)$$

and forms the likelihood function

$$\Lambda_N[y(t)] = \exp\left\{-\frac{2}{N_0} \int_0^{NT_s} [y(t) - s(t; \hat{\boldsymbol{\alpha}})]^2 \, dt\right\} \qquad (4\text{-}147)$$

As $N \to \infty$, the performance of this suboptimum detector approaches that of the optimum *maximum-likelihood sequence estimator* (MLSE) (since all possible sequences are assumed to be equally probable, this MLSE detector is, in fact, the MAP detector).

Now the exponent in (4-147) can be expanded and the likelihood ratio simplified to

$$\Lambda_N'[y(t)] = \exp\left\{\frac{4}{N_0} \int_0^{NT_s} y(t)s(t; \boldsymbol{\alpha}) \, dt\right\} \qquad (4\text{-}148)$$

by noting that $\int_0^{NT_s} y^2(t) \, dt$ and $\int_0^{NT_s} s^2(t; \boldsymbol{\alpha}) \, dt$ are independent of $\boldsymbol{\alpha}$ (the latter is just the energy of the constant-envelope signal in the interval $0 \leq t \leq NT_s$). The receiver can equally well choose $\hat{\boldsymbol{\alpha}}$ to maximize the logarithm of (4-148).

Now, let $\hat{\mathbf{a}}_{k,N}$ be defined as the sequence

$$\hat{\mathbf{a}}_{k,N} = \{k, \hat{a}_1, \hat{a}_2, \ldots, \hat{a}_{N-1}\} \tag{4-149}$$

where $k = \pm 1, \pm 3, \ldots, \pm(M - 1)$. There are M^N sequences $\hat{\mathbf{a}}$. Since the detector need only find an estimate $\hat{a}_0$ of a_0, these M^N sequences can be formed into M groups

$$\hat{\mathbf{a}}_{1,N}, \hat{\mathbf{a}}_{3,N}, \ldots, \hat{\mathbf{a}}_{M-1,N}$$

$$\hat{\mathbf{a}}_{-1,N}, \hat{\mathbf{a}}_{-3,N}, \ldots, \hat{\mathbf{a}}_{-(M-1),N}$$

The detector then finds the *group* of sequences that jointly maximize the logarithm of (4-148) and takes $\hat{a}_0$ as the first symbol in this group. For large signal-to-noise ratios and sufficiently long observation intervals the MLSE and this latter detector have equivalent performances.

The probability of an erroneous decision, by the union bound, is overbounded by

$$P(\epsilon) \le \frac{1}{M^{N-1}} \sum_{\substack{\text{all } k \text{ and } l \\ k \ne l}} Q\left[\frac{D(\mathbf{a}_{k,N}, \mathbf{a}_{l,N})}{\sqrt{2N_0}}\right] \tag{4-150}$$

where $Q(x)$ is the Q-function defined by (4-31) and $l, k = \pm 1, \pm 3, \ldots, \pm(M - 1)$, $l \ne k$. $D(\mathbf{a}_{k,N}, \mathbf{a}_{l,N})$ is the Euclidian distance between $s(t; \mathbf{a}_{k,N})$ and $s(t; \mathbf{a}_{l,N})$. Its square, $D^2(\cdot)$, can be written as

$$D^2(\mathbf{a}_{k,N}, \mathbf{a}_{l,N}) = \sum_{i=0}^{N} \int_{iT_s}^{(i+1)T_s} [s(t; \mathbf{a}_{k,N}) - s(t; \mathbf{a}_{l,N})]^2 \, dt$$

If $f_0 T_s \gg 1$, this can be simplified to

$$D^2(\cdot) = 2E_s \left\{ N - \frac{1}{T_s} \int_0^{NT_s} \cos\left[\varphi_e(\mathbf{a}_{k,N} - \mathbf{a}_{l,N})\right] dt \right\} \tag{4-151}$$

where $\varphi_e(\mathbf{a}_{k,N} - \mathbf{a}_{l,N})$ is the excess phase function corresponding to the *difference* sequence

$$\boldsymbol{\gamma}_N \triangleq \mathbf{a}_{k,N} - \mathbf{a}_{l,N} \tag{4-152}$$

That is, it is sufficient to consider a trellis of *phase differences* corresponding to (4-151).

For large signal-to-noise ratios, it is a good approximation to consider the term in (4-150) corresponding to the *closest* phase trajectories. Using this approximation, the error probability is approximated by

$$P(\epsilon) \simeq \Gamma_0 Q\left(\frac{D_{\min,N}}{\sqrt{2N_0}}\right) \tag{4-153}$$

where Γ_0 is a positive constant independent of E_s/N_0 and $D_{\min,N}$ is the minimum of $D(\mathbf{a}_{k,N}, \mathbf{a}_{l,N})$ with respect to a pair of sequences $\mathbf{a}_{k,N}$ and $\mathbf{a}_{l,N}$ with $k \ne l$. $D_{\min,N}$ can also be calculated by using

$$D_{\min,N}^2 = 2E_s \min_{\boldsymbol{\gamma}_N} \left\{ N - \frac{1}{T_s} \int_0^{NT_s} \cos\left[\varphi_e(\boldsymbol{\gamma}_N)\right] dt \right\} \tag{4-154}$$

with the restriction that

$$\gamma_0 \in \{2, 4, 6, \ldots, 2(M - 1)\}$$

$$\gamma_i \in \{0, \pm 2, \pm 4, \ldots, \pm 2(M - 1)\} \qquad i = 1, 2, \ldots, N - 1 \qquad (4\text{-}155)$$

It is convenient to consider a normalized distance $d^2 = D^2/2E_b$, $E_b = E_s/\log_2 M$. For MSK, BPSK, and QPSK, $d^2_{min} = 2$.

EXAMPLE 4-12

Consider the calculation of

$$d^2_{min} = \frac{D^2_{min}}{2E_b} = \min_{\gamma_N} \left\{ 2 - \frac{1}{T_b} \int_0^{T_b} \cos [\varphi_e(\gamma_N)] \, dt \right\} \qquad (4\text{-}156)$$

for CPFSK. The phase trellis of CPFSK is shown in Figure 4-22a. The minimum squared distance, (4-156), can be written in the form

$$d^2_{min} = 2 - \frac{1}{T_b} \int_0^{2T_b} \cos 2\pi h[2q(t) - 2q(t - T_b)] \, dt \qquad (4\text{-}157)$$

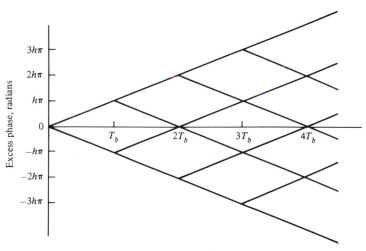

(a)   Phase trellis for CPFSK

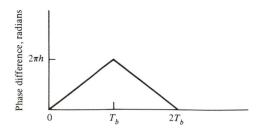

(b)   Phase difference function, $4\pi h [q(t-T_b)]$

FIGURE 4-22.   Excess phase functions for CPFSK.

where

$$q(t) = \begin{cases} 0 & t \leq 0 \\ \dfrac{t}{2T_b} & 0 \leq t \leq 2T_b \\ \dfrac{1}{2} & t > 2T_b \end{cases} \tag{4-158}$$

The phase difference function, $4\pi h[q(t) - q(t - T_b)]$, is shown in Figure 4-22b. From this figure it is seen that $d_{\min}^2$ can be expressed as

$$d_{\min}^2 = 2 - \frac{1}{T_b} \left\{ \int_0^{T_b} \cos \frac{2\pi h t}{T_b} \, dt + \int_{T_b}^{2T_b} \cos \left[2\pi h \left(\frac{2 - t}{T_b} \right) \right] dt \right\}$$

$$= 2[1 - \text{sinc} \, (2h)] \qquad h \leq 0.5 \tag{4-159}$$

where sinc $(u) = (\sin \pi u)/\pi u$. This result is an *upper bound* for $d_{\min}^2$ for all h.

Actual values for $d_{\min}^2$ may be less than $2[1 - \text{sinc} \, (2h)]$ for certain ranges of h because the phase trajectories must be viewed modulo 2π. ■

For further consideration of the calculation of $d_{\min}^2$, the reader is referred to Aulin and Sundberg. [12]

4-6.3 Calculation of Power Spectra for Multi-*h* CPM Signals

Because of the interdependence of the excess phase of a multi-*h* CPM signal between signaling intervals, the calculation of their power spectra is not as simple as for BPSK or quadrature modulation methods such as QPSK, OQPSK, or MSK. The three principal methods for calculating power spectra of multi-*h* CPM signals are (1) simulation, (2) the Markov chain approach, and (3) the direct method. For a summary of these methods and several references, the reader is referred to the special issue of the *IEEE Transactions on Communications* cited earlier [8].*

The direct approach utilizes the definition of the power spectral density of a signal, which is written as

$$G(f) = \lim_{N \to \infty} E \left[\frac{|S_{NT_s}(f)|^2}{NT_s} \right] \tag{4-160}$$

where $S_{NT}(f)$ is the Fourier transform of an NT-second interval of the signal under consideration which, for our purposes here, is given by (4-138). The expectation is taken with respect to the random data sequence and the random initial phase angle φ_0.

Closed-form results for the spectra of *M*-ary CPFSK, *M*-ary pulse-shaped FM, and partial-response FM signals have been obtained using the direct approach and appear in the literature. When this approach is used to obtain the spectrum of randomly modulated digital FM (see [13]) is applied to multi-*h* CPM signals, it is necessary to consider super-intervals of length $T' = PT_s$, where P is the period of the phase trellis (P is either K or $2K$ depending on Γ—see the discussion following

---

*In particular, see the article by Wilson and Gaus [10].

Example 4-11), because the phases are not identically distributed in each signaling interval due to the memory from phase state to phase state. Thus, if a PT_s segment of an M-ary, multi-h CPM signal is considered there are M^P distinct waveforms over which the expectation in (4-160) must be carried out. For the details and equations relating to this method, the reader is referred to Wilson and Gaus. The one-sided lowpass power spectra of the multi-h CPM signals considered in Example 4-10 are shown in Figure 4-23a. Also shown in Figures 4-23b and c for comparison

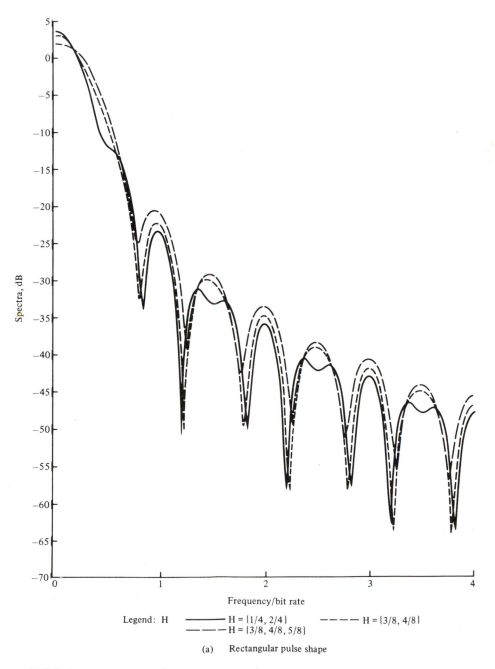

Legend: H ———— H = {1/4, 2/4}   − − − − H = {3/8, 4/8}
              — — — H = {3/8, 4/8, 5/8}

(a)   Rectangular pulse shape

FIGURE 4-23.   Spectra for the multi-h phase codes considered in Example 4-10.

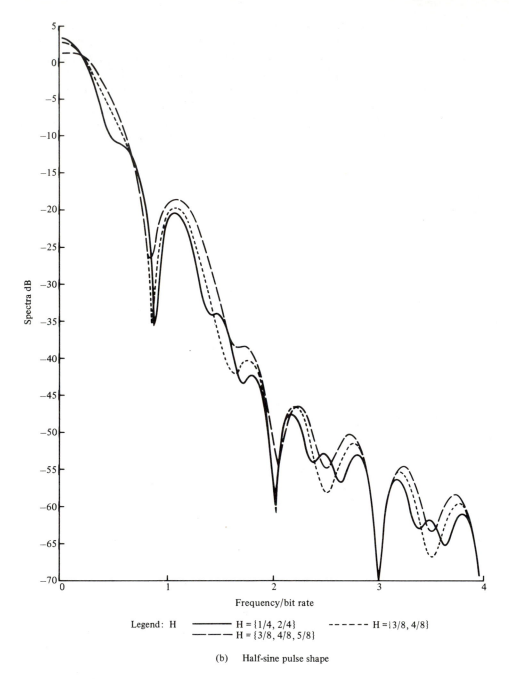

Legend: H ———— H = {1/4, 2/4}   - - - - - H = {3/8, 4/8}
———— H = {3/8, 4/8, 5/8}

(b)   Half-sine pulse shape

FIGURE 4-23.   continued.

are spectra for the same phase codes but with half-sine and raised-cosine pulse shapes. These were obtained through a computer implementation of the direct-method computation. Figure 4-24 shows out-of-band power for these phase codes for rectangular, half-sine, and raised-cosine pulse shapes. Note the faster roll-off of out-of-band power with frequency for the smoother pulse shapes.

Due to the complexity of the direct-method computations, it is beneficial to have

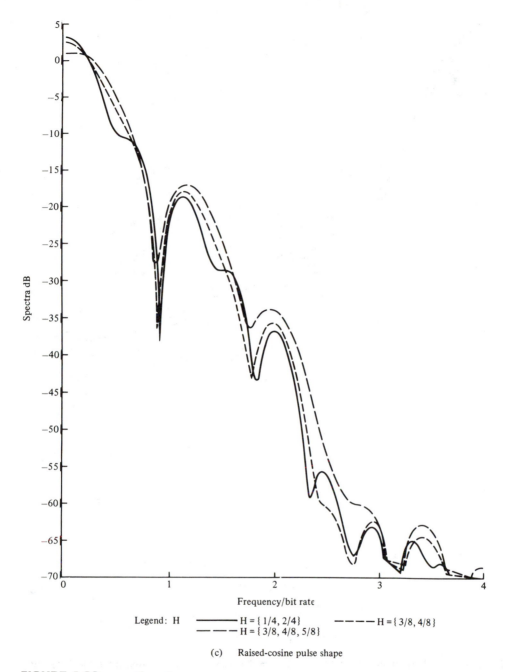

Spectra dB

Frequency/bit rate

Legend: H —————— H = { 1/4, 2/4 }   — — — — H = { 3/8, 4/8 }
— — — H = { 3/8, 4/8, 5/8 }

(c)   Raised-cosine pulse shape

FIGURE 4-23.   continued.

a simpler method of obtaining rough estimates for the power spectra of multi-h CPM signals. Two such approximations are suggested:

1. The multi-h CPM process is visualized as spending a fraction $1/K$ of symbols at each of the K modulation indices $h_1, h_2, \ldots, h_K$. The spectrum is computed as a weighted sum of constant $-h$ spectra as

$$G(f) \simeq \frac{1}{K} \sum_{i=1}^{K} G_{h_i}(f) \qquad (4\text{-}161)$$

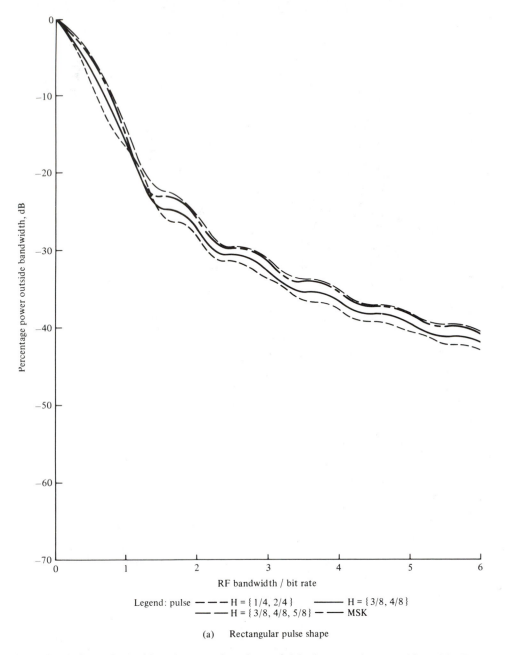

Legend: pulse — — — H = { 1/4, 2/4 }   ———— H = { 3/8, 4/8 }
— — H = { 3/8, 4/8, 5/8 } — — MSK

(a)   Rectangular pulse shape

FIGURE 4-24.   Out-of-band power for the multi-h phase codes considered in Example 4-10.

where $G_{h_i}(f)$ is the power spectrum of a constant-h signal with modulation index h_i. A closed-form expression is available for CPFSK and is given below by (4-163).

2. In the second method, the multi-h CPM signal is approximated by a constant-h signal with a modulation index which is the average over one cycle of the h_i's:

$$\bar{h} = \frac{1}{K} \sum_{i=1}^{K} h_i \qquad (4\text{-}162)$$

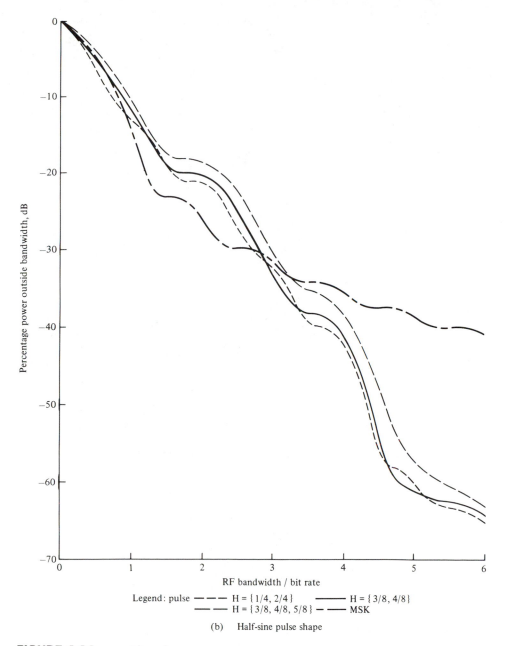

Legend: pulse — — — $H = \{1/4, 2/4\}$ ——— $H = \{3/8, 4/8\}$
— · — $H = \{3/8, 4/8, 5/8\}$ – – – MSK

(b)   Half-sine pulse shape

FIGURE 4-24.   continued.

In using either of these methods, an expression for the power spectrum of a constant-h signal is required. For CPFSK, the result is (see [14])

$$G(f) = G_+(f) + G_-(f) \tag{4-163}$$

where

$$G_\pm(f) = \frac{A^2 \sin^2[\pi(f \pm f_1)T_b] \sin^2[\pi(f \pm f_2)T_b]}{2\pi^2 T_b\{1 - 2\cos[2\pi(f \pm \alpha)T_b]\cos 2\pi\beta T_b + \cos^2(2\pi\beta T_b)\}}$$
$$\left[\frac{1}{f \pm f_1} - \frac{1}{f \pm f_2}\right]^2 \tag{4-164}$$

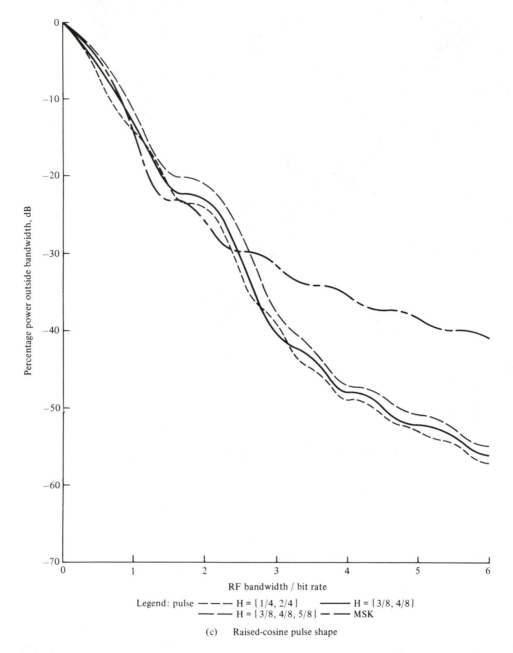

Legend: pulse — — — H = { 1/4, 2/4 }   ——— H = { 3/8, 4/8 }
— — H = { 3/8, 4/8, 5/8 } — · — MSK

(c)   Raised-cosine pulse shape

FIGURE 4-24.   continued.

In (4-164), the following definitions are used:

T_b = bit duration

A = signal amplitude

f_1, f_2 = signaling frequencies in hertz (i.e., for binary CPFSK,

$f_0 \pm h/2T_b$ hertz, where f_0 is the apparent carrier)

$\alpha = \frac{1}{2}(f_2 + f_1)$

$\beta = \frac{1}{2}(f_2 - f_1)$

Examples of the use of (4-161) and (4-162) together with (4-164) to approximate multi-h power spectra are reserved for the problems. In general, method 1 is the most accurate, with spectral computations using it being within ± 2 dB of the true spectrum for "good" multi-h codes. Good codes are defined to be those optimizing the power/bandwidth trade-off, and appear to be codes with the h-values closely grouped (e.g., $\{\frac{3}{16}, \frac{4}{16}\}$). Method 2 indicates the correct general shape of the spectrum, but results in spectral nulls where there should be only local minima and is therefore somewhat misleading. Both methods give fairly accurate results in computing out-of-band power.

4-6.4 Synchronization Considerations for Multi-h CPM Signals*

Three levels of synchronization are required in the demodulation of multi-h CPM signals. These are (1) carrier phase synchronization, (2) symbol or baud timing, and (3) interval synchronization, modulo K (sometimes referred to as *superbaud timing*). One technique for acquiring such timing information is to pass the multi-h CPM signal through a qth power-law device, where q is the denominator of the h-sequence. It can be shown that the spectrum of the output has discrete spectral components at the following frequencies:

$$(1)\ \Gamma\ =\ \sum_{i=1}^{K} L_i\ \text{even:}$$

$$f\ =\ qf_0\ +\ \frac{m-1}{KT_s} \tag{4-165a}$$

$$f\ =\ qf_0\ +\ \frac{m}{KT_s} \tag{4-165b}$$

m an integer

$(2)\ \Gamma$ odd:

$$f\ =\ qf_0\ +\ \frac{2m-1}{KT_s} \tag{4-166a}$$

$$f\ =\ qf_0\ +\ \frac{2m+1}{KT_s} \tag{4-166b}$$

If these adjacent spectral components are extracted from the spectrum by narrow-band filters or phase-locked loops, the resulting frequencies can be mixed to produce a phase coherent signal at the superbaud clock frequency $1/KT_s$. Frequency multiplication by K then produces a clock frequency at the symbol rate $1/T_s$.

Finally, reference frequencies at the signaling frequencies $f_0 \pm h_i/2qT_s$ are required. These are used to produce inphase and quadrature baseband components

*See [9]. The results of a simulation study on synchronization of multi-h is reported in [15].

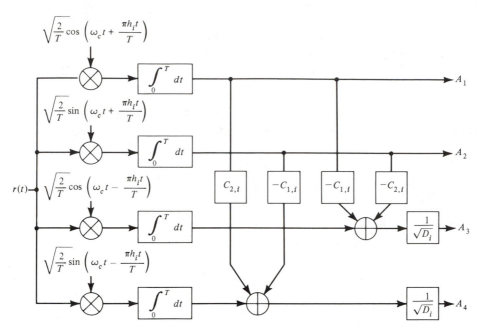

FIGURE 4-25. Resolution of multi-h CPM signals into signal-space coordinates.

from the received multi-h CPM signal as illustrated in Figure 4-25. Depending on whether Γ is even or odd, frequencies are present at the qth power-law device output symmetrically located about the frequency qf_0. For example, if Γ is even, two phase-locked loops may be employed with free-running frequencies $qf_0 + m/KT_s$ and $qf_0 - m/KT_s$ to produce a frequency at $2qf_0$ when these two frequencies are mixed. Frequency division by $2q$ then produces a reference frequency of the form $A \cos (2\pi f_0 t + \pi n/q)$, where the term $\pi n/q$, n integer, represents a $2q$-fold phase ambiguity due to the frequency multiplication by the qth power device which must be resolved.* A similar argument holds for Γ odd. To obtain the reference frequencies $f_0 \pm h_i/2qT_s$, the baud-rate clock can be frequency divided by $2q$ to obtain a reference of frequency $1/2qT_s$. Frequency multiplication by the appropriate h_i and mixing with f_0 then produces the frequencies $f_0 \pm h_i/2qT_s$.

One remaining task must be accomplished before optimal detection of the data sequence can be carried out, and that is to resolve the multi-h signal into a vector space spanned by an orthogonal basis-vector set. Since MSK can be represented in terms of the orthogonal references

$$ R_i = \cos 2\pi \left(f_0 \pm \frac{1}{4T_b} \right) t $$

the resolution process was accomplished easily in this case. The problem with multi-h in this regard is that the signals $\cos [2\pi(f_0 \pm h_i/2T_s)t], (i - 1)T_s \le t \le iT_s,$

---

*The Viterbi algorithm, to be discussed later in regard to detection, has computational symmetry to phase errors that are multiples of π/q, so this ambiguity does not affect the error performance.

are not necessarily orthogonal for arbitrary values of h_i.* Therefore, the reference signals

$$R_1(t) = \varphi_{1i}(t) = \sqrt{\frac{2}{T_s}} \cos\left[2\pi\left(\frac{f_c + h_i}{2T_s}\right)t\right]$$

$$R_2(t) = \varphi_{2i}(t) = \sqrt{\frac{2}{T_s}} \sin\left[2\pi\left(\frac{f_c + h_i}{2T_s}\right)t\right]$$

$$\begin{array}{l} 0 \le t \le T_s, \\ i = 1, 2, \ldots, K \end{array} \quad (4\text{-}167)$$

$$R_3(t) = \sqrt{\frac{2}{T_s}} \cos\left[2\pi\left(f_c - \frac{h_i}{2T_s}\right)t\right]$$

$$R_4(t) = \sqrt{\frac{2}{T_s}} \sin\left[2\pi\left(f_c - \frac{h_i}{2T_s}\right)t\right]$$

are defined, and an orthogonal reference set is obtained through the Gram–Schmidt orthogonalization procedure. Clearly, $R_1(t)$ and $R_2(t)$ are orthogonal and normalized for each i. Therefore, they are set equal to $\varphi_{1i}(t)$ and $\varphi_{2i}(t)$, respectively, which denote the orthonormal references. To find the remaining orthonormal references, form

$$v_3(t) = R_3(t) - (R_3, \varphi_{2i})\varphi_{2i}(t) - (R_3, \varphi_{1i})\varphi_{1i}(t) \qquad (4\text{-}168)$$

where

$$(R_j, \varphi_k) \triangleq \int_{(i-1)T_s}^{iT_s} R_j(t)\varphi_k(t)\, dt \qquad (4\text{-}169)$$

denotes the inner product of $R_j(t)$ and $\varphi_k(t)$. If $f_0 T_s \gg 1$, it can be shown that

$$(R_3, \varphi_{1i}) = \text{sinc } (2h_i) \triangleq C_{1i} \qquad (4\text{-}170\text{a})$$

$$(R_3, \varphi_{2i}) = \frac{1 - \cos 2\pi h_i}{2\pi h_i} \triangleq C_{2i} \qquad (4\text{-}170\text{b})$$

To normalize (4-168), the norm of (4-166) is required. The norm squared is

$$\|v_3\|^2 = (v_3, v_3) = 1 - C_{2i}^2 - C_{1i}^2 \triangleq D_i^2 \qquad (4\text{-}171)$$

Thus the third set (note that there is a different function for each h_i) of orthonormal functions is given by

$$\varphi_{3i}(t) = \frac{R_3(t) - C_{2i}\varphi_{2i}(t) - C_{1i}\varphi_{1i}(t)}{D_i} \qquad i = 1, 2, \ldots, K \quad (4\text{-}172)$$

The last orthonormal function set is found by forming

$$v_4(t) = R_4(t) - (R_4, \varphi_{3i})\varphi_{3i}(t) - (R_4, \varphi_{2i})\varphi_{2i}(t) - (R_4, \varphi_{1i})\varphi_{1i}(t) \qquad (4\text{-}173)$$

The required inner products can be shown to be

$$(R_4, \varphi_{1i}) = -C_{2i} \qquad (4\text{-}174\text{a})$$

$$(R_4, \varphi_{2i}) = C_{1i} \qquad (4\text{-}174\text{b})$$

$$(R_4, \varphi_{3i}) = 0 \qquad (4\text{-}174\text{c})$$

*This discussion can be generalized to arbitrary frequency pulse shapes, but we restrict attention to rectangular pulse shapes.

and the norm of $v_4(t)$ is again D_i as it was for $v_3(t)$. Therefore, the fourth set of orthonormal functions is

$$\varphi_{4i}(t) = \frac{R_4(t) + C_{2i}\varphi_{1i}(t) - C_{1i}\varphi_{2i}(t)}{D_i} \qquad i = 1, 2, \ldots, K \quad (4\text{-}175)$$

Any received multi-h CPM signal can be expressed a linear combination of these basis functions. Figure 4.25 shows the required operations on the correlations with reference signals (4-167) to project a received multi-h signal into the space formed by

$$(\varphi_{1i}, \varphi_{2i}, \varphi_{3i}, \varphi_{4i}) \qquad i = 1, 2, \ldots, K$$

In the next section we consider the use of the Viterbi algorithm for forming a maximum-likelihood estimate of the data sequence that modulates a multi-h signal received in the presence of AWGN.

4-6.5 Application of the Viterbi Algorithm to Detection of Multi-h CPM Signals

It was illustrated in Figure 4-25 how to resolve multi-h CPM signals into a finite-dimensional vector space. During the ith signaling interval and given data symbol α_j was sent, the received signal can be written as

$$s_{ij}(t) = \sqrt{\frac{2E_s}{T_s}} \cos \{2\pi[f_0 t + h_i \alpha_j q(t - iT_s)] + \varphi_{ij}\} \qquad iT_s \leq t \leq (i + 1)T_s$$

$$(4\text{-}176)$$

where E_s, T_s, f_0, h_i, and $q(t)$ are as defined previously, and $\alpha_j = 0, \pm 1, \pm 2, \ldots, \pm(M - 1)$ is the data symbol. The phase angle φ_{ij} represents the initial phase state for the ith signaling interval and takes on values in the set $\{n\pi/q: n = 0, \pm 1, \ldots, \pm q\}$.

We assume that AWGN of two-sided power spectral density $N_0/2$ is also present at the receiver input. When projected into the signal space spanned by the set of orthonormal functions $\varphi_{1i}(t)$, $\varphi_{2i}(t)$, $\varphi_{3i}(t)$, and $\varphi_{4i}(t)$ in the time interval $iT_s \leq t \leq (i + 1)T_s$, its effect can be fully represented, as far as detection is concerned, by the noise vector

$$\mathbf{N} = (N_1, N_2, N_3, N_4)^t \qquad (4\text{-}177)$$

where N_l, $l = 1, 2, 3, 4$, are independent Gaussian random variables with mean zero and variance $N_0/2$. They are also independent from one signaling interval to the next.

Thus, since the signal phase states from one signaling interval to the next can be represented by a trellis diagram, we may apply the VA, as summarized in Appendix D, to detect the data sequence. We make one slight change from that discussion in terms of computing the branch metrics. Due to the Gaussian noise, the branch metrics for transition ξ_{ij} at time i are of the form

$$\lambda(\xi_{ij}) = \|\mathbf{y}_i - \mathbf{s}_{ij}\|^2$$

$$= \|\mathbf{y}_i\|^2 - 2(\mathbf{y}_i, \mathbf{s}_{ij}) + \|\mathbf{s}_{ij}\|^2 \qquad (4\text{-}178)$$

where $\mathbf{y}_i$ is the data vector (signal plus noise) during the ith signaling interval.

Since the signals are of constant amplitude, $\|\mathbf{s}_{ij}\|^2 = $ constant. Also, $\|\mathbf{y}_i\|^2$ is not explicitly dependent on $\mathbf{s}_i$, so that minimization of the path lengths in the VA can be accomplished by computing only the transition correlations,

$$\beta_{ij} \triangleq (\mathbf{y}_i, \mathbf{s}_{ij})$$

$$= \int_{(i-1)T_s}^{iT_s} y(t)s_{ij}(t)\, dt \qquad i = 0, 1, 2, \ldots; \quad j = 1, 2, \ldots, qM \quad (4\text{-}179)$$

and choosing the data sequence that maximizes the path correlations.

If the projection of the received signal plus noise into the signal space during the ith interval is represented by the vector $(A_{1i}, A_{2i}, A_{3i}, A_{4i})^t$ as shown in Figure 4-25, it can be shown that the transition correlations are given by

$$\beta_{ij} = (D_i A_{3i} + C_{2i}A_{2i} + C_{1i}A_{1i})\cos\varphi_{ij}$$

$$- (D_i A_{4i} - C_{2i}A_{1i} + C_{1i}A_{2i})\sin\varphi_{ij} \qquad \text{if } \alpha_j = -1$$

and

$$\beta_{ij} = A_{1i}\cos\varphi_{ij} - A_{2i}\sin\varphi_{ij} \qquad \text{if } \alpha_j = +1 \qquad (4.180)$$

where binary signaling has been assumed. Similar expressions can be derived for M-ary signaling.

To explain further the use of the VA for detection of multi-h signals, we restrict our attention to a $\{\frac{2}{4}, \frac{3}{4}\}$ h-code. When fully developed, the phase trellis for this code is as shown in Figure 4-26, where we note that $q = 4$ but that the total number of phases is $2q$. The transitions for $\alpha_j = +1$ are shown by solid lines and those for $\alpha_j = -1$ are shown by dashed lines.

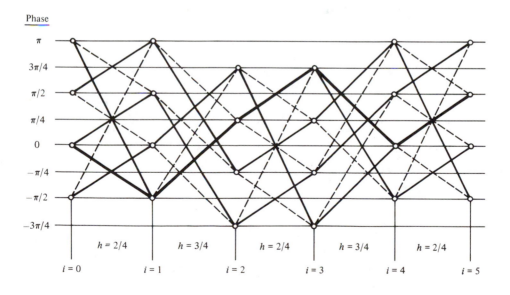

FIGURE 4-26. Fully developed phase trellis for the ($\frac{2}{4}$, $\frac{3}{4}$) multi-h phase code.

Now, at any given discrete-time instant, only q phase values are used, and therefore the implementation of the VA need only allow q storage registers for the correlations of the survivors. For illustrative purposes, we consider detection of the data sequence

$$\boldsymbol{\alpha} = (-1, 1, 1, -1, 1, -1, -1, 1, \ldots)$$

which is shown as the heavy path in Figure 4-26.

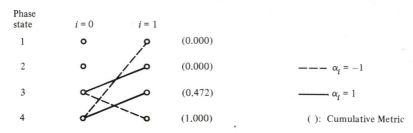

(a) Survivor phase state sequence and maximum cumulative metrics from time $i = 0$ to time $i = 1$

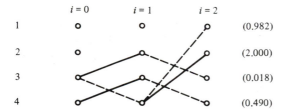

(b) Survivor phase state sequences and maximum cumulative metrics from time $i = 0$ to time $i = 2$

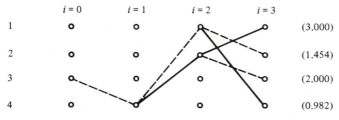

(c) Survivor phase state sequences and maximum cumulative metrics from time $i = 0$ to time $i = 3$

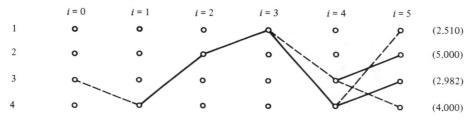

(d) Survivor phase state sequences and maximum cumulative metrics from time $i = 0$ to time $i = 5$

FIGURE 4-27. Development of the survivor phase state sequences for a $(\frac{2}{4}, \frac{3}{4})$ phase code and for the data sequence $(-1, 1, 1, -1, 1, -1, -1, 1, \ldots)$.

TABLE 4-1. Metric Computation for Viterbi Demodulation of a $(\frac{2}{4}, \frac{3}{4})$ Code

Phase State No.	Data Bit = +1 (Up Phase)			Data Bit = −1 (Down Phase)		
	Old Phase State No.	Phase $\times\, q/\pi$	Cumulative Metric	Old Phase State No.	Phase $\times\, q/\pi$	Cumulative Metric
			Time Interval: $i = 0$ to $i = 1$			
1	2	4	−0.472	4	4	0.000
2	3	2	−0.000	1	2	−1.000
3	4	0	0.472	2	0	0.000
4	1	−2	+0.000	3	−2	1.000
			Time Interval: $i = 1$ to $i = 2$			
1	3	3	0.472	4	3	0.982
2	4	1	2.000	1	1	−0.018
3	1	−1	−0.000	2	−1	0.018
4	2	−3	−1.000	3	−3	0.490
			Time Interval: $i = 2$ to $i = 3$			
1	2	3	3.000	4	3	0.490
2	3	1	0.018	1	1	1.454
3	4	−1	−0.510	2	−1	2.000
4	1	−3	0.982	3	−3	−0.454
			Time Interval: $i = 3$ to $i = 4$			
1	2	4	1.472	3	4	1.000
2	3	2	2.018	4	2	0.982
3	4	0	0.964	1	0	4.000
4	1	−2	2.982	2	−2	1.454
			Time Interval: $i = 4$ to $i = 5$			
1	2	4	2.018	4	4	2.510
2	3	2	5.000	1	2	1.472
3	4	0	2.982	2	0	2.490
4	1	−2	0.472	3	−2	4.000

The calculation of the metrics and the maximum correlation paths proceeds as illustrated in Figure 4-27 and Table 4-1. In Table 4-1, the phase states are labeled 1 to 4 starting from the top of the trellis, and the actual phase is π/q times the value shown in the figure. The metrics given are cumulative. No noise is present.

To explain the unfolding of the algorithm in time, let us suppose that it has been in progress for a long time (the metrics have been renormalized, let us suppose). To get to phase state 1 at time $i = 1$, the two possible paths are from old phase states 2 if the data bit is $a = 1$, or from old phase state 4 if the data bit is $a = -1$. Checking the first line of Table 4-1, we see that the maximum cumulative metric is 0.000, corresponding to a -1 data bit as opposed to a cumulative metric of -0.472 for a $+1$ data bit. To get to new phase state 2, the maximum cumulative metric is 0.000, which resulted in a transition from old phase state 3 due to a $+1$ data bit. Similarly, to get to phase state 3, the transition from old phase state 4 is chosen which corresponds to a maximum cumulative metric of 0.472 due to a $+1$ data bit. To get to phase state 4, the transition from old phase state 3 is chosen corresponding to a maximum cumulative metric of 1.000 due to a -1 data bit. The survivor paths and cumulative metrics are shown by the trellis diagram of Figure 4-26.

Going next to the phases at time $i = 2$ from time $i = 1$, we find from Table 4-1 that the maximum cumulative metric to phase state 1 originates from old phase state 4, to phase state 2 from old phase state 4, to phase state 3 from old phase state 2, and to phase state 4 from old phase state 3. The development of the trellis up to time $i = 3$ is shown in Figure 4-27b together with the maximum cumulative metrics for each node shown in parentheses.

In going from $i = 2$ to $i = 3$, the maximum cumulative metric to phase state 1 originates from old phase state 2, to phase state 2 from old phase state 1, to phase state 3 from phase state 2, and to phase state 4 from phase state 1. The development of the trellis up to $i = 3$ is shown in Figure 4-27c. Note that at time $i = 3$, the decision that the data bit at time $i = 1$ was a -1 can be made, since the only path left in the trellis in going from $i = 0$ to $i = 1$ is the transition from phase state 3 to phase state 4, which corresponds to a -1 data bit. Ordinarily, a decision depth would be chosen and decisions for time instants farther back than this decision depth would be forced by using some reasonable criterion such as maximum metric.

The development of the trellis proceeds as discussed above with the trellis at time $i = 5$ appearing as illustrated in Figure 4-27d. Note that a unique decision for the data sequence can be made up to time $i = 3$, which is $-1, 1, 1$. If the criterion of maximum cumulative metric is used in order to make decisions up to $i = 5$, the estimated data sequence is $-1, 1, 1, -1, 1$, which corresponds to the correct sequence. The reason this is possible is because no noise is present.

REFERENCES

[1] J. M. WOZENCRAFT and I. M. JACOBS, *Principles of Communication Engineering* (New York: Wiley, 1965).

[2] M. ABRAMOWITZ and I. STEGUN, eds., *Handbook of Mathematical Functions* (New York: Dover). Originally published in 1964 as NBS Applied Mathematics Series 55.

[3] W. C. LINDSEY and M. K. SIMON, *Telecommunication System Engineering* (Englewood Cliffs, N.J.: Prentice-Hall, 1973).

[4] R. W. LUCKY, J. SALZ, and E. J. WELDON, *Principles of Data Communication* (New York: McGraw-Hill, 1968).

[5] R. E. ZIEMER and W. H. TRANTER, *Principles of Communication* (Boston: Houghton Mifflin, 1976).

[6] J. H. PARK, "On Binary DPSK Reception," *IEEE Trans. Commun.*, Vol. COM-26, pp. 484–486, April 1978.

[7] V. K. PRABHU, "Error Rate Performance for Differential PSK," *IEEE Trans. Commun.*, Vol. COM-30, pp. 2547–2550, December 1982.

[8] Special Issue on Combined Modulation and Encoding, *IEEE Trans. Commun.*, Vol. COM-29, March 1981.

[9] J. B. ANDERSON and D. P. TAYLOR, "A Bandwidth-Efficient Class of Signal Space Codes," *IEEE Trans. Inf. Theory*, Vol. IT-24, pp. 703–712, November 1978.

[10] S. G. WILSON and R. C. GAUS, "Power Spectra of Multi-*h* Phase Codes," *IEEE Trans. Commun.*, Vol. COM-29, pp. 250–256, March 1981.

[11] A. T. LEREIM, "Spectral Properties of Multi-*h* Phase Codes," CRL Intern. Rept. CRL-57, McMaster University, Ontario, July 1978.

[12] T. AULIN and C. E. SUNDBERG, "Continuous Phase Modulation, Part I: Full Response Signaling," *IEEE Trans. Commun.*, Vol. COM-29, pp. 196–209, March 1981.

[13] J. E. MAZO and J. SALZ, "Spectra of Frequency Modulation with Random Waveforms," *Inf. Control*, Vol. 9, pp. 414–422, 1966.

[14] W. R. BENNETT and J. R. DAVEY, *Data Transmission* (New York: McGraw-Hill, 1965).

[15] B. A. MAZUR and D. P. TAYLOR, "Demodulation and Synchronization of Multi-h Phase Codes," *IEEE Trans. Commun.*, Vol. COM-29, pp. 257–266, March 1981.

[16] R. G. GALLAGER, "A Simple Derivation of the Coding Theorem and Some Applications," *IEEE Trans. Inf. Theory*, Vol. IT-11, pp. 3–18, January 1965.

[17] C. E. SHANNON, "Communication in the Presence of Noise," *Proc. IRE*, Vol. 37, pp. 10–21, January 1949.

ADDITIONAL READING

FORNEY, G. D., "The Viterbi Algorithm," *Proc. IEEE*, Vol. 61, pp. 268–278, March 1973.

PROBLEMS

5-1. A message source emitting one of q messages each $T_m = 1/R_m$ seconds, where R_m is the message rate, is used as the input to a modulator which associates a string of J messages with one of M signals (symbols) of duration $T_s = 1/R_s$ seconds, where R_s is the symbol rate. If no gaps are to exist between symbols, fill in the blanks in the following table.

q	R_m (messages/s)	M	R_s (symbols/s)
4		16	1,000
2	100,000	32	
	2,000	4	4,000
3	20,000		5,000

4-2. Consider a hypothesis-testing problem with the following conditional densities given the hypotheses H_1 and H_2, a priori probabilities, and costs:

$$f_z(Z|H_1) = \tfrac{1}{2} e^{-|Z|}$$

$$f_z(Z|H_2) = 10e^{-z/10}u(Z)$$

$$p_1 = \tfrac{1}{3} \qquad p_2 = \tfrac{2}{3} \qquad c_{11} = c_{22} = 0$$

$$c_{12} = c_{21}$$

Obtain the following:
(a) The likelihood ratio test based on a single sample. Simplify as much as possible.
(b) P_M and P_F.
(c) The average cost per decision. What else can the average cost be viewed as in this particular instance?

4-3. Given the signals

$$u_1(t) = e^{-t}u(t)$$

$$u_2(t) = e^{-2t}u(t)$$

$$u_3(t) = e^{-3t}u(t)$$

and the definition of scalar product (4-12) with $t_0 = 0$ and $T_s = \infty$, verify the properties of the scalar product as given below (4-12) for these signals.

4-4. (a) Using the Gram–Schmidt procedure, obtain an orthonormal basis set corresponding to the signals given in Problem 4-3.

(b) See if you can generalize this result to N such signals; that is,

$$u_n(t) = e^{-nt}u(t) \qquad n = 1, 2, \ldots$$

4-5. Verify Schwarz's inequality for $u_1(t)$ and $u_2(t)$ given in Problem 4-3.

4-6. Prove that (4-27) indeed provides the vector **a** which minimizes average energy.

4-7. Do a more rigious derivation of (4-55) using combinatorial principles.

4-8. For a waveform set of (4-58), obtain an expression for the peak-to-average power ratio and plot as a function of M.

4-9. Provide a derivation of (4-65).

4-10. Justify (4-66).

4-11. (a) Derive the relationship between a and E_s for 16-QASK given by (4-69).

(b) Generalize (4-69) to 4^n-QASK, where $n = 1, 2, 3, \ldots$.

4-12. Plot the probability of bit error for 16-QASK versus E_b/N_0. Compare with 16-ary PSK and 16-ary orthogonal signaling using bounds.

4-13. (a) Derive an expression for the symbol error probability for 64-QASK.

(b) Plot the probability of *bit error* versus E_b/N_0 for 64-QASK. Compare with 64-ary PSK and 64-ary orthogonal signaling.

4-14. Using Figure 3-15 and arguments like that given in Section 4-3.5, compare the bandwidth efficiencies in terms of bits/s/Hz for BPSK, QPSK, 8-PSK, 16-PSK, 32-PSK, 16-QASK, and 64-QASK using:

(a) A 90% energy containment bandwidth.

(b) A 99% energy containment bandwidth.

4-15. Referring to the NFSK analysis given in Section 4-4.1, generalize this to the case of a Rayleigh fading channel where

$$z(t) = \sqrt{\frac{2E_s}{T_s}}\, G \cos(\omega_i t + \alpha)$$

where G is a Rayleigh random variable with density function

$$f_G(x) = \frac{x}{\sigma^2} \exp\left(\frac{-x^2}{2\sigma^2}\right) \qquad x \geq 0$$

Note that $G \cos \alpha$ and $G \sin \alpha$ are independent, Gaussian random variables.

4-16. Obtain the closed-form result for $M = 2$ for (4-108).

4-17. Verify the phase trellis plots of Figure 4-21.

4-18. Do the following h-codes achieve the maximum possible value for constraint length?

(a) $\{\frac{3}{16}, \frac{4}{16}, \frac{5}{16}\}$

(b) $\{\frac{7}{16}, \frac{8}{16}\}$

(c) $\{\frac{8}{16}, \frac{9}{16}\}$

(d) $\{\frac{7}{16}, \frac{9}{16}\}$

4-19. What are the *periods* of the multi h-codes given in Problem 4-18?

4-20. Using the idea of an average modulation index for multi-h, as expressed by (4-162), derive and calculate the approximate spectra for the multi-h phase codes $\{\frac{1}{4}, \frac{2}{4}\}$, $\{\frac{3}{8}, \frac{4}{8}, \frac{5}{8}\}$, and $\{\frac{3}{8}, \frac{4}{8}, \frac{5}{8}\}$ for $fT_b = 0.5, 1, 2$, and 3. Compare with the exact results of Figure 4-23.

4-21. Go through the steps to verify (4-170) and (4-173).

4-22. Discuss how the block diagram of Figure 4-25 could serve as the basis of a computer simulation of multi-h phase modulation. Instead of adding channel noise, find the mean and variances of Gaussian random variables to be added to A_1, A_2, A_3, and A_4 to simulate the noise.

4-23. (a) Compare (4-133) and (4-137) by plotting both versus E_N/N_0 on the same graph. Recall that (4-137) gives an absolute bound for R_N, whereas (4-133) does not.

(b) Using N-dimensional signal sets constrained to have finite energy, a somewhat tighter bound parameter than (4-133) can be obtained (see [16] and [17]), which is

$$R_0^* = \frac{\log_2 e}{2}\left[1 + \frac{E_n}{N_0} - \sqrt{1 + \left(\frac{E_N}{N_0}\right)^2}\right] + \frac{1}{2}\log_2\left[\frac{1}{2}\left(1 + \sqrt{1 + \left(\frac{E_N}{N_0}\right)^2}\right)\right]$$

Compare this bound parameter with (4-133) and (4-137) as a function of E_N/N_0. Note that $C_n/2 < R_0^* < C_N$.

CHAPTER 5

Generation of Coherent References

5-1

INTRODUCTION

The implementation and performance of several types of digital data modems have just been considered in Chapters 3 and 4. Generation of stable carrier and clock frequencies is necessary at both the modulator and demodulator of a digital data communication system. In this chapter, means for generation of such reference signals are considered. Because the generation of reference signals involves the use of primary and secondary reference oscillators, the properties and statistical description of phase noise on signals produced by such reference sources will be summarized first. Phase-locked-loop structures and properties, as well as their application in establishing reference signals for carriers and clocks, are surveyed next. Finally, implementation of frequency synthesizers by the direct and phase-locked methods is described. Because of the brief treatment of these subjects, several references are recommended for further details [1–7].

DESCRIPTION OF PHASE NOISE AND ITS PROPERTIES

5-2.1 General Considerations

In order to design and characterize the behavior of systems for coherent reference generation, at least a casual understanding of the statistics of oscillator phase noise is necessary. To this end, a reference signal will be modeled as

$$r(t) = A[1 + a(t)] \cos\left[\omega_0 t + \varphi(t) + \frac{\alpha t^2}{2}\right] \qquad (5\text{-}1)$$

where $\omega_0 = 2\pi f_0$ is the nominal reference frequency of interest, $\varphi(t)$ is random phase jitter, and the term $\frac{1}{2}\alpha t^2$ is phase accumulation due to long-term frequency drift of the oscillator. Recalling that instantaneous frequency is the derivative with respect to time of instantaneous phase it is seen that α is the frequency drift in rad/s. In this introductory consideration, it is assumed that the unwanted amplitude noise, $a(t)$, is negligible, although it is important to note that the amplitude fluctuations, if sufficiently large, may have a nonneglible effect on the system operation.

The phase jitter, $\varphi(t)$, in (5-1) might occur at the output of a primary reference source, as in a transmitter synthesizer, or it might represent the cumulative effects of several oscillators in a large system such as a satellite communications link. The phase jitter may include both the effects of long-term phenomena such as component aging, temperature fluctuations, and power supply variations, as well as the effects of random noise. It is customary to focus attention on the random noise effects in $\varphi(t)$ and assume that any long-term effects, which result in nonstationary behavior of $\varphi(t)$, can be adequately modeled in the frequency drift term, $\frac{1}{2}\alpha t^2$.

5-2.2 Phase and Frequency Noise Power Spectra

Since $\varphi(t)$ is random, it is appropriate to discuss its properties in terms of statistical averages. If one is concerned with gross descriptions of phase noise effects, the standard deviation or variance of $\varphi(t)$ is perhaps adequate. Assuming that $\varphi(t)$ has zero mean, its variance can be obtained by integrating the phase-jitter power spectral density, $G_\varphi(f)$, with units of rad²/Hz, over all frequency. Sometimes the power spectral density of the instantaneous frequency fluctuations, $S_{\Delta f}(f)$, is used, which is related to the phase-jitter power spectral density by

$$S_\omega(f) = (2\pi f)^2 G_\varphi(f) \qquad (\text{rad/s})^2/\text{Hz} \qquad (5\text{-}2)$$

which follows by recalling that the Fourier transform of $dx(t)/dt$ is $(j2\pi f)X(f)$. The spectra $G_\varphi(f)$ and $S_\omega(f)$ are quite often plotted as single-sided (i.e., mean-square phase fluctuation is obtained by integrating over positive frequencies only) and single-sideband relative to the oscillator nominal frequency.

Typical (single-sideband) straight-line asymptote plots for $G_\varphi(f)$ and $S_\omega(f)$ are shown in Figure 5-1, where it is noted that the spectra consist of several regions. These regions are described as follows (f is frequency relative to the nominal carrier frequency of the oscillator):

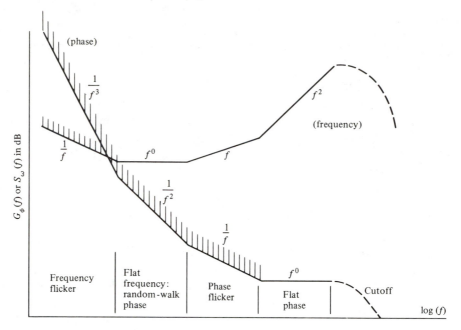

FIGURE 5-1.   Oscillator noise spectra; asymptotic approximations (From Ref. 1.)

1. Frequency flicker:

$$G_{ff}(f) = \frac{k_1}{f^3} \qquad \text{rad}^2/\text{Hz} \tag{5-3}$$

or

$$S_{ff}(f) = \frac{K_1}{f} \qquad (\text{rad/s})^2/\text{Hz} \tag{5-4}$$

2. Random-phase walk or white frequency noise:

$$G_{wf}(f) = \frac{k_2}{f^2} \qquad \text{rad}^2/\text{Hz} \tag{5-5}$$

or

$$S_{wf}(f) = K_2 \qquad (\text{rad/s})^2/\text{Hz} \tag{5-6}$$

3. Phase flicker:

$$G_{pf}(f) = \frac{k_3}{f} \qquad \text{rad}^2/\text{Hz} \tag{5-7}$$

or

$$S_{pf}(f) = K_3 f \qquad (\text{rad/s})^2/\text{Hz} \tag{5-8}$$

4. Flat phase:

$$G_{fp}(f) = k_4 \qquad \text{rad}^2/\text{Hz} \tag{5-9}$$

or

$$S_{fp}(f) = K_4 f^2 \qquad (\text{rad/s})^2/\text{Hz} \tag{5-10}$$

5. Cutoff:

$$G_c(f) = S_c(f) \simeq 0 \tag{5-11}$$

In all cases above, $K_i = 4\pi^2 k_i$, $i = 1, 2, 3, 4$.

From a physical standpoint, it is possible to explain the white frequency-noise and flat phase-noise portions of these spectra as being caused by the random motion of charge carriers. It is more difficult to explain the frequency-flicker and phase-flicker regions. Flicker noise is observed as a low-frequency disturbance in nearly all active electronic devices, but an acceptable explanation of how this low-frequency noise energy is translated into radio-frequency phase or frequency fluctuations is difficult to come up with. The existence of flicker-type phase-fluctuation spectra must simply be expected on the basis of experimental measurements, several examples of which are provided in Figure 5-2.

Mathematically, flicker-type spectra are difficult to work with. For example, integration of the phase fluctuation spectrum over the range of frequencies $(0, \infty)$ gives the variance of the phase fluctuations, which for spectra of the form k/f^n results in an undefined result if $n \geq 1$. Empirically, a lower cutoff frequency, f_x, can be determined such that

$$\hat{\sigma}_\varphi^2 = \int_{f_x}^{\infty} G_\varphi(f') \, df' << 1 \quad \text{rad}^2 \tag{5-12}$$

Fluctuations with frequencies lower than f_x can be viewed as contributing to broadening of the carrier frequency and those above f_x can be viewed as modulation on this broadened carrier.

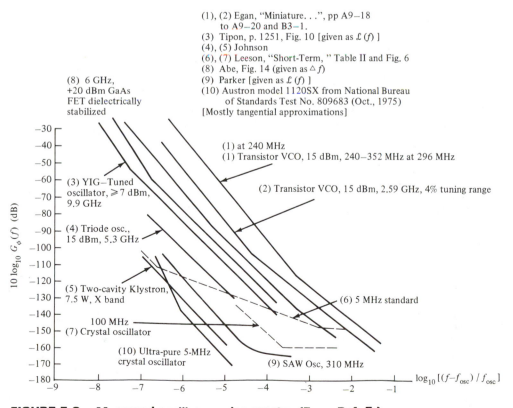

(1), (2) Egan, "Miniature. . .", pp A9–18 to A9–20 and B3–1.
(3) Tipon, p. 1251, Fig. 10 [given as $\mathcal{L}(f)$]
(4), (5) Johnson
(6), (7) Leeson, "Short-Term," Table II and Fig. 6
(8) Abe, Fig. 14 (given as $\triangle f$)
(9) Parker [given as $\mathcal{L}(f)$]
(10) Austron model 1120SX from National Bureau of Standards Test No. 809683 (Oct., 1975)
[Mostly tangential approximations]

(8) 6 GHz, +20 dBm GaAs FET dielectrically stabilized

(1) at 240 MHz
(1) Transistor VCO, 15 dBm, 240–352 MHz at 296 MHz

(3) YIG–Tuned oscillator, $\geqslant 7$ dBm, 9.9 GHz

(2) Transistor VCO, 15 dBm, 2.59 GHz, 4% tuning range

(4) Triode osc., 15 dBm, 5.3 GHz

(5) Two-cavity Klystron, 7.5 W, X band

(6) 5 MHz standard

100 MHz
(7) Crystal oscillator

(10) Ultra-pure 5-MHz crystal oscillator

(9) SAW Osc, 310 MHz

FIGURE 5-2. Measured oscillator noise spectra (From Ref. 5.).

EXAMPLE 5-1

Consider an oscillator with the phase fluctuation spectral density shown in Figure 5-3. Find k_1, k_2, k_3, and k_4 for the asymptotic approximations given by (5-3), (5-5), (5-7), and (5-9). Find f_x as defined by (5-12) such that $\hat{\sigma}_\varphi^2 = 0.1$ rad^2. Replot Figure 5-3 with the abscissa $\log_{10}[(f - f_0)/f_0]$ for $f_0 = 100$ MHz, where f_0 is the oscillator nominal frequency, and compare with Figure 5-3.

Solution: Using the relations

$$-3 \text{ dB/octave} = -10 \text{ dB/decade} \rightarrow 10 \log_{10}f^{-1}$$

$$-6 \text{ dB/octave} = -20 \text{ dB/decade} \rightarrow 10 \log_{10}f^{-2}$$

$$-9 \text{ dB/octave} = -30 \text{ dB/decade} \rightarrow 10 \log_1 f^{-3}$$

and a conveniently chosen set of frequencies, one obtains

$$k_1 = 10^{-3} \qquad (\text{rad-Hz})^2$$

$$k_2 = 10^{-5} \qquad \text{rad}^2\text{-Hz}$$

$$k_3 = 10^{-10} \qquad \text{rad}^2$$

$$k_4 = 10^{-16} \qquad \text{rad}^2/\text{Hz}$$

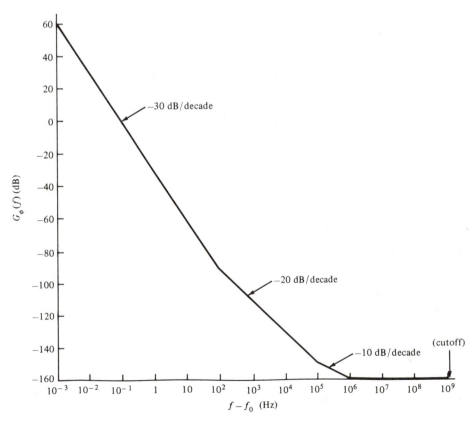

FIGURE 5-3. Oscillator phase-noise spectral density For Example 5-1.

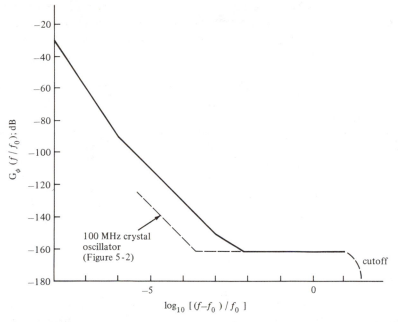

FIGURE 5-4. Oscillator phase-noise spectrum of Example 5-1 as a function of normalized frequency compared with spectrum of Figure 5-2.

Integration of $G_\varphi(f)$ from f_x to the cutoff frequency of 10 GHz* results in

$$\frac{5 \times 10^{-3}}{f_x^2} + \frac{10^{-5}}{f_x} + 10^{-10} \ln f_x + 10^{-16} f_x \simeq 0.1$$

where all upper limit evaluations are neglected relative to 0.1. Since the first term on the left-hand side dominates, it follows that

$$f_x \simeq 0.071 \text{ Hz}$$

The curve of Figure 5-3, with an abscissa normalized by the nominal oscillator frequency of 100 MHz, is replotted in Figure 5-4 and compared with the phase-noise spectrum of the 100-MHz crystal oscillator of Figure 5-2. It is seen that the hypothetical oscillator of this example has more phase noise than the 100-MHz crystal oscillator. ∎

5-2.3 Allan Variance

Although the phase- and frequency-noise power spectra will be used primarily in the discussion to follow, it is useful to point out the relationship of the phase-noise spectrum to frequency counter measurements of oscillator frequency variations due to noise. Define the fractional frequency stability of an oscillator as

$$\delta_j \triangleq \frac{\varphi[(j+1)T] - \varphi(jT)}{T\omega_0} \tag{5-13}$$

*Clearly, a cutoff frequency of 10 GHz has little physical meaning compared with a center frequency of 100 MHz. It is simply a frequency beyond which the phase-noise spectrum is negligible.

where $\varphi(t_s)$ is the phase at sampling instant t_s, T is the measurement interval, and ω_0 is the nominal oscillator frequency in rad/s. A conventional frequency counter measures $\{\varphi[(j + 1)T] - \varphi(jT)\}/2\pi$ in a T-second period. The Allan variance [7, p. 339] is defined as

$$\overline{\sigma^2(N, T)} \triangleq [1/(N - 1)] \sum_{n=1}^{N} (\delta_n - \delta_{sm})^2 \qquad (5\text{-}14)$$

where the overbar denotes the statistical average and δ_{sm} is the sample mean of δ_j, which is

$$\delta_{sm} \triangleq \frac{1}{N} \sum_{j=1}^{N} \delta_j \qquad (5\text{-}15)$$

It can be shown [5, 7] that the Allan variance is related to the frequency-noise power spectrum by

$$\overline{\sigma^2(N, T)} = \frac{N}{N - 1} \int_0^\infty \frac{S_{\triangle f}(f)}{(2\pi)^2} \operatorname{sinc}^2 (fT) \left(1 - \frac{\sin^2 \pi N f T}{N^2 \sin^2 \pi f T}\right) df \qquad (5\text{-}16)$$

where sinc $(u) = \sin(\pi u)/\pi u$. Given the phase- or frequency-noise power spectrum of an oscillator, (5-16) can be used to obtain the Allan variance. However, the reverse is not true. For further discussion of the Allan variance, see [5, Chap. 10].

5-2.4 Effect of Frequency Multipliers and Dividers on Phase-Noise Spectra

A frequency multiplier is a device whose output is a signal with instantaneous frequency that is an integer multiple of the instantaneous frequency of the signal at its input. Thus, for an input of the form

$$x(t) = A \cos [\omega_c t + \theta(t)] \qquad (5\text{-}17)$$

the output of a frequency multiplier is

$$y(t) = B \cos [n\omega_c t + n\theta(t)] \qquad (5\text{-}18)$$

where n is the multiplication factor and ω_c is the center frequency of the signal spectrum. A similar relation can be written for a frequency divider, except that the instantaneous frequency is divided by an integer, n.

The application of this result to phase-noise power spectra means that since the phase-noise spectrum is proportional to the *square* of the instantaneous phase, the phase-noise spectrum at the output of a frequency multiplier of multiplication factor n is n^2 *times the input phase-noise spectrum*. A similar statement holds for frequency-noise spectra since instantaneous frequency is proportional to the derivative of instantaneous phase. Similarly, the phase- or frequency-noise spectra at the output of a frequency divider is $1/n^2$ times the input phase- or frequency-noise spectra, where n is the divider ratio.

PHASE-LOCKED LOOP MODELS AND CHARACTERISTICS OF OPERATION

Since phase-locked loops play important roles in establishing coherent references in digital communication systems, their properties will be described in this section.

5-3.1 Synchronized Mode: Linear Operation

The block diagram for a phase-locked loop of arbitrary order is shown in Figure 5-5. It consists of a phase detector whose output is a monotonic function of the phase difference between the input signal and the reference input, a loop filter with transfer function $F(s)$, and a voltage-controlled oscillator which produces the reference signal, $e_o(t)$. The input signal is represented as

$$x_c(t) = A_c \cos (2\pi f_o t + \varphi) \tag{5-19}$$

and the voltage-controlled oscillator (VCO) output, or reference signal, is represented as*

$$e_o(t) = -A_v \sin (2\pi f_o t + \theta) \tag{5-20}$$

The frequency deviation of the VCO output is proportional to its input; that is,

$$\frac{d\theta}{dt} = K_v e_v(t) \tag{5-21}$$

where K_v is the VCO constant in rad/s/V.

If the phase detector is assumed to be an ideal multiplier followed by a lowpass filter whose sole effect is to remove the double-frequency component at the multiplier output, the phase detector output is

$$e_d(\psi) = K_d \sin \psi \tag{5-22}$$

where

$$\psi = \varphi - \theta \tag{5-23}$$

is the phase error and K_d is a proportionality constant. For the sinusoidal phase detector, $K_d = \frac{1}{2} A_c A_v K_m$, where K_m is the multiplier constant. The phase detector characteristic given by (5-22) is illustrated in Figure 5-6 together with several other possible phase detector characteristics. If the phase error is small, the sinusoidal phase detector characteristic shown in Figure 5-6a is linear to a good approximation. Therefore, if $|\psi| << 1$, all the phase detectors with characteristics shown in Figure 5-6 have approximately the same effect on loop operation. If the phase error is large, all impose nonlinear effects on system operation. Such nonlinear behavior will be discussed in more detail later, but for now loop operation is assumed to be entirely within one of the linear regions with positive slope. These can be shown to be stable lock-point regions for a first-order loop which has $F(s) = 1$ by employing phase-plane arguments [7].

---

*It is initially assumed that the loop is operating in the frequency synchronized mode; that is, only the phase of the VCO must be synchronized with the input signal phase.

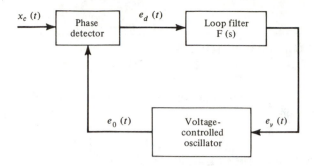

FIGURE 5-5.   General-order phase-locked-loop block diagram.

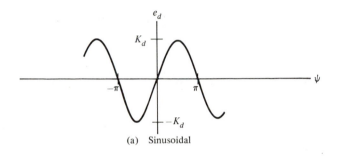

(a)   Sinusoidal

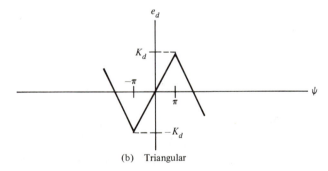

(b)   Triangular

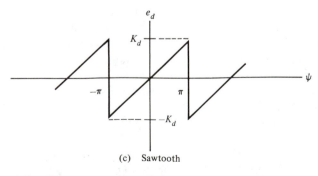

(c)   Sawtooth

FIGURE 5-6.   Phase detector characteristics.

With the definitions above and assuming operation in the linear mode, the equations describing loop operation will now be obtained. It is convenient to do so using Laplace transform notation and by considering the signal phase as the signal of interest. A loop model using Laplace transformed quantities and assuming linear operation is shown in Figure 5-7. The Laplace-transformed loop equations are

$$E_d(s) = K_d[\Phi(s) - \Theta(s)] = K_d\Psi(s) \tag{5-24}$$

$$E_v(s) = F(s)E_d(s) \tag{5-25}$$

$$\theta(s) = \frac{K_v E_v(s)}{s} \tag{5-26}$$

The following ratios of Laplace-transformed quantities, or transfer functions, relating to loop operation in the synchronized mode may be solved for, and are frequently used:

1. The closed-loop transfer function:

$$H(s) \triangleq \frac{\Phi(s)}{\Theta(s)} = \frac{K_v K_d F(s)}{s + K_v K_d F(s)} \tag{5-27}$$

2. The phase error transfer function:

$$H_e(s) \triangleq \frac{\Phi(s) - \Theta(s)}{\Phi(s)} = \frac{\Psi(s)}{\Phi(s)}$$

$$= 1 - \frac{\Theta(s)}{\Phi(s)}$$

$$= 1 - H(s) = \frac{s}{s + K_v K_d F(s)} \tag{5-28}$$

3. The VCO control-voltage/input-phase transfer function:

$$H_v(s) = \frac{V_c(s)}{\Phi(s)}$$

$$= \frac{sH(s)}{K_v}$$

$$= \frac{K_d s F(s)}{s + K_v K_d F(s)} \tag{5-29}$$

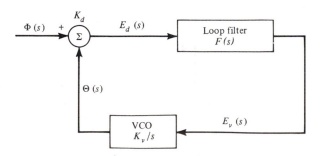

FIGURE 5-7. Laplace-transformed phase-locked-loop model for operation in linear mode.

It is convenient to write the closed-loop transfer function in terms of the open-loop transfer function, which is defined as

$$G(s) = K_v K_d F(s)/s \qquad (5\text{-}30)$$

Substituting (5-30) into (5-27) results in

$$H(s) = \frac{G(s)}{1 + G(s)} \qquad (5\text{-}31)$$

The open-loop dc gain is defined as

$$K = K_v K_d F(0) \qquad (5\text{-}32)$$

which is a generalization of the total effective loop gain of the first-order loop.

By appropriate choice of $F(s)$, any order closed-loop transfer function can be obtained. Consideration here will be restricted to first- and second-order loops. Various types of loop filters for second-order loops are employed. Circuit diagrams for two of these types are illustrated in Figure 5-8. For second-order loops, it is customary to express the denominator of the closed-loop transfer function in terms of the damping factor, ζ, and natural frequency, ω_n, as

$$D(s) = s^2 + 2\zeta\omega_n s + \omega_n^2 \qquad (5\text{-}33)$$

With these definitions, and the definition of noise equivalent bandwidth for a filter given by (A-114), the closed-loop transfer functions and noise equivalent bandwidths for the first- and second-order loops given in Table 5-1 result [1, 7].

The closed-loop frequency response for a second-order loop with active filter is shown in Figure 5-9 for several values of damping factor, ζ. The frequency response corresponding to its phase-error transfer function is shown in Figure 5-10 for $\zeta = 0.707$. In terms of its effect on input phase, Figure 5-9 shows that a phase-locked loop performs a lowpass filtering operation. In the application to FM demodulation, the loop bandwidth is made large in order that $\theta(t)$ closely tracks $\varphi(t)$, thus making the VCO input proportional (or nearly so) to the modulating signal. This follows from the defining equation for the VCO (5-21). Conversely, when establishing a coherent reference for digital data demodulation, it is desirable to have the loop bandwidth narrow to minimize the effects of input noise to the loop on $\varphi(t)$ in terms of phase jitter. The limitation on how narrow the loop bandwidth can be made is determined by the amount of phase-jitter noise on the carrier to which the loop is being locked. These points are examined further in the following subsections.

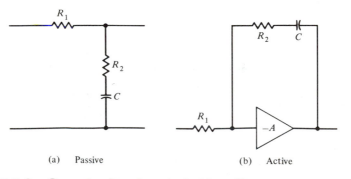

(a)   Passive   (b)   Active

FIGURE 5-8.   Second-order phase-locked-loop filters.

TABLE 5-1. Transfer Functions and Parameters for First- and Second-Order Phase-Locked Loops

Loop Filter, $F(s)$	Natural Frequency,[a] ω_n (rad/s)	Damping Factor	Closed-Loop Transfer Function, $H(s)$	Error Transfer Function, $1 - H(s)$	Single-sided Bandwidth (Hz)
1 (first order)	K	—	$\dfrac{K}{s + K}$	$\dfrac{s}{s + K}$	$\dfrac{K}{4}$
$\dfrac{s\tau_2 + 1}{s\tau_1 + 1}$ (passive, second order)	$\sqrt{\dfrac{K}{\tau_1}}$	$\dfrac{\omega_n}{2}\left(\tau_2 + K^{-1}\right)$	$\dfrac{(2\zeta\omega_n - \omega_n^2/K)s + \omega_n^2}{D(s)}$	$\dfrac{s^2 + \omega_n^2 s/K}{D(s)}$	$\dfrac{K\tau_2(1/\tau_2^2 + K/\tau_1)}{4(K + 1/\tau_2)}$
$\dfrac{s\tau_2 + 1}{s\tau_1}$ (active, second order)	$\sqrt{\dfrac{K}{\tau_1}}$	$\dfrac{\tau_2\omega_n}{2}$	$\dfrac{2\zeta\omega_n s + \omega_n^2}{D(s)}$	$\dfrac{s^2}{D(s)}$	$\dfrac{1}{2}\,\omega_n\left(\zeta + \dfrac{1}{4\zeta}\right)$ (note b)
$\dfrac{1}{s\tau + 1}$ (lag, second order)	$\sqrt{\dfrac{K}{\tau}}$	$\dfrac{1}{2\sqrt{K\tau}}$	$\dfrac{\omega_n^2}{D(s)}$	$\dfrac{s^2 + 2\zeta\omega_n}{D(s)}$	$\dfrac{K}{4}$

[a] $K = K_v K_d$

[b] For a second-order loop with $\zeta = 0.5$.. $B_L = 0.5\omega_n$; with $\zeta = 1/\sqrt{2}$, $B_L = 0.53\omega_n$. B_L is the single-sided noise bandwidth in hertz, and the dimensions of ω_n are rad/s.

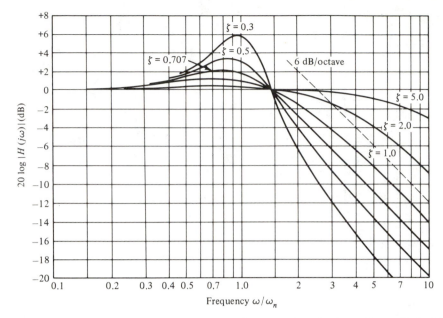

FIGURE 5-9.   Frequency response of a high-gain second-order loop (From Ref. 1.)

5-3.2   Effects of Noise

The input to the linear phase-locked loop of Figure 5-7 will now be assumed to be signal plus stationary, bandlimited, Gaussian noise:

$$x_r(t) = x_c(t) + n(t) \tag{5-34}$$

where $x_c(t)$ is given by (5-19) and $n(t)$ is represented in phase/quadrature form as

$$n(t) = n_c(t) \cos 2\pi f_o t + n_s(t) \sin 2\pi f_o t \tag{5-35}$$

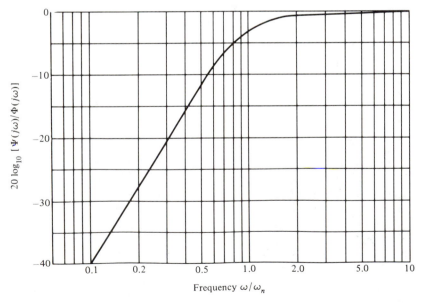

FIGURE 5-10.   Error response of high-gain loop, $\zeta = 0.707$. (From Ref. 1.)

Using the VCO output signal representation (5-20) and the loop parameters defined previously, it can be shown that the output of an ideal multiplier-type phase detector (sinusoidal characteristic as shown in Figure 5-6a) is [1]

$$e_d(t) = K_d[\sin (\varphi - \theta) - n'(t)] \tag{5-36}$$

where

$$n'(t) = \frac{n_c(t)}{A_c} \cos \theta + \frac{n_s(t)}{A_c} \sin \theta \tag{5-37}$$

Thus the noise equivalent model for an ideal multiplier-type phase detector is as shown in Figure 5-11. Furthermore, if the single-sided noise spectral density of $n(t)$ is N_0, the single-sided noise spectral density of $n'(t)$ can be shown to be

$$S_{n'}(f) = \frac{2N_0}{A_c^2} \qquad f \leq \frac{B}{2} \tag{5-38}$$

If the noise bandwidth of $n(t)$ is B hertz (single-sided), the variance of the input noise is

$$\sigma_n^2 = N_0 B \tag{5-39}$$

and that of $n'(t)$, with single-sided bandwidth $B/2$, is

$$\sigma_{n'}^2 = \frac{N_0 B}{A_C^2} = \frac{\sigma_n^2}{A_c^2} \tag{5-40}$$

The only assumption made in regard to deriving $n'(t)$ is that the VCO phase, $\theta(t)$, is very slowly varying (ideally time-invariant, but arbitrary). Linearity of the phase detector has not been imposed. Thus $n'(t)$ is Gaussian, assuming that θ is constant or very slowly varying.

If the input noise to the loop is sufficiently small, the $\sin (\varphi - \theta)$ operation in Figure 5-11 can be replaced by $(\varphi - \theta)$, and the appropriate closed-loop model with noise at the input is then as shown in Figure 5-12. Since the equivalent noise, $n'(t)$, is additive at the input, the variance of the VCO output phase is

$$\sigma_\theta^2 = S_{n'}(0)B_L$$

$$= \frac{2N_0 B_L}{A_c^2} \tag{5-41}$$

where B_L is the single-sided equivalent noise bandwidth of the closed loop. For the second-order loop with active filter, the ratio of equivalent noise bandwidth to natural frequency from Table 5-1 is

$$\frac{B_L}{\omega_n} = \frac{\zeta + \zeta/4}{2} \tag{5-42}$$

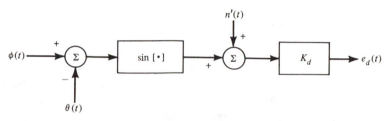

FIGURE 5-11. Noise equivalent model for sinusoidal phase detector.

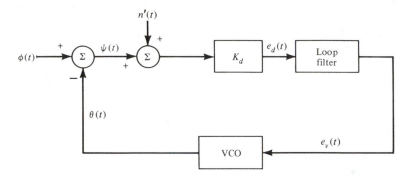

FIGURE 5-12. Linear model for phase-locked loop with additive noise present at the input.

which has a minimum value of $\frac{1}{2}$ for $\zeta = \frac{1}{2}$, giving a minimum VCO phase variance due to additive input noise of

$$\sigma_{\theta,min}^2 = \frac{N_0 \omega_n}{A_c^2} = \frac{2N_0 B_L}{A_c^2} \qquad \text{(second-order loop, active filter, } \zeta = \frac{1}{2}) \quad (5\text{-}43)$$

A damping factor of $\zeta = 1/\sqrt{2} = 0.707$, which is often used due to transient response considerations, gives a VCO phase variance due to noise which differs from (5-43) by only 6%.

The signal-to-noise ratio at the loop input, with noise measured in a loop bandwidth, is

$$\rho = (\text{SNR})_L = \frac{A_c^2}{2N_0 B_L} \qquad (5\text{-}44)$$

In terms of ρ, σ_θ^2 is given by

$$\sigma_\theta^2 = \frac{1}{\rho} \qquad (5\text{-}45)$$

a result that was derived assuming operation in the linear region of the phase detector characteristic.

An exact analysis for the variance of the VCO phase due to noise has been carried out only for the first-order phase-locked loop assuming no frequency offset and no modulation on the carrier. This result is shown in Figure 5-13 together with (5-43) from the linearized analysis. The method used to solve the nonlinear problem, known as the Fokker–Planck technique, gives a probability density function for the phase error of the form

$$p(\psi) = \frac{\exp (\rho \cos \psi)}{2\pi I_0(\rho)} \qquad |\psi| \leq \pi \qquad (5\text{-}46)$$

where $I_0(\cdot)$ is the modified Bessel function of order zero. For ρ large, it can be shown by using the asymptotic formula $I_0(\rho) \simeq \exp (\rho)/\sqrt{2\pi\rho}$ that $p(\psi)$ tends to a Gaussian density function with zero mean and variance σ_θ^2.

Other approximate noise analysis methods have been devised for analyzing the behavior of phase-locked loops operating into the nonlinear region. These methods deal only with second-order statistics such as variance, and do not take into account cycle slipping. Again, appealing to the Fokker-Planck analysis results for the first-

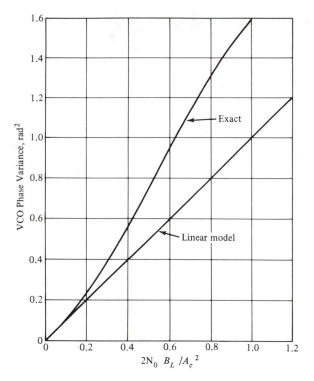

FIGURE 5-13. Comparison of exact and approximate for values for first-order PLL phase variance.

order phase-locked loop, it has been shown [2] that the average time between cycle slips (i.e., the average time for the loop phase error to reach $\pm 2\pi$ after starting initially at zero) is given by

$$T_{av} = \frac{2\pi\rho I_o^2(\rho)}{2B_L}$$

$$\simeq \frac{\pi}{4B_L} \exp(2\rho) \qquad \rho \gg 1 \qquad (5\text{-}47)$$

In addition, the probability distribution of the time between slips is exponential; that is,

$$P(T) = 1 - \exp\left(\frac{-T}{T_{av}}\right) \qquad (5\text{-}48)$$

where T is the time to a slip given the loop started with zero error.

These results apply exactly to a first-order loop with a sinusoidal phase detector characteristic operating in additive white Gaussian noise. Exact Fokker-Planck solutions for the second-order loop have not been obtained. However, experimental measurements and approximate nonlinear analyses for the second-order loop show that the exact nonlinear analysis results for the first-order loop are in close agreement with the second-order loop results for $(SNR)_L > 0$ dB.

5-3.3 Phase-Locked-Loop Tracking of Oscillators with Phase Noise

The effect of oscillator phase jitter on the phase error of a second-order phase-locked loop with active filter and damping factor $\zeta = 1/\sqrt{2}$ will now be analyzed. To find the variance of the loop phase error, σ_ψ^2, due to the phase-locked loop tracking an oscillator with phase-jitter power spectral density $G_\varphi(f)$, the relationship for the output noise variance of a linear system in terms of input noise spectral density and system frequency response function will be used. In the present context, however, the input noise spectral density is $G_\varphi(f)$ and the system frequency response function is the loop-phase-error frequency response, $1 - \tilde{H}(f)$, where the tilde denotes $H(s)|_{s=j2\pi f}$. Thus the loop-phase-error variance is*

$$\sigma_\psi^2 = \int_0^\infty G_\varphi(f)|1 - \tilde{H}(f)|^2 \, df \tag{5-49}$$

where the lower limit is zero since $G_\varphi(f)$ is a single-sided power spectral density. From Table 5-1 it follows that

$$|1 - \tilde{H}(f)|^2 = |1 - H(s)|^2_{s=j2\pi f, \zeta=1/\sqrt{2}}$$

$$= \left| \frac{(j\pi f)^2}{(j2\pi f)^2 + 2\zeta\omega_n(j2\pi f) + \omega_n^2} \right|^2_{\zeta=1/\sqrt{2}}$$

$$= \frac{(f/f_n)^4}{1 + (f/f_n)^4} \tag{5-50}$$

with $f_n \triangleq \omega_n/2\pi$. To carry out the evaluation of (5-49), $G_\varphi(f)$ is represented in terms of the asymptotic expression

$$G_\varphi(f) = \begin{cases} \dfrac{K_1}{f^3} + \dfrac{K_2}{f^2} + K_4 & f \leq f_m \\[3mm] \dfrac{K_1}{f^3} + \dfrac{K_2}{f^2} & f > f_m \end{cases} \tag{5-51}$$

which is based on the discussion centering on (5-3) through (5-11). The loop-phase-error variance due to phase jitter on the input signal can be written

$$\sigma_\psi^2 = \int_0^\infty \frac{f_n^{-2}K_1 x}{1 + x^4} \, dx + \int_0^\infty \frac{f_n^{-1}K_2 x^2}{1 + x^4} \, dx + \int_0^{f_m/f_n} \frac{K_1 x^4 f_n}{1 + x^4} \, dx \tag{5-52}$$

Carrying out the integration, the phase-error variance can be expressed as [7]

$$\sigma_\psi^2 = \frac{K_1 \pi^3}{\omega_n^2} + \frac{K_2 \pi^2}{\sqrt{2}\omega_n} + K_4 f_m \tag{5-53}$$

EXAMPLE 5-2
Evaluate σ_ψ^2 using the values for K_1, K_2, and K_4 found in Example 5-1 for a loop bandwidth of 1 Hz.

Solution: From Table 5-1, if $\zeta = 1/\sqrt{2}$, the loop bandwidth is

$$B_L = \frac{1}{2}\omega_n\left(\zeta + \frac{1}{4\zeta}\right) = 0.53\omega_n \tag{5-54}$$

---

*The phase-noise spectra are assumed to be single-sided and single-sideband in this chapter.

For $B_L = 1$ Hz, $\omega_n = 1.89$ rad/s. Therefore,

$$\sigma_\psi^2 = \frac{(10^{-3})\pi^3}{(1.89)^2} + \frac{(10^{-5})\pi^2}{\sqrt{2}(1.89)} + (10^{-16})(10^9)$$

$$= 8.7 \times 10^{-3} + 3.71 \times 10^{-5} + 10^{-7} = 8.8 \times 10^{-3}\ \text{rad}^2$$

or $\sigma_\psi = 0.094$ rad. ∎

5-3.4 Phase Jitter Plus Noise Effects

From (5-41) it is seen that the VCO phase variance due to additive noise at the loop input increases linearly with B_L. For a constant input phase, this translates directly to a linear increase in phase-error variance with B_L due to noise. On the other hand, (5-53) and (5-54) show that the phase-error variance due to phase jitter on the input signal decreases with increasing B_L. Therefore, an optimum value of B_L exists which provides a minimum in the phase-error variance due to both additive input noise and phase jitter on the input signal. This optimum value is illustrated by Figure 5-14.

EXAMPLE 5-3

Find the optimum loop bandwidth (in the sense of minimum phase-error variance) for an active filter second-order loop with $\zeta = 1/\sqrt{2}$ which is tracking an oscillator with the phase-noise characteristics given in Example 5-1 assuming that the loop operates in an additive noise background with

$$\frac{A_c^2}{2N_0} = 40\ \text{dB-Hz}$$

Solution: Total mean-square phase error, or variance, due to both phase noise and background noise from (5-41) and (5-53) is

$$\sigma_{\psi,T}^2 = \frac{2N_0 B_L}{A_c^2} + \frac{K_1 \pi^3}{\omega_n^2} + \frac{K_2 \pi^2}{\sqrt{2}\,\omega_n} + K_4 f_m$$

where $\omega_n = B_L/0.53 = 1.89 B_L$. Substituting previously obtained values for constants, this can be written as

$$\sigma_{\psi,T}^2 = 10^{-4} B_L + \frac{8.7 \times 10^{-3}}{B_L^2} + \frac{3.71 \times 10^{-5}}{B_L} + 10^{-7}$$

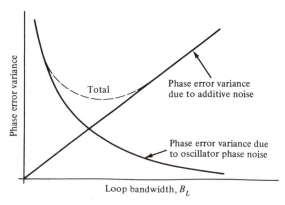

FIGURE 5-14. Optimization of phase error variance.

Differentiation of this with respect to B_L and setting the result equal to zero results in

$$10^{-4} - \frac{17.4 \times 10^{-3}}{B_{L,\text{opt}}^3} - \frac{3.71 \times 10^{-5}}{B_{L,\text{opt}}^2} = 0$$

or

$$B_{L,\text{opt}} \simeq 5.6 \text{ Hz}$$

For this optimum bandwidth, the total phase-error standard deviation is

$$\sigma_{\psi,T,\text{opt}} \simeq 0.03 \text{ rad}$$

Now consider a phase-locked loop with phase noise, $\theta_n(t)$, on the VCO output. It can be shown* that the phase-error variance due to the noisy VCO is

$$\sigma_{\psi,\text{VCO}}^2 = \int_0^\infty |1 - H(j2\pi f)|^2 G_{\theta_n}(f) \, df \qquad (5\text{-}55)$$

where G_{θ_n} is the *single-sided* power spectral density of the VCO phase noise. (Note that additive phase noise at the VOC output is no different from additive phase noise at the loop input.) Suppose that the VCO phase-noise spectral density is random phase walk with

$$G_{\theta_n}(f) = \frac{K_{\text{VCO}}}{f^2} \qquad (5\text{-}56)$$

For a second-order loop with $\zeta = 1/\sqrt{2}$ and natural frequency ω_n, it follows from (5-54) that

$$\sigma_{\psi,\text{VCO}}^2 = \frac{K_{\text{VCO}}\pi^2}{\sqrt{2}\omega_n} = \frac{K_{\text{VCO}}\pi^2}{\sqrt{2}(1.89)B_L} \qquad \left(\zeta = \frac{1}{\sqrt{2}}\right) \qquad (5\text{-}57)$$
■

If the loop also has additive white noise present at its input, an optimum bandwidth exists, as in Example 5-3, that will minimize the total phase-error variance due to VCO jitter and input noise.

One final comment needs stressing in regard to tracking an oscillator with phase jitter. If frequency multiplication or division by N is used prior to the loop, the phase noise is multiplied or divided by N and the phase-noise spectral density is multiplied or divided by N^2, respectively. This follows because the instantaneous output phase, $\theta_{\text{out}}(t)$, of a frequency multiplier is $\theta_{\text{out}}(t) = N\theta_{\text{in}}(t)$, where $\theta_{\text{in}}(t)$ is the input phase, with division by N for a divider.

5-3.5 Transient Response

The tracking error, $\psi(t)$, of the loop for various input phase functions, $\varphi(t)$, can be determined by obtaining the inverse Laplace transform of

$$\psi(s) = [1 - H(s)]\varphi(s) \qquad (5\text{-}58)$$

where $1 - H(s)$ is the phase-error transfer function [see (5-28)]. Typical transient input phase functions are:

---

*For example, see Blanchard [3]. Note that Blanchard's closed-loop VCO phase noise and instantaneous phase error due to VCO phase noise are the same.

1. A step, $\varphi(t) = \Delta\varphi\, u(t)$, for which

$$\varphi_s(s) = \frac{\Delta\varphi}{s} \tag{5-59}$$

2. A ramp (frequency step), $\varphi(t) = \Delta\omega\, tu(t)$, for which

$$\varphi_r(s) = \frac{\Delta\omega}{s^2} \tag{5-60}$$

3. A parabola (frequency ramp), $\varphi(t) = \frac{1}{2}\Delta\dot\omega\, t^2u(t)$, for which

$$\varphi_p(s) = \frac{\Delta\dot\omega}{s^3} \tag{5-61}$$

4. A parabola in frequency, $\varphi(t) = \frac{1}{6}\Delta\ddot\omega\, t^3u(t)$, for which

$$\varphi_{fp}(s) = \frac{\Delta\ddot\omega}{s^4} \tag{5-62}$$

The phase-error responses of a first-order loop to each of the first three inputs, respectively, are

$$(1)\quad \psi_s(t) = \Delta\varphi\, e^{-Kt}u(t) \tag{5-63}$$

$$(2)\quad \psi_r(t) = \frac{\Delta\omega}{K}(1 - e^{-Kt})u(t) \tag{5-64}$$

$$(3)\quad \psi_p(t) = \frac{\Delta\dot\omega}{K^2}(Kt + e^{-Kt} - 1)u(t) \tag{5-65}$$

Note that $\psi_r(t)$ is the indefinite integral of $\psi_s(t)$ with $\Delta\varphi$ replaced by $\Delta\omega$, and $\psi_p(t)$ is the indefinite integral of $\psi_r(t)$ with $\Delta\omega$ replaced by $\Delta\dot\omega$. Also note that only for the phase step is the steady-state VCO phase error zero. For the frequency step (phase ramp), the steady-state phase error is

$$\psi_{r,ss} = \frac{\Delta\omega}{K} \tag{5-66}$$

which approaches zero as the loop gain approaches infinity. However, the loop bandwidth also goes to infinity with increasing loop gain so that phase error variance due to additive noise at the input becomes progressively larger with decreasing steady-state phase error. From (5-65), it is seen that the frequency ramp (phase parabola) results in an essentially linearly increasing phase error.

All the comments above apply to the case where loop components do not saturate; that is, (5-58) is predicated under the assumption of linear loop operation. For a sinusoidal phase detector, the VCO input is really $K\sin\psi$, so that (5-66) becomes

$$\left|\frac{\Delta\omega}{K}\right| = |\sin(\psi_{r,ss})| \leq 1$$

This establishes the *hold-in range* of a first-order loop as

$$-K \leq \Delta\omega_H \leq K \qquad \text{rad/s} \tag{5-67}$$

For a high-gain second-order loop, the phase-error transfer function is given by

$$1 - H(s) = \frac{s^2}{s^2 + 2\zeta\omega_n s + \omega_n^2} \tag{5-68}$$

(see Table 5-1). The steady-state phase error can be found from

$$\lim_{t \to \infty} \psi(t) = \lim_{s \to 0} \{s[1 - H(s)]\varphi(s)\} \tag{5-69}$$

For the phase step and phase ramp (frequency offset) inputs it is seen that the steady-state phase error for a second-order loop is zero.

The Laplace transform inversion of (5-58) in response to a frequency ramp, (5-61), and parabola in frequency, (5-62), yields, respectively, the following transient response for $\zeta < 1$:

$$\psi_p(t) = \frac{\Delta\dot{\omega}}{\omega_n^2} \{1 - e^{-\zeta\omega_n t}[\cos(\omega_n\sqrt{1 - \zeta^2}t)$$

$$+ \frac{\zeta}{\sqrt{1 - \zeta^2}} \sin(\omega_n\sqrt{1 - \zeta^2}t)]\}u(t) \quad \text{(frequency ramp)} \tag{5-70}$$

$$\psi_{fp}(t) = \frac{\Delta\ddot{\omega}}{\omega_n^3} \{\omega_n t - 2\zeta + 2\zeta e^{-\zeta\omega_n t}[\cos(\omega_n\sqrt{1 - \zeta^2}t)$$

$$- \frac{1 - 2\zeta^2}{2\zeta\sqrt{1 - \zeta^2}} \sin(\omega_n\sqrt{1 - \zeta^2}t)]\}u(t) \quad \text{(frequency parabola)} \tag{5-71}$$

Figure 5-15 shows the transient phase error due to an input ramp in frequency, and Figure 5-16 shows the transient phase error due to a parabolic frequency input. For the frequency ramp, it is seen that the steady-state phase error is

$$\psi_{ss,p} = \frac{\Delta\dot{\omega}}{\omega_n^2} \tag{5-72}$$

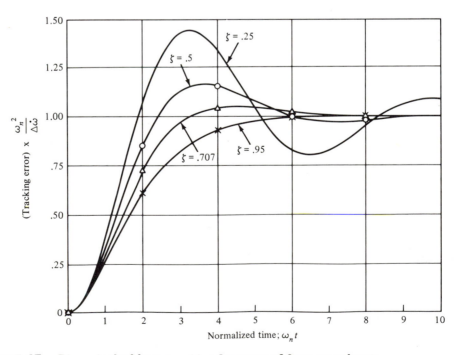

FIGURE 5-15.   Phase-locked-loop tracking for ramp of frequency input.

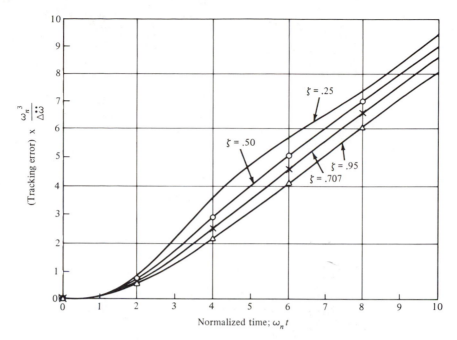

FIGURE 5-16. Phase-locked-loop tracking error for parabola of frequency input.

Again, use of a sinusoidal phase detector would have resulted in

$$\sin \psi_{ss,p} = \frac{\Delta\dot{\omega}}{\omega_n^2} \leq 1$$

which established that the *maximum permissible rate of change of input frequency to a second-order loop as*

$$\Delta\dot{\omega}_{max} = \omega_n^2 \text{ rad/s}^2 = 0.53B_L^2 \qquad \zeta = \frac{1}{\sqrt{2}} \qquad (5\text{-}73)$$

provided that no loop components saturate.

Theoretically, a second-order loop with infinite dc gain can never permanently lose lock. Its response to a large enough frequency offset will be a temporary loss of lock, resulting in cycle slipping, after which it will relock. The frequency-step limit below which the loop does *not* slip cycles is called the *pull-out frequency*. It has been established from phase-plane portraits for the second-order loop with sinusoidal phase detector that the pull-out frequency satisfies the empirical relation [1]

$$\Delta\omega_{po} = 1.8\omega_n(\zeta + 1) \qquad \text{rad/s} \qquad (5\text{-}74)$$

5-3.6 Phase-Locked-Loop Acquisition

The initial application of a sinusoidal signal to a phase-locked loop with carrier frequency different from the quiescent frequency of the VCO results in a beat frequency at the VCO output (i.e., the loop slips cycles as illustrated in Figure 5-17). The beat-frequency waveform has a nonzero average value, which tends to drive the VCO toward *frequency lock* (i.e., the VCO frequency matches that of

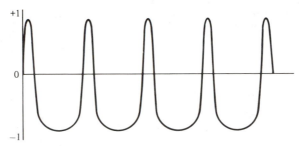

(a) Typical beat-note wave shape, first-order loop,
$\Delta\omega/K = 1.10$ (Gardner)

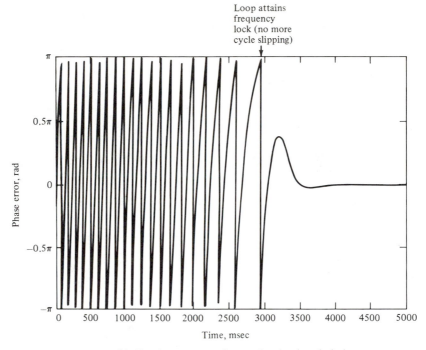

(b) Transient response of a second-order phase-locked
loop with an initial frequency offset of $\Delta f = 10$ Hz
and a Noise Bandwidth of $B_n = 5$ Hz (Spilker)

FIGURE 5-17.   Phase-error signals for phase-locked loops in acquisition.

the input carrier frequency). The dc component of the phase detector output is called the *pull-in voltage*. In a second-order loop, which includes an integrator prior to the VCO, the pull-in voltage is integrated and the loop will eventually reach frequency lock provided that saturation of a loop component does not occur first.

For a first-order loop, an integrator is not present, and the loop will acquire lock only if the frequency offset is within the *lock-in range* of the loop. The inequalities

$$-K < 2\pi(f_c - f_0) < K \qquad (5\text{-}75)$$

establish the frequency offset limit of VCO quiescent frequency from input carrier frequency within which a first-order loop may acquire lock. If the magnitude of the frequency offset in rad/s exceeds K, it is impossible for a first-order loop to have a static phase error which will drive the VCO frequency to match the input

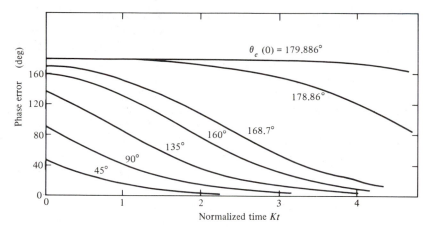

FIGURE 5-18. Transient phase errors in first-order PLL. [1]

frequency. If (5-75) is satisfied, the loop will acquire lock, but the time to lock depends on the initial phase error between the input and VCO. Figure 5-18 illustrates phase error transients for a first-order phase-locked loop for an initial frequency offset of zero. It is seen that if the initial phase error is near 180 degrees, an extremely long transient can take place. This phenomenon, known as *hang-up*, is not unique to the first-order loop.

Returning to the second-order loop, the duration of the initial beat frequency transient illustrated in Figure 5-17b is the *pull-in time*. If $\Delta\omega - 2\pi(f_c - f_0) >> K$, the pull-in time of a second-order loop is approximately [1]

$$T_p \simeq \frac{(\Delta\omega)^2}{2\zeta\omega_n^3} \qquad \text{seconds} \tag{5-76}$$

For a high-gain loop with $\zeta = 1/\sqrt{2}$, this becomes

$$T_p \simeq \frac{4(\Delta f)^2}{B_L^3} \qquad \text{seconds} \qquad \left(\zeta = \frac{1}{\sqrt{2}}\right) \tag{5-77}$$

where $\Delta f = f_c - f_0$ hertz and B_L is the single-sided loop bandwidth in hertz. Once $\Delta\omega \le K$, the loop ceases to skip cycles and quickly snaps into lock. The additional time required for the loop to *settle*, T_S, is approximately [7]

$$T_s \simeq \frac{1.5}{B_L} \tag{5-78}$$

where the final phase error is 0.1 rad or less.

Because the pull-in time for a second-order loop can be exceedingly long, an *acquisition aid* is often used. This usually takes the form of a ramp applied to the VCO input or a square wave applied to the integrator input. The maximum rate of change for the VCO frequency is given by (5-71) under *noise-free* conditions. (Note that there is no difference if a frequency ramp is placed on the input or the VCO output.) With noise present, the sweep rate must be reduced. Empirical data suggest that the maximum sweep rate should be limited to*

$$\Delta\dot{\omega}_{max} = \omega_n^2[1 - (SNR)_L^{-1/2}] \tag{5-79}$$

*See Gardner [1, p. 81]. Note that Gardner's $(SNR)_L$ is one-half of the definition used here.

where $(\mathrm{SNR})_L$ is the carrier-to-noise ratio with noise measured in a loop bandwidth. Once lock is acquired, the sweep can be removed. If injected as a square wave into the integrator of an active loop filter, the removal of the sweep does not have to be particularly rapid under normal conditions since once the loop is locked the sweep voltage is compensated by the phase detector output.

Removal of the sweep requires the use of a lock detector, which can be implemented by means of the *coherent amplitude detector* illustrated in Figure 5-19. The output of such a detector, once the loop is locked, is proportional to signal amplitude at the loop input. Therefore, it can also be used to control open-loop gain, which depends on input signal amplitude. Another way to remove the effect of variations of the input signal amplitude on loop gain is to precede the loop with a limiter.

Another way in which acquisition can be speeded up is to employ a wider loop bandwidth during acquisition. This is shown by (5-77). Narrowing the loop bandwidth once acquisition has been achieved is facilitated by means of a coherent amplitude detector.

Finally, note that the settling time for a phase-locked loop is improved with the addition of noise to the loop, either as external noise or as a *dithering signal*. This is illustrated by Figure 5-20 for a second-order loop. Note that an optimum value of the order of 20 dB apparently exists for loop signal-to-noise ratio to provide minimum settling time. The use of phase acquisition aids for reducing settling time has been studied recently with the conclusion that the phase acquisition time can be reduced for loop signal-to-noise ratios above 12 dB. Below 12 dB, no significant advantage from the acquisition aid was realized [8].

5-3.7 Other Configurations

Only the simplest phase-locked-loop configuration has been considered here. Many other phase-locked-loop structures perform functions other than locking onto a discrete spectral component. For example, in the spread-spectrum discussions to follow, delay-lock and tau-dither loops are described that can be used to track a pseudonoise digital sequence. Loop structures for tracking BPSK and QPSK, which contain no carrier component, are discussed in Chapter 6. Figure 5-21 shows

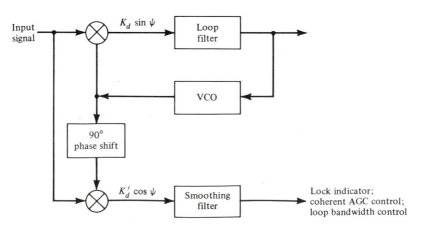

FIGURE 5-19. Coherent amplitude detector.

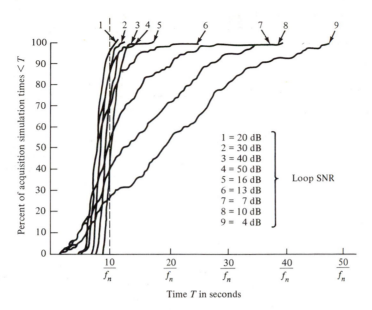

FIGURE 5-20. Computer simulations of second-order phase-locked-loop acquisition time for zero-frequency offset and an initial phase error of $\epsilon_T = \pi$ rad. (From Ref. 7.)

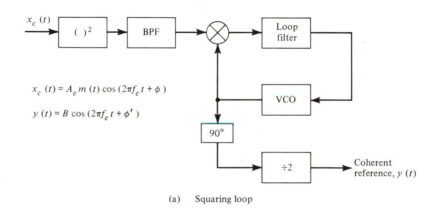

$x_c(t) = A_c m(t) \cos(2\pi f_c t + \phi)$

$y(t) = B \cos(2\pi f_c t + \phi')$

(a)   Squaring loop

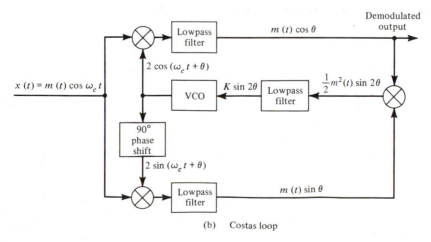

(b)   Costas loop

FIGURE 5-21.   Loops for deriving a coherent reference for demodulation of BPSK.

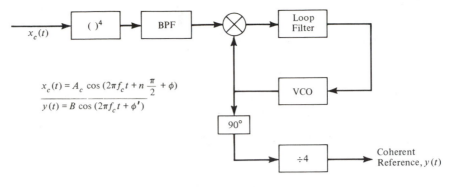

$$x_c(t) = A_c \cos\left(2\pi f_c t + n\frac{\pi}{2} + \phi\right)$$
$$\overline{y(t) = B \cos(2\pi f_c t + \phi')}$$

(a) Quadrupling loop

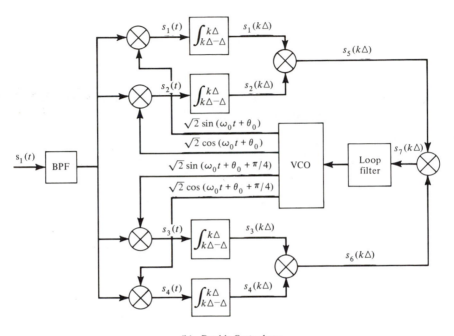

(b) Double Costas loop

FIGURE 5-22. Loops for demodulation of quadriphase.

squaring and Costas loops for demodulating BPSK. It has been shown that the Costas loop is equivalent to the squaring loop in terms of noise performance provided that the lowpass filters in the arms of the Costas loop are the lowpass equivalent of the bandpass filter of the squaring loop [9]. Similarly, the quadrupling and double Costas loop structures shown in Figure 5-22 perform equivalently, provided that the input filters are the same [10].

A disturbing phenomenon that occurs in biphase and quadriphase demodulators is that of false lock [11–15]. In a long loop, which includes hetrodyning to an intermediate frequency within the loop, it is shown in [4] that false lock is due to the accumulated delay in the loop. In short loops, as illustrated in Figure 5-21 and 5-22, false lock arises from filtering of the data in combination with the nonlinear operations in the loop, which produces spectral components at multiples of one-half the data rate.

5-3.8 Effects of Transport Delay

At sufficiently high frequencies, the delay associated with the phase-locked-loop layout can effectively add additional poles to the loop transfer function. Thus a loop designed to be a second-order loop is, in essence, a higher-than-second-order loop. Such delays may cause a loop to operate in a totally different manner from the one for which it was designed. In particular, a loop designed to be second order and therefore thought to be unconditionally stable may effectively be third order or higher and therefore by only conditionally stable.

5-4

FREQUENCY SYNTHESIS

A frequency synthesizer is a device for generating several possible output frequencies from a single, highly stable reference frequency. Systems applications to communications include HF radio, frequency-division multiple-access satellite communications, and spread-spectrum communications systems. There are three main techniques used for frequency synthesizer implementation, although combinations of these may be used as well as variants of these techniques. The three methods of frequency synthesis are referred to as

1. Digital (or table look-up).
2. Direct (or mix-and-divide).
3. Phase-locked (or indirect).

The function of a synthesizer is described mathematically by the equation

$$f_2 = \frac{n_2 f_1}{n_1} \tag{5-80}$$

where f_1 is the reference frequency and n_1 and n_2 are integers.

Recalling that frequency multiplication of a sinusoid multiplies both the nominal frequency and the phase deviation by the multiplication factor, n_2/n_1, it is seen from (5-1) that the long-term stability of a frequency synthesizer is that of the stable reference multiplied by n_2/n_1. That is, the term $\alpha t^2/2$ in the argument of (5-1), which reflects long-term drift, is multiplied by n_2/n_1, as are the terms $\omega_0 t$ and $\varphi(t)$. It is tempting at this point to say that the short-term stability term, or phase noise, of the synthesized frequency is determined from $n_2\varphi(t)/n_1$. However, short-term stability depends on the manner in which the output frequencies are synthesized. A discussion of short-term stability is therefore postponed until after the various types of synthesizers have been discussed.

Each of the synthesis techniques listed above has advantages and disadvantages. The reader is encouraged to consult more detailed discussion of synthesizer design, such as references [5, 6, 8], before embarking on any synthesizer design.

5-4.1 Digital Synthesizers

The basic idea of a digital synthesizer is illustrated by Figure 5-23. With each clock pulse, which occurs at frequency f_1, the accumulator increments a phase variable, θ, by the amount $a\Delta\theta$ where a is a proportionality constant. The value

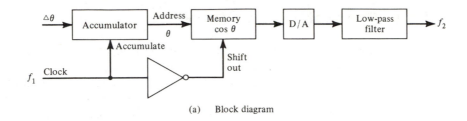

(a)   Block diagram

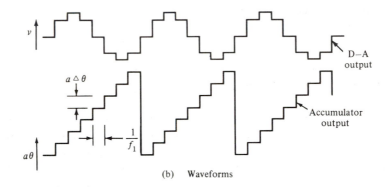

(b)   Waveforms

FIGURE 5-23. Principle of operation for a digital frequency synthesizer. (From Ref. 5.)

of the phase variable, θ, serves as the address to a memory containing N-bit numbers proportional to cos θ, quantized to 2^N levels. The memory output is converted to an analog voltage by a digital-to-analog (D/A) converter.

The capacity of the accumulator corresponds to one complete cycle of cos θ. Let n_1 be the capacity of the accumulator and let n_2 be the increment in the accumulator value for each clock cycle. Then the number of clock cycles required to cycle the accumulator is n_1/n_2, and the frequency of the accumulator cycle, which is also the frequency of the D/A converter output, is given by (5-80). The resolution of the synthesizer is the change in frequency that occurs when n_2 changes by one. This is

$$\Delta f = \left(\frac{n_2 + 1}{n_1} - \frac{n_2}{n_1} \right) f_1 = \frac{f_1}{n_1} \qquad (5\text{-}81)$$

Because the structure of the digital synthesizer implies no fewer than two phase values for each cycle of the output, the theoretical maximum value for f_2 is $f_1/2$. However,

$$f_{2,\text{max}} = \frac{f_1}{4} \qquad (5\text{-}82)$$

is more practical to allow reasonable lowpass output filters.

EXAMPLE 5-4

Consider a digital synthesizer for which the capacity of the accumulator is 2^8. Obtain the following:

(a) The clock frequency required to produce a 32-kHz resolution.

(b) The increment in accumulator contents at each clock pulse to produce a 160-kHz output frequency.

(c) The maximum synthesizer output frequency, $f_{2,\text{max}}$.

Solution: From (5-81) the clock frequency is

$$f_1 = n_1 \Delta f = (2^8)(2^5 \times 10^3)$$

$$= 2^{13} \text{ kHz}$$

$$= 8.192 \text{ MHz}$$

To produce $f_2 = 160$ kHz, n_2 is calculated from (5-80) to be

$$n_2 = \frac{n_1 f_2}{f_1}$$

$$= \frac{(2^8)(2^4 \times 10^4)}{2^{13} \times 10^3}$$

$$= 5$$

From (5-82) $f_{2,max} = 2^{11} \times 10^3$ Hz $= 2.048$ MHz. ■

Advantages of digital synthesizers are that frequencies can be changed very rapidly and that fine resolution is relatively easy to attain. A disadvantage is that the maximum synthesized frequency is limited, by the speed of the digital logic and memory, to a few megahertz. The spurious frequency components near the generated frequency depend on the quantization accuracy used to generate $\cos \theta$. Current limits on spurious sidelobe levels have been reported to be -50 to -60 dB relative to the desired spectral component [16].

5-4.2 Direct Synthesis

Configurations. In the direct frequency synthesis process the desired frequency is built up by multiplication, mixing (summation or subtraction of a reference), and division of a single reference frequency. Many combinations obviously can be used to produce a desired frequency in this way. For example, 7381 kHz can be produced as the 7381th harmonic of 1 kHz or as the 7th harmonic of 1000 kHz plus the 3rd harmonic of 100 kHz plus the 8th harmonic of 10 kHz plus the first harmonic of 1 kHz.

Figure 5-24 shows a direct synthesizer for producing frequencies over a 10-MHz

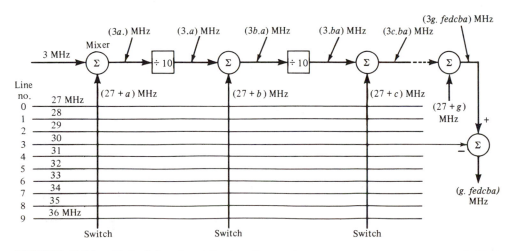

FIGURE 5-24. Principle of operation for a direct frequency synthesizer. (From Ref. 5.)

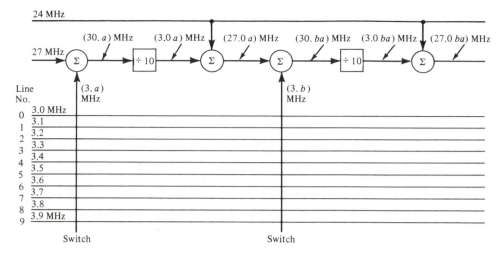

FIGURE 5-25. **Modified direct synthesizer that avoids mixer feedthrough problem.**

range with a resolution of 1 Hz. The reference frequencies of 3 MHz and 27 to 36 MHz could be derived by multiplication of a 1-MHz frequency, which in turn is derived from a stable oscillator, say, of 5 mHz by division. Note that at least one input to each mixer overlaps the output frequency range. Thus it is impossible to eliminate the mixer feed through of this frequency from the output by a fixed output filter. A more practical arrangement which avoids this problem is shown in Figure 5-25.

EXAMPLE 5-5
Synthesize the frequency 7.123456 MHz with the direct-synthesis scheme of Figure 5-25.

Solution: Figure 5-26 shows the mathematical construction of the desired frequency. ∎

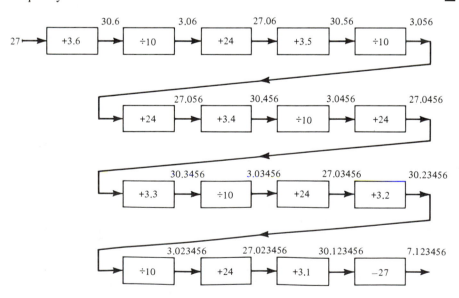

FIGURE 5-26. **Direct synthesis of the frequency 7.123456 MHz (all frequencies are in units of MHz).**

An advantage of direct synthesizers is that the output frequency can be changed rapidly—essentially at the speed of the switches, although some allowance must be made for delay through the system. Another advantage is that the output spectrum can be as clean as the reference oscillator spectrum with FM sidebands increased by the effective multiplication ratio from input to output. In addition, direct synthesizers can have very fine resolution. Disadvantages of direct synthesizers are that they require considerable power because of the large number of LO signals required, and they are bulky. Direct synthesizers with spurious sidelobe levels of -100 dB have been reported.

Spurious Frequency Component Generation in Direct Synthesizers. As an example of spurious frequency component generation in direct synthesizers, consider the synthesizer example of Figure 5-26 and the potential spurious responses (or "spurs") in the output of the top chain of mixers and dividers.

The spurious response or spurs at a mixer output, with inputs of frequencies f_S and f_L, are defined by the relationship

$$nf_s + mf_L = f_I \tag{5-83}$$

where f_I is the mixer output frequency of interest and m and n are integers. Equation (5-83) results from the fact that no mixer is a perfect product device but, rather, is more accurately modeled as producing cross-products of integer powers of each input at its output. Only one of the resultant output frequencies, usually

$$f_{I_1} = f_s + f_L \tag{5-84}$$

or

$$f_{I_2} = |f_s - f_L| \tag{5-85}$$

is of interest; others are referred to as *spurs*. The condition $m = 1$, $n = 0$ identifies signal port feedthrough and $m = 0$, $n = 1$ local oscillator feedthrough.

In the frequency synthesizer example of Figure 5-26, sum mixing is being used as expressed by (5-84). For simplicity, only spurs identified by carrier- and local-oscillator port feedthrough and the difference of first harmonics given by (5-85) are considered. Furthermore, only frequencies at each mixer output which are within an octave bandwidth of the frequency of the desired signal are included. Using this criterion, the first mix produces the following frequencies (MHz):

Signal	Signal-Port Feedthrough	Local Oscillator Port Feedthrough	Diff. Terms
30.6	3.6	27	23.4

The division by 10 results in the sum of these signals being hard-limited. Hence the spurs at 27 MHz and 23.4 MHz appear on the 30.6-MHz signal as frequency modulation. The result of the division by 10 produces the following frequency components:

Signal at 3.06 MHz:
   Spur at $|3.06 - 3.6| = 0.54$ MHz (folded back)
   Spur at $|3.06 - 7.2| = 4.14$ MHz (folded back)

The result of the second mix produces the following frequencies:

Signal	Signal-Port Feedthrough	Local Oscillator Port Feedthrough	Diff. Terms	Nonsignal Sum Terms
27.06	3.06	24	20.94	24.54
	0.54		19.86	28.54
	4.14		23.46	

After mix three, these are translated up 3.5 MHz to produce the following frequencies:

Signal	Signal-Port Feedthrough	Local Oscillator Port Feedthrough	Diff. Terms	Nonsignal Sum Terms
30.56	27.06	3.5	23.56	28.04
	24.54	(ignored)	21.04	32.04
	28.54		25.02	27.5
	24.		20.50	24.44
	20.94		17.44	23.36
	19.86		16.36	26.96
	23.46		19.96	
	(0.54 and 4.14 ignored)			

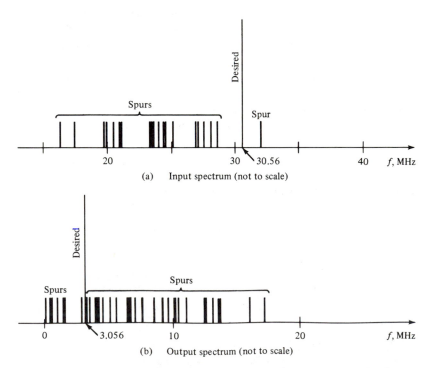

(a) Input spectrum (not to scale)

(b) Output spectrum (not to scale)

FIGURE 5-27. Spectra at divider input and output of a direct synthesizer.

Again, the frequencies far removed from the desired signal frequency of 30.56 MHz are ignored. Upon going through the limiter represented by the next divide-by-10 cycle, the 30.56-MHz signal is FM modulated by the spur frequency components. Spectra at the divider input and output might appear as shown in Figure 5-27. Although identification of spurs is straightforward with the aid of a sketch, the selection of signal and local oscillator frequencies is not particularly straightforward. Sometimes of help is a *spur chart*, which is a two-dimensional representation of (5-83), or a computer program.

5-4.3 Phase-Locked Frequency Synthesizers

Configurations. The block diagram of a simple phase-locked synthesizer is shown in Figure 5-28. The condition (5-80) is satisfied by virtue of the fact that the VCO output frequency divided by n_2 is locked to the reference frequency divided by n_1. Output frequency selection is provided by changing the divider integers, n_1 and n_2. Usually, digital counters are used to provide the desired divider integers.

The minimum increment in output frequency is given by (5-81). The loop bandwidth must be smaller than this minimum increment in order to suppress ripple and ensure loop stability. Therefore, small increments in output frequency demand small loop bandwidths. On the other hand, output phase jitter is dominated by the VCO if loop bandwidth is small. In addition, loop acquisition time is inversely proportional to loop bandwidth [see (5-77) and (5-78)], so that small loop bandwidth implies slow switching between synthesized frequencies.

These conflicting requirements present significant challenges in the design of phase-locked synthesizers. One simple solution to this problem is illustrated in Figure 5-29, where the final synthesizer output frequency is

$$f_2 = \frac{n_2 f_1}{n_1 m} \tag{5-86}$$

so that frequency increments of

$$\Delta f = \frac{f_1}{n_1 m} \tag{5-87}$$

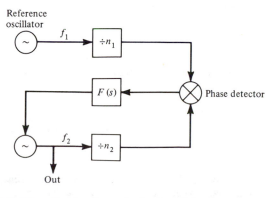

FIGURE 5-28. Basic phase-locked synthesizer.

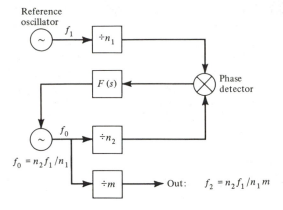

FIGURE 5-29. Modified basic phase-locked synthesizer with divider at output.

are obtained. However, phase comparison occurs at frequency f_1/n_1, which alleviates the loop bandwidth problem by a factor of m through operation of the dividers at the VCO output at m times the frequency required for the basic configuration of Figure 5-28.

Output Phase Noise. To consider the output phase noise of a phase-locked synthesizer, the model of Figure 5-30 will be used and following quantities are defined:

$G_\varphi(f)$ = single-sided phase-noise power spectrum of the reference oscillator
$G_\theta(f)$ = single-sided phase-noise power spectrum of the VCO
$\dfrac{2N_0}{A_c^2}$ = equivalent single-sided power spectral level of the additive white input noise *referred to the closed loop* (see Figure 5-12)

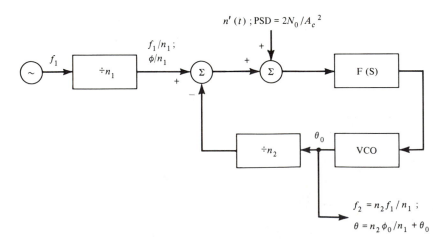

FIGURE 5-30. Model for computing phase noise at output of a phase-locked synthesizer (zero Subscripts on output phases denote loop filtered quantities).

With these definitions, the phase-noise variance of the phase-locked synthesizer output is

$$\sigma_0^2 = \frac{2N_0 B_L}{A_c^2} + \left(\frac{N_2}{N_1}\right)^2 \int_0^\infty G_\varphi(f)|H(f)|^2 \, df$$

$$+ \int_0^\infty G_\theta(f)|1 - H(f)|^2 \, df \qquad (5\text{-}88)$$

where $H(f)$ is the closed-loop frequency response. Thus, with negligible additive noise, the output phase noise variance is the same as the VCO for small loop bandwidths and that of the reference source multiplied by $(n_2/n_1)^2$ if loop bandwidth is large.

Spur Generation in Indirect Synthesizers.   The spur problem also exists in indirect synthesizers and is particularly troublesome in synthesizers where a large tuning range and fast acquisition is desired. The latter implies a wideband loop, although the problem can sometimes be alleviated by using preset tuning of the VCO, which permits a narrower bandwidth loop.

REFERENCES

[1] F. M. GARDNER, *Phaselock Techniques*, 2nd ed. (New York: Wiley, 1979).

[2] A. J. VITERBI, *Principles of Coherent Communication* (New York: McGraw-Hill, 1966).

[3] A. BLANCHARD, *Phase-Locked Loops* (New York: Wiley, 1976).

[4] W. C. LINDSEY, *Synchronization Systems in Communication and Control* (Englewood Cliffs, N.J.: Prentice-Hall, 1972).

[5] W. F. EGAN, *Frequency Synthesis by Phase Lock* (New York: Wiley, 1981).

[6] V. MANASSEWITSCH, *Frequency Synthesizers Theory and Design*, 2nd ed. (New York: Wiley, 1981).

[7] J. J. SPILKER, JR., *Digital Communications by Satellite* (Englewood Cliffs, N.J.: Prentice-Hall, 1977). Chap.12.

[8] H. MEYR, "Phase Acquisition Statistics for Phase-Locked Loops," *IEEE Trans. Commun.*, Vol. COM-28, pp. 1365–1372, August 1980.

[9] R. L. DIDDAY and W. C. LINDSEY, "Subcarrier Tracking Methods and Communication System Design," *IEEE Trans. Commun. Technol.*, Vol. COM-16, pp. 541–550, August 1968.

[10] L. C. PALMER and S. A. KLEIN, "Phase Slipping in Phase-Locked Loop Configurations That Track Biphase or Quadriphase Modulated Carriers," *IEEE Trans. Commun.*, Vol. COM-20, pp. 984–991, October 1972.

[11] G. HEDIN, J. K. HOLMES, W. C. LINDSEY, and K. T. WOO, "Theory of False Lock in Costas Loops," *IEEE Trans. Commun.*, Vol. COM-26, pp. 1–12, January 1978.

[12] M. K. SIMON, "The False Lock Performance of Costas Loops with Hard-Limited In-Phase Channel," *IEEE Trans. Commun.*, pp. 23–33, January 1978.

[13] M. K. SIMON, "False Lock Behavior of Quadriphase Receivers," *IEEE Trans. Commun.*, Vol. COM-27, pp. 1660–1669, November 1979.

[14] S. T. KLEINBERG and H. CHANG, "Sideband False-Lock Performance of Squaring, Fourth-Power and Quadriphase Costas Loops for NRZ Data Signals," *IEEE Trans. Commun.*, Vol. COM-28, pp. 1335–1342, August 1980.

[15] M. L. OLSON, "False-Lock Detection in Costas Demodulators," *IEEE Trans. Aerosp. Electron. Syst.*, Vol. AES-11, pp. 180–182, March 1975.

[16] J. GORSKI-POPIEL, ed., *Frequency Synthesis: Techniques and Applications* (New York: IEEE Press, 1975).

[17] M. ABRAMOWITZ and I. STEGUN, eds., *Handbook of Mathematical Functions*, (New York: Dover), originally published in 1964 as NBS Applied Mathematics Series 55.

PROBLEMS

5-1. Referring to Example 5-1, find the constants describing the various asymptotes for the *frequency* fluctuation spectral density. Sketch these asymptotes as a function of frequency, using a decibel scale.

5-2. For the phase-noise spectral density of Figure 5-3, compute the following:
  (a) The rms phase deviation between frequencies 1 to 10^6 Hz of the nominal oscillator center frequency.
  (b) The rms frequency deviation due to phase noise for frequencies from 1 to 10^6 Hz of the carrier.

5-3. Repeat Problem 5-2 with power spectra that are 10 dB lower and corner frequencies that are 100 times higher.

5-4. **(a)** A frequency counter measures the average frequency $\hat{f}$ of a sinusoidal source by counting the phase cycle increase in T seconds of the source output. Neglecting round-off errors and the drift term $\frac{1}{2}\alpha t^2$ in (5-1), the measurement for $\hat{f}$ can be expressed as

$$\hat{f} = \frac{\omega_0 T + \varphi(T) - \varphi(0)}{2\pi T}$$

where $\varphi(t)$ is the phase jitter in (5-1). Let the autocorrelation function of the phase jitter be $R_\varphi(\tau)$. Show that the variance of $\hat{f}$ is

$$\text{var }(\hat{f}) = \sigma_{\hat{f}}^2 = \frac{1}{(2\pi T)^2} E\{[\varphi(T) - \varphi(0)]^2\}$$

$$= \frac{2}{(2\pi T)^2} [R_\varphi(0) - R_\varphi(T)]$$

where it is assumed that $\varphi(t)$ is a zero-mean wide-sense stationary random process.
  (b) If $G_\varphi(f) = \mathcal{F}[R_\varphi(\tau)]$ is the power spectral density of $\varphi(t)$, show that

$$\sigma_{\hat{f}}^2 = 2\int_0^\infty G_\varphi(f) \frac{1 - \cos 2\pi fT}{2(\pi T)^2} \, df$$

  (c) Verify that the result found in part (b) checks with (5-16) if $N \gg 1$.

5-5. **(a)** Given a phase-jitter power spectrum of the form

$$G_\varphi(f) = \begin{cases} \dfrac{k}{f^2 f_1} + N_0 & f < f_1 \\[2mm] \dfrac{k}{f^3} + N_0 & f_1 < f < f_2 \\[2mm] N_0 & f_2 < f < f_3 \end{cases}$$

If $f_1 T \ll 1$ and $f_2 T \gg 1$, show that for this spectrum the frequency measurement variance $\sigma_{\hat{f}}^2$ as found in Problem 5-4 is approximately

$$\sigma_{\hat{f}}^2 \simeq k f_1 [\tfrac{5}{2} - \text{Ci}(2\pi f_1 T)] + [N_0 f_3 / 2(\pi T)^2][1 - \text{sinc}(2f_3 T)]$$

where $\text{Ci}(x) = \int_\infty^x [(\cos t)/t] \, dt$ is the cosine integral.

(b) The cosine integral can be expanded as

$$\text{Ci}(x) = \gamma + \ln x + \sum_{n=1}^{\infty} \frac{(-1)^n x^{2n}}{2n(2n)!}$$

where $\gamma = 0.577216$. Neglecting all the higher-order terms above $\ln x$ specialize the result found in part (a) to the case where $f_1 T \ll 1$. Note that σ_f^2 increases slowly as f_1 decreases, eventually approaching infinity. The Allan variance, on the other hand, remains finite.

5-6. Verify the transfer functions (5-27) through (5-29) for the linearized phase-locked loop.

5-7. Verify all the entries in Table 5-1.

5-8. Show that the noise component at the output of a multiplier-type (sinusoidal characteristic) phase detector of a phase-locked loop is given by (5-37) and that its spectral density is given by (5-38).

5-9. Show that the phase-error probability density function (5-46) for a first order phase-locked loop approaches a Gaussian density for large ρ.

5-10. **(a)** Verify (5-53) for the phase-error variance of a phase-locked loop tracking an oscillator with phase noise modeled by (5-51).
(b) Evaluate σ_ψ^2 for the phase-noise spectrum of Figure 5-3 as a function of loop bandwidth, B_L.

5-11. Plot $B_{L,\text{opt}}$ as defined in Example 5-3 as a function of $A_c^2/2N_0$. Plot the corresponding values of phase-error standard deviation.

5-12. A polynomial approximation for the modified Bessel function of order zero for $|x| \leq 3.75$ is given by

$$I_0(x) = 1 + 3.5156229t^2 + 3.0899424t^4$$

$$+ 1.2067492t^6 + 0.2659732t^8$$

$$+ 0.0360768t^{10} + 0.0045813t^{12} + \epsilon$$

where $t = x/3.75$ and $|\epsilon| < 1.6 \times 10^{-7}$ [18, p. 378]). Use this approximation and the asymptotic approximation

$$I_0(x) \simeq \frac{\exp(x)}{\sqrt{2\pi x}}$$

to plot (5-46) for $\rho = 0.1$, 1, and 10. Compare the result for $\rho = 10$ with a Gaussian density with mean zero and variance 0.1.

5-13. Use the approximations given in Problem 5-12 to evaluate the phase-error variance $\sigma_\psi^2 = \int_{-\pi}^{\pi} \psi^2 p(\psi) \, d\psi$ for $\rho = 0.1$, 1, and 2 with the aid of a programmable calculator. Check the results obtained for $\rho = 1$ and 2 with Figure 5-13. As $\rho \to 0$, $\sigma_\psi^2 \to \pi^2/3$, check the result for $\rho = 0.1$ with this limiting result. Explain why the phase-error variance approaches $\pi^2/3$ as $\rho \to 0$.

5-14. **(a)** Evaluate T_{av} as given by (5-47) for a loop bandwidth of $B_L = 20$ Hz and $\rho = 0.1$, 1, and 10. Comment on the meaning of the result for $\rho = 10$.
(b) Equation (5-47) is derived assuming a first-order loop using the Fokker–Planck equation. Experimental data have been used to obtain the empirical relation

$$B_L T_{\text{av}} = \exp(\pi\rho)$$

Compare T_{av} calculated from this equation with the results obtained in part (a).

5-15. Derive (5-70) and (5-71) for the transient response of a second-order phase-locked loop to a frequency ramp and a frequency parabola.

5-16. Consider the optimization of a second-order phase-locked loop with loop filter $F(s) = (s\tau_2 + 1)/s\tau_1$ in response to a frequency step of a specified value $\Delta\omega$ and noise. If the integral-square phase error is to be less than or equal to some value k, what is the optimum value of ζ to minimize the phase jitter due to noise?

5-17. Given a second-order phase-locked loop with active loop filter as in Problem 5-16. If its noise bandwidth is fixed, what value of damping factor will permit the largest frequency step $\Delta\omega$ without the loop being pulled out of lock, even temporarily?

5-18. (a) Show that the Costas loop signals shown in Figure 5-21b are correct.
(b) If $x(t) = -m(t) \cos \omega_0 t$, show that the same signal is obtained at the VCO input, thereby implying that the Costas loop will lock to either $x(t)$ or $-x(t)$. What implication does this have in regard to data detection?

5-19. (a) Verify that the loops shown in Figure 5-22 will provide coherent reference signals for the demodulation of QPSK.
(b) Referring to the ambiguity problem outlined in Problem 5-18(b) in regard to the Costas loop, what values of φ in $x_c(t)$ of Figure 5-22a provide the same output reference $y(t)$?

5-20. A digital synthesizer that produces a maximum output frequency of $f_{2,\max} = 512$ kHz with a frequency resolution of 2 kHz is desired. Compute the capacity of the accumulator and increment in accumulator value for each pulse cycle to produce the following frequencies: 6 kHz; 38 kHz; 40 kHz, 512 kHz.

5-21. (a) Using the symmetry of the cosine function, explain how the accumulator capacity for a digital synthesizer can be decreased from that specified by (5-81) for a given clock frequency and frequency increment.
(b) What accumulator capacity could be used in designing the synthesizer of Problem 5-20 if this symmetry is employed?

5-22. Consider the digital synthesizer of Problem 5-20. Sketch the output waveform for $f_2 = 512$ kHz. Compute the ratio of third harmonic to fundamental components in decibels for this waveform. What order Butterworth filter with a 3-dB cutoff frequency of 1024 kHz would suppress the third harmonic to 30 dB below the fundamental?

5-23. (a) Using the block diagram of Figure 5-24 to show how the frequency 678.132 kHz could be synthesized. Show both a flow diagram like Figure 5-26 and a diagram showing the switch closures for the frequencies 27 to 36 MHz (note that any convenient decade of frequencies separated by 1 MHZ would do).
(b) Revise the frequency plan found in part (a) to synthesize the frequency 678.132 kHz as shown in Figure 5-25. Discuss its merits over the scheme in part (a).

5-24. (a) Design a phase-locked-loop synthesizer to provide the maximum frequency and frequency resolution asked for in Problem 5-20. Assume that $f_1 = 8.192$ MHz and that a second-order phase-locked loop with $\zeta = 1/\sqrt{2}$ is used. Use the configuration of Figure 5-29. Choose a loop bandwidth that gives $T_p + T_s$ for the phase-locked loop of 1 ms or less when switching between adjacent frequencies. Make your selection of n_1 and m compatible with this bandwidth.
(b) If

$$G_\varphi(f) = \frac{10^{-3}}{f^3} + \frac{10^{-5}}{f^2}$$

and

$$G_\theta(f) = \frac{10^{-6}}{f^2}$$

compute the phase jitter on the synthesizer output.

CHAPTER 6

Synchronization of Digital Communication Systems

6-1

THE GENERAL PROBLEM OF SYNCHRONIZATION

The need for establishing various levels of synchronization in a digital communication system was discussed briefly in Section 1-3. The hierarchy of synchronization steps in a communication system begins with *carrier synchronization*, if a coherent modulation method is employed, followed by *symbol* or *bit synchronization*. Noncoherent modulation formats do not require carrier synchronization, but as shown in Chapter 4, this is achieved at the expense of an additional increment in signal-to-noise ratio for a given error probability over a corresponding coherent scheme. Levels of synchronization that may be required beyond symbol synchronization include *spreading code synchronization* in spread-spectrum systems, and *word* and *frame* synchronization. The problem of spreading code synchronization is addressed in Chapters 9 and 10. Theoretical approaches and methods for achieving carrier and symbol synchronization are addressed in this chapter.

Carrier and symbol synchronization schemes which derive the synchronization clock signal from the data-bearing signal itself will be considered. Another alternative is to divide the transmitted signal power between a data-bearing signal and a signal referred to as a pilot, expressly intended to provide a signal component for deriving the receiver reference clock. The latter method is analyzed in a straightforward fashion by considering the pilot signal as an input to a suitably designed phase-locked-loop circuit and using the theory given in Chapter 5. In the case of frame synchronization, the synchronization problem is usually solved by inserting a special symbol pattern, referred to as a *unique word*, into each frame. This unique word has properties especially chosen to facilitate the derivation of a frame-synchronization clock.

If a pilot signal is not included with the modulated signal and a modulation format is employed which contains no carrier component, the carrier synchronizer must, in effect, generate power at the carrier frequency through a nonlinear operation. The same is true at the symbol synchronization level if a symbol format is

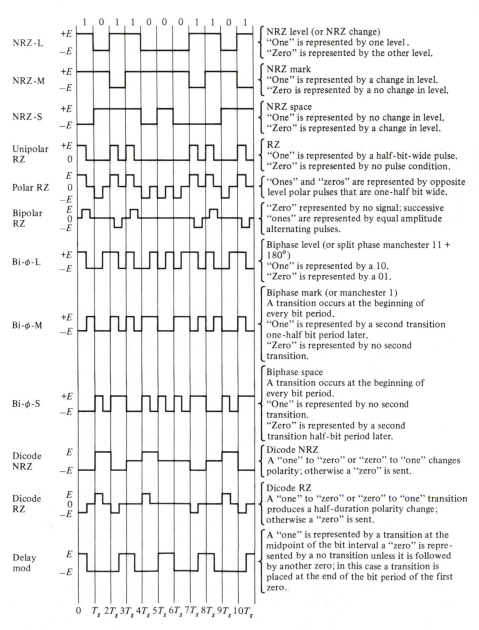

FIGURE 6-1. Binary data formats. (From Ref. 1.)

used which does not include a discrete spectral component at the symbol rate. Several binary symbol formats are illustrated in Figure 6-1.

A general approach for the derivation and analysis of synchronization structures involves the application of the *maximum a posteriori* (MAP) principle. Application of the MAP procedure results in synchronizer structures which seek a value for the parameter of interest so that the a posteriori probability, or the conditional probability (density), $p[A|y(t)]$, of the parameter of interest given the received signal plus interference, $y(t)$, is maximized. If the a priori probability density function of the parameter of interest is very broad, so that essentially no prior knowledge of the parameter is available, the MAP estimation procedure is equivalent to *maximum-likelihood* (ML) estimation. The ML estimate of a parameter is that value which makes the observed data most probable to have occurred.

A second approach that can be used in the estimation of synchronization parameters is the use of nonlinear operations on the signal received to produce spectral lines at frequencies which are rational multiples of the clock rates desired. For example, a spectral component at twice the carrier frequency of a DSB modulated signal can be generated by squaring the received signal. A special case of this approach is the use of cross-spectrum estimation, as illustrated in Figure 6-2. Analysis of this synchronization device makes use of the cyclostationary nature of the signal $z(t)$. Depending on the choices for the filter functions $H(f)$ and $G(f)$, many different types of synchronizers can be realized. Two examples will be mentioned here. If $G(f) = 1$, the familiar squaring loop for acquisition of a carrier reference from a BPSK modulated signal results. If $G(f) = e^{-j2\pi fT_d}$ (i.e., a pure delay), a delay-and-multiply circuit results which can be used to generate clock signals from random non-return-to-zero (NRZ) data streams. Applications include derivation of symbol clocks and extraction of spreading code clocks from direct-sequence spread-spectrum signals.

Because of the importance and breadth of the subject of synchronization, and because of the brevity of this chapter in relation to the subject, it is important at this point to provide references for further study. A recent issue of the *IEEE Transactions on Communications* [2] is devoted solely to the subject of synchronization. Books by Stiffler [3] and Lindsey and Simon [4] contain chapters dealing with various facets of the synchronization problem. Finally, the books by Bhargava et al. [5] and Holmes [1] provide single-chapter treatments of various methods for carrier and symbol synchronization. As already mentioned, Chapters 9 and 10 of this book provide an in-depth treatment of the spreading code acquisition and tracking problem in spread-spectrum communications.

In the next section, the MAP or ML approach to synchronization is considered and applied to the carrier, symbol, and frame synchronization cases. In Section 6-3, use of nonlinear operations to generate spectral lines at clock frequencies is considered. Finally, several practical implementations for carrier and symbol synchronizers are discussed in Section 6-4.

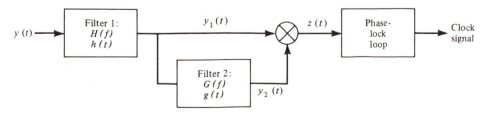

FIGURE 6-2. Typical cross-spectrum synchronizer.

APPLICATION OF THE MAP AND ML PRINCIPLES TO ESTIMATION OF SIGNAL PARAMETERS

6-2.1 Preliminary Definitions and Relationships

A general approach to parameter estimation is the Bayes procedure, in which the estimate of a random parameter is based on the minimization of a *cost function.** For the single-observation, single-parameter case, it can be shown that minimization of average cost with a square-well cost function (i.e., errors within Δ of the true parameter value cost zero, while those greater than Δ cost a fixed, positive amount with $\Delta \to 0$) result in the maximum a posteriori or MAP condition. In terms of the conditional density function of the parameter given the data, $f_{A|Z}(a|z)$, the MAP estimate, $\hat{a}_{\mathrm{MAP}}(Z)$, of the parameter is given by the solution(s) to the equation

$$f_{A|Z}(A|Z)\big|_{A = \hat{a}_{\mathrm{MAP}}(Z)} = \text{maximum} \tag{6-1}$$

Necessary, but not sufficient conditions that the MAP estimate must satisfy are

$$\frac{\partial}{\partial A} f_{A|Z}(A|Z)\bigg|_{A = \hat{a}_{\mathrm{MAP}}(Z)} = 0 \tag{6-2a}$$

and

$$\frac{\partial}{\partial A} \ln f_{A|Z}(A|Z)\bigg|_{A = \hat{a}_{\mathrm{MAP}}(Z)} = 0 \tag{6-2b}$$

where the derivatives are assumed continuous. The latter expression is convenient for exponential-type a posteriori density functions.

Although not as general as other estimators, such as the conditional mean estimator, the MAP criterion is relatively easy to apply. Furthermore, it can be interpreted as the maximum-likelihood (ML) estimate of a parameter if little is known about the parameter to be estimated so that its a priori density is nearly constant when compared with the conditional density function of the data given the parameter $f_{Z|A}(z|a)$. The ML estimate of A is defined as

$$f_{Z|A}(Z|A)\big|_{A = \hat{a}_{\mathrm{ML}}(Z)} = \text{maximum} \tag{6-3}$$

and can be found from the necessary, but not sufficient, conditions

$$\frac{\partial f_{Z|A}(Z|A)}{\partial A}\bigg|_{A = \hat{a}_{\mathrm{ML}}(Z)} = 0 \tag{6-4a}$$

and

$$l(A)\bigg|_{A = \hat{a}_{\mathrm{ML}}(Z)} \triangleq \frac{\partial \ln f_{Z|A}(Z|A)}{\partial A}\bigg|_{A = \hat{a}_{\mathrm{ML}}(Z)} = 0 \tag{6-4b}$$

where $l(A)$ is the likelihood function. From Bayes rule and (6-4b), it follows that the necessary condition for the MAP estimate can be expressed in terms of $l(A)$ as

*For a fuller discussion of estimation procedures applied to communication system design, see Ziemer and Tranter [6].

$$\left[l(A) + \frac{\partial}{\partial A} f_A(A) \right] \Bigg|_{A = \hat{a}_{\text{MAP}}(Z)} = 0 \tag{6-5}$$

where $f_A(A)$ is the a priori density function of A. If multiple parameters are to be estimated, (6-1) through (6-5), as the case may be, are written for each parameter. (Again, the necessary derivatives are assumed continuous.)

If multiple data observations are available, say $\mathbf{Z} = (Z_1, Z_2, \ldots, Z_K)$, on which to base the estimate of a parameter, the conditional density function $f_{Z|A}(z|A)$ in (6-3), (6-4), and (6-5) is simply replaced by the joint conditional density function, $f_{\mathbf{Z}|A}(\mathbf{z}|A)$. Furthermore, if the observations $Z_1, Z_2, \ldots, Z_K$ are independent, given A, then

$$f_{\mathbf{Z}|A}(\mathbf{z}|A) = \prod_{k=1}^{K} f_{Z_k|A}(z_k|A) \tag{6-6}$$

For example, observation of a signal $s(t, A)$ in white Gaussian noise can be cast into a multi-observation estimation problem with conditionally independent observations by using the signal-space representation introduced in Chapter 4. This will be done shortly. First, some additional properties regarding the quality of ML estimates will be given.

An estimate $\hat{a}(Z)$ is *unbiased* if

$$E\{\hat{a}(Z)|A\} = A \tag{6-7}$$

where, if A is random, the expectation is conditioned on A. The *bias*, B, of an estimate is given by

$$B = E[\hat{a}(Z)|A] - A \tag{6-8}$$

Clearly, unbiased estimates are to be preferred over biased estimates unless the bias is known.

A second measure of quality for an estimate is its variance, which may be difficult to compute. A lower bound for the variance of an unbiased ML estimate is provided by the Cramér–Rao inequality, which can be expressed in the two equivalent forms

$$\text{var } [\hat{a}(\mathbf{Z})] \geq \left\{ E\left[\left(\frac{\partial \ln f_{\mathbf{Z}|A}(\mathbf{z}|A)}{\partial A} \right)^2 \right] \right\}^{-1} \tag{6-9a}$$

or

$$\text{var } [\hat{a}(\mathbf{Z})] \geq \left\{ -E\left[\frac{\partial^2 \ln f_{\mathbf{Z}|A}(\mathbf{z}|A)}{\partial A^2} \right] \right\}^{-1} \tag{6-9b}$$

where, since A is regarded as nonrandom, the expectation is with respect to $\mathbf{Z}$. The proof of (6-9) requires that the partial derivatives exist and be absolutely integrable. For *efficient estimates*, (6-9a) and (6-9b) are satisfied with equality. A sufficient condition for equality in (6-9) is that

$$\frac{\partial \ln f_{\mathbf{Z}|A}(\mathbf{z}|A)}{\partial A} = [\hat{a}(\mathbf{Z}) - A]g(A) \tag{6-10}$$

where $g(\cdot)$ is a function of A independent of $\mathbf{Z}$. Further properties of ML estimates are the following:

1. If an efficient estimate of a parameter exists, it is the ML estimate.

2. In the limit as the number of independent observations becomes large, ML estimates are *Gaussian, unbiased,* and *efficient.*

6-2.2 Expressions for Estimation of Continuous Waveform Parameters

As mentioned above, the estimation of waveform parameters in AWGN backgrounds can be handled by using the signal-space approach introduced in Chapter 4. Equations relating to this problem are developed in this subsection.

Consider a signal $s(t, A)$, where A is a parameter of interest, observed in AWGN over a time interval of T seconds. Thus the observed waveform is

$$y(t) = s(t, A) + n(t) \qquad 0 \le t \le T \tag{6-11}$$

where the single-sided power spectral density of $n(t)$ is N_0. Let $\{\varphi_k(t) : k = 1, 2, 3, \ldots\}$ be a complete set of orthonormal functions defined on the interval $[0, T]$. Then, in the sense of limit in the mean [7, p. 179], the observed waveform can be expressed in terms of this orthonormal set as

$$y(t) = \sum_{k=1}^{\infty} S_k(A)\varphi_k(t) + \sum_{k=1}^{\infty} N_k\varphi_k(t) \tag{6-12}$$

where

$$S_k(A) = \int_0^T s(t, A)\varphi_k^*(t) \, dt \tag{6-13}$$

$$N_k = \int_0^T n(t)\varphi_k^*(t) \, dt \tag{6-14}$$

Because the noise is white, it follows that

$$E(N_k N_j^*) = \tfrac{1}{2} N_0 \delta_{kj} \tag{6-15}$$

The estimate of the parameter A, by virtue of (6-12), can be based on the coefficients

$$Y_k = S_k(A) + N_k = \int_0^T y(t)\varphi_k(t) \, dt \qquad k = 1, 2, \ldots \tag{6-16}$$

which, given A, are independent Gaussian random variables with means

$$E(Y_k|A) = S_k(A) \qquad k = 1, 2, \ldots \tag{6-17}$$

and variances $\tfrac{1}{2} N_0$. By considering the joint density function of the first K Y_k's given A and then taking the limit as $K \to \infty$, it can be shown that the likelihood function for A is (see [6, p. 411])

$$l(A) = \frac{2}{N_0} \sum_{k=1}^{\infty} Y_k S_k^*(A) - \frac{1}{N_0} \sum_{k=1}^{\infty} |S_k(A)|^2 \tag{6-18}$$

Using Parseval's theorem (Example 4-3), (6-18) can be written as

$$l(A) = \frac{2}{N_0} \int_0^T y(t)s^*(t, A) \, dt - \frac{1}{N_0} \int_0^T |s(t, A)|^2 \, dt \tag{6-19}$$

The necessary condition for the ML estimate is obtained by differentiating $l(A)$ with respect to A. The resulting equation is

$$\frac{1}{N_0} \int_0^T \left\{ [2y(t) - s(t, A)] \frac{\partial s^*(t, A)}{\partial A} - \frac{s(t, A)}{A} s^*(t, A) \right\} \Bigg|_{A = \hat{a}_{ML}} dt = 0 \quad (6\text{-}20)$$

If $s(t, A)$ is real, (6-20) simplifies to

$$\frac{2}{N_0} \int_0^T [y(t) - s(t, A)] \frac{\partial s(t, A)}{\partial A} \Bigg|_{A = \hat{a}_{ML}} dt = 0 \quad (6\text{-}21)$$

For estimation of multiple parameters, an equation such as (6-20) or (6-21) would be written for each parameter. If unwanted parameters are present which have a known probability density function, it is necessary to obtain an averaged conditional density function with respect to the unwanted parameter. One final note regarding (6-18) is that in many cases of interest the infinite sum is unnecessary because the dimensionality of the signal space may be finite. This is illustrated by the following example.

EXAMPLE 6-1

Consider an ML estimator for the phase of a sinusoidal signal observed over a T-second interval in AWGN. An appropriate signal model is

$$y(t) = B \cos (\omega_0 t + \theta) + n(t) \qquad 0 \le t \le T \quad (6\text{-}22)$$

where B and ω_0 are known constants (for convenience, assume that $\omega_0 T$ is an integer), θ is the unknown phase, and $n(t)$ is AWGN with single-sided power spectral density N_0. By expanding $B \cos (\omega_0 t + \theta)$ as

$$B \cos \theta \cos \omega_0 t - B \sin \theta \sin \omega_0 t$$

it is seen immediately that an appropriate orthonormal function set is

$$\varphi_1(t) = \sqrt{\frac{2}{T}} \cos \omega_0 t \qquad 0 \le t \le T$$

$$\varphi_2(t) = \sqrt{\frac{2}{T}} \sin \omega_0 t \qquad 0 \le t \le T$$

The estimate can be based solely on

$$z(t) = \sqrt{\frac{T}{2}} B \cos \theta \, \varphi_1(t) - \sqrt{\frac{T}{2}} B \sin \theta \, \varphi_2(t)$$

$$+ N_1 \varphi_1(t) + N_2 \varphi_2(t) \qquad 0 \le t \le T$$

where

$$N_i = \int_0^T n(t) \varphi_i(t) \, dt \qquad i = 1, 2$$

because $y(t) - z(t)$ is a noise component which is irrelevant data (Chapter 4). Thus (6-18) can be applied, where $A = \theta$, and

$$S_1(\theta) = \sqrt{\frac{T}{2}} \, B \cos \theta$$

$$S_2(\theta) = -\sqrt{\frac{T}{2}} \, B \sin \theta$$

$$S_k(\theta) = 0 \qquad k \geq 3$$

Thus (6-18) becomes

$$l(\theta) = \frac{\sqrt{2T}}{N_0} B[\cos(\theta)Y_1 - \sin(\theta)Y_2] - \frac{TB^2}{2N_0}$$

where $Y_k = \int_0^T y(t)\varphi_k(t) \, dt$, $k = 1, 2$. If $l(\theta)$ is differentiated with respect to θ, the resulting necessary condition for the ML estimate of θ is

$$\frac{2B}{N_0}\left[-\sin\hat{\theta}_{ML} \int_0^T y(t) \cos \omega_0 t \, dt - \cos\hat{\theta}_{ML} \int_0^T y(t) \sin \omega_0 t \, dt \right] = 0$$

or

$$-\frac{2B}{N_0} \int_0^T y(t) \sin(\omega_0 t + \hat{\theta}_{ML}) \, dt = 0 \qquad (6\text{-}23)$$

which can be interpreted as the feedback structure of Figure 6-3.

The Cramér–Rao lower bound on the variance of the estimate can be found by taking the second derivative of $l(\theta)$ and substituting it into (6-9b). The result is

$$\text{var}\{\hat{\theta}_{ML}\} \geq \left\{ \frac{2B}{N_0} \int_0^T E[y(t)] \cos(\omega_0 t + \theta) \, dt \right\}^{-1} \qquad (6\text{-}24)$$

where the left-hand side of (6-23) has been used for the first derivative and the expectation appearing in (6-9b) has been taken inside the integral. From (6-22), $E[y(t)] = B \cos(\omega_0 t + \theta)$, so that

$$\int_0^T E[y(t)] \cos(\omega_0 t + \theta) \, dt = \frac{BT}{2}$$

which, when substituted into (6-24), gives

$$\text{var}(\hat{\theta}_{ML}) \geq \frac{N_0}{P_s} \frac{1}{2T} \qquad (6\text{-}25)$$

where $P_s = \frac{1}{2}B^2$ is the signal power. Now, the equivalent noise bandwidth of an ideal integrator over the finite time interval T is $B_L = \frac{1}{2}T$, so that (6-25) can be

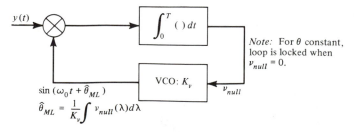

FIGURE 6-3. Feedback structure for estimating the phase of a sinusoid in AWGN.

written as

$$\text{var } (\hat{\theta}_{\text{ML}}) \geq \frac{N_0 B_I}{P_s} \qquad (6\text{-}26)$$

The numerator is the noise power passed by an ideal integrator—a fictitious quantity that allows the bound for the ML estimate to be put into a nice form. ∎

Example 6-1 is reassuring in that the optimum phase estimator is suggestive of a phase-locked loop, which is what one would use from practical considerations. The next example includes the effect of random binary data.

EXAMPLE 6-2

Consider the observation over one bit period of a BPSK-modulated signal with unknown phase, θ, in AWGN. The signal on which the estimate is to be based is

$$y(t) = \alpha_i B \cos (\omega_0 t + \theta) + n(t) \qquad 0 \leq t \leq T_b \qquad (6\text{-}27)$$

The data bit, α_i, is assumed equally likely to be $+1$ or -1. The orthonormal functions used in Example 6-1 apply here with $T = T_b$. Given θ, the data on which to base an observation are

$$z(t) = B \frac{T_b}{2} [\pm \cos \theta \; \varphi_1(t) \mp \sin \theta \; \varphi_2(t)] + N_1 \varphi_1(t) + N_2 \varphi_2(t)$$

where N_1 and N_2 are independent Gaussian random variables with mean zero and variance $N_0/2$. The ML estimate of θ is obtained as that value $\hat{\theta}_{\text{ML}}$ which maximizes the joint conditional density function of Z_1 and Z_2 given θ *averaged* with respect to the data bit distribution. Because N_1 and N_2 are Gaussian and independent, it follows that this conditional density function is

$$f_{\mathbf{Y}|\theta}(Y_1, Y_2|\theta) = \frac{1}{2} f_{\mathbf{Y}|\theta,\alpha_i=1} + \frac{1}{2} f_{\mathbf{Y}|\theta,\alpha_i=-1}$$

$$= \frac{1}{2(\pi N_0)} \left\{ \exp \left[-\frac{1}{N_0} \left(Y_1 - B\sqrt{\frac{T_b}{2}} \cos \theta \right)^2 - \frac{1}{N_0} \left(Y_2 + B\sqrt{\frac{T_b}{2}} \sin \theta \right)^2 \right] \right\}$$

$$+ \frac{1}{2(\pi N_0)} \left\{ \exp \left[-\frac{1}{N_0} \left(Y_1 + B\sqrt{\frac{T_b}{2}} \cos \theta \right)^2 - \frac{1}{N_0} \left(Y_2 - B\sqrt{\frac{T_b}{2}} \sin \theta \right)^2 \right] \right\}$$

$$= C \cosh \left[\frac{\sqrt{2T_b}}{N_0} B(Y_1 \cos \theta - Y_2 \sin \theta) \right] \qquad (6\text{-}28)$$

where C is a constant independent of θ, and

$$Y_i = \int_0^{T_b} y(t) \; \varphi_i(t) \; dt \qquad i = 1, 2 \qquad (6\text{-}29)$$

Because $\ln (\cdot)$ is monotonic, the ML estimate of θ can be found equally well from $\ln [f_{\mathbf{Y}|\theta} (Y_1, Y_2|\theta)]$. Differentiation of this likelihood function results in the necessary condition for $\hat{\theta}_{\text{ML}}$ given by

$$\tanh \left[\frac{\sqrt{2T_b}B}{N_0} (Y_1 \cos \hat{\theta}_{\text{ML}} - Y_2 \sin \hat{\theta}_{\text{ML}}) \right] x(-Y_1 \sin \hat{\theta}_{\text{ML}} - Y_2 \cos \hat{\theta}_{\text{ML}}) = 0$$

$$\qquad (6\text{-}30)$$

From (6-29) and the definitions of $\varphi_1(t)$ and $\varphi_2(t)$, it follows that

$$Y_1 \cos \hat{\theta}_{ML} - Y_2 \sin \hat{\theta}_{ML} = \sqrt{\frac{2}{T_b}} \int_0^{T_b} y(t) \cos(\omega_0 t + \hat{\theta}_{ML}) \, dt$$

$$-Y_1 \sin \hat{\theta}_{ML} - Y_2 \cos \hat{\theta}_{ML} = -\sqrt{\frac{2}{T_b}} \int_0^{T_b} y(t) \sin(\omega_0 t + \hat{\theta}_{ML}) \, dt$$

Thus (6-30) can be written as

$$\tanh\left[\frac{2B}{N_0} \int_0^{T_b} y(t) \cos(\omega_0 t + \hat{\theta}_{ML}) \, dt\right] \int_0^{T_b} y(t) \sin(\omega_0 t + \hat{\theta}_{ML}) \, dt = 0$$

$$(6\text{-}31)$$

This equation suggests the double-loop structure for estimating θ shown in Figure 6-4, which is reminiscent of a Costas loop. It is left as a problem to find the Cramér–Rao lower bound for the variance of this estimate. ∎

6-2.3 Generalization of the Estimator Equations to Multiple Symbol Intervals and Multiple Parameters

In this subsection the form of the received signal plus noise (6-11) will be written to reflect explicitly observation over several symbol intervals and dependence on several parameters. The dependence on one of these parameters—the data sequence—will be removed through averaging in the case of pulse-amplitude-modulated waveforms. By representing the data as a complex-valued sequence, parameter estimation in quadrature-modulated communication systems can be considered. Thus, in (6-11), let $s(t, \mathbf{A})$ be represented as

$$s(t, \mathbf{A}) = \sum_{l=1}^{L} s_l(t, \mathbf{A}) \tag{6-32}$$

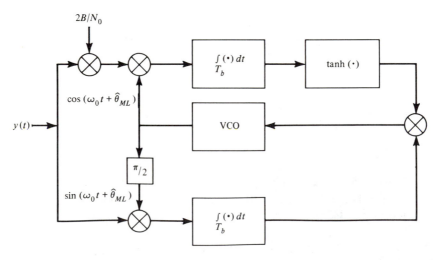

FIGURE 6-4. Feedback structure for estimating the phase of a BPSK-modulated signal.

where $\mathbf{A}$ is a vector of parameters and $s_l(t, \mathbf{A})$ represents the signal during the lth symbol interval of T seconds duration. For example,

$$s_l(t, \mathbf{A}) = \alpha_l p[t - (l - 1)T - \epsilon, \theta] \tag{6-33}$$

where $\alpha_l = \pm 1$ are the data during the lth symbol interval, $p(\cdot)$ is a basic pulse shape, which is nonzero over the interval $(0, T)$, and ϵ and θ are unknown parameters to be estimated that represent the symbol timing epoch and carrier phase, respectively.

Proceeding as in Section 6-2.2, a complete orthonormal set of basis functions over a T-second interval can be defined such that $s(t, \mathbf{A})$ is represented in terms of this orthonormal set by the series

$$y(t) = \lim_{K \to \infty} \sum_{l=1}^{L} y_{lK}(t) \tag{6-34}$$

in which

$$y_{lK}(t) = \sum_{k=1}^{K} [S_{lk}(\mathbf{A}) + N_{lk}]\varphi_k(t) \tag{6-35}$$

In (6-35), the expansion coefficients are given by

$$S_{lk}(\mathbf{A}) = \int_{(l-1)T-\epsilon}^{lT-\epsilon} s_l(t, \mathbf{A})\varphi_k(t) \, dt \tag{6-36}$$

for the signal, and

$$N_{lk} = \int_{(l-1)T-\epsilon}^{lT-\epsilon} n(t)\varphi_k(t) \, dt \tag{6-37}$$

for the noise. As before, it can be shown that the N_{lk} are zero-mean Gaussian random variables with variances $N_0/2$.

The ML estimates for the parameters to be estimated maximize the conditional probability density function $f(\mathbf{Y}|\mathbf{A})$.* Assuming that the data sequence symbols are independent from one interval to the next, this conditional density function can be written as

$$f(\mathbf{Y}|\mathbf{A}) = \prod_{l=1}^{L} f_l(\mathbf{Y}_l|\mathbf{A}', \alpha_l) \tag{6-38}$$

where $f_l(\mathbf{Y}_l|\mathbf{A}', \alpha_l)$ is the conditional density function of the observations during the lth symbol interval and $\mathbf{A}'$ is a new parameter vector for the lth interval, which excludes the data symbol. The remaining parameters do not change over the L symbol intervals used in making the ML estimate of $\mathbf{A}'$. Given $\mathbf{A}$, $\mathbf{Y}_l$ is a Gaussian L-component vector with each component having variance $N_0/2$ and mean $S_{lk}(\mathbf{A}', \alpha_l)$. During the lth symbol interval it therefore follows that the conditional probability density function is K-dimensional Gaussian of the form

$$f_l(\mathbf{Y}_l|\mathbf{A}', \alpha_l) = (\Pi N_0)^{-K/2} \exp \left\{ \frac{[\mathbf{Y}_l - \mathbf{S}_l(\mathbf{A})][\mathbf{Y}_l - \mathbf{S}_l(\mathbf{A})]^\dagger}{N_0} \right\} \tag{6-39}$$

where the superscript $\dagger$ denotes the conjugate transpose and

$$\mathbf{S}_l(\mathbf{A}) = [S_{l1}(\mathbf{A}), S_{l2}(\mathbf{A}), \ldots, S_{lK}(\mathbf{A})]^t$$

*See Lindsey and Simon [4] for a derivation that allows the possibility of MAP estimation. In actuality, their derivation is eventually specialized to the ML estimation.

is the vector of means for the K samples of the lth symbol interval. The exponent in (6-39) can be expanded as

$$\frac{1}{N_0} [\|\mathbf{Y}_l\|^2 - (\mathbf{S}_l, \mathbf{Y}_l) - (\mathbf{Y}_l, \mathbf{S}_l) + \|\mathbf{S}_l\|^2] \tag{6-40}$$

in which $\|\mathbf{Y}_l\|^2 = \mathbf{Y}_l\mathbf{Y}_l^\dagger$ and $\|\mathbf{S}_l\|^2 = \mathbf{S}_l\mathbf{S}_l^\dagger$ become the energies of the received data and signal in the lth symbol interval, respectively, as $K \to \infty$. If a signaling scheme is employed for which $\|\mathbf{S}_l\|^2$ is independent of l (e.g., a constant-envelope signal set), $\|\mathbf{S}_l\|^2$ is a constant independent of $\mathbf{A}$. Similarly, $\|\mathbf{Y}_l\|^2$ is not explicitly dependent on $\mathbf{A}$, and the estimate can be based on $(\mathbf{S}_l, \mathbf{Y}_l) + (\mathbf{Y}_l, \mathbf{S}_l)$. That is the ML estimate for the vector of parameters $\mathbf{A}'$ is that value for which

$$\lim_{K \to \infty} \prod_{l=1}^{L} C \exp \left\{ \frac{2}{N_0} \text{Re} \left[\overline{\sum_{k=1}^{K} S_{lk}(\mathbf{A}', \alpha_l) Y_{lk}^*} \right] \right\} \triangleq f(\mathbf{Y}|\mathbf{A}') \tag{6-41}$$

is maximum, where the overbar denotes an average with respect to the data sequence $(\alpha_1, \alpha_2, \ldots, \alpha_L)$. Without further assumptions, (6-41) cannot be simplified further.

Attention will now be focused in signaling schemes for which the coefficients $S_{lk}(\mathbf{A}', \alpha_l)$ are of the form

$$S_{lk}(\mathbf{A}', \alpha_l) = \alpha_l p_{lk}(\mathbf{A}') \tag{6-42}$$

where α_l and $p_{lk}(\mathbf{A}')$ may be complex. If $\alpha_l = a_l + jb_l$ is a complex data sequence, both its real and imaginary parts take on the values $+1$ or -1 with equal probability and are independent of each other. Let

$$p_{lk}(\mathbf{A}') = p_{lkr} + jp_{lki} \tag{6-43}$$

$$Y_{lk} = B_{lkr} + jB_{lki} \tag{6-44}$$

Then (6-41) can be expressed as

$$f(\mathbf{Y}|\mathbf{A}') = \lim_{K \to \infty} C' \prod_{l=1}^{L} \overline{\exp \left[\frac{1}{N_0} a_l \sum_{k=1}^{K} (p_{lkr} B_{lkr} + p_{lki} B_{lki}) \right]}$$

$$\times \prod_{l=1}^{L} \overline{\exp \left[\frac{1}{N_0} b_l \sum_{k=1}^{K} (p_{lkr} B_{lki} - p_{lki} B_{lkr}) \right]} \tag{6-45}$$

where C' is a constant and the overbars denote averaging with respect to the data sequences $\{a_l\}$ and $\{b_l\}$, respectively. Now $a_l = +1$ or -1 with probability $\frac{1}{2}$, so that the first exponential, which is of the form $\exp(a_l D)$, when averaged becomes

$$\overline{\exp(a_l D)} = \frac{1}{2} \exp(-D) + \frac{1}{2} \exp(D)$$

$$= \cosh D \tag{6-46}$$

A similar averaging operation can also be performed on the second exponential in (6-45). Also, in the limit as $K \to \infty$, the sums over K can be replaced by integrals. If the received signal plus noise envelope is written in terms of its real and imaginary components as

$$\tilde{y}(t) = y_r(t) + jy_i(t) \tag{6-47}$$

and similarly writing the signal envelope as

$$\bar{s}(t, \mathbf{A}') = s_r(t, \mathbf{A}') + js_i(t, \mathbf{A}') \qquad (6\text{-}48)$$

then the logarithm of the likelihood function (averaged with respect to the data) can be written as

$$\ln f(y(t)|\mathbf{A}') = \sum_{l=1}^{L} \ln \cosh \left\{ \frac{1}{N_0} \int_{T_{l\epsilon}} [s_r(t, \mathbf{A}')y_r(t) + s_i(t_i, \mathbf{A}')y_i(t)] \, dt \right\}$$

$$+ \sum_{l=1}^{L} \ln \cosh \left\{ \frac{1}{N_0} \int_{T_{l\epsilon}} [s_r(t, \mathbf{A}')y_i(t) - s_i(t, \mathbf{A}')y_r(t)] \, dt \right\}$$

$$(6\text{-}49)$$

where the constant term has been omitted since it makes no difference in the maximization and $T_{l\epsilon}$ is the interval $(l - 1)T - \epsilon \leq t \leq lT - \epsilon$. In (6-48), $\mathbf{A}'$ is a vector of all parameters to be estimated. Note that the timing epoch, ϵ, also appears explicitly in the integrals. The maximum-likelihood estimate for $\mathbf{A}'$ is found as the joint set of values, $\hat{\mathbf{A}}'$, which maximize (6-48). Alternatively, (6-48) can be differentiated partially with respect to each parameter and the resulting equations set equal to zero. The resulting set of equations form a necessary set of conditions to be satisfied by the ML parameter estimates. Application of the latter approach will be illustrated by example.

EXAMPLE 6-3

Consider an estimator for phase synchronization in QPSK data transmission. For QPSK, the transmitted signal during the symbol interval $(l - 1)T \leq t \leq lT$ can be written as

$$s(t, \theta) = \text{Re} \left[\frac{A}{\sqrt{2}} (a_l + jb_l) \exp(j\theta) \exp(j\omega_0 t) \right]$$

$$= \frac{A}{\sqrt{2}} [a_l \cos(\omega_0 t + \theta) - b_l \sin(\omega_0 t + \theta)] \qquad (l - 1)T \leq t \leq lT$$

$$(6\text{-}50)$$

The received signal plus noise is

$$y(t) = \text{Re}\{(y_r(t) + jy_i(t)] \exp(j\omega_0 t)\}$$

Now

$$\frac{\partial}{\partial \theta} \ln \cosh [F(\theta)] = \tanh [F(\theta)] \frac{\partial F(\theta)}{\partial \theta} \qquad (6\text{-}51)$$

The timing epoch ϵ is assumed known in this example. Differentiation of (6-49) with respect to θ produces the single necessary condition

$$\sum_{l=1}^{L} \tanh \left[\frac{2}{M_0} \int_{(l-1)T}^{lT} y(t) \cos(\omega_0 t + \hat{\theta}) \, dt \right] \int_{(l-1)T}^{lT} y(t) \sin(\omega_0 t + \hat{\theta}) \, dt$$

$$+ \sum_{l=1}^{L} \tanh \left[\frac{2}{M_0} \int_{(l-1)T}^{lT} y(t) \sin(\omega_0 t + \hat{\theta}) \, dt \right] \int_{(l-1)T}^{lT} y(t) \cos(\omega_0 t + \hat{\theta}) \, dt = 0$$

$$(6\text{-}52)$$

where appropriate trigonometric identities have been used to simplify (6-52). Figure 6-5 shows a block diagram of the feedback estimator structure for θ. If the tanh(·)

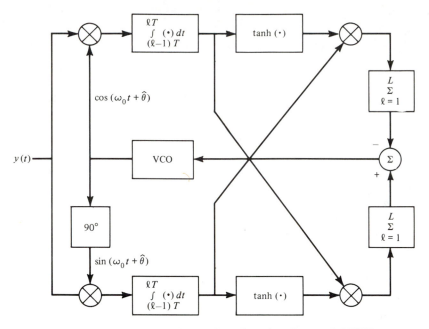

FIGURE 6-5. Estimator structure for estimating the phase of QPSK.

function in the upper leg is approximated as a hard limiter, it is suggestive of the data estimation loop for demodulation of QPSK. It is left to the problems to consider the tracking performance of the loop. ∎

EXAMPLE 6-4

In this example, the application of (6-49) to the estimation of the timing epoch of a pulse-amplitude-modulated (PAM) system will be considered. In this case, the data sequence is real. The signal is represented as a real baseband pulse function $p(t)$. The log-likelihood function (6-49) therefore becomes

$$\ln f(y(t)|\epsilon) = \sum_{l=1}^{L} \ln \cosh \left\{ \frac{1}{N_0} \int_{T_{l\epsilon}} p[t - (l - 1)T - \epsilon] y(t) \, dt \right\} \quad (6\text{-}53)$$

where $T_{l\epsilon}$ is the interval $(l - 1)T - \epsilon \leq t \leq lT - \epsilon$. The block diagram of a symbol synchronizer based on (6-53) is shown in Figure 6-6. The performance of this symbol synchronizer has been studied through Monte Carlo simulation on a computer [8]. Results for the average value of $|\epsilon - \hat{\epsilon}|/T$ are shown in Figure 6-7 as a function of $2E_s/N_0$, where E_s is the symbol energy with symbol periods of memory as a parameter. A raised-cosine pulse shape was employed for $p(t)$.

It is interesting to note that, since $\ln \cosh x \simeq \frac{1}{2}x^2$, the nonlinearities in Figure 6-6 can be replaced by square-law devices if the SNR is small. Thus the statistic to be maximized, assuming low SNR, is

$$\Lambda(\epsilon) = \frac{1}{2N_0} \sum_{l=1}^{L} \left\{ \int_{T_{l\epsilon}} p[t - (l - 1)T - \epsilon] y(t) \, dt \right\}^2 \quad (6\text{-}54)$$

A necessary condition for the optimum estimate of ϵ is provided by obtaining the derivative of $\Lambda(\epsilon)$ with respect to ϵ:

$$\frac{d\Lambda(\epsilon)}{d\epsilon}\bigg|_{\epsilon=\hat{\epsilon}} = -\frac{1}{N_0}\sum_{l=1}^{L}\int_{T_{l\epsilon}} p[t-(l-1)T-\hat{\epsilon}]y(t)\,dt$$

$$\times \int_{T_{l\epsilon}} \dot{p}[t-(l-1)T-\hat{\epsilon}]y(t)\,dt = 0 \qquad (6\text{-}55)$$

where $\dot{p}(\cdot)$ denotes the derivative of $p(\cdot)$ with respect to its argument. The statistic that the optimum estimate satisfies is formed by correlating the received signal plus

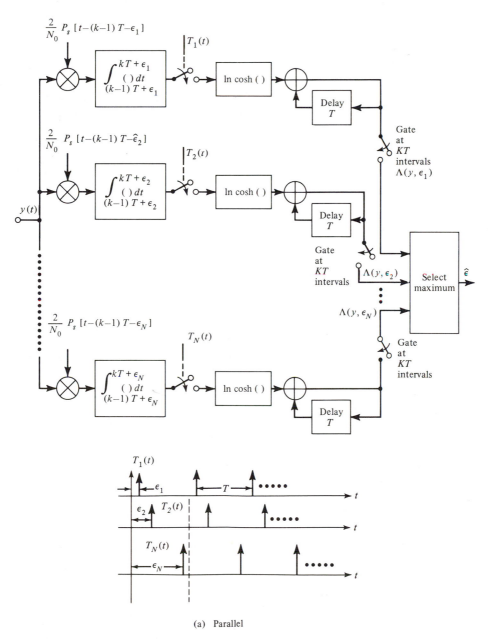

(a) Parallel

FIGURE 6-6. ML estimator of symbol timing epoch. (From Ref. 8.)

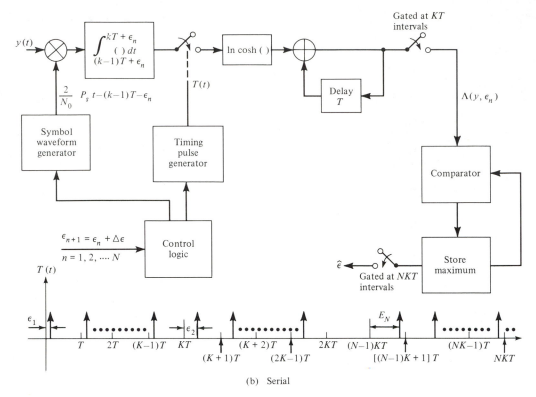

(b) Serial

FIGURE 6-6 continued.

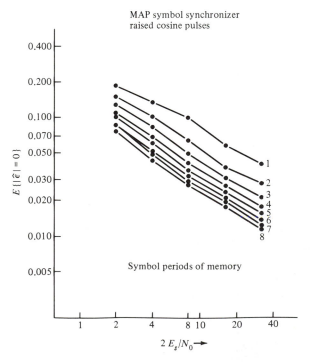

FIGURE 6-7. Performance of an ML symbol synchronization scheme. (From Ref. 8.)

noise with a replica of the pulse and its derivative in the lth interval, and the value of ϵ is sought that makes the product of these correlations zero. Alternatively, the correlation operations can be replaced by matched filter operations. A block diagram of the resulting feedback estimator structure is shown in Figure 6-8. ∎

6-2.4 Data-Aided Versus Non-Data-Aided Synchronization

The likelihood function (6-45) was averaged over the data that produced the cosh (·) in (6-49). The result is referred to as a non-data-aided (NDA) estimator. Another alternative is to assume that the data a_l and b_l are available or have been estimated. This results in a data-aided (DA) estimator. Thus the appropriate expression to be maximized through choice of $\mathbf{A}'$ is (6-45) without the overbars, which represent averaging over the data. In the limit as $K \to \infty$, the log-likelihood function becomes

$$\ln f(y(t)|\mathbf{A}', \hat{\boldsymbol{\alpha}}) = \sum_{l=1}^{L} \frac{\hat{a}_l}{N_0} \int_{T_{le}} [s_r(t, \mathbf{A}')y_r(t) + s_i(t, \mathbf{A}')y_i(t)] \, dt$$

$$+ \sum_{l=1}^{L} \frac{\hat{b}_l}{N_0} \int_{T_{le}} [s_r(t, \mathbf{A}')y_i(t) - s_i(t, \mathbf{A}')y_r(t)] \, dt \quad (6\text{-}56)$$

where $\hat{a}_l$ and $\hat{b}_l$ are the estimates of the lth data bits in the quadrature channels. These can be obtained through feeding back detections from an earlier signaling intervals, which, if the signal-to-noise ratio is not too small, will be essentially error-free (e.g., an error probability of 10^{-3} implies that only 1 bit in 1000 is in error on the average).

EXAMPLE 6-5
Consider the PAM system of Example 6-4. The DA estimator uses the statistic

$$\Lambda(\epsilon) = \sum_{l=1}^{L} \hat{a}_l \int_{T_{le}} p[t - (l - 1)T - \epsilon]y(t) \, dt \quad (6\text{-}57)$$

That is, the received signal plus noise in the lth interval is multiplied by the data estimate and then summed over the L symbol intervals used to make the estimate of the timing epoch. The correlation operation can be replaced by a matched filter. ∎

6-2.5 Joint Estimation of Parameters

To demonstrate the application of the ML technique for joint estimation of parameters, (6-56) will be specialized to QASK (QPSK) with

$$s(t, \theta, \epsilon) = \text{Re} \left\{ \frac{A}{\sqrt{2}} (a_l + jb_l)p(t - lT + \epsilon) \exp [j(\omega_0 t + \theta)] \right\} \quad (6\text{-}58)$$

where

$$p(t) = \begin{cases} 1 & -T \leq t \leq 0 \\ 0 & \text{otherwise} \end{cases}$$
$$(6\text{-}59)$$
$$y(t) = \text{Re} \{[y_r(t) + jy_i(t)] \exp (j\omega_0 t)\}$$

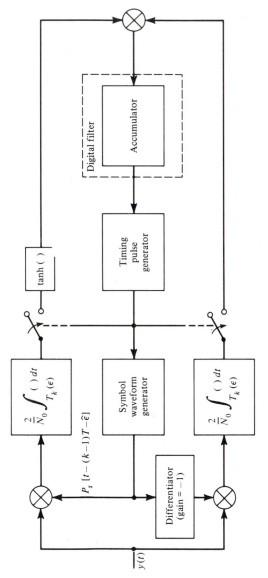

FIGURE 6-8. Feedback structure for the ML estimation of timing epoch.

The DA log-likelihood function (6-55) becomes

$$L(\epsilon, \theta, \hat{\alpha}) =$$

$$\sum_{l=1}^{L} \frac{A}{\sqrt{2N_0}} \left\{ \hat{a}_l \int_{-\infty}^{\infty} p(t - lT + \epsilon)[\cos(\omega_0 t + \theta)y_r(t) + \sin(\omega_0 t + \theta)y_i(t)] \, dt \right.$$

$$\left. + \hat{b}_l \int_{-\infty}^{\infty} p(t - lT + \epsilon)[\cos(\omega_0 t + \theta)y_i(t) - \sin(\omega_0 t + \theta)y_r(t)] \, dt \right\}$$

$$(6\text{-}60)$$

The necessary conditions for estimation of phase, θ, and timing epoch, ϵ, are obtained by differentiating $L(\epsilon, \theta, \hat{\alpha})$ with respect to θ and ϵ, respectively, and setting the resulting equations equal to zero. Using the approximations that

$$y_r(t) = \text{LPF}[2y(t) \cos \omega_0 t] \qquad (6\text{-}61\text{a})$$

and

$$y_i(t) = \text{LPF}[2y(t) \sin \omega_0 t] \qquad (6\text{-}61\text{b})$$

where $\text{LPF}(\cdot)$ denotes the lowpass-filtered part, these necessary conditions can be written as

$$\left. \frac{\partial L}{\partial \theta} \right|_{\substack{\theta = \hat{\theta} \\ \epsilon = \hat{\epsilon}}} = \sum_{l=1}^{L} \frac{A}{\sqrt{2N_0}} \left[\hat{a}_l \int_{-\infty}^{\infty} \left. \frac{\partial p(t - lT + \epsilon)}{\partial \epsilon} \right|_{\hat{\epsilon}} y(t) \cos(\omega_0 t + \hat{\theta}) \, dt \right.$$

$$\left. - \hat{b}_l \int_{-\infty}^{\infty} \left. \frac{\partial p(t - lT + \epsilon)}{\partial \epsilon} \right|_{\hat{\epsilon}} y(t) \sin(\omega_0 t + \hat{\theta}) \, dt \right] = 0$$

$$(6\text{-}62)$$

$$\left. \frac{\partial L}{\partial \epsilon} \right|_{\substack{\theta = \hat{\theta} \\ \epsilon = \hat{\epsilon}}} = \sum_{l=1}^{L} \frac{A}{\sqrt{2N_0}} \left[\hat{a}_l \int_{-\infty}^{\infty} p(t - lT + \epsilon) \right|_{\hat{\epsilon}} y(t) \sin(\omega_0 t + \hat{\theta}) \, dt$$

$$\left. + \hat{b}_l \int_{-\infty}^{\infty} p(t - lT + \epsilon) \right|_{\hat{\epsilon}} y(t) \cos(\omega_0 t + \hat{\theta}) \, dt \right] = 0$$

$$(6\text{-}63)$$

The tracking loop mechanization is shown in Figure 6-9. The NDA estimator structure can be obtained by specializing (6-48) to the appropriate signal structure.

6-2.6 Open-Loop Versus Closed-Loop Structures

Figures 6-6 and 6-8 illustrate two approaches to the implementation of parameter estimators. The first is an implementation that computes the log-likelihood function for a number of possible parameter values, and that value is selected which corresponds to an absolute maximum. Depending on whether the parallel or serial realization is used, the price paid in obtaining this estimate is either in terms of hardware for the parallel realization, or search time for the serial realization. If the parallel realization is used, the time required to make the estimate is KT plus the computation time. If the serial approach is taken, the time required is roughly N

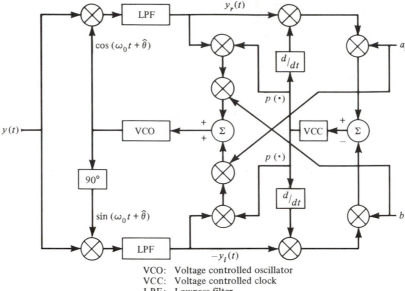

cos $(\omega_0 t + \hat{\theta})$

$y_r(t)$

d/dt

$p(\cdot)$

a_ℓ

$y(t)$

VCO

+
Σ
+

VCC

Σ

+

−

$p(\cdot)$

90°

d/dt

sin $(\omega_0 t + \hat{\theta})$

LPF

$-y_i(t)$

b_ℓ

VCO: Voltage controlled oscillator
VCC: Voltage controlled clock
LPF: Lowpass filter

FIGURE 6-9. DA feedback estimator structure for the joint estimation of carrier phase and timing epoch.

times the parallel search time, where N is the number of trial estimates made. In addition, the resulting log-likelihood function value must be saved after each trial for comparison with that obtained on the next trial in order to obtain the maximum.

The second basic implementation method, referred to as the *feedback approach*, trades hardware complexity for search time. In contrast to the serial method discussed above, however, a trial estimate is made, and then used to produce a new corrected estimate with the correction being proportional to the derivative of $p(\cdot)$ with respect to ϵ.

Yet a third way to mechanize the estimation algorithm is referred to as *sequential search*. In a sequential search procedure, the log-likelihood function (6-55) or an appropriate variation of it is compared with a threshold after an initial choice for ϵ_i has been made. If it is below the threshold, the decision is made that ϵ_i is not the proper value and a new value is tried; if the log-likelihood function is above the threshold, the decision is made that the chosen value of ϵ is the proper one. A more general version of the sequential search procedure makes use of two thresholds. If the log-likelihood function is below the lowest threshold or above the largest threshold, decisions for "out-of-synch" or "in synch" are made; if between the thresholds, more data are integrated.

6-2.7 Practical Timing Epoch Estimators

If the derivative of the log-likelihood function is approximated by the difference

$$\frac{\partial \Lambda(\epsilon)}{\partial \epsilon} \simeq \frac{\Lambda(\epsilon + \Delta\epsilon/2) - \Lambda(\epsilon - \Delta\epsilon/2)}{\Delta\epsilon} \qquad (6\text{-}64)$$

then the derivative of $\Lambda(\epsilon)$ (6-55) is approximately

$$\frac{d\Lambda(\epsilon)}{d\epsilon} \simeq \frac{1}{2N_0\Delta\epsilon} \sum_{l=1}^{L} \left\{ \left[\int_{T_{l\epsilon}} p\left[t - (l-1)T - \hat{\epsilon} - \frac{\Delta\epsilon}{2} \right] y(t)\, dt \right]^2 \right.$$
$$\left. - \left[\int_{T_{l\epsilon}} p\left[t - (l-1)T - \hat{\epsilon} + \frac{\Delta\epsilon}{2} \right] y(t)\, dt \right]^2 \right\} = 0$$

(6-65)

A block diagram of the synchronizer structure corresponding to (6-65) is illustrated in Fig. 6-10. The parameter $\Delta\epsilon$ can be chosen to optimize performance. If the signal-to-noise ratio is large, the square-law devices can be replaced by absolute-value functions in keeping with the approximation $\ln \cosh x \simeq |x|/2$, $|x| \gg 1$. Symbol synchronizers of this type are referred to as *early-late gate symbol synchronizers*.

The phase-noise performance of the absolute-value and square-law early-late gate symbol synchronizers has been analyzed in the literature [9]. For large signal-to-noise ratios, the timing variances normalized by the symbol duration, T, squared are

$$\sigma_{\epsilon,\text{AV}}^2 \simeq \frac{B_L T}{8(E_s/N_0)}$$

(6-66)

and

$$\sigma_{\epsilon,\text{SL}}^2 \simeq \frac{5B_L T}{32(E_s/N_0)}$$

(6-67)

for the absolute-value and square-law synchronizers, respectively, where E_s is the symbol energy, N_0 the single-sided noise power spectral density, and B_L is the noise bandwidth of the first-order tracking loop. The ratio of (6-67) to (6-66) expressed in decibels shows that the absolute-value synchronizer has a 0.97-dB advantage over the square-law synchronizer for large signal-to-noise ratios.

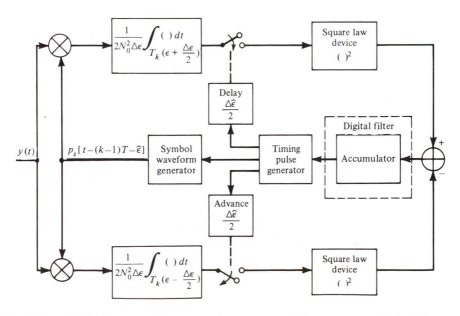

FIGURE 6-10. Block diagram of a square-law type early-late gate symbol timing estimator. (From Ref. 4.)

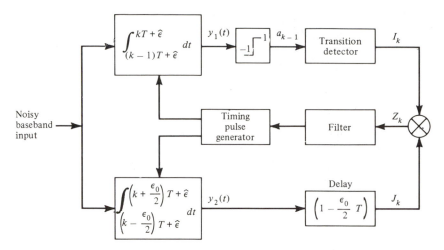

FIGURE 6-11. Block diagram of an in-phase/midphase symbol synchronizer. (From Ref. 5.)

Another type of closed-loop synchronizer, illustrated in Figure 6-11, is referred to as the *in-phase/midphase synchronizer* or *data-transition tracking loop* (DTTL). The upper channel, referred to as the in-phase channel, includes an integrate-and-dump of duration T followed by a limiter to make hard decisions on the data symbols. Following the limiter is a transition detector whose output is given by

$$I_k = \begin{cases} 0 & \text{two successive data symbols the same} \\ -1 & \text{minus-plus data transition} \\ +1 & \text{plus-minus data transition} \end{cases}$$

The lower channel, referred to as the *midphase channel*, also includes an integrate-and-dump operation of duration $\epsilon_0 T$ which is staggered relative to the in-phase channel. The parameter ϵ_0 is chosen to optimize performance. The samples from this integrate-and-dump are delayed by $(1 - \epsilon_0/2)T$ and multiplied by the transition detector outputs. The resulting product is filtered and used to control a voltage-controlled clock (VCC) generator to provide the proper phasing of the integrate-and-dump operations so that the control voltage is driven to zero. The timing variance for large signal-to-noise ratio (see [9]) is

$$\sigma_{\epsilon,\text{DTTL}}^2 = \frac{B_L T}{4(E_s/N_0)} \tag{6-68}$$

which is 3 dB inferior to the absolute-value early-late gate synchronizer.

6-3

SYNCHRONIZATION METHODS BASED ON PROPERTIES OF WIDE-SENSE CYCLOSTATIONARY RANDOM PROCESSES

A wide-sense cyclostationary random process is defined in Appendix A as one having a mean $E[s(t)]$ and autocorrelation function

$$R_s(t + \tau, t) = E[s(t)s(t + \tau)] \tag{6-69}$$

that are periodic functions of time. These properties are useful for deducing synchronization techniques for carrier and clock recovery. In this section these properties will be used to investigate synchronization circuits of various types. For a tutorial summary of this topic, see [10].

6-3.1 Carrier Recovery Circuits

The use of cyclostationarity in deducing carrier tracking circuits can be illustrated by considering a modulated signal of the form

$$s(t) = \text{Re}\ [d(t)e^{j\theta}e^{j\omega_0 t}] \tag{6-70}$$

where the frequency, ω_0, and phase, θ, are constants. The signal $d(t)$ is a random process related to the data. For example, for a BPSK-modulated signal,

$$d(t) = \sum_{-\infty}^{\infty} a_k p(t - kT_b - \Delta) \tag{6-71}$$

where $a_k \epsilon(-1, 1)$ is the kth bit of an independent, identically distributed (iid) data sequence, $p(t)$ is the pulse-shaping function, and Δ is a random variable uniformly distributed in $(0, T_b)$. Therefore, in this case $d(t)$ is a wide-sense stationary process.

In general, $d(t)$ will be assumed to be a complex-valued stationary random process. In this case, the autocorrelation function of $s(t)$ is

$$R_{ss}(t, t + \tau) = E\{\text{Re}\ [d(t)e^{j(\omega_0 t + \theta)}]\ \text{Re}\ [d(t + \tau)e^{j(\omega_0(t+\tau)+\theta)}]\}$$

$$= E\left\{\left[\frac{1}{2}d(t)e^{j(\omega_0 t + \theta)} + \frac{1}{2}d^*(t)e^{-j(\omega_0 t + \theta)}\right]\right.$$

$$\left. \times \left[\frac{1}{2}d(t + \tau)e^{j(\omega_0(t+\tau)+\theta)} + \frac{1}{2}d^*(t + \tau)e^{-j(\omega_0(t+\tau)+\theta)}\right]\right\}$$

$$= \frac{1}{2}\text{Re}\left[R_{dd*}(\tau)e^{j\omega_0\tau}\right] + \frac{1}{2}\text{Re}\left[R_{dd}(\tau)e^{j(2\omega_0 t + \omega_0\tau + 2\theta)}\right] \tag{6-72}$$

where

$$R_{dd}(\tau) = E[d(t)d(t + \tau)] \tag{6-73}$$

$$R_{dd*}(\tau) = E[d(t)d^*(t + \tau)] \tag{6-74}$$

If $s(t)$ is passed through a square-law device, the output has expected value

$$E[s^2(t)] = R_{ss}(t, t)$$

$$= \frac{1}{2}R_{dd*}(0) + \frac{1}{2}\text{Re}\left[R_{dd}(0)e^{j2(\omega_0 t + \theta)}\right] \tag{6-75}$$

which shows that the squarer output has a spectral component at twice the carrier frequency with phase that is twice that of the unknown signal phase. Therefore, a narrowband filter or phase-locked loop in combination with a frequency divider can be used to establish a local reference which is phase coherent with the phase of the received carrier. In considering the noise performance of such a combination there are, in general, three types of noise terms that must be considered. If the input to the squarer is

$$x(t) = s(t) + n(t) \qquad (6\text{-}76)$$

where $n(t)$ is a sample function of a wide-sense stationary random process independent of $s(t)$, the autocorrelation function of the squarer output, $y(t) = [s(t) + n(t)]^2$, is

$$R_{yy}(t, t + \tau) = R_{s \times s}(t, t + \tau) + R_{s \times n}(t, t + \tau) + R_{n \times n}(t, t + \tau) \qquad (6\text{-}77)$$

which is obtained under the assumption that both signal and noise have zero mean. In (6-77), the various correlation functions on the right-hand side are defined as follows:

$$R_{s \times s}(t, t + \tau) \triangleq E[s^2(t)s^2(t + \tau)] \qquad (6\text{-}78a)$$

$$R_{s \times n}(t, t + \tau) \triangleq E[s^2(t + \tau)]E[n^2(t)] + E[s^2(t)]E[n^2(t + \tau)]$$
$$+ 4E[s(t)s(t + \tau)]E[n(t)n(t + \tau)] \qquad (6\text{-}78b)$$

$$R_{n \times n}(t, t + \tau) \triangleq E[n^2(t)n^2(t + \tau)] \qquad (6\text{-}78c)$$

The noise components present at the filter or phase-locked-loop input then consist of the following:

1. The part of (6-78a) that does not contribute to the signal-related component $\frac{1}{2} \operatorname{Re} \{R_{dd*}(0) \exp [j2(\omega_0 t + \theta)]\}$ in (6-75). This noise term, which is dependent solely on the signal, is called *self-noise*.
2. The output components of the squarer due to the interaction of signal with noise which has the autocorrelation function given by (6-78b).
3. The squarer output components due to the interaction of noise with noise with corresponding autocorrelations given by (6-78c).

Quite often, the signal self-noise components are assumed negligible. They are discussed in Chapter 10, where tracking of spreading codes for spread-spectrum systems is considered.

If self-noise is negligible, noise influencing the phase estimate can be attributed to interaction of signal with noise and noise with noise. Assuming the signal to be stationary for the purposes of noise analysis and assuming $n(t)$ to be Gaussian allows (6-78b) and (6-78c) to be simplified to

$$R_{s \times n}(\tau) = 2\sigma_s^2 \sigma_n^2 + 4R_s(\tau)R_n(\tau) \qquad (6\text{-}79a)$$

$$R_{n \times n}(\tau) = \sigma_n^4 + 2R_n^2(\tau) \qquad (6\text{-}79b)$$

where

$$\sigma_s^2 = E[s^2(t)] \qquad (6\text{-}80a)$$

$$\sigma_n^2 = E[n^2(t)] \qquad (6\text{-}80b)$$

$$R_s(\tau) = E[s(t)s(t + \tau)] \qquad (6\text{-}80c)$$

$$R_n(\tau) = E[n(t)n(t + \tau)] \qquad (6\text{-}80d)$$

The power spectral density of the noise into the narrowband filter or phase-locked loop is

$$S_n(f) = (2\sigma_s^2 \sigma_n^2 + \sigma_n^4)\delta(f) + 4S_s(f) * S_n(f) + 2S_n(f) * S_n(f) \qquad (6\text{-}81)$$

where $S_s(f)$ and $S_n(f)$ are the power spectral densities of $s(t)$ and $n(t)$, respectively. The performance of the carrier reference acquisition device, be it a bandpass filter or phase-locked loop, can be characterized in terms of the phase-error variance, which is the square of the difference between the estimate, $\hat{\theta}$, and the true value, θ. Such a calculation is illustrated in the next example.

EXAMPLE 6-6

Consider the use of a squaring loop to estimate a phase reference from a BPSK signal in AWGN. Prior to the squaring device, the signal plus noise is passed through a bandpass filter to eliminate as much noise power as possible while leaving the signal essentially undistorted. A block diagram of the system is shown in Figure 6-12. Let

$$s(t) = A \sum_{k=-\infty}^{\infty} a_k \Pi \left(\frac{t - kT_b - \Delta}{T_b} \right) \cos 2\pi f_c t \qquad (6\text{-}82)$$

Then, from (5-64) with $Q(f) = 0$ and

$$P(f) = A^2 T_b \, \text{sinc}^2 \, (T_b f) \qquad (6\text{-}83)$$

the power spectral density of the signal is

$$S_s(f) = \frac{A^2 T_b}{4} \{ \text{sinc}^2 \, [T_b(f - f_0)] + \text{sinc}^2 \, [T_b(f + f_0)] \} \qquad (6\text{-}84)$$

Assuming the white noise, $n_w(t)$, to have double-sided power spectral density $N_0/2$, the power spectral density of the noise at the bandpass filter output is

$$S_n(f) = \frac{N_0}{2} \left[\Pi \left(\frac{f - f_0}{2W} \right) + \Pi \left[\left(\frac{f + f_0}{2W} \right) \right] \right] \qquad (6\text{-}85)$$

Because of the difficulty in convolving a rectangular function with $\text{sinc}^2(\cdot)$, the signal spectrum, $S_s(f)$, will be replaced by rectangular spectra of bandwidths equal to the equivalent noise bandwidth of the sinc functions composing the signal spectra, which is $1/T_b$ hertz. These spectra are shown in Figure 6-13a.

According to (5-43), the phase-error variance for a phase-locked loop tracking a sinusoidal signal in noise is

$$\sigma_\varphi^2 = \frac{N_0' B_L}{P_c} \qquad (6\text{-}86)$$

where P_c is the power of the signal being tracked, N_0' is the single-sided spectral density of the white input noise [primed here to distinguish it from N_0 in (6-85)], and B_L is the single-sided loop bandwidth. The latter is assumed to be very small compared with the bandwidths of the spectra shown in Figure 6-13. Therefore, as far as the phase-locked-loop tracking performance calculation is concerned, the noise spectrum at the phase-locked-loop input can be approximated as a constant throughout the loop passband. This maximum spectral level can be found by ap-

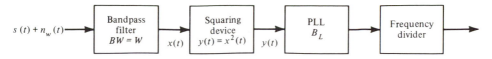

FIGURE 6-12. Block diagram of a carrier acquisition system for BPSK.

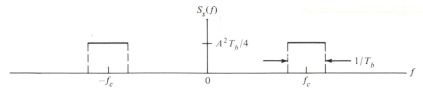

(a) Input signal spectrum (approximation)

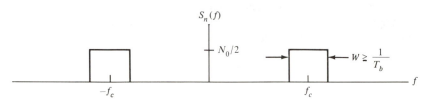

(b) Input noise spectrum

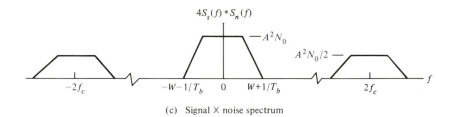

(c) Signal × noise spectrum

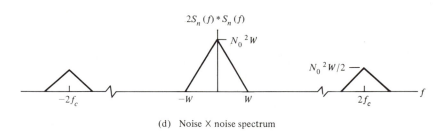

(d) Noise × noise spectrum

FIGURE 6-13. Signal and noise spectra for analyzing the performance of a cross-spectrum coherent reference estimator.

plying (6-81). The spectra $4S_s(f) * S_n(f)$ and $2S_n(f) * S_n(f)$ are shown in Figure 6-13c and d. Note that the first term of (6-81), being at dc, does not contribute to the noise power within the phase-locked-loop passband. From Figures 6-13c and d it follows that the double-sided noise spectral density, $N_0'/2$, at the loop input is approximated by

$$\frac{N_0'}{2} = A^2 N_0 + N_0^2 W \tag{6-87}$$

Furthermore, from (6-75), the power of the signal component tracked by the loop is

$$P_c = \frac{1}{2}\left[\frac{1}{2} R_{dd*}(0)\right]^2 = \frac{R_{dd*}^2(0)}{8} \tag{6-88}$$

But, from (6-83),

$$R_{dd*}(\tau) = \mathcal{F}^{-1}(A^2 T_b \ \text{sinc}^2 T_b f)$$

$$= A^2 \Lambda \left(\frac{t}{T_b} \right) \tag{6-89}$$

and therefore $P_c = A^4/8$. Consequently, (6-86) becomes

$$\sigma_\varphi^2 = 8 \ \frac{(2A^2 N_0 + N_0^2 W)B_L}{A^4}$$

$$= 4 \left(\frac{N_0 B_L}{P_s} \right) + 4 \left(\frac{N_0 B_L}{P_s} \right) \frac{N_0 W}{P_s} \tag{6-90}$$

where $P_s = A^2/2$ is the power of a sinusoidal signal of amplitude A. Since the angle being tracked is twice that desired (i.e., $\varphi = 2\theta$) it follows that

$$\sigma_\theta^2 = \frac{1}{4} \ \sigma_\varphi^2$$

$$= \left(\frac{N_0 B_L}{P_s} \right) \left(1 + \frac{N_0 W}{P_s} \right)$$

$$= \frac{S_L^{-1}}{\rho} \tag{6-91}$$

where ρ is the equivalent signal-to-noise ratio in the loop bandwidth as defined in Chapter 5 and S_L is known as the *squaring loss* [11], which for this particular case can be written as

$$S_L = \frac{1}{1 + 1/\rho\gamma} \tag{6-92}$$

where $\gamma = W/B_L$. The inverse of the squaring loss represents the factor by which the phase error variance of a squaring loop is increased over that of a normal phase-locked loop due to the noise $\times$ noise interaction in the loop. ■

6-3.2 Delay and Multiply Circuits for Symbol Clock Estimation

In this subsection, a specific form of the cross-spectrum synchronizer shown in Figure 6-2 will be considered. In particular, filter 2 of that circuit will be taken as an ideal delay of τ seconds, so that

$$G(f) = e^{-j2\pi f \tau} \tag{6-93}$$

As before, suppose that the input to the circuit is signal plus noise as expressed by (6-76) with the signal component a general modulated signal of the form (6-70). The filter with transfer function $H(f)$ is initially assumed to introduce negligible distortion to the signal $s(t)$, so that

$$y_1(t) = s(t) + n(t) \tag{6-94}$$

$$y_2(t) = s(t - \tau) + n(t - \tau) \tag{6-95}$$

The expectation of the output of the multiplier is

$$E[z(t)] = E[s(t)s(t - \tau)] + E[n(t)n(t - \tau)]$$

$$= E[s(t)s(t - \tau)] + R_n(\tau) \tag{6-96}$$

where $R_n(\tau)$ is the autocorrelation function of the noise. (Note that the cross-terms are zero becuase the average of the noise is zero.)

To see that spectral components are generated in the signal portion of the output $z(t)$ which can be used for clock recovery, consider the first term of (6-96), where $s(t)$ is given by (6-70). From (6-72) it follows that

$$E[s(t)s(t - \tau)] = R_{ss}(t, t - \tau)$$

$$= \frac{1}{2} \text{Re } [R_{dd*}(-\tau)e^{-j\omega_0\tau}] + \frac{1}{2} \text{Re } [R_{dd}(-\tau)e^{j(2\omega_0 t - \omega_0 \tau + 2\theta)}] \tag{6-97}$$

If, for example, $s(t)$ is a QPSK-modulated signal, then

$$d(t) = d_r(t) + jd_i(t) \tag{6-98}$$

where

$$d_r(t) = \frac{A}{\sqrt{2}} \sum_{k=-\infty}^{\infty} a_k p(t - kT_s) \tag{6-99}$$

$$d_i(t) = \frac{A}{\sqrt{2}} \sum_{k=-\infty}^{\infty} b_k p(t - kT_s) \tag{6-100}$$

in which $\{a_n\}$ and $\{b_n\}$ are independent iid sequences with each member taking on the values $+1$ or -1 with equal probability. In (6-100), A is the amplitude of the modulated carrier and $p(t)$ is a basic pulse shape. Since it is timing epoch that is being estimated there is no phase randomizing of the data sequences (6-99) and (6-100). Use of (6-98) and (6-99) to evaluate the correlation functions in (6-97) will indicate whether or not a clock-related spectral component is present. The correlation function $R_{dd}(-\tau)$ becomes

$$R_{dd}(-\tau) = E\left[\frac{A^2}{2} \sum_{k=-\infty}^{\infty} (a_k + jb_k)p(t - kT_s) \times \sum_{l=-\infty}^{\infty} (a_l + jb_l)p(t - \tau - lT_s)\right]$$

$$= \frac{A^2}{2} \sum_{k,l=-\infty}^{\infty} E[(a_k + jb_k)(a_l + jb_l) \times p(t - kT_s)p(t - \tau - lT_s)] \tag{6-101}$$

But by the independent and iid properties of the data sequences,

$$E[(a_k + jb_k)(a_l + jb_l)] = E(a_k a_l) - E(b_k b_l) = 0 \tag{6-102}$$

which means that $R_{dd}(-\tau) \equiv 0$. Now consider $R_{dd*}(-\tau)$, which is

$$R_{dd*}(-\tau) = E\left[\frac{A^2}{2} \sum_{k,l=-\infty}^{\infty} (a_k + jb_k)(a_l - jb_l)p(t - kT_s)p(t - \tau - lT_s)\right]$$

$$= \frac{A^2}{2} \sum_{k,l=-\infty}^{\infty} E[(a_k + jb_k)(a_l - jb_l)p(t - kT_s)p(t - \tau - lT_s)] \tag{6-103}$$

Again using the iid and independent properties of the data sequences, it follows that

$$E[(a_k + jb_k)(a_l - jb_l)] = E(a_k a_l) + E(b_k b_l)$$

$$= 2\delta_{kl} \qquad (6\text{-}104)$$

where δ_{kl} is the Kronecker delta. Thus (6-103) becomes

$$R_{dd*}(-\tau) = A^2 \sum_{k=-\infty}^{\infty} p(t - kT_s)p(t - \tau - kT_s) \qquad (6\text{-}105)$$

and the autocorrelation function of the signal, from (6-72), is

$$R_{ss}(t, t + \tau) = \frac{A^2}{2} \sum_{k=-\infty}^{\infty} p(t - kT_s)p(t - \tau - kT_s) \cos \omega_0\tau \qquad (6\text{-}106)$$

Two items regarding the synchronization can be noted from (6-106). First, relative maxima occur for $\omega_0\tau = 2n\pi$, where n is an integer. Second, $R_{ss}(t, t + \tau)$ is periodic in t, as can be seen by replacing t by $t + T_s$ [i.e., $s(t)$ is cyclostationary]. Thus the signal portion of the average output can be represented in terms of a Fourier series. For $\cos \omega_c\tau = 1$, this Fourier series is of the form

$$E[s(t)s(t - \tau)] = \sum_{n=-\infty}^{\infty} c_n e^{j2\pi nt/T_s} \qquad (6\text{-}107)$$

where

$$
\begin{aligned}
c_n &= \frac{1}{T_s} \int_{-T_s/2}^{T_s/2} \frac{A^2}{2} \sum_{k=-\infty}^{\infty} p(t - kT_s)p(t - \tau - kT_s)e^{-j2\pi nt/T_s}\, dt \\
&= \frac{A^2}{2T_s} \sum_{k=-\infty}^{\infty} \int_{-T_s/2}^{T_s/2} p(t - kT_s)p(t - \tau - kT_s)e^{-j2\pi nt/T_s}\, dt \\
&= \frac{A^2}{2T_s} \int_{-\infty}^{\infty} p(t)p(t - \tau)e^{-j2\pi nt/T_s}\, dt \qquad (6\text{-}108)
\end{aligned}
$$

The component at the clock frequency is given by

$$2\mathrm{Re}\,(c_1 e^{j2\pi nt/T_s})$$

with power $|c_1|^2$. If $\omega_c\tau = n\pi$ is assumed, it remains to adjust τ to maximize the clock component power for a given pulse shape. This is illustrated by the following example.

EXAMPLE 6-7

Optimum delay will be found for a rectangular pulse shape in this example.

Solution: Assume that $p(t) = \Pi(t/T_s)$ and $A = 1$, for which

$$
c_n = \begin{cases}
\dfrac{1}{2T_s} \displaystyle\int_{-T_s/2}^{\tau + T_s/2} \exp\left(\dfrac{-j2\pi nt}{T_s}\right) dt & \dfrac{-T_s}{2} \le \tau < 0 \\[4ex]
\dfrac{1}{2T_s} \displaystyle\int_{\tau - T_s/2}^{T_s/2} \exp\left(\dfrac{-j2\pi nt}{T_s}\right) dt & 0 \le \tau \le \dfrac{T_s}{2}
\end{cases}
$$

$$= (-1)^n \frac{|\tau|}{2T_s} \operatorname{sinc}\left(\frac{n\tau}{T_s}\right) \exp\left(\frac{-j\pi n|\tau|}{T_s}\right) \qquad |\tau| \le \frac{T_s}{2}$$

For $n = 1$, the value of $|c_1|^2$ is

$$|c_1|^2 = \left(\frac{\tau}{2T_s}\right)^2 \operatorname{sinc}^2\left(\frac{\tau}{T_s}\right)$$

which has a maximum value of 0.0253 for $\tau = T_s/2$. Thus if one is interested in acquiring a clock reference that is synchronous with the symbol rate, $1/T_s$, the optimum delay to use in a delay-and-multiply cross-spectrum synchronizer is one-half symbol period. ■

The question of performance of a cross-spectrum synchronizer in noise still remains. To illustrate the consideration of noise, consider the delay-and-multiply synchronizer for which $G(f) = e^{-j2\pi f\tau}$ in Figure 6-2. Assume the input to signal, $s(t)$, plus AWGN, $n_w(t)$, and assume that filter 1 leaves the signal undistorted while bandlimiting the noise. The signals $y_1(t)$ and $y_2(t)$ are, respectively, represented by

$$y_1(t) = s(t) + n_1(t) \tag{6-109}$$

$$y_2(t) = s(t - \tau) + n_2(t) \tag{6-110}$$

Both $n_1(t)$ and $n_2(t)$ are Gaussian random processes with identical power spectral densities since $|G(f)|^2 = 1$. To determine the noise spectrum of the multiplier output, however, it is useful to compute the autocorrelation function of $z(t)$, which can be written as

$$R_z(\lambda) = E[z(t)z(t + \lambda)]$$

$$= E\{[s(t) + n_1(t)][s(t - \tau) + n_2(t)]$$

$$\times [s(t + \lambda) + n_1(t + \lambda)][s(t + \lambda - \tau) + n_2(t + \lambda)]\}$$

$$= R_{s\times s}(\lambda) + R_{s\times n}(\lambda) + R_{n\times n}(\lambda) \tag{6-111}$$

where

$$R_{s\times s}(\lambda) = E[s(t)s(t - \tau)s(t + \lambda)s(t + \lambda - \tau)] \tag{6-112a}$$

$$R_{s\times n}(\lambda) = E[s(t + \lambda)s(t + \lambda - \tau)]E[n_1(t)n_2(t)]$$

$$+ E[s(t + \lambda)s(t - \tau)]E[n_1(t)n_2(t + \lambda)]$$

$$+ E[s(t + \lambda - \tau)s(t - \tau)]E[n_1(t)n_1(t + \lambda)]$$

$$+ E[s(t)s(t + \lambda - \tau)]E[n_2(t)n_1(t + \lambda)]$$

$$+ E[s(t)s(t - \tau)]E[n_1(t + \lambda)n_2(t + \lambda)] \tag{6-112b}$$

$$R_{n\times n}(\lambda) = E[n_1(t)n_2(t)n_1(t + \lambda)n_2(t + \lambda)] \tag{6-112c}$$

The first equation, (6-112a), represents the interaction of signal with signal, and contains both discrete spectral components of use in generating a symbol timing clock and a continuous part which represents self-noise. The latter component will be assumed to be negligible. The last correlation function, given by (6-112c), is due to the interaction of noise with noise and can be simplified by noting that $n_1(t) = n(t)$ and $n_2(t) = n(t - \tau)$ are Gaussian processes. Hence, by (A-43), $R_{n\times n}(\lambda)$ can be written as

$$R_{n\times n}(\lambda) = R_n^2(\tau) + R_n^2(\lambda) + R_n(\lambda - \tau)R_n(\lambda + \tau) \tag{6-113}$$

where $R_n(\lambda)$ is the autocorrelation function of the noise at the output of filter 1. Consequently, the spectrum of the noise × noise term is

$$S_{n \times n}(f) = R_n^2(\tau)\delta(f) + S_n(f) * S_n(f) + [S_n(f)e^{-j2\pi f\tau}] * [S_n(f)e^{j2\pi f\tau}] \qquad (6\text{-}114)$$

The final term to be considered is the signal × noise contribution with correlation function (6-112b). For purposes of simplification, the signal will be assumed stationary so that (6-112b) becomes

$$R_{s \times n}(\lambda) = 2R_s(\tau)R_n(\tau) + R_s(\lambda + \tau)R_n(\lambda - \tau)$$

$$+ R_s(\lambda)R_n(\lambda) + R_s(\lambda - \tau)R_n(\lambda + \tau) \qquad (6\text{-}115)$$

with the corresponding spectral density

$$S_{s \times n}(f) = 2R_s(\tau)R_n(\tau)\delta(f) + S_s(f) * S_n(f)$$

$$+ [S_s(f)e^{-j2\pi f\tau}] * [S_n(f)e^{j2\pi f\tau}]$$

$$+ [S_s(f)e^{j2\pi f\tau}] * [S_n(f)e^{-j2\pi f\tau}] \qquad (6\text{-}116)$$

The noise power within the phase-locked-loop passband can be approximated by these spectral densities evaluated at $f = 1/T_s$ times the loop bandwidth. The computations are similar to those of Example 6-6.

EXAMPLE 6-8

Consider a BPSK signal spectrum plus AWGN with a rectangular lowpass spectrum of bandwidth $2W$ and double-sided spectral density of $N_0/2$. In order to carry out the convolutions in (6-116), the signal will be approximated by a rectangular spectrum of lowpass bandwidth $\frac{1}{2}T_s$ hertz and double-sided spectral level A^2T_b. Carrying out the convolutions, the two-sided equivalent noise spectral densities are as shown in Figure 6-14. The equivalent noise spectral density at the phase-locked-loop input at $f = 1/T_s$ is

$$\frac{N_0'}{2} = A^2N_0 \cos\frac{2\pi\tau}{T_s} + \frac{A^2N_0}{2} + \frac{N_0^2W}{2}\left(1 - \frac{1}{2WT_s}\right)$$

$$= \frac{A^2N_0}{2} + \frac{N_0^2W}{4}$$

where the last equation results from assuming that $WT_s = 1$ and $\tau/T_s = \frac{1}{2}$. From Example 6-7, the power of the spectral component tracked by the loop is

$$P_c = 0.0253A^4$$

so that the tracking variance of the loop is

$$\sigma_\varphi^2 = \frac{N_0'B_L}{P_c} = \frac{N_0B_L}{0.0506A^2} + \frac{N_0^2WB_L}{0.1012A^4}$$

$$\simeq 10\frac{N_0B_L}{P_s}\left(1 + \frac{N_0B_L}{4P_s}\frac{W}{B_L}\right) \qquad \blacksquare$$

As in the case of tracking a carrier reference for a BPSK signal by means of a squaring loop, the performance of the delay and multiply circuit includes a squaring loss. Optimization of the filter bandwidths has been considered in a dissertation by McCallister [11].

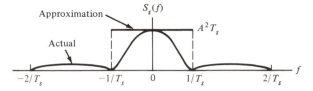

(a) Data spectrum and approximation

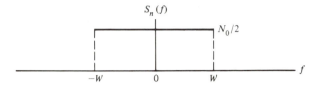

(b) Noise spectrum

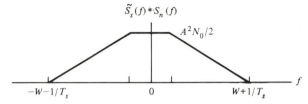

(c) Signal cross noise spectrum (approximation)

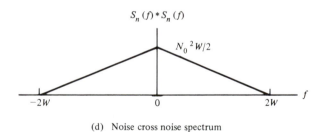

(d) Noise cross noise spectrum

FIGURE 6-14. Spectra pertinent to computing tracking error variance of a cross-spectrum symbol synchronizer.

6-4

SUMMARY

Two basic approaches to parameter estimation for synchronization purposes have been considered in this chapter. These are the MAP (or ML) estimation approach and application of the property of cyclostationarity. Applications to carrier phase and timing epoch estimation have been considered. Application to a third area, that of frame synchronization, will not be considered here; rather, the reader is referred to the literature [12, 13].

REFERENCES

[1] J. K. Holmes, *Coherent Spread Spectrum Systems* (New York: Wiley, 1982), Chap. 12.

[2] F. M. Gardner and W. C. Lindsey, eds., Special Issue on Synchronization, *IEEE Trans. Commun.*, Vol. COM-28, August 1980.

[3] J. J. Stiffler, *Theory of Synchronous Communications* (Englewood Cliffs, N.J.: Prentice-Hall, 1971), Part II.

[4] W. C. Lindsey and M. K. Simon, *Telecommunication System Engineering*, (Englewood Cliffs, N.J.: Prentice-Hall, 1973), Chap. 9.

[5] V. K. Bhargava, D. Haccoun, R. Matyas, and P. Nuspl, *Digital Commmunications by Satellite* (New York: Wiley, 1981), Chap. 5.

[6] R. E. Ziemer and W. H. Tranter, *Principles of Communications* (Second edition) (Boston: Houghton Mifflin, 1985), Chap. 7.

[7] H. L. Van Trees, *Detection, Estimation, and Modulation Theory*, Vol. I, (New York: Wiley, 1968), Chap. 4.

[8] P. A. Wintz and E. J. Luecke, "Performance of Optimum and Suboptimum Synchronizers," *IEEE Trans. Commun. Technol.*, Vol. COM-17, pp. 380–389, June 1969.

[9] M. K. Simon, "Nonlinear Analysis of an Absolute Value Type of an Early-Late Gate Bit Synchronizer," *IEEE Trans. Commun. Technol.*, Vol. COM-18, pp. 589–596, October 1970.

[10] L. E. Franks, "Carrier and Bit Synchronization in Data Communication—A Tutorial Review," *IEEE Trans. Commun.*, Vol. COM-28, pp. 1107–1129, August 1980.

[11] R. D. McCallister, "Generalized Cross-Spectrum Symbol Synchronization," Ph.D. dissertation, Arizona State University, December 1981.

[12] R. A. Scholtz, "Frame Synchronization Techniques," *IEEE Trans. Commun.*, Vol. COM-28, pp. 1204–1213, August 1980.

[13] C. R. Carter, "Survey of Synchronization Techniques for a TDMA Satellite—Switched System," *IEEE Trans. Commun.*, Vol. COM-28, pp. 1291–1301, August 1980.

PROBLEMS

6-1. Consider the maximum-likelihood estimation of a parameter A in independent, additive, Gaussian noise with mean zero and variance σ_n^2:

$$y(t) = A + n(t)$$

(a) The observed waveform is sampled at a single time instant, t_k, and an ML estimate made on the basis of this sample. Show that this estimate is

$$\hat{a}_{\mathrm{ML}} = y(t_k)$$

with variance σ_n^2.

(b) The observed waveform is sampled at N time instants spaced such that the samples are independent. Show that

$$\hat{a}_{\mathrm{ML}} = \frac{1}{N} \sum_{k=1}^{N} y(t_k)$$

and that

$$\mathrm{var}\,(\hat{a}_{\mathrm{ML}}) = \frac{\sigma_n^2}{N}$$

(c) Show that the Cramér–Rao inequality is satisfied with the equality in this case.

6-2. Obtain a lower bound for the variance for the phase estimate of a BPSK-modulated signal in AWGN considered in Example 6-2.

6-3. Starting with (6-45), show that (6-52) holds.

6-4. Derive an expression for the lower bound on the variance of the timing epoch estimator shown in Figure 6-8. Assume a raised-cosine pulse:

$$p(t) = \frac{1}{2}\left(1 + \cos\frac{\pi t}{\tau_0}\right)\Pi\left(\frac{t}{2\tau_0}\right)$$

Make any approximations necessary to obtain a closed-form result.

6-5. (a) Obtain the timing epoch estimate variance for the data-aided (DA) estimator of Example 6-5.
(b) Using a raised-cosine pulse, compare the variance of the DA estimator with the NDA estimator of Problem 6-4.

6-6. Draw suitable waveforms for each subsystem output of the early-late gate symbol timing estimator of Figure 6-10 to explain how it works.

6-7. Draw waveforms at the outputs of the subsystem blocks of the inphase/midphase symbol synchronizer of Figure 6-11 and explain how it works.

6-8. (a) Obtain the autocorrelation function of the square of QASK modulated signal of the form (6-58) by using (6-72). Simplify it as much as possible for the case of a rectangular pulse-shaping function.
(b) Identify the component of $R_{ss}(0, 0)$ suitable for deriving a carrier reference.

6-9. (a) Derive (6-77), showing that the terms defined by (6-78) are correct.
(b) Assuming the noise to be Gaussian, and using (A-43), show that (6-79) and (6-80) hold.
(c) Sketch power spectra corresponding to each term in (6-77) for a signal of the form

$$s(t) = A \cos(2\pi f_0 t + \theta)$$

where A and f_0 are constants and θ is a random variable uniformly distributed in $(0, 2\pi)$. Assume the noise to be bandlimited, Gaussian noise with power spectral density

$$S_n(f) = \begin{cases} \dfrac{N_0}{2} & f_0 - \dfrac{W}{2} \le |f| \le f_0 + \dfrac{W}{2} \\ 0 & \text{otherwise} \end{cases}$$

6-10. Verify the spectra shown in Figure 6-13 and justify (6-92).

6-11. Rework Example 6-7 for the half-cosine pulse shape

$$p(t) = \cos\frac{\pi t}{T_s}\Pi\left(\frac{t}{T_s}\right)$$

CHAPTER 7

Introduction to Spread-Spectrum Systems

7-1

INTRODUCTION

All of the modulation/demodulation techniques discussed so far have been designed to communicate digital information from one place to another as efficiently as possible in a stationary additive white Gaussian noise (AWGN) environment. The transmitted signals were selected to be relatively efficient in their use of the communication resources of power and bandwidth. The demodulators were designed to yield minimum bit error probability for the given transmitted signal in AWGN. Quantitative comparisons were made using the bandwidth and the E_b/N_0 required by the modem to achieve a specified bit error probability.

Although many real-world communication channels are accurately modeled as stationary AWGN channels, there are other important channels which do not fit this model. Consider, for example, a military communication system which might be jammed by a continuous wave (CW) tone near the modem's center frequency or by a distorted retransmission of the modem's own signal. The interference cannot be modeled as stationary AWGN in either of these cases. Another jammer may transmit AWGN, but the jamming signal may be pulsed and is therefore not stationary.

Another type of interference, which does not fit the stationary AWGN model, occurs when there are multiple propagation paths between the transmitter and receiver.

The modem then interferes with itself via a delayed reception of its own signal. This phenomenon is called *multipath reception* and is a problem in line-of-sight microwave digital radios such as those used for long-haul telephone transmission and in urban mobile radio, among other places.

The remainder of this text is devoted to discussing a modulation and demodulation technique that can be used as an aid in mitigating the deleterious effects of the types of interference described above. This modulation and demodulation technique is called *spread spectrum* because the transmission bandwidth employed is much greater than the minimum bandwidth required to transmit the digital information. To be classified as a spread-spectrum system, the modem must have the following characteristics:

1. The transmitted signal energy must occupy a bandwidth which is larger than the information bit rate (usually much larger) and which is independent of the information bit rate.
2. Demodulation must be accomplished, in part, by correlation of the received signal with a replica of the signal used in the transmitter to spread the information signal.

A number of modulation techniques use a transmission bandwidth much larger than the minimum required for data transmission but are not spread-spectrum modulations. Low-rate coding, for example, results in increased transmission bandwidth but does not satisfy either of the conditions above. Wideband frequency modulation also results in a large transmission bandwidth but is not spread spectrum.

Spread-spectrum techniques can be very useful in solving a wide range of communications problems. The amount of performance improvement that is achieved through the use of spread spectrum is defined as the *processing gain* of the spread-spectrum system. That is, processing gain is the difference between system performance using spread-spectrum techniques and system performance not using spread-spectrum techniques, all else being equal. An often used approximation for processing gain is the ratio of the spread bandwidth to the information rate. In fact, some authors define processing gain as this or a similar bandwidth ratio. The particular definition chosen is of little consequence as long as it is always understood that real system performance improvement is the primary concern of the spread-spectrum system designer.

This chapter is intended to provide further motivation for the study of spread-spectrum systems and to introduce the most widely used types of spread-spectrum systems. Two important communication problems which can be partially solved using spread-spectrum techniques are described in Section 7-2. The two fundamental types of spread-spectrum systems, direct-sequence (DS) and frequency-hop (FH), are described in sections 7-3 and 7-4, and hybrid forms are described in Section 7-5. The chapter concludes with a discussion of complex-envelope models used for spread spectrum.

7-2

TWO COMMMUNICATIONS PROBLEMS

7-2.1 Pulse-Noise Jamming

Consider a coherent binary phase-shift-keyed (BPSK) communication system which is being used in the presence of a pulse-noise jammer. A *pulse-noise jammer* transmits pulses of bandlimited white Gaussian noise having total average power J referred to the receiver front end. The jammer may choose the center frequency and bandwidth of the noise to be identical to the receiver's center frequency and bandwidth. In addition, the jammer chooses its pulse duty factor ρ to cause maximum degradation to the communication link while maintaining constant average transmitted power J.

It was shown earlier that the bit error probability of a coherent BPSK system is given by

$$P_E = Q\left(\sqrt{\frac{2E_b}{N_0}}\right) \tag{7-1}$$

The one-sided noise power spectral density N_0 in this expression represents receiver front end thermal noise. When transmitting, the noise jammer increases the receiver noise power spectral density from N_0 to $N_0 + N_J/\rho$, where $N_J = J/W$ is the one-sided average jammer power spectral density and W is the transmission bandwidth. The jammer transmits using duty factor ρ, so that the average bit error probability is

$$\overline{P}_E = (1 - \rho)Q\left(\sqrt{\frac{2E_b}{N_0}}\right) + \rho Q\left(\sqrt{\frac{2E_b}{N_0 + N_J/\rho}}\right) \tag{7-2}$$

The jammer, given this formula, chooses ρ to maximize $\overline{P}_E$.

When a system is being designed to operate in a jamming environment, the maximum possible transmitter power is generally used and thermal noise can be safely neglected. In this case, the first term in (7-2) vanishes and $\overline{P}_E$ can be approximated by

$$\overline{P}_E \simeq \rho Q\left(\sqrt{\frac{2E_b\rho}{N_J}}\right) \tag{7-3}$$

The Q-function can be bounded [1] by an exponential yielding

$$\overline{P}_E \leq \frac{\rho}{\sqrt{4\pi E_b\rho/N_J}}\, e^{-E_b\rho/N_J} \tag{7-4}$$

The maximum of this function over ρ can be found by taking the first derivative and setting it equal to zero. The maximizing ρ is found to be $\rho = N_J/2E_b$ and $\overline{P}_{E,\text{max}}$ is given by

$$\overline{P}_{E,\text{max}} \quad \frac{1}{\sqrt{2\pi e}}\, \frac{1}{2E_b/N_J} \tag{7-5}$$

Of course, the duty factor must be less than or equal to unity so that (7-5) applies only when $E_b/N_J \geq 0.5$. For $E_b/N_J < 0.5$, $\overline{P}_E$ is given by (7-3) with $\rho = 1.0$. Observe that the exponential dependence of bit error probability on signal-to-noise ratio of (7-1) has been replaced by an inverse linear relationship in (7-5). Equations (7-1) and (7-5) are plotted in Figure 7-1, where it can be seen that the pulse noise jammer causes a degradation of approximately 31.5 dB at a bit error probability of 10^{-5}.

The severe degradation in system performance caused by the pulse-noise jammer can be largely eliminated by using a combination of spread-spectrum techniques and forward error correction coding with appropriate interleaving. The effect of the spectrum spreading will be to change the abscissa from E_b/N_J to E_bK/N_J, where K is a constant about equal to W/R, where R is the data rate of the spread-

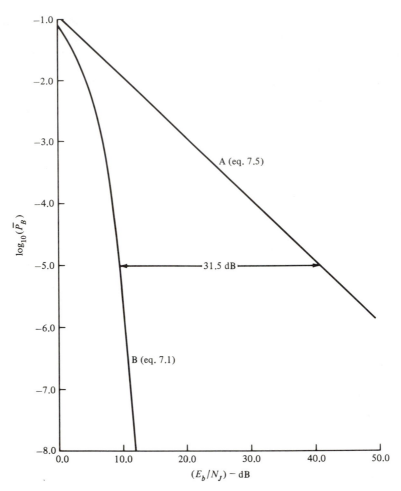

FIGURE 7-1. Bit error probability: (A) worst-case pulse noise jammer; (B) continuous noise jammer.

spectrum system. Error correction coding will be used to return from the inverse linear relation between error probability and signal-to-noise ratio to nearly the exponential relationship desired.

Finally, observe that in order to cause maximum degradation, the jammer must know the value of E_b/N_J at the receiver. This implies accurate knowledge of attenuation in both the transmitter-to-receiver path and jammer-to-receiver path. This knowledge would be difficult to obtain in a tactical environment, so that the results just described are worst-case. In addition, a real jammer would be limited in peak power output and would not be able to use an arbitrarily small duty factor. In spite of these limitations, the pulse jammer is a serious threat to military communications systems.

7-2.2 Low Probability of Detection

Situations exist where it is desirable that a communication link be operated without knowledge of certain parties. *Low probability of detection* (LPD) communication systems are designed to make their detection as difficult as possible by anyone but

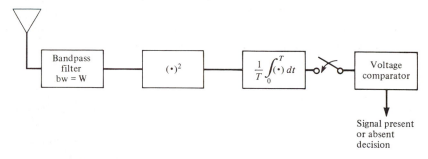

FIGURE 7-2. Energy detector or radiometer.

the intended receiver. This, of course, implies that the minimum signal power required to achieve a particular performance is used. The goal of the LPD system designer is to use a signaling scheme that results in the minimum probability of being detected within some time interval. Spread-spectrum techniques can significantly aid the system designer in achieving this goal.

Assume that the detector is using a radiometer. A radiometer detects energy received in a bandwidth W by filtering to this bandwidth, squaring the output of this filter, integrating the output of the squarer for time T, and comparing the output of the integrator at time T with a threshold as illustrated in Figure 7-2. If the integrator output is above a preset threshold at time T, the signal is declared present; otherwise, the signal is declared absent. The performance of the radiometer in detecting the desired communication signal is known if the probability density function of the integrator output at time T is known. This probability density function is used to calculate the probability, P_d, of detecting the signal if it is indeed present, and probability of falsely declaring a detection when noise alone is present, P_{fa}.

Two approximations are often used for the integrator output pdf. The first, which is used when the *time–bandwidth product TW* is large relative to the received energy to noise power spectral density ratio E/N_0, is to approximate the pdf as Gaussian. In this case, P_d, P_{fa}, TW, and E/N_0 are related by [2]

$$P_D = \Phi\left\{\left[\frac{P}{N_0}\sqrt{\frac{T}{W}} - \Phi^{-1}(1 - P_{fa})\right]\right\} \qquad (7\text{-}6)$$

where*

$$\Phi(y) = \frac{1}{\sqrt{2\pi}}\int_{-\infty}^{y}\exp\left(-\frac{1}{2}\zeta^2\right)d\zeta \qquad (7\text{-}7)$$

and $\Phi^{-1}(x)$ is equal to the variate y such that $\Phi(y) = x$. For a fixed P_{fa}, the probability of detection can be made smaller by reducing P/N_0 or increasing W. Integration time, T, is controlled by the detector, but W can be increased using spread-spectrum techniques. Thus another important application for spread spectrum is the reduction of signal detectability for fixed SNR, integration time, and detector false-alarm probability.

---

*The function $\Phi(y)$ is easily related to the Q-function, which can be calculated using the polynomial approximation given in Appendix E.

DIRECT-SEQUENCE SPREAD SPECTRUM

One method of spreading the spectrum of a data-modulated signal is to modulate the signal a second time using a very wideband spreading signal. This second modulation is usually some form of digital phase modulation, although analog amplitude or phase modulation is conceptually possible. The spreading signal is chosen to have properties which facilitate demodulation of the transmitted signal by the intended receiver, and which make demodulation by an unintended receiver as difficult as possible. These same properties will also make it possible for the intended receiver to discriminate between the communication signal and jamming. If the bandwidth of the spreading signal is large relative to the data bandwidth, the spread-spectrum transmission bandwidth is dominated by the spreading signal and is nearly independent of the data signal.

Bandwidth spreading by direct modulation of a data-modulated carrier by a wideband spreading signal or code is called *direct-sequence* (DS) *spread spectrum*. Other types of spread-spectrum systems exist in which the spreading code is used to control the frequency or time of transmission of the data-modulated carrier, thus indirectly modulating the data-modulated carrier by spreading code. These systems will be discussed later. The digital codes used for the spreading signal are discussed in detail in Chapter 8. The most common techniques used for direct-sequence spreading are discussed below.

7-3.1   BPSK Direct-Sequence Spread Spectrum

The simplest form of DS spread spectrum employs binary phase-shift keying (BPSK) as the spreading modulation. It was shown earlier that ideal BPSK modulation results in instantaneous phase changes of the carrier by 180 degrees and can be mathematically represented as a multiplication of the carrier by a function $c(t)$ which takes on the values ± 1. Consider a constant-envelope data-modulated carrier having power P, radian frequency ω_0, and data phase modulation $\theta_d(t)$ given by*

$$s_d(t) = \sqrt{2P} \cos \left[\omega_0 t + \theta_d(t) \right] \qquad (7\text{-}8)$$

This signal occupies a bandwidth typically between one-half and twice the data rate prior to DS spreading, depending on the details of the data modulation. BPSK spreading is accomplished by simply multiplying $s_d(t)$ by a function $c(t)$ representing the spreading waveform, as illustrated in Figure 7-3. The transmitted signal is

$$s_t(t) = \sqrt{2P} \, c(t) \cos \left[\omega_0 t + \theta_d(t) \right] \qquad (7\text{-}9)$$

The signal of (7-9) is transmitted via a distortionless path having transmission delay T_d. The signal is received together with some type of interference and/or Gaussian noise. Demodulation is accomplished in part by remodulating with the spreading code appropriately delayed as shown in Figure 7-4. This remodulation or correlation of the received signal with the delayed spreading waveform is called *despreading* and is a critical function in all spread-spectrum systems. The signal

---

*Hereafter, the carrier frequency will be denoted by ω_0 rather than ω_c so that ω_c may be used for the spreading code clock frequency.

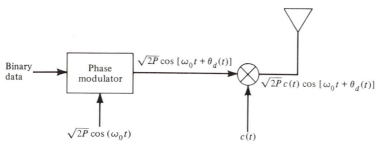

FIGURE 7-3. BPSK direct sequence spread spectrum transmitter.

component of the output of the despreading mixer is

$$\sqrt{2P}\ c(t - T_d)c(t - \hat{T}_d)\ \cos\ [\omega_0 t + \theta_d(t - T_d) + \varphi] \qquad (7\text{-}10)$$

where $\hat{T}_d$ is the receiver's best estimate of the transmission delay. Since $c(t) = \pm 1$, the product $c(t - T_d) \times c(t - \hat{T}_d)$ will be unity if $\hat{T}_d = T_d$, that is, if the spreading code at the receiver is synchronized with the spreading code at the transmitter. When correctly synchronized, the signal component of the output of the receiver despreading mixer is equal to $s_d(t)$ except for a random phase φ, and $s_d(t)$ can be demodulated using a conventional coherent phase demodulator.

Observe that the data modulation above does not also have to be BPSK; no restrictions have been placed on the form of $\theta_d(t)$. However, it is common to use the same type of digital phase modulation for the data and the spreading code. When BPSK is used for both modulators, one phase modulator (mixer) can be eliminated. The double-modulation process is replaced by a single modulation by the modulo-2 sum of the data and the spreading code.

Figure 7-5 illustrates the direct-sequence spreading and despreading operation when the data modulation and the spreading modulation are BPSK. In this case, the data modulation is represented by a multiplication of the carrier by $d(t)$, where $d(t)$ takes on values of ± 1. Thus

$$s_d(t) = \sqrt{2P}\ d(t)\ \cos\ \omega_0 t \qquad (7\text{-}11)$$

$$s_t(t) = \sqrt{2P}\ d(t)c(t)\ \cos\ \omega_0 t \qquad (7\text{-}12)$$

The data and spreading waveforms are illustrated in Figures 7-5a and b, and $s_d(t)$ and $s_t(t)$ are illustrated in Figures 7-5c and d. Figure 7-5e represents an incorrectly phased input to the receiver despreading mixer assuming zero propagation delay, and Figure 7-5f shows the output of this mixer. Observe that Figure 7-5f is not

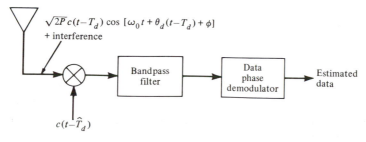

FIGURE 7-4. BPSK direct-sequence spread spectrum receivers.

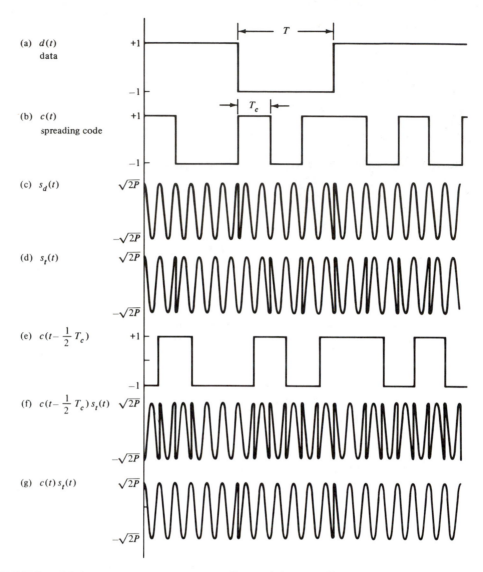

(a) $d(t)$
data

(b) $c(t)$
spreading code

(c) $s_d(t)$

(d) $s_t(t)$

(e) $c(t - \frac{1}{2} T_c)$

(f) $c(t - \frac{1}{2} T_c) s_t(t)$

(g) $c(t) s_t(t)$

FIGURE 7-5. BPSK direct-sequence spreading and despreading.

equivalent to $s_d(t)$, illustrating that the receiver must be synchronized with the transmitter. Finally, Figure 7-5g shows the despreading mixer output when the despreading code is correctly phased. In this case $c(t)s_t(t) = s_d(t)$ and the data-modulated carrier has been recovered.

It is also instructive to consider the power spectra of the signals of Figure 7-5. Recall that the two-sided power spectral density in W/Hz of a binary phase-shift-keyed carrier is given by

$$s_d(f) = \tfrac{1}{2} PT\{\text{sinc}^2[(f - f_0)T] + \text{sinc}^2[(f + f_0)T]\} \qquad (7\text{-}13)$$

which is plotted in Figure 7-6. Now observe that the signal $s_t(t)$ of Figure 7-5d is also a binary phase-shift-keyed carrier and therefore has a power spectral density which is given by (7-13) with T replaced by T_c, the duration of a spreading code symbol. The spreading code symbol duration T_c is often referred to as a spreading

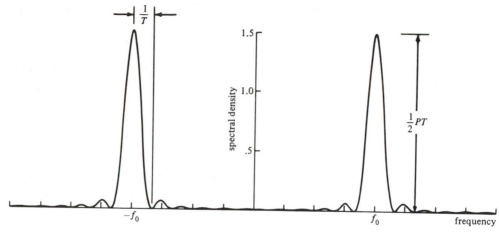

FIGURE 7-6. Power spectral density of data-modulated carrier.

code *chip*. Figure 7-7 shows the power spectral density of $s_t(t)$ in the case where $T_c = T/3$. Observe that the effect of the modulation by the spreading code is to spread the bandwidth of the transmitted signal by a factor of 3, and that this spreading operation reduces the level of the psd by a factor of 3. In actual systems, this spreading factor is typically much larger than 3.

Equation (7-13) applies only when both the data modulation and the spreading modulation are binary phase-shift keying, and when the data modulation and the spreading modulation are phase synchronous. In this case, since the data modulation is completely random, the signal $s_t(t)$ is a randomly biphase modulated signal and (7-13) applies. Consider again the case in which the data modulation is an arbitrary constant-envelope phase modulation. The data modulated carrier is represented by (7-8) and the transmitted signal is represented by (7-9). The power spectrum of the transmitted signal is calculated using the *Wiener–Khintchine theorem*, which states that the power spectrum and the autocorrelation function of a signal are a Fourier transform pair.

The data-modulated carrier is an ergodic random process, the spreading code is both deterministic and periodic, and their product, $s_t(t)$, is an ergodic random process. The signal $s_d(t)$ is independent of $c(t)$, so that the autocorrelation function

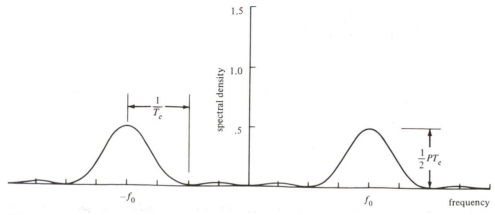

FIGURE 7-7. Power spectral density of data- and spreading code-modulated carrier.

$R_t(\tau)$ of the product $c(t)s_d(t)$ equals the product of the autocorrelation functions, that is,

$$R_t(\tau) = R_d(\tau)R_c(\tau) \tag{7-14}$$

Using the frequency convolution theorem of Fourier transform theory, the power spectral density of $s_t(t)$, which is the Fourier transform of $R_t(T)$, is

$$S_t(f) = \int_{-\infty}^{\infty} S_d(f')S_c(f - f') \, df' \tag{7-15}$$

EXAMPLE 7-1

Calculate the power spectrum of the direct-sequence spread spectrum transmitted signal when BPSK is used for both the data modulation and the spreading code modulation. Assume that the spreading code chip rate is 100 times the data rate, and that the period of the spreading code is infinite.

The power spectrum of the data-modulated carrier is given by (7-13). The power spectrum of the spreading code $c(t)$ is the Fourier transform of its autocorrelation function

$$R_c(\tau) = \lim_{A\to\infty} \frac{1}{2A} \int_{-A}^{A} c(t')c(t' - \tau) \, dt'$$

When $\tau = 0$, this integral is equal to 1.0 since $c^2(t) = 1.0$. When $\tau \geq T_c$, the integral is zero since the code has been modeled as an infinite sequence of independent random binary digits. For $0 < \tau < T_c$ the integral is equal to the fraction of the chip time for which $c(t' + T) = c(t')$, as illustrated in Figure 7-8. Therefore,

$$R_c(\tau) = \begin{cases} 1 - \dfrac{|\tau|}{T_c} & |\tau| < T_c \\ 0 & |\tau| \geq T_c \end{cases}$$

as illustrated in Figure 7-9. The Fourier transform of this triangular waveform is easily calculated. The result is

$$S_c(f) = T_c \operatorname{sinc}^2(fT_c) = \frac{T}{100} \operatorname{sinc}^2\left(\frac{fT}{100}\right)$$

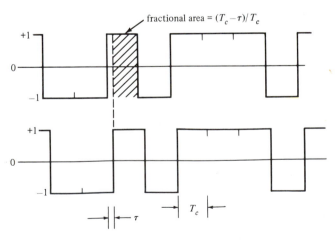

FIGURE 7-8. Calculation of autocorrelation function of an infinite sequence of random binary digits.

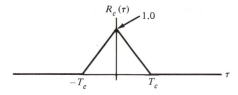

R_c(τ) 1.0

$-T_c$ T_c τ

FIGURE 7-9. Autocorrelation function of an infinite sequence of random binary digits.

The transmitted power spectrum is then

$$S_t(f) = \int_{-\infty}^{\infty} \frac{1}{2} PT \, \text{sinc}^2 \, [(f' - f_0)T] \, \frac{T}{100} \, \text{sinc}^2 \left[(f - f') \frac{T}{100} \right] df'$$

$$+ \int_{-\infty}^{\infty} \frac{1}{2} PT \, \text{sinc}^2 \, [(f' + f_0)T] \, \frac{T}{100} \, \text{sinc}^2 \left[(f - f') \frac{T}{100} \right] df'$$

Because the spreading code chip rate is much larger than the data rate, the second sinc function in each integral is approximately constant over the range of significant values of the first sinc function. Thus the convolution can be approximated by

$$S_t(f) \simeq \frac{PT^2}{200} \, \text{sinc}^2 \left[(f - f_0) \frac{T}{100} \right] \int_{-\infty}^{\infty} \text{sinc}^2 \, [(f' - f_0)T] \, df'$$

$$+ \frac{PT^2}{200} \, \text{sinc}^2 \left[(f + f_0) \frac{T}{100} \right] \int_{-\infty}^{\infty} \text{sinc}^2 \, [(f' + f_0)T] \, df'$$

$$= \frac{1}{2} \frac{PT}{100} \left\{ \text{sinc}^2 \left[(f - f_0) \frac{T}{100} \right] + \text{sinc}^2 \left[(f + f_0) \frac{T}{100} \right] \right\}$$

It was claimed earlier that one of the advantages of using spread spectrum is that it will enable the receiver to reject deliberate interference or jamming. Interference rejection is accomplished by the receiver despreading mixer, which *spreads* the spectrum of the interference at the same time that the desired signal is *despread*. If the interference energy is spread over a bandwidth much larger than the data bandwidth, most of the energy will be rejected by the data filter.

Suppose that BPSK is used for both the data modulation and the spreading modulation and that the interference is a single tone having power J. The jammer's best strategy is to place the jamming tone directly in the center of the modem's transmission bandwidth. If no spectrum spreading were employed, the ratio of jamming power to signal power in the data bandwidth would be J/P. The power spectrum of the received signal is approximately

$$S_r(f) \simeq \frac{1}{2} PT_c\{\text{sinc}^2 \, [(f - f_0)T_c] + \text{sinc}^2 \, [(f + f_0)T_c]\}$$

$$+ \frac{1}{2} J\{[\delta(f - f_0) + \delta(f + f_0)\} \qquad (7\text{-}16)$$

and the received signal is

$$r(t) = \sqrt{2P} \, d(t - T_d)c(t - T_d) \cos(\omega_0 t + \varphi)$$

$$+ \sqrt{2J} \cos(\omega_0 t + \varphi') \qquad (7\text{-}17)$$

Assume that the receiver despreading code is correctly phased so that the output of the despreading mixer is

$$y(t) = \sqrt{2P} \, d(t - T_d) \cos(\omega_0 t + \varphi)$$
$$+ \sqrt{2J} \, c(t - \hat{T}_d) \cos(\omega_0 t + \varphi') \tag{7-18}$$

and the power spectrum of $y(t)$ is

$$S_y(f) = \tfrac{1}{2} PT\{\text{sinc}^2 [(f - f_0)T] + \text{sinc}^2 [(f + f_0)T]\}$$
$$+ \tfrac{1}{2} JT_c\{\text{sinc}^2 [(f - f_0)T_c] + \text{sinc}^2 [(f + f_0)T_c]\} \tag{7-19}$$

Observe that the data signal has been despread to the data bandwidth, while the single-tone jammer has been spread over the full transmission bandwidth of the spread-spectrum system.

The power spectra of the signals discussed above are illustrated in Figure 7-10. The received power spectra are shown in Figure 7-10a, and the spectra after the despreading mixer are shown in Figure 7-10b. The despreading operation in spread-spectrum receivers is followed by a filtering operation to limit the bandwidth at the input to the data demodulator to approximately the data bandwidth. The power transfer function of an ideal filter accomplishing this is shown in Figure 7-10c and the output of this filter is shown in Figure 7-10d. This ideal filter represents the noise equivalent bandwidth of an actual intermediate-frequency (IF) filter whose noise bandwidth is equal to the data rate. Nearly all of the signal power is passed by the IF filter. A large fraction of the spread jammer power, on the other hand, is rejected by this filter. The magnitude of the jammer power passed by the IF filter is

$$J_0 = \int_{-\infty}^{\infty} S_J(f)|H(f)|^2 \, df \tag{7-20}$$

where $S_J(f)$ is the power spectrum of the jammer after the despreading mixer. If an ideal bandpass IF filter as shown in Figure 7-10c is assumed, then

$$J_0 = \int_{-f_0 - 1/2T}^{-f_0 + 1/2T} S_J(f) \, df + \int_{f_0 - 1/2T}^{f_0 + 1/2T} S_J(f) \, df$$

$$= \frac{1}{2} JT_c \int_{-f_0 - 1/2T}^{-f_0 + 1/2T} \text{sinc}^2 [(f + f_0)T_c] \, df$$

$$+ \frac{1}{2} JT_c \int_{f_0 - 1/2T}^{f_0 + 1/2T} \text{sinc}^2 [(f - f_0)T_c] \, df \tag{7-21}$$

For large ratios of data bandwidth to total spread bandwidth, that is, $T_c \ll T$, the sinc function is nearly constant over the range of the integration and

$$J_0 \simeq J \frac{T_c}{T} \tag{7-22}$$

Thus the jamming power at the input to the data demodulator has been reduced by a factor T_c/T over its value without the use of spread spectrum. The processing gain of this very simple spread-spectrum system is equal to the inverse of this jammer power reduction factor, or

$$G_p = \frac{T}{T_c} \tag{7-23}$$

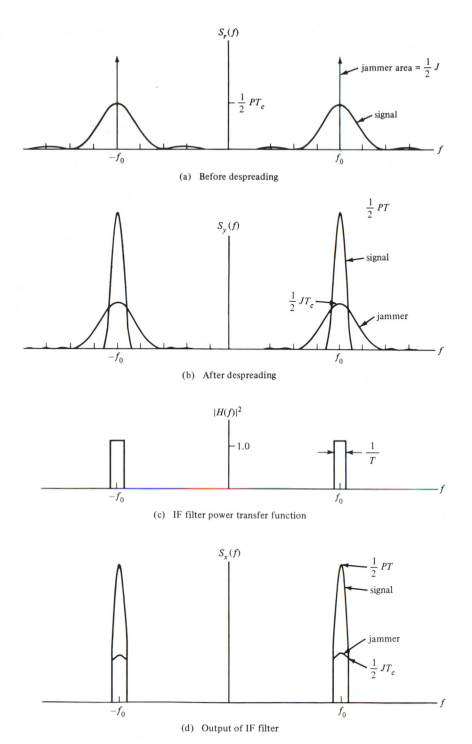

(a) Before despreading

(b) After despreading

(c) IF filter power transfer function

(d) Output of IF filter

FIGURE 7-10. Receiver power spectral densities with tone jamming.

Other equivalent definitions of processing gain are possible and are often used. Throughout this text a consistent definition of processing gain as an improvement factor is used. This results in different formulas for G_p, depending on the particular system being considered.

7-3.2 QPSK Direct-Sequence Spread Spectrum

Recall from the first part of this text that it is sometimes advantageous to transmit simultaneously on two carriers which are in phase quadrature. The principal reason for doing this is to conserve spectrum, since, for the same total transmitted power, the same bit error probability is achieved using one-half the transmission bandwidth. Bandwidth efficiency is not usually of primary importance in a spread-spectrum system, but quadrature modulations are still important. The reason for this is that quadrature modulations are more difficult to detect in low probability of detection applications, and quadrature modulations are less sensitive to some types of jamming.

Both the data modulation and the spreading modulation can be placed on quadrature carriers using a number of techniques. If no restriction is placed on the data phase modulator, QPSK spreading modulation can be added using the system of Figure 7-11a. Observe that the power at either output of the quadrature hybrid is one-half of the input power. The output of the QPSK modulator is

$$s(t) = \sqrt{P}\, c_1(t) \cos\left[\omega_0 t + \theta_d(t)\right] + \sqrt{P}\, c_2(t) \sin\left[\omega_0 t + \theta_d(t)\right]$$

$$\stackrel{\triangle}{=} a(t) + b(t) \tag{7-24}$$

where $c_1(t)$ and $c_2(t)$ are the in-phase and quadrature spreading waveforms. When written this way both spreading waveforms are assumed to take on only values of ± 1. These spreading waveforms are assumed to be chip synchronous but otherwise totally independent of one another.

The power spectrum of the QPSK spread spectrum signal of (7-24) can be calculated by observing that both terms of this equation are identical, except for amplitude and a possible phase shift, to (7-9) for BPSK spread spectrum. Thus, since the two signals are orthogonal, the power spectrum of the sum signal equals the algebraic sum of the two power spectra. That this is true is most conveniently illustrated by calculating the autocorrelation of $s(t)$, which is

$$R_s(\tau) = E[s(t)s(t + \tau)]$$

$$= E[a(t)a(t + \tau)] + E[b(t)b(t + \tau)]$$

$$\quad + E[a(t)b(t + \tau)] + E[b(t)a(t + \tau)]$$

$$= R_a(\tau) + R_b(\tau) + E[a(t)b(t + \tau)] + E[b(t)a(t + \tau)] \tag{7-25}$$

If the functions $a(t)$ and $b(t)$ are orthogonal [3, p. 298], the last two terms of (7-25) are equal to zero. This condition is satisfied in the present case since $c_1(t)$ and $c_2(t)$ are independent code waveforms.

The receiver for the transmitted signal of (7-24) is shown in Figure 7-11b. In this figure the bandpass filter is centered at frequency ω_{IF} and has a bandwidth sufficiently wide to pass the data-modulated carrier with negligible distortion. Using straightforward trigonometric identities, it can be shown that the components of $x(t)$ and $y(t)$ near the IF are given by

$$x(t) = \sqrt{\frac{P}{2}}\, c_1(t - T_d)c_1(t - \hat{T}_d) \cos\left[\omega_{\text{IF}} t - \theta_d(t)\right]$$

$$-\sqrt{\frac{P}{2}}\, c_2(t - T_d)c_1(t - \hat{T}_d) \sin\left[\omega_{\text{IF}} t - \theta_d(t)\right] \qquad (7\text{-}26)$$

$$y(t) = \sqrt{\frac{P}{2}}\, c_1(t - T_d)c_2(t - \hat{T}_d) \sin\left[\omega_{\text{IF}} t - \theta_d(t)\right]$$

$$+\sqrt{\frac{P}{2}}\, c_2(t - T_d)c_2(t - \hat{T}_d) \cos\left[\omega_{\text{IF}} t - \theta_d(t)\right] \qquad (7\text{-}27)$$

If the receiver-generated replicas of the spreading codes are correctly phased, then

$$c_1(t - T_d)c_1(t - \hat{T}_d) = c_2(t - T_d)c_2(t - \hat{T}_d) = 1.0 \qquad (7\text{-}28)$$

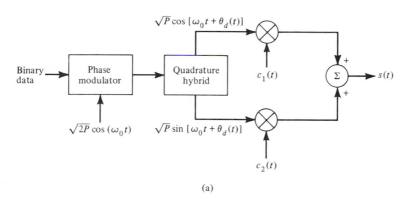

(a)

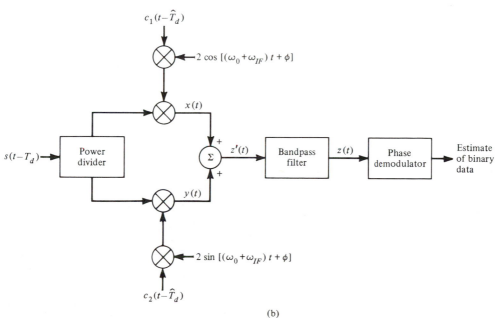

(b)

FIGURE 7-11. (a) QPSK spread-spectrum modulator with arbitrary data phase modulation; (b) QPSK spread-spectrum receiver for arbitrary data modulation.

and the desired signals have been despread. These despread signals will pass through the bandpass filter. The undesired terms of (7-26) and (7-27) cancel, so that

$$z(t) = \sqrt{2P} \cos [\omega_{\text{IF}}t - \theta_d(t)] \tag{7-29}$$

In deriving the results above, perfect receiver carrier phase tracking has been assumed. Observe in (7-29) that the data-modulated carrier has been completely recovered. That is, the QPSK spreading modulation added by the transmitter has been completely removed by the receiver despreading operation. The signal $z(t)$ is the input to a conventional phase demodulator where data is recovered. Other forms of the receiver are possible. The particular placement of the mixers and filter shown, however, is typical of an arrangement that might be found in actual hardware.

When the data modulation is binary phase-shift keying, the transmitter and receiver can be implemented as shown in Figure 7-12, where phase coherency has been assumed. The transmitted signal in this case is

$$s(t) = \sqrt{P} \, d(t) \, [c_1(t) \cos (\omega_0 t) + c_2(t) \sin (\omega_0 t)] \tag{7-30}$$

This type of modulation has been referred to as *balanced QPSK modulation* [4] since the data modulation is balanced between the in-phase and quadrature channels.

The receiver for the signal of (7-30) is shown in Figure 7-12b. The system is assumed to be coherent, so that the phase of the in-phase and quadrature local oscillators are equal to the received carrier phase. With this assumption, the difference frequency components of the mixer outputs are

$$x(t) = \sqrt{\frac{P}{2}} \, d(t - T_d) c_1(t - T_d) c_1(t - \hat{T}_d) \cos \omega_{\text{IF}}t$$

$$+ \sqrt{\frac{P}{2}} \, d(t - T_d) c_2(t - T_d) c_1(t - \hat{T}_d) \sin (-\omega_{\text{IF}}t) \tag{7-31a}$$

$$y(t) = \sqrt{\frac{P}{2}} \, d(t - T_d) c_2(t - T_d) c_2(t - \hat{T}_d) \cos \omega_{\text{IF}}t$$

$$+ \sqrt{\frac{P}{2}} \, d(t - T_d) c_1(t - T_d) c_2(t - \hat{T}_d) \sin \omega_{\text{IF}}t \tag{7-31b}$$

When the receiver spreading code replica is correctly phased, the output of the bandpass filter is

$$z(t) = \sqrt{2P} \, d(t - T_d) \cos \omega_{\text{IF}}t \tag{7-32}$$

This signal is the recovered data-modulated carrier, which is now demodulated by the BPSK data demodulator.

Another configuration for a QPSK spread-spectrum modem is shown in Figure 7-13. In this case, both the data modulation and the spreading code modulation are different for the in-phase and quadrature channels and the modulation is called [4] *dual-channel QPSK*. The transmitted waveform for dual-channel QPSK is

$$s(t) = \sqrt{P} d_1(t) c_1(t) \cos \omega_0 t + \sqrt{P} \, d_2(t) c_2(t) \sin \omega_0 t \tag{7-33}$$

which has total power P. The receiver for this waveform is shown in Figure 7-13b and is similar in operation to the balanced QPSK modem described above. A small but important variation on the signal of (7-33) yields one of the spread-spectrum

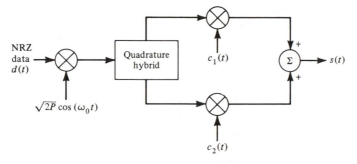

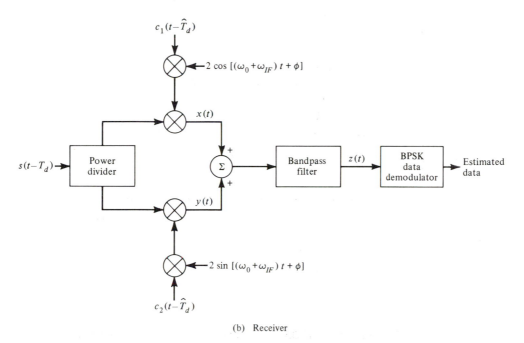

(b) Receiver

FIGURE 7-12. Balanced QPSK direct-sequence spread-spectrum modem.

signals used for the Tracking and Data Relay Satellite System (TDRSS). This variation is simply to permit the in-phase and quadrature channels to have unequal power. Thus the transmitted signal is

$$s(t) = \sqrt{2P_I}\, d_1(t)c_1(t)\, \cos\, \omega_0 t$$
$$+ \sqrt{2P_Q}\, d_2(t)c_2(t)\, \sin\, \omega_0 t \qquad (7\text{-}34)$$

Another variation applicable to all the QPSK spread-spectrum modems is to use offset QPSK for the spreading modulation. This variation is also employed on TDRSS.

The QPSK spread-spectrum modems discussed above are the principal types of QPSK modems either currently in use or widely discussed in the literature. Other variations, especially in the details of the implementations, are possible and may be more efficient under some conditions.

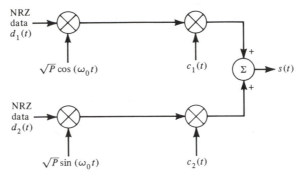

(a) Transmitter

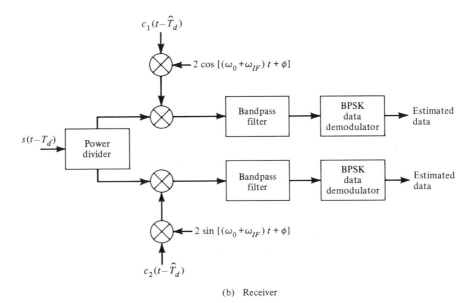

(b) Receiver

FIGURE 7-13. Dual-channel QPSK direct-sequence spread-spectrum modem: (a) transmitter; (b) receiver.

7-3.3 MSK Direct-Sequence Spread Spectrum

Another practical direct-sequence spread-spectrum modulation scheme is minimum shift keying (MSK). Although conventional schemes for generating MSK signals are more complex than the schemes for generating QPSK, the serial approach to MSK modulation is applicable to spread-spectrum systems and results in a system with the theoretical benefits of an in-phase and quadrature system together with hardware only slightly more complex than a BPSK system. The fundamental theory of both parallel and serial MSK implementations has been discussed earlier in this text and will not be reviewed here.

A conventional MSK spread-spectrum modem is illustrated in Figure 7-14. The transmitter output signal is

$$s(t) = \sqrt{2P}\, d(t) \left[c_1(t) \cos\left(\frac{\pi}{T_c} t\right) \cos \omega_0 t + c_2(t) \sin\left(\frac{\pi}{T_c} t\right) \sin \omega_0 t \right]$$

$$(7\text{-}35)$$

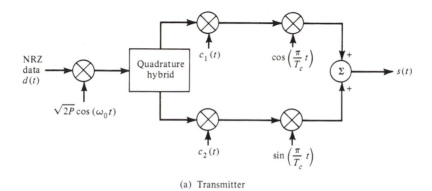

(a) Transmitter

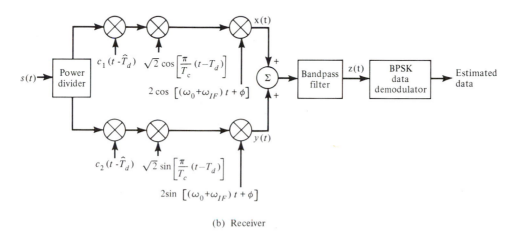

(b) Receiver

FIGURE 7-14. Minimum shift keying spread-spectrum modem: (a) transmitter; (b) receiver.

It is left as an exercise to show that the MSK modulation is removed by the receiver of Figure 7-14b when both the carrier tracking and code tracking loops are operating perfectly.

Although the modem of Figure 7-14 ideally provides proper MSK spreading and despreading modulation, the hardware is relatively complex. The serial technique for MSK modulation discussed earlier can be used to significantly simplify this hardware. Two serial MSK spread-spectrum modems are illustrated in Figure 7-15. In both the transmitters and the receiver, the MSK conversion filter is a passive, linear, time-invariant filter. Observe that a single spreading code is used in these modems rather than the two separate spreading codes that are required in the conventional MSK modem. For the same performance, however, the serial MSK spreading code is required to operate at twice the rate of the codes in the conventional modulator, and therefore may be more difficult to implement. The two modems differ only in the placement of the data modulator.

A convenient tool for understanding MSK modulation is the excess phase trellis introduced earlier. The excess phase trellis presents a graphical picture of the instantaneous phase difference between the transmitted signal and a carrier at the center frequency of the MSK spectrum. This excess phase is also, by definition, the phase of the complex envelope of the MSK signal. The excess phase trellis can

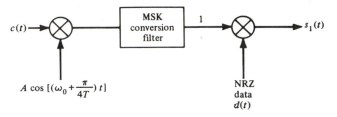

(a) Transmitter 1

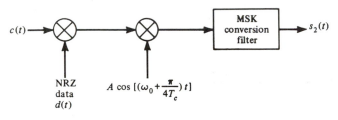

(b) Transmitter 2

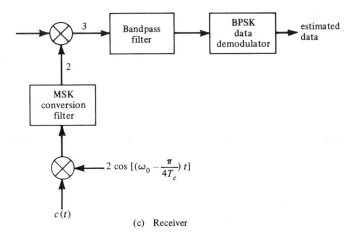

(c) Receiver

FIGURE 7-15. Serial MSK spread-spectrum modem: (a) transmitter 1; (b) transmitter 2; (c) receiver.

also be used to describe the transmitted phase in a spread-spectrum system using MSK spreading together with BPSK data modulation. Assume that the data modulation is phase synchronous with the spreading modulation. Figure 7-16 is an example excess phase trellis for both modems of Figure 7-15 when the spreading code rate is three times the data rate.

Observe that the data modulation in transmitter 1 generates abrupt $\pi/2$ phase transitions in the output signal. This is illustrated in Figure 7-16c, which is derived from Figure 7-16d by adding a $\pi/2$ phase shift whenever the data is a "mark." In the receiver, the received signal is despread by remodulating with a replica of the spreading code. The excess phase of the despreading mixer output is found by subtracting the excess phase of the reference from the excess phase of the received

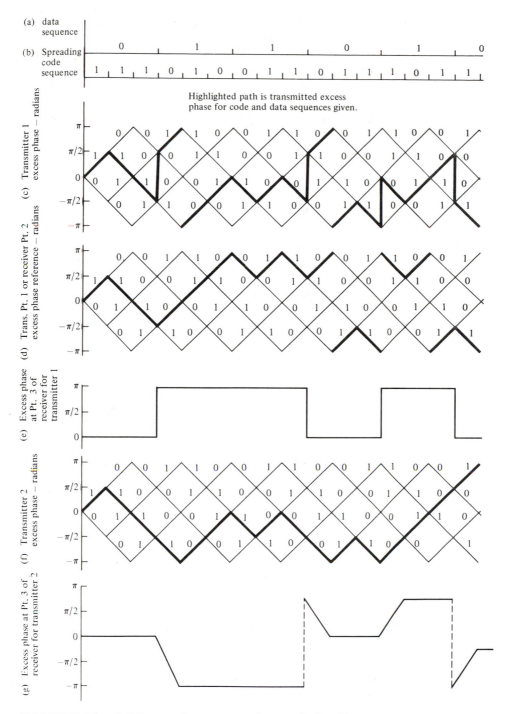

FIGURE 7-16. MSK spread-spectrum phase relationships.

signal. The result of this operation is illustrated in Figure 7-16e for transmitter 1. In this case, the BPSK data-modulated carrier has been recovered exactly.

In transmitter 2 of Figure 7-15b, the data is added modulo-2 to the spreading code prior to MSK modulation. The transmitted signal no longer has abrupt $\pi/2$ phase transitions as illustrated in Figure 7-16f, and the transmitted power spectrum is precisely that of a MSK modulated carrier at the spreading code rate. The excess phase of the despreading mixer output in the receiver in this case is shown in Figure 7-16g, where it is seen that the BPSK data-modulated carrier is only approximately recovered. The data phase transitions have been slowed by the transmitter conversion filter. In this slowing down of the data phase transitions, the direction that the phasor takes between zero and $\pm \pi$ also becomes apparent. Although modem 2 recovers the data modulation only approximately, for high processing gains the approximation is very good and this circuit is extremely practical.

7-4

FREQUENCY-HOP SPREAD SPECTRUM

A second method for widening the spectrum of a data-modulated carrier is to change the frequency of the carrier periodically. Typically, each carrier frequency is chosen from a set of 2^k frequencies which are spaced approximately the width of the data modulation spectrum apart, although neither condition is absolutely necessary. The spreading code in this case does not directly modulate the data-modulated carrier but is instead used to control the sequence of carrier frequencies. Because the transmitted signal appears as a data-modulated carrier which is hopping from one frequency to the next, this type of spread spectrum is called *frequency-hop* (FH) *spread spectrum*. In the receiver, the frequency hopping is removed by mixing (down-converting) with a local oscillator signal which is hopping synchronously with the received signal.

7-4.1 Coherent Slow-Frequency-Hop Spread Spectrum

Although in most cases the frequency hopping is done noncoherently, a fully coherent frequency-hop system is theoretically possible and is of pedagogical interest. Consider, for example, the FH system shown in Figure 7-17. The frequency synthesizer output is a sequence of tones of duration T_c, so $h_T(t)$ can be written

$$h_T(t) = \sum_{n=-\infty}^{\infty} 2p(t - nT_c) \cos (\omega_n t + \varphi_n) \qquad (7-36)$$

where $p(t)$ is a unit amplitude pulse of duration T_c starting at time zero, and ω_n and φ_n are the radian frequency and phase during the nth frequency-hop interval. The frequency ω_n is taken from a set of 2^k frequencies. In contrast to the DS system, where the spreading code was used one bit at a time, the spreading code here is used k bits at a time. The transmitted signal is the data-modulated carrier up-converted to a new frequency $(\omega_0 + \omega_n)$ on each FH chip,

$$s_t(t) = \left[s_d(t) \sum_{n=-\infty}^{+\infty} 2p(t - nT_c) \cos (\omega_n t + \varphi_n) \right]_{\substack{\text{sum freq.} \\ \text{components}}} \qquad (7-37)$$

Calculation of the transmitted power spectrum is accomplished using the frequency convolution theorem of Fourier theory. Define $S_d(f)$ to be the power spectral density of the data-modulated carrier and $S_h(f)$ to be the power spectral density of the hop carrier $h_T(t)$. These two signals are independent, so that the power spectrum

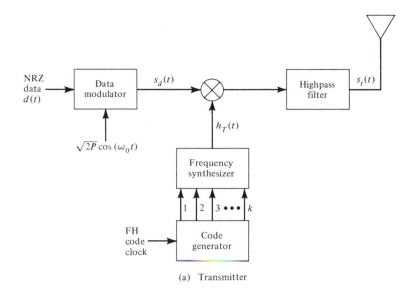

(a) Transmitter

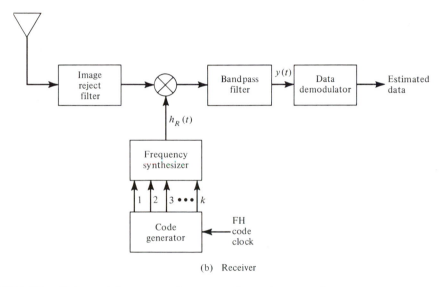

(b) Receiver

FIGURE 7-17. Coherent frequency hop spread-spectrum modem.

of the transmitted signal is the sum frequency term of the convolution of $S_d(f)$ with $S_h(f)$.

The signal $h_T(t)$ may or may not be periodic. In most cases, if $h_T(t)$ were periodic, its period would be sufficiently long that little error would be made in considering the period infinite. This assumption is made in the following. Thus $h_T(t)$ is considered a purely random sequence of frequencies. For the coherent frequency-hop system being considered, the same phase φ_m is used each time $h_T(t)$ returns to frequency ω_m, that is, $\varphi_n \in \{\varphi_m, m = 1, 2, \ldots, 2^k\}$. With these assumptions, $S_h(f)$ is given by [5, 6]

$$S_h(f) = \frac{1}{T_c^2} \sum_{n=-\infty}^{\infty} \left| \sum_{m=1}^{2k} p_m G_m(n/T_c) \right|^2 \delta\left(f - \frac{n}{T_c} \right)$$

$$+ \frac{1}{T_c} \sum_{m=1}^{2k} p_m(1 - p_m) |G_m(f)|^2$$

$$- \frac{2}{T_c} \sum_{\substack{m=1 \\ m \neq m' \\ m < m'}}^{2k} \sum_{m'=1}^{2k} p_m p_{m'} \, \text{Re} \, [G_m(f)G_{m'}^*(f)] \qquad (7\text{-}38)$$

where p_m is the probability that frequency m is selected, and $G_m(f)$ is the Fourier transform of $g_m(t)$, where

$$g_m(t) = \begin{cases} 2p(t) \cos (\omega_m t + \varphi_m) & 0 \leq t \leq T_c \\ 0 & \text{elsewhere} \end{cases} \qquad (7\text{-}39)$$

Observe that this psd has discrete components due to the assumption that the same phase is used each time $h_T(t)$ returns to frequency ω_m. The Fourier transform $G_m(f)$ is

$$G_m(f) = T_c \exp\{-j[\pi(f - f_m)T_c - \varphi_m]\} \, \text{sinc} \, [(f - f_m)T_c]$$

$$+ T_c \exp\{-j[\pi(f + f_m)T_c + \varphi_m]\} \, \text{sinc} \, [(f + f_m)T_c] \qquad (7\text{-}40)$$

Calculation of $S_h(f)$ can be simplified if the assumption is made that $G_m(f)$ and $G_{m'}(f)$ are nonoverlapping for $m \neq m'$. In this case, $G_m(f)G_{m'}^*(f) = 0$ and the third term of (7-38) vanishes. This assumption is very good whenever $1/T_c$ is small with respect to the minimum frequency spacing. Assuming also that all frequencies ω_m are equally likely leads to

$$S_h(f) \simeq \frac{1}{(T_c 2^k)^2} \sum_{n=-\infty}^{\infty} \sum_{m=1}^{2k} \left| G_m\left(\frac{n}{T_c} \right) \right|^2 \delta\left(f - \frac{n}{T_c} \right)$$

$$+ \frac{1}{T_c} \frac{1}{2^k} \left(1 - \frac{1}{2^k} \right) \sum_{m=1}^{2k} |G_m(f)|^2 \qquad (7\text{-}41)$$

The discrete components of $S_h(f)$ are negligible only when $T_c 2^k \gg 1$, which is not usually the case in systems of interest.

EXAMPLE 7-2

Calculate the transmitted power spectral density for a coherent frequency hop system. The system employs BPSK data modulation with a data rate of 1 Mbps. The hop rate is 100×10^3 hops per second, the frequency spacing equals the data rate, and four frequencies are employed.

Solution: Since the frequency spacing is considerably larger than the hop rate, $G_m(f)$ and $G_{m'}(f)$ are nearly nonoverlapping (orthogonal) and (7-41) applies. With $T_c = 10^{-5}$ and $2^k = 4$ and $G_m(f)$ given by (7-40), $S_h(f)$ is approximately

$$S_h(f) \simeq \frac{1}{2^{2k}} \sum_{n=-\infty}^{\infty} \sum_{m=1}^{2k} \{\text{sinc}^2 \, [(n - f_m T_c)] + \text{sinc}^2 \, [(n + f_m T_c)]\} \delta\left(f - \frac{n}{T_c} \right)$$

$$+ \frac{T_c}{2^k} \left(1 - \frac{1}{2^k} \right) \sum_{m=1}^{2k} \{\text{sinc}^2 \, [(f - f_m)T_c] + \text{sinc}^2[(f + f_m)T_c]\}$$

The psd of the BPSK data modulated carrier is given by (7-13).

Since the frequency spacing equals 1.0 MHz and the hop rate is 100×10^3, $f_m T_c$ is an integer so that the sinc function of the first term of $S_h(f)$ is sampled only at integer values and

$$S_h(f) \simeq \frac{1}{2^{2k}} \sum_{m=1}^{2^k} [\delta(f - f_m) + \delta(f + f_m)]$$

$$+ \frac{T_c}{2^k}\left(1 - \frac{1}{2^k}\right) \sum_{m=1}^{2^k} \{\text{sinc}^2[(f - f_m)T_c] + \text{sinc}^2[(f + f_m)T_c]\}$$

The convolution of $S_d(f)$ and $S_h(f)$ is the desired result. This convolution may be simplified by observing that one of the sinc functions within the convolution integral varies much more slowly than the other. Thus an approximate result is obtained by considering one sinc function a constant. After some manipulations the final approximate result is

$$S_t(f) \simeq \frac{PT}{2 \cdot 2^{2k}} \sum_{m=1}^{2^k} \{\text{sinc}^2[(f - f_m - f_0)T] + \text{sinc}^2[(f + f_m + f_0)T]\}$$

$$+ \left(1 - \frac{1}{2^k}\right)\frac{PT}{2 \cdot 2^k} \sum_{m=1}^{2^k} \{\text{sinc}^2[(f - f_m - f_0)T] + \text{sinc}^2[(f + f_m + f_0)T]\}$$

$$= \frac{1}{2}PT\frac{1}{2^k} \sum_{m=1}^{2^k} \{\text{sinc}^2[(f - f_m - f_0)T] + \text{sinc}^2[(f + f_m + f_0)T]\}$$

which is plotted in Figure 7-18.

This final expression for the transmitted power spectral density is usually written down by inspection since it is simply the sum of the data-modulated carrier psd translated to all hop frequencies and weighted by the probability of transmitting that frequency. The rather lengthy derivation of this result here is intended to provide an understanding of the approximations used in arriving at this result. ■

The result of (7-38) is based on the fact that the same phase being used each time the synthesizer of Figure 7-17 returns to frequency ω_m. If the frequency synthesizer phase is random for each successive time interval, the power spectral density of $h_T(t)$ is

$$S_h(f) = \frac{T_c}{2^k} \sum_{m=1}^{2^k} \{\text{sinc}^2[(f - f_m)T_c] + \text{sinc}^2[(f + f_m)T_c]\} \qquad (7\text{-}42)$$

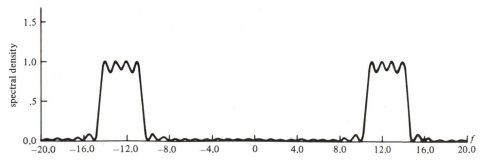

FIGURE 7-18. Transmitted power spectrum for a frequency hop system with $T = 1.0$ and $f_1 = 11$, $f_2 = 12$, $f_3 = 13$, $f_4 = 14$.

7-4 FREQUENCY-HOP SPREAD SPECTRUM

which is derived in Appendix F. When this result is convolved with $S_d(f)$, a transmitted power spectral density identical to that calculated in Example 7-2 is obtained.

Returning now to the problem of transmitting data with a coherent FH spread-spectrum modem, the received signal $s_t(t - T_d)$ for the transmitter of Figure 7-17 is

$$s_t(t - T_d) = \sqrt{2P} \sum_{n=-\infty}^{\infty} p(t - T_d - nT_c)$$

$$\cos\left[(\omega_0 + \omega_n)t + \varphi_n + \theta_d(t - T_d) - (\omega_0 + \omega_n)T_d\right] \quad (7\text{-}43)$$

In the receiver, this signal is down-converted using a locally generated reference

$$h_R(t) = 2 \sum_{n=-\infty}^{\infty} p(t - \hat{T}_d - nT_c) \cos\left[\omega_n t + \varphi_n - \omega_n \hat{T}_d\right] \quad (7\text{-}44)$$

After bandpass filtering to extract the difference frequency component of the down-conversion mixer, the received signal $y(t)$ is, assuming that $\hat{T}_d = T_d$,

$$y(t) = [s_t(t - T_d)h_R(t)]_{\text{LP}}$$

$$= \sqrt{2P} \sum_{n=-\infty}^{\infty} p(t - T_d - nT_c) \cos\left[\omega_0 t - \omega_0 T_d + \theta_d(t - T_d)\right]$$

$$= \sqrt{2P} \cos\left[\omega_0 t - \omega_0 T_d + \theta_d(t - T_d)\right] \quad (7\text{-}45)$$

and the data-modulated carrier has been completely recovered. If a tracking error exists, $\hat{T}_d \neq T_d$, the recovered carrier is phase modulated by terms having the form $\sum_n (T_d - \hat{T}_d)\omega_n$. This can be a serious problem in a fully coherent FH system unless a means is provided for coherent carrier tracking which is independent of the FH code tracking loop.

7-4.2 Noncoherent Slow-Frequency-Hop Spread Spectrum

Because of the difficulty of building truly coherent frequency synthesizers as well as the code tracking requirements alluded to above, most frequency-hop spread-spectrum systems use either noncoherent or differentially coherent data modulation schemes. The modem block diagram is unchanged from that illustrated in Figure 7-17. In the receiver, however, no effort is made to precisely recover the phase of the data-modulated carrier since it is not required by the demodulator.

A common data modulation for FH systems is M-ary frequency shift keying. Suppose, for example, that the data modulator outputs one of 2^L tones each LT seconds, where T is the duration of one information bit. Usually, these tones are spaced far enough apart so that the transmitted signals are orthogonal. This implies that the data modulator frequency spacing is at least $1/LT$ and that the data modulator output spectral width is approximately $2^L/LT$. Each T_c seconds, the data modulator output is translated to a new frequency by the frequency-hop modulator. When $T_c \geq LT$ the FH system is called a *slow-frequency-hop* system. The output of this spread-spectrum modulator is illustrated in Figure 7-19. In this figure the "instantaneous" transmitted spectrum is shown as a function of time for a system with $L = 2$ and $k = 3$. Two data bits are collected each $2T = T_s$ seconds and one of four frequencies is generated by the data modulator. This frequency is

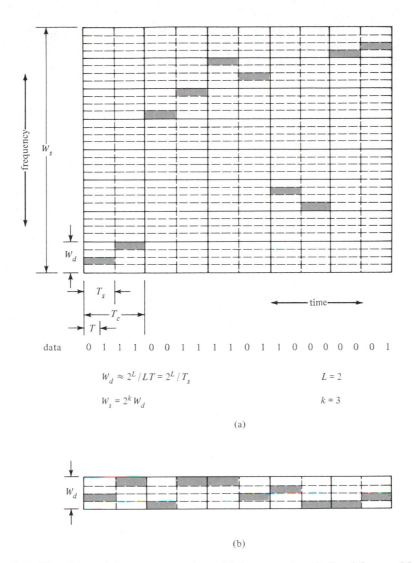

$W_d \approx 2^L / LT = 2^L / T_s$ \qquad\qquad $L = 2$

$W_s = 2^k W_d$ \qquad\qquad $k = 3$

(a)

(b)

FIGURE 7-19. Pictorial representation of (a) transmitted signal for an M-ary FSK slow-frequency hop spread-spectrum system; (b) receiver downconverter output.

translated to one of $2^k = 8$ frequency-hop bands by the FH modulator. In this example, a new frequency-hop band is selected after each group of 2 symbols or 4 bits is transmitted.

In the receiver, the transmitted signal is down-converted using a local oscillator which outputs the sequence of frequencies 0, $5W_d$, $6W_d$, $2W_d$, $7W_d$, . . . and the output of the down-converter is a sequence of tones in the first (lowest) FH band representing the data. The down-converter output is illustrated in Figure 7-19b. This signal can be demodulated using the conventional methods for noncoherent MFSK (i.e., a bank of bandpass filters with energy detectors at their outputs).

A very preliminary estimate of the processing gain of the FH system just described can be obtained by considering a noise jammer. In the absence of frequency hopping, the jammer chooses a bandwidth W_d centered on the proper carrier frequency and forces the receiver operating signal-to-noise ratio to $E_b/N_J = E_b W_d/J$, where J is the average jammer power. When frequency hopping is added, the jammer must place noise in all 2^k frequency-hop bands in order to cause the receiver to have the same performance as before. Thus the jammer requires a total power 2^k times as large as before and the processing gain is $2^k = W_s/W_d$.

7-4.3 Noncoherent Fast-Frequency-Hop Spread Spectrum

In contrast to the slow-FH system, where the hop-frequency band changes more slowly than symbols come out of the data modulator, the hop-frequency band can change many times per symbol in a *fast-frequency-hop* system. A significant benefit achieved when fast frequency hop is used is that frequency diversity gain is seen on each transmitted symbol. This is particularly beneficial in a partial-band jamming environment.

A representation of the transmitted signal for a fast-frequency-hop system is illustrated in Figure 7-20. The output of the MFSK modulator is one of 2^L tones

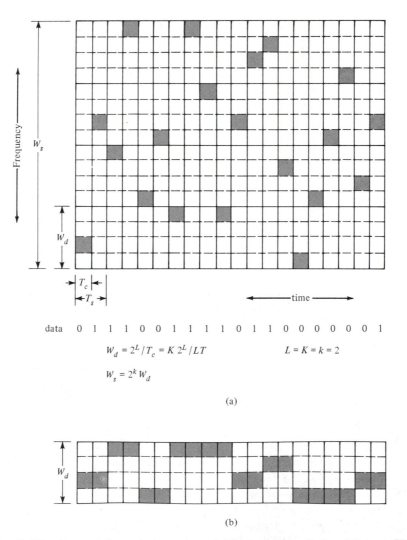

data   0   1   1   1   0   0   1   1   1   1   0   1   1   0   0   0   0   0   0   1

$$W_d = 2^L/T_c = K\,2^L/LT \qquad\qquad L = K = k = 2$$

$$W_s = 2^k\,W_d$$

(a)

(b)

FIGURE 7-20. Pictorial representation of (a) transmitted signal for an M-ary FSK fast frequency hop spread-spectrum system; (b) receiver downconverter output.

as before, but now this tone is subdivided into K chips. After each chip, the MFSK modulator output is hopped to a different frequency. Since the chip duration T_c is shorter than the data modulator output symbol duration T_s, the minimum tone spacing for orthogonal signals is now $1/T_c = K/LT$. The receiver frequency-dehopping operation functions in exactly the same way as before. The output of the down-conversion operation is shown in Figure 7-20b.

The data demodulator can operate in several different modes in a fast-frequency-hop system. One mode is to make a decision on each frequency-hop chip as it is received and to make an estimate of the data modulator output based on all K chip decisions. The decision rule could be a simple majority vote. Another mode would be to calculate the likelihood of each data modulator output symbol as a function of the total signal received over K chips and to choose the largest. A receiver which calculates the likelihood that each symbol was transmitted is optimum in the sense that minimum error probability is achieved for a given E_b/N_0. Each of these possible operating modes performs differently and has different complexity. The spread-spectrum system designer must choose the mode of operation that best solves the particular problem. It will be shown later that fast frequency hop is a very useful technique in either a fading-signal environment or in a partial band-jamming environment, and its use with error correction coding is particularly convenient.

7-5

HYBRID DIRECT-SEQUENCE/FREQUENCY-HOP SPREAD SPECTRUM

A third method for spectrum spreading is to employ both direct-sequence and frequency-hop spreading techniques in a hybrid direct-sequence/frequency-hop system. One reason for using hybrid techniques is that some of the advantages of both types of systems are combined in a single system. Hybrid techniques are widely used in military spread-spectrum systems and are currently the only practical way of achieving extremely wide spectrum spreading. Many methods of combining DS and FH spreading are possible. The method discussed here was selected because of its simplicity as an example of a hybrid system.

Figure 7-21 illustrates a hybrid DS/FH spread-spectrum modem which employs differential binary PSK data modulation. Because noncoherent frequency hopping is used, the data modulation must be either noncoherent or differentially coherent. As discussed earlier, DPSK modulation requires a differential data encoding prior to carrier modulation as shown. With this encoding the sampled output of the differential demodulator is the original data sequence. In Figure 7-21 the DPSK modulated carrier is first direct sequence spread by multiplication with the DS spreading waveform $c(t)$, and then frequency hopped through up-conversion using the sequence of FH tones $h_T(t)$. The signal $h_T(t)$ is defined by (7-36). The power spectral density of the transmitted signal $s_t(t)$ is

$$S_t(f) = [S_{ds}(f) * S_h(f)] \quad \text{(7-46)}$$
$$\underset{\substack{\text{sum freq.} \\ \text{terms}}}{}$$

where $S_h(f)$ is given by (7-42) and $S_{ds}(f)$ is approximately (see Example 7-1)

$$S_{ds}(f) \approx \tfrac{1}{2} PT_c\{\text{sinc}^2[(f - f_0)T_c] + \text{sinc}^2[(f + f_0)T_c]\} \quad \text{(7-47)}$$

where T_c is the DS spreading code chip duration. If the FH rate is slow relative to the DS bandwidth, the density $S_h(f)$ may be approximated by a sum of delta

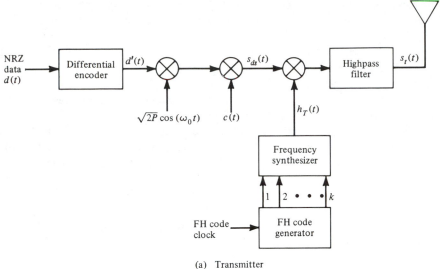

(a) Transmitter

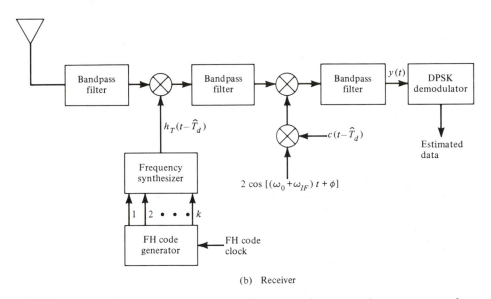

(b) Receiver

FIGURE 7-21. Hybrid direct-sequence/frequency-hop spread-spectrum modem.

functions and (7-46) becomes

$$S_t(f) \simeq \left[\int_{-\infty}^{\infty} \frac{1}{2} PT_c \{ \mathrm{sinc}^2 \left[(f' - f_0)T_c \right] + \mathrm{sinc}^2 \left[(f' + f_0)T_c \right] \} \right.$$

$$\left. \times \frac{1}{2^k} \sum_{m=1}^{2^k} \{ \delta(f - f' - f_m) + \delta(f - f' + f_m) \} \, df' \right]_{\substack{\text{sum freq.} \\ \text{terms}}}$$

$$= \frac{PT_c}{2^{k+1}} \sum_{m=1}^{2^k} \{ \mathrm{sinc}^2 \left[(f - f_m - f_0)T_c \right] + \mathrm{sinc}^2 \left[(f + f_m + f_0)T_c \right] \}$$

$$(7\text{-}48)$$

The receiver recovers the DPSK modulated carrier by first frequency dehopping the received signal and then DS despreading the signal. Both of these operations have been explained in detail above, so that it can be easily demonstrated that the recovered data-modulated carrier is

$$y(t) = \sqrt{2P}\, d'(t - T_d) \cos\left[\omega_{\text{IF}} t + \varphi(t)\right] \tag{7-49}$$

where $\varphi(t)$ is a function which accounts for the random phase changes of the receiver FH synthesizer at each hop frequency. Of course, both the DS and the FH code sequences must be correctly phased for despreading to occur.

7-6

COMPLEX-ENVELOPE REPRESENTATION OF SPREAD-SPECTRUM SYSTEMS

All of the spread-spectrum systems discussed above can be conveniently represented mathematically using complex-envelope notation. This common notation is useful not only as an aid in understanding the spreading/despreading process but also as an analytical and simulation tool. Recall from Chapter 2 that any bandpass signal $v(t)$ whose Fourier spectrum is centered at frequency ω_c can be expressed as

$$v(t) = \text{Re}\left[\tilde{v}(t) e^{j\omega_0 t}\right] \tag{7-50}$$

where $\tilde{v}(t)$ is the complex envelope of $v(t)$ and is a complex function of time, and Re $(\cdot)$ is the real part of the argument. All the signal processing steps required in spread-spectrum systems (i.e., linear filtering and mixing) can be mathematically modeled as operations on the complex envelope of the signal of interest.

A generic complex envelope model of a spread-spectrum modem is illustrated in Figure 7-22. In this figure, all signals are complex functions of time, double lines present real and imaginary (in-phase and quadrature) paths, and the mixers perform complex multiplications. The actual (real) transmitted signal is $s(t) = \text{Re}\left[\tilde{s}(t) e^{j\omega_0 t}\right]$, where power amplification has been ignored, and $\tilde{s}(t) = \tilde{d}(t)\tilde{c}(t)$. In writing this last expression, it has been assumed that the transmitter mixing operation is also part of an up-conversion process, so that $\omega_0 = \omega_1 + \omega_2$ where ω_1 and ω_2 are the actual center frequencies at the data modulator and spreading code generator outputs.

The receiver input $\tilde{r}(t)$ is the delayed transmitter output $\tilde{s}(t - T_d)$ plus interference $\tilde{u}(t)$ plus thermal noise $\tilde{n}(t)$. In the receiver, the mixing operations are assumed to be part of the receiver down-conversion chain. Since the difference frequency components of the mixer outputs are required, the complex conjugate of the reference signal envelopes are used [8] as the mixer inputs. The first receiver mixing operation accounts for all frequency and phase differences between the received carrier (includes doppler effects) and the local reference carrier. If the system is coherent, a carrier tracking loop will force $\hat{\omega}_0 = \omega_0$ and $\hat{\varphi} = \varphi$ so that this complex-envelope multiplication has no effect. The second receiver mixing operation is the spread-spectrum despreading operation. In general, the input to the data demodulator is

$$\hat{\tilde{d}}(t - T_d) = \tilde{d}(t - T_d)\tilde{c}(t - T_d)\tilde{c}^*(t - \hat{T}_d) \exp\left\{-j[(\omega_0 - \hat{\omega}_0)t + \varphi - \hat{\varphi}]\right\}$$

$$+ \tilde{u}(t)\tilde{c}^*(t - \hat{T}_d) \exp\left\{-j[(\omega_0 - \hat{\omega}_0)t + \varphi - \hat{\varphi}]\right\}$$

$$+ \tilde{n}(t)\tilde{c}^*(t - \hat{T}_d) \exp\left\{-j[(\omega_0 - \hat{\omega}_0)t + \varphi - \hat{\varphi}]\right\} \tag{7.51}$$

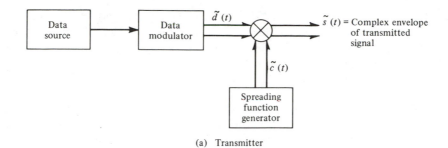

(a) Transmitter

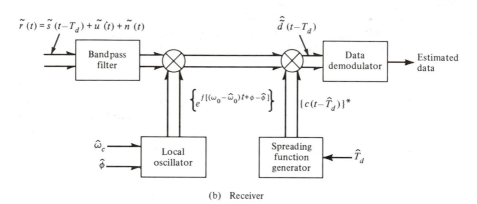

(b) Receiver

FIGURE 7-22. Generic complex-envelope model of spread-spectrum modem: (a) transmitter; (b) receiver. (From Ref. 7.)

where the input bandpass filter has been assumed to have a bandwidth sufficiently wide to pass all signals without distortion.

This representation is completely general and, with proper selection of $\tilde{d}(t)$ and $\tilde{c}(t)$, can be used for any direct-sequence or frequency-hop spread-spectrum system with any type of data modulation. Table 7-1 gives the complex envelope for a number of the most common digital modulation types which can be used for either the data modulation or the spreading modulation. In this table, $p_T(t)$ and $p_{2T}(t)$ are unit pulses of duration T and $2T$ seconds, respectively. The variable T is the information bit duration when these envelopes are used for data modulation and the spreading code chip duration T_c when they are used for the spreading modulation.

EXAMPLE 7-3

Calculate the complex envelope of the data demodulator input for a fully coherent spread-spectrum modem that uses BPSK data modulation and MSK spreading modulation when the only interference is a single-tone jammer which is not at the system carrier frequency.

Solution: The complex envelope of the data modulation is

$$\tilde{d}(t) = \sum_n d_n p_T(t - nT)$$

and the complex envelope of the spreading modulation is

$$\tilde{c}(t) = \sum_m p_{T_c}(t - mT_c) \exp\left[j\left(c_m \frac{\pi t}{2T_c} + x_m\right)\right]$$

TABLE 7-1. Complex Envelope of Common Digital Modulation Types

Modulation Type	Complex Envelope — $\tilde{v}(t)$
Binary phase-shift keying (BPSK)	$\tilde{v}(t) = \sum_{n} a_n p_T(t - nT)$ $a_n \in \{+1, -1\}$
Quaternary phase-shift keying (QPSK)	$\tilde{v}(t) = \sum_{n} p_{2T}(t - 2nT) \exp(j\beta_n)$ $\beta_n \in \left\{0, \dfrac{\pi}{2}, \pi, \dfrac{3\pi}{2}\right\}$
Offset quaternary phase-shift keying (OQPSK)	$\tilde{v}(t) = \sum_{n} p_T(t - nT) \exp(j\beta_n)$ $\beta_n = \beta_{n-1} + a_n \dfrac{\pi}{2}$ $a_n \in \{+1, 0, -1\}$
M-ary phase-shift keying (MPSK)	$\tilde{v}(t) = \sum_{n} p_{mT}(t - nmT) \exp(j\beta_n)$ $m = \log_2 M$ $\beta_n \in \{\beta_1, \beta_2, \ldots, \beta_m\}$
Binary frequency-shift keying (FSK)	$\tilde{v}(t) = \sum_{n} p_T(t - nT) \exp[j(\omega_n t + \varphi_n)]$ $\omega_n \in \{\omega_1, \omega_2\}$ $0 \le \varphi_n \le 2\pi$
Minimum-shift keying (MSK)	$\tilde{v}(t) = \sum_{n} p_T(t - nT) \exp\left[j\left(a_n \dfrac{\pi t}{2T} + x_n\right)\right]$ $a_n = \in \{+1, -1\}$ $x_n = x_{n-1} + (a_{n-1} - a_n)\dfrac{n\pi}{2}$
M-ary frequency-shift keying (MFSK)	$\tilde{v}(t) = \sum_{n} p_T(t - nT) \exp[j(\omega_n t + \varphi_n)]$ $\omega_n \in \{\omega_1, \omega_2, \ldots, \omega_m\}$ $0 \le \varphi_n \le 2\pi$

so that the transmitted signal is

$$\tilde{s}(t) = \sqrt{2P}\,\tilde{d}(t)\tilde{c}(t) = \sqrt{2P} \sum_{n} \sum_{m} d_n p_T(t - nT) p_{T_c}(t - mT_c)$$

$$\times \exp\left[j\left(c_m \frac{\pi t}{2T_c} + x_m\right)\right]$$

The transmitted signal power is P. The envelope of the jammer is

$$\tilde{u}(t) = \sqrt{2J} \exp(j\,\Delta\omega\, t)$$

where the offset frequency is $\Delta\omega$ and the jammer power is J.

Referring to Figure 7-22, and assuming perfect carrier tracking, the complex envelope at the input to the data demodulator is

$$\hat{\tilde{d}}(t - T_d) = \tilde{s}(t - T_d)\tilde{c}^*(t - \hat{T}_d)$$

$$+ \sqrt{2J}\,\tilde{c}^*(t - \hat{T}_d)\exp(j\Delta\omega\, t)$$

$$= \sqrt{2P}\sum_n d_n p_T(t - T_d - nT)$$

$$\times \sum_{n'}\sum_m p_{T_c}(t - T_d - n'T_c)p_{T_c}(t - \hat{T}_d - mT_c)$$

$$\times \exp\left\{j\left[c_{n'}\frac{\pi}{2T_c}(t - T_d) + x_{n'}\right] - j\left[c_m\frac{\pi}{2T_c}(t - \hat{T}_d) + x_m\right]\right\}$$

$$+ \sqrt{2J}\sum_{m'} p_{T_c}(t - \hat{T}_d - m'T_c)\exp\left\{-j\left[c_{m'}\frac{\pi}{2T_c}(t - \hat{T}_d)\right] + j\,\Delta\omega\, t\right\}$$

If perfect spreading code tracking is also assumed ($\hat{T}_d = T_d$), then

$$\hat{\tilde{d}}(t - T_d) = 2P\sum_n d_n p_T(t - T_d - nT)$$

$$+ \sqrt{2J}\sum_{m'} p_{T_c}(t - \hat{T}_d - m'T_c)\exp\left\{-j\left[c_{m'}\frac{\pi}{2T_c}(t - \hat{T}_d)\right] + j\,\Delta\omega\, t\right\}$$

and it is seen that the MSK spreading has been entirely removed from the desired signal and that the jamming tone has been MSK modulated via the despreading operation. ∎

EXAMPLE 7-4

Calculate the complex envelope of the data demodulator input for a slow-frequency-hop spread-spectrum modem that uses differential binary PSK data modulation.

Solution: The envelope of the data modulation is the same as in Example 7-3 except that the differentially encoded data sequence is used. The envelope of the spreading modulation is

$$\tilde{c}(t) = \sum_m p_{T_c}(t - mT_c)\exp[j(\omega_m t + \varphi_m)]$$

Assume that the receiver is able to perfectly track the frequency of the received signal using an AFC loop, but that no attempt is made to estimate the random phase changes associated with each frequency hop. Thus

$$\hat{\tilde{d}}(t - T_d) = \tilde{d}(t - T_d)\tilde{c}(t - T_d)\tilde{c}^*(t - T_d)\exp[j(\varphi - \hat{\varphi})]$$

$$= \sqrt{2P}\sum_n d_n p_T(t - T_d - nT)\sum_m\sum_{m'} p_{T_c}(t - T_d - mT_c)$$

$$\times p_{T_c}(t - \hat{T}_d - m'T_c)\exp\{+j[\omega_m(t - T_d) + \varphi_m]$$

$$- j[\omega_{m'}(t - \hat{T}_d) + \varphi_{m'}]\}$$

If perfect code tracking is assumed ($\hat{T}_d = T_d$),

$$\hat{\tilde{d}}(t - T_d) = \sqrt{2P}\sum_n d_n p_T(t - T_d - nT)\sum_m \exp(j\theta_m)$$

where $\theta_m = \varphi_m - \varphi_{m'}$ is the difference between the transmitter and receiver frequency-hop phases. ∎

REFERENCES

[1] J. M. WOZENCRAFT and I. M. JACOBS, *Principles of Communication Engineering* (New York: Wiley, 1965).

[2] R. A. DILLARD, "Detectability of Spread Spectrum Signals," *IEEE Trans. Aerosp. Electron. Syst.*, July 1979.

[3] A. PAPOULIS, *Probability, Random Variables, and Stocastic Processes* (New York: McGraw-Hill, 1965).

[4] B. K. LEVITT, "Effect of Modulation Format and Jamming Spectrum on Performance of Direct Sequence Spread Spectrum Systems," *Conf. Rec.*, IEEE Nat. Telecommun. Conf., 1980.

[5] R. C. TITSWORTH and L. R. WELCH, "Power Spectra of Signals Modulated by Random and Pseudorandom Sequences," Jet Propulsion Laboratory, Pasadena, Calif., Tech. Rep. 32-140, October 1961.

[6] W. C. LINDSEY and M. K. SIMON, *Telecommunication Systems Engineering* (Englewood Cliffs, N. J.: Prentice-Hall, 1973).

[7] R. A. SCHOLTZ, "The Spread Spectrum Concept," *IEEE Trans. Commun.*, August 1977.

[8] S. STEIN and J. J. JONES, *Modern Communication Principles* (New York: McGraw-Hill, 1967).

[9] J. K. HOLMES, *Coherent Spread Spectrum Systems* (New York: Wiley-Interscience, 1982).

[10] R. L. HARRIS, "Introduction to Spread Spectrum Techniques," in *Spread Spectrum Communications*, NATO AGARD Lecture Ser. 58, June 6, 1973 (AD 766 914).

PROBLEMS

7-1. Show that $z(t)$ of Figure 7-14b is the recovered data-modulated carrier at IF, that is,

$$z(t) = \sqrt{2P}\, d(t) \cos \omega_{IF} t$$

if perfect code and carrier tracking are assumed.

7-2. Consider a non-spread-spectrum communication system employing differential binary PSK modulation. Suppose that this system is jammed by a narrowband noise-pulse jammer having total average power J and duty factor ρ.
  (a) Find the optimum jammer duty factor ignoring thermal noise as a function of E_b/N_J and plot bit error probability versus E_b/N_J for the optimized jammer.
  (b) Plot bit error probability versus E_b/N_J for nonoptimum jammer duty factors $\rho = 0.25$, 0.50, 0.75, and 1.0.

7-3. Consider a BPSK direct-sequence spread-spectrum system using differential binary PSK data modulation. Suppose that this system is jammed by a narrowband pulse-noise jammer having a total average power J and duty factor ρ. Assume that the spreading code chip rate is 10 times the data bit rate.
  (a) Find the optimum jammer duty factor, ignoring thermal noise as a function of E_b/N_J and plot bit error probability versus E_b/N_J for the optimized jammer.
  (b) What is the processing gain of this spread-spectrum system?
  (c) Compare the optimum duty factors for Problems 7-2 and 7-3.

7-4. Consider a BPSK direct-sequence spread-spectrum system with arbitrary data modulation and continuous tone jamming. Assuming that the spreading code period is infinite, plot the processing gain of the system as a function of the jamming tone center frequency relative to the system carrier frequency. Assume that the spreading code rate is 100 times the data rate.

7-5. Repeat Problem 7-4 assuming MSK direct-sequence modulation with a null-to-null bandwidth equal to the BPSK null-to-null bandwidth used in Problem 7-4. Compare the plots of Problem 7-4 and 7-3.

7-6. Consider a direct-sequence BPSK spread-spectrum system which obtains its spreading code from the feedback shift register circuit shown below. Calculate the transmitter output power spectral density for arbitrary data modulation.

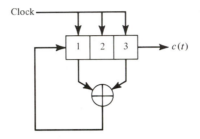

PROBLEM 7-6. Feedback shift register.

7-7. An elementary random process comprises four sample functions each of which is assigned equal probability.

$$x_1(t) = 1$$

$$x_2(t) = -2$$

$$x_3(t) = \sin \pi t$$

$$x_4(t) = \cos \pi t$$

(a) Is the process stationary?
(b) Calculate $E[x(t)]$ and $E[x(t_1)x(t_2)]$.
(Wozencraft and Jacobs [1, Prob. 3.1])

7-8. Let $x(t)$ and $y(t)$ be statistically independent, stationary random processes and define $z(t) = x(t)y(t)$. Is $z(t)$ stationary? Show that

$$S_z(f) = S_x(f) * S_y(f)$$

where, as usual, the symbol $*$ denotes convolution. (Wozencraft and Jacobs [1, Prob. 3.5])

7-9. Consider the random process

$$s(t, \mathbf{a}, T) = \sum_{n=-\infty}^{\infty} a_n p(t + T - nT_c)$$

where

$$p(t) = \sin\left(\frac{2\pi}{T_c} t\right) \qquad 0 \le t \le T_c$$

T is a random variable uniformly distributed on $(0, T_c)$, and $\mathbf{a} = (\ldots, a_{-1}, a_0, a_1, \ldots)$ is a doubly infinite sequence of equally likely binary random variables from the set $\{+1, -1\}$.
(a) Calculate the autocorrelation function of $s(t, \mathbf{a}, T)$.
(b) Calculate the power spectral density of $s(t, \mathbf{a}, T)$.

7-10. Consider a frequency-hop spread-spectrum system that uses binary FSK data modulation and 64 orthogonal frequency-hop bands. Suppose that the data rate is 1.0 Mbps and that the

frequency hop rate is 10^4 hops per second. Assume that the FH synthesizer is noncoherent from hop to hop and assume any convenient carrier frequency.
(a) Calculate the transmitter output power spectral density.
(b) Calculate the optimum number of frequency-hop bands for a partial band-noise jammer to jam assuming that the jammer has constant total power.

7-11. Consider a frequency-hop spread-spectrum system that uses binary FSK data modulation and 64 orthogonal frequency hop bands. Suppose that the data rate is 10 kbps and the frequency hop rate is 1.0×10^6 hops per second. Assume that the FH synthesizer is noncoherent from hop to hop and assume any convenient carrier frequency. Calculate the transmitter output power spectral density.

7-12. Consider a direct-sequence BPSK spread-spectrum receiver using arbitrary data modulation having the configuration illustrated below. The received thermal noise two-sided power spectral density is $N_0/2$ watts/Hz and the noise bandwidths of the input bandpass filter is B hertz. Assuming that the spreading code chip rate is very large relative to the data rate, calculate the thermal noise spectral density at the data demodulator input for $B = 2/T_c$, $3/T_c$, and $4/T_c$. T_c is the spreading code chip duration.

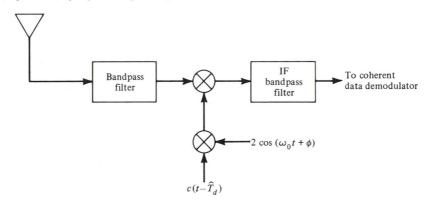

PROBLEM 7-12. Direct-sequence spread-spectrum receiver.

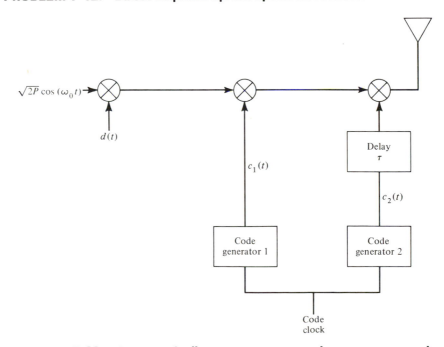

PROBLEM 7-13. Offset code direct-sequence spread-spectrum transmitter.

7-13. Consider the direct-sequence BPSK spread-spectrum transmitter illustrated on page 363. Assume that the two code generators operate synchronously and that each generates an independent sequence of totally random binary (± 1) symbols. Calculate the transmitted power spectral density as a function of τ, and plot results for $\tau = 0$, $T_c/4$, and $T_c/2$, where T_c is the code chip duration. (Harris [10])

7-14. In Chapter 3 it was demonstrated that a filter "matched" to the signaling waveform was optimum for data detection is an AWGN environment. Show that in a BPSK direct-sequence spread-spectrum system, the receiver despreading function is part of that optimum matched filtering operation.

7-15. Consider a direct-sequence BPSK spread-spectrum receiver that employs coherent BPSK data modulation as shown in Figure P7-15. Assume that a tone jammer is used against this receiver and that the jammer knows the correct carrier phase. Assume an infinite spreading code period, a jammer power J, and a spreading code chip rate that is N times the data rate. Determine the probability density function of the integrator output at the sampling instants. Write an expression for the bit error probability as a function of the jamming and signal noise powers, and N.

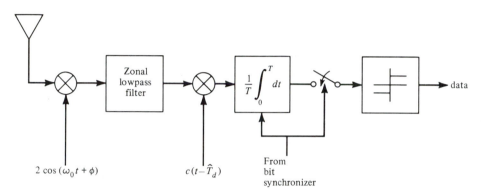

PROBLEM 7-15.   Coherent BPSK spread-spectrum receiver.

CHAPTER 8

Binary Shift Register Sequences for Spread-Spectrum Systems

8-1

INTRODUCTION

The waveform $c(t)$ used in the systems described in Chapter 7 to spread and despread the data-modulated carrier is usually generated using a shift register whose contents during each time interval is some linear or nonlinear combination of the contents of the register during the preceding time interval. In this chapter various techniques for generating $c(t)$ are described and analyzed.

For the spread-spectrum system to operate efficiently, the waveform $c(t)$ is selected to have certain desirable properties. For example, the phase of the received spreading code $c(t - T)$ must be initially determined and then tracked by the receiver. These functions are facilitated by choosing $c(t)$ to have a two-valued autocorrelation function as exhibited by the maximal-length sequences to be considered in detail later. It is often desirable to employ a $c(t)$ having a very wide bandwidth. This implies that electronically simple spreading code generators which can operate at very high speeds should be considered. When the spread-spectrum system is used for multiple access, sets of waveforms $c_1(t)$, $c_2(t)$, . . ., $c_m(t)$ must be found which have good cross-correlation properties. When jamming resistance is a concern, waveforms are used that have extremely long periods and are difficult for the jammer to generate. The spreading code generators discussed in this chapter have one or more of these properties. The codes discussed are the most commonly used codes in current spread-spectrum systems. Many other codes are possible and are discussed in the references.

In the following, the term *spreading code* is used to refer to the output of the binary shift register generator and the term *spreading waveform* is reserved for the function $c(t)$, which takes on values ± 1 and is used as the actual input to the spreading or despreading modulator. The ideal spreading code would be an infinite sequence of equally likely random binary digits. Unfortunately, the use of an infinite random sequence implies infinite storage in both the transmitter and receiver. This is clearly not possible, so that the periodic *pseudorandom codes* (*PN codes*) as described in this chapter are always employed. In this text the term "PN code" is used for any periodic spreading code with noise-like properties. Specific PN codes include the maximal-length codes and Gold codes, among others.

The chapter begins with definitions and a review of the mathematics that will be needed to analyze the spreading codes. This is followed by a comprehensive discussion of maximal-length codes, which are by far the most widely used spreading codes. Gold codes, which are combinations of maximal-length codes used in multiple-access systems, are discussed next. The chapter is concluded with a discussion of nonlinear codes and special sequences that can be used to assist the receiver in initial determination of the received code phase.

8-2

DEFINITIONS, MATHEMATICAL BACKGROUND, AND SEQUENCE GENERATOR FUNDAMENTALS

8-2.1 Definitions

All of the spreading codes to be discussed are periodic sequences of ones and zeros with period N. It is convenient to represent a sequence of binary digits $\ldots, b_{-2}$, $b_{-1}, b_0, b_1, b_2, \ldots$ by a polynomial $b(D) = \cdots + b_{-2}D^{-2} + b_{-1}D^{-1} + b_0 + b_1 D + b_2 D^2 + \cdots$. The delay operator D implies simply that the binary symbol which multiplies D^j occurs during the jth time interval of the sequence. Because the code is periodic, $b_n = b_{N+n}$ for any n. The spreading waveform $c(t)$ derived from this spreading code is also periodic with period $T = NT_c$ and is specified by

$$c(t) = \sum_{n=-\infty}^{\infty} a_n p(t - nT_c) \qquad (8\text{-}1)$$

where $a_n = (-1)^{b_n}$, and $p(t)$ is a unit pulse beginning at 0 and ending at T_c. The waveform $c(t)$ is deterministic, so that its *autocorrelation function* is defined by [1]

$$R_c(\tau) = \frac{1}{T} \int_0^T c(t)c(t + \tau)\, dt \qquad (8\text{-}2)$$

Since $c(t)$ is periodic with period T, it follows that $R_c(\tau)$ is also periodic with period T. Consider two different spreading waveforms $c(t)$ and $c'(t)$. The *cross-correlation function* of these two deterministic waveforms is

$$R_{cc'}(\tau) = \frac{1}{T} \int_0^T c'(t)c(t + \tau)\, dt \qquad (8\text{-}3)$$

where it has been assumed that both waveforms have the same period T. The cross-correlation function is also periodic with period T.

The variable τ in (8-2) and (8-3) can assume any value. That is, τ is not constrained to be an integral multiple of T_c. Substituting (8-1) into (8-3) yields

$$R_{cc'}(\tau) = \frac{1}{T} \sum_m \sum_n a_m a_n' \int_0^T p(t - mT_c)p(t + \tau - nT_c)\, dt \qquad (8\text{-}4)$$

The integral in (8-4) is nonzero only when $p(t - mT_c)$ and $p(t + \tau - nT_c)$ overlap. The delay τ can be expressed as $\tau = kT_c + \tau_\epsilon$, where $0 \le \tau_\epsilon < T_c$. Using this substitution, the pulses overlap only for $n = k + m$ and $n = k + m + 1$, so that (8-4) becomes

$$R_{cc'}(\tau) = R_{cc'}(k, \tau_\epsilon)$$

$$= \frac{1}{N} \sum_{m=0}^{N-1} a_m a_{k+m}' \frac{1}{T_c} \int_0^{T_c - \tau_\epsilon} p(\lambda)p(\lambda + \tau_\epsilon)\, d\lambda$$

$$+ \frac{1}{N} \sum_{m=0}^{N-1} a_m a_{k+m+1}' \frac{1}{T_c} \int_{T_c - \tau_\epsilon}^{T_c} p(\lambda)p(\lambda - T_c + \tau_\epsilon)\, d\lambda \qquad (8\text{-}5)$$

where the substitution $\lambda = t - mT_c$ has also been employed. The *discrete periodic cross-correlation function* of two codes $b(D)$ and $b'(D)$ is defined by [2]

$$\theta_{bb'}(k) = \frac{1}{N} \sum_{n=0}^{N-1} a_n a_{n+k}' \qquad (8\text{-}6)$$

where $a_n = (-1)^{b_n}$. Using this definition, the cross-correlation function $R_{cc'}(\tau)$ becomes

$$R_{cc'}(\tau) = R_{cc'}(k, \tau_\epsilon)$$

$$= \left(1 - \frac{\tau_\epsilon}{T_c}\right)\theta_{bb'}(k) + \frac{\tau_\epsilon}{T_c} \theta_{bb'}(k + 1) \qquad (8\text{-}7)$$

This expression is often convenient since the theory used to analyze code sequences yields results exclusively in terms of unit delays; that is, $\theta_{bb'}(k)$ is calculated rather than $R_{cc'}(\tau)$. The discrete periodic cross-correlation function can be calculated by representing the sequences $b(D)$ and $b'(D)$ as binary vectors $\mathbf{b}$ and $\mathbf{b}'$ of length N. A delay of k time units of the original sequence is represented as a cyclic shift of k time units of the vector representation. The kth cyclic shift of $\mathbf{b}$ is represented by $\mathbf{b}(k)$. Using this notation, the function $\theta_{bb'}(k) = (N_A - N_D)/N$, where N_A is the number of places in which $\mathbf{b}(0)$ agrees and N_D the number of places in which $\mathbf{b}(0)$ disagrees with $\mathbf{b}'(k)$. Equivalently, N_A is the number of zeros and N_D the number of ones in the modulo-2 sum of $\mathbf{b}(0)$ and $\mathbf{b}(k)$. The *discrete periodic autocorrelation function* is denoted by $\theta_b(k)$ and is defined by (8-6) with $a_n' = a_n$. When the periodic autocorrelation function is used in place of the periodic cross-correlation function in (8-5), the result is the autocorrelation function

$$R_c(\tau) = \left(1 - \frac{\tau_\epsilon}{T_c}\right)\theta_b(k) + \frac{\tau_\epsilon}{T_c} \theta_b(k + 1) \qquad (8\text{-}8)$$

8-2.2 Finite-Field Arithmetic

Some of the manipulations that will be performed on the code sequences introduced later require an understanding of the mechanics of finite-field arithmetic. In particular, determining the initial load of a shift register generator which will produce a particular known delay of the code requires a knowledge of the various ways of representing the elements of a finite field. This same knowledge is required to determine what phases of a maximal-length sequence to add together to obtain a specific known third phase. This discussion is intended to provide adequate information to perform these basic manipulations. A complete introductory discussion of this subject can be found in Lin and Costello [3] and a comprehensive treatment can be found in Birkhoff and MacLane [4].

Consider a set $S = \{e_0, e_1, e_2, \ldots, e_{M-1}\}$ having M elements. A finite field is constructed by defining two binary operations on the set called *addition* and *multiplication* such that certain conditions are satisfied. Addition and multiplication of two elements e_j and e_k are denoted $e_j + e_k$ and $e_j \cdot e_k$, respectively. The conditions that must be satisfied for S and the two operations to be a field are:

1. The addition or multiplication of any two elements of S must yield an element of S. That is, the set is *closed* under both addition and multiplication.
2. Both addition and multiplication must be *commutative*.
3. There must exist an *additive identity element* which will always be denoted by 0.
4. The set S must contain an *additive inverse element* $-e_j$ for every element e_j.
5. The set must contain a *multiplicative identity element* which will always be denoted by 1.
6. Excluding the element 0, the set must contain a *multiplicative inverse element* e_j^{-1} for every element e_j.
7. Multiplication is *distributive* over addition.
8. Both addition and multiplication must be *associative*.

EXAMPLE 8-1

Consider the set $S = \{0, 1, 2\}$ with addition and multiplication defined in Tables 8-1 and 8-2. It can easily be verified that this set, with the operations defined in Tables 8-1 and 8-2, satisfies all the conditions above and is therefore a field. By inspection it is seen that the set is closed under both operations. The symmetry of the tables shows that both operations are commutative. The additive inverse ele-

TABLE 8-1.   Modulo-3 Addition

+	0	1	2
0	0	1	2
1	1	2	0
2	2	0	1

TABLE 8-2.   Modulo-3 Multiplication

·	0	1	2
0	0	0	0
1	0	1	2
2	0	2	1

TABLE 8-3.	Modulo-2 Addition		TABLE 8-4.	Modulo-2 Multiplication	
+	0	1	·	0	1
0	0	1	0	0	0
1	1	0	1	0	1

ments are $-0 = 0$, $-1 = 2$, and $-2 = 1$. The multiplicative inverse elements are $1^{-1} = 1$ and $2^{-1} = 2$. All other conditions can be verified similarly. ■

EXAMPLE 8-2

Consider the set $S = \{0, 1\}$ with addition and multiplication defined in Tables 8-3 and 8-4. It can easily be verified that this set is a field of two elements. This is the binary number field that will be used extensively in what follows. Observe that addition can be accomplished electronically using an Exclusive-OR gate and multiplication can be accomplished using an AND gate. ■

It can be shown that the set of integers $\{0, 1, 2, \ldots, M - 1\}$, where M is prime and addition and multiplication are carried out modulo-M, is a field [3]. These fields are called *prime fields*. The operations of subtraction and division are also easily defined for any field using the addition and multiplication tables, just as is done with the real-number field. Subtraction is defined as the addition of the additive inverse and division is defined as multiplication by the multiplicative inverse. For example, 1-2 in the ternary field of Example 8-1 is defined by $1 + (-2) = 1 + 1 = 2$. Similarly, $1 \div 2 = 1 \cdot (2^{-1}) = 1 \cdot 2 = 2$. Note that nonprime fields do not necessarily employ modulo-M arithmetic.

Fields can be constructed having any prime number of elements p or any integral power of a prime number p^m of elements. A field having p^m elements is called an *extension field* of the field having p elements. Finite fields are often referred to as *Galois fields*, using the notation GF(M) for the field having M elements. The remainder of this discussion will be concerned exclusively with the binary number field GF(2) and its extensions GF(2^m). The reason for this is that the electronics used to implement the code generators is binary, and some of the shift register generators will be shown to generate the elements of GF(2^m).

Consider next the arithmetic of polynominals in D whose coefficients are elements of GF(2). A polynomial "of degree m over GF(2)" has the form $f(D) = f_0 + f_1 D + f_2 D^2 + \cdots + f_m D^m$, where f_j is an element of GF(2), that is, $f_j = 0$ or 1. The operations of addition, subtraction, multiplication, and division are defined for these polynomials in exactly the same way as for polynomials with real coefficients except that binary arithmetic is used. For example, the addition of $f(D)$ and $g(D) = g_0 + g_1 D + g_2 D + \cdots + g_m D^m$ yields $h(D) = h_0 + h_1 D + h_2 D^2 + \cdots + h_m D^m$ where $h_0 = f_0 + g_0$, $h_1 = f_1 + g_1$, $h_2 = f_2 + g_2$, $\ldots, h_m = f_m + g_m$ and modulo-2 addition of coefficients is used. The multiplication of these same polynomials yields a product $h(D)$ with the following coefficients:

$$h_0 = f_0 g_0$$

$$h_1 = f_0 g_1 + f_1 g_0$$

$$h_2 = f_0 g_2 + f_1 g_1 + f_2 g_0$$

$$\vdots$$

$$h_m = f_0 g_m + f_1 g_{m-1} + \cdots + f_m g_0$$

$$h_{m+1} = f_1 g_m + f_2 g_{m-1} + \cdots + f_m g_1$$

$$\vdots$$

$$h_{2m} = f_m g_m \tag{8-9}$$

where all additions and multiplications are modulo-2. Since modulo-2 arithmetic is commutative, associative, and distributive, it is easy to show that polynomial arithmetic, as just defined, is also commutative, associative, and distributive.

The division of one polynomial over GF(2) by another yields a quotient $q(D)$ and a remainder $r(D)$ just as with ordinary long division of two polynomials. For example, suppose that $f(D) = 1 + D^5$ and $g(D) = 1 + D + D^3 + D^4$. The long division of $f(D)$ by $g(D)$ yields

$$\tag{8-10}$$

$$
\begin{array}{r}
D + 1 \\
D^4 + D^3 + D + 1\overline{\smash{\big)}\,D^5 \qquad\qquad\qquad\quad + 1} \\
\underline{D^5 + D^4 \qquad + D^2 + D} \\
D^4 \qquad + D^2 + D + 1 \\
\underline{D^4 + D^3 \qquad + D + 1} \\
D^3 + D^2
\end{array}
$$

so that $q(D) = 1 + D$ and $r(D) = D^2 + D^3$. This result can be easily verified by calculating $f(D) = q(D)g(D) + r(D)$. When the remainder $r(D) = 0$, $f(D)$ is said to be *divisible* by $g(D)$.

EXAMPLE 8-3

Let $f(D) = 1 + D + D^{10} + D^{19}$ and $g(D) = D^2 + D^{10}$. Then $h(D) = f(D) + g(D) = (1 + 0) + (1 + 0)D + (0 + 1)D^2 + (1 + 1)D^{10} + (1 + 0)D^{19} = 1 + D + D^2 + D^{19}$. Now subtract $f(D)$ from $h(D)$. Recall that subtraction is defined as addition of the additive inverse. In the binary number field, each element is its own additive inverse since $0 + 0 = 0$ and $1 + 1 = 0$, so that addition and subtraction are identical. Therefore, $h(D) - f(D) = h(D) + f(D) = (1 + 1) + (1 + 1)D + (1 + 0)D^2 + (0 + 1)D^{10} + (1 + 1)D^{19} = D^2 + D^{10} = g(D)$, as expected. ∎

EXAMPLE 8-4

Divide $f(D) = 1 + D^6$ by $g(D) = 1 + D + D^3 + D^4$.

Solution:

$$
\begin{array}{r}
D^2 + D + 1 \qquad\qquad\qquad\qquad\quad (8\text{-}11) \\
D^4 + D^3 + D + 1 \,\overline{\big)\, D^6 \qquad\qquad\qquad\qquad\quad + 1} \\
\underline{D^6 + D^5 \qquad\quad + D^3 + D^2} \\
D^5 \qquad\quad + D^3 + D^2 \qquad + 1 \\
\underline{D^5 + D^4 \qquad\qquad + D^2 + D} \\
D^4 + D^3 \qquad\qquad + D + 1 \\
\underline{D^4 + D^3 \qquad\qquad + D + 1} \\
0
\end{array}
$$

Thus $g(D)$ divides $f(D)$. This result is verified by multiplying the quotient $q(D) = 1 + D + D^2$ by $g(D)$.

$$
\begin{array}{ll}
(1 + D + D^2)(1 + D + D^3 + D^4) = 1 + D + \quad + D^3 + D^4 & (8\text{-}12) \\
\qquad\qquad\qquad\qquad\qquad\qquad\quad + D + D^2 \qquad + D^4 + D^5 \\
\qquad\qquad\qquad\qquad\qquad\qquad\qquad\quad + D^2 + D^3 \qquad + D^5 + D^6 \\
\qquad\qquad\qquad\qquad\qquad\qquad\quad \overline{} \\
\qquad\qquad\qquad\qquad\qquad\qquad\qquad 1 \qquad\qquad\qquad\qquad\qquad + D^6 = f(D)
\end{array}
$$

Polynomials over GF(2) have roots and may be factorable. Substituting $D = 1$ into $g(D)$ of Example 8-4 will demonstrate that 1 is a root, so that $D - 1 = D + 1$ should be a factor of $g(D)$. To verify this, divide $g(D)$ by $D + 1$:

$$
\begin{array}{r}
D^3 \qquad\qquad\quad + 1 \qquad\qquad (8\text{-}13) \\
D + 1 \,\overline{\big)\, D^4 + D^3 \qquad + D + 1} \\
\underline{D^4 + D^3} \\
D + 1 \\
\underline{D + 1} \\
0
\end{array}
$$

showing that $g(D) = (D + 1)(D^3 + 1)$. The roots of a polynomial over GF(2) may not be elements of GF(2), just as a real polynomial may not have real roots. For example, $D^4 + D^3 + 1$ does not have 0 or 1 as a root.

At this point, all the tools are in place to construct the extension field GF(2^m) of the binary field GF(2) for $m > 1$. The extension field has 2^m elements. Consider all the polynomials of degree $m - 1$ over GF(2). There are 2^m such polynomials, so each polynomial can be used to represent a single element of the extension field GF(2^m). For example, if $m = 2$, the extension field contains $2^2 = 4$ elements, and there are four polynomials of degree $m - 1 = 1$, which are 0, 1, D, and $1 + D$. The operations of addition and multiplication must now be defined such that all the properties of a field are satisfied. Suppose that addition of two elements of the field is defined as the normal modulo-2 polynomial addition of the two polynomials representing the field elements. It can be easily verified that, using this addition rule, the addition of any two elements of the field yields another element of the field so that the field is closed under addition. The additive identity element is 0 and the additive inverse of any element is the element itself. It has already been noted that addition of polynomials over GF(2) is commutative.

Multiplication of two elements of GF(2^m) is defined using special polynomials of degree m which are called *primitive polynomials*. A polynomial $h(D)$ of degree

m is said to be primitive if the smallest integer n for which $h(D)$ divides $D^n + 1$ is $n = 2^m - 1$. Primitive polynomials are said to be *irreducible* since they are not the product of any two polynomials of lower degree (i.e., they cannot be factored). Primitive polynomials of any degree m are known to exist and have been tabulated in [5, 6]. The product of two elements of GF(2^m) is defined as the remainder when the normal polynomial product is divided by the primitive polynomial chosen to define multiplication. This multiplication is referred to as modulo-$h(D)$ multiplication. Observe that more than one primitive polynomial exists for most $m > 1$ and different multiplication rules will be obtained depending on which primitive polynomial is chosen. The polynomial product of two polynomials of degree $m - 1$ or less has a degree of at most $2m - 2$, and the remainder when dividing by a polynomial of degree m has a degree of at most $m - 1$. Since the remainder has degree at most $m - 1$, it is another element of GF(2^m), so that the field is closed under multiplication. Polynomial multiplication is associative and commutative, so that multiplication of elements of GF(2^m) is also associative and commutative.

If it can be demonstrated that a multiplicative inverse element exists for each nonzero element of GF(2^m), all the conditions defining a field will have been satisfied. To determine the multiplicative inverse elements, consider the sequence of nonzero elements of GF(2^m) beginning with 1 and such that each element is the modulo-$h(D)$ product of D and the preceding element [3]:

$$1$$

$$D$$

$$D \cdot D = D^2$$

$$D^2 \cdot D = D^3$$

$$\vdots \tag{8-14}$$

For all products where the normal polynomial product has degree less than m (the degree of the primitive polynomial), the remainder obtained when dividing by $h(D)$ is the normal polynomial product. That is, the normal polynomial product equals the modulo-$h(D)$ product. At some point in this sequence the normal polynomial product D^m will appear. At this point the remainder or modulo-$h(D)$ product is $r(D) = 1 + h_1 D + h_2 D^2 + \cdots + h_{m-1} D^{m-1}$, as can be seen by long division

$$
D^m + h_{m-1}D^{m-1} + \cdots + h_2 D^2 + h_1 D + 1 \overline{\big)
\begin{array}{l}
1 \\
D^m \\
\underline{D^m + h_{m-1}D^{m-1} + \cdots + h_2 D^2 + h_1 D + 1} \\
\phantom{D^m + {}}h_{m-1}D^{m-1} + \cdots + h_2 D^2 + h_1 D + 1
\end{array}}
\tag{8-15}
$$

Thus $D^{m-1} \cdot D = D^m = h_{m-1}D^{m-1} + \cdots + h_2 D^2 + h_1 D + 1$ and the sequence of powers of D can be written as polynomials of degree less than or equal to $m - 1$.

EXAMPLE 8-5 [3]

Let $m = 4$ and consider the multiplication law defined by $h(D) = D^4 + D + 1$. Performing the long division of D^4 by $h(D)$ yields a remainder of $r(D) = 1 + D$. Thus $D^4 = 1 + D$ and the sequence of nonzero elements of GF(2^4) written in the order described above becomes

D^0:   1

D^1: $1 \cdot D = D$

D^2: $D \cdot D = D^2$

D^3: $D^2 \cdot D = D^3$

D^4: $D^3 \cdot D = D^4 = 1 + D$

D^5: $(1 + D) \cdot D = D + D^2$

D^6: $(D + D^2) \cdot D = D^2 + D^3$

D^7: $(D^2 + D^3) \cdot D = D^3 + D^4 = 1 + D + D^3$

D^8: $(1 + D + D^3) \cdot D = D + D^2 + D^4 = 1 + D^2$

D^9: $(1 + D^2) \cdot D = D + D^3$

D^{10}: $(D + D^3)D = D^2 + D^4 = 1 + D + D^2$

D^{11}: $(1 + D + D^2)D = D + D^2 + D^3$

D^{12}: $(D + D^2 + D^3)D = D^2 + D^3 + D^4 = 1 + D + D^2 + D^3$

D^{13}: $(1 + D + D^2 + D^3)D = D + D^2 + D^3 + D^4 = 1 + D^2 + D^3$

D^{14}: $(1 + D^2 + D^3)D = D + D^3 + D^4 = 1 + D^3$

D^{15}: $(1 + D^3)D = D + D^4 = 1$

D^{16}: $1 \cdot D = D$

D^{17}: $D \cdot D = D^2$

$\vdots$

Observe that the sequence is periodic and there are 15 distinct elements. ■

A primitive polynomial of degree m divides $D^{2^m-1} + 1$ so that $D^{2^m-1} + 1 = q(D)h(D)$, which implies that $D^{2^m-1} = q(D)h(D) + 1$, and the remainder when dividing D^{2^m-1} by $h(D)$ is 1. Thus the sequence of elements of GF(2^m) written such that each element is D times the preceding element and using the proposed multiplication rule repeats after at most $2^m - 1$ elements. Since the smallest n for which a primitive polynomial divides $D^n + 1$ is $n = 2^m - 1$, the sequence of elements cannot repeat sooner than after $2^m - 1$ elements, so that the sequence always contains exactly $2^m - 1$ distinct elements. Thus there are two useful ways of representing the elements of the extension field GF(2^m). The first representation is as polynomials of degree $m - 1$ over GF(2). The second representation is as power of D. The two representations are related through the primitive polynomial used to define the multiplication law. Using the second representation, the fact that $D^{2^m-1} = 1$ leads immediately to the multiplicative inverse of any element of GF(2^m). The multiplicative inverse of any element D^j is $(D^j)^{-1} = D^{2^m-1-j}$ since $D^j \cdot D^{2^m-1-j} = D^{2^m-1} = 1$.

EXAMPLE 8-6
Derive a table of the multiplicative inverse elements in polynomial form for the nonzero elements of GF(2^4) using the primitive polynomial $h(D) = D^4 + D + 1$.

Solution: The nonzero elements of this field are the first 15 terms of the sequence derived in Example 8-5. These elements, along with their inverses, are as follows:

element			inverse element	
1	1		1	1
D	D		D^{14}	$1 + D^3$
D^2	D^2		D^{13}	$1 + D^2 + D^3$
D^3	D^3		D^{12}	$1 + D + D^2 + D^3$
D^4	$1 + D$		D^{11}	$D + D^2 + D^3$
D^5	$D + D^2$		D^{10}	$1 + D + D^2$
D^6	$D^2 + D^3$		D^9	$D + D^3$
D^7	$1 + D + D^3$		D^8	$1 + D^2$	(8-16)
D^8	$1 + D^2$		D^7	$1 + D + D^3$
D^9	$D + D^3$		D^6	$D^2 + D^3$
D^{10}	$1 + D + D^2$		D^5	$D + D^2$
D^{11}	$D + D^2 + D^3$		D^4	$1 + D$
D^{12}	$1 + D + D^2 + D^3$		D^3	D^3
D^{13}	$1 + D^2 + D^3$		D^2	D^2
D^{14}	$1 + D^3$		D	D

Several products of an element and its inverse will verify that the inverses are correct. Using the multiplication law defined by $h(D)$, $(D^7)(D^7)^{-1} = (D^7)(D^8) = (1 + D + D^3)(1 + D^2)$. The normal polynomial multiplication yields

$$
\begin{array}{l}
1 + D \qquad\quad + D^3 \\
\qquad\quad + D^2 + D^3 + D^5 \\
\hline
1 + D + D^2 \qquad\quad + D^5
\end{array}
$$

(8-17)

and long division by $h(D)$ yields

(8-18)

$$
\begin{array}{r}
D \\
D^4 + D + 1 \overline{\smash{\big)}\, D^5 + \qquad\quad D^2 + D + 1} \\
\underline{D^5 \qquad\quad + D^2 + D} \\
1
\end{array}
$$

so that $(D^7)(D^8) = 1$, as expected. ∎

The preceding paragraphs have presented a brief introduction to finite-field arithmetic. The purpose of the discussion was to give the student sufficient proficiency with finite fields to enable certain manipulations with maximal-length codes to be defined later. In summary, a finite field or Galois field is a set of elements upon which two operations (addition and multiplication) are defined. The set and these operations must satisfy certain very precise rules. Integer fields having a prime number of elements and using modulo-M multiplication and addition exist. The most commonly used integer field used in this text is the binary number field with elements 0 and 1 and using modulo-2 addition and multiplication. Extension fields of the binary field having 2^m elements also exist. Two representations for the elements of these extension fields are convenient. One representation associates one element of GF(2^m) with each possible binary polynomial of degree less than or equal to $m - 1$. Addition is defined as normal polynomial addition. Multiplication is defined as the modulo-$h(D)$ product of the polynomials, where $h(D)$ is primitive. A second representation of the elements of GF(2^m) is as powers of D.

It was shown that $D^{2^m-1} = 1$, which resulted in easy identification of the multiplicative inverse for any element of GF(2^m). Facility with finite-field arithmetic will be useful in the study of maximal-length codes. Error correction coding also makes extensive use of the concepts discussed here.

8-2.3 Sequence Generator Fundamentals

In this section the actual shift registers that are used to generate PN codes are examined. It will be shown that shift registers with feedback and/or feedforward connections can be used to multiply and divide polynomials over GF(2). Since these shift registers execute the mathematical operations described in the preceding section, that same mathematics becomes a powerful tool in the design and analysis of the generators. In what follows, modulo-2 polynomial arithmetic will be used to calculate the output of a shift register generator given a particular input and feedback/feedforward connections, to determine the maximum possible period of a linear feedback shift register, and to determine several different feedback schemes which provide identical outputs but which are different electronically. It is convenient now to limit the binary sequences under consideration to semi-infinite sequences rather than the doubly infinite sequences discussed previously. Thus all sequences are assumed to begin at time zero, and a code sequence $a(D)$ contains only positive powers of the delay operator D. That is,

$$a(D) = \sum_{j=0}^{\infty} a_j D^j \qquad (8\text{-}19)$$

Consider the logic circuits illustrated in Figure 8-1. In this figure, the boxes represent unit delays or shift registers, circles containing subscripted letter coefficients represent a connection if the coefficient is a 1 or no connection if the coefficient is a 0, and circles containing a " + " represent modulo-2 adders or Exclusive-OR gates. The circles containing subscripted coefficients may also be viewed as a modulo-2 multiplication of the input by the coefficient. The output of any shift register stage of the upper circuit is the delayed input sequence $D^j a(D)$, where j is the number of the shift register as shown in Figure 8-1. The output of the upper circuit is the modulo-2 sum of certain of the current and delayed input symbols selected by the transfer function polynomial $h(D)$, where

$$h(D) = \sum_{j=0}^{r} h_j D^j \qquad (8\text{-}20)$$

The output $b(D)$ is

$$b(D) = \sum_{k=0}^{r} [a(D)D^k] h_k$$

$$= h(D)a(D) \qquad (8\text{-}21)$$

Thus the upper circuit of Figure 8-1 performs the modulo-2 polynomial multiplication of the input sequence $a(D)$ and the transfer function $h(D)$.

Equation (8-21) can be expanded to obtain*

*This development was suggested by a reviewer.

$$b(D) = \sum_{k=0}^{r} h_k \left(\sum_{j=0}^{\infty} a_j D^j \right) D^k$$

$$= \sum_{j=0}^{\infty} \left(\sum_{l=0}^{\min\{j,r\}} a_{j-l} h_l \right) D^j$$

Now consider the lower circuit of Figure 8-1. The output of the jth modulo-2 adder, denoted $b_j(D)$, is

$$b_j(D) = b_{j-1}(D)D + a(D)h_{r-j}$$

for $j = 2, \ldots, r$, and for $j = 1$,

$$b_1(D) = a(D)Dh_r + a(D)h_{r-1}$$

Thus, by iteration

$$b(D) = b_r(D)$$

$$= b_{r-1}(D)D + a(D)h_0$$

$$= b_{r-2}(D)D^2 + a(D)Dh_1 + a(D)h_0$$

$$\vdots$$

$$= b_1(D)D^{r-1} + a(D)D^{r-2}h_{r-2} + \cdots + a(D)h_0$$

$$= a(D)D^r h_r + a(D)D^{r-1}h_{r-1} + \cdots + a(D)h_0$$

$$= \sum_{k=0}^{r} [a(D)D^k]h_k$$

$$(8-22)$$

and it is seen that both circuits of Figure 8-1 produce the same output sequences. An advantage of the lower circuit over the upper circuit for high-speed applications is that the Exclusive-OR gates are not cascaded, and therefore there is less delay from input to output.

The lower circuit of Figure 8-1 can also be configured as a two-input multiplier as illustrated in Figure 8-2. The output of this circuit is

$$b(D) = a_1(D)h(D) + a_2(D)k(D) \qquad (8-23)$$

where $k(D)$ is the transfer function for the second input $a_2(D)$. Suppose now that $k_0 = 0$ in the second transfer function $k(D)$, and suppose further that $a_2(D)$ is taken from the output of Figure 8-2, that is, $a_2(D) = b(D)$. Since $k_0 = 0$, define

$$k(D) = g(D) + 1 \qquad (8-24)$$

where $g(D)$ is a transfer function with $g_0 = 1$. The modified circuit is illustrated in Figure 8-3. The input–output relationship of this circuit is

$$b(D) = a_1(D)h(D) + b(D)[g(D) + 1] \qquad (8-25)$$

Adding $b(D)g(D) + b(D)$ to both sides of this equation yields

$$b(D)g(D) - a_1(D)h(D) \qquad (8-26)$$

which can be solved for $b(D)$ if a polynomial $c(D)$ can be found which satisfies the relationship $g(D)c(D) = 1$. With $c(D)$ so defined, (8-26) can be written

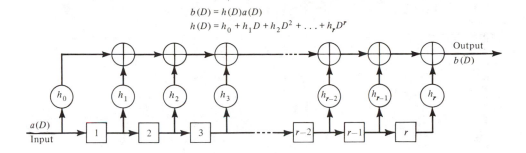

$$b(D) = h(D)a(D)$$
$$h(D) = h_0 + h_1 D + h_2 D^2 + \ldots + h_r D^r$$

$$b(D) = h(D)a(D)$$
$$h(D) = h_0 + h_1 D + h_2 D^2 + \ldots + h_r D^r$$

Key: $\oplus$ = modulo - 2 adder

h_j = modulo - 2 multiplication of input by h_j

$\square$ = single-stage shift register

FIGURE 8-1. Two equivalent circuits for multiplying polynomials. (From [5])

$$b(D) = a_1(D)h(D)c(D) \tag{8-27}$$

The coefficients of $c(D)$ must satisfy the following relationships:

$$g_0 c_0 = 1 \tag{8-28a}$$

$$\sum_{l=0}^{\min\{j,r\}} g_l c_{j-l} = 0 \qquad j = 1, 2, \ldots \tag{8-28b}$$

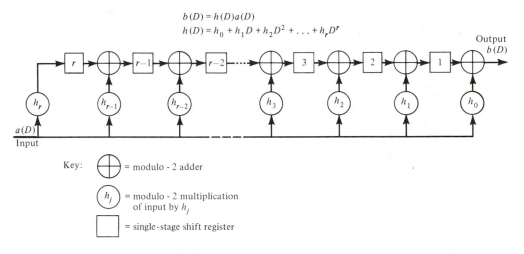

$$b(D) = h(D)a_1(D) + k(D)a_2(D)$$
$$h(D) = h_0 + h_1 D + h_2 D^2 + \ldots + h_r D^r$$

$$k(D) = k_0 + k_1 D + k_2 D^2 + \ldots + k_r D^r$$

FIGURE 8-2. Two-input modulo-2 polynomial multiplier. (From [5])

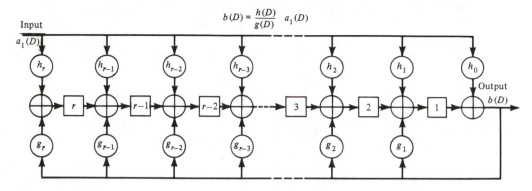

$$b(D) = \frac{h(D)}{g(D)} \, a_1(D)$$

Input
$a_1(D)$

Output
$b(D)$

FIGURE 8-3. Circuit that simultaneously multiplies by h(D) and divides by g(D).

in order for the product $g(D)c(D)$ to equal unity. Note that the polynomial multiplication being considered here is *not* modulo a primitive polynomial but is normal multiplication using modulo-2 arithmetic to combine coefficients. Using (8-28), all of the coefficients of $c(D)$ can be found and then used in (8-27) to determine $b(D)$.

Consider the polynomial long division of 1 by $g(D)$, where $g(D)$ is written with low-order terms on the left:

$$
\begin{array}{r}
1 + g_1 D + (g_2 + g_1^2)D^2 + \cdots \hspace{4cm} (8\text{-}29)
\end{array}
$$

$$
1 + g_1 D + g_2 D^2 + \cdots + D^r \,|\, 1
$$

$$
\begin{array}{l}
\quad 1 + g_1 D + \quad\quad g_2 D^2 + \quad\quad\quad g_3 D^3 + \cdots + \quad\quad\quad\quad\quad D^r \\
\quad\quad\quad g_1 D + \quad\quad g_2 D^2 + \quad\quad\quad g_3 D^3 + \cdots + \quad\quad\quad\quad\quad D^r \\
\quad\quad\quad g_1 D + \quad\quad g_1^2 D^2 + \quad\quad\quad g_1 g_2 D^3 + \cdots + \quad g_1 g_{r-1} D^r + g_1 D^{r+1} \\
\hline
\quad\quad\quad\quad\quad (g_2 + g_1^2)D^2 + (g_3 + g_1 g_2)D^3 + \cdots + (1 + g_1 g_{r-1})D^r + g_1 D^{r+1} \\
\quad\quad\quad\quad\quad (g_2 + g_1^2)D^2 + g_1(g_2 + g_1^2)D^3 + \cdots \\
\hline
\quad\quad\quad\quad\quad\quad\quad\quad\quad\quad \cdots
\end{array}
$$

Solving for the first few coefficients of $c(D)$ from (8-28) yields

$$c_0 = g_0 = 1$$

$$c_1 = g_1 c_0 = g_1$$

$$c_2 = g_1 c_1 + g_2 c_0 = g_1^2 + g_2$$

$$\vdots$$

$$(8\text{-}30)$$

Comparing (8-29) and (8-30) it can be seen that $c(D) = 1/g(D)$, where the long division is carried out in the manner of (8-29). Thus the circuit of Figure 8-3 has been shown to multiply an arbitrary input polynomial by $h(D)$ and divide it by $g(D)$ simultaneously, that is,

$$b(D) = a_1(D) \, \frac{h(D)}{g(D)} \qquad\qquad (8\text{-}31)$$

EXAMPLE 8-7

Suppose that $h(D) = D^6$ and $g(D) = 1 + D + D^2 + D^3 + D^6$. The multiplier/divider shift register for these polynomials is illustrated in Figure 8-4a. What is the output of this circuit given that the input is $a(D) = 1$ (i.e., the input is a 1 at time zero followed by an infinite string of zeros)?

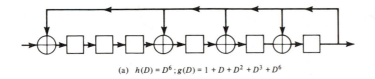

(a) $h(D) = D^6 \, ; g(D) = 1 + D + D^2 + D^3 + D^6$

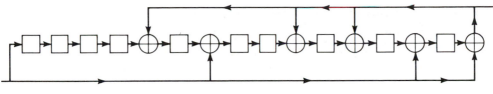

(b) $h(D) = 1 + D + D^5 + D^{10} \, ; g(D) = 1 + D^2 + D^3 + D^6$

FIGURE 8-4. Multiplier/divider circuits: (a) $h(D) = D^6$; $g(D) = 1 + D + D^2 + D^3 + D^6$. (b) $h(D) = 1 + D + D^5 + D^{10}$; $g(D) = 1 + D^2 + D^3 + D^6$.

Solution: From the previous discussion, the output is

$$b(D) = a(D) \frac{h(D)}{g(D)}$$

$$= \frac{D^6}{1 + D + D^2 + D^3 + D^6}$$

Performing the long division with high-order coefficients of the divisor on the right yields

$$
\begin{array}{r}
D^6 + D^7 \qquad\qquad\quad + D^{10} + D^{11} + D^{12} + \cdots \\
\hline
1 + D + D^2 + D^3 + D^6 \,\big|\, D^6 \qquad\qquad\qquad\qquad\qquad\qquad \\
D^6 + D^7 + D^8 + D^9 + \qquad\quad + D^{12} \\
\hline
D^7 + D^8 + D^9 \qquad\quad + D^{12} \\
D^7 + D^8 + D^9 + D^{10} \qquad\quad + D^{13} \\
\hline
D^{10} \qquad + D^{12} + D^{13} \\
D^{10} + D^{11} + D^{12} + D^{13} \qquad\qquad\qquad + D^{16} \\
\hline
D^{11} \qquad\qquad\qquad + D^{16} \\
D^{11} + D^{12} + D^{13} + D^{14} \qquad\qquad\qquad D^{17} \\
\hline
D^{12} + D^{13} + D^{14} \qquad + D^{16} + D^{17} \\
\vdots
\end{array}
$$

This output sequence can be verified by manually calculating the contents of the shift register for the first few shifts. ∎

EXAMPLE 8-8

The circuit for simultaneously multiplying by $h(D) = 1 + D + D^5 + D^{10}$ and dividing by $g(D) = 1 + D^2 + D^3 + D^6$ is illustrated in Figure 8-4b. ∎

The input to the circuit of Figure 8-3 can be any binary sequence, including a sequence that ends at some finite time. After the input sequence ends, the circuit is equivalent to the feedback shift register shown in Figure 8-5. Suppose that the input sequence ends at time j. Then the highest power of D in $a_1(D)$ is D^j and the highest power of D in the product $a_1(D)h(D)$ in (8-26) is D^{j+r}, since $h(D)$ had degree r. Therefore, the coefficient of any power of D greater than $j + r$ on the left side of (8-26) must be zero, and the coefficients of $b(D)$ and $g(D)$ must satisfy

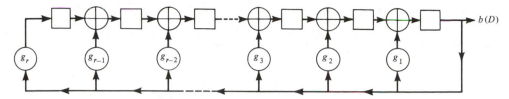

FIGURE 8-5. High-speed linear feedback shift register generator.

$$\sum_{m=0}^{r} g_m b_{i-m} = 0 \tag{8-32}$$

for $i > j + r$. Since $g_0 = 1$, this relationship can be written

$$b_i = \sum_{m=1}^{r} g_m b_{i-m} \tag{8-33}$$

This recurrence relation must be satisfied at all times subsequent to the end of the input sequence.

A second circuit configuration which satisfies the recurrence relationship of (8-33) and therefore generates an identical output sequence is illustrated in Figure 8-6. This can be verified by considering the output of the generator to be the input to the leftmost shift register stage as shown. By inspection the output is

$$b(D) = g_1 D b(D) + g_2 D^2 b(D) + \cdots + g_r D^r b(D) \tag{8-34}$$

Equating coefficients of D^i on both sides of this equation will show that (8-33) is satisfied. This feedback configuration is, in fact, a commonly used configuration. The equivalence of the circuits of Figures 8-5 and 8-6 is not at all surprising given the equivalence of the multipliers of Figure 8-1. The configuration chosen for a particular application depends on such things as the speed at which the hardware must operate and whether delayed outputs are also required. Delayed outputs for all delays up to r are available from the configuration of Figure 8-6 but not from the configuration of Figure 8-5. The configuration of Figure 8-5, however, can function at higher speeds than that of Figure 8-6 since there is less propagation delay in the feedback path.

It is of interest to be able to determine the output $b(D)$ of either the circuit of Figure 8-5 or 8-6 given the initial contents of the shift register. This is easily accomplished for the circuit of Figure 8-5. Its output $b(D)$ is identical to the output $b'(D)$ beginning at time r of Figure 8-3 with $h(D) = D^r$ and with an input

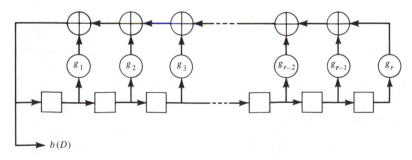

$b(D)$

FIGURE 8-6. Linear feedback shift register whose output satisfies the same recurrence relationship as the generator of Figure 8-5.

$$a_1(D) = a_0 + a_1D + a_2D^2 + \cdots + a_{r-1}D^{r-1} \tag{8-35}$$

This input is nonzero just long enough to load the shift register. The output, from (8-31), is

$$b'(D) = \frac{D^r a_1(D)}{g(D)} \tag{8-36}$$

The problem being considered is to find the output beginning at the time that the shift register is completely loaded. The loading process just described consumes r time units, so the desired result is $b'(D)$ of (8-36) beginning at time r. Observe that the output $b'(D)$ is zero for the first r time units while the shift register loads so that the output beginning at time r is simply $b'(D)$ shifted by r time units or

$$b(D) = \frac{a_1(D)}{g(D)} \tag{8-37}$$

EXAMPLE 8-9

Find the output of the circuit of Figure 8-5 with $g(D) = 1 + D + D^3 + D^4$ and an initial shift register load of 0001. The circuit and the initial load are illustrated in Figure 8-7.

Solution: The initial load is described by

$$a_1(D) = 1$$

and the output is

$$b(D) = \frac{1}{1 + D + D^3 + D^4}$$

Performing the polynomial long division yields

$$
\begin{array}{r}
1 + D + D^2 \qquad\qquad\qquad + D^6 + D^7 + D^8 \qquad\qquad\qquad + \cdots \\
\hline
1 + D + D^3 + D^4\,\big|\,1 \\
\underline{1 + D \qquad\quad + D^3 + D^4} \\
D \qquad\quad + D^3 + D^4 \\
\underline{D + D^2 \qquad\quad + D^4 + D^5} \\
D^2 + D^3 \qquad\quad + D^5 \\
\underline{D^2 + D^3 \qquad\quad + D^5 + D^6} \\
D^6 \\
\underline{D^6 + D^7 \qquad\quad + D^9 + D^{10}} \\
D^7 \qquad\quad + D^9 + D^{10} \\
\underline{D^7 + D^8 \qquad\quad + D^{10} + D^{11}} \\
D^8 + D^9 \qquad\quad + D^{11} \\
\underline{D^8 + D^9 \qquad\quad + D^{11} + D^{12}} \\
D^{12} \\
\vdots
\end{array}
$$

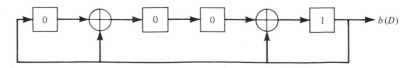

FIGURE 8-7. Circuit configuration for Example 8-9.

Observe that the output sequence is periodic with a period of six. This can be verified with a manual calculation of the contents of the shift register as a function of time. ∎

The procedure just described applies *only* to the configuration of Figure 8-5. Although the two shift register configurations generate identical output sequences, the sequence of shift register states each goes through is different. A simple means of finding the output sequence for the shift register of Figure 8-6 is to find the equivalent initial state of the circuit of Figure 8-5 and then to use the results just described. Suppose that the initial state for the circuit of Figure 8-6 is

$$a(D) = a_0 + a_1D + a_2D^2 + \cdots + a_{r-1}D^{r-1} \tag{8-38}$$

This means that a_0 is in the rightmost shift register stage, a_1 in the second from the right, and so on. Define $c(D)$ to be the output of the rightmost shift register of Figure 8-6, that is, $c(D) = D^r b(D)$. The first r elements of $c(D)$ are a_0, a_1, a_2, . . ., a_{r-1}. Since the circuits of Figures 8-5 and 8-6 are equivalent, the initial load of the circuit of Figure 8-5 can be chosen such that its output $b'(D) = c(D)$. Let the initial load that accomplishes this be

$$a'(D) = a_0' + a_1'D + a_2'D^2 + \cdots + a_{r-1}'D^{r-1} \tag{8-39}$$

Since the two output sequences are equal, (8-37) becomes

$$a_0 + a_1D + \cdots + a_{r-1}D^{r-1} + b_r'D^r + b_{r+1}'D^{r+1} + \cdots$$
$$= \frac{a_0' + a_1'D + a_2'D^2 + \cdots + a_{r-1}'D^{r-1}}{g_0 + g_1D + g_2D^2 + \cdots + g_rD^r} \tag{8-40}$$

Thus the initial state of the configuration of Figure 8-5, which produces the same output sequence as the configuration of Figure 8-6 with the initial load of $a(D)$, is found by equating the first r coefficients of

$$\{a(D) + b_r'D^r + b_{r+1}'D^{r+1} + \cdots\}g(D) = a'(D) \tag{8-41}$$

The entire output sequence can be found using (8-37) and the fact that $c(D) = b'(D) = D^r b(D)$.

EXAMPLE 8-10

Find the output of the circuit of Figure 8-6 with $g(D) = 1 + D + D^3 + D^4$ and an initial shift register load of 0001. The circuit and the initial load are illustrated in Figure 8-8.

Solution: The initial load in polynomial form is $a(D) = 1$ and the product on the left side of (8-41) is

$$g(D) + (b_r'D^r + b_{r+1}'D^{r+1} + \cdots)g(D) = a_0' + a_1'D + a_2'D^2 + a_3'D^3$$

so that

$$a_0' = g_0 = 1$$
$$a_1' = g_1 = 1$$
$$a_2' = g_2 = 0$$
$$a_3' = g_3 = 1$$

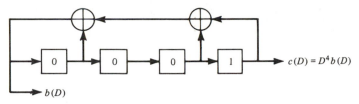

FIGURE 8-8. Circuit configuration for example 8-10.

Thus $a'(D) = 1 + D + D^3$ and the complete output sequence is $b'(D) = a'(D)/g(D)$ or

$$
\begin{array}{r}
1 \qquad\qquad\qquad + D^4 + D^5 + D^6 \qquad\qquad\qquad + D^{10} + D^{11} + D^{12} + \cdots \\
\hline
1 + D + D^3 + D^4 \, \big|\, 1 + D \qquad + D^3 \\
\underline{1 + D \qquad + D^3 + D^4} \\
D^4 \\
\underline{D^4 + D^5 \qquad\quad + D^7 + D^8} \\
D^5 \qquad\quad + D^7 + D^8 \\
\underline{D^5 + D^6 \qquad\qquad + D^8 + D^9} \\
D^6 + D^7 \qquad\qquad + D^9 \\
\underline{D^6 + D^7 \qquad\qquad + D^9 + D^{10}} \\
D^{10} \\
\vdots
\end{array}
$$

Multiplying by $D^{-r} = D^{-4}$ yields the output $b(D) = D^{-4} + 1 + D + D^2 + D^6 + D^7 + D^8 + \cdots$. Observe that the output is identical to the output of Example 8-9 except for beginning with the D^{-4} term. This means simply that the output sequence beginning at time -4 is known. These additional known output symbols are the load of the shift register. ∎

Several observations about the circuits of Figures 8-5 and 8-6 are now made. First, given nonzero initial conditions, neither of the registers will ever reach an all-zeros state. This can be seen from the circuit of Figure 8-5, which would reach the state where all registers except the rightmost contain zeros just prior to reaching the all-zeros state. The single 1 in the rightmost register would be fed back to some other register. Since all g's cannot be zero, the all-zeros state is never reached. Second, since the register contains r stages and an r-stage shift register has at most $2^r - 1$ nonzero states, the output must be periodic with a period of *at most* $2^r - 1$. The period can be significantly less than $2^r - 1$.

Consider the determination of the maximum period of either of the shift register circuits. Note that the same circuit may generate many different output sequences; the particular output sequence generated depends on the initial state of the register. For example, consider the linear feedback shift register generator illustrated in Figure 8-9. Four different sets of shift register states are possible depending on the initial state. These four possible cycles have periods of 1, 1, 2, and 4 as shown. Since all possible shift register states are included in one of the four cycles, there are no other cycles. The maximum possible period for an arbitrary feedback shift register connection defined by $g(D)$ can be found [5] by defining the reciprocal polynomial of $g(D)$ by

$$
g_r(D) = D^r g\left(\frac{1}{D}\right) \tag{8-42}
$$

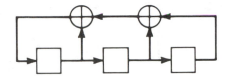

Cycle 1	Cycle 2	Cycle 3	Cycle 4
0 0 0	1 1 1	0 1 0	1 0 0
0 0 0	1 1 1	1 0 1	1 1 0
0 0 0	1 1 1	0 1 0	0 1 1
•	•	1 0 1	0 0 1
•	•	•	1 0 0
•	•	•	1 1 0
		•	0 1 1
			0 0 1
			•
			•
			•

FIGURE 8-9. Linear feedback shift register cycles for four different initial conditions. (From [7])

It can be shown [5] that the maximum possible period of the shift register generator is the smallest possible integer N for which $D^N + 1$ is divisible by $g_r(D)$. That is, the maximum period is the smallest N for which a polynomial $h_r(D)$ exists such that

$$g_r(D)h_r(D) = D^N + 1 \qquad (8\text{-}43)$$

EXAMPLE 8-11

Suppose that $g(D) = 1 + D + D^3 + D^4$. Then

$$g_r(D) = D^4 g\left(\frac{1}{D}\right)$$

$$= D^4(1 + D^{-1} + D^{-3} + D^{-4})$$

$$= D^4 + D^3 + D + 1$$

By performing all the necessary long divisions, it can be shown that the smallest integer N for which $D^N + 1$ is evenly divisible by $g_r(D)$ is $N = 6$.

$$
\begin{array}{r}
D^2 + D + 1 \\
\hline
D^4 + D^3 + D + 1 \overline{\smash{\big)} D^6 \qquad\qquad\qquad\qquad + 1} \\
\underline{D^6 + D^5 \qquad + D^3 + D^2} \\
D^5 \qquad + D^3 + D^2 \qquad + 1 \\
\underline{D^5 + D^4 \qquad + D^2 + D} \\
D^4 + D^3 \qquad + D + 1 \\
\underline{D^4 + D^3 \qquad + D + 1} \\
0
\end{array}
$$

Thus the maximum possible period is 6. Observe that this period is less than the maximum possible period for a four-stage shift register, which is $2^4 - 1 = 15$.

8-3

MAXIMAL-LENGTH SEQUENCES

All of the discussion in Section 8-1 was general in that, except for requiring that $g_0 = 1$, no restrictions were placed on the generator functions. In this section discussion is limited to linear feedback shift register generators having the form of Figure 8-5 or 8-6 with $g(D)$ a primitive polynomial. Recall that the maximum possible period of a shift register generator is the smallest N for which the reciprocal $g_r(D)$ of the generator polynomial $g(D)$ divides $D^N + 1$. It can be demonstrated [5] that the reciprocal of a primitive polynomial is also primitive. Thus the smallest N for which a primitive polynomial $g(D)$ of degree r divides $D^N + 1$ is $N = 2^r + 1$. This means that a shift register initial condition exists which results in a cycle with period $N = 2^r - 1$. Since an r-stage shift register has a total of $2^r - 1$ nonzero states, all states are passed through in this cycle having period $N = 2^r - 1$, and there is only one possible cycle. Shift register sequences having the maximum possible period for an r-stage shift register are called *maximal-length sequences* or *m-sequences*. Since the shift register passes through all possible states, each different initial condition results in a different phase of the same *m*-sequence.

8-3.1 Properties of *m*-Sequences

Maximal-length sequences have a number of properties which are useful in their application to spread-spectrum systems. Some of these properties are given here.

Property I. *A maximal-length sequence contains one more one than zero. The number of ones in the sequence is $\frac{1}{2}(N + 1)$.*

Proof: *Consider the generator of Figure 8-5, where the rightmost symbol of the shift register state is the output symbol. The shift register passes through all possible nonzero states. Of these states, $2^{r-1} = \frac{1}{2}(N + 1)$ have a one in the rightmost position, and $2^{r-1} - 1$ have a zero in the rightmost position. Thus there is one more one than zero in the output sequence.*

Property II. *The modulo-2 sum of an m-sequence and any phase shift of the same sequence is another phase of the same m-sequence (shift-and-add property).*

Proof: *Consider the shift register generator of Figure 8-5. The output is given by (8-37) for any initial condition. Since any different initial condition results in a different phase of the same m-sequence, two phases b(D) and b'(D) of the same sequence can be written $b(D) = a(D)/g(D)$ and $b'(D) = a'(D)/g(D)$, where a(D) and a'(D) are distinct initial conditions. The modulo-2 sum b(D) + b'(D) = [a(D) + a'(D)]/g(D) = a''(D)/g(D). Since the modulo-2 sum of any two distinct initial conditions is a third distinct initial condition, a''(D)/g(D) = b''(D) is a third distinct phase of the original sequence b(D).*

Property III. *If a window of width r is slid along the sequence for N shifts, each r-tuple except the all zero r-tuple will appear exactly once.*

Proof: *Consider the shift register generator of Figure 8-6. The sequence b(D) passes through the shift register of this generator so that the window of width r is simply the state of the shift register. Since the shift register passes through all nonzero states exactly once, all possible r-tuples appear in the window exactly once.*

Property IV. *The periodic autocorrelation function $\theta_b(k)$ is two-valued and is given by*

$$\theta_b(k) = \begin{cases} 1.0 & k = lN \\ -\dfrac{1}{N} & k \neq lN \end{cases} \tag{8-44}$$

where l is any integer and N is the sequence period.

Proof: *The value of the periodic autocorrelation function $\theta_b(k)$ is $(N_A - N_D)/N$, where N_A is the number of zeros and N_D is the number of ones in the modulo-2 sum of the sequence **b** and the kth cyclic shift of **b**. For $k = lN$, the kth cyclic shift of **b** is identical to **b**, since the sequence period is N, so that the modulo-2 sum contains all zeros and $N_A = N$, $N_D = 0$, and $\theta_b(lN) = N/N = 1.0$. For $k \neq lN$, the modulo-2 sum is some phase of the original sequence by Property II. Then, by Property I, there is one more one than zero in the modulo-2 sum, so that $N_A - N_D = -1$ and $\theta_b(k) = -1/N$.*

Property V. *Define a* run *as a subsequence of identical symbols within the m-sequence. The length of this subsequence is the length of the run. Then, for any m-sequence, there is*

1. *1 run of ones of length r.*
2. *1 run of zeros of length r − 1.*
3. *1 run of ones and 1 run of zeros of length r − 2.*
4. *2 runs of ones and 2 runs of zeros of length r − 3.*
5. *4 runs of ones and 4 runs of zeros of length r − 4.*

$$\vdots$$

r. *2^{r-3} runs of ones and 2^{r-3} runs of zeros of length 1.*

Proof [8]: *Consider the shift register of Figure 8-6. There can be no run of ones having length $l \geq r$ since this would require that the all-ones shift register state be followed by another all-ones state. This cannot occur since each shift register state occurs once and only once during N cycles. Thus there is a single run of r consecutive ones, and this run is preceded by a zero and followed by a zero.*

A run of r − 1 ones must be preceded by and followed by a zero. This requires that the shift register state which is r − 1 ones followed by a 0 be followed immediately by the state which is a 0 followed by r − 1 ones. These two states are also passed through in the generation of the run of r ones, where they are separated by the all-ones state. Since each state occurs only once, there can be no run of r − 1 ones. A run of r − 1 zeros must be preceded by and followed by 1's. Thus the shift register must pass through the state which is a 1 followed by $r - 1$ zeros. This state occurs only once, so there is a single run of $r - 1$ zeros.

Now consider a run of k ones where $1 \leq k < r - 1$. Each run of k ones must be preceded by and followed by a 0. Thus the shift register must pass through the state which is a 0 followed by k ones followed by a 0, with the $r - k - 2$ remaining positions taking on arbitrary values. There are 2^{r-k-2} possible ways to complete these remaining positions in the shift register, so there are 2^{r-k-2} runs of k ones. Similarly, there are 2^{r-k-2} runs of k zeros.

8-3.2   Power Spectrum of *m*-Sequences

The power spectrum of the spreading waveform $c(t)$ is frequently used in the analysis of the performance of spread-spectrum systems. This power spectrum is easily calculated using the Wiener–Khintchine theorem and Property IV above for maximal-length spreading waveforms. The power spectrum of $c(t)$ is the Fourier transform of the autocorrelation function $R_c(\tau)$, which is given by (8-8). For $0 \leq \tau \leq T_c$, $k = 0$ $\tau_\epsilon = \tau$, and (8-8) becomes

$$R_c(\tau) = \left(1 - \frac{\tau}{T_c}\right) - \frac{1}{N}\left(\frac{\tau}{T_c}\right) \qquad 0 \leq \tau \leq T_c$$

$$= 1 - \frac{\tau}{T_c}\left(1 + \frac{1}{N}\right) \tag{8-45a}$$

where (8-44) has been used to evaluate $\theta_b(k)$. For $T_c < \tau < (N - 1)T_c$, $k \neq lN$ and $(k + 1) \neq lN$ for any integer l, so that

$$R_c(\tau) = \left(1 - \frac{\tau_\epsilon}{T_c}\right) - \frac{1}{N} + \left(\frac{\tau_\epsilon}{T_c}\right) - \frac{1}{N} \qquad T_c < \tau < (N - 1)T_c$$

$$= -\frac{1}{N} \tag{8-45b}$$

For $(N - 1)T_c \leq \tau < NT_c$, $k = N - 1$ and $k + 1 = N$, so that (8-8) becomes

$$R_c(\tau) = \left(1 - \frac{\tau_\epsilon}{T_c}\right) - \frac{1}{N} + \frac{\tau_\epsilon}{T_c} \qquad (N - 1)T_c \leq \tau < NT_c$$

$$= \frac{\tau_\epsilon}{T_c}\left(1 + \frac{1}{N}\right) - \frac{1}{N} \tag{8-45c}$$

where $\tau_\epsilon = \tau - (N - 1)T_c$, so that $0 < \tau_\epsilon \leq T_c$. Since $\theta_b(k)$ of (8-44) is periodic, $R_c(\tau)$ is also periodic and has a period $T = NT_c$. Thus (8-45) defines one complete cycle of $R_c(\tau)$. The autocorrelation function $R_c(\tau)$ is illustrated in Figure 8-10.

The power spectrum is found by taking the Fourier transform of (8-45). The result is

$$S_c(f) = \sum_{m=-\infty}^{\infty} P_m \, \delta(f - mf_0) \tag{8-46}$$

where $P_0 = 1/N^2$, $P_m = [(N + 1)/N^2]\text{sinc}^2(m/N)$, and $f_0 = 1/NT_c$. The power spectrum as a function of f is illustrated in Figure 8-11. This power spectrum consists of discrete spectral lines at all harmonics of $1/NT_c$. The envelope of the amplitude of these lines is given by $[(N + 1)/N^2] \, \text{sinc}^2(fT_c)$ except for the dc

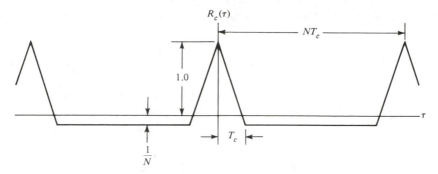

FIGURE 8-10. Autocorrelation function for a maximal-length sequence with chip duration T_c and period NT_c.

term, which has an amplitude $1/N^2$. Note that the ordinate in Figure 8-11 is absolute and is not decibels.

Suppose that the m-sequence $c(t)$ is used to biphase modulate a sinusoidal carrier having power P and frequency f_0. The modulated carrier is

$$s(t) = \sqrt{2P}\, c(t) \cos 2\pi f_0 t \qquad (8\text{-}47)$$

The power spectrum of this modulated carrier is the convolution of the power spectrum of the carrier and the power spectrum of the spreading code. Thus

$$S_s(f) = S_c(f) * \frac{P}{2}\, \delta(f - f_0) + S_c(f) * \frac{P}{2}\, \delta(f + f_0) \qquad (8\text{-}48)$$

and the resultant power spectrum is a translation of the discrete spectrum $S_c(f)$ upward and downward by a frequency f_0. In most spread-spectrum systems the carrier is randomly modulated by data as well as the spreading code. For this reason, the transmitted spectrum is continuous and not discrete.

8-3.3 Tables of Polynomials Yielding m-Sequences

It is often necessary to design circuits that generate m-sequences having a particular number of stages. Since finding the primitive polynomials used to generate these sequences is difficult, a number of authors have generated tables of primitive

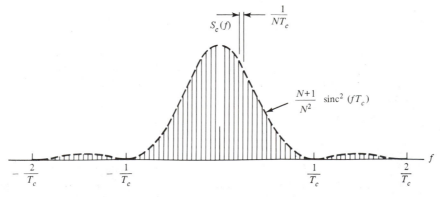

FIGURE 8-11. Power spectrum of a maximal-length sequence with chip duration T_c and period NT_c.

polynomials for quick reference. In particular, Peterson and Weldon [5] have an extensive table of polynomials in their Appendix C. The use of this table is described here since it provides a large selection of primitive polynomials which will be useful to the spread-spectrum system designer.

In the table, all polynomials are specified by an octal number which defines the coefficients of $g(D)$. The table also defines some polynomials which are not primitive and therefore will not yield a maximal-length sequence. An example entry in the table is

DEGREE 7   1 211E   3 217E   5 235E   7 367H   9 277E

11 325G   13 203F   19 313H   21 345G.

The entry for each polynomial consists of an integer whose use will be described later, an octal number defining $g(D)$, and a letter designating the type of polynomial. The letters E, F, G, and H designate primitive polynomials. The octal number gives the coefficients of $g(D)$ beginning with g_0 on the right and proceeding to g_r in the last nonzero position on the left.

EXAMPLE 8-12

The table contains the entry 7 367H. The letter H means that the entry is a primitive polynomial. Expanding the octal entry 367 into binary form yields

$$
\begin{array}{ccc}
\underbrace{\hspace{1.2em}3\hspace{1.2em}} & \underbrace{\hspace{1.2em}6\hspace{1.2em}} & \underbrace{\hspace{1.2em}7\hspace{1.2em}}
\end{array} \quad \text{octal}
$$

0 1 1   1 1 0   1 1 1   binary

↓ ↓ ↓   ↓ ↓ ↓   ↓ ↓ ↓

$g_7\ g_6\ g_5\ g_4\ g_3\ g_2\ g_1\ g_0$ coefficient

so that

$$g(D) = 1 + D + D^2 + D^4 + D^5 + D^6 + D^7$$

The shift register generator can be either the form of Figure 8-5 or 8-6. ■

Table 8-5 is a list of primitive polynomials of all degrees up to 40 and degrees 61 and 89. All polynomials in Table 8-5 are primitive so that the letter designation has been dropped as well as the number preceding the octal generator specification. Each entry in brackets represents one primitive polynomial as a series of octal numbers, exactly as explained above. The entries followed by an asterisk correspond to circuit implementation with only two feedback connections. Two feedback connection implementations are very useful for high-speed applications. No reciprocal polynomials are listed in Table 8-5. Since the reciprocal polynomial of a primitive polynomial is also primitive, each entry of Table 8-5 can be used to generate two distinct m-sequences. It can be demonstrated that the sequences generated by the reciprocal polynomial $g_r(D)$ is equivalent to the reverse of the sequence generated by $g(D)$.

EXAMPLE 8-13

Consider the sequence generated by the polynomial corresponding to the entry [13] of Table 8-1. The primitive polynomial is $g(D) = 1 + D + D^3$ and its reciprocal is $D^3g(1/D) = D^3 + D^2 + 1$. Using the configuration of Figure 8-5 with an initial load of $a(D) = 1$, the output sequence for $g(D)$ is, by the polynomial long division, $1 + D + D^2 + D^4 + D^7 + \cdots$. The output corresponding to

TABLE 8-5. Primitive Polynomials Having Degree $r \le 34$

Degree	Octal Representation of Generator Polynomial (g_0 on right to g_r on left)
2	[7]*
3	[13]*
4	[23]*
5	[45]*,[75],[67]
6	[103]*,[147],[155]
7	[211]*,[217],[235],[367],[277],[325],[203]*,[313],[345]
8	[435],[551],[747],[453],[545],[537],[703],[543]
9	[1021]*,[1131],[1461],[1423],[1055],[1167],[1541], [1333] ,[1605],[1751],[1743],[1617],[1553],[1157]
10	[2011]*,[2415],[3771],[2157],[3515],[2773],[2033], [2443] ,[2461],[3023],[3543],[2745],[2431],[3177]
11	[4005]*,[4445],[4215],[4055],[6015],[7413],[4143], [4563] ,[4053],[5023],[5623],[4577],[6233],[6673]
12	[10123],[15647],[16533],[16047],[11015],[14127], [17673],[13565],[15341],[15053],[15621],[15321], [11417],[13505]
13	[20033],[23261],[24623],[23517],[30741],[21643], [30171],[21277],[27777],[35051],[34723],[34047], [32535],[31425]
14	[42103],[43333],[51761],[40503],[77141],[62677], [44103],[45145],[76303],[64457],[57231],[64167], [60153],[55753]
15	[100003]*,[102043],[110013],[102067],[104307],[100317], [177775] ,[103451],[110075],[102061],[114725],[103251], [100021]*,[100201]*
16	[210013],[234313],[233303],[307107],[307527],[306357], [201735],[272201],[242413],[270155],[302157],[210205], [305667],[236107]
17	[400011]*,[400017],[400431],[525251],[410117],[400731], [411335] ,[444257],[600013],[403555],[525327],[411077], [400041]*,[400101]*
18	[1000201]*,[1000247],[1002241],[1002441],[1100045], [1000407] ,[1003011],[1020121],[1101005],[1000077], [1001361] ,[1001567],[1001727],[1002777]
19	[2000047],[2000641],[2001441],[2000107],[2000077], [2000157],[2000175],[2000257],[2000677],[2000737], [2001557],[2001637],[2005775],[2006677]
20	[4000011]*,[4001051],[4004515],[6000031],[4442235]
21	[10000005]*,[10040205],[10020045],[10040315],[10000635], [10103075] ,[10050335],[10002135],[17000075]
22	[20000003]*,[20001043],[2222223],[25200127],[20401207], [20430607] ,[20070217]

TABLE 8-5. **Primitive Polynomials Having Degree** $r \le 34$ **(cont.)**

Degree	Octal Representation of Generator Polynomial (g_0 on right to g_r on left)
23	[40000041]*,[40404041],[40000063],[40010061],[50000241], [40220151] ,[40006341],[40405463],[40103271],[41224445], [4043561]
24	[100000207],[125245661],[113763063]
25	[200000011]*,[200000017],[204000051],[200010031], [200402017] ,[252001251],[201014171],[204204057], [200005535] ,[200014731]
26	[400000107],[430216473],[402365755],[426225667], [510664323],[473167545],[411335571]
27	[1000000047],[1001007071],[1020024171],[1102210617], [1250025757],[1257242631],[1020560103],[1112225171], [1035530241]
28	[2000000011]*,[2104210431],[2000025051],[2020006031], [2002502115] ,[2001601071]
29	[4000000005]*,[4004004005],[4000010205],[4010000045], [4400000045] ,[4002200115],[4001040115],[4004204435], [4100060435] ,[4040003075],[4004064275]
30	[10,040,000,007],[10,104,264,207],[10,115,131,333],[11,362,212,703], [10,343,244,533]
31	[20,000,000,011]*,[20,000,000,017],[20,000,020,411],[21,042,104,211] [20,010,010,017] ,[20,005,000,251],[20,004,100,071],[20,202,040,217] [20,000,200,435] ,[20,060,140,231],[21,042,107,357]
32	[40,020,000,007],[40,460,216,667],[40,035,532,523],[42,003,247,143], [41,760,427,607]
33	[100,000,020,001]*,[100,020,024,001],[104,000,420,001], [100,020,224,401] ,[111,100,021,111],[100,000,031,463], [104,020,466,001] ,[100,502,430,041],[100,601,431,001]
34	[201,000,000,007],[201,472,024,107],[377,000,007,527], [225,213,433,257],[227,712,240,037],[251,132,516,577], [211,636,220,473],[200,000,140,003]
35	[400,000,000,005]*
36	[1,000,000,004,001]*
37	[2,000,000,012,005]
38	[4,000,000,000,143]
39	[10,000,000,000,021]*
40	[20,000,012,000,005]
61	[200,000,000,000,000,000,047]
89	[400,000,000,000,000,000,000,000,000,151]

Source: Refs. 5, 6, and 9.

the same initial state for the reciprocal polynomial is, again by long division, $1 + D^2 + D^3 + D^4 + D^7 + \cdots$.

The two output sequences are

$$g(D) \rightarrow 1\ 1\ 1\ 0\ 1\ 0\ 0\ ,\ 1\ 1\ 1\ 0\ 1\ 0\ 0\ ,\ \ldots$$

$$g_r(D) \rightarrow 1\ 0\ 1\ 1\ 1\ 0\ 0\ ,\ 1\ 0\ 1\ 1\ 1\ 0\ 0\ ,\ \ldots$$

which, except for a phase shift, are simply the reverse of one another. ◼

8-3.4 Partial Autocorrelation Properties of *m*-Sequences

The autocorrelation properties of maximal-length sequences are defined over a complete cycle of the sequence. That is, the two-valued autocorrelation of Property IV can be guaranteed only when the integration of (8-2) is over a full period of the waveform $c(t)$. In Chapter 10, where the code synchronization problem of spread spectrum communications is addressed, it will be shown that rapid synchronization of long codes often requires an estimate of the correlation between the received code and the receiver despreading code be made in less than a full code period. Thus the correlation estimate is based on a correlation over a partial period and is related to the partial autocorrelation properties of the code.

Since the partial autocorrelation is associated with an integration over a fraction of the code period, the partial autocorrelation function is dependent on the size of this fraction and the starting time of the integration. The *partial autocorrelation function* of the spreading waveform $c(t)$ is defined by

$$R_c(\tau, t, T_w) = \frac{1}{T_w} \int_t^{t+T_w} c(\lambda)c(\lambda + \tau)\, d\lambda \qquad (8\text{-}49)$$

where T_w is the duration of the correlation and t is the starting time of the correlation. Using (8-1) for $c(t)$ and the substitution $\gamma = \lambda - t$ yields

$$R_c(\tau, t, T_w) = \frac{1}{T_w} \sum_{n=-\infty}^{\infty} \sum_{m=-\infty}^{\infty} a_m a_n \int_0^{T_w} p(\gamma + t - mT_c)p(\gamma + t + \tau - nT_c)\, d\gamma \qquad (8\text{-}50)$$

Now let $\tau = kT_c + \tau_\epsilon$, $T_w = WT_c$, and assume that $t = k'T_c$. Then the integral is nonzero only for $n = m + k$ or $n = m + k + 1$ and the autocorrelation can be written

$$R_c(\tau_\epsilon, k, k', W) = \frac{1}{WT_c} \sum_{m=-\infty}^{\infty} a_m a_{m+k} \int_0^{WT_c} p[\gamma - (m - k')T_c]p[\gamma - (m - k')T_c + \tau_\epsilon]\, d\gamma$$

$$+ \frac{1}{WT_c} \sum_{m=-\infty}^{\infty} a_m a_{m+k+1} \int_0^{WT_c} p[\gamma - (m - k')T_c]p[\gamma - (m - k' + 1)T_c + \tau_\epsilon]\, d\gamma \qquad (8\text{-}51)$$

The integrand of (8-51) is nonzero within the limits of integration only when $0 \le (m - k')T_c \le (W - 1)T_c$, which implies that the limits on the summations can be reduced to $k' \le m \le w + k' - 1$. For any fixed value of m, the integrand of the first integral is nonzero only for $(m - k')T_c \le \gamma \le (m + 1 - k')T_c - \tau_\epsilon$, and the integrand of the second integral is nonzero only for $(m + 1 - k')T_c - \tau_\epsilon \le \gamma \le (m + 1 - k')T_c$, so that (8-51) can be written

$$R_c(\tau_\epsilon, k, k', W) = \frac{1}{WT_c} \sum_{m=k'}^{W+k'-1} a_m a_{m+k} \int_{(m-k')T_c}^{(m+1-k')T_c - T_\epsilon} p[\gamma - (m - k')T_c]$$

$$\times\ p[\gamma - (m - k')T_c + \tau_\epsilon]\ d\gamma$$

$$+ \frac{1}{WT_c} \sum_{m=k'}^{W+k'-1} a_m a_{m+k+1} \int_{(m+1-k')T_c - \tau_\epsilon}^{(m+1-k')T_c} p[\gamma - (m - k')T_c]$$

$$\times\ p[\gamma - (m + 1 - k')T_c + \tau_\epsilon]\ d\gamma$$

$$= \frac{1}{W} \sum_{m=k'}^{W+k'-1} a_m a_{m+k} \left(1 - \frac{\tau_\epsilon}{T_c}\right)$$

$$+ \frac{1}{W} \sum_{m=k'}^{W+k'-1} a_m a_{m+k+1} \left(\frac{\tau_\epsilon}{T_c}\right) \tag{8-52}$$

The *discrete partial autocorrelation function* of a sequence $b(D)$ is defined by [2]

$$\theta_b(k, k', W) = \frac{1}{W} \sum_{m=k'}^{k'+W-1} a_m a_{m+k} \tag{8-53}$$

so that the partial autocorrelation function can be written

$$R_c(\tau_\epsilon, k, k', W) = \left(1 - \frac{\tau_\epsilon}{T_c}\right) \theta_b(k, k', W) + \frac{\tau_\epsilon}{T_c} \theta_b(k + 1, k', W) \tag{8-54}$$

Thus $R_c(\tau, t, T_w)$ can easily be calculated from knowledge of $\theta_b(k, k', W)$. The discrete partial autocorrelation function can be calculated in the same manner as the discrete periodic autocorrelation function. The value of $\theta_b(k, k', W)$ is the number of agreements N_A minus the number of disagreements between $\mathbf{b}(0)$ and $\mathbf{b}(k)$ over the window beginning at k' and ending at $k' + W$ divided by W. This is equivalent to the difference between the number of zeros and the number of ones over the same window of the modulo-2 sum of $\mathbf{b}(0)$ and $\mathbf{b}(k)$.

EXAMPLE 8-14
Evaluate $\theta_b(6, k', 7)$ for the 15 bit m-sequence generated using the generator of Figure 8-5 with an initial condition $a(D) = 1$ and the primitive polynomial $g(D) = 1 + D + D^4$.

Solution: One cycle of $\mathbf{b}(0)$ and $\mathbf{b}(6)$ and their modulo-2 sum are

$$\mathbf{b}(0) \qquad = 1\ 1\ 1\ 1\ 0\ 1\ 0\ 1\ 1\ 0\ 0\ 1\ 0\ 0\ 0$$

$$\mathbf{b}(6) \qquad = 0\ 1\ 1\ 0\ 0\ 1\ 0\ 0\ 0\ 1\ 1\ 1\ 1\ 0\ 1$$

$$\mathbf{b}(0) + \mathbf{b}(6) = 1\ 0\ 0\ 1\ 0\ 0\ 0\ 1\ 1\ 1\ 1\ 0\ 1\ 0\ 1$$

Then $\theta_b(6, k', 7)$ is the number of zeros minus the number of ones in $\mathbf{b}(0) + \mathbf{b}(6)$ in a seven-unit window beginning at k'. The value of $\theta_b(6, k', 7)$ is plotted in Figure 8-12 as a function of k'.

Observe that the partial autocorrelation function is not well behaved as was the full-period autocorrelation function. The partial-period autocorrelation is not two-valued and its variation as a function of window size and window placement can cause serious difficulties if not taken into account in the system design. The mean and variance over k' of $\theta_b(k, k', W)$ are useful quantities for the spread-spectrum

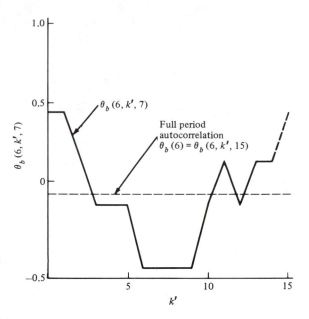

FIGURE 8-12. Discrete partial autocorrelation function $\theta_b(6,k',7)$ for 15-symbol *m*-sequence generated by $g(D) = 1 + D + D^4$.

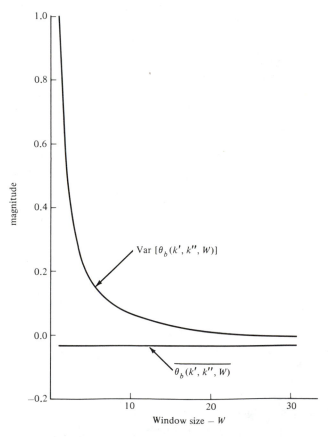

FIGURE 8-13. Mean and variance of the discrete partial autocorrelation function as a function of window size for 31-bit *m*-sequences.

system designer. Because of Property II of maximal-length sequences, the modulo-2 sum of $\mathbf{b}(0)$ and $\mathbf{b}(k)$ is another phase $\mathbf{b}(q)$ of the same m-sequence. Then

$$\theta_b(k, k', W) = \frac{1}{W} \sum_{i=0}^{W-1} a_{i+q+k'}$$

where $a_i = (-1)^{b_i}$; and the average of $\theta_b(k, k', W)$ over all k' becomes

$$\overline{\theta_b(k, k', W)} = \frac{1}{N} \sum_{k'=0}^{N-1} \frac{1}{W} \sum_{i=0}^{W-1} a_{i+q+k'}$$

$$= \frac{1}{WN} \sum_{i=0}^{W-1} \sum_{k'=0}^{N-1} a_{i+q+k'} \tag{8-55}$$

Because of Property I of the m-sequence, the inner summation in the last line of (8-55) is equal to -1 for all i and q so that [10]

$$\overline{\theta_b(k, k', W)} = -\frac{1}{N} \tag{8-56}$$

The second moment of the discrete partial autocorrelation function is [10]

$$\overline{\theta_b^2(k, k', W)} = \frac{1}{N} \sum_{k'=0}^{N-1} \theta_b^2(k, k', W)$$

$$= \frac{1}{W}\left(1 - \frac{W-1}{N}\right) \tag{8-57}$$

The variance over k' of $\theta_b(k, k', W)$ is then

$$\mathrm{var}\,[\theta_b(k, k', W)] = \overline{\theta_b^2(k, k', W)} - [\overline{\theta_b(k, k', W)}]^2$$

$$= \frac{1}{W}\left(1 - \frac{W-1}{N}\right) - \frac{1}{N^2} \tag{8-58}$$

Observe that for $W = N$ the variance equals zero as expected. ∎

EXAMPLE 8-15

Plot the mean and variance as a function of window size W for the family of 31-bit maximal-length sequences.

Solution: For any window size,

$$\overline{\theta_b(k, k', W)} = \frac{1}{31}$$

Equation (8-58) for the variance is

$$\mathrm{var}\,[\theta_b(k, k', W)] = \frac{1}{W}\left(1 - \frac{W-1}{31}\right) - \left(\frac{1}{31}\right)^2$$

These relationships are plotted in Figure 8-13

The results just derived can be used to determine approximate thresholds when correlations are being performed in order to determine whether or not two code phases agree. Higher-order moments of $\theta_b(k, k', W)$ are calculated in [10]. These higher-order moments are especially useful when a lowpass filtered m-sequence is being used as a noise source. ∎

8-3.5 Power Spectrum of $c(t)c(t + \epsilon)$

In Chapter 7 the despreading operation in all the receivers was accomplished by correlating the received signal with a replica of the spreading waveform $c(t)$. When the receiver-generated code replica is at exactly the correct phase, despreading occurs and the data modulation can then be extracted using a conventional data demodulator. In Chapter 9 it will be shown that one method of maintaining or tracking the correct receiver code phase will involve correlating the received signal with a replica of the code waveform which is offset in phase by some fraction of a code period. The power spectrum of the output $b(t, \epsilon) = c(t)c(t + \epsilon)$ of the despreading correlator is calculated in Appendix F. The result is

$$S_b(f, \epsilon) = \left[1 - \left(1 + \frac{1}{N} \right) \frac{|\epsilon|}{T_c} \right]^2 \delta(f)$$

$$+ \left(1 + \frac{1}{N} \right) \left(\frac{|\epsilon|}{T_c} \right)^2 \sum_{\substack{n=-\infty \\ n \neq 0}}^{\infty} \text{sinc}^2 (nf_c |\epsilon|) \delta(f - nf_c)$$

$$+ \frac{N + 1}{N^2} \left(\frac{|\epsilon|}{T_c} \right)^2 \sum_{\substack{m=-\infty \\ m \neq 0}}^{\infty} \text{sinc}^2 \left(\frac{mf_c}{N} |\epsilon| \right) \delta \left(f - \frac{mf_c}{N} \right) \quad (8\text{-}59)$$

where $N = m$-sequence period
$T_c = 1/f_c =$ code clock period

The power spectrum $S_b(f, \epsilon)$ of $b(t, \epsilon)$ is illustrated in Figure 8-14 for $\epsilon = 0$, $0.1T_c$, $0.5T_c$, and T_c. For $\epsilon = 0$ observe that all the spectral lines collapse into a single spectral line at zero frequency. This corresponds to complete despreading of the spread-spectrum signal. For $\epsilon = T_c$ the function $b(t, \epsilon)$ is simply a phase-shifted replica of $c(t)$ by the shift-and-add property, so $S_b(f, \epsilon) = S_c(f)$. Finally, observe that the power spectrum for any $\epsilon \neq 0$ or T_c is significantly wider than the spectrum of the spreading waveform $c(t)$.

8-3.6 Generation of Specific Delays of m-Sequence

It is sometimes useful to be able to generate two different phases of an m-sequence for phase differences that would be impractical to generate using a shift register or a delay line. For example, suppose that a spreading code chip rate of 100 MHz were being used in a system that transmits over a range of 150 km. In the spread-spectrum receiver, a reference code having a relative delay of approximately 0.5 ms or 5×10^4 chips must be generated. A trivial means of achieving this delay would be to start both the receiver and transmitter code generators from a common initial condition and to then inhibit the receiver clock for the anticipated delay. This trivial method may, however, not be sufficiently accurate for some applications. Another method of achieving this delay is to use a shift register that is 5×10^4 bits long. This method is obviously impractical. The distortion and loss in a delay line that would generate the required delay may make delay lines un-usable also.

Two techniques are discussed in this section for generating specific delays of an m-sequence. One of these methods is simply to calculate the correct shift register initial conditions required to generate a sequence delayed by k chips from the

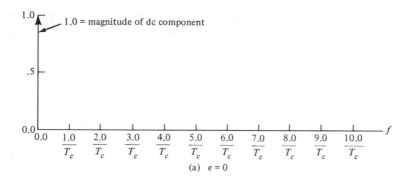

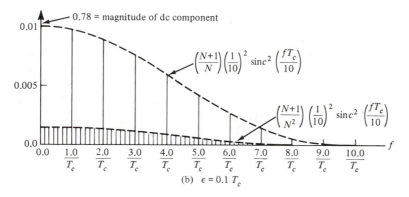

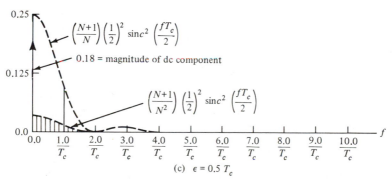

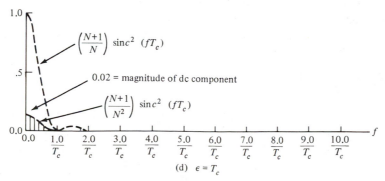

FIGURE 8-14. Power spectrum of $c(t)\, c(t + \epsilon)$ for various ϵ and $N = 7$: (a) $\epsilon = 0$; (b) $\epsilon = 0.1T_c$; (c) $\epsilon = 0.5T_c$; (d) $\epsilon = T_c$.

sequence generated from another specific initial condition. The other method makes use of the shift-and-add property of m-sequence generators. Both methods require knowledge of the mechanics of finitie-field arithmetic already discussed.

Determining the Initial Condition Yielding a Specific Delay. Consider the sequence generator of Figure 8-5. Given an initial condition $a(D)$, the output of this generator is $b(D) = a(D)/g(D)$. Another initial condition $a'(D)$ will produce another output sequence $b'(D)$, which is equal to $D^k b(D)$; that is, $b'(D)$ is the sequence $b(D)$ delayed by k chips. The task being addressed here is determining $a'(D)$ for a specific k.

This problem is most easily solved if the states $a(D)$ of the m-sequence generator are associated with elements of the extension field $GF(2^m)$ defined by the primitive polynomial $g(D)$, where m is the number of elements in the shift register and $m - 1$ is the degree of $a(D)$. The degree of $g(D)$ is m. The state $a(D)$ will sequence through all elements of $GF(2^m)$. Recall from an earlier discussion that each element in the sequence of elements of $GF(2^m)$ in polynomial form is generated by multiplying the preceding element by D and taking the remainder when dividing this product by $g(D)$. Let $q(D)$ represent an element of $GF(2^m)$. Then the element l units later, denoted by $q'(D)$, satisfies

$$D^l q(D) = p(D)g(D) + q'(D) \qquad (8\text{-}60)$$

The contents of the shift register at time zero are specified by $a(D)$. From Figure 8-5 it is seen that the contents of the shift register at time 1, denoted by $a'(D)$, satisfies

$$a'(D) = D^{-1}a(D) + D^{-1}a_0 g(D) \qquad (8\text{-}61)$$

or equivalently,

$$D^{-1}a(D) = D^{-1}a_0 g(D) + a'(D) \qquad (8\text{-}62)$$

Since the contents of the shift register represent elements of $GF(2^m)$, it is appropriate to consider D^{-1} as the multiplicative inverse element of D in $GF(2^m)$ so that $D^{-1} = D^{2^m - 2}$ and (8-62) becomes

$$D^{2^m - 2}a(D) = a_0 D^{2^m - 2}g(D) + a'(D) \qquad (8\text{-}63)$$

This equation has the same form as (8-60) with $l = 2^m - 2$, so that it is concluded that the contents of the shift register advance $2^m - 2$ steps through the sequence of elements of $GF(2^m)$ on each cycle. Since there are only $2^m - 1$ nonzero elements of $GF(2^m)$, the register may also be considered to cycle through the elements of $GF(2^m)$ in reverse order.

EXAMPLE 8-16
The elements of $GF(2^4)$ are given in Example 8-6 for the primitive polynomial $g(D) = 1 + D + D^4$. The shift register generator corresponding to $g(D)$ is illustrated in Figure 8-15 together with the contents of the register $a(D)$ beginning with state $a(D) = 1$. The element of $GF(2^m)$ corresponding to $a(D)$ is found by comparing $a(D)$ with the polynomials of Example 8-6. ■

At this point the problem of finding the shift register initial conditions corresponding to a particular advance or delay reduces to a problem of manipulating elements of $GF(2^m)$ using the techniques developed earlier. With $a(D)$ defining one initial condition and $a'(D)$ defining the initial condition corresponding to an *ad-*

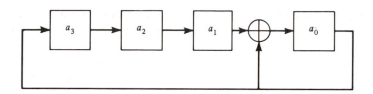

Cycle	Register state	$a(D)$	Element of $GF(2^4)$
0	0 0 0 1	1	D^0
1	1 0 0 1	$1 \qquad\quad + D^3$	D^{14}
2	1 1 0 1	$1 \quad + D^2 + D^3$	D^{13}
3	1 1 1 1	$1 + D + D^2 + D^3$	D^{12}
4	1 1 1 0	$D + D^2 + D^3$	D^{11}
5	0 1 1 1	$1 + D + D^2$	D^{10}
6	1 0 1 0	$D \qquad\quad + D^3$	D^9
7	0 1 0 1	$1 \quad + D^2$	D^8
8	1 0 1 1	$1 + D \qquad + D^3$	D^7
9	1 1 0 0	$D^2 + D^3$	D^6
10	0 1 1 0	$D + D^2$	D^5
11	0 0 1 1	$1 + D$	D^4
12	1 0 0 0	D^3	D^3
13	0 1 0 0	D^2	D^2
14	0 0 1 0	D	D^1

FIGURE 8-15. Comparison of shift register states with elements of **GF(2⁴)** for a typical *m*-sequence.

vance of k units, the discussion above implies that $a'(D)$ is the remainder found when dividing $D^{k(2^m - 2)}a(D)$ by $g(D)$. The product $D^{k(2^m - 2)}$ can be reduced using the fact that $D^{2^m - 1} = 1$. The load corresponding to a *delay* of k units is the remainder found when dividing $D^k a(D)$ by $g(D)$.

EXAMPLE 8-17

Consider the generator of Example 8-16 with an initial condition of $a(D) = 1$. What initial condition $a'(D)$ will produce an advance of 20 units? What $a'(D)$ will produce a delay of 20 units?

Solution: The period of the *m*-sequence is 15 units, so that an advance of 20 units is equivalent to an advance of 5 units. $D^{k(2^m - 2)} = D^{5 \cdot 14} = D^{70} = D^{10}$. Thus $a(D)D^{k(2^m - 2)} = D^{10} \cdot 1 = D^{10}$. From Example 8-6, $D^{10} = 1 + D + D^2$, and since the degree of $1 + D + D^2$ is less than the degree of $g(D) = 1 + D + D^4$, the remainder desired is just $1 + D + D^2$. Thus $a'(D) = 1 + D + D^2$. The same remainder is found if D^{10} itself is divided by $g(D)$. The result can be verified by comparison with the cycle 5 shift register state of Figure 8-15.

A delay of 20 units is equivalent to a delay of 5 units, so that $a'(D)$ is the remainder when dividing $D^5 a(D)$ by $g(D)$. This remainder is $D^2 + D = a'(D)$. This result can also be verified by comparison with the table within Figure 8-15.

∎

The technique just described works only for the shift register configuration of Figure 8-5. The shift register configuration of Figure 8-6 unfortunately does not cycle through the elements of $GF(2^m)$. The most convenient technique for finding the desired initial condition $a'(D)$ is to find the corresponding initial conditions for the configuration of Figure 8-5. Recall that this same technique was employed when the output sequence of this shift register configuration was calculated earlier. The procedure for generating the initial condition for one configuration from the other is described in (8-38) through (8-41).

EXAMPLE 8-18

Consider the shift register $g(D) = 1 + D + D^4$ and initial condition $a_1(D) = 1$. The generator is illustrated in Figure 8-16 together with all register states. Denote the corresponding initial condition for the configuration of Figure 8-5 by $e(D)$. Equation (8-41) becomes

$$e(D) = g(D)(a(D) + b_r D^r + \cdots)$$

and equating coefficients of D yields

$$e_0 = a_0 \qquad = 1$$

$$e_1 = a_1 + a_0 = 1$$

$$e_2 = a_2 + a_1 = 0$$

$$e_3 = a_3 + a_2 = 0$$

Thus $e(D) = 1 + D$. Suppose that $k = 20$ as in Example 8-17. Then $D^{k(2^m - 2)} = D^{10}$ and $D^{10}e(D) = D^{10} + D^{11} = 1 + D + D^2 + D + D^2 + D^3 = 1 + D^3 = e'(D)$. Now (8-45) can be used to return to the original shift register

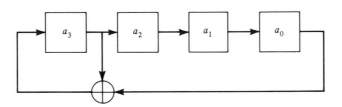

Cycle	Register state	$a(D)$
0	0 0 0 1	1
1	1 0 0 0	D^3
2	1 1 0 0	$D^2 + D^3$
3	1 1 1 0	$D + D^2 + D^3$
4	1 1 1 1	$1 + D + D^2 + D^3$
5	0 1 1 1	$1 + D + D^2$
6	1 0 1 1	$1 + D \qquad + D^3$
7	0 1 0 1	$1 \qquad + D^2$
8	1 0 1 0	$D \qquad + D^3$
9	1 1 0 1	$1 \qquad + D^2 + D^3$
10	0 1 1 0	$D + D^2$
11	0 0 1 1	$1 + D$
12	1 0 0 1	$1 \qquad + D^3$
13	0 1 0 0	D^2
14	0 0 1 0	D

FIGURE 8-16. Shift register states for Example 8-18.

configuration. Thus

$$e'(D) = g(D)(a'(D) + b_r D^r + \cdots)$$

or

$$1 + D^3 = (1 + D + D^4)(a_0 + a_1 D + a_2 D^2 + a_3 D^3 + \cdots)$$

which implies

$$a_0 = 1$$

$$a_0 + a_1 = 0$$

$$a_2 + a_1 = 0$$

$$a_3 + a_2 = 1$$

or $a(D) = 1 + D + D^2$. This result can be verified by comparison with cycle 5 of Figure 8-16. ∎

Determining the Phases of an *M*-sequence that Add to Produce a Particular Third Phase. Consider the sequence generator configuration with output shift register as illustrated in Figure 8-17, where there are *m* stages in the generator. The output $b(D)$ can be written $b(D) = a(D)/g(D)$. The present task is to determine the correct set of delayed outputs to add which will yield $b'(D)$ such that $b'(D) = D^k b(D)$. If the output shift register were k units in length, the problem would be trivial; however, the shift register is only $r - 1$ units long. The output $b'(D)$ is defined by

$$b'(D) = s_0 b(D) + s_1 D b(D) + s_2 D^2 b(D) + \cdots + s_{r-1} D^{r-1} b(D)$$

$$= s(D)b(D) \qquad (8\text{-}64)$$

The polynomial $s(D)$ is a connection polynomial with binary coefficients representing whether or not a particular delay of $b(D)$ is included in the modulo-2 sum used to obtain $b'(D)$. The polynomial $s(D)$ must be found such that $b'(D) = D^k b(D)$. Making use of the results of the preceding section, both $b(D)$ and $b'(D)$ can be written as a function of the generator polynomial and a known initial condition. That is,

$$b'(D) = \frac{a'(D)}{g(D)} = s(D)b(D) = s(D)\frac{a(D)}{g(D)} \qquad (8\text{-}65)$$

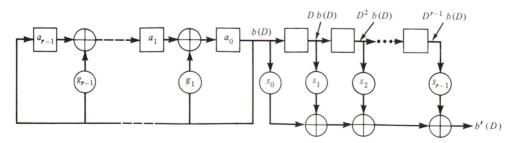

FIGURE 8-17. Generation of alternate phases of an *m*-sequence from short delays of *b(D)*.

so that the initial conditions are related by

$$a'(D) = s(D)a(D) \qquad (8\text{-}66)$$

Since the problem being addressed is concerned with the relative phase of two sequences and not the absolute phase of either, $a(D)$ can be arbitrarily chosen to be $a(D) = 1$. Then

$$s(D) = a'(D) \qquad (8\text{-}67)$$

and the problem has been solved provided that $a'(D)$ can be determined. The initial conditions $a'(D)$ are found using the procedure described in the preceding section. The final output of this calculation is a connection polynomial $s(D)$ which can be used with either shift register configuration (Figure 8-5 or 8-6).

EXAMPLE 8-19

Determine the proper phases of the output of the shift register generator of Figure 8-16 which may be added to produce another sequence which is delayed from the original by 12 symbols.

Solution: Using the results of the preceding section, $a'(D)$ is the remainder after dividing $D^{12}a(D)$ by $g(D)$. Thus, with $g(D) = 1 + D + D^4$ and $a(D) = 1$,

$$a'(D) = 1 + D + D^2 + D^3 = s(D)$$

and all four delays of $b(D)$ must be added. Observe that in Figure 8-17 it was presumed that the output $b(D)$ is taken from the rightmost shift register of the generator. For the alternative shift register configuration, the output is conveniently taken to be the input to the leftmost shift register stage, as shown in Figure 8-6. With this convention, the delays of the sequence $b(D)$ required for the sum to produce $b'(D)$ are available within the generator itself and no external delays are needed. Figure 8-18 shows the final configuration and associated outputs for a delay of 12 units. A single cycle of each sequence has been illustrated. ∎

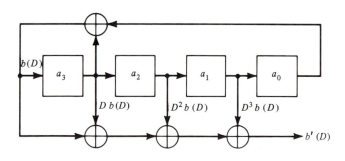

time	$=$	0 1 2 3 4 5 6 7 8 9 10 11 12 13 14
$b\,(D)$	$=$	0 1 1 1 1 0 1 0 1 1 0 0 1 0 0 $\cdots$
$D\,b(D)$	$=$	0 0 1 1 1 1 0 1 0 1 1 0 0 1 0 $\cdots$
$D^2\,b(D)$	$=$	0 0 0 1 1 1 1 0 1 0 1 1 0 0 1 $\cdots$
$D^3\,b(D)$	$=$	1 0 0 0 1 1 1 1 0 1 0 1 1 0 0 $\cdots$
$b'\,(D)$	$=$	1 1 0 1 0 1 1 0 0 1 0 0 0 1 1 $\cdots$

FIGURE 8-18.   Shift register configuration yielding a 12-symbol delay.

Security of Maximal-Length Sequences. Spread-spectrum systems are often used to protect digital transmissions from being jammed or to preclude unintended reception of the signal. Both of these objectives can only be met if the jammer or unintended receiver does not have knowledge of the spreading waveform $c(t)$. Unfortunately, when the jammer or interceptor can receive a relatively noise-free copy of the transmitted signal, the spreading code feedback connections and initial phase can be determined in a straightforward manner. For this reason, maximal-length sequences are a poor choice for the spreading code when a high level of security is required.

Suppose that the unintended party has access to an uncorrupted version of the transmitted spreading code. Thus the unintended party knows the sequence b_0, b_1, b_2, b_3, . . ., and would like to determine the shift register feedback connections used to generate this sequence. The party knows that an m-sequence is being transmitted and can easily determine the period of the sequence by measuring the received power spectrum accurately. Each symbol of the m-sequence must satisfy the recursion relationship of (8-33), so that the unintended party can write the following series of equations:

$$b_i = b_{i-1}g_1 + b_{i-2}g_2 + \cdots + b_{i-m}g_m$$
$$b_{i+1} = b_i g_1 + b_{i-1}g_2 + \cdots + b_{i-m+1}g_m$$
$$b_{i+2} = b_{i+1}g_1 + b_i g_2 + \cdots + b_{i-m+2}g_m$$
$$\cdot$$
$$\cdot$$
$$\cdot$$

(8-68)

After m such equations have been written, the unintended party will have m equations in the m unknowns g_1 through g_m which can be solved. Massey [11] has provided an efficient technique for solving this system of equations. The purpose of the present discussion is merely to make the student aware that algorithms exist for determining the shift register generator feedback connections so that the details of Massey's algorithm are left as a reference. The system of (8-68) can also be solved by brute force, as demonstrated in the following example.

EXAMPLE 8-20

Suppose that the sequence 0 1 1 0 0 1 0 0 is received and that the known period of the m-sequence is 15. Thus $m = 4$ and the set of equations to solve is

$$(1) \quad 0 = 0 \cdot g_1 + 1 \cdot g_2 + 1 \cdot g_3 + 0 \cdot g_4$$

$$(2) \quad 1 = 0 \cdot g_1 + 0 \cdot g_2 + 1 \cdot g_3 + 1 \cdot g_4$$

$$(3) \quad 0 = 1 \cdot g_1 + 0 \cdot g_2 + 0 \cdot g_3 + 1 \cdot g_4$$

$$(4) \quad 0 = 0 \cdot g_1 + 1 \cdot g_2 + 0 \cdot g_3 + 0 \cdot g_4$$

Adding (1) and (4) yields $0 = g_3$. Substituting $g_3 = 0$ into (1) yields $g_2 = 0$. Substituting $g_2 = g_3 = 0$ into (2) yields $g_4 = 1$; then substituting $g_2 = g_3 = 0$ and $g_4 = 1$ into (3) yields $g_1 = 1$. Therefore,

$$g(D) = 1 + D + D^4$$

This is the generator used in Example 8-19 and the received sequence is a subsequence of $b(D)$ in Figure 8-18. ∎

Finally, note that the number of symbols which must be received is $2m$, where m is the degree of $g(D)$ and $2m$ is much shorter than the period $N = 2^m - 1$ of the m-sequence. The assumption that the $2m$ symbols be received without error can only be made in special circumstances so that the security of m-sequences may in fact be slightly better than is implied in this discussion.

8-4
GOLD CODES

One of the applications of spread-spectrum systems is to provide a means other than frequency-division multiple access or time-division multiple access of sharing the scarce channel resources. When channel resources are shared using spread-spectrum techniques, all users are permitted to transmit simultaneously using the same band of frequencies. Users are each assigned a different spreading code so that they can be separated in the receiver despreading process. A goal of the spread-spectrum system designer for a multiple-access system is to find a set of spreading codes or waveforms such that as many users as possible can use a band of frequencies with as little mutual interference as possible.

Recall that the receiver despreading operation is a correlation operation with the spreading code of the desired transmitter. Ideally, a received signal that has been spread using a different spreading code will not be despread and will cause minimal interference in the desired signal. The specific amount of interference from a user employing a different spreading code is related to the cross-correlation between the two spreading codes. The Gold codes introduced in this section were invented in 1967 at the Magnavox Corporation specifically for multiple-access applications of spread spectrum. Relatively large sets of Gold codes exist which have well controlled cross-correlation properties. The treatment here is intended only to familiarize the student with the fact that this code family exists and with some of its most fundamental properties. A considerably more thorough discussion of these codes can be found in Sarwate and Pursley [2], from which the following information has been extracted. The reference also provides a large bibliography for further study.

The cross-correlation between two spreading codes was defined in (8-6), which is repeated here:

$$\theta_{bb'}(k) = \frac{1}{N} \sum_{n=0}^{N-1} a_n a'_{n+k} \tag{8-6}$$

Although the detailed correlation could be evaluated for multiple-access code sets, in many cases adequate information for system analysis can be obtained from the cross-correlation spectrum. The *cross-correlation spectrum* is a list of all possible values of $\theta_{bb'}(k)$ and the number of values of k which yield that particular cross-correlation. The cross-correlation spectrum is denoted by $\theta_{bb'}$. When $b = b'$ the cross-correlation spectrum becomes the *autocorrelation spectrum*. The autocorrelation spectrum for an m-sequence is

$$1.0 \quad \text{occurs} \quad 1 \text{ time}$$

$$-\frac{1}{N} \quad \text{occurs} \quad N - 1 \text{ times}$$

The Gold code sets to be defined shortly have a cross-correlation spectrum which is three-valued.

Consider an m-sequence that is represented by a binary vector **b** of length N, and a second sequence **b**′ obtained by sampling every qth symbol of **b**. The second sequence is said to be a *decimation* of the first, and the notation **b**′ = **b**[q] is used to indicate that **b**′ is obtained by sampling every qth symbol of **b**. The decimation of an m-sequence may or may not yield another m-sequence. When the decimation does yield an m-sequence, the decimation is said to be a *proper decimation*. The table of irreducible polynomials in Peterson and Weldon [5] can be used to determine whether a particular decimation of a particular m-sequence is proper. One entry from this table is:

$$\text{DEGREE 6} \quad 1 \ \ 103\text{F} \quad 3 \ \ 127\text{B} \quad 5 \ \ 147\text{H} \quad 7 \ \ 111\text{A}$$

$$9 \ \ 015 \quad 11 \ \ 155\text{E} \quad 21 \ \ 007$$

Recall that each octal number represents a polynomial and those numbers followed by an E, F, G, or H are primitive polynomials which generate m-sequences. Let **b** denote the m-sequence generated by 103 in the table. The decimal number q preceding the octal entry indicates that the sequence generated by that polynomial is the qth decimation of the sequence generated by the first entry in the table. Thus **b**′ = **b**[3] is generated by the polynomial 127, which is not primitive, so that **b**′ is not an m-sequence and this decimation is not proper. It has also been proven (Sarwate and Pursley [2]) that **b**′ = **b**[q] has period N if and only if gcd(N, q) = 1, where "gcd" denotes the greatest common divisor. Since $N = 2^6 - 1 = 63$ for the degree 6 polynomials, gcd(63, 3) = 3 and the period of **b**′ does not equal N. Sarwate and Pursley [2] have also shown that proper decimation by odd integers q will give *all* of the m-sequences of period N. Thus any pair of m-sequences having the same period N can be related by **b**′ = **b**[q] for some q.

The cross-correlation spectrum of pairs of m-sequences can be three-valued, four-valued, or possibly many-valued. Certain special pairs of m-sequences whose cross-correlation spectrum is three-valued, where those three values are

$$-\frac{1}{N} t(n)$$

$$-\frac{1}{N}$$

$$\frac{1}{N} [t(n) - 2] \tag{8-69}$$

where

$$t(n) = \begin{cases} 1 + 2^{0.5(n+1)} & \text{for } n \text{ odd} \\ 1 + 2^{0.5(n+2)} & \text{for } n \text{ even} \end{cases}$$

where the code period $N = 2^n - 1$, are called *preferred pairs* of m-sequences. Finding preferred pairs of m-sequences is necessary in defining sets of Gold codes. The following conditions are sufficient to define a preferred pair **b** and **b**′ of m-sequences:

1. $n \neq 0 \bmod 4$; that is, n is odd or $n = 2 \bmod 4$
2. **b**′ = **b**[q] where q is odd and either

$$q = 2^k + 1$$

or

$$q = 2^{2k} - 2^k + 1$$

3. $\gcd(n, k) = \begin{cases} 1 & \text{for } n \text{ odd} \\ 2 & \text{for } n = 2 \mod 4 \end{cases}$

EXAMPLE 8-21

Find a preferred pair of m-sequences having a period of 31 units and evaluate their cross-correlation spectrum.

Solution: Since $N = 31$, $n = 5$. The referenced (Peterson and Weldon [5]) table of irreducible polynomials contains the following entry:

DEGREE 5   1 45 E   3 75G   5 67H.

Arbitrarily choose **b** as the m-sequence generated by the primitive polynomial 45. The decimation $\mathbf{b}' = \mathbf{b}[3]$ is proper, so that the pair $(\mathbf{b}, \mathbf{b}[3])$ is a candidate pair. The first condition is satisfied since $n = 1 \mod 4$. The second condition is satisfied also since q is odd and $q = 2^k + 1$ for $k = 1$. Finally, $\gcd(5, 1) = 1$, so that all three conditions are satisfied and a preferred pair has been found. The m-sequences **b** and **b**[3] are

$$\mathbf{b} = \frac{0\ 1\ 2\ 3\ 4\ 5\ 6\ 7\ 8\ 9\ 10\ 11\ 12\ 13\ 14\ 15\ 16\ 17\ 18\ 19\ 20\ 21\ 22\ 23\ 24\ 25\ 26\ 27\ 28\ 29\ 30}{1\ 0\ 1\ 0\ 1\ 1\ 1\ 0\ 1\ 1\ \ 0\ \ 0\ \ 0\ \ 1\ \ 1\ \ 1\ \ 1\ \ 0\ \ 0\ \ 1\ \ 1\ \ 0\ \ 1\ \ 0\ \ 0\ \ 1\ \ 0\ \ 0\ \ 0\ \ 0\ \ 0}$$

$\mathbf{b}' = 1\ 0\ 1\ 1\ 0\ 1\ 0\ 1\ 0\ 0\ \ 0\ \ 1\ \ 1\ \ 1\ \ 0\ \ 1\ \ 1\ \ 1\ \ 1\ \ 1\ \ 0\ \ 0\ \ 1\ \ 0\ \ 0\ \ 1\ \ 1\ \ 0\ \ 0\ \ 0\ \ 0$

A straightforward but tedious manual calculation of the cross-correlation function will show that for any phase shift the cross-correlation takes on one of the three values $-9/31$, $-1/31$, or $+7/31$. ∎

Let $b(D)$ and $b'(D)$ represent a preferred pair of m-sequences having period $N = 2^n - 1$. The family of codes defined by $\{b(D), b'(D), b(D) + b'(D), b(D) + Db'(D), b(D) + D^2b(D), \ldots, b(D) + D^{N-1}b'(D)\}$ is called the set of *Gold codes* for this preferred pair of m-sequences. In this definition, the notation $D^j b'(D)$ represents a phase shift of the m-sequence $b'(D)$ by j units. Gold codes sets have the property that any pair of codes in the set, say **y** and **z**, have a three-valued cross-correlation spectrum which takes on the values defined in (8-69). A typical shift register configuration used to generate a family of Gold codes is illustrated in Figure 8-19. The particular preferred pair of m-sequences used in this figure are those found in Example 8-21. The complete family of Gold codes for this generator is obtained using different initial loads of the shift register. The code $b(D)$ is obtained by choosing some nonzero $a(D)$ for the upper generator and setting $a'(D) = 0$ for the lower generator. Similarly, the code $b'(D)$ is obtained with $a(D) = 0$ and $a'(D)$ an arbitrary nonzero initial condition. Thirty-one other codes of the family are obtained using the same $a(D)$ used for $b(D)$ with all possible nonzero $a'(D)$. There are a total of $N + 2$ codes in any family of Gold codes.

In summary, Gold codes are families of codes with well-behaved cross-correlation properties which are constructed by a modulo-2 addition of specific relative phases of a preferred pair of m-sequences. The period of any code in the family is N, which is the same as the period of the m-sequences. These codes are important since they have been selected by NASA for use on the Tracking and Data Relay Satellite System (TDRSS). The particular set of Gold codes employed in TDRSS are defined in STDN 108 by the Goddard Space Flight Center. This brief description

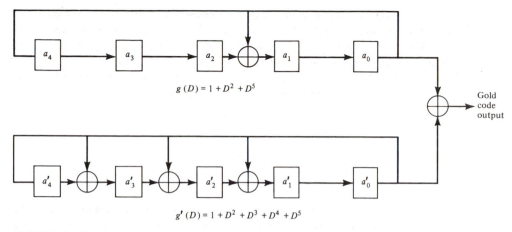

$$g(D) = 1 + D^2 + D^5$$

$$g'(D) = 1 + D^2 + D^3 + D^4 + D^5$$

FIGURE 8-19. Typical Gold code generator.

of Gold codes has not been rigorous and the student is referred to Sarwate and Pursley [2], Holmes [9], or Spilker [12] for further information on this interesting subject.

RAPID ACQUISITION SEQUENCES

Initial determination of the correct receiver spreading code phase is one of the difficult problems in spread-spectrum communications. Determining this code phase is often referred to as *code acquisition* or simply as *acquisition* and Chapter 10 is devoted to this problem. The most common acquisition technique is to serially correlate the received signal with a sequence of despreading waveform phases until the correct waveform phase is detected. The correct phase is detected using an energy detector at the output of a filter following the despreading operation. This technique will be discussed in detail later. The correlation time required at each test phase is related to the received signal-to-noise ratio and is longer for lower SNRs. In the absence of any a priori information about the received phase, analysis of a very simplified model of the acquisition process indicates that, on the average, $\frac{1}{2}N$ correlations are required to achieve synchronization. This result is based on the assumption that the received code phase is uniformly distributed over the full code period N. Each of these $\frac{1}{2}N$ test correlations lasts T_1 seconds, where T_1 is a function of the required probability of a false detection when the phase is incorrect, the probability of a missed detection when the phase is correct, and the received signal-to-noise ratio. Thus the average acquisition time is $\overline{T}_{acq} = \frac{1}{2}NT_1$, and can be extremely long for large N.

In the early 1960s, the Jet Propulsion Laboratory began to address the acquisition-time problem, which was becoming an issue on planetary missions using extremely long ranging codes. A particular code was found which enabled the average number of correlations to achieve acquisition to be reduced to $\log_2 N$ rather than $\frac{1}{2}N$. Although the correlation time for each step of the acquisition process is longer than that required in a conventional system, a significant savings in acquisition time resulted from the use of these codes. The following discussion is based on a very readable paper [13], which is recommended for further reading.

Suppose that the received spreading code phase uncertainty is equivalent to no more than N chips. That is, the time of arrival of the code epoch is known only to within $\pm\frac{1}{2}NT_c$, where T_c is, again, the code chip interval. Thus, to avoid ambiguity, an acquisition code which is a minimum of N bits long is required. If N is not a power of 2, choose a code length which is the smallest power of 2 greater than N. Let $n = \log_2 N$ and construct the N binary n-tuples

$$b_j = (\sigma_j^1, \sigma_j^2, \sigma_j^3, \ldots, \sigma_j^n) \qquad j = 1, 2, \ldots, N \qquad (8\text{-}70)$$

where $\sigma_j^k = \pm 1$ such that

$$\sum_{i=1}^{n} \left(\frac{1 - \sigma_j^i}{2}\right) 2^{i-1} = j - 1 \qquad (8\text{-}71)$$

These n-tuples b_j are simply the binary representations of the numbers 0 through $N - 1$ with the MSB placed on the right. For example, if $N = 16$, $n = 4$ and the n-tuples b_j are

$$b_1 = (1, 1, 1, 1) = (\sigma_1^1, \sigma_1^2, \sigma_1^3, \sigma_1^4)$$

$$b_2 = (-1, 1, 1, 1) = (\sigma_2^1, \sigma_2^2, \sigma_2^3, \sigma_2^4)$$

.
.
.

$$b_{16} = (-1, -1, -1, -1) = (\sigma_{16}^1, \sigma_{16}^2, \sigma_{16}^3, \sigma_{16}^4)$$

Given these vectors, the *rapid acquisition binary sequence* is

$$\chi = (\xi_1, \xi_2, \xi_3, \ldots, \xi_N) \qquad (8\text{-}72)$$

where

$$\xi_j = \begin{cases} 1 & \text{if } \sum_{i=1}^{n} \sigma_j^i \geq 0 \\ \\ -1 & \text{if } \sum_{i=1}^{n} \sigma_j^i < 0 \end{cases}$$

The rapid acquisition sequence for $N = 16$ is

$$\chi = (1, 1, 1, 1, 1, 1, 1, -1, 1, 1, 1, -1, 1, -1, -1, -1)$$

This sequence is transmitted periodically. The task given the receiver is to determine the correct starting phase of the rapid acquisition N-bit sequence. The receiver is able to accomplish this task with n correlations.

To reduce the number of correlations from $\frac{1}{2}N$ to $\log_2 N = n$, the acquisition sequence has been cleverly selected so that it has a relatively high cross-correlation with square-wave sequences which are in phase with χ. Square-wave sequences are sequences such as $(1, -1, 1, -1, \ldots, 1, -1)$, $(1, 1, -1, -1, 1, 1, \ldots, 1, 1, -1, -1)$, and so on. Specifically, the sequences which the receiver correlates with the received sequence are

$$s_i = (\sigma_1^i, \sigma_2^i, \ldots, \sigma_N^i) \qquad i = 1, 2, 3, \ldots, n \qquad (8\text{-}73)$$

These sequences are simply the transpose of the columns of a $(N \times n)$ matrix whose rows are b_j. In addition to high cross-correlation between χ and s_i, the cross-correlation for all s_i, $i = 1, 2, \ldots, n$, are equal. For $N = 16$, the cross-correlation between χ and s_i is

$$\frac{1}{N} \sum_{j=1}^{N} \xi_j \sigma_j^i = 0.375 \tag{8-74}$$

which can be simply verified by counting the number of agreements and disagreements between χ and any s_i. Notice also that the cross-correlation between the inverse $\bar{s}_i$ (change all $+1$'s to -1's, and vice versa) of any s_i and χ is equal to the negative of the in-phase correlation. Denote these normalized cross-correlations by $\pm\rho$.

The first correlation performed in the receiver is between the received sequence and s_1. The s_1 sequence has period two and thus only two possible phases, $(1, -1, 1, -1, \ldots, -1)$ and $(-1, 1, -1, 1, \ldots, 1)$. If the received sequence is in phase with s_1, the normalized cross-correlation (in the absence of noise) will equal $+\rho$. If the received sequence is not in phase with s_1, the cross-correlation will be $-\rho$. Thus the receiver can make a binary decision as to whether the starting phase of the received acquisition sequence occurs on a $+1$ or a -1 of the s_1 sequence. This decision eliminates 50% of the possible starting phases. The ability to eliminate 50% of all possible starting phases with a single decision is the reason for the reduction in acquisition time for this scheme.

Suppose that the first receiver decision is correct. The next cross-correlation performed by the receiver is between the received sequence and s_2, where the starting phase of s_2 occurs on one of the possible acquisition sequence starting phases determined in the first cross-correlation. The result of this second cross-correlation is positive if s_2 is in phase with the received acquisition sequence and negative if the acquisition sequence is in phase with $\bar{s}_2$. Again, the receiver makes a binary decision which eliminates 50% of the possible acquisition sequence starting phases. This process continues until the received sequence has been correlated with all n s_i's. The last binary decision determines the exact acquisition sequence starting phase from the two possible phases determined by the next-to-last step.

EXAMPLE 8-22

Consider the received sequence

$$(-\overset{1}{1}, \overset{2}{1}, \overset{3}{1}, \overset{4}{1}, -\overset{5}{1}, \overset{6}{1}, -\overset{7}{1}, -\overset{8}{1}, -\overset{9}{1}, \overset{10}{1}, \overset{11}{1}, \overset{12}{1}, \overset{13}{1}, \overset{14}{1}, \overset{15}{1}, \overset{16}{1})$$

which is a cyclic shift of the 16-bit rapid acquisition sequence defined above. In a conventional acquisition system, the receiver would correlate with one possible phase shift of acquisition sequence after another until the correct phase is found. The fast acquisition receiver would first correlate with the sequence

$$s_1 = (\overset{1}{1}, -\overset{2}{1}, \overset{3}{1}, -\overset{4}{1}, \overset{5}{1}, -\overset{6}{1}, \overset{7}{1}, -\overset{8}{1}, \overset{9}{1}, -\overset{10}{1}, \overset{11}{1}, -\overset{12}{1}, \overset{13}{1}, -\overset{14}{1}, \overset{15}{1}, -\overset{16}{1})$$

The result is $\frac{1}{16}(5 - 11) = -0.375$, indicating that correct acquisition sequence phase is one of the even-numbered positions above. Thus the odd-numbered positions have been eliminated. Next, the receiver correlates with the sequence

$$s_2 = (-\overset{1}{1}, \overset{2}{1}, \overset{3}{1}, -\overset{4}{1}, -\overset{5}{1}, \overset{6}{1}, \overset{7}{1}, -\overset{8}{1}, -\overset{9}{1}, \overset{10}{1}, \overset{11}{1}, -\overset{12}{1}, -\overset{13}{1}, \overset{14}{1}, \overset{15}{1}, -\overset{16}{1})$$

which has been phased so that it begins on an even numbered position. The result is $\frac{1}{16}(11 - 5) = +0.375$, indicating that the correct acquisition sequence phase is position 2, 6, 10, or 14. That is, 50% of the phases remaining after step 1 have been eliminated. The phases eliminated are the even-numbered positions which do not fall on one of the four starting phases of the $1, 1, -1, -1, \ldots$ sequence.

Next, the receiver correlates with

$$s_3 = (-\overset{1}{1}, \overset{2}{1}, \overset{3}{1}, \overset{4}{1}, \overset{5}{1}, -\overset{6}{1}, -\overset{7}{1}, -\overset{8}{1}, -\overset{9}{1}, \overset{10}{1}, \overset{11}{1}, \overset{12}{1}, \overset{13}{1}, -\overset{14}{1}, -\overset{15}{1}, -\overset{16}{1})$$

which again has been shifted to begin on position 2, 6, 10, or 14. The result is $\frac{1}{16}(11 - 5) = +0.375$, so that the correct acquisition sequence phase is either 2 or 10.

Finally, the receiver correlates with

$$s_4 = (-\overset{1}{1}, \overset{2}{1}, \overset{3}{1}, \overset{4}{1}, \overset{5}{1}, \overset{6}{1}, \overset{7}{1}, \overset{8}{1}, \overset{9}{1}, -\overset{10}{1}, -\overset{11}{1}, -\overset{12}{1}, -\overset{13}{1}, -\overset{14}{1}, -\overset{15}{1}, -\overset{16}{1})$$

which is phased to start on position 2 or 10 (two has been arbitrarily chosen). The result is $\frac{1}{16}(5 - 11) = -0.375$, indicating that 2 is not the correct starting phase but that position 10 is correct. The correct starting phase has been determined in four steps rather than an average of eight in a conventional system. The savings are more dramatic in cases with larger N. ∎

To determine average acquisition time, it is important that the cross-correlation between the received sequence and s_i be constant for all $i = 1, \ldots, n$. It is shown in [13] that the cross-correlation is constant and is given by

$$\rho = \begin{cases} \dfrac{1}{2^{n-1}} \dbinom{n-1}{\dfrac{n-1}{2}} & \text{for } n \text{ odd} \\[2em] \dfrac{1}{2^n} \dbinom{n}{\dfrac{n}{2}} & \text{for } n \text{ even} \end{cases} \tag{8-75}$$

where $\dbinom{a}{b}$ is the usual binomial coefficient. For $N = 16$, $n = 4$ and

$$\rho = \frac{1}{2^4} \binom{4}{2} = 0.375$$

as was calculated in Example 8-22.

For large n, Stirling's formula can be used to evaluate the binomial coefficients and to establish that

$$\rho \simeq \left(\frac{2}{\pi}\right)^{1/2} (\log_2 N)^{-1/2} \tag{8-76}$$

Since $\rho < 1$ for these codes, it must be remembered that the integration time for each binary decision must be longer than the integration time used in a conventional acquisition system. In spite of this longer integration time, significant acquisition time improvements can be achieved with these codes. The price paid for this reduced acquisition time is reduced jamming resistance and increased detectability for the spread-spectrum system. Both of these factors are the result of the additional structure (i.e., decreased randomness) of the code.

NONLINEAR CODE GENERATORS

In spread-spectrum applications where security is important, it is necessary that an eavesdropper and/or jammer not be able to obtain complete knowledge of the spreading code being employed. If the spreading code is known, the eavesdropper can demodulate the transmitted signal just as the desired receiver does, and the jammer can transmit the spreading code as a highly effective jamming signal. Thus the difficulty of determining the specific code generator configuration from knowledge of the transmitted signal is an important issue for the spread-spectrum system designer. Recall that the feedback connections for an n-stage maximal-length code can be easily determined from knowledge of $2n$ successive code symbols. For this reason, m-sequences are never used when a high degree of security is required. The overall subject of security is beyond the scope of this text. Tutorial articles on communications privacy can be found in a special issue of the IEEE *Communications Society Magazine* (November 1978) devoted to this subject and a discussion of encryption techniques can be found in [14]. The discussion of nonlinear spreading codes that follows indicates one way in which increased security through increased complexity can be achieved. One possible tact for the system designer is to develop codes that cannot be described by the simple linear recurrence relationship of (8-33) or codes for which r in (8-33) is so large that the solution of the system of (8-68) is computationally impossible. Fortunately, other methods of increasing complexity exist (Groth [15]) which are significantly more practical than either of the methods above. These methods make use of modulo-2 multiplications in addition to modulo-2 addition in order to increase effective complexity.

It is interesting to note that *any periodic sequence* of binary digits of period $2^r - 1$ can be generated by a *linear* feedback shift register. In the worst case, this generator is simply a recirculating shift register of length $2^r - 1$. In most cases, combinations of several short shift register generators can be used. For example, consider the periodic sequence [15]

$$1\ 1\ 1\ 1\ 1\ 0\ 1\ 0\ 1\ 0\ 0\ 1\ 1\ 0\ 0\ 0\ 1\ 0\ 0\ 0\ 0\ \ldots$$

having a period of 21. This sequence can obviously be generated using a 21-stage recirculating shift register. The output sequence of this recirculating shift register can be represented in polynomial form by

$$b(D) = \frac{1 + D + D^2 + D^3 + D^4 + D^6 + D^8 + D^{11} + D^{12} + D^{16}}{1 + D^{21}} \quad (8\text{-}77)$$

This ratio has the same form as (8-37) with $g(D) = 1 + D^{21}$, so that the shift register feedback connections are simply one connection from the last stage of a 21-stage shift register to the input to the first stage. The numerator above is the initial load of the shift register $a(D)$. The denominator of (8-77) factors into two terms, one of which is identical to the numerator and can be canceled to obtain

$$b(D) = \frac{1}{1 + D + D^5}$$

Thus, this 21-bit sequence can be more simply generated by a five-stage shift register with appropriate feedback connections. The denominator of the preceding expression can be further factored into

$$b(D) = \frac{1}{(1 + D + D^2)(1 + D^2 + D^3)} \qquad (8\text{-}78)$$

indicating that the sequence can also be generated by two separate sequence generators whose outputs are modulo-2 summed. Not all sequences yield polynomials that can be factored as conveniently as those above. The important point of this discussion is that *any periodic sequence* can be generated by a *linear* feedback shift register. The number of stages in the generator, however, may be excessive and the sequence may be much more conveniently generated if the use of nonlinear elements is allowed.

The major contribution of [15] is the demonstration of a technique that enables the construction of a linear equivalent of many nonlinear generators. This enables evaluation of the "complexity" of the nonlinear circuit. Consider only sequences of length $2^r - 1$ which are generated by linear and non-linear operations on an r-stage shift register. The *complexity of a sequence* is defined as the number of stages in the linear equivalent feedback shift register. Maximal-length sequences have minimum possible complexity ($= r$). In secure applications, the use of sequences which have very high complexity is desired so that the number of equations in (8-68) is large. Thus much computational power is required to solve (8-68).

One type of circuit used to generate high-complexity sequences is illustrated in Figure 8-20. In this figure, nonlinear feedforward logic has been added to a conventional linear feedback shift register generator. For analytical simplicity, all binary multipliers have only two inputs and the $e_{i,j}$ coefficients indicate which shift register stages are connected to the multiplier. As before, $e_{i,j} = 0$ indicates no connections or that stage i and stage j are not multiplied. The outputs of all multipliers are added to form the final output sequence. A specific example of a high-complexity sequence generator is illustrated in Figure 8-21a. In Figure 8-21 all $e_{i,j} = 0$ except $e_{1,2}$, $e_{3,4}$, and $e_{5,6}$, which equal 1. Note that the period of the sequence of Figure 8-20 is $2^r - 1$; that is, the period has not been increased by using nonlinear feedforward logic.

Groth [15] gives specific procedures for determining the linear equivalent of a circuit such as that shown in Figure 8-20. This procedure is long and complicated but not conceptually difficult. As a specific example, the linear circuit of Figure 8-21b produces exactly the same output as the circuit of Figure 8-21a. Notice that Figure 8-21b contains 21 shift register stages, so that 42 equations must be solved to determine the linear feedback connections. The original shift register contains only six shift registers and, if connected in a linear manner, would require 12 equations to specify.

This brief discussion of nonlinear sequence generators for secure applications is in no way comprehensive. Other nonlinear connections are possible, but significant analytical results are not available in the open literature. The basic concept, however, is always the same: Generate sequences that are very difficult to regenerate. Not considered above are schemes where the nonlinearities are placed in the feedback path rather than just in the feedforward path. Nonlinear generators of many types are discussed by Golomb [8].

Further Reading. Much has been written on the subject of codes for spread-spectrum systems. Comprehensive treatment of the mathematical background can be found in Birkhoff and MacLane [4] or Jacobson [16], while concise summaries of this material can be found in Peterson and Weldon [5] or Lin and Costello [3]. The text by Peterson and Weldon also contains a considerable amount of material

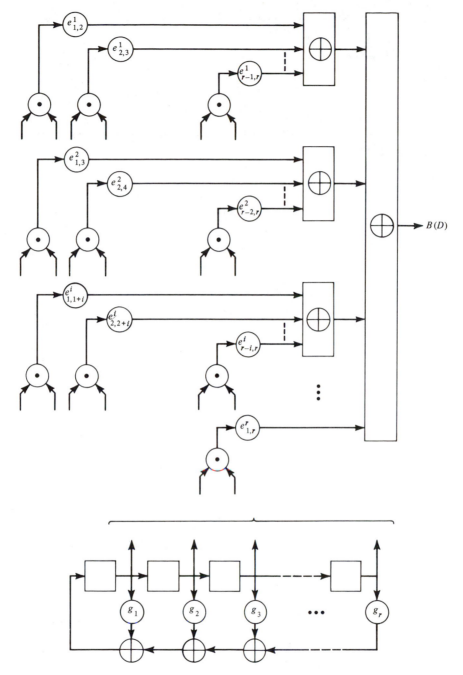

FIGURE 8-20. Generalized feedforward nonlinear Logic attached to linear generator. (From Ref. 15.)

on basic shift register structures and their use in error correction coding. A comprehensive treatment of both linear and nonlinear codes can be found in the text by Golomb [8], which is the only text available devoted entirely to shift register sequences. Concise summaries of the properties and mathematical structure of maximal-length sequences can be found in MacWilliams and Sloane [17] or Sarwate and Pursley [2]; the first reference also discusses pseudorandom arrays, while

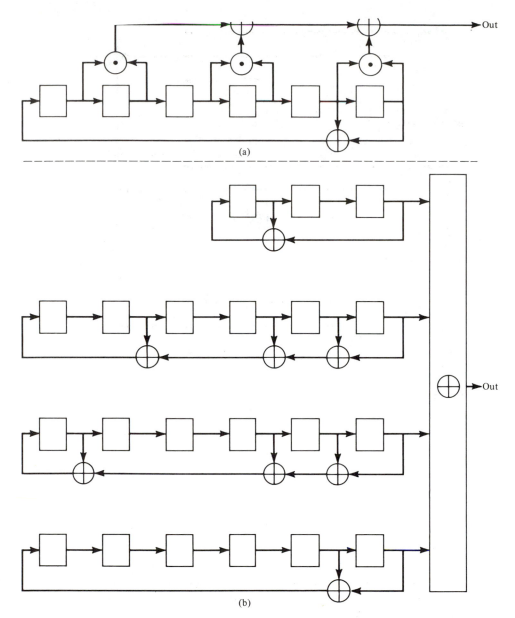

FIGURE 8-21. (a) Nonlinear generator; (b) linear equivalent. (From Ref. 15.)

the second includes a great deal of information on the correlation properties of other code families as well as m-sequences. The partial correlation properties of m-sequences are analyzed by Lindholm [10]. The power spectrum of m-sequences and products of m-sequences is discussed in Gill [18] and Gill and Spilker [19]. Most of the references contain rather extensive reference lists which will lead the student to a rich variety of research material on code sequences.

The use of special codes for ranging and acquisition is discussed in Titsworth [20] and Stiffler [13]. Both of these references provide a straightforward treatment of their subject which does not depend on finite-field mathematics. The nonlinear theory is discussed in Golomb [8] as well as in two more recent papers (Groth [15] and Key [21]). The latter two papers depend heavily on finite mathematics. Unfortunately, many of the current research results in the nonlinear theory remain in the classified literature.

Finally, two recent texts on spread-spectrum communications (Holmes [9] and Simon [22]) contain chapters on spreading codes as well as further reference lists.

REFERENCES

[1] R. E. ZIEMER and W. H. TRANTER, *Principles of Communications: Systems, Modulation, and Noise* (Boston: Houghton Mifflin, 1976).

[2] D. V. SARWATE and M. B. PURSLEY, "Crosscorrelation Properties of Pseudorandom and Related Sequences," *Proc. IEEE,* May 1980.

[3] S. LIN and D. J. COSTELLO, *Error Control Coding: Fundamentals and Applications* (Englewood Cliffs, N.J.: Prentice-Hall, 1983).

[4] G. BIRKHOFF and S. MACLANE, *A Survey of Modern Algebra* (New York: Macmillan, 1965).

[5] W. W. PETERSON and E. J. WELDON, *Error Correction Codes* (Cambridge, Mass.: MIT Press, 1972).

[6] K. METZGER and R. J. BOUWENS, "An Ordered Table of Primitive Polynomials over GF(2) of Degrees 2 Through 19 for Use with Linear Maximal Sequence Generators," TM107, Cooley Electronics Laboratory, University of Michigan, Ann Arbor, July 1972 [AD 746876].

[7] D. L. SCHILLING, B. H. BATSON, and R. PICKHOLTZ, *Spread Spectrum Communications*, Short Course Notes, Nat. Telecomm. Conf., 1980.

[8] S. W. GOLOMB, *Shift Register Sequences* (San Francisco: Holden-Day, 1967).

[9] J. K. HOLMES, *Coherent Spread Spectrum Systems* (New York: Wiley-Interscience, 1982).

[10] J. H. LINDHOLM, "An Analysis of the Pseudo-randomness Properties of Subsequences of Long *m*-Sequences," *IEEE Trans. Inf. Theory*, July 1968.

[11] J. L. MASSEY, "Shift-Register Synthesis and BCH Decoding," *IEEE Trans. Inf. Theory,* January 1969.

[12] J. J. SPILKER, JR., *Digital Communications by Satellite* (Englewood Cliffs, N.J.: Prentice-Hall, 1977).

[13] J. J. STIFFLER, "Rapid Acquisition Sequences," *IEEE Trans. Inf. Theory*, March 1968.

[14] W. DIFFIE and M. E. HELLMAN, "New Directions in Cryptography," *IEEE Trans. Inf. Theory*, November 1976.

[15] E. J. GROTH, "Generation of Binary Sequences with Controllable Complexity," *IEEE Trans. Inf. Theory*, May 1971.

[16] N. JACOBSON, *Basic Algebra 1* (San Francisco: W. H. Freeman, 1974).

[17] F. J. MACWILLIAMS and N. J. A. SLOANE, "Pseudo-random Sequences and Arrays," *Proc. IEEE*, December 1976.

[18] W. J. GILL, "Effect of Synchronization Error in Pseudo-random Carrier Communications," *Conf. Rec.,* First Annual IEEE Commun. Conf., June 1965.

[19] W. J. GILL and J. J. SPILKER, "An Interesting Decomposition Property for the Self-Products of Random or Pseudorandom Binary Sequences," *IEEE Trans. Commun. Syst.,* June 1963.

[20] R. C. TITSWORTH, "Optimal Ranging Codes," *IEEE Trans. Space Electron. and Telem.,* March 1964.

[21] E. L. KEY, "An Analysis of the Structure and Complexity of Nonlinear Binary Sequence Generators," *IEEE Trans. Inf. Theory*, November 1976.

[22] M. K. SIMON et al., *A Unified Approach to Spread Spectrum Communications* (Rockville, Md.: Computer Science Press, 1985).

PROBLEMS

8-1. Demonstrate that the set of integers $\{0, 1, 2, 3\}$ with addition and multiplication defined modulo-4 is not a field. Define the addition and multiplication tables which, together with this set, define a field of four elements. The polynomial $1 + D + D^2$ is primitive.

8-2. Construct a Galois field having 32 elements and list all elements in polynomial format together with its multiplicative inverse element. The polynomial $1 + D + D^5$ is primitive.

8-3. Prove that the polynomial $g(D) = 1 + D + D^3$ is primitive.

8-4. Prove that the additive and multiplicative identity elements of a finite field are unique.

8-5. Factor the polynomial $g(D) = 1 + D^2 + D^3 + D^4$, and evaluate $[g(D)]^4$.

8-6. Let D^{10} and D^{12} represent elements of the Galois field having 16 elements and defined using the primitive polynomial $h(D) = 1 + D + D^4$.
(a) Evaluate the product $D^{10} \cdot D^{12}$ using both the polynomial representation of these elements and the power-of-D representation.
(b) Evaluate $D^{10} + D^{12}$.
(c) Evaluate D^{10}/D^{12}.
(d) Evaluate $D^{10} - D^{12}$.

8-7. Let $h(D) = 1 + D + D^4$, $g(D) = 1 + D + D^2$, and $a(D) = 1 + D + D^2 + D^3$. Develop a shift register configuration which generates the function

$$b(D) = \frac{g(D)}{h(D)} a(D)$$

and determine the circuits output as a function of time. Can this circuit be used to automate the calculation of Problem 8-6?

8-8. Consider the feedback-shift-register configuration shown below. Determine the output of this circuit with the initial condition $a(D) = 1 + D + D^2 + D^3$.

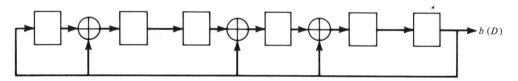

PROBLEM 8-8.   Feedback shift register.

8-9. Consider the feedback-shift-register configuration shown below. Determine the output of this circuit with initial condition $a(D) = 1 + D + D^2 + D^3$ and compare with the result of Problem 8-8.

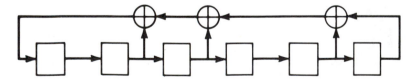

PROBLEM 8-9.   Feedback shift register.

8-10. What is the maximum possible period of the binary sequence generated by the linear feedback-shift-register configuration below? Find all possible output cycles of this circuit.

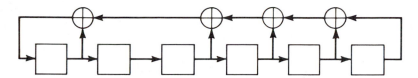

PROBLEM 8-10. Non-maximal-length linear feedback shift register.

8-11. Consider the maximal-length sequence generated using the primitive polynomial $g(D) = 1 + D + D^2 + D^4 + D^5$. Demonstrate that Properties I through V for m-sequences are satisfied for this particular m-sequence.

8-12. Plot the autocorrelation function and the detailed power spectrum for the maximal-length sequence specified by $g(D) = 1 + D + D^4$ using a shift register clock rate of 1.0 kHz.

8-13. The code of Problem 8-12 is used to BPSK modulate a 1.0-MHz carrier. Plot the power spectrum of the modulated carrier.

8-14. The code of Problem 8-12 is used to MSK modulate a 1.0-MHz carrier. Plot the power spectrum of the modulated carrier. (Hint: Consider serial MSK implementations.)

8-15. Two maximal-length shift register generators having periods N_1 and N_2 are run off the same clock. A third sequence is generated by modulo-2 adding the outputs of these generators. What is the period of the third output sequence?

8-16. Consider the m-sequence generator specified by the primitive polynomial 155_8. Define a shift register configuration that will generate this sequence in the forward direction and a shift register configuration that will generate this sequence in the reverse direction.

8-17. The correlation-filter arrangement illustrated below is one part of a code tracking loop in a direct-sequence spread-spectrum modem. Suppose that the spreading code is an m-sequence having period N and that the bandpass filter is ideal and is specified by

$$H(\omega) = \begin{cases} 1.0 & |\omega \pm \omega_0| \leq 0.2\pi/T_c \\ 0.0 & \text{elsewhere} \end{cases}$$

Plot the ratio of the filter output power at the carrier frequency to all other power at the filter output as a function of the code period N. All of the filter output power except that at the carrier frequency is called *code self-noise*.

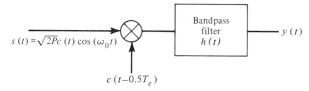

PROBLEM 8-17. Correlator illustrating spreading code self-noise.

8-18. A direct-sequence spread-spectrum modem is used to communicate between a base station and a mobile station separated by a distance which yields a propagation delay of 0.15 μs. The spreading code is a maximal-length code specified by the generator $g(D) = 1 + D^3 + D^{10}$ and the code generator clock rate is 100 MHz. Determine a set of initial conditions for the transmitter and receiver code generators such that, if the generators are started simultaneously, the received signal will be despread.

8-19. The receiver despreading code of Problem 8-18 is generated by modulo-2 addition of some number of delays of a code generator which is phase synchronous with the transmitter code generator. What delays should be added to achieve despreading in the receiver?

8-20. The following sequence of symbols from a spread-spectrum transmitter are received:

$$1\ 0\ 1\ 1\ 1\ 1\ 1\ 0\ 1\ 1\ 0\ 0\ 1$$

Spectral analysis of the received signal indicates that the power spectrum consists of discrete lines which are spaced at 322.6 kHz and that the spreading code rate is 10 MHz. What code generator is being used in the transmitter?

8-21. The following sequence of 32 symbols is received from a transmitter using a rapid acquisition sequence:

1	2	3	4	5	6	7	8	9	10	11	12	13	14	15	16
−1	1	1	1	−1	1	−1	−1	−1	1	1	1	−1	1	−1	−1

17	18	19	20	21	22	23	24	25	26	27	28	29	30	31	32
−1	1	−1	−1	−1	−1	−1	−1	−1	1	1	1	1	1	1	1

What is the starting phase of the received sequence?

8-22. Solve the following set of simultaneous equations over GF(2):

$$
\begin{aligned}
1 &= x + z \\
1 &= w + y \\
0 &= w + x + z \\
0 &= x + y
\end{aligned}
$$

8-23. Calculate the discrete periodic cross-correlation function for the pair of m-sequences defined by $g_1(D) = 1 + D^2 + D^5$ and

$$g_2(D) = 1 + D^2 + D^3 + D^4 + D^5$$

8-24. Calculate the discrete partial autocorrelation function for an m-sequence defined by $g(D) = 1 + D^2 + D^5$ using a window size of five chips and a phase difference of two chips.

CHAPTER 9

Code Tracking Loops

INTRODUCTION

Spread-spectrum communication requires that the transmitter and receiver spreading waveforms be synchronized. If the two waveforms are out of synchronization by as little as one chip, insufficient signal energy will reach the receiver data demodulator for reliable data detection. The task of achieving and maintaining code synchronization is always delegated to the receiver. There are two components of the synchronization problem. The first component is the determination of the initial code phase from whatever a priori information is available. This part of the problem is called *code acquisition* and is addressed in Chapter 10. The second component is the problem of maintaining code synchronization after initial acquisition. This problem is called code tracking and is addressed in this chapter.

Code tracking is accomplished using phase-lock techniques very similar to those discussed earlier for generation of coherent carrier references. The principal difference between the phase-locked loops discussed earlier and those discussed here is in the implementation of the phase discriminator. For carrier tracking, the discriminator is a simple multiplier, whereas for modern code tracking loops, several multipliers and usually pairs of filters and envelope detectors will be employed in the phase discriminator. Since code tracking loops are phase-locked loops (PLLs), the goal of the analysis to follow is to develop models of the various code tracking loops which are identical to the conventional PLL model and then to draw on the vast store of PLL results.

Code tracking loops for spread-spectrum systems can be categorized in several ways. First, there are coherent and noncoherent loops. Coherent loops make use of received carrier phase information whereas noncoherent loops do not. All but one of the phase discriminators to be discussed make use of correlation operations between the received signal and two different phases (early and late) of the receiver-generated spreading waveform. These two correlation operations can be accomplished using two independent channels or using a single channel which is time shared. A tracking loop that makes use of two independent correlators is called a *full-time early-late tracking loop*, and a tracking loop that time shares a single correlator is called a *tau-dither early-late tracking loop*. All the categories of tracking loops mentioned above are discussed in this chapter.

The tracking loops discussed are designed to achieve low rms tracking jitter in the presence of AWGN while tracking the dynamics of the received spreading waveform. The transmission delay, T_d, is actually a function of time, $T_d(t)$, when there is relative motion between the transmitter and the receiver. This function $T_d(t)$ must be tracked by the code-tracking loop. The same issues arise in the design of a code-tracking loop which arise in the design of a carrier-tracking phase-locked loop. The loop bandwidth will be selected to be a compromise between a wide bandwidth, which facilitates tracking the dynamics of $T_d(t)$, and a narrow bandwidth, which minimizes the tracking jitter due to AWGN. The problem of cycle skipping is more serious for code tracking than it is for carrier tracking. A cycle skip in a code-tracking loop may cause the system to re-enter the acquisition mode.

The chapter begins with a discussion of the optimum tracking loop for direct-sequence spread-spectrum systems. The baseband full-time early-late tracking loop is discussed next since many important concepts can be introduced with this simple code-tracking loop. The baseband early-late tracking loop operates directly on the spreading waveform itself rather than on a modulated carrier. Thus it is assumed that a demodulator preceding the tracking loop has accurately stripped the spreading waveform from the modulated carrier. Next, the full-time early-late noncoherent tracking loop is discussed. This loop is similar to the baseband loop except for a more complex phase discriminator. The tau-dither early-late noncoherent loop is discussed following the full-time loop because, even though it is electronically simpler than the full-time loop, it is analytically more complicated. Tracking loops for frequency-hop spread-spectrum systems are discussed last.

Some of the earliest work on code tracking discriminators was done by Spilker and Magill [1]. This early work was very general in nature and showed that the proper error signal for tracking could be derived from a correlation of the received signal with a receiver generated first derivative (with respect to time) of the spreading waveform. Later work [2–6] specialized Spilker and Magill's early work to the types of tracking loops currently employed. Code tracking loops are also discussed in [7–9].

9-2

OPTIMUM TRACKING OF WIDEBAND SIGNALS

The transmitted waveform $s(t)$ in a spread-spectrum communication system is a wideband signal. It has been stated [1] that the optimum tracking discriminator for

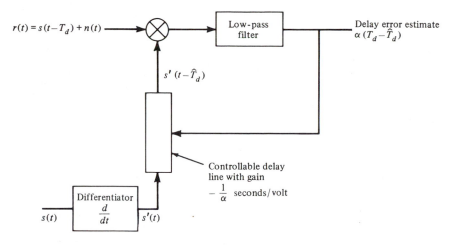

FIGURE 9-1. Tracking loop for arbitrary wideband signal. (From Ref. 1.)

an arbitrary wideband signal is a multiplier which forms the product of the received signal plus noise and the first derivative with respect to time of the receiver generated replica of the transmitted signal. This discriminator is optimum in that its output is a maximum likelihood estimate of the phase difference between the two wideband signals in an AWGN environment. This means that the output phase error estimate is the most probable phase error, given the available received information. A tracking loop making use of this optimum discriminator is illustrated in Figure 9-1. The received signal $r(t) = s(t - T_d) + n(t)$ is multiplied by a differentiated and delayed receiver-generated replica of $s(t)$. The multiplier output contains a dc component related to the delay error $(T_d - \hat{T}_d)$, where $\hat{T}_d$ is the receiver estimate of the transmission delay. This dc component is extracted by the lowpass filter and used to correct the delay of the controllable delay line.

This tracking loop configuration will not be analyzed in detail here since it is not commonly used in modern spread-spectrum systems. However, because it is optimum and because modern tracking loop configurations are usually discrete approximations of this loop, the operation of the loop will be described briefly for a single special case. Suppose that the received signal is the baseband spreading waveform $c(t - T_d)$ itself,* and suppose that thermal noise is ignored. Let the spreading waveform be derived from an m-sequence which has a period of seven symbols. The received signal is shown in Figure 9-2a. The function of the tracking loop is to produce the signal $c(t - \hat{T}_d)$ with $\epsilon = T_d - \hat{T}_d$ as small as possible. The signal $c(t - \hat{T}_d)$ is shown in Figure 9-2b and its first derivative with respect to time is shown in Figure 9-2c. The derivative is a series of impulse functions which are mathematically convenient but are difficult to handle electronically. The occurrence of these impulse functions is the primary reason that other types of tracking loops are used in practice. The tracking loop multiplier output is shown in Figure 9-2d. Since $\hat{T}_d > T_d$ and $|T_d - \hat{T}_d| < T_c$, all the impulse functions at the multiplier output are positive. The dc component of the multiplier output is the time average of

$$ c(t - T_d) \frac{d}{dt} [c(t - \hat{T}_d)] $$

---

*The technique described here will not work if the received signal is data modulated since the discriminator slope will change sign with the data modulation.

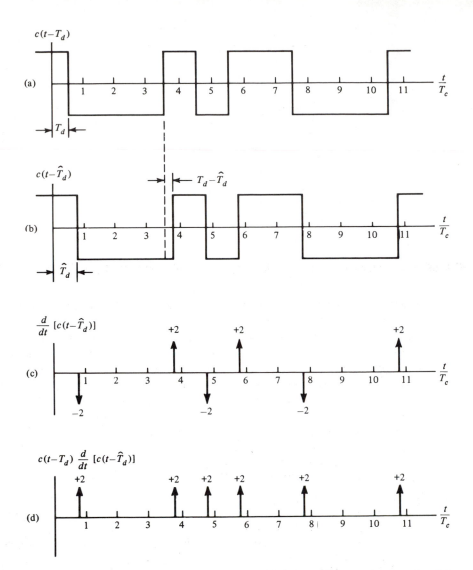

FIGURE 9-2. Waveforms of optimum code tracking loop: (a) received code; (b) receiver generated replica of spreading code; (c) derivative of b; (d) multiplier output.

which is $(N + 1)/NT_c$. When $\hat{T}_d < T_d$ and $|T_d - \hat{T}_d| < T_c$, all of the impulses at the multiplier output are negative and the dc component is $-(N + 1)/NT_c$. When $|T_d - \hat{T}_d| \geq T_c$, there is an equal number of positive and negative multiplier output impulses, and the dc level is zero. The dc output of the multiplier is shown in Figure 9-3 as a function of the delay difference $\delta = T_d - \hat{T}_d$. Observe that when $\delta < 0$, the delay line input will be positive and that this will decrease $\hat{T}_d$ as required to drive δ to zero. Whenever $|T_d - \hat{T}_d| < T_c$, a correct voltage exists which pushes the delay $\hat{T}_d$ in the correct direction.

The optimum delay discriminator operates similarly for any wideband signal. The reader is referred to Spilker [1, 7] for a complete analysis of this discriminator. Finally, observe that the conventional phase-locked loop employs exactly the discriminator just described. In a conventional PLL, the input signal $\sin(\omega t + \varphi)$ is

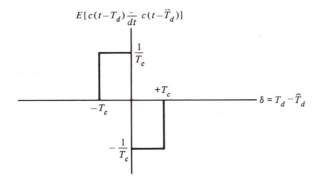

$$E[c(t-T_d)\tfrac{-}{dt} c(t-\hat{T}_d)]$$

FIGURE 9-3. Optimum delay discriminatior output dc component for *m*-sequence baseband tracking loop.

correlated with a signal cos $(\omega t + \hat{\varphi})$, which is, within a constant, the derivative of the received signal.

9-3

BASEBAND FULL-TIME EARLY-LATE TRACKING LOOP

The function of a baseband full-time loop is to track the time-varying phase of the received spreading waveform $c(t - T_d)$. The function $\hat{T}_d(t)$ will denote the receiver estimate of $T_d(t)$, and T_d and $\hat{T}_d$ are always functions of time, whether or not this dependence is written explicitly. The received signal consists of the spreading waveform $\sqrt{P}\, c(t - T_d)$ with power P and additive white Gaussian noise $n(t)$ with two-sided power spectral density $N_0/2$ W/Hz. That is,

$$s_r(t) = \sqrt{P}\, c(t - T_d) + n(t) \tag{9-1}$$

Figure 9-4 is a conceptual block diagram of the tracking loop. It consists of a phase discriminator, a loop filter, a voltage-controlled oscillator, and a spreading-waveform generator.

The received signal is input to the delay-lock discriminator where, after power division, it is correlated with an early spreading waveform $c(t - \hat{T}_d + (\Delta/2)T_c)$, and a late spreading waveform, $c(t - \hat{T}_d - (\Delta/2)T_c)$. The parameter Δ is the total normalized time difference between the early and late discriminator channels.

Consider the operation of the delay-lock discriminator as a static phase-measuring device in a noiseless environment. That is, let T_d and $\hat{T}_d$ be fixed and determine the output of the discriminator. This output will contain a component which is a function of $\delta = (T_d - \hat{T}_d)/T_c$ and is suitable for driving the VCO just as the phase-locked-loop multiplier output contained a component sin $(\theta_i - \theta_0)$. In the static case, it is convenient to write $y_1(t)$, $y_2(t)$, and $\epsilon(t, \delta)$ as explicit functions of T_d, $\hat{T}_d$, and t. Thus the early-correlator output is

$$y_1(t, T_d, \hat{T}_d) = K_1 \sqrt{\frac{P}{2}}\, c(t - T_d)c\!\left(t - \hat{T}_d + \frac{\Delta}{2} T_c \right) \tag{9-2a}$$

and the late-correlator output is

$$y_2(t, T_d, \hat{T}_d) = K_1 \sqrt{\frac{P}{2}}\, c(t - T_d)c\!\left(t - \hat{T}_d - \frac{\Delta}{2} T_c \right) \tag{9-2b}$$

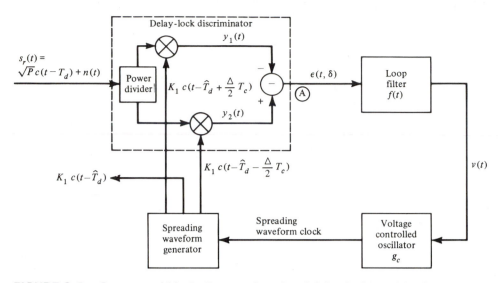

FIGURE 9-4. Conceptual block diagram: baseband delay-lock tracking loop.

In these expressions, K_1 is the multiplier gain and is dependent on the particular multiplier hardware implementation. The input signal has been divided by $\sqrt{2}$ to account for the power division, and noise has been ignored. The delay-lock discriminator output is the difference of $y_2(t)$ and $y_1(t)$ and is

$$\epsilon(t, T_d, \hat{T}_d) = y_2(t, T_d, \hat{T}_d) - y_1(t, T_d, \hat{T}_d)$$

$$= K_1 \sqrt{\frac{P}{2}} c(t - T_d) \left[c\left(t - \hat{T}_d - \frac{\Delta}{2} T_c\right) - c\left(t - \hat{T}_d + \frac{\Delta}{2} T_c\right) \right]$$

(9-3)

The dc component of this signal is used for code tracking. The time-varying component, which is also a function of δ, is called *code self-noise*.

The dc component of $\epsilon(t, T_d, \hat{T}_d)$ is denoted $K_1 \sqrt{P/2} \, D_\Delta(T_d, \hat{T}_d)$ and is the time average of $\epsilon(t, T_d, \hat{T}_d)$. Thus

$$K_1 \sqrt{\frac{P}{2}} D_\Delta(T_d, \hat{T}_d) = \frac{1}{NT_c} \int_{-NT_c/2}^{NT_c/2} K_1 \sqrt{\frac{P}{2}} c(t - T_d)$$

$$\left[c\left(t - \hat{T}_d - \frac{\Delta}{2} T_c\right) - c\left(t - \hat{T}_d + \frac{\Delta}{2} T_c\right) \right] dt \quad (9\text{-}4)$$

where NT_c is the period of $c(t)$. Recalling the definition of the autocorrelation function of $c(t)$ from (8-2),

$$D_\Delta(T_d, \hat{T}_d) = R_c\left(T_d - \hat{T}_d - \frac{\Delta}{2} T_c\right) - R_c\left(T_d - \hat{T}_d + \frac{\Delta}{2} T_c\right)$$

$$= R_c\left[\left(\delta - \frac{\Delta}{2}\right) T_c\right] - R_c\left[\left(\delta + \frac{\Delta}{2}\right) T_c\right]$$

$$\triangleq D_\Delta(\delta) \quad (9\text{-}5)$$

This function is plotted in Figure 9-5 for four values of Δ, where $c(t)$ is the waveform derived from a maximal-length sequence and $R_c(\tau)$ is defined in (8-45).

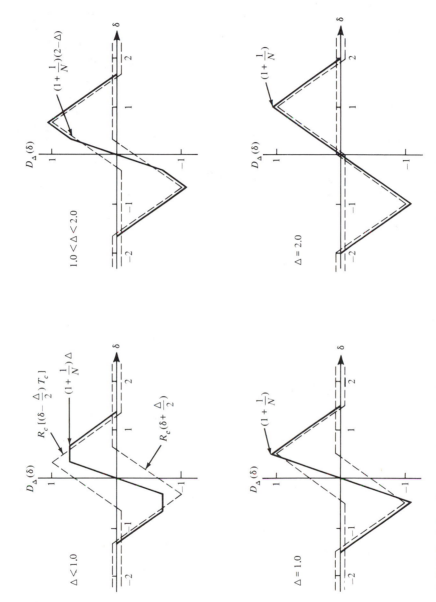

FIGURE 9-5. Delay-lock discriminator dc outputs for maximal-length sequence spreading codes and various values of Δ.

Observe from Figure 9-5 that there is a range of δ near zero for which $D_\Delta(\delta)$ is linearly related to δ. This region is always selected as the normal operating region for the tracking loop. The slope of the discriminator s-curve near $\delta = 0$ is $2(1 + 1/N)$ for all $0 < \Delta < 2.0$. The range of δ for which the discriminator characteristic has the slope $2(1 + 1/N)$ is $|\delta| < \Delta/2$ for $\Delta \leq 1.0$ and $|\delta| < 1 - \Delta/2$ for $1 \leq \Delta < 2$. This range decreases to zero at $\Delta = 2.0$, and is maximum for $\Delta = 1.0$.

In most analyses of code tracking loops the time-varying component of $\epsilon(t, T_d, \hat{T}_d)$, which is denoted $K_1 \sqrt{P/2} N_\Delta(t, T_d, \hat{T}_d)$, can be safely ignored since most of this self-noise power is at frequencies which are well outside the bandwidth of the tracking loop. A detailed proof of this statement requires calculation of the power spectrum of $\epsilon(t, T_d, \hat{T}_d)$. This calculation is tedious and will not be given here. An overbound on the magnitude of the power spectrum of $\epsilon(t, T_d, \hat{T}_d)$ is easily obtained, however, using results developed previously. This overbound is sufficient in most instances to prove that the code self-noise can be ignored. The overbound is obtained by observing that both terms of (9-3) have identical power spectra. The worst-case power spectrum of the sum is obtained by assuming that all components add phase synchronously. Thus the worst-case power spectrum of $\epsilon(t, T_d, \hat{T}_d)$ is four times the magnitude of the power spectrum of

$$\epsilon'(t, T_d, \hat{T}_d) = K_1 \sqrt{\frac{P}{2}} c(t - T_d) c\left(t - \hat{T}_d - \frac{\Delta}{2} T_c \right) \tag{9-6}$$

The power spectrum of the product of two m-sequences is calculated in Appendix F and is given in (8-59). The code self-noise can be ignored whenever the magnitude of the bound just calculated is sufficiently below the magnitude of the thermal noise power-spectral density calculated at the same point. A lower bound on code self-noise power spectrum is obtained by assuming that the two terms of (9-3) are uncorrelated. When this is true, the power spectrum of $K_1 \sqrt{P/2} N_\Delta(t, T_d, \hat{T}_d)$ is twice the magnitude of the power spectrum of $\epsilon'(t, T_d, \hat{T}_d)$. When $\Delta \geq 1.0$ and the spreading code is an m-sequence, the two terms of (9-3) are nearly uncorrelated.

Consider now the operation of the phase discriminator in an AWGN environment. With noise included,

$$y_1(t, T_d, \hat{T}_d) = K_1 \sqrt{\frac{P}{2}} c(t - T_d) c\left(t - \hat{T}_d + \frac{\Delta}{2} T_c \right)$$

$$+ \frac{K_1}{\sqrt{2}} c\left(t - \hat{T}_d + \frac{\Delta}{2} T_c \right) n(t) \tag{9-7a}$$

$$y_2(t, T_d, \hat{T}_d) = K_1 \sqrt{\frac{P}{2}} c(t - T_d) c\left(t - \hat{T}_d - \frac{\Delta}{2} T_c \right)$$

$$+ \frac{K_1}{\sqrt{2}} c\left(t - \hat{T}_d - \frac{\Delta}{2} T_c \right) n(t) \tag{9-7b}$$

The discriminator output is

$$\epsilon(t, T_d, \hat{T}_d) = K_1 \sqrt{\frac{P}{2}} [D_\Delta(\delta) + N_\Delta(t, T_d, \hat{T}_d)]$$

$$+ \frac{K_1}{\sqrt{2}} n(t) \left[c\left(t - \hat{T}_d - \frac{\Delta}{2} T_c \right) - c\left(t - \hat{T}_d + \frac{\Delta}{2} T_c \right) \right] \tag{9-8}$$

Assuming that code self-noise can be ignored, this can be written

$$\epsilon(t, T_d, \hat{T}_d) = K_1 \sqrt{\frac{P}{2}} \left[D_\Delta(\delta) + \frac{1}{\sqrt{P}} n'(t) \right] \tag{9-9}$$

where

$$n'(t) = n(t) \left[c\left(t - \hat{T}_d - \frac{\Delta}{2} T_c \right) - c\left(t - \hat{T}_d + \frac{\Delta}{2} T_c \right) \right] \tag{9-10}$$

To evaluate the noise performance of the tracking loop, the power spectrum of $n'(t)$ must be calculated. This calculation is accomplished by first evaluating the autocorrelation function of $n'(t)$. Care must be exercised in defining the autocorrelation function since, with $\hat{T}_d$ and Δ fixed, the random process $n'(t)$ defined above is not wide-sense stationary and therefore does not possess a power spectrum in the normal sense. With $\hat{T}_d$ interpreted as a random variable, stationarity is achieved and

$$R_{n'}(\tau) = E\left\{ n(t)n(t + \tau) \left[c\left(t - \hat{T}_d - \frac{\Delta}{2} T_c \right) - c\left(t - \hat{T}_d + \frac{\Delta}{2} T_c \right) \right] \right.$$

$$\left. \times \left[c\left(t + \tau - \hat{T}_d - \frac{\Delta}{2} T_c \right) - c\left(t + \tau - \hat{T}_d + \frac{\Delta}{2} T_c \right) \right] \right\} \tag{9-11}$$

The white noise $n(t)$ is independent of the spreading code $c(t)$, so that the expected value can be factored to obtain

$$R_{n'}(\tau) = E[n(t)n(t + \tau)]E\left\{ \left[c\left(t - \hat{T}_d - \frac{\Delta}{2} T_c \right) - c\left(t - \hat{T}_d + \frac{\Delta}{2} T_c \right) \right] \right.$$

$$\left. \left[c\left(t + \tau - \hat{T}_d - \frac{\Delta}{2} T_c \right) - c\left(t + \tau - \hat{T}_d + \frac{\Delta}{2} T_c \right) \right] \right\} \tag{9-12}$$

The autocorrelation function of the noise is a delta function, that is,

$$E[n(t)n(t + \tau)] = \frac{N_0}{2} \delta(\tau) \tag{9-13}$$

Substituting (9-13) into (9-12) and setting $\tau = 0$, since $\delta(\tau)$ is zero for all $\tau \neq 0$, results in

$$R_{n'}(\tau) = \frac{N_0}{2} \delta(\tau)E\left\{ \left[c\left(t - \hat{T}_d - \frac{\Delta}{2} T_c \right) - c\left(t - \hat{T}_d + \frac{\Delta}{2} T_c \right) \right]^2 \right\}$$

$$= \frac{N_0}{2} \delta(\tau) \left\{ E\left[c^2\left(t - \hat{T}_d - \frac{\Delta}{2} T_c \right) \right] \right.$$

$$- 2E\left[c\left(t - \hat{T}_d - \frac{\Delta}{2} \right) c\left(t - \hat{T}_d + \frac{\Delta}{2} T_c \right) \right]$$

$$\left. + E\left[c^2\left(t - \hat{T}_d + \frac{\Delta}{2} T_c \right) \right] \right\} \tag{9-14}$$

The spreading waveforms take on only values of ± 1, so that $c^2(t) = 1.0$. Assume that the spreading waveforms are derived from maximal-length sequences. Because of the shift-and-add property of m-sequences, the function $c(t - \hat{T}_d - (\Delta/2)T_c)$

$c(t - \hat{T}_d + (\Delta/2)T_c)$ may be viewed as a signal that switches between two phases of the same m-sequence for $\Delta \geq 1.0$ or between the all-ones sequence and some phase of the m-sequence for $\Delta < 1.0$. For $\Delta > 1.0$, the expected value of the product signal is $-1/N$. For $\Delta < 1.0$, the all-ones sequence is output for $(1 - \Delta)T_c$ seconds and some phase of the m-sequence is output for ΔT_c seconds of every chip. The expected value of this signal is $(1 - \Delta) - \Delta(1/N)$. Thus

$$R_{n'}(\tau) = \frac{N_0}{2} \delta(\tau) \left\{ 2 - 2E\left[c\left(t - \hat{T}_d - \frac{\Delta}{2}T_c \right) c\left(t - \hat{T}_d + \frac{\Delta}{2}T_c \right) \right] \right\}$$

$$= \begin{cases} N_0 \, \delta(\tau) \left(1 + \dfrac{1}{N} \right) & \text{for } \Delta \geq 1.0 \\[4mm] N_0 \, \delta(\tau) \, \Delta \left(1 + \dfrac{1}{N} \right) & \text{for } \Delta \leq 1.0 \end{cases} \qquad (9\text{-}15)$$

The two-sided power spectral density of $n'(t)$ is the Fourier transform of $R_{n'}(\tau)$ and is

$$S_{n'}(f) = \begin{cases} N_0 \left(1 + \dfrac{1}{N} \right) & \text{for } \Delta \geq 1.0 \\[4mm] N_0 \Delta \left(1 + \dfrac{1}{N} \right) & \text{for } \Delta < 1.0 \end{cases} \qquad (9\text{-}16)$$

Thus the random process $n'(t)$ is also white. The process is not, however, Gaussian since the second term in the product of (9-10) is periodic and takes on only values of ± 2 and 0.

At this point the delay-lock discriminator of Figure 9-4 has been adequately characterized to enable the development of a linear model of the entire tracking loop which will be valid for small tracking errors δ. Consider the voltage-controlled oscillator. The output frequency of this oscillator is $f_0 + g_c v(t)$, where f_0 is the rest or quiescent frequency and g_c is the VCO gain in Hz/V. The instantaneous output phase of this oscillator is the integral of frequency and is

$$2\pi f_0 t + 2\pi \frac{\hat{T}_d(t)}{T_c} = 2\pi f_0 t + 2\pi \int_0^t g_c v(\lambda) \, d\lambda + \theta_0 \qquad (9\text{-}17)$$

where θ_0 is the phase at time zero. The initial phase is set equal to zero in all that follows. The left side of the equation is the oscillator output phase written directly as a function of $\hat{T}_d(t)$. Subtracting $2\pi f_0 t$ from both sides of (9-17) and dividing by 2π yields

$$\frac{\hat{T}_d(t)}{T_c} = g_c \int_0^t v(\lambda) \, d\lambda \qquad (9\text{-}18)$$

and taking the Laplace transform yields

$$\frac{\hat{T}_d(s)}{T_c} = \frac{g_c V(s)}{s} \qquad (9\text{-}19)$$

where $\hat{T}(s)$ represents the Laplace transform of $\hat{T}(t)$.

The tracking loop filter is assumed to be passive, linear, and time invariant, so that its input–output relationship can be described by a differential equation having the form

$$a_m \frac{d^m\epsilon}{dt^m} + a_{m-1} \frac{d^{m-1}\epsilon}{dt^{m-1}} + \cdots + a_0\epsilon$$

$$= b_n \frac{d^n v}{dt^n} + b_{n-1} \frac{d^{n-1}v}{dt^{n-1}} + \cdots + b_0 v \tag{9-20}$$

where $m \leq n$ and ϵ represents $\epsilon(t, \delta)$. This linear differential equation can be solved using classical techniques. Using Laplace transform techniques, the ratio of the Laplace transform of the output, $V(s)$, to the Laplace transform of the input $E(s, \delta)$ is found to be

$$\frac{V(s)}{E(s, \delta)} = \frac{a_m s^m + a_{m-1}s^{m-1} + \cdots + a_0}{b_n s^n + b_{n-1}s^{n-1} + \cdots + b_0}$$

$$\triangleq F(s) \tag{9-21}$$

where $F(s)$ is the transfer function of the filter. This filter can also be described using its impulse response function $f(t)$. The filter output is the convolution of the input signal and the impulse response. That is,

$$v(t) = \int_{-\infty}^{t} \epsilon(\lambda, \delta)f(t - \lambda) \, d\lambda \tag{9-22}$$

The impulse response $f(t)$ is the inverse Laplace transform of the transfer function $F(s)$.

Using (9-9) to represent the delay-lock discriminator output, and (9-18) and (9-22) to represent the VCO and loop filter characteristics, the nonlinear integral equation representing the operation of the tracking loop of Figure 9-4 is

$$\frac{\hat{T}_d(t)}{T_c} = g_c \int_0^t v(\lambda) \, d\lambda$$

$$= g_c \int_0^t \int_{-\infty}^{\lambda} \epsilon(\alpha, \delta)f(\lambda - \alpha) \, d\alpha \, d\lambda$$

$$= g_c \int_0^t \int_{-\infty}^{\lambda} \left\{ K_1 \sqrt{\frac{P}{2}} D_\Delta[\delta(\alpha)] + \frac{K_1}{\sqrt{2}} n'(\alpha) \right\} f(\lambda - \alpha) \, d\alpha \, d\lambda \tag{9-23}$$

where $n'(t)$ is defined in (9-10). In this equation, the dependence of the normalized phase error $\delta(t) = [T_d(t) - \hat{T}_d(t)]/T_c$ on time has been shown explicitly. Equation (9-23) suggests the equivalent circuit illustrated in Figure 9-6. The equation that describes this circuit can be written down by inspection, and will be identical to (9-23). This circuit is the nonlinear [because of the function $D_\Delta(\delta)$] equivalent circuit for the baseband full-time early-late tracking loop.

For small tracking error, the phase discriminator output is a linear function of the tracking error and (9-23) can be written

$$\frac{\hat{T}_d(t)}{T_c} = g_c \int_0^t \int_{-\infty}^{\lambda} \left[K_1 \sqrt{\frac{P}{2}} 2 \left(1 + \frac{1}{N}\right) \frac{T_d(\alpha) - \hat{T}_d(\alpha)}{T_c} + \frac{K_1}{\sqrt{2}} n'(\alpha) \right]$$

$$\times f(\lambda - \alpha) \, d\alpha \, d\lambda \tag{9-24}$$

Consider the system illustrated in Figure 9-7. The equation that describes the operation of this system can be written down by inspection, and is identical to (9-24), where K_d is defined in Figure 9-7. Therefore the system of Figure 9-7

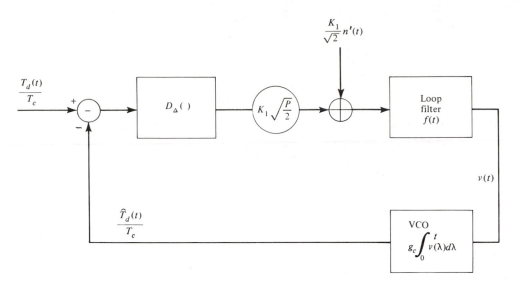

FIGURE 9-6. Nonlinear equivalent circuit for the baseband full-time early-late tracking loop.

describes the operation of the baseband full-time early-late tracking loop for small tracking errors. The noise function at the input to the circuit of Figure 9-7 will produce noise at point A, which is identical to the noise at point A of Figure 9-4. The power spectral density of the white noise function $n'(t)$ was given in (9-16). The power spectral density of $K_1 n'(t)/\sqrt{2}K_d$ is $(K_1/\sqrt{2}K_d)^2$ times the result of (9-16).

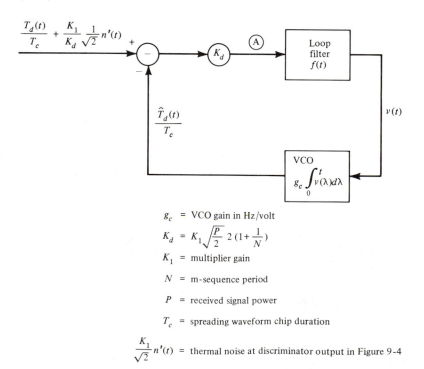

g_c = VCO gain in Hz/volt

$K_d = K_1 \sqrt{\dfrac{P}{2}} \, 2 \left(1 + \dfrac{1}{N}\right)$

K_1 = multiplier gain

N = m-sequence period

P = received signal power

T_c = spreading waveform chip duration

$\dfrac{K_1}{\sqrt{2}} n'(t)$ = thermal noise at discriminator output in Figure 9-4

FIGURE 9-7. Linear equivalent circuit for the baseband full-time early-late tracking loop.

The systems of Figures 9-4 and 9-7 are equivalent, so that either can be analyzed to find the tracking loop's noise and dynamic tracking performance. To illustrate the equivalence of the two systems, the loop filter was characterized by its impulse response and the VCO was characterized by an integral of its control voltage. These functions can also be characterized by the Laplace transform relationships of (9-19) and (9-21). The Laplace transform is a linear operation, so that the input difference circuit can be characterized by the difference of the Laplace transforms of $T_d(t)/T_c$ and $\hat{T}_d(t)/T_c$. Using transform notation, the system of Figure 9-7 can be represented by the system of Figure 9-8, which is identical to the linear model for the PLL of Figure 5-12. Although these circuits are equivalent, the student is cautioned that all of the analysis of the PLL assumed white Gaussian input noise. The noise process $n'(t)$ is not Gaussian. The pdf of $n'(t)$ contains an impulse function at zero amplitude. Since noise of zero amplitude has no effect on tracking, and since the remaining pdf is Gaussian, negligible errors will be obtained if the Gaussian analysis is used.

The tracking loop output $\hat{T}_d(s)$ can be related directly to its input $T_d(s)$ using Figure 9-8. This relationship is written by inspection and is

$$\frac{\hat{T}_d(s)}{T_c} = \frac{T_d(s) - \hat{T}_d(s)}{T_c}\left[K_d g_c \frac{F(s)}{s}\right] \tag{9-25}$$

Solving this equation for $\hat{T}_d(s)/T_d(s)$ yields

$$\frac{\hat{T}_d(s)}{T_d(s)} = \frac{K_d g_c F(s)}{s + K_d g_c F(s)} \triangleq H(s) \tag{9-26}$$

and solving for $[\hat{T}_d(s) - T_d(s)]/T_c$, which is the code tracking error, yields

$$\frac{T_d(s) - \hat{T}_d(s)}{T_c} = \frac{T_d(s)}{T_c}\left[\frac{s}{s + K_d g_c F(s)}\right] \tag{9-27}$$

These equations are the classical closed-loop servomechanism equations and from this point on the analysis of this tracking loop is identical to the analysis of any other servo loop. In particular, the response of the tracking loop to steps, ramps, and parabolas of input phase can be determined for various loop filter configurations. In addition, the model of Figure 9-8 is the same, except for the definition of some of the constants, as the PLL model, so that much PLL analysis also applies to the code tracking loop.

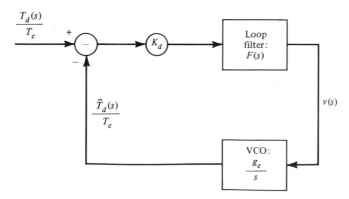

FIGURE 9-8. Laplace transform model of linear equivalent circuit for baseband full-time early-late tracking loop.

A result that is particularly important to the spread-spectrum system designer is the relationship between the mean-square tracking error or tracking jitter and the received signal-to-noise ratio in the loop bandwidth. The power spectrum of the tracking jitter is given by

$$S_\delta(f) = |H(j2\pi f)|^2 S_{n''}(f) \tag{9-28}$$

where $H(s)$ is the closed-loop transfer function defined by (9-26), and $S_{n''}(f)$ is the two-sided power spectrum of the Gaussian noise process at the input to the loop model of Figure 9-7. The power spectrum of $n''(t)$ is $K_1^2/2K_d^2$ times the result of (9-16) and is

$$S_{n''}(f) = \begin{cases} \dfrac{1}{2} \left(\dfrac{K_1}{K_d}\right)^2 N_0 \left(1 + \dfrac{1}{N}\right) & \text{for } \Delta \geq 1.0 \\[4mm] \dfrac{1}{2} \left(\dfrac{K_1}{K_d}\right)^2 N_0\, \Delta\left(1 + \dfrac{1}{N}\right) & \text{for } \Delta < 1.0 \end{cases} \tag{9-29}$$

Then the variance of δ which is denoted σ_δ^2 is

$$\sigma_\delta^2 = \int_{-\infty}^{\infty} S_{n''}(f)|H(j2\pi f)|^2\, df \tag{9-30}$$

Since $S_{n''}(f)$ is constant over all f [i.e., $n''(t)$ is a white noise process],

$$\sigma_\delta^2 = S_{n''}(0) \int_{-\infty}^{\infty} |H(j2\pi f)|^2\, df \tag{9-31}$$

The integral on the right side of this equation is defined as the two-sided noise bandwidth

$$W_L = \int_{-\infty}^{\infty} |H(j2\pi f)|^2\, df \tag{9-32}$$

The units of W_L are hertz. The final result for tracking jitter is then

$$\sigma_\delta^2 = \begin{cases} \dfrac{1}{2} \left(\dfrac{K_1}{K_d}\right)^2 N_0\left(1 + \dfrac{1}{N}\right) W_L & \text{for } \Delta \geq 1.0 \\[4mm] \dfrac{1}{2} \left(\dfrac{K_1}{K_d}\right)^2 N_0\, \Delta\left(1 + \dfrac{1}{N}\right) W_L & \text{for } \Delta < 1.0 \end{cases} \tag{9-33}$$

Observe that the tracking error variance is reduced as Δ decreases for $\Delta \leq 1.0$. This can be explained by noticing that as Δ decreases, the slope of the discriminator characteristic remains constant while some of the input noise cancels in (9-10) when the early and late codes overlap. Evaluating K_d as a function of K_1, P, and N in (9-33), and defining $\rho_L = 2P/N_0 W_L$ yields

$$\sigma_\delta^2 = \begin{cases} \dfrac{1}{2\rho_L} \left(\dfrac{N}{N+1}\right) & \text{for } \Delta \geq 1.0 \\[4mm] \dfrac{1}{2\rho_L} \left(\dfrac{N}{N+1}\right) \Delta & \text{for } \Delta < 1.0 \end{cases} \tag{9-34}$$

where ρ_L is the signal-to-noise ratio in the loop bandwidth.

EXAMPLE 9-1

A commonly used tracking loop filter is the simple lead-lag filter whose transfer function is

$$F(s) = \frac{1 + \tau_2 s}{\tau_1 s} \tag{9-35}$$

With this loop filter, the closed-loop transfer function of (9-26) becomes

$$H(s) = \frac{[K_d g_c(\tau_2/\tau_1)]s + \dfrac{K_d g_c}{\tau_1}}{s^2 + [K_d g_c(\tau_2/\tau_1)]s + \dfrac{K_d g_c}{\tau_1}}$$

$$= \frac{2\zeta\omega_n s + \omega_n^2}{s^2 + 2\zeta\omega_n s + \omega_n^2} \tag{9-36}$$

where the usual loop natural frequency and damping factor have been defined by

$$\omega_n = \sqrt{\frac{K_d g_c}{\tau_1}} \tag{9-37}$$

$$\zeta = \frac{\tau_2}{2}\omega_n \tag{9-38}$$

Since K_d is a function of the input signal level, both K_d and ζ are functions of P. This implies that automatic gain control is an important issue in tracking loop design. The two-sided noise bandwidth for a second-order loop with this loop filter is given by

$$W_L = \omega_n\left(\zeta + \frac{1}{4\zeta}\right) \tag{9-39}$$

This result is derived using contour integration to evaluate (9-32). The units of W_L are hertz. ∎

This completes the analysis of the baseband full-time early-late tracking loop. Fundamentally the same techniques will be used in the following sections to analyze other types of tracking loops. The two major components of this analysis were the derivation of the phase discriminator characteristic or S-curve, and the characterization of the thermal noise at the discriminator output. These two analyses were used to develop a linear model of the tracking loop which then enables the application of the vast store of control loop theory to the code tracking problem.

9-4

FULL-TIME EARLY-LATE NONCOHERENT TRACKING LOOP

Two difficulties arise when the baseband loop of Section 9-3 is applied to actual spread-spectrum communication systems. First, since the tracking loop input is the spreading waveform $c(t)$, this spreading waveform must be recovered from the carrier prior to code tracking. That is, the received signal must be demodulated prior to code tracking. Since spread-spectrum systems typically operate at very low signal-to-noise ratios in the transmission bandwidth, this demodulation will be difficult. In addition, the modulation is coherent and therefore a coherent carrier reference must be generated prior to demodulation. The generation of this coherent reference at extremely low signal-to-noise ratios is also difficult. The second difficulty stems from the fact that any communication system must convey information from the transmitter to the receiver. This implies that the carrier is in some way modulated with this information. The baseband signals analyzed in Section 9-3 conveniently ignored any data modulation principally because the baseband loop would not function properly had the received signal been $d(t - T_d)c(t - T_d)$ rather than $c(t - T_d)$.

Neither of these difficulties are present for the full-time early-late noncoherent tracking loop discussed in this section. The noncoherent loop employs a phase discriminator which is significantly different from that used in the baseband loop. The discriminator contains two energy detectors, which are, of course, not sensitive to data modulation or carrier phase, and thus enable the discriminator to ignore data modulation and carrier phase. Figure 9-9 is a conceptual block diagram of the tracking loop for the special case where the spreading modulation is binary phase-shift keying. With minor modifications, which will be discussed later, this tracking loop can be used for any direct-sequence spreading modulation. BPSK is used to introduce the student to this tracking loop because the analysis is simplified.

The received signal in this case is a data and spreading code-modulated carrier in bandlimited AWGN. This signal is represented by

$$r(t) = \sqrt{2P}\ c(t - T_d)\ \cos\ [\omega_0 t + \theta_d(t - T_d) + \varphi] + n(t) \qquad (9\text{-}40)$$

where P is the received signal power, $\theta_d(t - T_d)$ is the arbitrary data phase modulation, T_d is the transmission delay, φ is the random received carrier phase, ω_0 is the carrier radian frequency, and

$$n(t) = \sqrt{2}\ n_I(t)\ \cos\ \omega_0 t - \sqrt{2}\ n_Q(t)\ \sin\ \omega_0 t \qquad (9\text{-}41)$$

represents the bandlimited zero-mean white Gaussian noise process. The received noise is assumed to have a two-sided power spectral density of $N_0/2$ watts/hertz. The functions $n_I(t)$ and $n_Q(t)$ are independent zero-mean lowpass white Gaussian

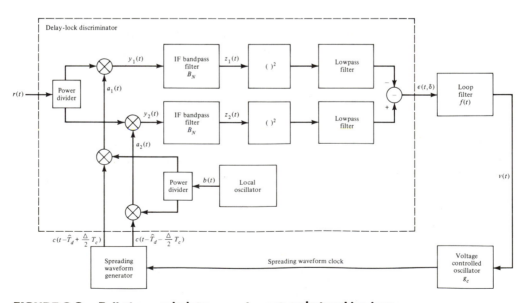

FIGURE 9-9. Full-time early-late noncoherent code tracking loop.

noise processes each having a two-sided power spectral density of $N_0/2$. Other authors choose to omit the $\sqrt{2}$ factors in (9-41). When these factors are omitted the baseband processes have a two-sided power spectral density of N_0 watts/hertz.

The received signal is power divided and then correlated with early and late spreading waveform modulated local oscillator signals. The reference local oscillator output is

$$b(t) = 2\sqrt{2K_1}\cos\left[(\omega_0 - \omega_{\text{IF}})t + \varphi'\right] \qquad (9\text{-}42)$$

so that

$$a_1(t) = 2\sqrt{K_1}c\left(t - \hat{T}_d + \frac{\Delta}{2}T_c\right)\cos\left[(\omega_0 - \omega_{\text{IF}})t + \varphi'\right] \qquad (9\text{-}43a)$$

$$a_2(t) = 2\sqrt{K_1}c\left(t - \hat{T}_d - \frac{\Delta}{2}T_c\right)\cos\left[(\omega_0 - \omega_{\text{IF}})t + \varphi'\right] \qquad (9\text{-}43b)$$

where ω_{IF} is the intermediate radian frequency, and φ' is the random local oscillator phase. The amplitude of these signals has been chosen so that the difference frequency outputs of the early and late correlator mixers due to signal will have power $PK_1/2$. Thus K_1 may be interpreted as the RF-to-IF conversion loss of the mixers. Any gain in an actual system between the input and the input to the squaring circuit may also be included in K_1. The IF bandpass filters are assumed to have center frequency ω_{IF} rad/s and a one-sided noise bandwidth of B_N hertz, so that the sum frequency mixer output components are rejected by these filters. Therefore, only the difference frequency mixer outputs need to be calculated. These components are

$$y_1(t) = \sqrt{K_1P}\; c(t - T_d)c\left(t - \hat{T}_d + \frac{\Delta}{2}T_c\right)\cos\left[\omega_{\text{IF}}t + \varphi - \varphi' + \theta_d(t - T_d)\right]$$

$$+ \sqrt{K_1}\; n_I(t)c\left(t - \hat{T}_d + \frac{\Delta}{2}T_c\right)\cos(\omega_{\text{IF}}t - \varphi')$$

$$- \sqrt{K_1}\; n_Q(t)c\left(t - \hat{T}_d + \frac{\Delta}{2}T_c\right)\sin(\omega_{\text{IF}}t - \varphi') \qquad (9\text{-}44a)$$

$$y_2(t) = \sqrt{K_1P}\; c(t - T_d)c\left(t - \hat{T}_d - \frac{\Delta}{2}T_c\right)\cos\left[\omega_{\text{IF}}t + \varphi - \varphi' + \theta_d(t - T_d)\right]$$

$$+ \sqrt{K_1}\; n_I(t)c\left(t - \hat{T}_d - \frac{\Delta}{2}T_c\right)\cos(\omega_{\text{IF}}t - \varphi')$$

$$- \sqrt{K_1}\; n_Q(t)c\left(t - \hat{T}_d - \frac{\Delta}{2}T_c\right)\sin(\omega_{\text{IF}}t - \varphi') \qquad (9\text{-}44b)$$

These correlator outputs are the sum of a component due to signal and a component due to noise. Define the noise components by

$$n_{1,\text{in}}(t) = \sqrt{\frac{K_1}{2}}\;c\left(t - \hat{T}_d + \frac{\Delta}{2}T_c\right)n'(t) \qquad (9\text{-}45a)$$

$$n_{2,\text{in}}(t) = \sqrt{\frac{K_1}{2}}\;c\left(t - \hat{T}_d - \frac{\Delta}{2}T_c\right)n'(t) \qquad (9\text{-}45b)$$

where

$$n'(t) = \sqrt{2}\;n_I(t)\cos(\omega_{\text{IF}}t - \varphi') - \sqrt{2}\;n_Q(t)\sin(\omega_{\text{IF}}t - \varphi') \qquad (9\text{-}46)$$

The signal component is again the sum of a desired signal, which will be used for code tracking, and a code self-noise term, which is negligible when the processing gain is sufficiently high. Thus there are three components of interest at the correlator output.

To determine the signal and self-noise components of the discriminator output, consider the noiseless case. Assume that the spreading code is independent of the data modulation and local oscillator carrier phase so that the power spectrum of $y_1(t)$ is the convolution of the power spectrum of $c(t - T_d)c(t - \hat{T}_d + (\Delta/2)T_c)$ and the power spectrum of $\sqrt{K_1P} \cos [\omega_{IF}t + \varphi - \varphi' + \theta_d(t - T_d)]$. The power spectrum of the product of two maximal length codes is derived in Appendix G and the result is given in (8-59), which is repeated here for convenience using radian frequency and defining $\alpha = T_d - \hat{T}_d + \dfrac{\Delta}{2} T_c$:

$$S_{cc'}(\omega) = \left[1 - \left(1 + \frac{1}{N} \right) \frac{|\alpha|}{T_c} \right]^2 2\pi \, \delta(\omega)$$

$$+ \left(1 + \frac{1}{N} \right) \left(\frac{|\alpha|}{T_c} \right)^2 \sum_{\substack{n=-\infty \\ n \neq 0}}^{\infty} \text{sinc}^2 \left(\frac{n\omega_c}{2\pi} |\alpha| \right) 2\pi \, \delta(\omega - n\omega_c)$$

$$+ \frac{N+1}{N^2} \left(\frac{|\alpha|}{T_c} \right)^2 \sum_{\substack{m=-\infty \\ m \neq 0}}^{\infty} \text{sinc}^2 \left(\frac{m\omega_c}{2\pi N} |\alpha| \right) 2\pi \, \delta\left(\omega - \frac{m\omega_c}{N} \right) \quad (8\text{-}59)$$

In this expression cc' denotes the maximal-length sequence product, and α is the phase difference in seconds between c and c'. The power spectrum, $S_d(\omega)$, of the data-modulated IF carrier can be calculated using the techniques described in Chapter 3. The convolution of $S_{cc'}(\omega)$ and $S_d(\omega)$ is

$$S_{y1}(\omega) = \frac{1}{2\pi} \int_{-\infty}^{\infty} S_d(\omega) S_{cc'}(\omega - \omega') \, d\omega'$$

$$= \left[1 - \left(1 + \frac{1}{N} \right) \frac{|\alpha|}{T_c} \right]^2 S_d(\omega)$$

$$+ \left(1 + \frac{1}{N} \right) \left(\frac{|\alpha|}{T_c} \right)^2 \sum_{\substack{n=-\infty \\ n \neq 0}}^{\infty} \text{sinc}^2 \left(\frac{n\omega_c}{2\pi} |\alpha| \right) S_d(\omega - n\omega_c)$$

$$+ \frac{N+1}{N^2} \left(\frac{|\alpha|}{T_c} \right)^2 \sum_{\substack{m=-\infty \\ m \neq 0}}^{\infty} \text{sinc}^2 \left(\frac{m\omega_c}{2\pi N} |\alpha| \right) S_d\left(\omega - \frac{m\omega_c}{N} \right) \quad (9\text{-}47)$$

Observe that this is a sum of frequency translations of $S_d(\omega)$ by all frequency components of cc'. The bandwidth of the IF filter has been selected to be just wide enough to pass the data-modulated IF carrier. In all spread-spectrum systems, the spreading waveform chip frequency ω_c is much greater than the maximum significant data modulation frequency. Therefore, all translations of $S_d(\omega)$ by harmonics of ω_c [i.e., the second line of (9-47)], will lie outside the IF filter passband and will be rejected. The components of the third line of (9-47) are translated, not by harmonics of ω_c but by harmonics of ω_c/N. The code period N is usually large, so that translations of $S_d(\omega)$ by ω_c/N may result in components within the

IF passband. Fortunately, these components have a magnitude proportional to $(N + 1)/N^2 \approx 1/N$, so that their effect on the discriminator may be minimal.

To gain a preliminary idea of the magnitude relative to the desired term of the terms on the third line of (9-47), assume that $S_d(\omega)$ has a uniform density over the frequency range $0 \leq |\omega \pm \omega_{IF}| < \pi B_N$, and calculate the power spectrum of the third term at $\omega = \pm\omega_{IF}$ for the limiting case where N approaches infinity. Denote this limiting value at $\omega = +\omega_{IF}$ by β, where

$$\beta = \lim_{N\to\infty} \frac{N + 1}{N^2} \left(\frac{|\alpha|}{T_c}\right)^2 \sum_{\substack{m = -\infty \\ m \neq 0}}^{\infty} \mathrm{sinc}^2\left(\frac{m\omega_c}{2\pi N}|\alpha|\right) S_d\left(\omega_{IF} - \frac{m\omega_c}{N}\right) \quad (9\text{-}48)$$

It can be shown that β is approximately given by

$$\beta = \left(\frac{\Delta}{2}\right)^2 \frac{B_N}{f_c} S_d(\omega_{IF}) \quad (9\text{-}49)$$

An indentical expression can be derived for the value of the power spectral density at $\omega = -\omega_{IF}$. The most important characteristic of (9-49) is that β decreases inversely with increasing f_c, so that, with sufficiently high code rate, β can be safely ignored. Unless otherwise stated, the code rate of the spread-spectrum system will be assumed adequately high that both the second and third lines of (9-47) can be ignored.

The preceding arguments have shown that for high processing gains, the only component of interest in

$$y_1(t) = \sqrt{K_1 P} c(t - T_d)c\left(t - \hat{T}_d + \frac{\Delta}{2}T_c\right) \cos\left[\omega_{IF}t + \varphi - \varphi' + \theta_d(t - T_d)\right]$$

is the component resulting from the product of the dc component of $c(t - T_d)c(t - \hat{T}_d + (\Delta/2)T_c)$ with the cosine. By definition, the dc component of the spreading waveform product is the autocorrelation function $R_c(\tau)$ of the spreading waveform evaluated at $\tau = T_d - \hat{T}_d + (\Delta/2)T_c$. Thus, for high processing gains,

$$y_1(t) \approx \sqrt{K_1 P}\, R_c\left[\left(\delta + \frac{\Delta}{2}\right)T_c\right] \cos\left[\omega_{IF}t + \varphi - \varphi' + \theta_d(t - T_d)\right]$$

$$\triangleq x_1(t) \quad (9\text{-}50a)$$

where $\delta = (T_d - \hat{T}_d)/T_c$. An identical development for the late channel will yield

$$y_2(t) \approx \sqrt{K_1 P}\, R_c\left(\delta - \frac{\Delta}{2}T_c\right) \cos\left[\omega_{IF}t + \varphi - \varphi' + \theta_d(t - T_d)\right]$$

$$\triangleq x_2(t) \quad (9\text{-}50b)$$

If IF bandpass filters have been designed to pass these signals with negligible distortion so that, in the noiseless case being considered, the square-law device inputs are exactly these signals.

The input to the square-law devices of Figure 9-9 is a narrowband signal centered at $\omega = \omega_{IF}$. The output of the squaring circuit has a component at baseband and a component centered at $\omega = 2\omega_{IF}$. The lowpass filters reject all components near $2\omega_{IF}$. Therefore, the signal component of the delay-lock discriminator output is

$$\epsilon(t, \delta) = [x_2^2(t) - x_1^2(t)]_{\text{lowpass}}$$

$$= \frac{1}{2} K_1 P \left\{ R_c^2 \left[\left(\delta - \frac{\Delta}{2} \right) T_c \right] - R_c^2 \left[\left(\delta + \frac{\Delta}{2} \right) T_c \right] \right\}$$

$$\triangleq \frac{1}{2} K_1 P D_\Delta(\delta) \tag{9-51}$$

where

$$D_\Delta(\delta) \triangleq R_c^2 \left[\left(\delta - \frac{\Delta}{2} \right) T_c \right] - R_c^2 \left[\left(\delta + \frac{\Delta}{2} \right) T_c \right] \tag{9-52}$$

For maximal-length sequence spreading waveforms, the autocorrelation function is given by (8-45) and, after some straightforward algebraic simplification,

$$D_\Delta(\delta) = \begin{cases} & \text{for } -N + 1 + \frac{\Delta}{2} < \delta \le -\left(1 + \frac{\Delta}{2}\right) \\[8pt] \frac{1}{N^2} - \left[1 + \left(1 + \frac{1}{N}\right)\left(\delta + \frac{\Delta}{2}\right) \right]^2 & \text{for } -\left(1 + \frac{\Delta}{2}\right) < \delta \le -\frac{\Delta}{2} \\[8pt] \frac{1}{N^2} - \left[1 - \left(1 + \frac{1}{N}\right)\left(\delta + \frac{\Delta}{2}\right) \right]^2 & \text{for } -\frac{\Delta}{2} < \delta \le -\left(1 - \frac{\Delta}{2}\right) \\[8pt] 2\left(1 + \frac{1}{N}\right)\left[2 - \left(1 + \frac{1}{N}\right)\Delta \right]\delta & \text{for } -\left(1 - \frac{\Delta}{2}\right) < \delta \le \left(1 - \frac{\Delta}{2}\right) \\[8pt] \left[1 + \left(1 + \frac{1}{N}\right)\left(\delta - \frac{\Delta}{2}\right) \right]^2 - \frac{1}{N^2} & \text{for } \left(1 - \frac{\Delta}{2}\right) < \delta \le \frac{\Delta}{2} \\[8pt] \left[1 - \left(1 + \frac{1}{N}\right)\left(\delta - \frac{\Delta}{2}\right) \right]^2 - \frac{1}{N^2} & \text{for } \frac{\Delta}{2} < \delta \le 1 + \frac{\Delta}{2} \end{cases} \tag{9-53a}$$

for $\Delta \ge 1.0$ and

$$D_\Delta(\delta) = \begin{cases} 0 & \text{for } -N + 1 + \frac{\Delta}{2} < \delta < -\left(1 + \frac{\Delta}{2}\right) \\[8pt] \frac{1}{N^2} - \left[1 + \left(\delta + \frac{\Delta}{2}\right)\left(1 + \frac{1}{N}\right) \right]^2 & \text{for } -\left(1 + \frac{\Delta}{2}\right) < \delta < \left(\frac{\Delta}{2} - 1\right) \\[8pt] -2\left(1 + \frac{1}{N}\right)\Delta\left[1 + \left(1 + \frac{1}{N}\right)\delta \right] & \text{for } \left(\frac{\Delta}{2} - 1\right) < \delta < -\frac{\Delta}{2} \\[8pt] 2\left(1 + \frac{1}{N}\right)\left[2 - \left(1 + \frac{1}{N}\right)\Delta \right]\delta & \text{for } -\frac{\Delta}{2} < \delta < +\frac{\Delta}{2} \\[8pt] 2\left(1 + \frac{1}{N}\right)\Delta\left[1 - \left(1 + \frac{1}{N}\right)\delta \right] & \text{for } \frac{\Delta}{2} < \delta < \left(1 - \frac{\Delta}{2}\right) \\[8pt] 1 - \left(1 + \frac{1}{N}\right)\left(\delta - \frac{\Delta}{2}\right) \right]^2 - \frac{1}{N^2} & \text{for } \left(1 - \frac{\Delta}{2}\right) < \delta < \left(1 + \frac{\Delta}{2}\right) \end{cases} \tag{9-53b}$$

for $\Delta \le 1.0$. This function is periodic with period N. Equation (9-53) defines a single cycle of $D_\Delta(\delta)$. Observe that in the area near $\delta = 0$, $D_\Delta(\delta)$ is a linear function of δ despite the fact that nonlinear operations have been used in the discriminator. The slope of $D_\Delta(\delta)$ near $\delta = 0$ is a function of Δ and equals zero when $\Delta = 2.0$. Because of this, the noncoherent delay-lock discriminator is never used with an early-to-late delay difference of two code chip times. Figure 9-10 illustrates $D_\Delta(\delta)$ for the same four values of Δ which were used in Figure 9-5 for the baseband tracking loop. This completes the analysis of the signal component of the discriminator output $\epsilon(t, \delta)$.

Consider next the noise component of $\epsilon(t, \delta)$. The noise process at the correlator outputs have already been defined in (9-45) and (9-46). These noise processes, $n_{1,in}(t)$ and $n_{2,in}(t)$, are the products of a bandlimited white Gaussian noise process $n'(t)$ centered at $\omega = \omega_{IF}$ and the spreading waveforms $\sqrt{K_1/2}\ c(t - \hat{T}_d \pm (\Delta/2)\,T_c)$. Since the noise and the spreading functions are independent, the power spectrum of $n_{1,in}(t)$ or $n_{2,in}(t)$ can be calculated by convolving the spectrum of $n'(t)$ and the spectrum of the spreading waveform $\sqrt{K_1/2}\ c(t)$. That is

$$S_{n_{j,in}}(\omega) = \frac{K_1}{4\pi}\,S_{n'}(\omega) * S_c(\omega) \qquad j = 1, 2 \qquad (9\text{-}54)$$

The effect of this convolution is to spread the noise power over a bandwidth wider than its original bandwidth and thereby reduce the power spectral density in the frequency range passed by the IF filter. The IF filter is narrowband relative to the result of (9-54), so that the convolution needs to be evaluated only for frequencies near ω_{IF}. In the case of interest, the power spectrum of $n'(t)$ has a much wider bandwidth than the power spectrum of $c(t)$, so that the convolution has little effect

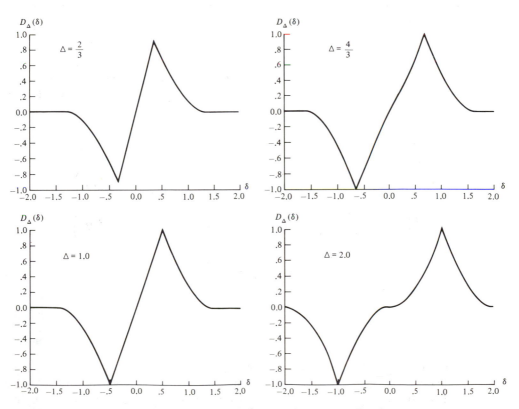

FIGURE 9-10. Full-time early-late noncoherent delay-lock discriminator S-curves for various Δ.

and $S_{n_{j,\text{in}}}(\omega_{\text{IF}})$ may be approximated by

$$S_{n_{j,\text{in}}}(\omega_{\text{IF}}) = \begin{cases} \dfrac{K_1}{2}\left(\dfrac{N_0}{2}\right) & 0 \le |\omega \pm \omega_{\text{IF}}| < \pi B \\ 0 & \text{elsewhere} \end{cases} \tag{9-55}$$

where B is the one-sided bandwidth of the received noise process. The $K_1/2$ factor accounts for the conversion gain of the mixer and the fact that noise power has been divided equally between the early and late correlator channels.

The noise processes $n_{1,\text{in}}(t)$ and $n_{2,\text{in}}(t)$ are Gaussian since $n'(t)$ is zero-mean Gaussian and the multiplying function takes on values of ± 1 with equal probability. The two processes are not, however, independent since one can always be derived from the other through multiplication by $+1$ or -1. Consider the noise output of the IF bandpass filters, which will be denoted by $n_1(t)$ and $n_2(t)$. The bandpass filters will be characterized by their impulse response $h(t)$, so that

$$n_j(t) = \int_{-\infty}^{\infty} n_{j,\text{in}}(\lambda)h(t - \lambda)\,d\lambda \qquad j = 1, 2 \tag{9-56}$$

Therefore,

$$n_j(t) = \int_{-\infty}^{\infty} \sqrt{\frac{K_1}{2}}\, c\left(\lambda - \hat{T}_d \pm \frac{\Delta}{2}T_c\right)n'(\lambda)h(t - \lambda)\,d\lambda \tag{9-57}$$

The processes $n_1(t)$ and $n_2(t)$ are Gaussian since the filter inputs are Gaussian. If they are also uncorrelated, they are independent. The cross-correlation between $n_1(t)$ and $n_2(t)$ is

$$R_{n_1 n_2}(\tau) \triangleq E[n_1(t)n_2(t + \tau)] \tag{9-58}$$

This function is difficult to calculate for the general case where the spreading waveform is assumed to be a deterministic function. The calculation is considerably simplified if $c(t)$ is assumed to be random. In that case

$$\begin{aligned}
R_{n_1 n_2}(\tau) &= E\left[\frac{K_1}{2}\int_{-\infty}^{\infty}\int_{-\infty}^{\infty} c\left(\gamma - \hat{T}_d + \frac{\Delta}{2}T_c\right)n'(\gamma)h(t - \gamma)\right. \\
&\qquad\qquad\qquad \left. \times\, c\left(\lambda - \hat{T}_d - \frac{\Delta}{2}T_c\right)n'(\lambda)h(t + \tau - \lambda)\,d\gamma\,d\lambda\right] \\
&= \frac{K_1}{2}\int_{-\infty}^{\infty}\int_{-\infty}^{\infty} h(t - \gamma)h(t + \tau - \lambda)E\left[c\left(\gamma - \hat{T}_d + \frac{\Delta}{2}T_c\right)\right. \\
&\qquad\qquad\qquad \left. \times\, c\left(\lambda - \hat{T}_d - \frac{\Delta}{2}T_c\right)\right]E[n'(\gamma)n'(\lambda)]\,d\gamma\,d\lambda \\
&= \frac{K_1}{2}\int_{-\infty}^{\infty}\int_{-\infty}^{\infty} h(t - \gamma)h(t + \tau - \lambda)R_c(\gamma - \lambda + \Delta T_c)R_{n'}(\gamma - \lambda)\,d\gamma\,d\lambda
\end{aligned} \tag{9-59}$$

The evaluation of this autocorrelation function is further simplified by assuming that the bandwidth of $n'(t)$ is significantly wider than the bandwidth of $c(t)$. With this assumption $R_c(\gamma - \lambda + \Delta T_c)$ is nearly constant over the range of significant values of $R_{n'}(\gamma - \lambda)$ and the desired cross-correlation can be approximated by

$$R_{n_1 n_2}(\tau) = \frac{K_1}{2} R_c(\Delta T_c) \int_{-\infty}^{\infty} \int_{-\infty}^{\infty} h(t - \gamma)h(t + \tau - \lambda)R_{n'}(\gamma - \lambda)\, d\gamma\, d\lambda$$

$$= \frac{K_1}{2} R_c(\Delta T_c) \int_{-\infty}^{\infty} \int_{-\infty}^{\infty} h(\gamma')h(\lambda')R_{n'}(\lambda' - \gamma' - \tau)\, d\gamma'\, d\lambda'$$

$$= \frac{K_1}{2} R_c(\Delta T_c) \int_{-\infty}^{\infty} \int_{-\infty}^{\infty} \int_{-\infty}^{\infty} S_{n'}(f)e^{j2\pi f(\lambda' - \gamma' - \tau)}h(\gamma')h(\lambda')\, df\, d\lambda'\, d\gamma'$$

$$= \frac{K_1}{2} R_c(\Delta T_c) \int_{-\infty}^{\infty} S_{n'}(f)e^{-j2\pi f\tau} \int_{-\infty}^{\infty} h(\gamma')e^{-j2\pi f\gamma'} \int_{-\infty}^{\infty} h(\lambda')e^{+j2\pi f\lambda'}\, d\lambda'\, d\gamma'\, df$$

$$= \frac{K_1}{2} R_c(\Delta T_c) \int_{-\infty}^{\infty} |H(j2\pi f)|^2 S_{n'}(f)e^{-j2\pi f\tau}\, df \tag{9-60}$$

Recall that the spreading code was assumed to be random for this calculation, so that $R_c(\Delta T_c)$ is defined by (F-6). In particular, $R_c(0) = 1.0$ and $R_c(T_c) = 0.0$. Thus the cross-correlation function of $n_1(t)$ and $n_2(t)$ is equal to zero when $\Delta = 1.0$.

The point of the preceding analysis is that $R_{n_1 n_2}(\tau)$ can be made small with the proper choice of the delay difference between the early and late correlator channels. A similar result is conjectured for maximal-length spreading codes, with the slight difference that the minimum cross-correlation would not be zero but would instead be related to $1/N$. The IF filter output-noise processes may or may not be strictly independent. In either case, they are narrowband white Gaussian noise processes and can be represented by

$$n_j(t) = \sqrt{2}n_{jI}(t) \cos \omega_{IF}t - \sqrt{2}n_{jQ}(t) \sin \omega_{IF}t \tag{9-61}$$

with $j = 1, 2$. The power transfer function of either of the IF filters is $|H(f)|^2$, where $H(f)$ is the Fourier transform of $h(t)$. Therefore, the power spectrum of $n_j(t)$ is

$$S_{nj}(f) = S_{nj,\text{in}}(f)|H(f)|^2 \tag{9-62}$$

and $S_{nj,\text{in}}(f)$ is defined by (9-54) or (9-55). Suppose, for example, that $H(f)$ is an ideal bandpass filter with bandwidth B_N. Then $S_{nj}(f)$ at the output of the IF filter has magnitude $\frac{1}{4}K_1 N_0$ and bandwidth B_N centered on $\pm f_{IF}$, and the power spectra of $n_{1I}(t)$, $n_{1Q}(t)$, $n_{2I}(t)$, or $n_{2Q}(t)$ have the same magnitude and bandwidth as $S_{nj}(f)$ but centered on zero frequency.

The output of the delay-lock discriminator $\epsilon(t, \delta)$ is given by

$$\epsilon(t, \delta) = [x_2(t) + n_2(t)]_{\text{LP}}^2 - [x_1(t) + n_1(t)]_{\text{LP}}^2 \tag{9-63}$$

where $x_1(t)$ and $x_2(t)$ are defined in (9-50), and the subscript LP denotes the lowpass components of the squared function. Expanding this equation and eliminating all double-frequency terms yields

$$\epsilon(t, \delta) = \frac{1}{2} K_1 P \left\{ R_c^2 \left[\left(\delta - \frac{\Delta}{2} \right) T_c \right] - R_c^2 \left[\left(\delta + \frac{\Delta}{2} \right) T_c \right] \right\}$$

$$+ \sqrt{2K_1 P} \left\{ R_c \left[\left(\delta - \frac{\Delta}{2} \right) T_c \right] n_{2I}(t) - R_c \left[\left(\delta + \frac{\Delta}{2} \right) T_c \right] n_{1I}(t) \right\}$$

$$\times \cos [\varphi - \varphi' + \theta_d(t - T_d)]$$

$$+ \sqrt{2K_1 P} \left\{ R_c \left[\left(\delta - \frac{\Delta}{2} \right) T_c \right] n_{2Q}(t) - R_c \left[\left(\delta + \frac{\Delta}{2} \right) T_c \right] n_{1Q}(t) \right\}$$

$$\times \sin \left[\varphi - \varphi' + \theta_d(t - T_d) \right]$$

$$+ [n_{2I}(t)]^2 + [n_{2Q}(t)]^2 - [n_{1I}(t)]^2 - [n_{1Q}(t)]^2 \qquad (9\text{-}64)$$

The first term of this equation is the desired signal defined in (9-51). The second and third terms are (signal $\times$ noise) terms, and the last four terms are (noise)^2 terms. The power spectrum of $\epsilon(t, \delta)$ is denoted $S_\epsilon(f)$ and is the Fourier transform of

$$R_\epsilon(\tau) = E[\epsilon(t, \delta)\epsilon(t + \tau, \delta)] \qquad (9\text{-}65)$$

where the expectation is over all sample functions of the Gaussian noise random process and over all received carrier phase angles φ and reference phase angles φ'. Equation (9-65) is evaluated in detail in Appendix H; the result is

$$R_\epsilon(\tau) = \frac{1}{4} K_1^2 P^2 \left\{ R_c^2 \left[\left(\delta - \frac{\Delta}{2} \right) T_c \right] - R_c^2 \left[\left(\delta + \frac{\Delta}{2} \right) T_c \right] \right\}^2$$

$$+ 2K_1 P \left\{ R_c^2 \left[\left(\delta - \frac{\Delta}{2} \right) T_c \right] + R_c^2 \left[\left(\delta + \frac{\Delta}{2} \right) T_c \right] \right\} E[n_b(t)n_b(t + \tau)]$$

$$\times E\{\cos [\theta_d(t) - \theta_d(t + \tau)]\}$$

$$+ 4E\{[n_b(t)n_b(t + \tau)]^2\} - 4E^2\{[n_b(t)]^2\} \qquad (9\text{-}66)$$

In this expression, $n_b(t)$ has been used to represent any one of the four baseband processes n_{1I}, $n_{1Q}(t)$, $n_{2I}(t)$, or $n_{2Q}(t)$. Since all four have identical statistics, the statistics of any one can be used for calculations of the autocorrelation, power, and so on. In deriving (9-66) it has been assumed that $n_{1I}(t)$, $n_{1Q}(t)$, $n_{2I}(t)$, and $n_{2Q}(t)$ are all independent. This assumption results in accurate results when the input noise bandwidth is large relative to the signal bandwidth and simultaneously $\Delta \geq 1.0$.

The purpose of calculating $R_\epsilon(\tau)$ is to enable the calculation of the delay-lock discriminator output power spectrum with signal and noise at its input. The desired power spectrum is the Fourier transform of $R_\epsilon(\tau)$. This Fourier transform is also calculated in Appendix H. The result is

$$S_\epsilon(f) = \frac{1}{4} K_1^2 P^2 \left\{ R_C^2 \left[\left(\delta - \frac{\Delta}{2} \right) T_c \right] - R_C^2 \left[\left(\delta + \frac{\Delta}{2} \right) T_c \right] \right\}^2 \delta(f)$$

$$+ 2K_1 P \left\{ R_C^2 \left[\left(\delta - \frac{\Delta}{2} \right) T_c \right] + R_C^2 \left[\left(\delta + \frac{\Delta}{2} \right) T_c \right] \right\} S_{n_b}(f) * S_{\theta_d}(f)$$

$$+ 8 S_{n_b}(f) * S_{n_b}(f) \qquad (9\text{-}67)$$

where $S_{\theta_d}(f)$ is the Fourier transform of twice the real part of the complex autocorrelation of the complex envelope of a carrier having random phase β and phase modulation $\theta_d(t)$. That is,

$$S_{\theta_d}(f) = \int_{-\infty}^{\infty} 2 \operatorname{Re}[R_A(\tau)] e^{-j2\pi f\tau} \, d\tau \qquad (9\text{-}68)$$

where

$$R_A(\tau) = \tfrac{1}{2} E[A^*(t)A(t + \tau)] \qquad (9\text{-}69)$$

$$A(t) = \exp[j\theta_d(t) + j\beta] \qquad (9\text{-}70)$$

Equation (9-67) can be evaluated for any data modulation to obtain the desired signal and noise power spectra at the phase discriminator output. The result has been derived assuming BPSK direct-sequence spreading modulation and is therefore not valid for arbitrary spreading modulation.

To derive a simple model for the tracking loop, consider the special case where the received carrier is not modulated with data. This special case, although somewhat unrealistic for a communication system, will result in the maximum possible noise component in the power spectral density, $S_\epsilon(f)$, within the tracking loop bandwidth. To verify this, observe that with no data modulation $S_{\theta_d}(f)$ is a delta function at $f = 0$ and the convolution does no spectral widening of the (signal $\times$ noise) term. Specifically, with no data modulation

$$A(t) = \exp(j\beta) \tag{9-71}$$

$$R_A(\tau) = \tfrac{1}{2}E[A^*(t)A(t + \tau)] = \tfrac{1}{2} \tag{9-72}$$

From (9-68), then

$$S_{\theta_d}(f) = 2\int_{-\infty}^{\infty} \tfrac{1}{2}e^{-j2\pi f\tau}\,d\tau = \delta(f) \tag{9-73}$$

Assume also that the IF filters are perfect brick wall filters. Then $S_{n_b}(f)$ has magnitude $\tfrac{1}{4}K_1N_0$, two-sided bandwidth B_N, and the convolution of (9-67) can be performed by inspection to obtain the final result for $S_\epsilon(f)$ illustrated in Figure 9-11.

The calculation of the discriminator output power spectrum gives no information about the probability density function of this output. Although the input noise is Gaussian, the nonlinear devices used in the detector assure that the output statistics are not Gaussian. Fortunately, the loop filter following the discriminator has a bandwidth which is much smaller than the bandwidth of the discriminator output random process, so that the central limit theorem (see, e.g., [10]) may be applied to show that the filter output is nearly Gaussian. Therefore, little error is made by assuming that the input is Gaussian. Also because the loop filter bandwidth is small, the discriminator output power spectral density may be assumed to be white with a value equal to the zero-frequency noise component of Figure 9-11.

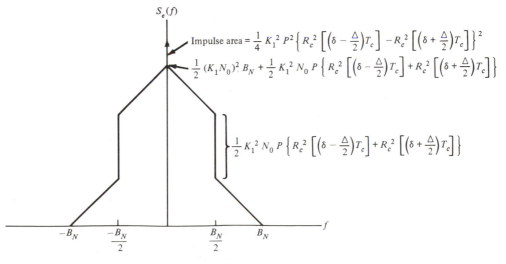

FIGURE 9-11. Complete full-time early-late noncoherent delay discriminator output power spectrum.

The remainder of the analysis of the noncoherent tracking loop is identical to the analysis of the baseband tracking loop performed earlier. Specifically, the VCO is characterized in the time domain by (9-18) and in the frequency domain by (9-19). The loop filter is again assumed to be a lumped linear time-invariant passive filter which is characterized by a differential equation having the form of (9-20). This filter can also be described by its Laplace transform transfer function $F(s)$ as in (9-21), or by its impulse response $f(t)$ in the time domain. Using (9-18), with $v(t)$ defined in Figure 9-9, the VCO can be described by

$$\frac{\hat{T}_d(t)}{T_c} = g_c \int_0^t v(\lambda)\, d\lambda \tag{9-74}$$

where the units of g_c are Hz/V. The VCO input is the convolution of the loop filter input and the filter impulse response, so that

$$\frac{\hat{T}_d(t)}{T_c} = g_c \int_0^t \int_{-\infty}^{\lambda} \epsilon(\lambda,\, \delta) f(\lambda - \alpha)\, d\alpha\, d\lambda \tag{9-75}$$

In this expression, $\epsilon(t,\, \delta)$ is the complete signal plus noise discriminator output waveform. The noise component of this signal is assumed to be a lowpass white Gaussian noise process having a two-sided power spectral density $\eta/2$ given by the last two terms of (9-67) evaluated at $f = 0$. In the special (worst) case where there is no data modulation, the noise power spectrum is illustrated in Figure 9-11 and

$$\frac{\eta}{2} = \frac{1}{2}(K_1 N_0)^2 B_N + \frac{1}{2} K_1^2 N_0 P \left\{ R_c^2 \left[\left(\delta - \frac{\Delta}{2} \right) T_c \right] + R_c^2 \left[\left(\delta + \frac{\Delta}{2} \right) T_c \right] \right\} \tag{9-76}$$

The noise component of $\epsilon(t,\, \delta)$ will be denoted by $n_\epsilon(t)$. The signal component of $\epsilon(t,\, \delta)$ was given by (9-51). Combining these results yields

$$\epsilon(t,\, \delta) = \tfrac{1}{2} K_1 P D_\Delta(\delta) + n_\epsilon(t) \tag{9-77}$$

and (9-75) becomes

$$\frac{\hat{T}_d(t)}{T_c} = g_c \int_0^t \int_{-\infty}^{\lambda} [\tfrac{1}{2} K_1 P D_\Delta(\delta) + n_\epsilon(\alpha)] f(\lambda - \alpha)\, d\alpha\, d\lambda \tag{9-78}$$

where $\delta = [T_d(\alpha) - \hat{T}_d(\alpha)]/T_c$. The model of Figure 9-12 is also described by (9-78), so that it is a correct model for the full-time early-late noncoherent code tracking loop.

For small δ the discriminator S-curve $D_\Delta(\delta)$ can be approximated by a linear function of δ. From (9-53), and for small δ,

$$D_\Delta(\delta) = 4 \left(1 + \frac{1}{N} \right) \left[1 - \left(1 + \frac{1}{N} \right) \frac{\Delta}{2} \right] \delta \tag{9-79}$$

This linear model for the phase discriminator permits moving the noise process $n_\epsilon(t)$ to the loop input with appropriate gain adjustment. Letting

$$K_d = 4 \left(1 + \frac{1}{N} \right) \left[1 - \left(1 + \frac{1}{N} \right) \frac{\Delta}{2} \right] \left(\frac{1}{2} K_1 P \right) \tag{9-80}$$

the linearized equivalent circuit for the tracking loop can be represented as illustrated in Figure 9-13. The Laplace transform model for this tracking loop is iden-

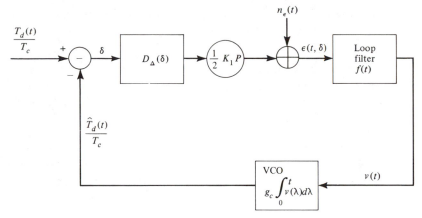

Notes: 1) $n_\epsilon(t)$ is lowpass white Gaussian noise process
with two-sided psd $\eta/2$ given by Equation (9-76).
2) $D_\Delta(\delta)$ is given by Equation (9-52).

FIGURE 9-12. Nonlinear equivalent circuit for the full-time early-late noncoherent tracking loop.

tical to the model shown in Figure 9-8 for the baseband tracking loop with appropriate changes in the definition of K_d.

The derivation of the linear model for this tracking has been long and many assumptions have been made. The two fundamental calculations that led to the linear model were the calculation of the noiseless discriminator output characteristic

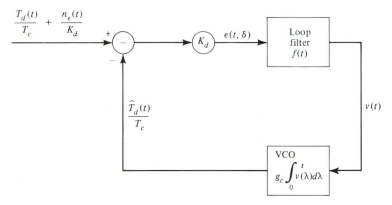

g_c = VCO gain in Hz / volt

$K_d = \frac{1}{2} K_1 P \left[\frac{d}{d\delta} D_\Delta(\delta) \right]_{\delta = 0}$

$K_d = \left(1 + \frac{1}{N} \right) \left[1 - \left(1 + \frac{1}{N} \right) \frac{\Delta}{2} \right] (2 K_1 P)$ for BPSK only

K_1 = mixer conversion gain/loss

Δ = normalized difference between early and late correlator channels

N = m−sequence spreading code period

P = received signal power

T_c = spreading waveform chip duration

$n_\epsilon(t)$ = thermal noise component at discriminator output (see Note on Figure 9.12)

FIGURE 9-13. Linear equivalent circuit for the full-time early-late noncoherent tracking loop.

and the modeling of the discriminator output noise process. The significant assumptions made in this development are:

1. The code self-noise and clock components of (8-59) have been ignored. As a preliminary step in including the effects of these terms, their spectra can be modeled as continuous spectra and added to the thermal noise spectrum.
2. The thermal noise processes at the two IF bandpass filter outputs have been assumed to be independent. This assumption is quite accurate for $\Delta \geq 1.0$. For $\Delta < 1.0$, the assumption is not accurate, however, it is conjectured that the correlated components will cancel one another in the difference circuit at the discriminator output.
3. The discriminator output noise has been assumed to be Gaussian. Because of the small bandwidth of the loop filter relative to the discriminator output bandwidth, this assumption is quite good.
4. The effect of the data modulation in the (signal $\times$ noise) term at the discriminator output has been ignored. The effect of data modulation can be included by using (9-67) to calculate the power spectrum of $n_\epsilon(t)$. Ignoring data modulation results in pessimistic performance estimates.

These assumptions are consistent with those made in the existing literature on this subject.

Using the linear model, the closed-loop transfer function $H(s)$ is given by (9-26), which is repeated here for convenience:

$$H(s) = \frac{K_d g_c F(s)}{s + K_d g_c F(s)} = \frac{\hat{T}_d(s)}{T_d(s)} \tag{9-26}$$

and the Laplace transform of the tracking error is given by (9-27):

$$\frac{T_d(s) - \hat{T}_d(s)}{T_c} = \frac{T_d(s)}{T_c}\left[\frac{s}{s + K_d g_c F(s)}\right] \tag{9-27}$$

The two-side loop noise bandwidth W_L is defined by (9-32) and the rms tracking jitter is given by

$$\sigma_\delta^2 = \int_{-\infty}^{\infty} S_{n''}(f)|H(j2\pi f)|^2 \, df \tag{9-30}$$

where $S_{n''}(f)$ is the noise power spectrum at the input to the model of Figure 9-13. This power spectrum is approximately flat over the loop bandwidth. Its magnitude is given by $\eta/2K_d^2$, where $\eta/2$ is defined in (9-76). Thus

$$\sigma_\delta^2 = \frac{\eta}{2K_d^2} W_L \tag{9-81}$$

EXAMPLE 9-2

Find an expression for the tracking jitter for the noncoherent tracking loop for the special case where $\Delta = 1.0$, and $N \gg 1$.

Solution: When $\Delta = 1.0$, (9-76) simplifies to

$$\frac{\eta}{2} = \frac{1}{2}(K_1 N_0)^2 B_N + \frac{1}{2}K_1^2 N_0 P\left(\frac{1}{4} + \frac{1}{4}\right) = \frac{1}{2}K_1^2 N_0\left(N_0 B_N + \frac{P}{2}\right) \tag{9-82}$$

and for $N >> 1$ and $\Delta = 1.0$,

$$K_d \approx K_1 P \qquad (9\text{-}83)$$

Therefore,

$$\sigma_\delta^2 = \frac{\frac{1}{2}K_1^2 N_0 \left(N_0 B_N + \frac{P}{2} \right)}{(K_1 P)^2} W_L = \frac{1}{2\rho_L} \left(1 + \frac{2}{\rho_{IF}} \right) \qquad (9\text{-}84)$$

where

$$\rho_L = \frac{2P}{N_0 W_L} \qquad (9\text{-}85)$$

$$\rho_{IF} = \frac{P}{N_0 B_N} \qquad (9\text{-}86)$$

The variables ρ_L and ρ_{IF} are the signal-to-noise ratios in the loop bandwidth and in the IF filter bandwidth, respectively. Observe that (9-84) is similar to (9-34). The difference in the two expressions is the second term within the parentheses of (9-84). This term is the result of the squaring operation used in the noncoherent loop. ■

This completes the analysis of the full-time early-late noncoherent code tracking loop. This tracking loop is often used in modern spread-spectrum systems.

9-5

TAU-DITHER EARLY-LATE NONCOHERENT TRACKING LOOP

The code tracking loop discussed in Section 9-4 performs the task of code tracking efficiently and is widely used. It has two problems, however, which led spread-spectrum researchers to invent the tau-dither early-late code tracking loop discussed in this section. The first problem with the full-time loop is that the early and late IF channels must be precisely amplitude balanced. When they are not properly balanced, the discriminator characteristic is offset and does not produce zero volts out when the tracking error is zero. The other problem is that the full-time loop uses costly components somewhat freely. Both of these problems are solved in the tau-dither early-late tracking loop by time sharing a single correlation channel for both early and late channel use. The price paid to solve these problems is slightly worse noise performance and considerably more difficult analysis. The application of time-shared tracking discriminators to noncoherent spread-spectrum code tracking was first proposed by Hartmann [5], whose analysis is followed closely in this section. The tau-dither tracking loop is also analyzed by Simon [6].

A functional block diagram of the tau-dither tracking loop is illustrated in Figure 9-14. Except for the discriminator, the loop is the same as the full-time tracking loop of Figure 9-9. The discriminator has a single channel which is switched between use as an early correlator and a late correlator by a switching signal $q(t)$. The signal $q(t)$ is a square wave of frequency f_q which takes on values of ± 1. When $q(t) = -1$, the switch at the spreading waveform generator is in the position shown and the correlator functions as a late correlator. When $q(t) = +1$, the

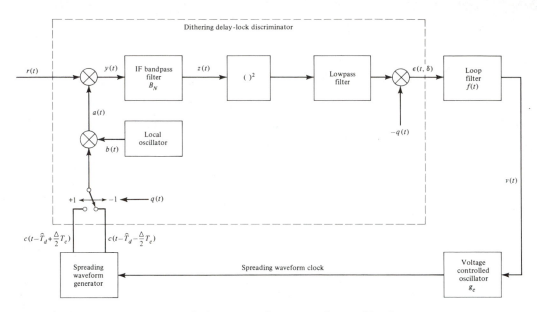

FIGURE 9-14.   Tau-dither early-late noncoherent code tracking loop.

correlator functions as an early correlator. The signal $q(t)$ is also used to multiply the squaring circuit output. This multiplication provides the sign inversion necessary to generate the discriminator S-curve from the early and late autocorrelation functions. The analysis of the dithering loop is similar to the tracking loop analyses performed in Sections 9-3 and 9-4. The discriminator output signal component will be modeled as a linear function of the normalized tracking error δ, and the discriminator output noise power spectral density will be calculated. These calculations are facilitated by developing a discriminator model which is very similar to the full-time loop model, so that some of the previous analysis can be used directly.

The input to the tau-dither loop $r(t)$ is a data- and direct-sequence-modulated carrier plus bandlimited AWGN with two-sided power spectral density $N_0/2$. That is,

$$r(t) = \sqrt{2P}\, c(t - T_d) \cos\left[\omega_0 t + \theta_d(t - T_d) + \varphi\right] + n(t) \qquad (9\text{-}87)$$

where

$$n(t) = \sqrt{2}\, n_I(t) \cos \omega_0 t - \sqrt{2}\, n_Q(t) \sin \omega_0 t \qquad (9\text{-}88)$$

and the baseband in-phase and quadrature noise functions are independent lowpass white Gaussian noise processes with two-sided power spectral densities $N_0/2$. The random received carrier phase is denoted by φ. Consideration has been limited to BPSK direct-sequence spreading modulation, but the data modulation $\theta_d(t)$ is an arbitrary phase modulation. The reference local oscillator output is

$$b(t) = 2\sqrt{K_1} \cos\left[(\omega_0 - \omega_{IF})t + \varphi'\right] \qquad (9\text{-}89)$$

The signal $a(t)$ depends on $q(t)$ and is

$$a(t) = 2\sqrt{K_1}\, c\left(t - \hat{T}_d + q(t)\frac{\Delta}{2} T_c\right) \cos\left[(\omega_0 - \omega_{IF})t + \varphi'\right] \qquad (9\text{-}90)$$

Assume now that the frequency of $q(t)$ is sufficiently low relative to the IF filter bandwidth that filter transients can be ignored. In this case, an equivalent discriminator is a two-channel discriminator as shown in Figure 9-15, whose output is switched between the early and late channels by functions $q_1(t)$ and $q_2(t)$. The functions $q_1(t)$ and $q_2(t)$ are defined by

$$q_1(t) = \tfrac{1}{2}[1 + q(t)] \tag{9-91a}$$

$$q_2(t) = \tfrac{1}{2}[1 - q(t)] \tag{9-91b}$$

and are plotted in Figure 9-16. In Figure 9-15 the functions $a_1(t)$ and $a_2(t)$ are

$$a_1(t) = 2\sqrt{K_1}\, c\!\left(t - \hat{T}_d + \frac{\Delta}{2}\, T_c\right) \cos\,[(\omega_0 - \omega_{IF})t + \varphi'] \tag{9-92a}$$

$$a_2(t) = 2\sqrt{K_1}\, c\!\left(t - \hat{T}_d - \frac{\Delta}{2}\, T_c\right) \cos\,[(\omega_0 - \omega_{IF})t + \varphi'] \tag{9-92b}$$

Observe that the power dividers that were shown in the full-time discriminator of Figure 9-9 are not shown in Figure 9-15 since this figure is used only as an analytical model. In this model, the received signal is always correlated with the early code in the upper channel, and always correlated with the late code in the lower channel. Because $q_1(t)$ and $q_2(t)$ take on values of $+1$ or 0, the discriminator output is alternately switched between the early and late channels. Thus, except for the transient which occurs in the IF filter of Figure 9-14 when the early/late transition occurs, the discriminator output is identical in Figures 9-14 and 9-15. Except for the multipliers for $q_1(t)$ and $q_2(t)$ and a factor of 2 in input power, Figure 9-15 is analytically identical to the full-time loop of Figure 9-9.

Calculation of the discriminator output in the noiseless case follows the same steps described earlier for the full-time loop. In particular, code clock and self-noise components can be ignored under the same conditions as those described earlier. The mixer output signals are given by (9-50) except for a factor of 2 in

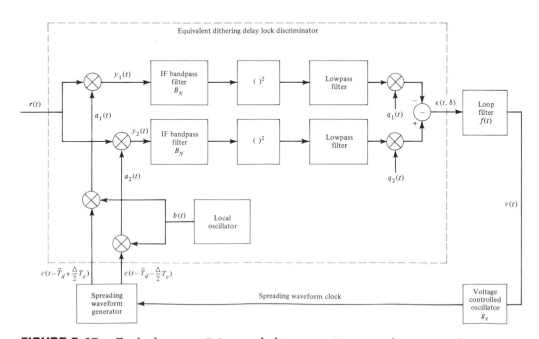

FIGURE 9-15. Equivalent tau-dither early-late noncoherent code tracking loop.

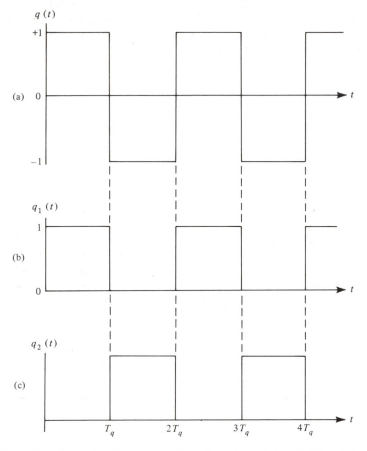

FIGURE 9-16. Dithering loop switching functions: (a) $q(t)$; (b) $q_1(t)$; and (c) $q_2(t)$.

power; thus

$$y_1(t) = \sqrt{2K_1P}\, R_c\left[\left(\delta + \frac{\Delta}{2}\right)T_c\right] \cos\left[\omega_{IF}t + \varphi - \varphi' + \theta_d(t - T_d)\right] \triangleq x_1(t)$$

$$(9\text{-}93a)$$

$$y_2(t) = \sqrt{2K_1P}\, R_c\left[\left(\delta - \frac{\Delta}{2}\right)T_c\right] \cos\left[\omega_{IF}t + \varphi - \varphi' + \theta_d(t - T_d)\right] \triangleq x_2(t)$$

$$(9\text{-}93b)$$

Again, the IF bandpass filters are assumed to pass the data-modulated IF carrier without distortion, so that the discriminator output signal component is

$$\epsilon(t, \delta) = [x_2^2(t)q_2(t) - x_1^2(t)q_1(t)]_{\text{lowpass}} \qquad (9\text{-}94)$$

Writing $q_1(t)$ and $q_2(t)$ as functions of $q(t)$ and simplifying (9-94) yields

$$\epsilon(t, \delta) = \{\tfrac{1}{2}[x_2^2(t) - x_1^2(t)] - \tfrac{1}{2}q(t)[x_2^2(t) + x_1^2(t)]\}_{\text{lowpass}} \qquad (9\text{-}95)$$

Substituting for $x_1(t)$ and $x_2(t)$ from (9-93) and simplifying by eliminating all double-frequency terms results in

$$\epsilon(t, \delta) = \frac{1}{2}K_1P\left\{R_c^2\left[\left(\delta - \frac{\Delta}{2}\right)T_c\right] - R_c^2\left[\left(\delta + \frac{\Delta}{2}\right)T_c\right]\right\}$$

$$- \frac{1}{2}q(t)K_1P\left\{R_c^2\left[\left(\delta + \frac{\Delta}{2}\right)T_c\right] + R_c^2\left[\left(\delta - \frac{\Delta}{2}\right)T_c\right]\right\} \quad (9\text{-}96)$$

The first term of this equation is identical to (9-51) and is the desired tracking error. The second term consists of harmonics of the dithering frequency. At this point the assumption is made that the dithering frequency is significantly higher than the bandwidth of the loop filter. When this is true, the second term of (9-96) is eliminated by this filter and

$$\epsilon(t, \delta) \simeq \frac{1}{2}K_1P\left\{R_c^2\left[\left(\delta - \frac{\Delta}{2}\right)T_c\right] - R_c^2\left[\left(\delta + \frac{\Delta}{2}\right)T_c\right]\right\}$$

$$\triangleq \frac{1}{2}K_1PD_\Delta(\delta) \quad (9\text{-}97)$$

Thus, in the noiseless case, the discriminator output is the same as the discriminator output of the full-time loop. A factor of 2 decrease in output does not occur since the input power divider used in the full-time loop is unnecessary in the dithering loop. The discriminator S-curve is defined by (9-53) and is illustrated in Figure 9-10.

The next step in the analysis is to calculate the power spectral density of the noise component at the discriminator output. The noise analysis up to the output of the IF bandpass filter is identical to the noise analysis for the full-time tracking loop. The noise at the output of these filters is specified in (9-61).

Analysis in Section 9-4 showed that the noise processes of (9-61) are not independent in general but are very nearly independent when $\Delta \geq 1.0$ and the system input noise bandwidth is wide relative to the received signal bandwidth. In this section it will be assumed that these processes are independent. To simplify the following analysis, the IF bandpass filter is assumed to be an ideal brick wall filter. Then, the noise power spectrum is given by (9-55) and (9-62):

$$S_{n_1}(f) = S_{n_2}(f) = \begin{cases} 2\frac{K_1}{2}\left(\frac{N_0}{2}\right) = \frac{K_1N_0}{2} & 0 \leq |f \pm f_{\text{IF}}| < \frac{B_N}{2} \\ 0 & \text{elsewhere} \end{cases} \quad (9\text{-}98)$$

where the density of (9-55) has been increased by a factor of 2 to account for the absence of a power divider in the tau-dither loop. In summary, the noise components at the IF bandpass filter outputs are independent, bandlimited, white Gaussian noise processes with two-sided power spectral densities given by (9-98).

With signal components defined in (9-93) and noise components defined in (9-61), the discriminator output signal is

$$\epsilon(t, \delta) = [x_2(t) + n_2(t)]_{\text{LP}}^2 q_2(t) - [x_1(t) + n_1(t)]_{\text{LP}}^2 q_1(t) \quad (9\text{-}99)$$

where the subscript LP denotes that the double-frequency terms will be ignored. Expanding (9-99) and then simplifying yields

$$\epsilon(t, \delta) = K_1 P \left\{ R_c^2 \left[\left(\delta - \frac{\Delta}{2} \right) T_c \right] q_2(t) - R_c^2 \left[\left(\delta + \frac{\Delta}{2} \right) T_c \right] q_1(t) \right\}$$

$$+ 2\sqrt{K_1 P} \cos \left[\varphi - \varphi' + \theta_d(t - T_d) \right] \left\{ n_{2I}(t) R_c \left[\left(\delta - \frac{\Delta}{2} \right) T_c \right] q_2(t) \right.$$

$$\left. - n_{1I}(t) R_c \left[\left(\delta + \frac{\Delta}{2} \right) T_c \right] q_1(t) \right\}$$

$$+ 2\sqrt{K_1 P} \sin \left[\varphi - \varphi' + \theta_d(t - T_d) \right] \left\{ n_{2Q}(t) R_c \left[\left(\delta - \frac{\Delta}{2} \right) T_c \right] q_2(t) \right.$$

$$\left. - n_{1Q}(t) R_c \left[\left(\delta + \frac{\Delta}{2} \right) T_c \right] q_1(t) \right\}$$

$$+ [n_{2I}(t)]^2 q_2(t) + [n_{2Q}(t)]^2 q_2(t) - [n_{1I}(t)]^2 q_1(t) - [n_{1Q}(t)]^2 q_1(t)$$

$$\tag{9-100}$$

To calculate the discriminator output power spectrum, the autocorrelation function

$$R_\epsilon(\tau) = E[\epsilon(t, \delta)\epsilon(t + \tau, \delta)] \tag{9-101}$$

must be evaluated. This is done in Appendix H, where it is shown that

$$R_\epsilon(\tau) = K_1^2 P^2 \left\{ R_c^2 \left[\left(\delta - \frac{\Delta}{2} \right) T_c \right] + R_c^2 \left[\left(\delta + \frac{\Delta}{2} \right) T_c \right] \right\}^2 R_{q1}(\tau)$$

$$- K_1^2 P^2 R_c^2 \left[\left(\delta - \frac{\Delta}{2} \right) T_c \right] R_c^2 \left[\left(\delta + \frac{\Delta}{2} \right) T_c \right]$$

$$+ 8K_1 P \sigma_n^2 \left\{ R_c^2 \left[\left(\delta - \frac{\Delta}{2} \right) T_c \right] + R_c^2 \left[\left(\delta + \frac{\Delta}{2} \right) T_c \right] \right\} R_{q1}(\tau)$$

$$- 2K_1 P \sigma_n^2 \left\{ R_c^2 \left[\left(\delta - \frac{\Delta}{2} \right) T_c \right] + R_c^2 \left[\left(\delta + \frac{\Delta}{2} \right) T_c \right] \right\}$$

$$+ 4K_1 P \left\{ R_c^2 \left[\left(\delta - \frac{\Delta}{2} \right) T_c \right] + R_c^2 \left[\left(\delta + \frac{\Delta}{2} \right) T_c \right] \right\} R_{n_b}(\tau) R_{q1}(\tau)$$

$$\times E\{\cos [\theta_d(t - T_d) - \theta_d(t + \tau - T_d)]\}$$

$$+ 4E\{[n_b(t)n_b(t + \tau)]^2\} R_{q1}(\tau)$$

$$+ 12\sigma_n^4 R_{q1}(\tau) - 4\sigma_n^4 \tag{9-102}$$

In this equation, $n_b(t)$ represents any one of the four signals $n_{1I}(t)$, $n_{1Q}(t)$, $n_{2I}(t)$, $n_{2Q}(t)$, and σ_n^2 is the mean square of $n_b(t)$.

The Fourier transform of $R_\epsilon(\tau)$ is the power spectrum needed to complete the analysis of the tau-dither tracking loop. The Fourier transform is also calculated in Appendix H; the result is†

---

†It has been assumed in deriving this result that only the first three harmonics of the dither frequency cause significant noise components within the tracking loop bandwidth.

$$S_\epsilon(f) = \frac{1}{4} K_1^2 P^2 \left\{ R_c^2 \left[\left(\delta - \frac{\Delta}{2} \right) T_c \right] - R_c^2 \left[\left(\delta + \frac{\Delta}{2} \right) T_c \right] \right\}^2 \delta(f)$$

$$+ \left[K_1 P \left\{ R_c^2 \left[\left(\delta - \frac{\Delta}{2} \right) T_c \right] + R_c^2 \left[\left(\delta + \frac{\Delta}{2} \right) T_c \right] \right\} + 4\sigma_n^2 \right]^2$$

$$\times \left[\frac{1}{\pi^2} \delta(f - f_q) + \frac{1}{\pi^2} \delta(f + f_q) + \frac{1}{9\pi^2} \delta(f + 3f_q) + \frac{1}{9\pi^2} \delta(f - 3f_q) \right]$$

$$+ \left[K_1 P \left\{ R_c^2 \left[\left(\delta - \frac{\Delta}{2} \right) T_c \right] + R_c^2 \left[\left(\delta + \frac{\Delta}{2} \right) T_c \right] \right\} \right]$$

$$\times \left[S_{n_b}(f) + \frac{4}{\pi^2} S_{n_b}(f - f_q) + \frac{4}{\pi^2} S_{n_b}(f + f_q) + \frac{4}{9\pi^2} S_{n_b}(f - 3f_q) \right.$$

$$\left. + \frac{4}{9\pi^2} S_{n_b}(f + 3f_q) \right]$$

$$+ 2S_T(f) + \frac{8}{\pi^2} S_T(f - f_q) + \frac{8}{\pi^2} S_T(f + f_q)$$

$$+ \frac{8}{9\pi^2} S_T(f - 3f_q) + \frac{8}{9\pi^2} S_T(f + 3f_q) \tag{9-103}$$

where $S_T(f) = S_{n_b}(f) * S_{n_b}(f)$. The first term of this equation is the square of (9-97). The square is expected since this is a power spectrum. This term is the desired error correction signal for code tracking. The remainder of the equation is undesired dither frequency clock components plus (signal × noise) and (noise × noise) components. The dither clock frequency must be selected to be large enough that the clock components of (9-103) are outside the tracking loop bandwidth. The power spectrum of $n_b(f)$ is

$$S_{n_b}(f) = \begin{cases} \dfrac{K_1 N_0}{2} & |f| < \frac{1}{2} B_N \\ 0 & \text{elsewhere} \end{cases} \tag{9-104}$$

The power spectra $S_{n_b}(f)$ and $S_T(f)$ are illustrated in Figure 9-17, and the complete power spectrum $S_\epsilon(f)$ is illustrated in Figure 9-18 for the particular case where $f_q = \frac{1}{4} B_N$.

Denote the total noise at the output of the tau-dither delay discriminator by $n_\epsilon(t)$. The signal component at the discriminator output was given in (9-97), and the total discriminator output is

$$\epsilon(t, \delta) = \frac{1}{2} K_1 P D_\Delta(\delta) + n_\epsilon(t) \tag{9-105}$$

Although $n_\epsilon(t)$ is considerably different, (9-74), (9-75), and (9-78), which were derived for the full-time tracking loop, also apply directly to the tau-dither loop. Therefore, the nonlinear model of Figure 9-12 is also an accurate model for the tau-dither loop with $D_\Delta(\delta)$ defined in (9-97). Since $D_\Delta(\delta)$ for the full-time and the tau-dither loop are identical, the linearization of the model for $\delta \ll 1$ is the same and Figure 9-13 is also a valid model for the tau-dither loop. It follows that (9-26) and (9-27) also apply to the tau-dither loop.

The code tracking jitter for the tau-dither loop is calculated using (9-30):

$$\sigma_\delta^2 = \int_{-\infty}^{\infty} S_{n''}(f) |H(j2\pi f)|^2 \, df \tag{9-30}$$

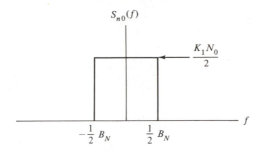

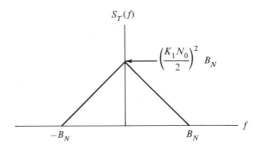

FIGURE 9-17. Power spectra $S_{no}(f)$ and $S_T(f) = S_{no}(f) * S_{no}(f)$.

where $n''(t)$ is the noise at the input to the linear model of Figure 9-13. That is, $n''(t) = n_\epsilon(t)/K_d$. In all cases of interest, the tracking loop noise bandwidth is very much smaller than the IF filter noise bandwidth or the dither frequency, so that $n''(t)$ may be assumed to be a lowpass white Gaussian noise process having a two-sided power spectral density $\eta/2$ given by the (signal × noise) and (noise × noise) components of (9-103) evaluated at $f = 0$ and divided by K_d^2. For $B_N/6 < f_q < B_N/3$ evaluation of these components of (9-103) yields

$$\frac{\eta}{2} = \frac{K_1^2 N_0 P}{2K_d^2} \left\{ R_c^2\left[\left(\delta - \frac{\Delta}{2}\right)T_c \right] + R_c^2\left[\left(\delta + \frac{\Delta}{2}\right)T_c \right] \right\}\left(1 + \frac{8}{\pi^2} \right)$$

$$+ \frac{1}{2}(K_1 N_0)^2 \frac{B_N}{K_d^2}\left[1 + \frac{8}{\pi^2}\left(1 - \frac{f_q}{B_N} \right) + \frac{8}{9\pi^2}\left(1 - \frac{3f_q}{B_N} \right) \right] \qquad (9\text{-}106)$$

The phase detector gain K_d is given by (9-80). For $N \gg 1$,

$$K_d \simeq \left(1 - \frac{\Delta}{2} \right)2K_1 P$$

so that

$$\sigma_\delta^2 = \frac{\eta}{2}W_L$$

$$\simeq \frac{N_0 W_L}{8P(1 - \Delta/2)^2}\left\{ R_c^2\left[\left(\delta - \frac{\Delta}{2}\right)T_c \right] + R_c^2\left[\left(\delta + \frac{\Delta}{2}\right)T_c \right] \right\}\left(1 + \frac{8}{\pi^2} \right)$$

$$+ \frac{N_0^2 B_N W_L}{8P^2(1 - \Delta/2)^2}\left[1 + \frac{8}{\pi^2}\left(1 - \frac{f_q}{B_N} \right) + \frac{8}{9\pi^2}\left(1 - \frac{3f_q}{B_N} \right) \right]$$

$$(9\text{-}107)$$

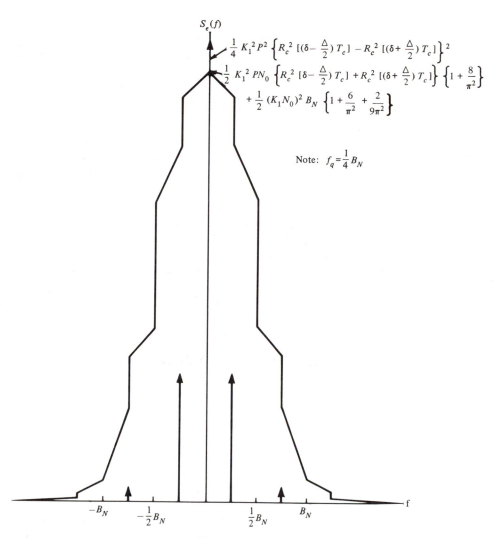

$$\frac{1}{4} K_1^2 P^2 \left\{ R_c^2 \left[(\delta - \frac{\Delta}{2}) T_c \right] - R_c^2 \left[(\delta + \frac{\Delta}{2}) T_c \right] \right\}^2$$

$$\frac{1}{2} K_1^2 P N_0 \left\{ R_c^2 \left[\delta - \frac{\Delta}{2}) T_c \right] + R_c^2 \left[(\delta + \frac{\Delta}{2}) T_c \right] \right\} \left\{ 1 + \frac{8}{\pi^2} \right\}$$

$$+ \frac{1}{2} (K_1 N_0)^2 B_N \left\{ 1 + \frac{6}{\pi^2} + \frac{2}{9\pi^2} \right\}$$

Note: $f_q = \frac{1}{4} B_N$

FIGURE 9-18. Typical power spectrum of noise and clock components at output of tau-dither discriminator.

EXAMPLE 9-3

Consider a tau-dither tracking loop for which $\Delta = 1.0$, $N \gg 1$, and $f_q = B_N/4$. In this case (9-107) becomes

$$\sigma_\delta^2 = \frac{N_0 W_L}{2P} \left(\frac{1}{4} + \frac{1}{4} \right) \left(1 + \frac{8}{\pi^2} \right)$$

$$+ \frac{N_0^2 B_N W_L}{2P^2} \left[1 + \frac{8}{\pi^2} \left(\frac{3}{4} \right) + \frac{8}{9\pi^2} \left(\frac{1}{4} \right) \right]$$

$$= \frac{1}{2\rho_L} \left(1.811 + \frac{3.261}{\rho_{IF}} \right) \qquad (9\text{-}108)$$

where

$$\rho_L = \frac{2P}{N_0 W_L} \tag{9-109}$$

$$\rho_{IF} = \frac{P}{N_0 B_N} \tag{9-110}$$

are the signal-to-noise ratios in the loop and IF bandwidth, respectively. Comparing (9-108) to the comparable result (9-84) for the full-time tracking loop shows that noise performance has indeed been sacrificed to attain the benefits of using the tau-dither tracking loop. ∎

This completes the rather lengthy discussion of the tau-dither early-late code tracking loop. The assumptions that have been made in this analysis are the same as those made in the analysis of the full-time tracking loop. Spreading code clock and self-noise components have been ignored in the discriminator and the noise processes at the output of the two IF filters in the model of Figure 9-15 have been assumed to be independent. The latter assumption is very good for the tau-dither loop since the two outputs affect the error signal at different times.

9-6

DOUBLE-DITHER EARLY-LATE NONCOHERENT TRACKING LOOP

In certain applications, the noise performance degradation of the tau-dither tracking loop relative to the full-time loop is unacceptable. The double-dither early-late noncoherent tracking loop proposed in Hopkins [11] can be used in these cases to solve the gain-imbalance problem of the full-time tracking loop. The double-dither loop-noise performance is the same as the full-time loop-noise performance. The price paid for simultaneously achieving good noise performance and solving the amplitude balance problem is increased hardware complexity.

The hardware configuration of the double-dither loop is illustrated in Figure 9-19. Observe the similarity between the full-time loop of Figure 9-9 and the double-dither loop. The only difference is that the use of the two channels in the double-dither loop alternates between early and late channel correlation. When $q(t) = +1$ in Figure 9-19, the configuration is identical to Figure 9-9. When $q(t) = -1$, the switches are reversed, as is the sign of the differencing circuit output, so that the output $\epsilon(t, \delta)$ is unchanged. If the transient that occurs when $q(t)$ changes states can be ignored, and if the two channels are identical, the double-dither loop performance is identical to the full-time loop performance.

To understand the mechanism used by the double-dither loop to solve the gain-imbalance problem, consider the operation of the discriminator in the noiseless case. Code self-noise and code clock components at the IF bandpass filter inputs can be ignored under the same conditions that they are ignored in the full-time and tau-dither discriminators. Attributing all the gain imbalance between the two channels to the mixer conversion gain, analysis identical to that which resulted in (9-50) can be used to show that

$$y_1(t) = \sqrt{K_1 P} \, R_c \left[\left(\delta + q(t) \frac{\Delta}{2} \right) T_c \right] \cos \left[\omega_{IF} t + \varphi - \varphi' + \theta_d(t - T_d) \right]$$

$$\tag{9-111a}$$

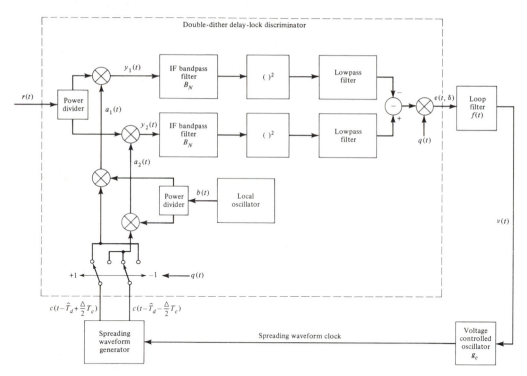

FIGURE 9-19. Double-dither early-late noncoherent code tracking loop. (From Ref. 11.)

$$y_2(t) = \sqrt{K_2 P}\, R_c\left[\left(\delta - q(t)\frac{\Delta}{2}\right)T_c\right]\cos\left[\omega_{\mathrm{IF}}t + \varphi - \varphi' + \theta_d(t - T_d)\right]$$

(9-111b)

where K_1 is the upper channel mixer conversion gain and K_2 is the lower channel mixer conversion gain. It immediately follows that the noiseless discriminator output is

$$\epsilon(t, \delta) = \frac{P}{2}\left\{K_2 R_c^2\left[\left(\delta - q(t)\frac{\Delta}{2}\right)T_c\right] - K_1 R_c^2\left[\left(\delta + q(t)\frac{\Delta}{2}\right)T_c\right]\right\}q(t)$$

(9-112)

This expression can also be written

$$\epsilon(t, \delta) = \frac{P}{2}\left\{[K_2 q_1(t) + K_1 q_2(t)]R_c^2\left[\left(\delta - \frac{\Delta}{2}\right)T_c\right]\right.$$
$$\left. - [K_1 q_1(t) + K_2 q_2(t)]R_c^2\left[\left(\delta + \frac{\Delta}{2}\right)T_c\right]\right\}$$

(9-113)

where $q_1(t)$ and $q_2(t)$ are defined by (9-91). If the two channels are identical (i.e., $K_1 = K_2$), then (9-113) reduces to (9-51) showing that the double-dither loop is identical to the full-time loop. If $K_1 \neq K_2$ and there is no dither [i.e., $q(t) = 1$ for all t], (9-112) becomes

$$\epsilon(t, \delta) = \frac{P}{2}\left\{K_2 R_c^2\left[\left(\delta - \frac{\Delta}{2}\right)T_c\right] - K_1 R_c^2\left[\left(\delta + \frac{\Delta}{2}\right)T_c\right]\right\}$$

(9-114)

Discriminator S-curves for several values of (K_1, K_2) and $\Delta = 1.0$ are illustrated in Figure 9-20. Observe that the curves do not cross the origin. This is the problem that is solved by the double-dither discriminator. When $q(t)$ is a square wave whose frequency is well above the bandwidth of the loop filter, the time average of (9-113) is the desired tracking error,

$$\langle \epsilon(t, \delta) \rangle = \frac{P}{2} \left(\frac{K_1 + K_2}{2} \right) \left\{ R_c^2 \left[\left(\delta - \frac{\Delta}{2} \right) T_c \right] - R_c^2 \left[\left(\delta + \frac{\Delta}{2} \right) T_c \right] \right\}$$

(9-115)

In this case, the discriminator S-curve crosses the origin as in Figure 9-10; however, the magnitude of the S-curve is the average magnitude over the two gains K_1 and K_2.

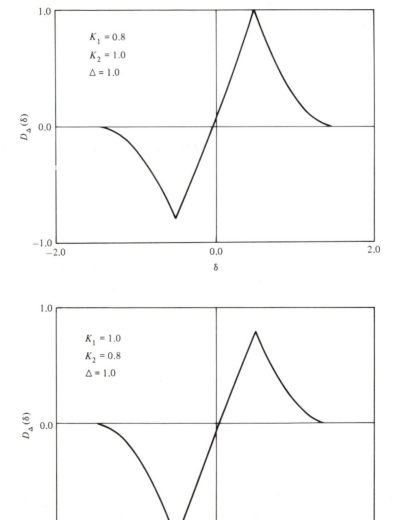

FIGURE 9-20. Full-time early-late discriminator *S*-curve with gain imbalance.

Space does not permit derivation of the noise performance of this loop here. The student, at this point, has all the tools necessary to do this derivation. The principal task of this derivation is to generate the complete power spectrum at the double-dither discriminator output. Finally, note that the double-dither technique can also be applied to other dual-channel phase detectors. The student is referred to Hopkins [11] for a complete analysis of this code tracking loop.

9-7

FULL-TIME EARLY-LATE NONCOHERENT TRACKING LOOP WITH ARBITRARY DATA AND SPREADING MODULATION

The discussion of Section 9-4 was limited to systems that employ BPSK direct-sequence spreading modulation. In this section it will be shown that the same results apply to a system having an arbitrary constant-envelope phase modulation when the BPSK autocorrelation function is replaced by the autocorrelation function for the modulation being considered. The code tracking loop configuration for arbitrary constant-envelope direct-sequence modulation is illustrated in Figure 9-21. This configuration is identical to that of Figure 9-9 except that the two mixers which were used as BPSK modulators have been replaced by more general phase modulators.

The received signal $r(t)$ is the sum of the data-modulated and direct-sequence spread carrier and bandlimited white Gaussian noise

$$r(t) = \sqrt{2P} \cos \left[\omega_0 t + \theta_{ss}(t - T_d) + \theta_d(t - T_d) + \varphi \right] + n(t) \quad (9\text{-}116)$$

where $n(t)$ is the AWGN, $\theta_{ss}(t - T_d)$ is the direct-sequence spreading modulation, $\theta_d(t - T_d)$ is the information modulation, and φ is the random received carrier

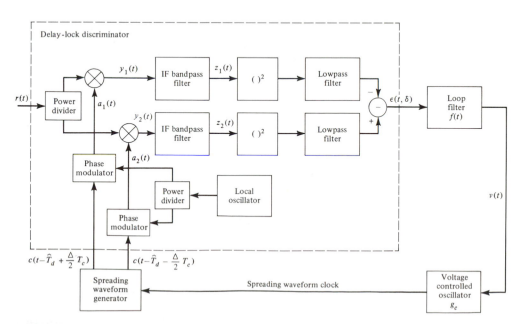

FIGURE 9-21. Full-time early-late noncoherent tracking loop for arbitrary direct-sequence spreading modulation.

phase. The received signal has power P. The early and late reference signals are

$$a_1(t) = 2\sqrt{K_1}\, \cos\left[(\omega_0 - \omega_{\text{IF}})t + \theta_{ss}\left(t - \hat{T}_d + \frac{\Delta}{2}T_c\right) + \varphi'\right] \quad (9\text{-}117a)$$

and

$$a_2(t) = 2\sqrt{K_1}\, \cos\left[(\omega_0 - \omega_{\text{IF}})t + \theta_{ss}\left(t - \hat{T}_d - \frac{\Delta}{2}T_c\right) + \varphi'\right] \quad (9\text{-}117b)$$

where K_1 is the small-signal conversion loss of the mixers, φ' is the random reference oscillator phase, and Δ is the normalized delay difference between the early and late correlator channels.

The mixer output signals are the products of (9-117) and (9-116) with a gain adjustment to account for the power divider. Because of the bandpass filters, only the difference frequency component of this product is of interest. The desired outputs are

$$
\begin{aligned}
y_1(t) = \sqrt{PK_1}\, \cos\Bigg[&\omega_{\text{IF}}t + \theta_{ss}(t - T_d) - \theta_{ss}\left(t - \hat{T}_d + \frac{\Delta}{2}T_c\right) \\
&+ \theta_d(t - T_d) + \varphi - \varphi'\Bigg] \\
+ \sqrt{K_1}\, n_I(t)\, &\cos\left[\omega_{\text{IF}}t - \theta_{ss}\left(t - \hat{T}_d + \frac{\Delta}{2}T_c\right) - \varphi'\right] \\
- \sqrt{K_1}\, n_Q(t)\, &\sin\left[\omega_{\text{IF}}t - \theta_{ss}\left(t - \hat{T}_d + \frac{\Delta}{2}T_c\right) - \varphi'\right]
\end{aligned}
$$

$$(9\text{-}118a)$$

$$
\begin{aligned}
y_2(t) = \sqrt{PK_1}\, \cos\Bigg[&\omega_{\text{IF}}t + \theta_{ss}(t - T_d) - \theta_{ss}\left(t - \hat{T}_d - \frac{\Delta}{2}T_c\right) \\
&+ \theta_d(t - T_d) + \varphi - \varphi'\Bigg] \\
+ \sqrt{K_1}\, n_I(t)\, &\cos\left[\omega_{\text{IF}}t - \theta_{ss}\left(t - \hat{T}_d - \frac{\Delta}{2}T_c\right) - \varphi'\right] \\
- \sqrt{K_1}\, n_Q(t)\, &\sin\left[\omega_{\text{IF}}t - \theta_{ss}\left(t - \hat{T}_d - \frac{\Delta}{2}T_c\right) - \varphi'\right]
\end{aligned}
$$

$$(9\text{-}118b)$$

It is convenient at this point to express $y_1(t)$ and $y_2(t)$ using complex-envelope notation. Thus

$$
\begin{aligned}
\tilde{y}_1(t) = \sqrt{PK_1}\, \exp&\left[j\theta_{ss}(t - T_d) - j\theta_{ss}\left(t - \hat{T}_d + \frac{\Delta}{2}T_c\right)\right] \\
&\times \exp\left[j\theta_d(t - T_d)\right] \exp\left[j(\varphi - \varphi')\right] \\
+ \sqrt{K_1}\, \exp&\left[-j\theta_{ss}\left(t - \hat{T}_d + \frac{\Delta}{2}T_c\right)\right] \exp\left[-j\varphi'\right][n_I(t) + jn_Q(t)]
\end{aligned}
$$

$$(9\text{-}119a)$$

$$\tilde{y}_2(t) = \sqrt{PK_1} \, \exp\left[j\theta_{ss}(t - T_d) - j\theta_{ss}\left(t - \hat{T}_d - \frac{\Delta}{2} T_c\right)\right]$$

$$\times \exp\left[j\theta_d(t - T_d)\right] \exp\left[j(\varphi - \varphi')\right]$$

$$+ \sqrt{K_1} \, \exp\left[-j\theta_{ss}\left(t - \hat{T}_d - \frac{\Delta}{2} T_c\right)\right] \exp\left[-j\varphi'\right][n_I(t) + jn_Q(t)]$$

$$(9\text{-}119\text{b})$$

Adding and subtracting appropriate quantities from (9-119a), $\tilde{y}_1(t)$ can be expressed as

$$\tilde{y}_1(t) = \sqrt{PK_1} \, E\left\{ \exp\left[j\theta_{ss}(t - T_d) - j\theta_{ss}\left(t - \hat{T}_d + \frac{\Delta}{2} T_c\right)\right]\right\}$$

$$\times \exp\left[j\theta_d(t - T_d)\right] \exp\left[j(\varphi - \varphi')\right]$$

$$+ \sqrt{PK_1} \left[\exp\left[j\theta_{ss}(t - T_d) - j\theta_{ss}\left(t - \hat{T}_d + \frac{\Delta}{2} T_c\right)\right]\right.$$

$$\left. - E\left\{ \exp\left[j\theta_{ss}(t - T_d) - j\theta_{ss}\left(t - \hat{T}_d + \frac{\Delta}{2} T_c\right)\right]\right\}\right]$$

$$\times \exp\left[j\theta_d(t - T_d)\right] \exp\left[j(\varphi - \varphi')\right]$$

$$+ \sqrt{K_1} \, \exp\left[-j\theta_{ss}\left(t - \hat{T}_d + \frac{\Delta}{2} T_c\right)\right] \exp\left(-j\varphi'\right)[n_I(t) + jn_Q(t)]$$

$$(9\text{-}120)$$

where the expected value is over all possible transmission delays T_d and delay estimates $\hat{T}_d$. For any useful spreading modulation, the power in the second term of this equation is approximately uniformly spread over a bandwidth which is one or two times the bandwidth of the transmitted spread-spectrum signal. When the system processing gain is large, the data modulation bandwidth is very much smaller than the spreading bandwidth. The IF filter bandwidth is approximately equal to the data modulation bandwidth, so that the majority of the power in the second term of (9-120) will be rejected by the IF bandpass filter. This term contains the code clock and self-noise components which were shown to be negligible for BPSK spreading in Section 9-4. In the following, these terms are ignored. Thus

$$\tilde{y}_1(t) \approx \sqrt{K_1 P} \, E\left\{ \exp[j\theta_{ss}(t - T_d) - j\theta_{ss}\left(t - \hat{T}_d + \frac{\Delta}{2} T_c\right)]\right\}$$

$$\exp\left[j\theta_d(t - T_d)\right] \exp\left[j(\varphi - \varphi')\right]$$

$$+ \sqrt{K_1} \, \exp\left[-j\theta_{ss}\left(t - \hat{T}_d + \frac{\Delta}{2} T_c\right)\right] \exp\left(-j\varphi'\right)[n_I(t) + jn_Q(t)]$$

$$(9\text{-}121)$$

The complex autocorrelation function of the spreading modulation envelope is [12]

$$R_{\theta_{ss}}(\tau) = \tfrac{1}{2} E\{\exp[j\theta_{ss}(t) - j\theta_{ss}(t + \tau)]\} \qquad (9\text{-}122)$$

so that

$$\tilde{y}_1(t) \approx \sqrt{K_1 P} \, 2R_{\theta_{ss}} \left[\left(\delta + \frac{\Delta}{2} \right) T_c \right] \exp\left[j\theta_d(t - T_d) \right] \exp\left[j(\varphi - \varphi') \right]$$

$$+ \sqrt{K_1} \exp\left[-j\theta_{ss}\left(t - \hat{T}_d + \frac{\Delta}{2} T_c \right) \right] \exp(-j\varphi')[n_I(t) + jn_Q(t)]$$

$$(9\text{-}123\text{a})$$

where $\delta = (T_d - \hat{T}_d)/T_c$. Similarly,

$$\tilde{y}_2(t) \approx \sqrt{K_1 P} \, 2R_{\theta_{ss}} \left[\left(\delta - \frac{\Delta}{2} \right) T_c \right] \exp\left[j\theta_d(t - T_d) \right] \exp\left[j(\varphi - \varphi') \right]$$

$$+ \sqrt{K_1} \exp\left[-j\theta_{ss}\left(t - \hat{T}_d - \frac{\Delta}{2} T_c \right) \right] \exp(-j\varphi')[n_I(t) + jn_Q(t)]$$

$$(9\text{-}123\text{b})$$

Denote the complex noise components of $\tilde{y}_1(t)$ and $\tilde{y}_2(t)$ by $\tilde{n}_1(t)$ and $\tilde{n}_2(t)$, respectively. The power spectra of $\tilde{n}_1(t)$ and $\tilde{n}_2(t)$, denoted $S_{n1}(f)$ and $S_{n2}(f)$, are calculated from their complex autocorrelation functions. Consider the early channel. The complex noise autocorrelation function is

$$R_{n1}(\tau) = \tfrac{1}{2}E[\tilde{n}_1(t)\tilde{n}_1^*(t + \tau)]$$

$$= \tfrac{1}{2}K_1 E\left\{ \exp\left[-j\theta_{ss}\left(t - \hat{T}_d + \frac{\Delta}{2} T_c \right) + j\theta_{ss}\left(t + \tau - \hat{T}_d + \frac{\Delta}{2} T_c \right) \right] \right\}$$

$$\times E\{[n_I(t) + jn_Q(t)][n_I(t + \tau) - jn_Q(t + \tau)]\}$$

$$(9\text{-}124)$$

where the expectation factors because the spreading modulation and the noise are independent. In (9-124) the expected value is over all $\hat{T}_d$ and all sample functions of the noise processes. Using the fact that $n_I(t)$ and $n_Q(t)$ are independent and identically distributed, and using (9-122), equation (9-124) becomes

$$R_{n1}(\tau) = 2K_1 R_{\theta_{ss}}(\tau) R_{n_I}(\tau) \qquad (9\text{-}125)$$

The power spectrum of $\tilde{n}_1(t)$ is the Fourier transform of this autocorrelation. Using the Fourier transform convolution theorem,

$$S_{n1}(f) = 2K_1 S_{\theta_{ss}}(f) * S_{n_I}(f) \qquad (9\text{-}126)$$

When the bandwidth of the noise is much larger than the spreading modulation bandwidth, $R_{\theta_{ss}}(\tau)$ may be approximated by its value at $\tau = 0$,

$$R_{\theta_{ss}}(\tau) \approx \tfrac{1}{2} E[\exp(j0)] = \tfrac{1}{2} \qquad (9\text{-}127)$$

and

$$S_{n1}(f) \approx 2K_1 \tfrac{1}{2} S_{n_I}(f) = \frac{K_1 N_0}{2} \qquad |f| < B_{in}$$

$$= 0 \qquad (9\text{-}128)$$

where $N_0/2$ is the input noise two-sided power spectral density, and an input bandwidth of $2B_{in}$ has been assumed. An identical expression can be derived for $S_{n2}(f)$.

Equation (9-123) takes the IF bandpass filter partially into account since code self-noise has been ignored. The remaining terms are also processed by the IF filters. As in Section 9-4 it is assumed that the data modulation passes through the IF filter without distortion. The noise components, on the other hand, are wideband relative to the IF filter bandwidth, so that the filters have a significant effect on these signals. The IF bandpass filters are characterized by their impulse response $h(t)$ or the lowpass equivalent complex envelope impulse response $h_l(t)$, where

$$h(t) = 2 \, \text{Re} \, [h_l(t) \exp (j\omega_{\text{IF}}t)] \tag{9-129}$$

Expressions for the complex envelope filter output noise which are analogous to (9-57) for BPSK spreading are then

$$\tilde{n}_1^0(t) = \sqrt{K_1} \exp (-j\varphi') \int_{-\infty}^{+\infty} \exp \left[-j\theta_{\text{ss}} \left(\lambda - \hat{T}_d + \frac{\Delta}{2} T_c \right) \right]$$

$$[n_I(\lambda) + jn_Q(\lambda)]h_l(t - \lambda)d\lambda \tag{9-130a}$$

$$\tilde{n}_2^0(t) = \sqrt{K_1} \exp (-j\varphi') \int_{-\infty}^{\infty} \exp \left[-j\theta_{\text{ss}} \left(\lambda - \hat{T}_d - \frac{\Delta}{2} T_c \right) \right]$$

$$[n_I(\lambda) + jn_Q(\lambda)]h_l(t - \lambda) \, d\lambda \tag{9-130b}$$

These filter outputs are Gaussian since the filter inputs are Gaussian. Thus, if $\tilde{n}_1^0(t)$ and $\tilde{n}_2^0(t)$ are uncorrelated, they are also independent. The complex cross-correlation of the two noise components is

$$R_{n,\text{out}}(\tau) \triangleq \frac{1}{2} E\{[\tilde{n}_1^0(t)][\tilde{n}_2^0(t + \tau)]^*\}$$

$$= \frac{1}{2} K_1 E \left\{ \int_{-\infty}^{\infty} \int_{-\infty}^{\infty} \exp \left[-j\theta_{\text{ss}} \left(\lambda - \hat{T}_d + \frac{\Delta}{2} T_c \right) + j\theta_{\text{ss}} \left(\gamma - \hat{T}_d - \frac{\Delta}{2} T_c \right) \right] \right.$$

$$\left. \times \, [n_I(\lambda) + jn_Q(\lambda)][n_I(\gamma) - jn_Q(\gamma)]h_l(t - \lambda)h_l^*(t + \tau - \gamma) \, d\lambda \, d\gamma \right\}$$

$$= \frac{1}{2} K_1 \int_{-\infty}^{\infty} \int_{-\infty}^{\infty} E \left\{ \exp \left[-j\theta_{\text{ss}} \left(\lambda - \hat{T}_d + \frac{\Delta}{2} T_c \right) + j\theta_{\text{ss}}(\gamma - \hat{T}_d - \frac{\Delta}{2} T_c) \right] \right\}$$

$$\times \, \{E[n_I(\lambda)n_I(\gamma)] + E[n_Q(\lambda)n_Q(\gamma)]\}h_l(t - \lambda)h_l^*(t + \tau - \gamma) \, d\lambda \, d\gamma \tag{9-131}$$

where the expected value is over all estimated propagation delays $\hat{T}_d$ and all sample functions of the random processes $n_I(t)$ and $n_Q(t)$. Using (9-122) and remembering that $n_I(t)$ and $n_Q(t)$ are identically distributed, (9-131) becomes

$$R_{n,\text{out}}(\tau) = 2K_1 \int_{-\infty}^{\infty} \int_{-\infty}^{\infty} R_{\theta_{\text{ss}}}(\lambda - \gamma + \Delta T_c)R_{n_I}(\lambda - \gamma)h_l(t + \tau - \gamma)h_l^*(t - \gamma) \, d\lambda \, d\gamma \tag{9-132}$$

This equation is comparable to (9-59) for BPSK spreading. Using similar steps to those following (9-59), it may be shown that with the assumption that $R_{\theta_{\text{ss}}}(\tau)$ is slowly varying relative to $R_{n_I}(\tau)$, $R_{n,\text{out}}(\tau)$ is directly proportional to $R_{\theta_{\text{ss}}}(\Delta T_c)$. For $\Delta = 0$, it is clear that the two filter output noise processes are identical and therefore not independent. The two processes may be assumed independent only when Δ is sufficiently large. Although the analysis is more complicated when the filter output noise processes are dependent, it is conjectured that loop noise performance is unaffected since the correlated components will cancel at the discriminator output. In the following it will be assumed that the noise processes at the IF filter outputs are independent.

The complex envelopes of the filter output noise processes will be represented by

$$\tilde{n}_j^0(t) = \sqrt{2}\, n_{jI}^0(t) + j\sqrt{2}\, n_{jQ}^0(t) \tag{9-133}$$

where $n_{jI}^0(t)$ and $n_{jQ}^0(t)$, $j = 1, 2$, are assumed independent lowpass Gaussian noise processes. The power spectrum of $\tilde{n}_j^0(t)$ is the product of the power transfer function of the filter and the power spectrum of $\tilde{n}_j(t)$ given in (9-128). Assuming that the IF filter bandwidth is B_N,

$$S_{n,\text{out}}(f) = \begin{cases} \frac{1}{2}\, K_1 N_0 & |f| < \frac{1}{2} B_N \\ 0 & \text{elsewhere} \end{cases} \tag{9-134}$$

This expression is valid for both $\tilde{n}_1^0$ and $\tilde{n}_2^0$. The power spectrum of any one of the in-phase or quadrature noise components of (9-133) is one-half the magnitude of (9-134). Combining all of the above, the filter output complex envelopes are given by

$$\tilde{z}_1(t) = 2\sqrt{K_1 P}\, R_{\theta_{ss}}\left[\left(\delta + \frac{\Delta}{2}\right)T_c\right] \exp\left[j\theta_d(t - T_d)\right] \exp\left[j(\varphi - \varphi')\right]$$
$$+ \sqrt{2}\, n_{1I}^0(t) + j\sqrt{2}\, n_{1Q}^0(t) \tag{9-135a}$$

$$\tilde{z}_2(t) = 2\sqrt{K_1 P}\, R_{\theta_{ss}}\left[\left(\delta - \frac{\Delta}{2}\right)T_c\right] \exp\left[j\theta_d(t - T_d)\right] \exp\left[j(\varphi - \varphi')\right]$$
$$+ \sqrt{2}\, n_{2I}^0(t) + j\sqrt{2}\, n_{2Q}^0(t) \tag{9-135b}$$

These signals are input to the square-law detectors. The desired output of the square-law detector is the difference frequency component whose envelope is $\frac{1}{2}\tilde{z}_1(t)\tilde{z}_1^*(t)$ or $\frac{1}{2}\tilde{z}_2(t)\tilde{z}_2^*(t)$. The delay-lock discriminator output is then

$$\epsilon(t, \delta) = \frac{1}{2}\{\tilde{z}_2(t)\tilde{z}_2^*(t) - \tilde{z}_1(t)\tilde{z}_1^*(t)\}$$

$$= 2K_1 P\left\{\left|R_{\theta_{ss}}\left[\left(\delta - \frac{\Delta}{2}\right)T_c\right]\right|^2 - \left|R_{\theta_{ss}}\left[\left(\delta + \frac{\Delta}{2}\right)T_c\right]\right|^2\right\}$$

$$+ 2\sqrt{2K_1 P}\left\{\text{Re}\left[R_{\theta_{ss}}\left[\left(\delta - \frac{\Delta}{2}\right)T_c\right]\right]n_{2I}^0(t) + \text{Im}\left[R_{\theta_{ss}}\left[\left(\delta - \frac{\Delta}{2}\right)T_c\right]\right]n_{2Q}^0(t)\right.$$

$$\left. - \text{Re}\left[R_{\theta_{ss}}\left[\left(\delta + \frac{\Delta}{2}\right)T_c\right]\right]n_{1I}^0(t) - \text{Im}\left[R_{\theta_{ss}}\left[\left(\delta + \frac{\Delta}{2}\right)T_c\right]\right]n_{1Q}^0(t)\right\}$$

$$\times \cos\left[\theta_d(t - T_d) + \varphi - \varphi'\right]$$

$$+ 2\sqrt{2K_1 P}\left\{\text{Re}\left[R_{\theta_{ss}}\left[\left(\delta - \frac{\Delta}{2}\right)T_c\right]n_{2Q}^0(t)\right] + \text{Im}\left[R_{\theta_{ss}}\left[\left(\delta - \frac{\Delta}{2}\right)T_c\right]\right]n_{2I}^0(t)\right.$$

$$\left. - \text{Re}\left[R_{\theta_{ss}}\left[\left(\delta + \frac{\Delta}{2}\right)T_c\right]n_{1Q}^0(t)\right] - \text{Im}\left[R_{\theta_{ss}}\left[\left(\delta + \frac{\Delta}{2}\right)T_c\right]\right]n_{1I}^0(t)\right\}$$

$$\times \sin\left[\theta_d(t - T_d) + \varphi - \varphi'\right]$$

$$+ [n_{2I}^0(t)]^2 + [n_{2Q}^0(t)]^2 - [n_{1I}^0(t)]^2 - [n_{1Q}^0(t)]^2 \tag{9-136}$$

The first term of $\epsilon(t, \delta)$ is the desired phase correcting component, the second and third terms are the $s \times n$ components, and the remaining terms are the $n \times n$ components. The discriminator S-curve is given by the first term. The S-curve is a function of the specific spreading modulation employed.

For all constant-envelope phase modulations of interest, the complex autocorrelation is real. In this special case, (9-136) is identical to (9-64) with the exception that $R_c[(\delta \pm \Delta/2)T_c]$ has been replaced by $2 \, \mathrm{Re} \, [R_{\theta_{ss}}[(\delta \pm \Delta/2)T_c]]$, and $R_c^2[(\delta \pm \Delta/2)T_c]$ has been replaced by $4|R_{\theta_{ss}}[(\delta \pm \Delta/2)T_c]|^2$. This difference is expected since the complex envelope autocorrelation function for BPSK is

$$R_{\theta_{ss}}(\tau) \triangleq \tfrac{1}{2} E[c(t)c^*(t + \tau)] = \tfrac{1}{2} R_c(\tau) \qquad (9\text{-}137)$$

The similarity of (9-64) and (9-136) enables direct application of the analysis for the BPSK system to the more general case. In particular, the autocorrelation function of the delay-lock discriminator output is derived from (9-66) and is

$$
\begin{aligned}
R_\epsilon(\tau) = \; & \frac{1}{4} K_1^2 P^2 \left\{ 4 \left| R_{\theta_{ss}}\left[\left(\delta - \frac{\Delta}{2}\right)T_c\right] \right|^2 - 4 \left| R_{\theta_{ss}}\left[\left(\delta + \frac{\Delta}{2}\right)T_c\right] \right|^2 \right\}^2 \\
& + 2K_1 P \left\{ 4 \, \mathrm{Re}\left[R_{\theta_{ss}}\left[\left(\delta - \frac{\Delta}{2}\right)T_c\right] \right]^2 \right. \\
& \left. + 4 \, \mathrm{Re}\left[R_{\theta_{ss}}\left[\left(\delta + \frac{\Delta}{2}\right)T_c\right] \right]^2 \right\} E[n^0(t)n^0(t + \tau)] \\
& \times E\{\cos[\theta_d(t) - \theta_d(t + \tau)]\} \\
& + 4E\{[n^0(t)n^0(t + \tau)]^2\} - 4E^2\{[n^0(t)]^2\} \qquad (9\text{-}138)
\end{aligned}
$$

where $n^0(t)$ has been used to represent any one of the four IF filter output processes $n_{1I}^0(t)$, $n_{1Q}^0(t)$, $n_{2I}^0(t)$, or $n_{2Q}^0(t)$. The Fourier transform of (9-138) is the discriminator output power spectrum. This spectrum is given by a simple modification of (9-67); the result is

$$
\begin{aligned}
S_\epsilon(f) = \; & \frac{1}{4} K_1^2 P^2 \left\{ 4 \left| R_{\theta_{ss}}\left[\left(\delta - \frac{\Delta}{2}\right)T_c\right] \right|^2 - 4 \left| R_{\theta_{ss}}\left[\left(\delta + \frac{\Delta}{2}\right)T_c\right] \right|^2 \right\}^2 \delta(f) \\
& + 2K_1 P \left\{ 4 \, \mathrm{Re}\left[R_{\theta_{ss}}\left[\left(\delta - \frac{\Delta}{2}\right)T_c\right] \right]^2 \right. \\
& \left. + 4 \, \mathrm{Re}\left[R_{\theta_{ss}}\left[\left(\delta + \frac{\Delta}{2}\right)T_c\right] \right]^2 \right\} S_{n^0}(f) * S_{\theta_d}(f) \\
& + 8 S_{n^0}(f) * S_{n^0}(f) \qquad (9\text{-}139)
\end{aligned}
$$

where $S_{\theta_d}(f)$ is the Fourier transform of the complex-envelope autocorrelation function $R_A(\tau)$ defined in (9-69). Under the assumption that the data modulation is narrowband relative to the noise, the discriminator output power spectrum is illustrated in Figure 9-11 with the proper substitution for the spreading modulation autocorrelation.

Define $D_\Delta(\delta)$ and $n_\epsilon(t)$ as the desired and undesired components of the discriminator output as in (9-77). That is,

$$D_\Delta(\delta) = 4 \left| R_{\theta_{ss}}\left[\left(\delta - \frac{\Delta}{2}\right)T_c\right] \right|^2 - 4 \left| R_{\theta_{ss}}\left[\left(\delta + \frac{\Delta}{2}\right)T_c\right] \right|^2 \qquad (9\text{-}140)$$

$$\epsilon(t, \delta) = \frac{1}{2} K_1 P D_\Delta(\delta) + n_\epsilon(t) \tag{9-141}$$

The discriminator S-curve $D_\Delta(\delta)$ is dependent only on the spreading phase modulation characteristics. With these definitions, the nonlinear loop model is given in Figure 9-12 and the linear model is given in Figure 9-13. Specific results for tracking jitter are dependent on the particular spreading modulation employed.

EXAMPLE 9-4

Consider the full-time early-late code tracking loop with MSK spreading modulation and a normalized early-late difference of $\Delta = 1.0$. Assume that the spreading code is a totally random binary sequence. The complex autocorrelation function for MSK phase modulation is

$$R_{\theta_{ss}}(\tau) = \frac{1}{2\pi} \left[\pi \left(1 - \frac{|\tau|}{2T_c} \right) \cos \left(\frac{\pi|\tau|}{2T_c} \right) + \sin \left(\frac{\pi|\tau|}{2T_c} \right) \right] \tag{9-142}$$

The discriminator S-curve is calculated from (9-142) and (9-140) and the result is illustrated in Figure 9-22. Using straightforward algebraic manipulations, the slope of $D_\Delta(\delta)$ at $\delta = 0$ can be shown to be

$$\left[\frac{d}{d\delta} D_\Delta(\delta) \right]_{\substack{\delta = 0 \\ \Delta = 1.0}} = \frac{3}{2} \left(\frac{3\pi}{4} \cos \frac{\pi}{4} + \sin \frac{\pi}{4} \right) \sin \frac{\pi}{4}$$

$$= 2.517 \tag{9-143}$$

so that

$$K_d = 1.259 K_1 P \tag{9-144}$$

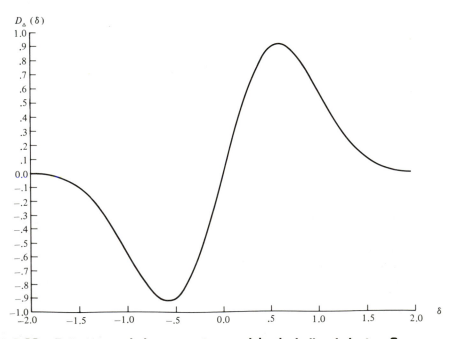

FIGURE 9-22. Full-time early-late noncoherent delay-lock discriminator S-curve using MSK spreading modulation.

The comparable result for BPSK spreading is

$$\left[\frac{d}{d\delta}D_\Delta(\delta)\right]_{\substack{\delta=0\\\Delta=1.0}} = 2.0 \tag{9-145}$$

$$K_d = K_1 P \tag{9-146}$$

The magnitude of the delay-lock discriminator output noise power spectrum is given in (9-76) for the case where data modulation has been ignored. Thus

$$S_{n_\epsilon}(f) = \frac{1}{2}K_1^2 N_0^2 B_n$$

$$+ \frac{1}{2}K_1^2 N_0 P\left\{4R_{\theta_{ss}}^2\left[\left(\delta - \frac{\Delta}{2}\right)T_c\right] + 4R_{\theta_{ss}}^2\left[\left(\delta + \frac{\Delta}{2}\right)T_c\right]\right\}$$

$$= \frac{1}{2}K_1^2 N_0^2 B_N$$

$$+ \frac{1}{2}K_1^2 N_0 P \frac{2}{\pi^2}\left(\frac{3\pi}{4}\cos\frac{\pi}{4} + \sin\frac{\pi}{4}\right)^2$$

$$= \frac{1}{2}K_1^2 N_0(N_0 B_N + 1.141P) \tag{9-147}$$

and the equivalent white noise power spectral density at the input to the linear equivalent circuit is

$$\frac{S_{n_\epsilon}(t)}{K_d^2} = \frac{1}{2}\left(\frac{N_0}{2P}\right)\left[1.44 + 1.26\left(\frac{N_0 B_N}{P}\right)\right] \tag{9-148}$$

The tracking jitter variance is equal to the product of this noise density and the two-sided loop noise bandwidth W_L, so that

$$\sigma_\delta^2 = \frac{1}{2}\left(\frac{N_0 W_L}{2P}\right)\left[1.44 + 1.26\left(\frac{N_0 B_N}{P}\right)\right]$$

$$= \frac{1}{2\rho L}\left(1.44 + \frac{1.26}{\rho_{IF}}\right) \tag{9-149}$$

Comparing the last expression to (9-84) for BPSK spreading modulation, it is observed that, for this special case where $\Delta = 1.0$, the optimum choice of modulation for minimum tracking jitter depends on whether or not the squaring loss component is dominant. Note, however, that the choice may be more clear for other delay differences. ■

9-8

CODE TRACKING LOOPS FOR FREQUENCY-HOP SYSTEMS

The full-time early-late noncoherent code tracking loop can also be used in a frequency-hopping spread-spectrum system. In particular, the conceptual block diagram of Figure 9-21 can be used if the phase modulators are replaced with frequency synthesizers. The tracking loop configuration for a frequency-hopping system is illustrated in Figure 9-23. Since frequency synthesizers are complex, and

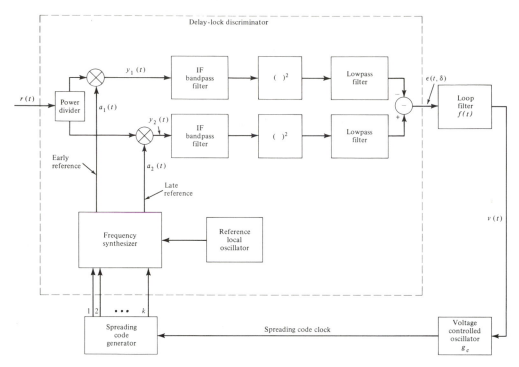

FIGURE 9-23. Full-time early-late noncoherent tracking loop for frequency-hopping spread-spectrum system.

therefore costly, a single synthesizer is used and the early-late delay difference is generated using a delay line. Observe that the "spreading waveform generator" of Figure 9-21 has been replaced by a "spreading code generator," indicating that its output is not a waveform $c(t) = \pm 1$ but a digital signal which controls the frequency of the synthesizer. The code generator outputs k binary digits at a time, implying that the synthesizer generates 2^k frequencies. As seen in all the other tracking loops analyzed in this chapter, the delay-lock discriminator output consists of a desired phase correction signal and noise. The most difficult part of the loop analysis is the characterization of these components. Although the analysis of Section 9-7 was general, an assumption was made that the transmitter and receiver spreading code phase modulators were identical. In the present case, this assumption implies that the transmitter and receiver frequency synthesizers are precisely phase matched. Because of the hardware difficulties that this implies, that assumption is not made in the analysis that follows. Discussion is limited to tracking loops having an early-late channel difference of one chip, and systems that employ a slow hop relative to the data modulation.

The received signal is the sum of the frequency-hopped and data-modulated carrier having power P and bandlimited additive white Gaussian noise having two-sided power spectral density $N_0/2$. That is,

$$r(t) = \sqrt{2P} \cos \left[\omega_0 t + \sum_{n=-\infty}^{\infty} (\omega_n t + \varphi_n) p_{T_c}(t - T_d - nT_c) + \theta_d(t - T_d) \right]$$
$$+ \sqrt{2}\, n_I(t) \cos \omega_0 t - \sqrt{2}\, n_Q(t) \sin \omega_0 t \tag{9-150}$$

In this equation, $(\omega_0 + \omega_n)$ is the transmission frequency during time interval n, φ_n is the frequency synthesizer random phase during this same interval, $p_{T_c}(t)$ is a unit pulse beginning at $t = 0$ and ending at $t = T_c$, and $\theta_d(t)$ is the arbitrary data phase modulation. The reference signals $a_1(t)$ and $a_2(t)$ are frequency hopped using the same hop pattern as used in the transmitter, but using frequencies that are offset from the transmitter frequencies by the IF. The hopping patterns of the reference signals are offset in phase from the receiver estimate of the transmission delay $\hat{T}_d$ by $\pm\frac{1}{2}$ chip, where a chip is the frequency-hop dwell time T_c. Thus

$$a_1(t) = 2\sqrt{K_1}\cos\left[(\omega_0 + \omega_{\text{IF}})t + \sum_{n=-\infty}^{\infty}(\omega_n t + \varphi_n')p_{T_c}\left(t - \hat{T}_d + \frac{\Delta}{2}T_c - nT_c\right)\right]$$

$$(9\text{-}151\text{a})$$

$$a_2(t) = 2\sqrt{K_1}\cos\left[(\omega_0 + \omega_{\text{IF}})t + \sum_{n=-\infty}^{\infty}(\omega_n t + \varphi_n')p_{T_c}\left(t - \hat{T}_d - \frac{\Delta}{2}T_c - nT_c\right)\right]$$

$$(9\text{-}151\text{b})$$

where K_1 is the mixer conversion loss, φ_n' is the receiver synthesizer random phase during the nth time interval, and all other terms have been defined earlier.

Consider the early channel mixer output signal $y_1(t)$. The sum frequency terms of this mixing operation will be rejected by the IF bandpass filter, so that the signal of interest is

$$
\begin{aligned}
y_1(t) = \sqrt{K_1 P}\cos\Bigg[& \omega_{\text{IF}}t - \sum_{n=-\infty}^{\infty}(\omega_n t + \varphi_n)p_{T_c}(t - T_d - nT_c) - \theta_d(t - T_d) \\
& + \sum_{m=-\infty}^{\infty}(\omega_m t + \varphi_m')p_{T_c}\left(t - \hat{T}_d + \frac{\Delta}{2}T_c - mT_c\right)\Bigg] \\
& + \sqrt{K_1}\,n_I(t)\cos\left[\omega_{\text{IF}}t + \sum_{n=-\infty}^{\infty}(\omega_n t + \varphi_n')p_{T_c}\left(t - \hat{T}_d + \frac{\Delta}{2}T_c - nT_c\right)\right] \\
& - \sqrt{K_1}\,n_Q(t)\sin\left[\omega_{\text{IF}}t + \sum_{n=-\infty}^{\infty}(\omega_n t + \varphi_n')p_{T_c}\left(t - \hat{T}_d + \frac{\Delta}{2}T_c - nT_c\right)\right]
\end{aligned}
$$

$$(9\text{-}152)$$

The bandwidth of the received noise covers the entire received signal spectrum. During each hop dwell, a different portion of the received noise spectrum is translated to near the IF and will pass through the IF filter. It will be assumed that the frequency hop rate is slow enough that hop transients may be neglected. Thus the mixer output noise may be approximated by a narrowband white Gaussian noise process centered at the IF (i.e., not hopped), whose two-sided power spectral density is $K_1 N_0/4$.

The signal component of (9-152) is centered on the IF only when the two summations within the argument of the cosine are generating the same frequency. This occurs whenever $p_{T_c}(t - T_d - nT_c)$ and $p_{T_c}(t - \hat{T}_d + (\Delta/2)T_c - mT_c)$ overlap. If any overlap of these pulses occurs, it will occur on every hop interval. During that portion of time where no overlap occurs, the signal component is translated to a frequency that will not pass through the IF filter and the signal may be ignored during this time. Thus the signal component may be replaced by an equivalent (as far as the tracking loop is concerned) pulsed signal and (9-152) may be approximated by

$$y_1'(t) = \sqrt{K_1 P} \sum_{n=-\infty}^{\infty} p_{T_c}(t - T_d - nT_c) p_{T_c}\left(t - \hat{T}_d + \frac{\Delta}{2}T_c - nT_c\right)$$

$$\times \cos\left[\omega_{IF}t - (\varphi_n - \varphi_n') - \theta_d(t - T_d)\right]$$

$$+ \sqrt{K_1}\, n_{1I}(t) \cos \omega_{IF}t - \sqrt{K_1}\, n_{1Q}(t) \sin \omega_{IF}t \qquad (9\text{-}153a)$$

Figure 9-24 illustrates typical early channel mixer input and output spectra. The portions of the early mixer output spectrum marked with the letter "A" will not pass through the IF filter and are ignored in (9-153). Similarly, the late channel

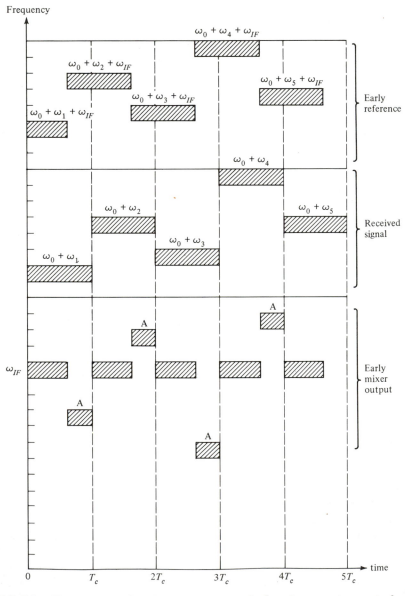

FIGURE 9-24. Illustration of typical early channel mixer input and output signal spectra.

mixer equivalent output is

$$
y_2'(t) = \sqrt{K_1 P} \sum_{n=-\infty}^{\infty} p_{T_c}(t - T_d - nT_c) p_{T_c}\left(t - \hat{T}_d - \frac{\Delta}{2}T_c - nT_c\right)
$$

$$
\times \cos\left[\omega_{IF}t - (\varphi_n - \varphi_n') - \theta_d(t - T_d)\right]
$$

$$
+ \sqrt{K_1}\, n_{2I}(t) \cos \omega_{IF}t - \sqrt{K_1}\, n_{2Q}(t) \sin \omega_{IF}t \tag{9-153b}
$$

The discussion in this section is limited to systems with $\Delta = 1.0$. Thus the early and late reference signals are never simultaneously at the same frequency and, at any instant of time, the early and late channel noise processes come from different portions of the received signal band and are therefore independent.

The frequency-hop rate is assumed to be very slow relative to the IF filter bandwidth so that a quasi-static analysis similar to that used in the dithering loop analysis of Section 9-5 can be used. With this assumption, the pulse modulation of (9-153) passes through the filter without distortion. The filter output noise processes are Gaussian and independent since the inputs are Gaussian and independent. Thus the filter outputs are approximately

$$
z_1(t) = \sqrt{K_1 P} \sum_{n=-\infty}^{\infty} p_{T_c}(t - T_d - nT_c) p_{T_c}\left(t - \hat{T}_d + \frac{\Delta}{2}T_c - nT_c\right)
$$

$$
\times \cos\left[\omega_{IF}t - (\varphi_n - \varphi_n') - \theta_d(t - T_d)\right]
$$

$$
+ \sqrt{2}\, n_{1I}^0(t) \cos \omega_{IF}t - \sqrt{2}\, n_{1Q}^0(t) \sin \omega_{IF}t \tag{9-154a}
$$

$$
z_2(t) = \sqrt{K_1 P} \sum_{n=-\infty}^{\infty} p_{T_c}(t - T_d - nT_c) p_{T_c}\left(t - \hat{T}_d - \frac{\Delta}{2}T_c - nT_c\right)
$$

$$
\times \cos\left[\omega_{IF}t - (\varphi_n - \varphi_n') - \theta_d(t - T_d)\right]
$$

$$
+ \sqrt{2}\, n_{2I}^0(t) \cos \omega_{IF}t - \sqrt{2}\, n_{2Q}^0(t) \sin \omega_{IF}t \tag{9-154b}
$$

The noise signals $n_{1I}^0(t)$, $n_{1Q}^0(t)$, $n_{2I}^0(t)$, and $n_{2Q}^0(t)$ are all independent, lowpass, white Gaussian noise processes with two-sided power spectral densities of $K_1 N_0/4$. Their two-sided bandwidth equals the IF filter one-sided bandwidth.

The slow hop assumption also enables simplification of the calculation of the output of the lowpass filters which follow the squaring devices. The outputs are calculated for the signal present (i.e., pulses overlap) and signal absent cases independently and combined at the output. As usual, the sum frequency terms at the squaring device outputs are rejected by the lowpass filters. The (signal × signal) and (signal × noise) output terms are present only when the pulse signals overlap; however, the (noise × noise) term is always present. The early channel lowpass filter output is

$$
[z_1^2(t)]_{LP} = \{\tfrac{1}{2} K_1 P + n_{1I}^0 \sqrt{2K_1 P} \cos\left[\varphi_n - \varphi_n' - \theta_d(t - T_d)\right]
$$

$$
- n_{1Q}^0(t)\sqrt{2K_1 P} \sin\left[\varphi_n - \varphi_n' - \theta_d(t - T_d)\right]\}
$$

$$
\times \left\{\sum_{n=-\infty}^{\infty} p_{T_c}(t - T_d - nT_c) p_{T_c}\left(t - \hat{T}_d + \frac{\Delta}{2}T_c - nT_c\right)\right\}
$$

$$
+ [n_{1I}^0(t)]^2 + [n_{1Q}^0(t)]^2 \tag{9-155}
$$

A similar expression is found for $[z_2^2(t)]_{LP}$.

Straightforward algebraic and trigonometric manipulation on $[z_2^2(t)]_{LP} - [z_1^2(t)]_{LP}$ yields the following result for the discriminator output $\epsilon(t, \delta)$:

$$\epsilon(t, \delta) = \frac{1}{2} K_1 P \left[\sum_{n=-\infty}^{\infty} p_{T_c}(t - T_d - nT_c) p_{T_c}\left(t - \hat{T}_d - \frac{\Delta}{2} T_c - nT_c \right) \right.$$

$$\left. - \sum_{m=-\infty}^{\infty} p_{T_c}(t - T_d - mT_c) p_{T_c}\left(t - \hat{T}_d + \frac{\Delta}{2} T_c - mT_c \right) \right]$$

$$+ \sqrt{2K_1 P} \cos \left[\theta_d(t - T_d)\right] [n_{2I}^0(t)\beta_1(t) - n_{1I}^0(t)\beta_2(t)]$$

$$+ \sqrt{2K_1 P} \sin \left[\theta_d(t - T_d)\right] [n_{2I}^0(t)\beta_3(t) - n_{1I}^0(t)\beta_4(t)]$$

$$+ \sqrt{2K_1 P} \sin \left[\theta_d(t - T_d)\right] [n_{2Q}^0(t)\beta_1(t) - n_{1Q}^0(t)\beta_2(t)]$$

$$- \sqrt{2K_1 P} \cos \left[\theta_d(t - T_d)\right] [n_{2Q}^0(t)\beta_3(t) - n_{1Q}^0(t)\beta_4(t)]$$

$$+ [n_{2I}^0(t)]^2 + [n_{2Q}^0(t)]^2 - [n_{1I}^0(t)]^2 + [n_{1Q}^0(t)]^2 \qquad (9\text{-}156)$$

where $\beta_1(t) = \sum_{n=-\infty}^{\infty} \cos (\varphi_n - \varphi_n') p_{T_c}(t - T_d - nT_c) p_{T_c}\left(t - \hat{T}_d - \frac{\Delta}{2} T_c - nT_c \right)$

$$\beta_2(t) = \sum_{n=-\infty}^{\infty} \cos (\varphi_n - \varphi_n') p_{T_c}(t - T_d - nT_c) p_{T_c}\left(t - \hat{T}_d + \frac{\Delta}{2} T_c - nT_c \right)$$

$$\beta_3(t) = \sum_{n=-\infty}^{\infty} \sin (\varphi_n - \varphi_n') p_{T_c}(t - T_d - nT_c) p_{T_c}\left(t - \hat{T}_d - \frac{\Delta}{2} T_c - nT_c \right)$$

$$\beta_4(t) = \sum_{n=-\infty}^{\infty} \sin (\varphi_n - \varphi_n') p_{T_c}(t - T_d - nT_c) p_{T_c}\left(t - \hat{T}_d + \frac{\Delta}{2} T_c - nT_c \right)$$

Consider the term of this equation defined by

$$e(t) = \sqrt{2K_1 P} \cos \left[\theta_d(t - T_d)\right] \left[n_{2I}^0(t) \sum_{n=-\infty}^{\infty} \cos (\varphi_n - \varphi_n') p_{T_c}(t - T_d - nT_c) \right.$$

$$\times p_{T_c}\left(t - \hat{T}_d - \frac{\Delta}{2} T_c - nT_c \right)$$

$$\left. - n_{1I}^0(t) \sum_{m=-\infty}^{\infty} \cos (\varphi_m - \varphi_m') p_{T_c}(t - T_d - mT_c) p_{T_c}\left(t - \hat{T}_d + \frac{\Delta}{2} T_c - mT_c \right) \right]$$

$$(9\text{-}157)$$

Discussion is limited to cases where $\Delta = 1.0$. Thus $p_{T_c}(t - T_d - nT_c)$ $p_{T_c}(t - \hat{T}_d - (\Delta/2)T_c - nT_c)$ is nonzero only where $p_{T_c}(t - T_d - mT_c)$ $p_{T_c}(t - \hat{T}_d + (\Delta/2)T_c - mT_c)$ is zero, and vice versa, and the factor within parentheses switches between its first term and its second term twice each T_c seconds. Assume that the frequency-hop rate is slow enough that the transient due to the switching may be neglected. Then, since $n_{2I}^0(t)$ and $n_{1I}^0(t)$ have identical statistics, and since the arguments of the cosines are identical, the function $e(t)$ may be approximated by

$$e(t) \approx \sqrt{2K_1 P} \cos \left[\theta_d(t - T_d)\right] n_{1I}^0(t) \cos \left[\varphi(t)\right] \qquad (9\text{-}158)$$

where $\varphi(t)$ represents the phase modulation due to the noncoherence of the frequency synthesizers. That is,

$$\varphi(t) = \sum_{n=-\infty}^{\infty} (\varphi_n - \varphi'_n) p_{T_c}(t - T_d - nT_c) \tag{9-159}$$

Identical arguments may be used to simplify three other terms of $\epsilon(t, \delta)$, resulting in

$$\epsilon(t, \delta) \approx \tfrac{1}{2} K_1 P \Bigg[\sum_{n=-\infty}^{\infty} p_{T_c}(t - T_d - nT_c) p_{T_c}(t - \hat{T}_d - \tfrac{1}{2}T_c - nT_c)$$

$$- \sum_{m=-\infty}^{\infty} p_{T_c}(t - T_d - mT_c) p_{T_c}(t - \hat{T}_d + \tfrac{1}{2}T_c - mT_c) \Bigg]$$

$$+ \sqrt{2K_1 P} \cos [\theta_d(t - T_d)] n_{1I}^0(t) \cos \varphi(t)$$

$$+ \sqrt{2K_1 P} \sin [\theta_d(t - T_c)] n_{1I}^0(t) \sin \varphi(t)$$

$$+ \sqrt{2K_1 P} \sin [\theta_d(t - T_d)] n_{1Q}^0(t) \cos \varphi(t)$$

$$- \sqrt{2K_1 P} \cos [\theta_d(t - T_d)] n_{1Q}^0(t) \sin \varphi(t)$$

$$+ [n_{2I}^0(t)]^2 + [n_{2Q}^0(t)]^2 - [n_{1I}^0(t)]^2 - [n_{1Q}^0(t)]^2 \tag{9-160}$$

The first term of this equation contains the desired phase correction information in its dc component. Calculating the dc component and denoting the remaining signal by $n_s(t)$ yields

$$\epsilon(t, \delta) = \tfrac{1}{2} K_1 P\{R[(\delta - \tfrac{1}{2})T_c] - R[(\delta + \tfrac{1}{2})T_c]\} + n_s(t)$$

$$+ \sqrt{2K_1 P} \cos [\theta_d(t - T_d) - \varphi(t)] n_{1I}^0(t)$$

$$+ \sqrt{2K_1 P} \sin [\theta_d(t - T_d) - \varphi(t)] n_{1Q}^0(t)$$

$$+ [n_{2I}^0(t)]^2 + [n_{2Q}^0(t)]^2 - [n_{1I}^0(t)]^2 - [n_{1Q}^0(t)]^2 \tag{9-161}$$

where

$$R(\tau) = \begin{cases} 0 & \text{for } \tau < -T_c \\[2mm] \dfrac{\tau}{T_c} + 1.0 & \text{for } -T_c \le \tau < 0 \\[4mm] -\dfrac{\tau}{T_c} + 1.0 & \text{for } 0 \le \tau < T_c \\[4mm] 0 & \text{for } T_c \le \tau \end{cases} \tag{9-162}$$

The immediate goal of this analysis is the calculation of the power spectrum of $\epsilon(t, \delta)$. This power spectrum is calculated in the usual manner through calculation of the autocorrelation function. In this calculation, $n_s(t)$ will be ignored since its power spectrum consists of impulses at harmonics of the hop frequency, which is assumed to be much higher than the bandwidth of the loop filter. The calculation of the autocorrelation function is simplified by taking advantage of the fact that $\theta_d(t)$, $\varphi(t)$, and all noise components are independent. Thus the expectations factor and many terms are immediately seen to equal zero. The calculation is very similar to the calculation detailed in Appendix H. The result, after some manipulation, is

$$R_\epsilon(\tau) = \tfrac{1}{4}K_1^2 P^2\{R[(\delta - \tfrac{1}{2})T_c] - R[(\delta + \tfrac{1}{2})T_c]\}^2$$

$$+ 2K_1 PR_{n^0}(\tau)\, E\{\cos [\varphi(t) - \varphi(t + \tau)]\}E\{\cos [\theta_d(t) - \theta_d(t + \tau)]\}$$

$$- 2K_1 PR_{n^0}(\tau)E\{\sin [\varphi(t) - \varphi(t + \tau)]\}E\{\sin [\theta_d(t) - \theta_d(t + \tau)]\}$$

$$+ 4E\{[n^0(t)]^2[n^0(t + \tau)]^2\} - 4E^2\{[n^0(t)]^2\} \tag{9-163}$$

In this equation, $n^0(t)$ represents any one of the four independent processes $n_{1I}^0(t)$, $n_{1Q}^0(t)$, $n_{2I}^0(t)$, or $n_{2Q}^0(t)$.

Consider the expectations $E\{\cos [\varphi(t) - \varphi(t + \tau)]\}$ and $E\{\sin [\varphi(t) - \varphi(t + \tau)]\}$. The expected value is over all phase angles φ_n and φ_n' of (9-159) as well as all transmission delays T_d. From (9-159),

$$\varphi(t) - \varphi(t + \tau) = \sum_{n=-\infty}^{\infty} (\varphi_n - \varphi_n')p_{T_c}(t - T_d - nT_c)$$

$$- \sum_{m=-\infty}^{\infty} (\varphi_m - \varphi_m')p_{T_c}(t + \tau - T_d - mT_c) \tag{9-164}$$

For $|\tau| < T_c$, the unit pulses overlap for a fraction $(T_c - |\tau|)/T_c$ of all T_d. For these T_d, $\varphi(t) - \varphi(t + \tau) = 0$, and for all other T_d, $\varphi(t) - \varphi(t + \tau)$ is uniformly distributed over $(0, 2\pi)$. Thus

$$E\{\cos [\varphi(t) - \varphi(t + \tau)]\} = \frac{T_c - |\tau|}{T_c} \qquad \text{for } |\tau| < T_c \tag{9-165}$$

$$E\{\sin [\varphi(t) - \varphi(t + \tau)]\} = 0 \qquad \text{for } |\tau| < T_c \tag{9-166}$$

For $T_c < |\tau|$, the unit pulses do not overlap, $\varphi(t) - \varphi(t + \tau)$ is uniformly distributed over $(0, 2\pi)$ for all T_d, and both expectations are zero. Thus

$$R_\epsilon(\tau) = \tfrac{1}{4} K_1^2 P^2\{R[(\delta - \tfrac{1}{2})T_c] - R[(\delta + \tfrac{1}{2})T_c]\}^2$$

$$+ 2K_1 PR_{n^0}(\tau)R_h(\tau)2\,\text{Re}\,[R_A(\tau)]$$

$$+ 4E\{[n^0(t)]^2[n^0(t + \tau)]^2\} - 4E^2\{[n^0(t)]^2\} \tag{9-167}$$

where $R_A(\tau)$ is the data modulation autocorrelation, and

$$R_h(\tau) = \begin{cases} 1 - \dfrac{|\tau|}{T_c} & |\tau| \le T_c \\ 0 & \text{elsewhere} \end{cases} \tag{9-168}$$

The Fourier transform of (9-167) yields the desired delay-lock discriminator output power spectral density. The first term is a constant and its Fourier transform is a delta function at zero frequency. The second term is the product of three autocorrelations and its Fourier transform is the convolution of the Fourier transforms of the individual terms. The transform of $R_{n^0}(\tau)$ is the power spectrum of any one of the lowpass white noise processes given by

$$S_{n^0}(f) = \begin{cases} \tfrac{1}{4} K_1 N_0 & |f| < \tfrac{1}{2} B_N \\ 0 & \text{elsewhere} \end{cases} \tag{9-169}$$

The transform of $R_h(\tau)$ is

$$S_h(f) = T_c\, \text{sinc}^2\, (fT_c) \tag{9-170}$$

and the transform of $2 \operatorname{Re} \{R_A(\tau)\}$ is denoted by $S_{\theta_d}(f)$ and is given by Equation (9-68). The transform of the last two terms of (9-167) are evaluated in Appendix H; the result is $8S_{n^0}(f) * S_{n^0}(f)$. Combining these results yields

$$S_\epsilon(f) = \tfrac{1}{4} K_1^2 P^2 \{R[(\delta - \tfrac{1}{2})T_c] - R[(\delta + \tfrac{1}{2})T_c]\}^2 \, \delta(f) \tag{9-171}$$
$$+ \, 2K_1 P S_{n^0}(f) * S_h(f) * S_{\theta_d}(f) + 8S_{n^0}(f) * S_{n^0}(f)$$

In order to generate specific code tracking jitter results for comparison with the results of the other analyses of this chapter, consider the case where there is no data modulation. In this case, $S_{\theta_d}(f) = \delta(f)$. It was assumed earlier that the frequency-hop rate is slow relative to the IF filter bandwidth. Therefore, $S_{n^0}(f)$ is nearly constant over the range of significant values of $S_h(f)$ and

$$S_{n^0}(f) * S_h(f) = \int_{-\infty}^{\infty} S_{n^0}(\lambda) S_h(f - \lambda) \, d\lambda$$

$$\approx S_{n^0}(f) \int_{-\infty}^{\infty} S_h(f - \lambda) \, d\lambda = S_{n^0}(f) \tag{9-172}$$

Thus

$$S_\epsilon(f) \approx \tfrac{1}{4} K_1^2 P^2 \{R[(\delta - \tfrac{1}{2})T_c] - R[(\delta + \tfrac{1}{2})T_c]\}^2 \, \delta(f)$$
$$+ \, 2K_1 P S_{n^0}(f) + 8S_{n^0}(f) * S_{n^0}(f) \tag{9-173}$$

which is illustrated in Figure 9-25. This figure should be compared to Figure 9-11 for the full-time early-late direct sequence code tracking loop.

The discriminator S-curve is defined by the discriminator dc output component. The normalized S-curve, $D_\Delta(\delta)$, is plotted in Figure 9-26. It is defined by

$$D_\Delta(\delta) = R[(\delta - \tfrac{1}{2})T_c] - R[(\delta + \tfrac{1}{2})T_c]$$

with $R(\tau)$ defined by (9-162). The total discriminator output is then

$$\epsilon(t, \delta) = \tfrac{1}{2} K_1 P D_\Delta(\delta) + n_\epsilon(t) \tag{9-174}$$

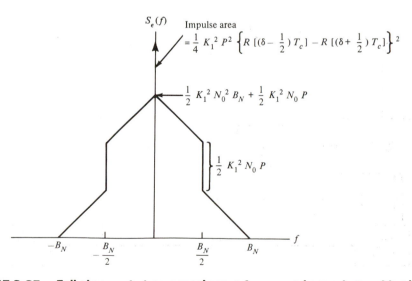

FIGURE 9-25. Full-time early-late noncoherent frequency-hop code tracking loop discriminator output power spectrum.

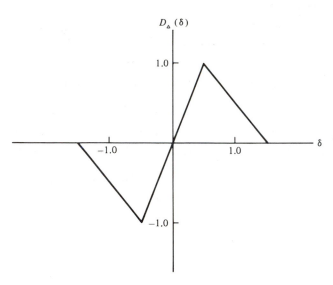

FIGURE 9-26.   Normalized *S*-curve for full-time early-late noncoherent code tracking loop discriminator.

where $n_\epsilon(t)$ represents all output noise. From this point on the loop analysis is identical to that of Section 9-4. The linearized model of the loop is given in Figure 9-13 with

$$K_d = \tfrac{1}{2} K_1 P \left[\frac{d}{d\delta} D_\Delta(\Delta) \right]_{\delta=0} = K_1 P \qquad (9\text{-}175)$$

and where the input noise power spectrum has a magnitude given by the zero-frequency value of the continuous component of Figure 9-25 multiplied by $1/K_d^2$.

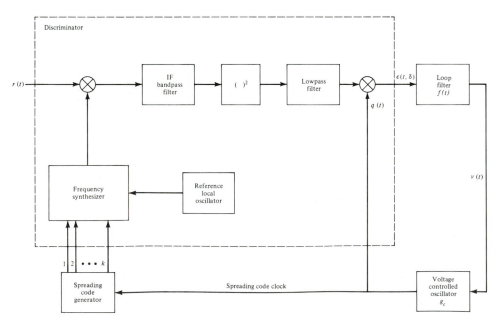

FIGURE 9-27.   Time-shared early-late noncoherent code tracking loop configuration for slow-frequency hop. (From Ref. 13.)

Finally, the variance of the code tracking jitter is calculated using (9-81); the result is

$$\sigma_\delta^2 = \frac{\frac{1}{2} K_1^2 N_0^2 B_N + \frac{1}{2} K_1^2 N_0 P}{(K_1 P)^2} W_L$$

$$= \frac{1}{\rho_L}\left(1 + \frac{1}{\rho_{IF}}\right) \tag{9-176}$$

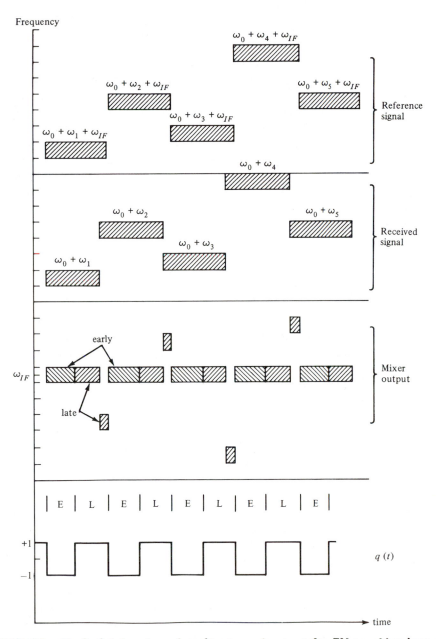

FIGURE 9-28. Typical dehopping mixer inputs and output for **FH** tracking loop configuration of **Figure 9-27.**

This result should be compared to (9-84). The increased tracking jitter is the result of an increased (signal × noise) term in the power spectrum of Figure 9-25, which is, in turn, the result of the fact that slow frequency hop is used.

This completes the analysis of the full-time frequency hop code tracking loop. Other tracking loop configurations are possible. A configuration that is similar to the direct-sequence dithered tracking loop is illustrated in Figure 9-27. This configuration can be derived from the loop of Figure 9-23 by observing that, when δ is small, one-half of the early or late channel mixer output power is out of band and is rejected by the IF bandpass filter. In addition, the two channels never translate the received signal to IF at the same time so that time sharing a single filter and square law detector appears possible. Figure 9-28 illustrates typical de-hopping mixer input and output spectra. The reference signal may be thought of as an early reference for one-half time and a late reference for the other one-half time. This is illustrated by dividing the output spectrum into two regions denoted by different crosshatching in Figure 9-28. When the reference is late as illustrated, the duration of the late half of the mixer output is reduced and a dc correction is generated at the output of the second mixer. Similarly, when the reference is early, the early portion of the mixer output is reduced. Comparison of Figures 9-24 and 9-28 will show that the slope of the discriminator characteristic is larger by a factor of 2 for the full-time loop. This is seen by observing that whereas the early or late channel output is reduced by a tracking error in the time-shared loop, in the full-time loop, the output power lost in one channel is added to the other. It is conjectured that the performance of the time-shared loop is slightly inferior to that of the full-time tracking loop.

9-9

SUMMARY

The primary goal of this chapter has been to present a method of analysis that can be applied to all types of code tracking loops. In all cases, the analysis was aimed at developing a linear model of the tracking loop identical to that of Figure 9-7. This model was developed through characterization of the signal and noise components of the discriminator output. Accomplishing this task required numerous assumptions to be made about the independence of certain noise processes and the relative bandwidths of certain signals. The analysis is sufficiently detailed that the point of departure for extensions should be clear. A factor that has been consistently ignored is the effect of distortions due to the IF bandpass filter for the noncoherent loops. The student is referred to Simon [6] for an analysis that does not make this assumption.

Table 9-1 summarizes a few of the most significant results of this chapter. For noncoherent systems, the full-time loops using either BPSK or MSK spreading clearly provide the best noise performance. MSK spreading appears to be a good choice for high data rates where the IF bandwidth is large, resulting in low IF signal-to-noise ratios. The tau-dither tracking loop provides inferior noise performance but is significantly simpler in hardware and also solves the gain-balance problem of the full-time loops. In any system design, many parameters must be optimized which are not explicitly shown in Table 9-1. For example, early-late delay differences other than $\Delta = 1.0$ may be desirable in some cases. For the dithering loop, other dithering frequencies may provide superior performance under some conditions.

TABLE 9-1. Summary of Direct-Sequence Code Tracking Loop Noise Performance Results for $\Delta = 1.0$

Type of Loop	Discriminator Gain, K_d	Equivalent Input Noise PSD	Code Tracking Jitter Variance
Full-time baseband coherent loop	$K_1\sqrt{2P}$	$\dfrac{N_0}{4P}$	$\sigma_\delta^2 = \dfrac{1}{2\rho_L}$
Full-time early-late noncoherent loop with BPSK spreading	$K_1 P$	$\dfrac{N_0}{4P} + \dfrac{B_N N_0^2}{2P^2}$	$\sigma_\delta^2 = \dfrac{1}{2\rho_L}\left(1 + \dfrac{2}{\rho_{IF}}\right)$
Tau-dither early-late noncoherent loop with BPSK spreading and $f_q = B_N/4$	$K_1 P$	$\dfrac{N_0}{4P}(1.81) + \dfrac{B_N N_0^2}{2P^2}(1.63)$	$\sigma_\delta^2 = \dfrac{1}{2\rho_L}\left(1.811 + \dfrac{3.261}{\rho_{IF}}\right)$
Full-time early-late noncoherent loop with MSK spreading	$1.26 K_1 P$	$\dfrac{N_0}{4P}(1.44) + \dfrac{B_N N_0^2}{2P^2}(0.63)$	$\sigma_\delta^2 = \dfrac{1}{2\rho_L}\left(1.44 + \dfrac{1.26}{\rho_{IF}}\right)$

This concludes the discussion of code tracking loops for spread-spectrum systems. The references provide a large selection of material for further reading.

REFERENCES

[1] J. J. SPILKER and D. T. MAGILL, "The Delay-Lock Discriminator—An Optimum Tracking Device," *Proc. IRE*, September 1961.

[2] J. J. SPILKER, "Delay-Lock Tracking of Binary Signals," *IEEE Trans. Space Electron. Telem.*, March 1963.

[3] W. J. GILL, "A Comparison of Binary Delay-Lock Tracking Loop Implementations," *IEEE Trans. Aerosp. Electron. Syst.*, July 1966.

[4] R. B. WARD, "Digital Communications on a Pseudonoise Tracking Link Using Sequence Inversion Modulation," *IEEE Trans. Commun. Technol.*, February 1967.

[5] H. P. HARTMAN, "Analysis of a Dithering Loop for PN Code Tracking," *IEEE Trans. Aerosp. Electron. Syst.*, January 1974.

[6] M. K. SIMON, "Noncoherent Pseudonoise Code Tracking Performance of Spread Spectrum Receivers," *IEEE Trans. Commun.*, March 1977.

[7] J. J. SPILKER, *Digital Communications by Satellite* (Englewood Cliffs, N.J.: Prentice-Hall, 1977).

[8] J. K. HOLMES *Coherent Spread Spectrum Systems* (New York: Wiley-Interscience, 1982).

[9] R. C. DIXON, *Spread Spectrum Systems* (New York: Wiley-Interscience, 1976).

[10] A. PAPOULIS, *Probability Random Variables and Stochastic Processes* (New York: McGraw-Hill, 1965).

[11] P. M. HOPKINS, "Double Dither Loop for Pseudonoise Code Tracking," *IEEE Trans. Aerosp. Electron. Syst.*, November 1977.

[12] S. STEIN and J. JONES, *Modern Communications Principles* (New York: McGraw-Hill, 1967).

[13] R. L. PICKHOLTZ, D. L. SCHILLING, and L. B. MILSTEIN, "Theory of Spread Spectrum Communications—A Tutorial," *IEEE Trans. Commun.*, May 1982.

PROBLEMS

9-1. Calculate the variance of the code tracking error for a full-time early-late noncoherent code tracking loop in a system that utilizes offset-QPSK spreading modulation. Express the result as a function of the signal-to-noise ratios in the loop bandwidth and in the IF filter bandwidth.

9-2. Consider a spread-spectrum communications system which is used between two platforms that move relative to one another. Suppose that BPSK spreading modulation is used and that the tracking loop is tracking the correct code phase. At time $t = 0$ the platforms begin accelerating at a rate of 1.5 g's relative to one another. This acceleration continues until the relative velocity is 3 mach, at which time the acceleration ceases. Derive a complete expression for the normalized tracking error as a function of time with tracking loop bandwidth as a parameter. Assume that the carrier frequency is 10 GHz, the spreading code rate is 100 MHz, and that the tracking loop is second order and critically damped.

9-3. Derive an expression for the variance of the tracking jitter for a system using BPSK spreading modulation and a tau-dither early-late noncoherent tracking loop using a dither frequency $f_q << B_N$. Assume that $\Delta = 1.0$. Compare the result with the result derived in the text for $f_q = B_N/4$.

9-4. Consider a spread-spectrum communication system using a full-time early-late noncoherent code tracking loop having a noise bandwidth of 1 kHz and an IF filter noise bandwidth of 100 kHz. Plot the variance of the tracking jitter as a function of loop signal-to-noise ratio for $\Delta = 1.0, 0.5,$ and 0.1.

9-5. A tone jammer is being used to attempt to disrupt communications over a link which employs a tau-dither tracking loop with $f_q = B_N/4$ and $\Delta = 1.0$. Assume that the spreading code rate is 100 MHz, the IF filter noise bandwidth is 100 kHz, and that the tracking loop two-sided noise bandwidth is 1 kHz. Derive an expression for the variance of the tracking error as a function of the ratio of received signal power to received jammer power.

9-6. Calculate the sensitivity of the loop bandwidth and damping to changes in received power level for a second-order full-time noncoherent tracking loop and a second-order tau-dither tracking loop. Assume that the design point damping is $\delta = 1/\sqrt{2}$.

9-7. A system designer has a choice between BPSK and MSK spreading modulation. The available transmission bandwidth and the data modulation bandwidth are fixed. What is the best modulation for a system that must achieve minimum absolute tracking error variance?

9-8. Plot the normalized discriminator S-curve for a full-time noncoherent tracking loop using MSK spreading modulation for $\Delta = 0.5$, 1.0, 1.5, and 2.0.

9-9. Consider a received signal

$$r(t) = \sqrt{2P} \cos [\omega_0 t + \theta(t)]$$
$$+ \sqrt{2}\, n_I(t) \cos \omega_0 t - \sqrt{2}\, n_Q(t) \sin \omega_0 t$$

where $n_I(t)$ and $n_Q(t)$ are independent lowpass white Gaussian noise processes with two-sided bandwidth B_N and two-sided noise power spectral density $N_0/2$. Derive an alternative representation for the noise using $\cos [\omega_0 t + \theta(t)]$ and $\sin [\omega_0 t + \theta(t)]$ as quadrature carriers and determine what conditions are necessary for $n_I'(t)$ and $n_Q'(t)$ of this representation to be independent.

9-10. Define the sum of two random processes $y_1(t)$ and $y_2(t)$ by

$$z(t) = y_1(t) + y_2(t)$$

Derive an expression for the power spectrum of $z(t)$ as a function of the individual and cross power spectra of $y_1(t)$ and $y_2(t)$.

9-11. Consider a linear second-order tracking loop as shown in Figure 9-8 with a loop filter defined by

$$F(s) = \frac{1 + \tau_2 s}{1 + \tau_1 s}$$

Derive expressions for the dynamic tracking error when the input is defined by (a) $T_d(t)/T_c = AU(t)$; (b) $T_d(t)/T_c = BtU(t)$; (c) $T_d(t)/T_c = Ct^2U(t)$. In these expressions, $U(t)$ is the unit step function.

9-12. Repeat Problem 9.11 for

$$F(s) = \frac{1 + \tau_2 s}{\tau_1 s}$$

9-13. A BPSK transmitter is moving at a constant velocity V_r relative to its intended receiver. The transmitted carrier frequency is f_t and the transmitted bit rate is R_t. The receiver carrier synchronization and bit synchronization loops operate perfectly. What is the receiver output bit rate R_r? Prove your result.

9-14. The figure below illustrates three different mechanizations for a second-order loop filter.
(a) Derive the filter transfer function $F(s)$ for each filter shown.
(b) Derive the closed-loop transfer function for each filter shown.
Assume that the operational amplifier has infinite input impedance, zero output impedance, but finite gain.

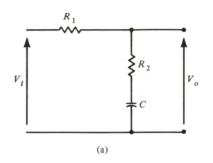

(a)

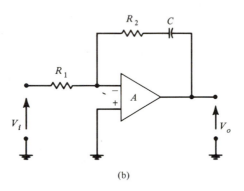

(b)

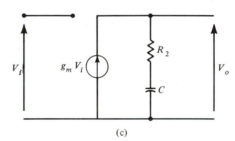

(c)

PROBLEM 9-14.   Loop filters for second-order tracking loops.

9-15.   Two bandpass signals $x_1(t)$ and $x_2(t)$ have complex envelopes $\tilde{x}_1(t)$ and $\tilde{x}_2(t)$ and different carrier frequencies f_1 and f_2. Let $y(t) = x_1(t)x_2(t)$. Prove that the complex envelope of the difference frequency component is

$$\tilde{y}_d(t) = \tfrac{1}{2}\tilde{x}_1(t)\tilde{x}_2^*(t)$$

and that the complex envelope of the sum frequency component is

$$\tilde{y}_s(t) = \tfrac{1}{2}\tilde{x}_1(t)\tilde{x}_2(t)$$

9-16.   Consider the *RLC* bandpass filter illustrated below. Derive the impulse response of the equivalent lowpass filter under the assumption of high-*Q* components.

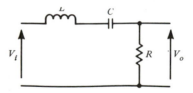

PROBLEM 9-16.   RLC bandpass filter.

9-17. Reconsider the bandpass filter of Problem 9-16. The input to this circuit is a rectangular pulse of a carrier at frequency f_c whose complex envelope is

$$p_T(t) = \begin{cases} A & 0 < t < T \\ 0 & \text{elsewhere} \end{cases}$$

Derive and plot an expression for the filter output.

9-18. The receiver front end illustrated in the figure below is used in a spread-spectrum communication system. At point 1, the received signal power is -116.3 dBm and the received noise single-sided noise power spectral density is -174 dBm per hertz.
   (a) Using the component gains and noise figures shown, calculate the value of E_b/N_0 at the input to the delay-lock discriminator. Assume a data rate of 10 kbps.
   (b) Assume a filter noise bandwidth of 10 MHz and calculate the signal-to-noise ratio P_s/P_N at the input to the delay-lock discriminator.

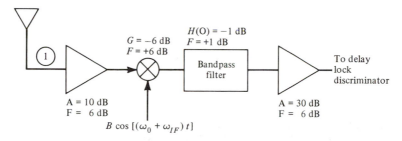

PROBLEM 9-18.   Typical receiver front end.

9-19. Consider a full-time early-late noncoherent code tracking loop designed so that the ratio of the signal power to thermal noise power at the IF bandpass filter output is $+10$ dB. When spreading code self-noise is taken into account, calculate the additional signal power required to achieve $+10$ dB SNR as a function of the ratio of spreading code chip rate to IF filter bandwidth. Assume that $\Delta = 1.0$ and that the data modulation spectrum is uniform over the IF filter bandwidth.

9-20. In actual spread-spectrum systems, the squaring device in the energy detectors of the phase discriminators are usually crystal detectors characterized by their sensitivity in units of millivolts of output per milliwatt of input. These devices can be modeled by an ideal squaring device either preceded by or followed by an ideal amplifier. Calculate the gain of the amplifier in both cases as a function of the detector sensitivity K.

9-21. A BPSK spread-spectrum system is to be used for determining the range between a satellite whose position is accurately known and a vehicle on the surface of the earth. This range is measured by measuring the propagation time of the spread spectrum signal using the spreading code epoch as a time reference. Select a spreading code, spreading code rate, tracking loop configuration, and estimate the required signal-to-noise ratio in the tracking loop bandwidth to achieve a variance of 3 m for the range measurement. The system should be designed to have a minimum range ambiguity of 10,000 km.

CHAPTER 10

Initial Synchronization of the Receiver Spreading Code

10-1

INTRODUCTION

The analysis of the code tracking loops in Chapter 9 presumed that the received spreading waveform and receiver-generated replica of the spreading waveform are initially synchronized in both phase and frequency. This initial synchronization of the spreading waveforms is a significant problem in spread-spectrum system design and is the subject of the present chapter. Phase/frequency synchronization is difficult because typical spreading waveform periods are long and bandwidths are large. Thus uncertainty in the estimated propagation delay $\hat{T}_d$ translates into a large number of symbols of code phase uncertainty. Oscillator instabilities and doppler frequency shifts result in frequency uncertainties which must also be resolved. The correct code phase/frequency must be found quickly using the minimum amount of hardware possible. In many cases this process must be accomplished at very low signal-to-noise ratios or in the presence of jamming.

A widely used technique for initial synchronization is to search serially through all potential code phases and frequencies until synchronization is achieved. Each reference phase/frequency is evaluated by attempting to despread the received signal. If the estimated code phase and frequency are correct, despreading will occur and will be sensed. If the code phase or frequency is incorrect, the received signal will not be despread, and the reference waveform will be stepped to a new phase/frequency for evaluation. This technique is called *serial search*. Because of the widespread use of serial search techniques, a large fraction of this chapter is devoted to the analysis of this technique. Specifically, strategies for rapidly evaluating the correctness of the trial phase are discussed in detail. In addition, a method for calculating mean synchronization time is given. This calculation requires knowledge of the probability of false alarm P_{fa} and the probability of detection P_d as a function of evaluation (integration) time and signal-to-noise ratio for various methods of detecting a sine wave in white Gaussian noise. Methods of calculating P_d and P_{fa} are described. The effects of modified sweep patterns are also briefly discussed.

A highly efficient method of initial synchronization is to matched filter detect the received signal directly. The matched filter is designed to output a pulse when a particular sequence of code symbols is received. When this pulse is sensed, the receiver code generator is started using an initial condition corresponding to the received code stream and synchronization is complete. This technique requires matched filters with extremely large time–bandwidth products. The difficulty of implementing these filters is the principal reason this synchronization technique is not more widely used. In the usual case where the spreading code period is long, the matched filter must also be easily programmable. For example, the spreading code period may be longer than the entire mission being carried out. Thus a matched filter designed for a single sequence of code symbols may never receive these symbols. A programmable matched filter can be programmed for a particular code phase dependent on an estimate of the propagation delay. The requirement for programmability has led researchers to investigate high-speed digital correlators as well as programmable convolvers for this application.

Consider a direct-sequence spread-spectrum system employing BPSK spreading. When the spreading code is an *m*-sequence that is generated using the generator of Figure 8-6 a convenient method of synchronization is to demodulate the received symbol stream directly and to load the shift register generator with this symbol stream. At high signal-to-noise ratios it is likely that the demodulated symbols are correct, so that the shift register initial condition is also correct. At reduced signal-to-noise ratios, the shift register load may not be correct and additional loads will have to be attempted. Each load is evaluated by attempting to despread the received signal just as was described for serial search techniques. This technique is called *rapid acquisition by sequential estimation* (RASE) and was first described by Ward [1]. A slightly modified version of this technique performs a preliminary evaluation of the shift register load by comparing its output with several received symbols prior to attempting the full correlation despreading evaluation. This modified technique is called *recursion-aided RASE* (RARASE) and is described in Ward and Yiu [2]. Both of these techniques are described in this chapter.

Yet another synchronization technique is to use special-purpose synchronization codes such as described in Section 8-5. These codes are designed to have good correlation with all square-wave sequences. The received signal is correlated with a series of square waves having increasing periods. Each correlation eliminates one-half of the possible code phase positions so that a total of $\log_2 N$ correlations will completely identify the phase of a received code with a period of N symbols. Unfortunately, the use of spreading codes with very specific correlation properties implies an increased vulnerability to detection and to jamming. For this reason, the use of special-purpose synchronization codes is limited to nonmilitary systems.

The number of code phases and frequencies that must be evaluated to obtain initial synchronization is proportional to propagation delay uncertainty expressed in spreading code chips and the relative dynamics of the transmitter and receiver. Since the code chip duration is inversely proportional to the chip rate, synchronization time is also directly proportional to the clock rate used for the spreading code generators. Since a frequency-hop spread-spectrum system may employ a clock rate which is much lower than the transmission bandwidth, synchronization time for frequency-hop systems is typically much lower than for direct-sequence systems having the same transmission bandwidth. This factor is a principal reason why frequency hopping has been selected for some current spread-spectrum systems. Initial synchronization techniques for frequency-hop spread-spectrum systems are the same as those used for direct-sequence systems except that RASE techniques and the special synchronization codes of Section 8-5 are not applicable.

The techniques to be described in this chapter are generally capable of determining the received spreading code phase to within an accuracy of $\pm \frac{1}{2}$ to $\pm \frac{1}{4}$ of a chip. When the code tracking loop is closed, there may therefore be a phase error of $\pm \frac{1}{2}$ chip which the tracking loop is expected to eliminate. This transition from the completion of the initial synchronization function to fine code tracking is called *tracking loop pull-in* and is important because it affects the required loop bandwidth selection. The study of the tracking loop pull-in characteristic is a nonlinear analysis problem. A method for evaluating the loop pull-in trajectories is discussed.

10-2

PROBLEM DEFINITION AND THE OPTIMUM SYNCHRONIZER

In general, both the phase and frequency of the received spread-spectrum signal will be unknown to the receiver. In this chapter it will be assumed that a priori information is available to the receiver which bounds the phase uncertainty to a range of ΔT seconds and the frequency uncertainty to a range $\Delta \Omega$ radians/second. The method almost universally employed to evaluate the receiver reference phase and frequency is to attempt to despread the received signal using that phase and frequency. An energy detector at the despreader output measures the signal plus noise energy in a narrow bandwidth at a known frequency. If the phase and frequency of the receiver-generated replica of the spreading waveform are correct, the received signal will be collapsed in bandwidth, translated to the center frequency of the bandpass filter, and the energy detector will detect the presence of signal. Figure 10-1 is a simplified conceptual block diagram of the functions needed to evaluate a single reference phase and frequency. The control logic shown in dashed lines may be added to select values of $\hat{T}_d$ and $\hat{\omega}$ for evaluation.

The system of Figure 10-1 will yield a "correct phase/frequency" or "hit" decision over a range of phase/frequency values near the correct values. The size of this range is calculated from the spreading waveform autocorrelation function and the bandwidth of the IF filter. For example, suppose that the received signal $r(t)$ is a carrier which is BPSK modulated by an m-sequence waveform $c(t - T_d)$ plus AWGN. Thus

$$r(t) = c(t - T_d) \cos (\omega_0 t + \varphi) + n(t) \tag{10-1}$$

where ω_0 is the carrier frequency and φ is the random received carrier phase. The reference waveform is

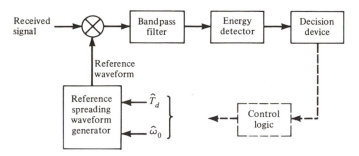

FIGURE 10-1. System used to evaluate a single spreading waveform phase and frequency.

$$a(t) = c(t - \hat{T}_d) \cos [(\hat{\omega}_0 + \omega_{IF})t] \qquad (10\text{-}2)$$

and the despreader output difference frequency term is

$$x(t) = c(t - T_d)c(t - \hat{T}_d) \cos [\omega_{IF}t + (\hat{\omega}_0 - \omega_0)t - \varphi] + n'(t) \qquad (10\text{-}3)$$

The power spectrum of $c(t - T_d)c(t - \hat{T}_d)$ is given in (8-59) and plotted in Figure 8-14 for m-sequence spreading codes. This power spectrum contains an impulse at zero frequency which corresponds to a single spectral line at $\omega = \omega_{IF} + \hat{\omega}_0 - \omega_0$ at the output of the despreader. The power in this component is $R_c^2(\tau)$, where $R_c(\tau)$ is the autocorrelation of the spreading waveform and $\tau = \hat{T}_d - T_d$. The power in this component will be sensed by the energy detector if τ is sufficiently small so that $R_c(\tau)$ is near unity and if $\hat{\omega}_0 - \omega_0$ is not so large that the desired component will lie outside the passband of the bandpass filter. The phase/frequency uncertainty region may be graphically depicted as a rectangle with dimensions $\Delta\Omega \times \Delta T$ as in Figure 10-2. This rectangle can be subdivided into smaller rectangles whose dimensions Δt and $\Delta\omega$ are the range around the correct phase/frequency over which the system of Figure 10-1 will yield a "hit." Thus a single test in each cell of Figure 10-2 will be sufficient to determine the correct received phase and frequency.

An initial synchronization system which is optimum in the sense that it achieves synchronization with a given probability in the minimum possible time is one that evaluates all cells of Figure 10-2 simultaneously. This system requires a subsystem similar to that shown in Figure 10-1 for every phase/frequency cell of

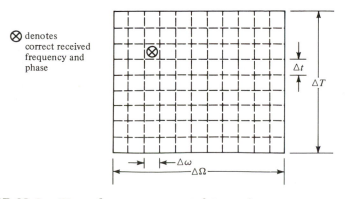

FIGURE 10-2. Phase frequency uncertainty region.

Figure 10-2 and is thus not optimum in a minimum-hardware sense. This minimum-acquisition-time system is never implemented because of its excessive hardware complexity. All the systems discussed in this chapter are designed to achieve a compromise between acquisition time and reasonable complexity, without compromising jamming resistance or any other important system characteristic. Synchronization time is proportional to the total number of cells that must be evaluated and it makes no difference whether these cells were all generated from frequency uncertainty or all from phase uncertainty or a combination. Thus, in some of the following analysis, frequency uncertainty will be ignored with the understanding that phase uncertainty is scaled appropriately.

10-3

SERIAL SEARCH SYNCHRONIZATION TECHNIQUES

Serial search techniques are by far the most commonly used spread-spectrum synchronization techniques. Any synchronization system that evaluates the phase/frequency cells of Figure 10-2 serially (i.e., one after another) until the correct cell is found is said to use serial search. In this section the mean and variance of the synchronization time are calculated for a very general serial search system. The result is simple enough that the system designer will be able to use it to make design trade-off studies to minimize mean synchronization time. Inputs to the mean synchronization time result are the probability of detection when the correct cell is being evaluated, P_d, and the probability of false alarm when an incorrect cell is being evaluated, P_{fa}, as a function of the integration time T_i and SNR. The calculation of P_d and P_{fa} is also discussed in detail.

10-3.1 Calculation of the Mean and Variance Synchronization Time

For simplicity assume that no frequency uncertainty exists and that the correct phase is uniformly distributed over the region ΔT. Based on the spreading waveform autocorrelation function, a phase step size of Δt seconds has been chosen. For convenience assume that $\Delta T / \Delta t$ is an integer C. Because it is equally probable that the correct phase is in any cell, the serial search can begin anywhere. Let the search begin at one boundary of the uncertainty region. The search will advance through one cell at a time until C cells have been evaluated. If synchronization has not been achieved at that time, a retrace will start the search over again at the starting position.

The mean synchronization time is calculated in a straightforward manner by considering all possible sequences of events leading to a correct synchronization. An event in the probability space being considered is defined by a particular location n for the correct phase cell, a particular number of missed detections j of the correct phase cell, and a particular number of false alarms k in all incorrect phase cells evaluated. The total synchronization time for a particular event defined by (n, j, k) is

$$T(n, j, k) = nT_i + jCT_i + kT_{fa} \tag{10-4}$$

where T_i is the (fixed) integration time for evaluation of each cell, and T_{fa} is the time required to reject an incorrect cell when a false alarm occurs. The false-alarm penalty T_{fa} may be many times larger than T_i, so that false alarms are undesirable events. The total number of cells evaluated for this event is $(n + jC)$, the total number of correct cells is $(j + 1)$, and the total number of incorrect cells is $(n + jC - j - 1)$. The probability of the correct cell being the nth cell is $1/C$ and the probability of j missed detections followed by a correct detection is $P_d(1 - P_d)^j$. The k false alarms can occur in any order within the $(n + jC - j - 1) \triangleq K$ incorrect cells. The probability of a particular ordering is $P_{\text{fa}}^k(1 - P_{\text{fa}})^{K-k}$, and there are $\binom{K}{k}$ orderings of k false alarms in K cells.

Combining all of the above, the probability of the event (n, j, k) is

$$\Pr(n, j, k) = \frac{1}{C} P_d(1 - P_d)^j \binom{K}{k} P_{\text{fa}}^k(1 - P_{\text{fa}})^{K-k} \tag{10-5}$$

The mean synchronization time is

$$\overline{T}_s = \sum_{n,j,k} T(n, j, k) \Pr(n, j, k) \tag{10-6}$$

The correct cell number can range over $(1, C)$, the number of missed detections can range from zero to infinity, and the number of false alarms ranges over $(0, K)$. Thus the mean synchronization time is

$$\overline{T}_s = \frac{1}{C} \sum_{n=1}^{C} \sum_{j=0}^{\infty} \sum_{k=0}^{K} [(n + jC)T_i + kT_{\text{fa}}] \binom{K}{k} P_{\text{fa}}^k(1 - P_{\text{fa}})^{K-k} P_d(1 - P_d)^j$$

$$= \frac{1}{C} \sum_{n=1}^{C} \sum_{j=0}^{\infty} (n + jC)T_i \left[\sum_{k=0}^{K} \binom{K}{k} P_{\text{fa}}^k(1 - P_{\text{fa}})^{K-k} \right] P_d(1 - P_d)^j$$

$$+ \frac{1}{C} \sum_{n=1}^{C} \sum_{j=0}^{\infty} T_{\text{fa}} \left[\sum_{k=0}^{K} \binom{K}{k} k P_{\text{fa}}^k(1 - P_{\text{fa}})^{K-k} \right] P_d(1 - P_d)^j \tag{10-7}$$

The first summation over k is evaluated using the identity

$$\sum_{k=0}^{K} \binom{K}{k} a^k b^{K-k} = \binom{K}{0} b^K + \binom{K}{1} ab^{K-1} + \cdots + \binom{K}{K} a^K$$

$$= (b + a)^K$$

With $a = P_{\text{fa}}$ and $b = 1 - P_{\text{fa}}$, $b + a = 1.0$ and the first summation over k equals unity. The second sum over k is the mean of a discrete random variable which has a binomial distribution. The mean value is KP_{fa} [3]. Thus (10-7) simplifies to

$$\overline{T}_s = \frac{1}{C} \sum_{n=1}^{C} \sum_{j=0}^{\infty} [(n + jC)T_i + KT_{\text{fa}}P_{\text{fa}}]P_d(1 - P_d)^j \tag{10-8}$$

Expanding this expression using $K = n + jC - j - 1$ and simplifying yields

$$\overline{T}_s = \frac{1}{C} \sum_{n=1}^{C} \sum_{j=0}^{\infty} [(n - 1)T_{\text{da}} + (j + 1)T_i + j(C - 1)T_{\text{da}}]P_d(1 - P_d)^j \tag{10-9}$$

where

$$T_{\text{da}} = T_i + T_{\text{fa}}P_{\text{fa}} \tag{10-10}$$

is the average dwell time at an incorrect phase cell. Equation (10-9) can be evaluated completely using the identities

$$\sum_{i=1}^{L} i = L\left(\frac{L+1}{2}\right) \tag{10-11a}$$

$$\sum_{i=0}^{\infty} m^i = \frac{1}{1-m} \tag{10-11b}$$

$$\sum_{i=0}^{\infty} im^i = \frac{m}{(1-m)^2} \tag{10-11c}$$

After some straightforward algebraic manipulation, the result is

$$\overline{T}_s = (C-1)T_{da}\left(\frac{2-P_d}{2P_d}\right) + \frac{T_i}{P_d} \tag{10-12}$$

This expression for mean synchronization time is a function of P_{fa} through the definition of T_{da}. This result has been derived in Holmes and Chen [4] using signal flow graph techniques.

Synchronization time for any serial search strategy is a random variable. The mean of this random variable has been calculated above. The second moment or variance will now be calculated. Recall that the variance of a random variable x with probability distribution $p_x(\alpha)$ is given by

$$\sigma_x^2 = E[(x - \bar{x})^2] = E[x^2] - E^2[x] \tag{10-13}$$

Since the mean of the synchronization time has already been calculated, the variance will be known if the mean-square value can be calculated. The synchronization time for a received phase which lies in the nth cell and which is detected after j missed detections and k false alarms is given by (10-4). The mean-square value is calculated using

$$\overline{T_s^2} = \sum_{n,j,k} T^2(n, j, k) \Pr[n, j, k] \tag{10-14}$$

The probability distribution $\Pr[n, j, k]$ is the same as used to calculate the mean. Substituting (10-4) and (10-5) into (10-14) yields

$$\overline{T_s^2} = \frac{1}{C} \sum_{n=1}^{C} \sum_{j=0}^{\infty} \sum_{k=0}^{K} [(n + jC)T_i + kT_{fa}]^2 \binom{K}{k} P_{fa}^k (1 - P_{fa})^{K-k} P_d (1 - P_d)^j \tag{10-15}$$

Expanding the square and grouping like powers of k results in

$$\overline{T_s^2} = \frac{P_d}{C} \sum_{n=1}^{C} \sum_{j=0}^{\infty} \sum_{k=0}^{K} (A_2 k^2 + A_1 k + A_0) \binom{K}{k} P_{fa}^k (1 - P_{fa})^{K-k} (1 - P_d)^j \tag{10-16}$$

where $A_2 = T_{fa}^2$
$A_1 = 2(n + jC)T_i T_{fa}$
$A_0 = (n + jC)^2 T_i^2$

This expression simplifies by recognizing the sums over k are the moments of a binomially distributed random variable. The first sum is [3]

$$\sum_{k=0}^{K} A_2 k^2 \binom{K}{k} P_{fa}^k (1 - P_{fa})^{K-k} = A_2[K^2 P_{fa}^2 + K P_{fa}(1 - P_{fa})] \tag{10-17}$$

and the other two terms were evaluated above. Thus (10-16) simplifies to

$$\overline{T_s^2} = \frac{P_d}{C} \sum_{n=1}^{C} \sum_{j=0}^{\infty} \{A_2[K^2 P_{\text{fa}}^2 + KP_{\text{fa}}(1 - P_{\text{fa}})] + A_1[KP_{\text{fa}}] + A_0\}(1 - P_d)^j$$

$$(10\text{-}18)$$

This expression is evaluated by grouping like powers of j after substituting $K = n + jC - j - 1$, which results in

$$\overline{T_s^2} = \frac{P_d}{C} \sum_{n=1}^{C} \sum_{j=0}^{\infty} (B_2 j^2 + B_1 j + B_0)(1 - P_d)^j \qquad (10\text{-}19)$$

where $B_2 = C^2 T_i^2 + 2C(C - 1)T_i T_{\text{fa}} P_{\text{fa}} + T_{\text{fa}}^2 P_{\text{fa}}^2 (C - 1)^2$

$\qquad B_1 = 2nC T_i^2 + 2(2Cn - n - C)T_i T_{\text{fa}} P_{\text{fa}}$
$\qquad\qquad + 2T_{\text{fa}}^2 P_{\text{fa}}^2 (n - 1)(C - 1) + T_{\text{fa}}^2 (C - 1)P_{\text{fa}}(1 - P_{\text{fa}})$
$\qquad B_0 = n^2 T_i^2 + 2n(n - 1)T_i T_{\text{fa}} P_{\text{fa}} + T_{\text{fa}}^2 (n - 1)^2 P_{\text{fa}}^2$
$\qquad\qquad + (n - 1)P_{\text{fa}}(1 - P_{\text{fa}})T_{\text{fa}}^2$

The summations over j can be evaluated using (10-11) together with

$$\sum_{i=0}^{\infty} i^2 m^i = \frac{m(1 + m)}{(1 - m)^3} \qquad (10\text{-}11\text{d})$$

The result is

$$\overline{T_s^2} = \frac{1}{C} \sum_{n=1}^{C} \left[B_2 \frac{(1 - P_d)(2 - P_d)}{P_d^2} + B_1 \frac{1 - P_d}{P_d} + B_0 \right] \qquad (10\text{-}20)$$

The summation over n is evaluated by grouping like powers of n, which yields

$$\overline{T_s^2} = \frac{1}{C} \sum_{n=1}^{C} (D_2 n^2 + D_1 n + D_0) \qquad (10\text{-}21)$$

where $D_2 = (T_i + T_{\text{fa}} P_{\text{fa}})^2$
$\qquad D_1 = -2T_i T_{\text{fa}} P_{\text{fa}} - 2T_{\text{fa}}^2 P_{\text{fa}}^2 + T_{\text{fa}}^2 P_{\text{fa}}(1 - P_{\text{fa}})$
$\qquad\qquad + [2C T_i^2 + 2(2C - 1)T_i T_{\text{fa}} P_{\text{fa}} + 2(C - 1)T_{\text{fa}}^2 P_{\text{fa}}^2]\left(\dfrac{1 - P_d}{P_d} \right)$

$\qquad D_0 = T_{\text{fa}}^2 P_{\text{fa}}^2 - T_{\text{fa}}^2 P_{\text{fa}}(1 - P_{\text{fa}})$
$\qquad\qquad + [-2C T_i T_{\text{fa}} P_{\text{fa}} - 2(C - 1)T_{\text{fa}}^2 P_{\text{fa}}^2 + (C - 1)T_{\text{fa}}^2 P_{\text{fa}}^2(1 - P_{\text{fa}})]\left(\dfrac{1 - P_d}{P_d} \right)$
$\qquad\qquad + [C^2 T_i^2 2C(C - 1)T_i T_{\text{fa}} P_{\text{fa}} + (C - 1)^2 T_{\text{fa}}^2 P_{\text{fa}}^2] \dfrac{(1 - P_d)(2 - P_d)}{P_d^2}$

The summations over n are evaluated using [5]

$$\sum_{n=1}^{C} n = \frac{C(C + 1)}{2} \qquad (10\text{-}22\text{a})$$

$$\sum_{n=1}^{C} n^2 = \frac{C(C + 1)(2C + 1)}{6} \qquad (10\text{-}22\text{b})$$

The remainder of the calculation is simply a bookkeeping exercise. Using (10-12) to obtain the square of the mean value of T_s, the final result is

$$\sigma_{T_s}^2 = \overline{T_s^2} - (\overline{T}_s)^2$$

$$= \left[\frac{C^2 - 1}{12} - \frac{(C - 1)^2}{P_d} + \frac{(C - 1)^2}{P_d^2} \right] T_{da}^2$$

$$+ (2C - 1) \frac{1 - P_d}{P_d^2} T_i^2 + 2(C - 1) \frac{1 - P_d}{P_d^2} T_i T_{fa} P_{fa}$$

$$- (C - 1) \frac{2 - P_d}{2 P_d} T_{fa}^2 P_{fa}^2 + (C - 1) \frac{2 - P_d}{2 P_d} T_{fa}^2 P_{fa} \qquad (10\text{-}23)$$

This result is similar to the result derived in Holmes and Chen [4] using signal flow graph techniques. For $C \gg 1$, $1 - P_d \ll 1$, and $P_{fa} \ll 1.0$, the variance is approximated by

$$\sigma_{T_s}^2 \approx T_{da}^2 C^2 \left(\frac{1}{12} - \frac{1}{P_d} + \frac{1}{P_d^2} \right) \qquad (10\text{-}24)$$

which agrees exactly with Holmes and Chen [4]. For $P_d = 1.0$, $P_{fa} = 0.0$, and $C \gg 1$, the variance is

$$\sigma_{T_s}^2 = T_{da}^2 C^2 (\tfrac{1}{12}) \qquad (10\text{-}25)$$

which is equal to the variance of a random variable which is uniformly distributed over the range $(0, CT_{da})$. The variance increases as P_d falls below unity.

One of the goals of the spread-spectrum system designer will be to design a synchronization system that minimizes mean synchronization time. Equation (10-12) indicates that mean synchronization is a function of P_d, P_{fa}, T_i, T_{fa}, and C. The designer has some degree of control over all these variables, including C. Even though the phase uncertainty region in seconds is fixed by system requirements, the region can be subdivided into any number of cells C by the system designer. The remaining four variables are not independent of one another and therein lies the difficulty in the design. It will be seen quantitatively later that high P_d together with low P_{fa} implies large T_i. Thus there will be an optimum set of P_d, P_{fa}, T_i, T_{fa} which minimizes mean synchronization time. It is not correct to assume that minimum average synchronization time will always be achieved with $1 - P_d \ll 1.0$, so that the correct phase cell is detected on the first sweep. In some cases the selection of a moderate P_d will result in a much lower T_i than a high P_d, and will thus result in reduced average T_s, even though several sweeps of the uncertainty region may be required.

10-3.2 Modified Sweep Strategies

All the results up to this point have been derived assuming that the received spreading waveform phase is uniformly distributed over a particular uncertainty region. Suppose now that the received phase distribution is defined by $p(n)$, the probability that the received phase is within the nth phase cell. When $p(n)$ is known, the sweep strategy should be modified to search the most likely phase cells first and then the less likely cells. For example, if the received phase distribution were Gaussian, a reasonable search strategy would be (Braun [6]) to search the cells within one standard deviation of the most likely cell first and then expand the search to cells within two standard deviations, and so on. Note that $p(n)$ is a discrete distribution and therefore cannot be Gaussian. The distribution of the received phase is, however, continuous; the function $p(n)$ is easily derived from the continuous density and the cell boundaries. Figure 10-3 illustrates the Gaussian distribution and the search strategy proposed by Braun [6]. Observe that the received phase density has been truncated at three standard deviations.

The average synchronization time for a system based on the assumptions illustrated in Figure 10-3 is calculated using the same technique used above for the uniform distribution. Equation (10-6) applies directly. The proper limits for all summations as well as $T(n, j, k)$ and Pr $[n, j, k]$ must be determined. The limits for the first summation are the same as before except that the phase cells will be numbered differently. The limits are $-C/2 \leq n \leq +C/2$ where cell number zero is at the center of the uncertainty region. The discrete probability of the nth phase cell being correct is

$$p(n) = A \exp\left(-\frac{n^2}{2T^2}\right) \qquad -\tfrac{1}{2} C \leq n \leq +\tfrac{1}{2} C \qquad (10\text{-}26)$$

where the constant A is calculated from

$$\sum_{n=-(1/2)C}^{(1/2)C} p(n) = 1.0$$

$$T = \frac{1}{6} C$$

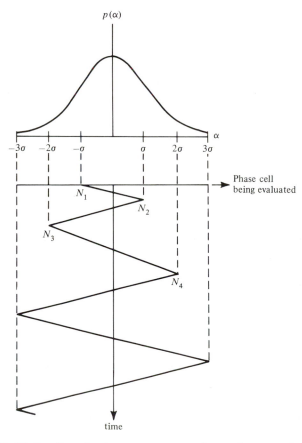

FIGURE 10-3. Received spreading waveform phase probability density and possible sweep strategy. (From Ref. 6.)

Since all phase cells are not evaluated on the first sweep, the probability of finding the correct cell on the first sweep is zero for some n. Each pass (partial or complete) through the uncertainty region will be numbered using the variable b. Let N_b and N_{b+1} denote the starting and ending cell number for the bth pass. For the strategy of Figure 10-3, $N_1 = -C/6$, $N_2 = +C/6$, $N_3 = -C/3$, $N_4 = +C/3$, and so on, where all fractions are rounded to the nearest integer. Let $f(n)$ denote the number of the pass that evaluates cell n for the first time. For example, if $C/6 < n < C/3$, the cell will not be evaluated until the third sweep. Recall that j denotes the number of missed detections prior to a correct detection of the correct cell. The limits of the sum over j are unchanged from above (i.e., $0 \leq j \leq \infty$). The variable k is the total number of false alarms in all incorrect phase cells evaluated prior to synchronization. The total of all incorrect cells is denoted $K(n, j)$, and is given by

$$K(n, j) = \sum_{b=1}^{f(n)+j-1} |N_{b+1} - N_b| + |n - N_{f(n)+j}| - j \qquad (10\text{-}27)$$

As before, the false alarms can occur in any order within the $K(n, j)$ incorrect phase cells and the number of false alarms is binomially distributed. Combining all of the above yields

$$\Pr[n, j, k] = A \exp\left(-\frac{n^2}{2T^2}\right) P_d (1 - P_d)^j \binom{K(n, j)}{k} P_{\text{fa}}^k (1 - P_{\text{fa}})^{K(n, j) - k} \qquad (10\text{-}28)$$

$$\overline{T}_s = \sum_{n=-C/2}^{C/2} \sum_{j=0}^{\infty} \sum_{k=0}^{K(n,j)} \{[K(n, j) + j]T_i + kT_{\text{fa}}\} A \exp\left(-\frac{n^2}{2T^2}\right)$$

$$\times P_d (1 - P_d)^j \binom{K(n, j)}{k} P_{\text{fa}}^k (1 - P_{\text{fa}})^{K(n,j) - k} \qquad (10\text{-}29)$$

This expression has been evaluated in Braun [6] using the characteristic function. For reasonably high P_d, the expression can be evaluated on a digital computer to obtain specific numerical results.

Modified sweep strategies of the type described here have been used for many years in radar. Their use results in reduced synchronization time when the distribution of the received phase is nonuniform. The magnitude of synchronization time savings is a function of the variance of the received phase distribution.

10-3.3 Continuous Linear Sweep of Uncertainty Region

Most current spread-spectrum systems employ a sweep system which moves from one uncertainty cell to the next in discrete steps. An obvious variation of this strategy is to offset the clock frequency of the reference waveform generator slightly so that the phase of the waveform slips linearly past the received waveform phase. Recall that phase is the integral of frequency. The output of the despreading mixer is then a wideband signal except when the received and reference waveform phases have slipped sufficiently close to one another that despreading occurs. When despreading occurs, the despread energy can be detected and the sweep terminated. This technique was analyzed in Sage [7]. Although more sophisticated search strategies have been developed since this early paper, Sage's results are a reasonable baseline to which other strategies may be compared.

Consider a system that employs BPSK direct spreading modulation. The received waveform is

$$r(t) = \sqrt{2P}\, c(t - T_d)\cos \omega_0 t + n(t) \qquad (10\text{-}30)$$

where $n(t)$ is the usual bandlimited white Gaussian noise. The receiver configuration for synchronization is illustrated in Figure 10-4. For simplicity, the center frequency of the bandpass filter is approximately equal to the received carrier frequency. The reference waveform is then simply the spreading waveform, that is, $a(t) = c(t - \hat{T}_d)$. The despreading mixer output is

$$y'(t) = \sqrt{2P}\, c(t - T_d)c(t - \hat{T}_d)\cos(\omega_0 t + \theta) + n(t)c(t - \hat{T}_d) \quad (10\text{-}31)$$

where θ is a random phase that is assumed unknown. The reference waveform clock is offset in frequency so that the propagation delay estimate is varying linearly with time, that is, $\hat{T}_d = \hat{T}_d(t) = \hat{T}_{do} + Kt$, where $\hat{T}_{do}$ is an arbitrary fixed initial condition, $K = T_c/T_{ss}$, and T_{ss} is the time required to search one chip.

Recalling previous analysis of waveforms of the type of (10-31), the signal component at the mixer output is wideband when $T_d - \hat{T}_d(t) > T_c$ and has a sinusoidal component at frequency ω_0 whose magnitude is $\sqrt{2P}\, R_c(\tau)$ when $T_d - \hat{T}_d(t) = \tau < T_c$. $R_c(\tau)$ is the autocorrelation function of $c(t)$ and may be approximated by the function of Figure 7-9 for the present discussion. Since $\hat{T}_d$ is varying linearly with t, so is τ. Therefore, the sinusoidal component at frequency ω_0 at the mixer output is amplitude modulated by the autocorrelation function and is

$$x'(t) = \sqrt{2P}\, R_c(T_d - \hat{T}_{do} - Kt)\cos(\omega_0 t + \theta) \qquad (10\text{-}32)$$

which is illustrated in Figure 10-5.

The purpose of the bandpass filter, envelope detector, and threshold detector is to detect this signal approximately when the peak amplitude is reached. The filter is selected to have an impulse response which is matched to the signal $x'(t)$. The matched impulse response $h(t)$ is a time-reversed and delayed replica of the input waveform [8]. The delay is included so that the filter will be realizable. In this case,

$$h(t) = bR_c(Kt - T_c)\cos \hat{\omega}_0 t \qquad (10\text{-}33)$$

Since the frequency of the received signal is unknown during synchronization, $\hat{\omega}_0$ has been used in (10-33).

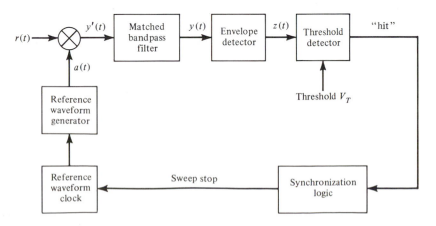

FIGURE 10-4. Synchronization system for linear sweep of uncertainty region.

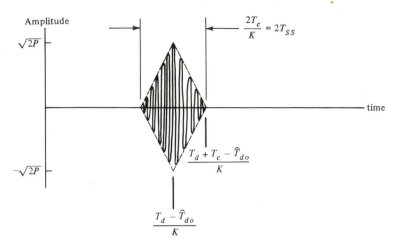

FIGURE 10-5. Signal component at despreading mixer output.

The matched filter output due to signal is the convolution of $h(t)$ and $x'(t)$. This output is

$$x(t) = \int_{-\infty}^{\infty} h(\alpha)x'(t - \alpha)\, d\alpha$$

$$= \sqrt{\frac{P}{2}}\, b \int_{-\infty}^{\infty} R_c(K\alpha - T_c)R_c(T_d - \hat{T}_{do} - Kt + K\alpha)$$

$$\times \cos(\Delta\omega_0\, \alpha + \omega_0 t + \theta)\, d\alpha$$

$$+ \sqrt{\frac{P}{2}}\, b \int_{-\infty}^{\infty} R_c(K\alpha - T_c)R_c(T_d - \hat{T}_{do} - Kt + K\alpha)$$

$$\times \cos[(\omega_0 + \hat{\omega}_0)\alpha - \omega_0 t - \theta]\, d\alpha \qquad (10\text{-}34)$$

where $\Delta\omega_0 = \hat{\omega}_0 - \omega_0$. The carrier frequency is assumed to be large relative to the maximum rate of change of the autocorrelation function product above, so that the second integral in the last line of (10-34) may be ignored. Expanding the cosine within the first integral yields

$$x(t) \approx \sqrt{\frac{P}{2}}\, b \cos(\omega_0 t + \theta) \int_{-\infty}^{\infty} R_c(K\alpha - T_c)R_c(T_d - \hat{T}_{do} - Kt + K\alpha)$$

$$\times \cos(\Delta\omega_0\alpha)\, d\alpha$$

$$- \sqrt{\frac{P}{2}}\, b \sin(\omega_0 t + \theta) \int_{-\infty}^{\infty} R_c(K\alpha - T_c)R_c(T_d - \hat{T}_{do} - Kt + K\alpha)$$

$$\times \sin(\Delta\omega_0\alpha)\, d\alpha \qquad (10\text{-}35)$$

At this point the assumption is made that the frequency error $\Delta f_0 = 2\pi\, \Delta\omega_0$ is small enough that the sine and cosine within the integrals of (10-35) are approximately constant over the range of nonzero values of the autocorrelation products. Using $\Delta\omega_0\alpha = \Delta\omega_0 T_c/K$ for the arguments of the sine and cosine, (10-35) becomes

$$x(t) \approx \sqrt{\frac{P}{2}} \, b \cos \left(\omega_0 t + \theta + \Delta\omega_0 \frac{T_c}{K} \right)$$

$$\times \int_{-\infty}^{\infty} R_c(K\alpha - T_c)R_c(T_d - \hat{T}_{\text{do}} - Kt + K\alpha) \, d\alpha \qquad (10\text{-}36)$$

This signal is a cosine with frequency ω_0 and arbitrary phase and with an amplitude given by $\sqrt{P/2} \, b$ times the integral of the autocorrelation function product. The integral is the output of a filter matched to the envelope of the signal of Figure 10-5. Although the absolute time of the maximum output of this filter is not known, it is known that the maximum occurs at the end of the triangular pulse which occurs at $t = (T_d + T_c - \hat{T}_{\text{do}})/K$. Thus the maximum filter output is

$$x_{\max}(t) = \sqrt{\frac{P}{2}} \, b \cos \left(\omega_0 t + \theta + \Delta\omega_0 \frac{T_c}{K} \right) \int_{-\infty}^{\infty} R_c^2(K\alpha - T_c) \, d\alpha$$

$$= \frac{\sqrt{2P}}{3} \, bT_{\text{ss}} \cos (\omega_0 t + \theta + \Delta\omega_0 T_{\text{ss}}) \qquad (10\text{-}37)$$

for the autocorrelation function of Figure 7-9. The search rate for this system is $1/T_{\text{ss}}$ chips per second.

The noise power at the output of the matched filter is calculated using the baseband equivalent filter transfer function. The impulse response $h_I(t)$ of the baseband equivalent filter is calculated from (10-33) and the relationship

$$h(t) = 2 \, \text{Re} \, [\tilde{h}(t)e^{j\hat{\omega}_0 t}] \qquad (10\text{-}38)$$

The result is

$$\tilde{h}(t) = \frac{b}{2} R_c(Kt - T_c) \qquad (10\text{-}39)$$

The Fourier transform of $\tilde{h}(t)$ is the filter transfer function $\tilde{H}(f)$. This Fourier transform was calculated earlier and the result is, ignoring the delay term,

$$\tilde{H}(f) = \frac{b}{2} T_{\text{ss}} \, \text{sinc}^2(fT_{\text{ss}}) \qquad (10\text{-}40)$$

The narrowband noise at the input to the matched filter is

$$n_i(t) = n(t)c(t - \hat{T}_d)$$

$$= \sqrt{2} \, n_I(t)c(t - \hat{T}_d) \cos \hat{\omega}_0 t - \sqrt{2} \, n_Q(t)c(t - \hat{T}_d) \sin \hat{\omega}_0 t \qquad (10\text{-}41)$$

and the complex envelope of this noise process is

$$\tilde{n}(t) = c(t - \hat{T}_d)[\sqrt{2} \, n_I(t) + j\sqrt{2} \, n_Q(t)] \qquad (10\text{-}42)$$

The power spectrum of $\tilde{n}(t)$ is the Fourier transform of its autocorrelation function. Since $n_I(t)$ and $n_Q(t)$ are independent of one another and both are independent of $c(t)$,

$$R_{\tilde{n}}(\tau) = \tfrac{1}{2} E[\tilde{n}(t)\tilde{n}^*(t + \tau)]$$

$$= 2R_c(\tau)R_{n_I}(\tau) \qquad (10\text{-}43)$$

The Fourier transform of (10-43) is

$$S_{\tilde{n}}(f) = 2S_c(f) * S_{n_I}(f) \qquad (10\text{-}44)$$

Assume now that the input noise spectrum is wideband relative to the spectrum of $c(t)$ so that

$$S_{\tilde{n}}(f) \approx 2S_{n_I}(f) = \begin{cases} N_0 & |f| < B \\ 0 & \text{elsewhere} \end{cases} \qquad (10\text{-}45)$$

where $1/T_c << B << f_0$. The noise power at the filter output is

$$\begin{aligned} N &= \int_{-\infty}^{\infty} |\tilde{H}(f)|^2 S_{\tilde{n}}(f) \, df \\ &\approx \tfrac{1}{4} \, b^2 T_{ss}^2 N_0 \int_{-B}^{+B} \text{sinc}^4 \, (fT_s) \, df \\ &\approx \frac{b^2}{6} \, T_{ss} N_0 \end{aligned} \qquad (10\text{-}46)$$

The last equality is from Gradshteyn and Ryzhik [5]. Finally, the maximum signal-to-noise power ratio at the matched filter output is calculated from (10-37) and (10-46) and is

$$\text{SNR}_{\text{max}} = \frac{\tfrac{1}{2}[(\sqrt{2P}/3) \, b \, T_{ss}]^2}{\tfrac{1}{6} \, b^2 T_{ss} N_0} = \tfrac{2}{3}\left(\frac{PT_{ss}}{N_0}\right) \qquad (10\text{-}47)$$

Observe that signal-to-noise ratio increases with increasing T_{ss} or, equivalently, with decreasing sweep rate.

The matched filter output calculated above is the sum of a sine wave and Gaussian noise. The peak signal-to-noise ratio occurs at the end of the pulse of Figure 10-5 and has a magnitude given by (10-47). The function of the envelope detector and threshold comparator in Figure 10-4 is to detect the presence of the sine wave. Since the phase of the sine wave is unknown, the envelope detector is the optimum detection device. Both the probability of detecting the signal P_d and the probability of falsely declaring signal present P_{fa} can be calculated if the probability density function of the envelope detector output is known. This density function is the well-known Ricean pdf (Rice [9]) of the envelope of a sine wave in Gaussian noise. Denote the envelope detector output by $z(t)$; then

$$p_z(\alpha) = \begin{cases} \dfrac{\alpha}{N} \exp\left(-\dfrac{\alpha^2 + A^2}{2N}\right) I_0\left(\dfrac{\alpha A}{N}\right) & \text{for } \alpha \geq 0 \\ 0 & \text{elsewhere} \end{cases} \qquad (10\text{-}48)$$

where N is the noise power or variance, A is the amplitude of the sine wave, and $I_0(\cdot)$ is the zeroth-order modified Bessel function. The probability of detection is the integral from V_T to infinity of this pdf evaluated for the maximum signal-to-noise ratio. Probability of false alarm is the same integral evaluated at the minimum signal-to-noise ratio (i.e., for $A = 0$). These calculations are depicted graphically in Figure 10-6, which shows the pdf for SNRs of zero and eight. The area which is crosshatched is P_d and the area which is double crosshatched is P_{fa}. Both P_d and P_{fa} are functions of SNR and V_T and cannot be independently selected. For a specific SNR, the selection of P_{fa} implies a particular threshold V_T which then sets P_d. Figure 10-7 illustrates these relationships.

EXAMPLE 10-1

Consider a spread-spectrum system using a spreading code clock frequency $f_c = 3$ MHz. Suppose that the received carrier power-to-noise power spectral

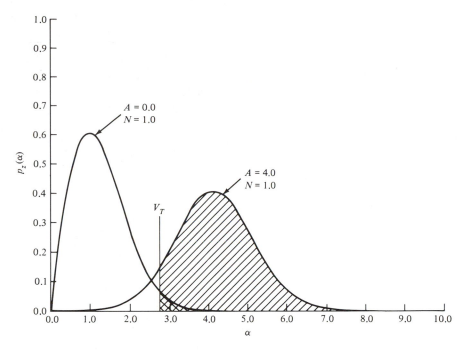

FIGURE 10-6. Example of the calculation of P_d and P_{fa}.

density ratio is 46.25 dB $= 10 \log (P/N_0)$ and that the propagation delay uncertainty is ± 1.2 ms. What sweep rate is required to yield a single sweep (P_d, P_{fa}) pair equal to $(0.9, 10^{-6})$, $(0.8, 10^{-6})$, $(0.9, 10^{-3})$, and $(0.8, 10^{-3})$? What is the average synchronization time for each case assuming a false alarm penalty of $100T_i$?

Solution: From Figure 10-7, the required signal-to-noise ratios for these four (P_d, P_{fa}) pairs are 13.4, 12.8, 11.0, and 10.3 dB, respectively. From (10-47)

$$10 \log (\text{SNR}_{max}) = 10 \log \left(\frac{2}{3}\right) + 10 \log \left(\frac{P}{N_0}\right) + 10 \log T_{ss}$$

Solving for T_{ss} for each of the four cases yields

$$T_{ss} = \begin{cases} 778 \ \mu s & \text{for } (0.9, 10^{-6}) \\ 678 \ \mu s & \text{for } (0.8, 10^{-6}) \\ 448 \ \mu s & \text{for } (0.9, 10^{-3}) \\ 381 \ \mu s & \text{for } (0.8, 10^{-3}) \end{cases}$$

Equation (10-12) for mean synchronization time applies to the discrete phase step analysis. At low P_{fa}, it also applies approximately to the present case. In (10-12) use

$$T_i = T_{ss}$$

$$T_{da} = T_{ss} + P_{fa}(100T_{ss}) = T_{ss}(1 + 100P_{fa})$$

$$C = 2(1.2 \text{ ms})(3 \times 10^6 \text{ chips/s}) = 7.2 \times 10^3$$

With these substitutions, (10-12) becomes

$$\overline{T}_s = (7199)T_{ss}(1 + 100P_{fa})\left(\frac{2 - P_d}{2P_d}\right) + \frac{T_{ss}}{P_d}$$

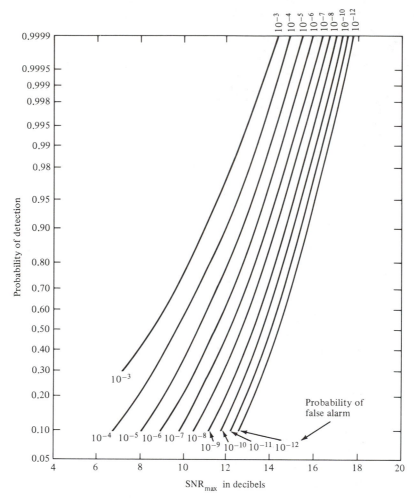

FIGURE 10-7. Probability of detection for a sine wave in Gaussian noise as a function of signal-to-noise ratio and the probability of false alarm. (From Ref. 10.)

and the final result is

$$
\overline{T}_s = \begin{cases} 3.42 \text{ s} & \text{for } (0.9, 10^{-6}) \\ 3.66 \text{ s} & \text{for } (0.8, 10^{-6}) \\ 2.17 \text{ s} & \text{for } (0.9, 10^{-3}) \\ 2.26 \text{ s} & \text{for } (0.8, 10^{-3}) \end{cases}
$$

Observe that permitting P_{fa} to increase by three orders of magnitude can decrease synchronization time significantly. ■

Finally, a method of calculation of additional points on the curves of Figure 10-7 is desired. These points are calculated by integrating the Ricean density function over the limits of V_T to infinity using the desired signal-to-noise ratio. This integral is called the Marcum Q-function defined by

$$
Q(a, b) \triangleq \int_b^\infty \alpha \exp\left[-\tfrac{1}{2}(\alpha^2 + a^2)\right] I_0(a\alpha) \, d\alpha \tag{10-49}
$$

An efficient algorithm for calculating $Q(a, b)$ has been given in Shnidman [11]. Equation (10-49) is related to the desired integral through straightforward normalization.

10-3.4 Detection of a Signal in Additive White Gaussian Noise

All synchronization systems that employ a discrete step serial search evaluate a particular phase/frequency cell by estimating whether or not signal energy is present at the output of a filter following the despreading mixer. Calculation of the mean and variance of the synchronization time requires knowledge of P_d and P_{fa} for this estimate as a function of evaluation time T_i and received signal-to-noise ratio. Three somewhat different methods of detecting signal energy are discussed in this section. Each method results in a different relationship between P_d, P_{fa}, T_i, and SNR. For each method, this relationship will also depend on other factors, such as detection thresholds. In the first case, a single integration of the output of a square-law envelope detector for T_i seconds will be used to detect signal energy. In the second case, multiple integrations together with some logic will be used to detect the signal. This will result in an evaluation time that is a random variable. The final method considered will integrate the square-law envelope detector output for a variable length of time until a selected P_d and P_{fa} are achieved. All these methods are used in current spread-spectrum systems.

Fixed Integration Time Detection. The simplest method of detecting the presence of signal energy at the output of a narrowband filter is illustrated in Figure 10-8. The input to the bandpass filter is taken from the output of the despreading mixer. When the reference waveform phase is correct, the received signal will be despread and $s(t)$ will appear at the filter output. The AWGN process $n(t)$ will always be present at the filter output. The filter output is squared and then lowpass filtered to eliminate the double-frequency terms that result from the squaring operation. The lowpass squared output is then integrated for T_i seconds, and the output of this operation is compared to a fixed threshold V_T. If the integrator output is above the threshold, the signal is declared present. Otherwise, the signal is declared absent. This system has been analyzed in detail in Urkowitz [12].

To evaluate this energy detector, the probability density function of the integrator output must be calculated. The exact calculation of this pdf is difficult and systems analysts always settle for approximate results. Several different approximations will be discussed in the following. Consider first the output due to noise alone.

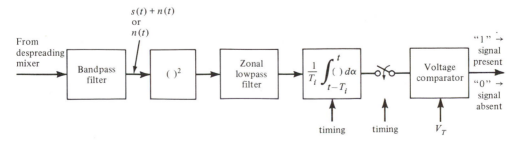

FIGURE 10-8. Fixed integration time energy detector.

The filter output noise process is represented by

$$n(t) = \sqrt{2}\, n_I(t) \cos \omega_{IF} t - \sqrt{2}\, n_Q(t) \sin \omega_{IF} t \tag{10-50}$$

Assume that the IF filter bandwidth is B so that both $n_I(t)$ and $n_Q(t)$ are baseband white Gaussian noise processes with two-sided power spectral densities $N_0/2$ over the frequency range $|f| < B/2$. Squaring (10-50) and retaining only the baseband components of the result yields

$$[n^2(t)]_{LP} = n_I^2(t) + n_Q^2(t) \tag{10-51}$$

and the integrator output at the sampling instant is

$$V = \frac{1}{T_i} \int_0^{T_i} [n^2(t)]_{LP}\, dt = \frac{1}{T_i} \int_0^{T_i} n_I^2(t)\, dt + \frac{1}{T_i} \int_0^{T_i} n_Q^2(t)\, dt \tag{10-52}$$

Consider the baseband noise processes $n_I(t)$ and $n_Q(t)$. These signals are bandlimited to the frequency range $-\frac{1}{2} B < f < +\frac{1}{2} B$. The sampling theorem [13] states that these bandlimited signals can be exactly represented by a sum of orthonormal sampling functions appropriately weighted and time translated. In particular,

$$n_I(t) = \sum_{k=-\infty}^{\infty} \frac{1}{\sqrt{B}}\, n_I\!\left(\frac{k}{B}\right) \sqrt{B}\, \text{sinc}\,(Bt - k) \tag{10-53}$$

A similar expression is obtained for $n_Q(t)$. A finite number of terms of this infinite sum can be used as approximation to the baseband noise process over the range $0 \le t \le T_i$. For example, let $n_I'(t)$ be defined by

$$n_I'(t) = \begin{cases} n_I(t) & 0 \le t \le T_i \\ 0 & \text{elsewhere} \end{cases} \tag{10-54}$$

as illustrated in Figure 10-9. Formally applying the sampling theorem to the signal $n_I'(t)$ yields

$$n_I'(t) = \sum_{k=1}^{BT_i} \frac{1}{\sqrt{B}}\, n_I'\!\left(\frac{k}{B}\right) \sqrt{B}\, \text{sinc}\,(Bt - k) \tag{10-55}$$

since samples of $n_I'(t)$ outside of the indicated range of k are zero. Unfortunately, the signal defined by (10-54) is not bandlimited, since it is time limited, and the sampling theorem does not apply. For large BT it can be argued that the range of significant frequency components of $n_I'(t)$ is not much greater than $-\frac{1}{2} B \le f \le +\frac{1}{2} B$, so that the result of (10-55) is approximately correct. With this understanding, the first integral of the right side of (10-52) can be approximately represented by

$$\frac{1}{T_i} \int_0^{T_i} n_I^2(t)\, dt = \frac{1}{T_i} \int_0^{T_i} n_I'^2(t)\, dt = \frac{1}{T_i} \int_{-\infty}^{\infty} n_I'^2(t)\, dt$$

$$\approx \frac{1}{T_i} \int_{-\infty}^{\infty} \left[\sum_{k=1}^{BT_i} \frac{1}{\sqrt{B}}\, n_I'\!\left(\frac{k}{B}\right) \sqrt{B}\, \text{sinc}\,(Bt - k) \right]$$

$$\times \left[\sum_{k'=1}^{BT_i} \frac{1}{\sqrt{B}}\, n_I'\!\left(\frac{k'}{B}\right) \sqrt{B}\, \text{sinc}\,(Bt - k') \right] dt \tag{10-56}$$

The sampling functions $\sqrt{B}\, \text{sinc}\,(Bt - k)$ are orthonormal over the limits $-\infty < t < +\infty$, that is,

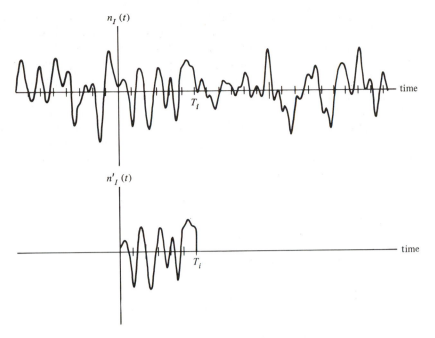

FIGURE 10-9. Baseband noise processes $n_I(t)$ and $n'_I(t)$.

$$\int_{-\infty}^{\infty} \sqrt{B} \text{ sinc } (Bt - k) \sqrt{B} \text{ sinc } (BT - k') \, dt = \begin{cases} 1.0 & k = k' \\ 0 & k \neq k' \end{cases} \quad (10\text{-}57)$$

Using (10-57), equation (10-56) simplifies to

$$\frac{1}{T_i} \int_0^{T_i} n_I^2(t) \, dt \approx \frac{1}{T_i B} \sum_{k=1}^{BT_i} n'^2_I \left(\frac{k}{B} \right) \quad (10\text{-}58)$$

A similar expression results for the second integral on the right side of (10-52) and V can be approximately represented by

$$V = \frac{1}{T_i} \int_0^{T_i} [n^2(t)]_{\text{LP}} \, dt \approx \frac{1}{T_i B} \sum_{k=1}^{BT_i} \left[n_I^2 \left(\frac{k}{B} \right) + n_Q^2 \left(\frac{k}{B} \right) \right] \quad (10\text{-}59)$$

which is a constant times the sum of the squares of $2BT_i$ independent Gaussian random variables. The fact that the samples of the noise processes are independent is deduced directly from the autocorrelation function of $n_I(t)$ or $n_Q(t)$, which is

$$R_n(\tau) = \tfrac{1}{2} N_0 B \text{ sinc } (B\tau) \quad (10\text{-}60)$$

and indicates that time samples spaced $1/B$ apart are uncorrelated and therefore independent.

The probability density function of the sum of the squares of $n = 2BT_i$ independent Gaussian random variables is chi-square with n degrees of freedom [3]. The variance σ^2 of the samples is $N_0 B/2T_i B = N_0/2T_i$, where the normalization factor $1/T_i B$ of (10-59) has been included. The chi-square density of V is [3]

$$p_c(\alpha) = \begin{cases} \dfrac{1}{2^{n/2} \, \sigma^n \Gamma(n/2)} \, \alpha^{(n/2)-1} \exp\left(-\dfrac{\alpha}{2\sigma^2} \right) & \text{for } \alpha > 0 \\ 0 & \text{for } \alpha \leq 0 \end{cases} \quad (10\text{-}61)$$

The exact result for the probability density of V has been calculated in Grenander et al. [14]. This calculation is well beyond the scope of this text; however, it is instructive to compare the exact result with the chi-square approximation. Figure 10-10 compares the exact result to the chi-square result for time–bandwidth products of $8/\pi$ and $16/\pi$. Observe that even for these small time–bandwidth products, the approximation is quite good. The approximation improves as time–bandwidth product increases.

When the input to the squaring device of Figure 10-8 is signal plus noise, indicating that the correct spreading waveform phase is being evaluated, the analysis is somewhat more complicated. Assume for simplicity that the signal is a sine wave with amplitude $\sqrt{2P}$ and arbitrary phase θ which can be represented by

$$s(t) = \sqrt{2P}\cos\theta\cos\omega_{\text{IF}}t - \sqrt{2P}\sin\theta\sin\omega_{\text{IF}}t \qquad (10\text{-}62)$$

The integrator output is

$$V = \frac{1}{T_i}\int_0^{T_i} [s(t) + n(t)]_{\text{LP}}^2\, dt$$

$$= \frac{1}{T_i}\int_0^{T_i} \{[\sqrt{2P}\cos\theta + \sqrt{2}\,n_I(t)]\cos\omega_{\text{IF}}t$$

$$- [\sqrt{2P}\sin\theta + \sqrt{2}\,n_Q(t)]\sin\omega_{\text{IF}}t\}_{\text{LP}}^2\, dt$$

$$= \frac{1}{T_i}\int_0^{T_i} [\sqrt{P}\cos\theta + n_I(t)]^2\, dt + \frac{1}{T_i}\int_0^{T_i} [\sqrt{P}\sin\theta + n_Q(t)]^2\, dt \qquad (10\text{-}63)$$

Using identical reasoning to that used above to develop an approximation to the integrator output in the noise-alone case, the two integrals in the last line of

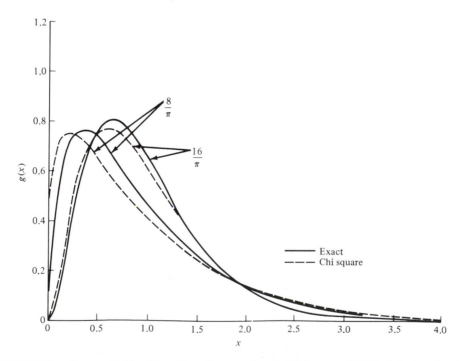

FIGURE 10-10. Comparison of exact and approximate (chi-square) integrator output probability density functions. (From Ref. 14.)

(10-63) are approximately given by

$$\frac{1}{T_i} \int_0^{T_i} [\sqrt{P} \cos \theta + n_I(t)]^2 \, dt \approx \frac{1}{T_iB} \sum_{k=1}^{BT_i} \left[\sqrt{P} \cos \theta + n_I\left(\frac{k}{B}\right) \right]^2 \quad (10\text{-}64a)$$

$$\frac{1}{T_i} \int_0^{T_i} [\sqrt{P} \sin \theta + n_Q(t)]^2 \, dt \approx \frac{1}{T_iB} \sum_{k=1}^{BT_i} \left[\sqrt{P} \sin \theta + n_Q\left(\frac{k}{B}\right) \right]^2 \quad (10\text{-}64b)$$

Thus the integrator output V may be approximated by the sum of $2BT_i$ independent Gaussian random variables all with the same variance $\sigma^2 = N_0/2T_i$. Half of the random variables have mean $\sqrt{P} \cos \theta / \sqrt{T_iB}$ and half have a mean $\sqrt{P} \sin \theta / \sqrt{T_iB}$. The phase θ is also a random variable and is uniformly distributed over the range $(0, 2\pi)$. For a particular integration, however, this phase may be considered constant. It will be seen shortly that the probability density of the integrator output is independent of θ. The probability density function of this sum has a noncentral chi-square density [15] with $2T_iB$ degrees of freedom and noncentrality parameter

$$\lambda = \frac{1}{T_iB} \sum_{k=1}^{BT_i} [(\sqrt{P} \cos \theta)^2 + (\sqrt{P} \sin \theta)^2] = P \quad (10\text{-}65)$$

Other authors (e.g., Urkowitz [12]) choose to normalize the integrator output such that the noncentrality parameter is equal to the signal energy-to-noise psd ratio and so that the variance of the summed random variables is unity. The noncentral chi-square probability density function is [15]

$$p_{nc}(\alpha) = \frac{1}{2^{n/2}} \exp\left(-\frac{\lambda}{2}\right) \exp\left(-\frac{\alpha}{2\sigma^2}\right) \sum_{j=0}^{\infty} \frac{(\alpha/\sigma^2)^{1/2n+j-1} \lambda^j}{\Gamma(\frac{1}{2}n + j)2^{2j}j!} \quad (10\text{-}66)$$

Observe that the first term of this expression is the chi-square density of (10-61).

Knowledge of the probability density function of the integrator output theoretically permits evaluation of P_d and P_{fa} for a particular fixed T_i, B, P, N_0, and V_T. This calculation is difficult because of the complexity of (10-61) and (10-66). The desired P_d is given by

$$P_d = \text{Pr } (V > V_T | \text{signal present})$$

$$= \int_{V_T}^{\infty} p_{nc}(\alpha) \, d\alpha \quad (10\text{-}67a)$$

and the desired P_{fa} is given by

$$P_{fa} = \text{Pr } (V > V_T | \text{no signal present})$$

$$= \int_{V_T}^{\infty} p_c(\alpha) \, d\alpha \quad (10\text{-}67b)$$

Tables of both the central [16] and noncentral [17] chi-square probability integrals are available. In addition, selected receiver operating characteristics (ROC) are plotted in Urkowitz [12], which give sets of (P_d, P_{fa}, λ) for time–bandwidth products of 2, 10, 20, 30, and 50. These ROCs are similar to the curves of Figure 10-7. When using the curves of Urkowitz [12], attention must be paid to the definitions used, which are different than those used above. A third alternative for calculating the integrals for P_d and P_{fa} is through the use of a nomogram presented in Urkowitz [12], which was derived from a nomogram in Smirnov and Potapov

[18]. This nomogram is reproduced here as Figure 10-11. In this nomogram, the integrator output voltage V' is normalized, that is,

$$V' = \frac{2T_i}{N_0} V \qquad (10\text{-}68)$$

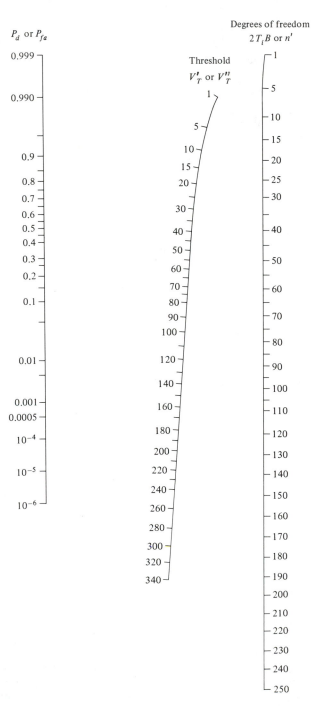

FIGURE 10-11. Nomogram for evaluating P_d and P_{fa} for the detector of Figure 10-8. (From Ref. 12.)

Probability of false alarm is calculated from Figure 10-11 using the given time–bandwidth product $2T_iB$ and the selected normalized threshold V_T'. The probability of detection is calculated using a modified number of degrees of freedom n' given by

$$n' = \frac{(2T_iB + \lambda')^2}{2T_iB + 2\lambda'}$$

(10-69)

and a modified threshold V_T'' given by

$$V_T'' = V_T'\left(\frac{2T_iB + \lambda'}{2T_iB + 2\lambda'}\right)$$

(10-70)

where $\lambda' = 2\lambda T_i/N_0$ is the normalized noncentrality parameter.

EXAMPLE 10-2 [12]

Suppose that $n = 2T_iB = 40$, $\lambda' = 30$, and $P_{fa} = 10^{-4}$ are given. The normalized threshold is $V_T' = 81$, which is read directly from the nomogram. For the given normalized noncentrality parameter λ', the modified number of degrees of freedom is calculated from (10-69) and the modified threshold from (10-70). Thus

$$n' = \frac{(40 + 30)^2}{40 + 60} = 49$$

$$V_T'' = 81\left(\frac{40 + 30}{40 + 60}\right) = 56.7$$

Using n' and V_T'', $P_d \simeq 0.23$ is read directly from the nomogram. ∎

A fourth alternative for calculating P_d and P_{fa} is to use numerical integration techniques. For the central chi-square density numerical integration is not difficult; however, for P_d close to unity or P_{fa} very small, numerical accuracy becomes a problem. For the noncentral chi-square density numerical integration is difficult due to the infinite summation in (10-66). When highly accurate values for the integral of the noncentral chi-square density are required, it is recommended that the reader refer to Robertson [19], where an infinite series is developed for this integral.

Another approximation to the integrator output probability density valid for all time–bandwidth products has been proposed in Baer [20]. This approximation was specifically derived for the particular case where the predetection filter of Figure 10-8 consists of a single resonator. The proposed approximation to (10-61) and (10-66) is [20]

$$p_B(\alpha) \approx \left(\frac{E_V}{\sigma_V^2}\right)^{E_V/\sigma_V^2} \frac{1}{\Gamma(E_V^2/\sigma_V^2)} \alpha^{(E_V^2/\sigma_V^2)-1} \exp\left(-\frac{E_V}{\sigma_V^2}\alpha\right) \qquad \alpha \geq 0$$

$$= 0 \qquad\qquad\qquad\qquad\qquad\qquad\qquad\qquad\qquad\qquad \alpha < 0 \quad (10\text{-}71)$$

where E_V and σ_V^2 are the mean and variance, respectively, of the integrator output and $\Gamma(\cdot)$ is the gamma function. The mean of the integrator output is

$$E(V) = E\left\{\frac{1}{T_i}\int_0^{T_i} [s(t) + n(t)]_{LP}^2\, dt\right\}$$

(10-72)

where the expected value is over all sample functions of $n(t)$ and all received signal phases. Replacing $s(t)$ and $n(t)$ with their narrowband representations of (10-50) and (10-62) and simplifying yields

$$E(V) = \frac{1}{T_i} \int_0^{T_i} E\{[\sqrt{P} \cos \theta + n_I(t)]^2\} \, dt$$

$$+ \frac{1}{T_i} \int_0^{T_i} E\{[\sqrt{P} \sin \theta + n_Q(t)]^2\} \, dt$$

$$= E\{[\sqrt{P} \cos \theta + n_I(t)]^2\} + E\{[\sqrt{P} \sin \theta + n_Q(t)]^2\}$$

$$= P + E[n_I^2(t)] + E[n_Q^2(t)]$$

$$= P + 2 \sigma_n^2 \tag{10-73}$$

where the subscript n represents either $n_I(t)$ or $n_Q(t)$.

Calculations of the variance of the integrator output is given in Papoulis [3]. The result is

$$\sigma_V^2 = \frac{2}{T_i} \int_0^{T_i} \left(1 - \frac{\lambda}{T_i}\right) \{R_w(\lambda) - E^2[w(t)]\} \, d\lambda \tag{10-74}$$

where $w(t)$ is the integrator (baseband) input and $R_w(\tau)$ is the autocorrelation function of $w(t)$. In the present case,

$$w(t) = [s(t) + n(t)]_{\text{LP}}^2$$

$$= P + 2\sqrt{P} \cos \theta \, n_I(t) + 2\sqrt{P} \sin \theta \, n_Q(t)$$

$$+ n_I^2(t) + n_Q^2(t) \tag{10-75}$$

The mean of $w(t)$ is

$$E[w(t)] = P + E[n_I^2(t)] + E[n_Q^2(t)] = P + 2\sigma_n^2 \tag{10-76}$$

The autocorrelation function is

$$R_w(\tau) = E[w(t)w(t + \tau)]$$

$$= E\{[P + 2\sqrt{P} \cos \theta \, n_I(t) + 2\sqrt{P} \sin \theta \, n_Q(t) + n_I^2(t) + n_Q^2(t)]$$

$$[P + 2\sqrt{P} \cos \theta \, n_I(t + \tau) + 2\sqrt{P} \sin \theta \, n_Q(t + \tau)$$

$$+ n_I^2(t + \tau) + n_Q^2(t + \tau)]\}$$

$$= P^2 + 4PE[n_I^2(t)] + 4PE[\cos^2\theta]E[n_I(t)n_I(t + \tau)]$$

$$+ 4PE[\sin^2\theta]E[n_I(t)n_I(t + \tau)]$$

$$+ 2E[n_I^2(t)n_I^2(t + \tau)] + 2E^2[n_I^2(t)] \tag{10-77}$$

The last step follows from the independence of $n_I(t)$, $n_Q(t)$ and θ and the fact that $E[n_I(t)] = E[n_Q(t)] = E[\cos \theta] = E[\sin \theta] = 0$. It has also been recognized that the statistics of $n_I(t)$ and $n_Q(t)$ are identical and both processes are stationary, so that terms such as $E[n_Q^2(t)]$ can be equated to $E[n_I^2(t)]$. The expected values over θ are

$$E(\cos^2\theta) = E(\tfrac{1}{2} + \tfrac{1}{2} \cos 2\theta) = \tfrac{1}{2} \tag{10-78a}$$

and

$$E(\sin^2\theta) = E(\tfrac{1}{2} - \tfrac{1}{2}\cos 2\theta) = \tfrac{1}{2} \qquad (10\text{-}78b)$$

so that (10-77) simplifies to

$$R_w(\tau) = P^2 + 4P\sigma_n^2 + 4PR_n(\tau) + 2R_{n^2}(\tau) + 2\sigma_n^4 \qquad (10\text{-}79)$$

The autocorrelation function $R_{n^2}(\tau)$ is [21]

$$R_{n^2}(\tau) = 2R_n^2(\tau) + \sigma_n^4$$

so that

$$R_w(\tau) = (P + 2\sigma_n^2)^2 + 4PR_n(\tau) + 4R_n^2(\tau) \qquad (10\text{-}80)$$

Substituting (10-76) and (10-80) into (10-74) yields

$$\sigma_V^2 = \frac{2}{T_i} \int_0^{T_i} \left(1 - \frac{\lambda}{T_i}\right)[4PR_n(\lambda) + 4R_n^2(\lambda)] \, d\lambda \qquad (10\text{-}81)$$

The autocorrelation function of $n_I(t)$ or $n_Q(t)$ is

$$R_n(\tau) = \tfrac{1}{2}N_0 B \text{ sinc } (B\tau) \qquad (10\text{-}82)$$

so that

$$\sigma_V^2 = \frac{4N_0 B}{T_i} \int_0^{T_i} \left(1 - \frac{\lambda}{T_i}\right)[P \text{ sinc } (B\lambda) + \tfrac{1}{2}N_0 B \text{ sinc}^2(B\lambda)] \, d\lambda \qquad (10\text{-}83)$$

For small BT_i products, the variance may be evaluated numerically. For large BT_i, (10-83) can be approximated by

$$\sigma_V^2 \approx \frac{4N_0 B}{T_i} \int_0^{\infty} [P \text{ sinc } (B\lambda) + \tfrac{1}{2}N_0 B \text{ sinc}^2(B\lambda)] \, d\lambda \qquad (10\text{-}84)$$

The integrals are evaluated in Gradshteyn and Ryzhik [5]. The final approximate result is

$$\sigma_V^2 \approx \frac{2N_0}{T_i} (P + \tfrac{1}{2}N_0 B)$$

$$= \frac{1}{BT_i} [2PN_0 B + (N_0 B)^2] \qquad (10\text{-}85)$$

For time–bandwidth products larger than about 3, the approximate result is within 10% of the correct result.

With both E_V and σ_V^2 determined, (10-71) can be easily evaluated. The form of this equation is simple enough that numerical integration is now possible to determine rather accurate values for P_d and P_{fa} for any time–bandwidth product. The accuracy of (10-71) in representing the actual pdf of the integrator output is important in systems calculation. Baer [20] does not give specific numerical results which can be checked. However, several plots of the approximation and a numerically calculated pdf are presented which show good agreement.

All the techniques described above for evaluation of P_d and P_{fa} are difficult to use and are impractical when a large number of values need to be calculated, as, for example, would be required in a system optimization. Fortunately, yet another approximation exists which makes calculation of P_d and P_{fa} fairly efficient. This approximation is the well-known Gaussian approximation to the density function of the integrator output. This approximation applies *only* to large time–bandwidth products. The justification for applying a Gaussian distribution to the integrator output is the central limit theorem [3], which states that the probability density of the sum of n independent random variables tends to a Gaussian density as n becomes large. This result is valid for random variables having any probability density. A Gaussian density function is specified once its mean and variance are defined. The mean E_V is defined in (10-73) and the variance is defined in (10-85) and the approximate density of the integrator output voltage is

$$p_G(\alpha) = \frac{1}{\sqrt{2\pi\sigma_V^2}} \exp\left[-\frac{(\alpha - E_V)^2}{2\sigma_V^2} \right] \tag{10-86}$$

The integral of this density from V_T to infinity yields the desired probabilities P_d and P_{fa}. Specifically, both P_d and P_{fa} are calculated from

$$\frac{1}{\sqrt{2\pi\sigma_V^2}} \int_{V_T}^{\infty} \exp\left[-\frac{(\alpha - E_V)^2}{2\sigma_V^2} \right] d\alpha = \frac{1}{\sqrt{2\pi}} \int_{(V_T - E_V/\sigma_V)}^{\infty} \exp\left(-\tfrac{1}{2}\alpha^2\right) d\alpha$$

$$= Q\left(\frac{V_T - E_V}{\sigma_V}\right) \tag{10-87}$$

where the Q-function was defined earlier. Appendix E gives a convenient and quite accurate rational approximation to $Q(\cdot)$, which can be used when many calculations are to be made.

A feeling for the behavior of the density function of the integrator output can be gained by example. Figure 10-12 is a plot of $p_B(\alpha)$ of Equation (10-71) for various signal-to-noise ratios. In order to obtain specific numerical results, the following parameters were selected:

$$B = 2.0 \text{ MHz}$$

$$T_i = 2.5 \text{ μs}$$

$$N_0 = 10^{-7} \text{ W/Hz}$$

The signal-to-noise ratio used is the ratio of carrier power P to one-sided noise power spectral density N_0 expressed in decibels, that is, SNR $= 10 \log_{10}(P/N_0)$. Observe that both the mean and variance increase as signal level P increases as

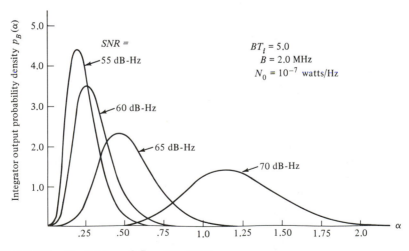

FIGURE 10-12. Variation of P_B with SNR.

expected from (10-73) and (10-85). Figure 10-13 is also a plot of $p_B(\alpha)$ from (10-71), but now only two signal-to-noise ratios are considered and BT_i is allowed to vary between 1 and 15. Notice that increasing BT_i does not change the mean appreciably; however, the variance decreases with increasing BT_i. This fact explains crudely why increasing BT_i makes the achievement of simultaneously high P_d and low P_{fa} possible. The different approximations to the integrator output density give different accuracies in the final calculation of P_d and P_{fa}. The choice of approximation will depend on the particular range of BT_i of interest and the required accuracy for P_d and P_{fa}.

In summary, the performance of the energy detector of Figure 10-8 is difficult to calculate exactly. A number of approximations to the probability density function of the integrator output exist with varying degrees of accuracy and computational difficulty. The central and noncentral chi-square approximation is the most accurate but evaluation of the integral of the pdf must be via tables, nomograms, or numerical integration. The approximation proposed by Baer results in simplified numerical integration and is also quite accurate. For large time–bandwidth products, the integrator output pdf is nearly Gaussian. In this case, polynomial approximations for the probability integral exist and are convenient for systems analysis. The choice of which approximation to use requires a trade between required accuracy and computational effort.

EXAMPLE 10-3

Reconsider the spread-spectrum system of Example 10-1, which uses a spreading code rate of $f_c = 3$ MHz. The design point SNR = 46.25 dB-Hz and the propagation delay uncertainty is ± 1.2 ms, which corresponds to ± 3600 chips. Suppose that the serial search step size is $\frac{1}{2}$ chip, so that the total number of phase cells which can be searched is $C = 14,400$. Evaluate the average synchronization time and comparator threshold for $P_d = 0.9$ and 0.7 and $BT_i = 10$ and 50. Assume that the IF bandpass filter bandwidth is 24 kHz and that the false alarm penalty is $100T_i$.

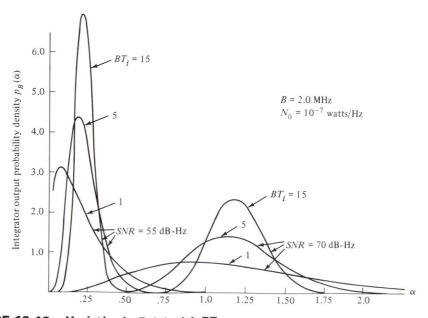

FIGURE 10-13. Variation in $P_B(\alpha)$ with BT.

Solution: The integration times for the BT_i products of 10 and 50 are $T_i = 10/24 \times 10^3 = 417$ μs and $T_i = 50/24 \times 10^3 = 2.08$ ms. The normalized noncentrality parameter is calculated from $\lambda' = 2\lambda T_i/N_0 = 2PT_i/N_0$. Substituting the design point $P/N_0 = 4.22 \times 10^4$ into this equation yields $\lambda' = 35.1$ for $BT_i = 10$ and $\lambda' = 176$ for $BT_i = 50$. The modified number of degrees of freedom is calculated from (10-69). For $BT_i = 10$, $n' = 33.7$ and for $BT_i = 50$, $n' = 169$. Using n' with the nomogram of Figure 10-11 yields the modified thresholds V_T'' given in the following table:

P_d	BT_i	V_T''	V_T'	P_{fa}
0.9	10	24	39.3	0.007
0.9	50	144	236	$<10^{-6}$
0.7	10	29	47.5	0.0006
0.7	50	159	260	$<10^{-6}$

The normalized thresholds V_T' are calculated using (10-70) and the results are also given in the table. Finally, P_{fa} is found using the nomogram with the normalized thresholds V_T' and unmodified number of degrees of freedom $n = 20$ and 100. The nomogram stops at $P_{fa} = 10^{-6}$, so these values are used in the synchronization-time calculations.

Substituting C, P_d, P_{fa}, T_i, and $T_{fa} = 100T_i$ into (10-10) and (10-12) yields the mean synchronization times given in the following table:

P_d	BT_i	$\overline{T}_s$	V_T
0.9	10	6.24 s	4.71 mV
0.9	50	18.3 s	5.67 mV
0.7	10	5.91 s	5.70 mV
0.7	50	27.8 s	6.25 mV

The actual threshold voltage can be calculated from (10-68) only after an absolute voltage reference is established at the filter output. If $N_0 = 10^{-7}$ W/Hz is arbitrarily chosen, the V_T values of the table above result. ∎

Comparing the results of Examples 10-1 and 10-3 points to a conclusion that a linear sweep with matched filter detection is superior to the scheme of this example. This conclusion is incorrect, however, because no attempt has been made to optimize the selection of P_d, P_{fa}, and BT_i for the stepped search. In addition, the matched filter approach ignores the fact that the despread signal is modulated by a data signal. This fact will certainly degrade performance of the matched filter approach. The IF filter bandwidth of the energy detector for the stepped phase serial search is selected to be wide enough to pass the data-modulated despread carrier. It should be clear from the results of this example that mean synchronization time can vary over a wide range as P_d, P_{fa}, and BT_i are varied, so that some optimization procedure is necessary.

Multiple-Dwell Detection. Fixed integration time detection is limited in that only two parameters, T_i and V_T, can be varied to reduce synchronization time. With signal-to-noise ratio and the number of uncertainty cells fixed, the selection of T_i and V_T completely determine P_d, P_{fa}, T_{da} and therefore average synchronization time. There is only one correct cell within the uncertainty region. Thus most of the cells evaluated by the energy detector are noise alone. An energy detection scheme that is capable of rejecting these incorrect phase cells rapidly while not letting P_{fa} become so large that false-alarm penalty time dominates synchronization time is desirable. The multiple-dwell detection scheme, discussed in this section, accomplishes this using multiple evaluations of the same phase cell. The first evaluation is very short and results in immediate rejection of many incorrect cells. The short integration time of this first evaluation also results in a high false-alarm probability. When a false alarm occurs on the first evaluation, a second evaluation of the same cell begins, using a longer integration time. This second evaluation is capable of rejecting most of the false alarms of the first evaluation. The second evaluation may be followed by a third, fourth, or as many as desired to achieve a particular performance level. Since a particular phase cell may be rejected after one or more integrations, the time required to evaluate a cell is a random variable. In this section a very general technique for evaluating the mean of this random variable is described. All integration times and thresholds as well as the logic followed by the multiple-dwell detection scheme are chosen to yield minimum average synchronization time. Fixed integration time detection is a special case (single-dwell) of the multiple-dwell systems described here.

Figure 10-14 is a simplified conceptual block diagram of a multiple-dwell energy detector. The detector is similar to the detector in Figure 10-8 except for the addition of a detection logic function which selects integration times T_j and thresholds V_j for each integration, and which outputs signal present/absent signals.

Figure 10-15 is a typical flow diagram for a multiple-dwell system. The system begins by selecting a code phase for evaluation using whatever a priori information available. Received signal despreading is attempted with this code phase, resulting

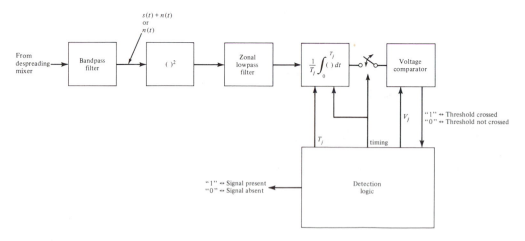

FIGURE 10-14. Simplified conceptual block diagram of multiple-dwell energy detector.

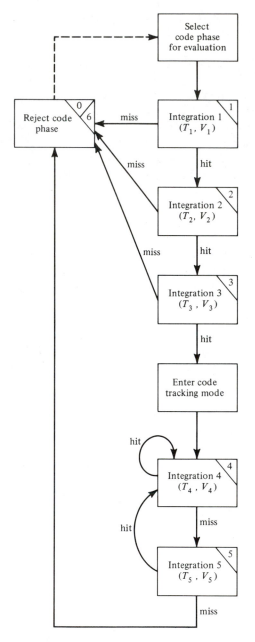

FIGURE 10-15. Typical logic flow diagram for a multiple-dwell detector.

in noise alone at the bandpass filter output if the phase is incorrect or signal plus noise at the filter output if the phase is correct. The filter output is squared and lowpass filtered, and the result is integrated for T_1 seconds. At the end of the integration, the integrator output V is compared with a threshold V_1. If $V < V_1$, a "miss" (incorrect phase) is declared and the phase cell is rejected. If $V \geq V_1$ a "hit" (potential correct phase) is declared and a second integration begins. This second integration is T_2 seconds long. At the end of the second integration, the integrator output V is compared with a threshold V_2. If $V < V_2$, the phase is rejected and, if $V \geq V_2$, the phase remains a potentially correct phase. When $V \geq V_2$, a third integration begins, after which the phase is either rejected or declared correct. The "hit" output from integration 3 generates the "signal present" output from the detection logic of Figure 10-14. The "signal absent" output is generated when a "miss" occurs on any of the first three integrations.

The logic of Figure 10-15 also includes a provision for continuing to evaluate the spreading code phase during code tracking. Integrations 4 and 5 provide a means for detecting when the code tracking loop has lost lock. During normal code tracking, the logic continues to cycle around the loop connecting integration 4 with itself. When a false alarm occurs, integrations 4 and 5 must occur before the code phase can be rejected. These two integration times may be quite large, since an incorrect dismissal of the correct code phase is very serious. The average time required to go between block 4 and block 6, after a false alarm has occurred, is the false-alarm penalty referred to earlier. The following analysis will include the false-alarm event explicitly. The flow diagram of Figure 10-15 is one example of a multiple-dwell system. Many other examples are possible. Other systems may differ in the total number of integrations and also in the interconnections between states. For example, a "miss" on integration 2 may return the system to integration 1 rather than to the "reject code phase" condition.

The goal of the analysis that follows is to calculate the average dwell time T_{da} at an incorrect phase cell, the probability of detecting the correct phase cell P_d when it is evaluated, and the average time required to evaluate the correct phase cell T_i. Knowledge of these variables will permit calculations of the mean synchronization time using (10-12).

Consider first the calculation of T_{da}. This average dwell time is the average time required by the system to cycle through the logic diagram of Figure 10-15 from the "integration 1" function to the "reject code phase" function with noise alone present. For convenience, some of the blocks of Figure 10-15 have been numbered and will denote the system states. The blocks that are not numbered are those blocks which are included to clarify the logic flow but which do not enter into the analysis to follow. Thus T_{da} is the average time required for the system to progress from state 1 to state 0/6. The reject code phase block has been numbered as both state 0 and state 6 because, in what follows, it will be convenient to distinguish between routes going between state 1 and 0/6 which do or do not pass through states 4 and 5. The flow diagram of Figure 10-15 can be represented by a state transition diagram as shown in Figure 10-16. In this figure, the functions that do not affect the analysis to follow have been deleted and the 0/6 state has been split into two separate states. Each transition of the diagram has been labeled with the probability of occurrence of the transition and z raised to a power equal to the integration time associated with the "from" state. At this point it may be

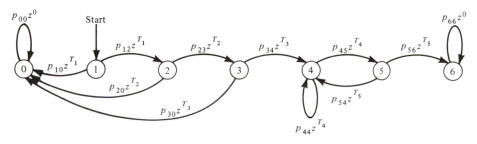

FIGURE 10-16. State transition diagram of a typical multiple-dwell detector.

recognized that the multiple-dwell system is exactly represented by a finite state Markov chain with absorbing boundaries [22]. The transition probabilities p_{ij} are the probabilities of passing from state i to state j and are calculated from the integration time T_i, the threshold V_i, and the SNR using the technique described in Section 10-3.4.

There are an infinite number of paths which can be followed through the state transition diagram between state 1 and state 0 or 6. The mean time required to reject an incorrect phase is given by

$$T_{da} = \sum_{l \in L} \Pr(l)T_l \tag{10-88}$$

where L denotes the set of all paths beginning at state 1 and ending at either state 0 or 6, l denotes a particular path in L, $\Pr(l)$ is the probability of following path l, and T_l is the time required to traverse path l. Consider any specific path through the state diagram of Figure 10-16, say path 1-2-3-0, which will be referred to as path l_0. The probability of following this path is the product of the state transition probabilities encountered while traversing this path, that is,

$$\Pr(l_0) = p_{12}p_{23}p_{30} \tag{10-89}$$

The time required to traverse this path is

$$T_{l_0} = T_1 + T_2 + T_3 \tag{10-90}$$

and the term of (10-88) corresponding to path l_0 is

$$\Pr(l_0)T_{l_0} = p_{12}p_{23}p_{30}(T_1 + T_2 + T_3) \tag{10-91}$$

Let $B(l, z)$ denote the product of the path labels encountered as path l is traversed. Thus, for path l_0,

$$B(l_0, z) = p_{12}z^{T_1}p_{23}z^{T_2}p_{30}z^{T_3}$$

$$= p_{12}p_{23}p_{30}z^{T_1+T_2+T_3} \tag{10-92}$$

The sum of integration times in the exponent can be transformed into a product term by taking the derivative with respect to z. The result is

$$\frac{d}{dz}B(l_0, z) = p_{12}p_{23}p_{30}(T_1 + T_2 + T_3)z^{T_1+T_2+T_3-1} \tag{10-93}$$

This equation is identical to (10-91) if $z = 1.0$. That is,

$$\Pr(l_0)T_{l_0} = \left[\frac{d}{dz}B(l_0, z)\right]_{z=1} \tag{10-94}$$

This result is true in general for any path l, so that

$$T_{da} = \sum_{l \in L} \Pr(l)T_l = \sum_{l \in L}\left[\frac{d}{dz}B(l, z)\right]_{z=1} \tag{10-95}$$

The state transition diagram can also be described by a transition matrix whose rows correspond to starting states, whose columns correspond to ending states, and whose elements are the path labels on the transition from the row state to the column state. The transition matrix Q' corresponding to the system described in Figure 10-16 is

$$Q' = \begin{array}{c} \\ 0 \\ 6 \\ 1 \\ 2 \\ 3 \\ 4 \\ 5 \end{array} \begin{bmatrix} z^0 & 0 & 0 & 0 & 0 & 0 & 0 \\ 0 & z^0 & 0 & 0 & 0 & 0 & 0 \\ p_{10}z^{T_1} & 0 & 0 & p_{12}z^{T_1} & 0 & 0 & 0 \\ p_{20}z^{T_2} & 0 & 0 & 0 & p_{23}z^{T_2} & 0 & 0 \\ p_{30}z^{T_3} & 0 & 0 & 0 & 0 & p_{34}z^{T_3} & 0 \\ 0 & 0 & 0 & 0 & 0 & p_{44}z^{T_4} & p_{45}z^{T_4} \\ 0 & p_{56}z^{T_5} & 0 & 0 & 0 & p_{54}z^{T_5} & 0 \end{bmatrix}$$

$$\triangleq \left[\begin{array}{c|c} \mathbf{U} & \mathbf{0} \\ \hline \mathbf{R} & \mathbf{Q} \end{array} \right] \tag{10-96}$$

Following Hopkins [23], the end states 0 and 6 have been assigned to the first two rows and columns, and the matrix has been partitioned into four submatrices $\mathbf{U}$, $\mathbf{R}$, $\mathbf{Q}$, and $\mathbf{0}$. The submatrix $\mathbf{Q}$ contains information about all the internal states of the system. These definitions are similar to those in Hopkins [23] except for the use of the delay operator z which permits the selection of different integration times for each decision. Define a new matrix $\mathbf{X} = \mathbf{Q}^n\mathbf{R}$, where $\mathbf{Q}^n$ indicates the nth power of $\mathbf{Q}$. Since $\mathbf{Q}$ is a square matrix, this new matrix is proper. The elements of $\mathbf{X}$ are denoted x_{ij}. The rows of $\mathbf{X}$ correspond to the same states as the rows of $\mathbf{Q}$, that is, the inner states, and the columns of $\mathbf{X}$ correspond to the same states as the columns of $\mathbf{R}$, that is, the end states. Each term of each element x_{ij} will correspond to a path of length $n + 1$ through the state transition diagram from state i to state j. If there are no paths of length $n + 1$ from state i to state j, then $x_{ij} = 0$. If there is one path, then x_{ij} is equal to the product of path labels encountered when traversing this path. If there are two paths, x_{ij} will have two terms each equal to the product of path labels encountered on the associated path.

EXAMPLE 10-4

Consider the state transition diagram of Figure 10-17. The matrix Q' is

$$Q' = \begin{array}{c} \\ 0 \\ 4 \\ 1 \\ 2 \\ 3 \end{array} \begin{bmatrix} 1 & 0 & 0 & 0 & 0 \\ 0 & 1 & 0 & 0 & 0 \\ p_{10}z^{T_1} & 0 & 0 & p_{12}z^{T_1} & 0 \\ 0 & 0 & p_{21}z^{T_2} & 0 & p_{23}z^{T_2} \\ 0 & p_{34}z^{T_3} & 0 & 0 & p_{33}z^{T_3} \end{bmatrix}$$

Therefore,

$$\mathbf{R} = \begin{bmatrix} p_{10}z^{T_1} & 0 \\ 0 & 0 \\ 0 & p_{34}z^{T_3} \end{bmatrix}$$

and

$$\mathbf{Q} = \begin{bmatrix} 0 & p_{12}z^{T_1} & 0 \\ p_{21}z^{T_2} & 0 & p_{23}z^{T_2} \\ 0 & 0 & p_{33}z^{T_3} \end{bmatrix}$$

10-3 SERIAL SEARCH SYNCHRONIZATION TECHNIQUES **517**

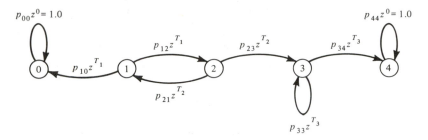

FIGURE 10-17. State transition diagram for Example 10-4.

The product $\mathbf{Q}^3\mathbf{R}$ is easily calculated to be

$$\mathbf{X} = \mathbf{Q}^3\mathbf{R} = \begin{matrix} & & 0 & & 4 \\ 1 \\ 2 \\ 3 \end{matrix} \begin{bmatrix} 0 & p_{12}p_{23}p_{33}p_{34}z^{T_1+T_2+2T_3} \\ p_{21}p_{12}p_{21}p_{10}z^{2T_1+2T_2} & p_{21}p_{12}p_{23}p_{34}z^{T_1+2T_2+T_3} + p_{23}p_{33}^2p_{34}z^{T_2+3T_3} \\ 0 & p_{33}^3p_{34}z^{4T_3} \end{bmatrix}$$

From this matrix it is seen that there are no paths of length exactly four between state 1 and state 0. This can be easily verified with a study of Figure 10-17. There is, however, one path of length 4 between state 2 and state 0; the path followed is 2-1-2-1-0. There are two paths of length 4 between state 2 and state 4; the paths followed are 2-1-2-3-4 and 2-3-3-3-4. ∎

Example 10-4 leads to the conjecture that the products of the path labels on all possible paths from inner states to end states through the state transition diagram are enumerated using matrix products of the form $\mathbf{Q}^n\mathbf{R}$. In addition, the infinite matrix sum

$$\mathbf{Y} = \mathbf{R} + \mathbf{Q}\mathbf{R} + \mathbf{Q}^2\mathbf{R} + \mathbf{Q}^3\mathbf{R} + \cdots \qquad (10\text{-}97)$$

enumerates all paths of all lengths between inner states and end states. Specifically,

$$y_{ij} = \sum_{l \in L(i,j)} B(l, z) \qquad (10\text{-}98)$$

where $L(i, j)$ is the set of all paths beginning at state i and ending at state j.

Return now to (10-95) for the average time to reject an incorrect phase cell. In this equation, L denotes all paths between state 1 and state 0 or 6; thus

$$L = L(1, 0) + L(1, 6) \qquad (10\text{-}99)$$

and (10-95) becomes

$$\begin{aligned} T_{da} &= \sum_{l \in L(1,0)} \left[\frac{d}{dz} B(l, z) \right]_{z=1} + \sum_{l \in L(1,6)} \left[\frac{d}{dz} B(l, z) \right]_{z=1} \\ &= \left\{ \frac{d}{dz} \left[\sum_{l \in L(1,0)} B(l, z) \right] + \frac{d}{dz} \left[\sum_{l \in L(1,6)} B(l, z) \right] \right\}_{z=1} \\ &= \left[\frac{d}{dz} (y_{10}) + \frac{d}{dz} (y_{16}) \right]_{z=1} \qquad (10\text{-}100) \end{aligned}$$

The derivative of a matrix is the matrix of the derivatives of the elements of the original matrix. Thus, each of the terms on the last line of (10-100) can be calculated from $\frac{d}{dz}\{\mathbf{Y}\}$. Equation (10-97) can be rewritten as

$$\begin{aligned}\mathbf{Y} &= (\mathbf{1} + \mathbf{Q} + \mathbf{Q}^2 + \cdots)\mathbf{R} \\ &= (\mathbf{1} - \mathbf{Q})^{-1}\mathbf{R}\end{aligned} \tag{10-101}$$

The last equality is derived by defining

$$\mathbf{A} = \mathbf{1} + \mathbf{Q} + \mathbf{Q}^2 + \cdots \tag{10-102}$$

Then

$$\mathbf{AQ} = \mathbf{Q} + \mathbf{Q}^2 + \mathbf{Q}^3 + \cdots \tag{10-103}$$

and

$$\mathbf{A} - \mathbf{AQ} = \mathbf{A}(\mathbf{1} - \mathbf{Q}) = \mathbf{1} \tag{10-104}$$

so that

$$\mathbf{A} = (\mathbf{1} - \mathbf{Q})^{-1} \tag{10-105}$$

The derivative with respect to z of (10-101) is

$$\begin{aligned}\frac{d}{dz}(\mathbf{Y}) &= (\mathbf{1} - \mathbf{Q})^{-1}\left(\frac{d}{dz}\mathbf{R}\right) + \left[\frac{d}{dz}(\mathbf{1} - \mathbf{Q})^{-1}\right]\mathbf{R} \\ &= (\mathbf{1} - \mathbf{Q})^{-1}\left(\frac{d}{dz}\mathbf{R}\right) - (\mathbf{1} - \mathbf{Q})^{-1}\left[\frac{d}{dz}(\mathbf{1} - \mathbf{Q})\right](\mathbf{1} - \mathbf{Q})^{-1}\mathbf{R} \end{aligned} \tag{10-106}$$

where the last equality follows from the matrix identity

$$\frac{d}{dz}\mathbf{A}^{-1} = -\mathbf{A}^{-1}\left(\frac{d}{dz}\mathbf{A}\right)\mathbf{A}^{-1} \tag{10-107}$$

At this point it is convenient to define a diagonal $n \times n$ matrix of integration times by

$$\mathbf{T} = \begin{bmatrix} T_1 & & & \\ & T_2 & & 0 \\ & & \ddots & \\ 0 & & & T_n \end{bmatrix} \tag{10-108}$$

where n is the number of internal states in the state transition diagram. With this definition it is easily seen that

$$\left(\frac{d}{dz}\mathbf{R}\right)_{z=1} = (\mathbf{TR})_{z=1} \tag{10-109}$$

and that

$$\left[\frac{d}{dz}(\mathbf{1} - \mathbf{Q})\right]_{z=1} = (-\mathbf{TQ})_{z=1} \tag{10-110}$$

Using these equations in (10-106) with $z = 1$,

$$\left(\frac{d}{dz}\mathbf{Y}\right)_{z=1} = \left.[(\mathbf{I} - \mathbf{Q})^{-1}\mathbf{TR} + (\mathbf{I} - \mathbf{Q})^{-1}\mathbf{TQ}(\mathbf{I} - \mathbf{Q})^{-1}\mathbf{R}\right]_{z=1}$$

$$= \{(\mathbf{I} - \mathbf{Q})^{-1}\mathbf{T}[\mathbf{R} + \mathbf{Q}(\mathbf{I} - \mathbf{Q})^{-1}\mathbf{R}]\}_{z=1}$$

$$= \{(\mathbf{I} - \mathbf{Q})^{-1}\mathbf{T}[\mathbf{I} + \mathbf{Q}(\mathbf{I} - \mathbf{Q})^{-1}]\mathbf{R}\}_{z=1}$$

$$= \{(\mathbf{I} - \mathbf{Q})^{-1}\mathbf{T}[\mathbf{I} + \mathbf{Q}(\mathbf{I} + \mathbf{Q} + \mathbf{Q}^2 + \cdots)]\mathbf{R}\}_{z=1}$$

$$= [(\mathbf{I} - \mathbf{Q})^{-1}\mathbf{T}(\mathbf{I} - \mathbf{Q})^{-1}\mathbf{R}]_{z=1} \qquad (10\text{-}111)$$

Finally, combining (10-100) and (10-111), the average time required to reject an incorrect phase cell is the sum of the elements on the first row of the last line above. Knowledge of all transition probabilities and integration times is required to evaluate this matrix equation. Since incorrect phase cells are being considered, the transition probabilities are calculated using the techniques discussed in Section 10-3.4 with noise alone as the detector input. It is assumed that the thresholds and integration times for each state are given.

The probability of detecting the correct phase cell when it is tested, P_d, is evaluated using the same techniques used to calculate T_{da}. In the example system of Figure 10-15, the probability of detection is the probability of passing from state 1 to the "enter code tracking mode" function. In any case except the noiseless case, arrival in the code tracking mode also implies that at some future time the system will pass through states 4 and 5 and will arrive at state 6. The probability of this event is unity since there is no other path to an end state and the system is guaranteed to eventually reach an end state [22]. Therefore, P_d is the probability of passing from state 1 to state 6. This probability is

$$P_d = \sum_{l \in L(1,6)} \Pr\,(l) = \sum_{l \in L(1,6)} B(l,\,z)_{z=1} \qquad (10\text{-}112)$$

Using (10-98),

$$P_d = (y_{16})_{z=1}$$

and from (10-101), P_d is the element of the first row and second column of

$$(\mathbf{Y})_{z=1} = [(\mathbf{I} - \mathbf{Q})^{-1}\mathbf{R}]_{z=1} \qquad (10\text{-}113)$$

When P_d is evaluated, the transition probabilities must be calculated for the signal plus noise input to the energy detector. Thus two sets of transition probabilities must be calculated in order to evaluate average synchronization time. One set is calculated for the noise alone case and is used to evaluate T_{da}. The other set is calculated for the signal plus noise case and is used to calculate P_d.

To complete the calculation of average synchronization time, T_i, the average time used in evaluating the correct phase cell must be determined. This calculation is identical to the calculation of T_{da} except that the transition probabilities for signal plus noise input to the detector are used, and the integration times in all states following the "enter tracking mode" block are set to zero. These integration times are set to zero so that "drop lock" time will not be included in T_i. The "drop lock" time is the average time that the system will remain cycling through the states following the tracking mode block when the signal is indeed present. Equation (10-111) yields the average times required to pass from state 1 to states 0 or 6. The path from state 1 to state 6 includes the "drop lock" time unless the integration times mentioned are set to zero.

All the tools are now in place to calculate mean synchronization time for a spread-spectrum system using stepped serial search with multiple-dwell detection. First, the transition probabilities are calculated for the noise alone and signal plus noise cases. Integration times and thresholds are assumed known. Next, T_{da} is evaluated using (10-111) and (10-100), P_d is evaluated using (10-113), and T_i is evaluated using (10-111). Finally, T_{da}, P_d, and T_i are substituted into (10-12) to calculate $\bar{T}_s$. Note that average values for T_{da} and T_i can be used in (10-12) since both terms appear as linear factors. The student is cautioned that the same substitution cannot be used to calculate the variance of the synchronization time.

EXAMPLE 10-5

Reconsider the system of Examples 10-1 and 10-3. The spreading code chip rate is $f_c = 3$ MHz, the design point SNR $= 46.25$ dB-Hz, and the propagation delay uncertainty is ± 1.2 ms. Assume again that the serial search step size is $1/2$ chip, so that $C = 14{,}400$. Assume that a double-dwell detection system is used which has a state transition diagram as shown in Figure 10-17. The IF filter bandwidth is 24 kHz. Arbitrarily choose time–bandwidth products for integrations 1, 2, and 3 of 4, 10, and 50, respectively. Also, arbitrarily choose signal plus noise transition probabilities of $p_{12} = p_{23} = 0.9$ and $p_{33} = 0.99$. Calculate the mean synchronization time for this system.

Solution: The integration times for the stated time–bandwidth products are $T_1 = 167$ μs, $T_2 = 417$ μs, and $T_3 = 2.08$ ms. The signal-to-noise ratio is $P/N_0 = 4.22 \times 10^4$. The normalized noncentrality parameters are calculated from $\lambda'_j = 2PT_j/N_0$; the results are $\lambda'_1 = 14.07$, $\lambda'_2 = 35.1$, and $\lambda'_3 = 176$. The modified number of degrees of freedom, calculated from (10-69) are $n'_1 = 13.5$, $n'_2 = 33.7$, and $n'_3 = 169$. Using these values along with the specified P_d's in the nomogram of Figure 10-11 yields normalized modified thresholds $V''_{T1} = 7.7$, $V''_{T2} = 24$, and $V''_{T3} = 129$. Using (10-70) the normalized thresholds are found to be $V'_{T1} = 12.6$, $V'_{T2} = 39.3$, and $V'_{T3} = 211.5$. The transition probabilities with noise alone at the detector input will be distinguished with a prime. The transition probabilities p'_{12}, p'_{23}, and p'_{33} are read from the nomogram using the given time–bandwidth products and normalized thresholds. The results are $p'_{12} = 0.12$, $p'_{23} = 0.007$, and $p'_{33} < 10^{-6}$. Using all of these values, the state transition diagrams for the noise alone input and the signal plus noise input are shown in Figure 10-18.

Equation (10-111) is evaluated with $z = 1.0$ to find T_{da}, the result is

$$\left(\frac{d}{dz}\mathbf{Y}\right)_{z=1} = \begin{bmatrix} 1.000 & -0.120 & 0 \\ -0.993 & 1.000 & -0.007 \\ 0 & 0 & 0.999999 \end{bmatrix}^{-1} \begin{bmatrix} 167\ \mu s & 0 & 0 \\ 0 & 417\ \mu s & 0 \\ 0 & 0 & 2.08\ ms \end{bmatrix}$$

$$\times \begin{bmatrix} 1.000 & -0.120 & 0 \\ -0.993 & 1.000 & -0.007 \\ 0 & 0 & 0.999999 \end{bmatrix}^{-1} \begin{bmatrix} 0.880 & 0 \\ 0 & 0 \\ 0 & 0.999999 \end{bmatrix}$$

$$= \begin{bmatrix} 246\ \mu s & 2.62\ \mu s \\ 489\ \mu s & 20.3\ \mu s \\ 0 & 2.08\ ms \end{bmatrix}$$

The average dwell time at an incorrect phase cell is 246 μs $+ 2.62$ μs ≈ 249 μs.

The probability of detection of the correct phase cell is calculated using (10-113) with $z = 1$. The result is

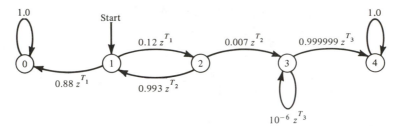

(a) Noise alone input

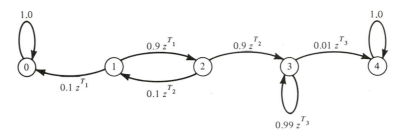

(b) Signal plus noise input

FIGURE 10-18.   State transition diagram for Example 10-5.

$$(\mathbf{Y})_{z=1} = \begin{bmatrix} 1.0 & -0.90 & 0 \\ -0.10 & 1.0 & -0.90 \\ 0 & 0 & 0.01 \end{bmatrix}^{-1} \begin{bmatrix} 0.1 & 0 \\ 0 & 0 \\ 0 & 0.01 \end{bmatrix}$$

$$= \begin{bmatrix} 0.11 & 0.89 \\ 0.0011 & 0.99 \\ 0 & 1.0 \end{bmatrix}$$

In this calculation the matrix $(\mathbf{I} - \mathbf{Q})^{-1}\mathbf{R}$ has been evaluated using the transition probabilities of Figure 10-18b. The detection probability is the element in the first row and last column of this matrix; thus $P_d = 0.89$.

Next, T_i is evaluated using (10-111). The transition probabilities of Figure 10-18b are used, and T_3 is set to zero. The result is

$$\left(\frac{d}{dz}\mathbf{Y}\right)_{z=1} = \begin{bmatrix} 1.0 & -0.90 & 0 \\ -0.10 & 1.0 & -0.90 \\ 0 & 0 & 0.01 \end{bmatrix}^{-1} \begin{bmatrix} 167 \ \mu s & 0 & 0 \\ 0 & 417 \ \mu s & 0 \\ 0 & 0 & 0 \end{bmatrix}$$

$$\times \begin{bmatrix} 1.0 & -0.90 & 0 \\ -0.10 & 1.0 & -0.90 \\ 0 & 0 & 0.01 \end{bmatrix}^{-1} \begin{bmatrix} 0.10 & 0 \\ 0 & 0 \\ 0 & 0.01 \end{bmatrix}$$

$$= \begin{bmatrix} 24.7 \ \mu s & 572 \ \mu s \\ 7.1 \ \mu s & 470 \ \mu s \\ 0 & 0 \end{bmatrix}$$

The time T_i is the sum of terms in the first row of the final result; thus $T_i = 572 \ \mu s + 24.7 \ \mu s \approx 597 \ \mu s$.

The final result for mean synchronization time from (10-12) is

$$\overline{T}_s = (14{,}399)(249 \ \mu s)\left[\frac{2 - 0.89}{2(0.89)}\right] + \frac{597 \ \mu s}{0.89}$$

$$= 2.24 \ s + 671 \ \mu s$$

$$\approx 2.24 \ s$$

The relative magnitude of the two terms of (10-12) seen above is typical when a large number of cells are being evaluated. For this reason, the last term is usually not calculated. ∎

Comparing the final results of Examples 10-3 and 10-5, it is concluded that a significant saving in average synchronization time is possible using multiple-dwell detection. Note, however, that neither example used optimum integration times and that the false-alarm penalty in Example 10-3 is larger than the equivalent penalty in Example 10-5. The conclusion that multiple-dwell systems do result in reduced synchronization times with respect to a fixed-integration-time system is valid.

Finally, the state transition diagrams for multiple-dwell detectors discussed above are typical, but many others are possible. Several types of transition diagrams are illustrated in Figure 10-19. Still other types are possible. It appears from some of these diagrams that time could by saved by beginning several integrations at the same time or, equivalently, using a single integrator which is sampled at appropriate times. This is not possible, however, because the analysis above assumes that all integrations are independent. Solution of this problem with dependent transition probabilities remains a challenging research problem.

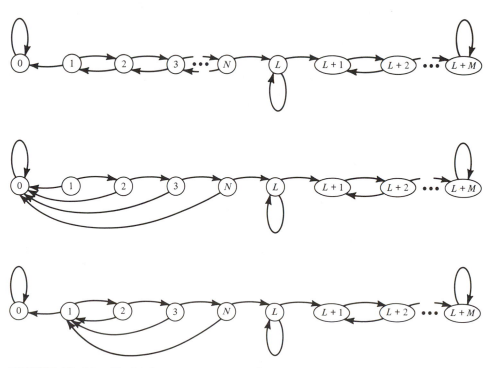

FIGURE 10-19. Typical state transition diagram types for multiple-dwell detection.

Sequential Detection. Consider the system illustrated in Figure 10-20 for detecting whether the phase of the receiver generated spreading waveform is correct. Once again, if the phase is correct, the received waveform will be despread, and signal power will appear at the output of the bandpass filter. The purpose of the envelope detector and the sequential detection processor is to detect this signal power reliably but without generating excessive false alarms when no signal is present. For any $K = 1, 2, \ldots$, let $x_K = (x_1, x_2, x_3, \ldots, x_K)$ denote a sequence of samples of the envelope detector output, and assume that the joint pdf of these samples is known. This pdf is denoted $p_s(\mathbf{x}_K)$ when signal is present and $p_n(\mathbf{x}_K)$ when noise alone is present.

During the evaluation of a single spreading waveform phase, samples of the detector output are taken one at a time and input to the sequential detection processor. After each sample (including the first), the likelihood ratio calculator computes

$$\Lambda(x_1, x_2, \ldots, x_K) = \frac{p_s(\mathbf{x}_K)}{p_n(\mathbf{x}_K)} \tag{10-114}$$

This likelihood ratio is input to the sequential detection logic. The logic function compares $\Lambda(\mathbf{x}_K)$ with an upper threshold A and a lower threshold B where $A > B$. If $\Lambda(\mathbf{x}_K) > A$, the signal is declared present and the test ends. If $\Lambda(\mathbf{x}_K) < B$, the signal is declared absent and the test ends. However, if $B < \Lambda(\mathbf{x}_K) < A$, no decision is made about the presence or absence of the signal and the test continues by taking another sample, calculating another likelihood ratio, and so on.

The likelihood ratio indicates whether the sequence of samples is more likely to have resulted from signal and noise at the filter output, $\Lambda(\mathbf{x}_K) > 1.0$, or from noise alone, $\Lambda(\mathbf{x}_K) < 1.0$. When $\Lambda(\mathbf{x}_K) >> 1.0$, the sequence of samples is much more likely to have come from signal and noise and a decision that the spreading waveform phase is correct is very reliable. Similarly, when $\Lambda(\mathbf{x}_K) << 1.0$ it is much more likely that the sequence of samples are the result of noise alone. The thresholds A and B are selected so that a specific reliability is achieved on the signal present/absent decision. Figure 10-21 illustrates $p_s(x_1)$ and $p_n(x_1)$ for three possible first sample values. For sample S_1, $\Lambda(x_1) << 1.0$; for sample S_2, $\Lambda(x_1) = 1.0$;

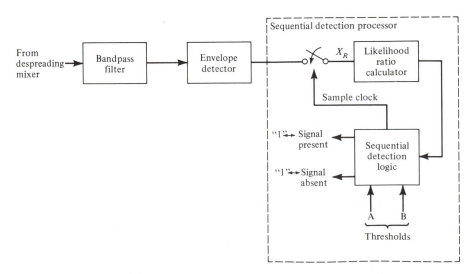

FIGURE 10-20. Sequential detector.

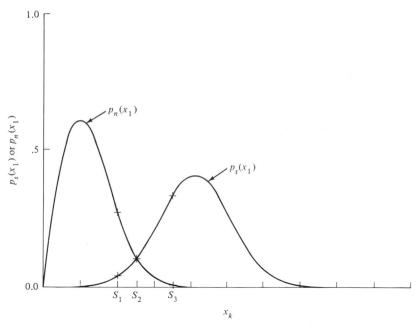

FIGURE 10-21. Envelope detector output probability density function.

for sample S_3, $\Lambda(x_1) \gg 1.0$. The probability density function for signal plus noise shown in this figure is calculated for a signal-to-noise ratio of about $+3$ dB.

Assume that the samples of the envelope detector output are spaced sufficiently in time that they may be considered independent. Then the joint pdf's are products of single sample pdf's, that is,

$$p_s(\mathbf{x}_K) = \prod_{k=1}^{K} p_s(x_k) \qquad (10\text{-}115a)$$

$$p_n(\mathbf{x}_K) = \prod_{k=1}^{K} p_n(x_k) \qquad (10\text{-}115b)$$

and

$$\Lambda(\mathbf{x}_K) = \prod_{k=1}^{K} \lambda(x_k) \qquad (10\text{-}116)$$

where

$$\lambda(x_k) \triangleq \frac{p_s(x_k)}{p_n(x_k)} \qquad (10\text{-}117)$$

When signal is present at the design point SNR, envelope detector samples usually result in $\lambda(x_k) > 1.0$ and the product $\Lambda(\mathbf{x}_K)$ grows with increasing K. When no signal is present, the envelope detector samples usually result in $\lambda(x_k) < 1.0$ and the product $\Lambda(\mathbf{x}_K)$ approaches zero as K increases. In the limit as $K \to \infty$, $\Lambda(\mathbf{x}_K)$ will always cross the upper threshold when signal is present at the design point SNR or the lower threshold when no signal is present. For finite K, errors can be made. A missed detection occurs whenever $\Lambda(\mathbf{x}_K)$ crosses the lower threshold before crossing the upper threshold when signal is indeed present. The probability of this event is $1.0 - P_d$. A false alarm occurs whenever $\Lambda(\mathbf{x}_K)$ crosses the upper threshold prior to crossing the lower threshold when no signal is present. The probability of this event is P_{fa}. Typical sequences of values of $\Lambda(\mathbf{x}_K)$ for signal plus noise and noise alone are illustrated in Figure 10-22. No errors occur in the cases shown. Observe that the sample values move toward one of the thresholds with a step size that is a random variable. The statistics of this random variable and the absolute value of the thresholds will determine the average number of samples that must be processed to cross one of the thresholds. The average sample number (ASN) is directly related to the quantities T_{da} and T_i, which must be evaluated to calculate the average synchronization time. The ASN is a function of the received signal-to-noise ratio, the detector characteristic, and the thresholds A and B.

Detectors having the general form shown in Figure 10-20 are called sequential detectors and were first discussed in Wald [24]. A great deal of literature is available on sequential detection. Two particularly good discussions are found in Helstrom [25] and Hancock and Wintz [26]. Sequential detection is sometimes referred to as the *sequential probability ratio test* (SPRT). It has been proven that sequential detection is optimum in the sense that it yields the minimum average detection time for a specified P_d and P_{fa} [27].

Relationship Between Thresholds and Error Probabilities. To design systems that use sequential detection, it is necessary to be able to relate the thresholds and the probabilities P_d and P_{fa} for any signal-to-noise ratio. Consider the set of all real vectors $\mathbf{x}_K$ for all $K = 1, 2, \ldots$, and assume that the functions $p_s(\mathbf{x}_K)$ and $p_n(\mathbf{x}_K)$ are known for the particular SNR. Let Γ_1 denote the set of all vectors that result in a decision that signal is present at the output of the IF bandpass filter. Thus, for every $\mathbf{x}_K \in \Gamma_1$,

$$A < \frac{p_s(\mathbf{x}_K)}{p_n(\mathbf{x}_K)} \qquad (10\text{-}118)$$

Note that all vectors in Γ_1 do not have the same dimensionality since different sequences $\mathbf{x}_K$ may result in a threshold crossing after different numbers of samples

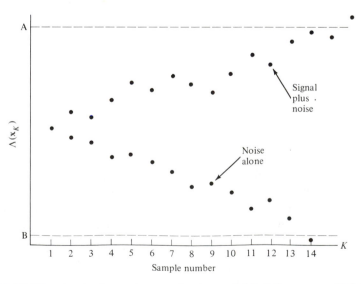

FIGURE 10-22. Typical sequences of values of $\Lambda(x_K)$ for sequential detection.

are taken. Assume now that the bandpass filter output is noise alone. A false alarm occurs for any $\mathbf{x}_K \in \Gamma_1$ and the probability of this event is

$$P_{\text{fa}} = \int_{\mathbf{x}_K \in \Gamma_1} p_n(\mathbf{x}_K) \, d\mathbf{x}_K \qquad (10\text{-}119a)$$

Similarly, when the filter output is signal plus noise, a correct detection occurs whenever $\mathbf{x}_K \in \Gamma_1$, and the probability of this event is

$$P_d = \int_{\mathbf{x}_K \in \Gamma_1} p_s(\mathbf{x}_K) \, d\mathbf{x}_K \qquad (10\text{-}119b)$$

Multiplying both sides of (10-118) by $p_n(\mathbf{x}_K)$, integrating over Γ_1, and using the results of (10-119) yields

$$P_d > A P_{\text{fa}} \qquad (10\text{-}120)$$

Let Γ_2 denote the set of all vectors which result in a decision that noise alone is present at the output of the filter. For every $\mathbf{x}_K \in \Gamma_2$,

$$\frac{p_s(\mathbf{x}_K)}{p_n(\mathbf{x}_K)} < B \qquad (10\text{-}121)$$

Assume that the bandpass filter output is noise alone. A correct dismissal occurs for all $\mathbf{x}_K \in \Gamma_2$, and this event occurs with probability

$$1 - P_{\text{fa}} = \int_{\mathbf{x}_K \in \Gamma_2} p_n(\mathbf{x}_K) \, d\mathbf{x}_K \qquad (10\text{-}122a)$$

When the filter output is signal plus noise, a missed detection occurs when $\mathbf{x}_K \in \Gamma_2$. This event occurs with probability

$$1 - P_d = \int_{\mathbf{x}_K \in \Gamma_2} p_s(\mathbf{x}_K) \, d\mathbf{x}_K \qquad (10\text{-}122b)$$

Combining (10-121) and (10-122) results in

$$1 - P_d < B(1 - P_{\text{fa}}) \qquad (10\text{-}123)$$

If (10-120) and (10-123) were equalities, the thresholds could be solved for as functions of P_d and P_{fa}. For very low signal-to-noise ratios, these equalities can be assumed with little error. To see this, reconsider the set Γ_1 and (10-118). Suppose that Γ_1' is defined as the set of all $\mathbf{x}_K$ which lead to a decision that signal is present and for which

$$A < \frac{p_s(\mathbf{x}_K)}{p_n(\mathbf{x}_K)} < A + \Delta A \qquad (10\text{-}124)$$

The set Γ_1' is a subset of Γ_1. For small signal-to-noise ratios, there is a correspondingly small difference in the pdf's $p_n(x_1)$ and $p_s(x_1)$, so that samples of the likelihood ratio are typically near unity, although large or small values do occur from time to time. Thus, from one sample to the next, small changes in $\Lambda(\mathbf{x}_K)$ are typical. Since the changes in $\Lambda(\mathbf{x}_K)$ are small, it is likely that the value of $\Lambda(\mathbf{x}_K)$ after the sample, which results in the threshold crossing is very close to the threshold. This in turn implies that the most likely $\mathbf{x}_K$ which result in an upper threshold crossing are in Γ_1'. Let $\overline{\Gamma}_1'$ denote the set of all $\mathbf{x}_K$ in Γ_1 but not in Γ_1', that is, $\Gamma_1' + \overline{\Gamma}_1' = \Gamma_1$. Equation (10-119) can be rewritten as

$$P_{\text{fa}} = \int_{\mathbf{x}_K \in \Gamma_1'} p_n(\mathbf{x}_K)\, d\mathbf{x}_K + \int_{\mathbf{x}_K \in \overline{\Gamma}_1'} p_n(\mathbf{x}_K)\, d\mathbf{x}_K \qquad (10\text{-}125a)$$

$$P_d = \int_{\mathbf{x}_K \in \Gamma_1'} p_s(\mathbf{x}_K)\, d\mathbf{x}_K + \int_{\mathbf{x}_K \in \overline{\Gamma}_1'} p_s(\mathbf{x}_K)\, d\mathbf{x}_K \qquad (10\text{-}125b)$$

Based on the discussion above, it is reasonable to assume that the rightmost term in (10-125) vanishes as SNR approaches zero. Multiplying both sides of (10-124) by $p_n(\mathbf{x}_K)$ and integrating over Γ_1' yields

$$AP_{\text{fa}} < P_d < (A + \Delta A)P_{\text{fa}} \qquad (10\text{-}126)$$

As SNR approaches zero, this equation remains valid for vanishingly small ΔA and

$$AP_{\text{fa}} = P_d \qquad (10\text{-}127)$$

is valid in the limit. A similar argument leads to

$$1 - P_d = B(1 - P_{\text{fa}}) \qquad (10\text{-}128)$$

These equations can be solved for P_d and P_{fa}, the result is

$$P_{\text{fa}} = \frac{1 - B}{A - B} \qquad (10\text{-}129a)$$

$$P_d = A\left(\frac{1 - B}{A - B}\right) \qquad (10\text{-}129b)$$

It cannot be overemphasized that the (10-127) and (10-128) are valid only for SNR approaching zero. Inaccuracies occur when the value of $\Lambda(\mathbf{x}_K)$ that first crosses the threshold is significantly different from the threshold. This problem has been discussed in detail in Wald [24], Helstrom [25], and Hancock and Wintz [26] and is called the *excess over boundary problem*.

When the SNR is not small and the excess over boundary problem cannot be ignored, the designer normally resorts to simulations to characterize the system. Simulations of these systems are straightforward and provide the only reasonable method of performance characterization when SNR is medium to high. The computer simulation also enables the designer to include the effects of different detector types as well as the effects of dependent samples. Additional discussion of simulators for sequential detection can be found in Cobb and Darby [28].

Operating Characteristic Function. In the discussion above it was assumed that the SNR was fixed and known. It is usually necessary to be able to calculate the performance of the system at SNRs other than the design point. A quick look at (10-129) indicates that the SNR does not enter in the relationship between the thresholds and P_d and P_{fa}. The test itself, however, is a function of SNR since $p_n(\mathbf{x}_K)$ and $p_s(\mathbf{x}_K)$ are functions of SNR. Thus, the relationships of (10-129) are valid only when the likelihood ratio calculation and the received SNR are matched. Suppose that the system is designed for a particular design point SNR and that the actual SNR is different from the design point. When noise-alone tests are being performed, the system will operate as designed; that is, P_{fa} is not affected by the change in SNR. When signal is present, however, modified performance will be observed and P_d will not necessarily equal the design point P_d. The actual P_d as a function of SNR for a sequential test designed for SNR_0 is related to the *operating characteristic function* (OCF), which is usually denoted $L(\text{SNR})$.

The OCF has been derived in Wald [24] and is also discussed in Helstrom [25] and Hancock and Wintz [26]. The OCF is defined as the probability that the sequential test will result in a decision that noise alone is present when actually signal is present at a specified SNR. Suppose that the SPRT is designed using $p_n(x_1)$ and $p_{s_0}(x_1)$ but that the actual detector output pdf with signal present is $p_s(x_1)$. Consider a new pdf $p(x_1)$, which is defined by

$$p(x_1) = \left[\frac{p_{s_0}(x_1)}{p_n(x_1)} \right]^h p_s(x_1) \tag{10-130}$$

where the parameter h is selected so that

$$\int_{-\infty}^{\infty} p(x_1) \, dx_1 = 1.0 \tag{10-131}$$

With this choice of h, $p(x_1)$ is a valid probability density function. An SPRT is designed to distinguish between the density functions $p(x_1)$ and $p_s(x_1)$. Assuming independent samples, this SPRT employs the likelihood ratio

$$\Lambda(\mathbf{x}_K) = \prod_{k=1}^{K} \frac{p(x_k)}{p_s(x_k)} \tag{10-132}$$

and compares the result with two thresholds $A' = A^h$ and $B' = B^h$, where A and B are thresholds used in the original SPRT. The new SPRT estimates that $p_s(x_1)$ is the correct pdf if $\Lambda(\mathbf{x}_k) < B'$ and that $p(x_1)$ is the correct pdf if $\Lambda(\mathbf{x}_K) > A'$. The likelihood ratio for the original test is

$$\Lambda_0(\mathbf{x}_K) = \prod_{k=1}^{K} \frac{p_{s_0}(x_k)}{p_n(x_k)} \tag{10-133}$$

From (10-130), (10-131), and (10-132) it is found that $[\Lambda_0(\mathbf{x}_K)]^h = \Lambda(\mathbf{x}_K)$. Thus, whenever $\Lambda(\mathbf{x}_K)$ crosses the threshold A' or B', $\Lambda_0(\mathbf{x}_K)$ crosses the threshold A or B, respectively.

The value of the operating characteristic function is the probability that the original test estimates that $p_n(x_1)$ is the correct pdf when $p_s(x_1)$ is correct. This probability is equal to the probability that threshold B is crossed by the original test when $p_s(x_1)$ is true, which is equivalent to the probability that threshold B' is crossed by the new test when $p_s(x_1)$ is true. The latter probability is the probability that $p_s(x_1)$ is estimated to be the correct pdf by the new test when $p_s(x_1)$ is true. The derived test is matched to its input statistics so that (10-129a) can be used to determine the probability of its estimating that $p_s(x_1)$ is correct when $p_s(x_1)$ is indeed correct. This probability is

$$1 - P_{\text{fa}} = 1 - \frac{1 - B^h}{A^h - B^h} = \frac{A^h - 1}{A^h - B^h} = L(\text{SNR}) \tag{10-134}$$

The probability of detecting the signal at the received SNR is the probability of $\Lambda_0(\mathbf{x}_K)$ crossing the upper threshold when $p_s(x_1)$ is true. This probability is

$$P_d(\text{SNR}) = 1 - L(\text{SNR}) = \frac{1 - B^h}{A^h - B^h} \tag{10-135}$$

Observe that when the received SNR is zero, $p_s(x_1) = p_n(x_1)$ and $h = +1$. From (10-135)

$$P_d(0) = \frac{1 - B}{A - B} \tag{10-136}$$

which is seen to equal P_{fa} for the original test as expected. Similarly, when the received SNR is equal to the design point SNR, $p_s(x_1) = p_{s_0}(x_1)$, and $h = -1$. From (10-135)

$$P_d(\mathrm{SNR}_0) = \frac{1 - B^{-1}}{A^{-1} - B^{-1}} = A\left(\frac{1 - B}{A - B}\right) \qquad (10\text{-}137)$$

which is equal to the design point P_d of (10-129b).

Unfortunately, calculation of the operating characteristic function and $P_d(\mathrm{SNR})$ is often extremely difficult because of the difficulty of determining h from (10-130) and (10-131). When the detector output pdf's are Gaussian, however, a simple solution exists [26]. In most other cases simulations can be used to evaluate $P_d(\mathrm{SNR})$ with little difficulty. An example operating characteristic function is shown in Figure 10-23. This example is the OCF for the SPRT of the mean of a Gaussian distribution. The design point mean is denoted μ and the actual mean is denoted θ. The test is designed to determine whether the detector output samples have a mean of zero (noise alone) or a mean of θ (signal plus noise). Observe that, at the design point $(\theta/\mu = 1.0)$, $L(\theta) = 1 - P_d$.

SPRT Using a Linear Envelope Detector and the Log-Likelihood Ratio. Although the sequential test can be analyzed in its most basic form where the actual likelihood ratio is calculated, the multiplications of the likelihood ratios as in (10-116) are inconvenient and not necessary. The performance of the SPRT is unchanged if, instead of using the likelihood ratio, any monotonic strictly increasing function of the likelihood ratio is used. Of course, the thresholds are also modified by this same monotonic function. Since the likelihood ratio is positive, the logarithm is a convenient function to use since it also converts the undesirable product functions to summations. At this point it is also convenient to limit attention to the specific detector and associated probability densities of most interest for application to spread-spectrum systems. The detector is a linear envelope detector

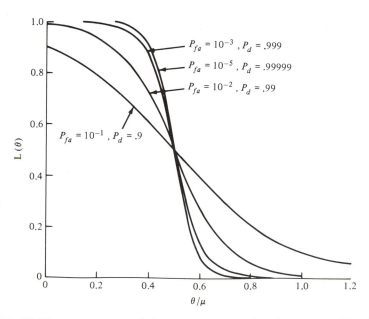

FIGURE 10-23. Operating characteristic function for the SPRT of the mean of a Gaussian distribution. (From Ref. 26.)

and the synchronization hardware is trying to detect the presence/absence of a sine wave at the bandpass filter output. Therefore, the pdf of a single sample of the envelope detector output is given by (10-48), which is due to Rice. The likelihood ratio, using (10-48) and (10-117) with slightly modified notation, is

$$
\lambda(x_k) = \frac{p_s(x_k)}{p_n(x_k)}
$$

$$
= \frac{(x_k/N)\exp{(-(x_k^2 + 2P)/2N)}I_0(x_k\sqrt{2P}/N)}{(x_k/N)\exp{(-x_k^2/2N)}} \qquad (10\text{-}138)
$$

where $p_n(x_k)$ is found from (10-48) by setting the signal power $P = 0$. The output of the bandpass filter has been assumed to be AWGN with power N or to be AWGN with power N plus a sine wave with power P and amplitude $\sqrt{2P}$. Simplifying (10-138) and taking the logarithm yields the log-likelihood ratio for a single sample

$$
\ln[\lambda(x_k)] = -\frac{P}{N} + \ln\left[I_0\left(\frac{x_k\sqrt{2P}}{N}\right)\right] \qquad (10\text{-}139)
$$

The log-likelihood ratio for a sequence of samples is the logarithm of (10-116), which is

$$
\ln[\Lambda(\mathbf{x}_K)] = \ln\left[\prod_{k=1}^{K}\lambda(x_k)\right]
$$

$$
= \sum_{k=1}^{K}\ln[\lambda(x_k)]
$$

$$
= \sum_{k=1}^{K}\left[\ln\left[I_0\left(\frac{x_k\sqrt{2P}}{N}\right)\right] - \frac{P}{N}\right] \qquad (10\text{-}140)
$$

Figure 10-24 is a conceptual block diagram of a sequential detector that uses the simplifications discussed above. Samples of the linear envelope detector output are scaled by $\sqrt{2P}/N$ and input to the $\ln[I_0(\cdot)]$ calculator. The output of this function is input to the adder after the bias P/N is subtracted. The adder adds the new value $\ln[\lambda(x_k)]$ to the sum of all previous values to obtain $\ln[\Lambda(\mathbf{x}_K)]$. The final result is compared with an upper threshold $\ln A$ and a lower threshold $\ln B$. If the result is above the upper threshold, the signal is declared present. If the result is below the lower threshold, signal is declared absent. If the result is between the two thresholds, another sample is taken and the process repeats. This sequential detector is a special case of the sequential detector described earlier.

When the log-likelihood ratio is used in place of the likelihood ratio, the upper threshold, which was larger than unity is positive, but the lower threshold, which was less than unity, is negative. Figure 10-25 illustrates typical sequences of values of $\ln[\Lambda(\mathbf{x}_K)]$ leading to upper and lower threshold crossings for the log-likelihood SPRT. With this detector, sequences corresponding to signal plus noise typically have a positive slope and sequences corresponding to noise alone typically have a negative slope. It will be seen later that at SNRs lower than the design point SNR, the positive slope sequence of Figure 10-25 may have nearly a zero slope with the effect that many samples are required to reach either threshold.

Average Sample Number. The last performance measure that needs to be calculated in order to evaluate the mean synchronization time for the spread-spectrum

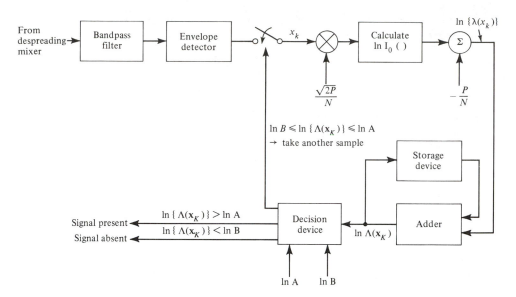

FIGURE 10-24. Sequential detector using linear envelope detector and log-likelihood ratio.

system is the average sample number (ASN). The ASN is the average number of samples taken by the detector prior to crossing one of the thresholds. With noise alone at the filter output, ASN determines the average dwell time T_{da} at an incorrect spreading waveform phase. Denote the ASN for noise alone by $\overline{K}_n$ and the sampling interval by T_s. Then

$$T_{da} = \overline{K}_n T_s + T_{fa} P_{fa} \qquad (10\text{-}141)$$

where the false-alarm penalty time T_{fa} is assumed known. With signal and noise at the filter output, ASN determines the average time to evaluate the correct spread-

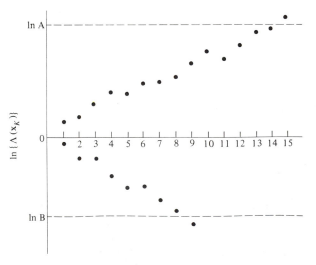

FIGURE 10-25. Typical sequences of values of ln $\{ \Lambda(x_K) \}$ leading to upper and lower threshold crossing for SPRT using log-likelihood ratio.

ing waveform phase, T_i. Denote the ASN for signal plus noise by $\overline{K}_s$; then

$$T_i = \overline{K}_s T_s \qquad (10\text{-}142)$$

In the calculation of either $\overline{K}_N$ or $\overline{K}_s$ the possibilities of the test ending with either an upper or a lower threshold crossing must be taken into account.

The derivation of the ASN begins with a short development from Kendall and Stuart [29], which is also presented in Helstrom [25]. Define a random variable y_k which is equal to 1 if the SPRT has not finished prior to the kth sample and thus observes x_k, and is equal to 0 otherwise. Let P_i denote the probability that a SPRT finishes at exactly the ith stage. The mean value of y_k is $E(y_k)$, where the expected value is over all possible runs of the sequential test. The value of $y_k = 1$ for all those tests which finish at or after the kth sample. The value of $y_k = 0$ for all other tests. Therefore,

$$E(y_k) = \sum_{i=1}^{k-1} 0 \cdot P_i + \sum_{i=k}^{\infty} 1 \cdot P_i$$

$$= \sum_{i=k}^{\infty} P_i \qquad (10\text{-}143)$$

For a test that finishes on the Kth sample, the log-likelihood ratio at the end of the test is

$$\ln\,[\Lambda(\mathbf{x}_K)] = \sum_{k=1}^{K} \ln\,[\lambda(x_k)]$$

$$= \sum_{k=1}^{\infty} y_k \ln\,[\lambda(x_k)] \qquad (10\text{-}144)$$

since $y_k = 1$ for $1 \le k \le K$ and $y_k = 0$ for $K < k$. The random variable y_k depends only on the values of the detector output x_1 through x_{k-1} since they determine whether or not a test finishes on or before the $(k-1)$th stage and therefore whether a kth sample is required. Therefore, the random variables y_k and x_k are independent and the expected value of (10-144) over all tests can be written

$$E\{\ln\,[\Lambda(\mathbf{x}_K)]\} = E\left\{ \sum_{k=1}^{\infty} y_k \ln\,[\lambda(x_k)] \right\}$$

$$= \sum_{k=1}^{\infty} E(y_k)E\{\ln\,[\lambda(x_k)]\} \qquad (10\text{-}145)$$

Since the detector output pdf is the same for every sample x_k, the right expected value above is not a function of k and may be factored, resulting in

$$E\{\ln\,[\Lambda(\mathbf{x}_K)]\} = E\{\ln\,[\lambda(x_k)]\} \sum_{k=1}^{\infty} E(y_k)$$

$$= E\{\ln\,[\lambda(x_k)]\} \sum_{k=1}^{\infty} \sum_{i=k}^{\infty} P_i$$

$$= E\{\ln\,[\lambda(x_k)]\} \sum_{k=1}^{\infty} kP_k$$

$$= E\{\ln\,[\lambda(x_k)]\}E(k) \qquad (10\text{-}146)$$

This equation is Kendall's result stating that the mean value of the log-likelihood ratio at the end of the SPRT is equal to the product of the mean value of the log-likelihood ratio for each sample and the expected number of samples taken to complete the test. The expected value $E[k]$ is the average sample number being calculated.

Once again, it is necessary to make the assumption that excess over boundaries can be neglected. This implies tests using a large number of samples on the average are being considered which in turn implies low SNR and/or $1 - P_d << 1$ and $P_{fa} << 1$. With this assumption, the value of the log-likelihood ratio at the end of the test is either $\ln A$ or $\ln B$. When the SPRT is evaluating incorrect phase cells (i.e., noise alone at the filter output), the value is $\ln A$ with probability P_{fa} and $\ln B$ with probability $1 - P_{fa}$. In this case,

$$E\{\ln [\Lambda(\mathbf{x}_k)]\} = P_{fa} \ln A + (1 - P_{fa}) \ln B \tag{10-147}$$

and the average sample number is

$$\overline{K}_n = \frac{P_{fa} \ln A + (1 - P_{fa}) \ln B}{E\{\ln[\lambda(x_k)]\}} \tag{10-148}$$

When the SPRT is evaluating the correct phase cell, the final log-likelihood ratio is $\ln A$ with probability P_d and $\ln B$ with probability $1 - P_d$. Recall that the probability of detection is given, in general, by the operating characteristic function; specifically $P_d = 1 - L(\text{SNR})$. Using the OCF,

$$E\{\ln [\Lambda(\mathbf{x}_k)]\} = L(\text{SNR}) \ln B + [1 - L(\text{SNR})] \ln A \tag{10-149}$$

and the ASN is

$$\overline{K}_s = \frac{L(\text{SNR}) \ln B + [1 - L(\text{SNR})] \ln A}{E\{\ln [\lambda(x_k)]\}} \tag{10-150}$$

At the design point SNR,

$$\overline{K}_s = \frac{(1 - P_d) \ln B + P_d \ln A}{E\{\ln [\lambda(x_k)]\}} \tag{10-151}$$

The equations above are all based on the assumption that excess over boundaries can be ignored.

The average sample numbers of (10-148), (10-150), and (10-151) can be evaluated if the average of the log-likelihood ratio for a single sample can be determined. The single sample log-likelihood ratio is given by (10-139). Since the ASN equations apply principally at low SNR, it is appropriate to attempt to evaluate the mean of (10-139) at low SNR. The zeroth-order modified Bessel function may be represented by the series [30]

$$I_0(z) = \sum_{m=0}^{\infty} \frac{z^{2m}}{2^{2m}(m!)^2}$$

$$= 1 + \frac{z^2}{4} + \frac{z^4}{64} + \frac{z^6}{2304} + \cdots \tag{10-152}$$

For small z, $I_0(z)$ can be approximated by the first three terms of the series. The logarithm of (10-152) can be simplified with the well-known series expansion

$$\ln (1 + \alpha) = \alpha - \tfrac{1}{2}\alpha^2 + \tfrac{1}{3}\alpha^3 - \cdots \tag{10-153}$$

Let $\alpha = \frac{1}{4}z^2 + \frac{1}{64}z^4$ and combine (10-152) and (10-153) to obtain

$$\ln[I_0(z)] \approx \left(\frac{z^2}{4} + \frac{z^4}{64}\right) - \frac{1}{2}\left(\frac{z^2}{4} + \frac{z^4}{64}\right)^2 + \frac{1}{3}\left(\frac{z^2}{4} + \frac{z^4}{64}\right)^3 - \cdots$$

$$\approx \frac{z^2}{4} - \frac{z^4}{64} \tag{10-154}$$

where all powers of z greater than 4 have been ignored. Substituting $z = x_k\sqrt{2P}/N$ yields

$$\ln\left[I_0\left(x_k\frac{\sqrt{2P}}{N}\right)\right] \approx \left(\frac{P}{2N^2}\right)x_k^2 - \left(\frac{P^2}{16N^4}\right)x_k^4 \tag{10-155}$$

The problem being addressed is the calculation of the mean of (10-139). Substituting (10-155) into (10-139) and taking the expected value results in

$$E\{\ln[\lambda(x_k)]\} \approx E\left[-\frac{P}{N} + \left(\frac{P}{2N^2}\right)x_k^2 - \left(\frac{P^2}{16N^4}\right)x_k^4\right]$$

$$= -\frac{P}{N} + \frac{P}{2N^2}E(x_k^2) - \left(\frac{P^2}{16N^4}\right)E(x_k^4) \tag{10-156}$$

The expected values in this equation are over all noise processes at the bandpass filter output. The random variables x_k are samples of the output of the envelope detector and have a pdf given by (10-48). This is the well-known Ricean pdf, and the expected values above are the second and fourth moments of this pdf. These moments have been calculated in Rice [9] and the results are

$$E(x_k^2) = 2N\Gamma(2)\,_1F_1\left(-1;\, 1;\, -\frac{P'}{N}\right) \tag{10-157a}$$

and

$$E(x_k^4) = (2N)^2\Gamma(3)\,_1F_1\left(-2;\, 1;\, -\frac{P'}{N}\right) \tag{10-157b}$$

where $_1F_1(a;\, b;\, z)$ is the confluent hypergeometric function, which has a series representation [9]

$$_1F_1(a, b, z) = 1 + \frac{az}{b} + \frac{a(a+1)}{b(b+1)}\frac{z^2}{2!} + \frac{a(a+1)(a+2)}{b(b+1)(b+2)}\frac{z^3}{3!} + \cdots \tag{10-158}$$

$\Gamma(n) = (n-1)!$ is the gamma function and P'/N is the SNR being considered. When the expected values are being evaluated for noise-alone phase cells, $P'/N = 0$, and when the expected values are being evaluated for the correct phase cells, $P'/N =$ (operating point SNR). The SNR used in (10-156) is the design point SNR, which may or may not equal the operating point SNR. Observe that the infinite series is truncated for $a = -1, -2$, so that

$$E(x_k^2) = 2N\left(1 + \frac{P'}{N}\right) = 2(N + P') \tag{10-159a}$$

$$E(x_k^4) = (2N)^2 2\left[1 + \frac{2P'}{N} + \frac{1}{2}\left(\frac{P'}{N}\right)^2\right] = 8(N^2 + 2P'N + \tfrac{1}{2}[P']^2) \tag{10-159b}$$

Substituting these relationships into (10-156) and simplifying yields

$$E\{\ln\,[\lambda(x_k)]\} \approx -\frac{1}{2}\left(\frac{P}{N}\right)^2 + \left(\frac{P}{N}\right)\left(\frac{P'}{N}\right) - \left(\frac{P}{N}\right)^2\left(\frac{P'}{N}\right) - \frac{1}{4}\left(\frac{P}{N}\right)^2\left(\frac{P'}{N}\right)^2$$

$$\approx -\frac{1}{2}\left(\frac{P}{N}\right)^2 + \left(\frac{P}{N}\right)\left(\frac{P'}{N}\right) \tag{10-160}$$

Observe that the average increment for incorrect phase cells is $-P^2/2N^2$ and that at the design point SNR (i.e., $P' = P$) the average increment is $+P^2/2N^2$. Combining this result with (10-148) and substituting into (10-141) gives the average dwell time at an incorrect phase cell for low SNR,

$$T_{da} = \frac{P_{fa}\ln A + (1 - P_{fa})\ln B}{-\frac{1}{2}(P/N)^2}T_s + T_{fa}P_{fa} \tag{10-161}$$

where P'/N is set equal to zero since noise alone phase cells are being considered. Similarly, the average time required to evaluate a correct phase cell is calculated from (10-160), (10-150), and (10-142)

$$T_i = \frac{L(\text{SNR})\ln B + [1 - L(\text{SNR})]\ln A}{-\frac{1}{2}(P/N)^2 + (P/N)(P'/N)}T_s \tag{10-162}$$

For $P'/N = P/2N$ this equation is indeterminate. The average evaluation time in this special case is calculated in Helstrom [25]. In this equation $L(\text{SNR})$ is evaluated at the operating point SNR.

At this point the student has all the tools necessary to calculate the mean synchronization time for a spread-spectrum system using stepped serial search with sequential detection. The sequential test is designed for a particular operating point SNR which is assumed to be low. P_d and P_{fa} are assumed to be given, although their selection is part of a long optimization process. The thresholds A and B are calculated using (10-127) and (10-128). The sampling interval T_s is given but must be sufficiently long that independent samples result. The dwell times T_{da} and T_i are calculated from (10-161) and (10-162), where $L(\text{SNR}) = 1 - P_d$ at the design point. Average synchronization time is calculated by substituting P_d, T_{da}, and T_i into (10-12) together with the given C.

A typical plot of the normalized average sample number versus SNR when signal plus noise is being evaluated is illustrated in Figure 10-26. The particular curve shown is from Helstrom [25] and is calculated from using a SPRT designed to distinguish two Gaussian probability density functions. The design point SNR is 1.0 and the ASN is normalized to the ASN at the design point. Observe that ASN increases significantly in the midrange of SNR. At midrange SNRs the mean of the single sample log-likelihood ratio of (10-156) is near zero and the cumulative log-likelihood ratio wanders near zero. For SNRs in region 1, (10-156) is negative and mostly missed detections occur. For SNRs in region 2, (10-156) is positive and the signal is usually detected.

Practical Considerations. Calculation of the log-likelihood ratio in a system using the SPRT is a difficult practical problem. The calculation can be made using a microprocessor that implements (10-152) and (10-153). Even with fast microprocessors, the time consumed by this calculation may dominate the overall synchronization time. A system using a microprocessor implementation of the SPRT is reported on in Carson [31]. An alternative digital implementation of the log-

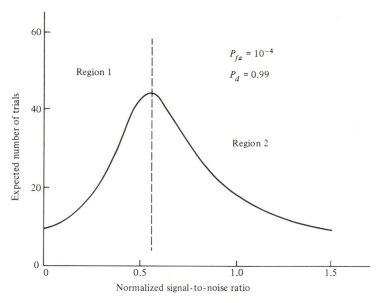

FIGURE 10-26. Typical plot of average sample number versus normalized signal-to-noise ratio. (From Ref. 25.)

likelihood ratio calculation is a table look-up using a high-speed read-only memory (ROM). This technique permits higher processing speed and has been implemented.

For very low SNR, the same approximations that were used in the calculation of ASN can be used to simplify the calculation. In particular,

$$\ln\left[\lambda(x_k)\right] \approx -\frac{P}{N} + \left(\frac{P}{2N^2}\right)x_k^2 - \left(\frac{P^2}{16N^4}\right)x_k^4 \tag{10-163}$$

for small SNR. It was seen earlier that the x_k^4 term contributes to the mean of this function and therefore cannot be ignored. However, common practice is to include the mean of this term at the design point SNR with the bias term $(-P/N)$ resulting in

$$\ln\left[\lambda(x_k)\right] \approx \left[-\frac{P}{N} - \frac{1}{2}\left(\frac{P}{N}\right)^2 - \frac{1}{8}\left(\frac{P}{N}\right)^3 - \frac{1}{32}\left(\frac{P}{N}\right)^4\right] + \frac{P}{2N^2}x_k^2 \tag{10-164}$$

With this simplification, the linear envelope detector and log-likelihood calculator can be replaced by a *square-law* envelope detector and a bias circuit. The student is cautioned that terms up to at least the $(P/N)^2$ terms of the bias must be retained for proper operation [32]. The detailed performance of this simplified detector is discussed in Kendall [33].

Finally, some implementations of sequential detection replace the upper threshold comparator with a time-out mechanism on the total number of samples. After each sample is taken and processed, the log-likelihood ratio is compared with a lower threshold and the total number of samples is compared with a fixed limit. If the lower threshold is exceeded, the signal is declared absent. If the sample count is exceeded before the lower threshold is crossed, the signal is declared present. Discussion of this sequential detector can be found in Holmes [34].

SYNCHRONIZATION USING A MATCHED FILTER

In all the synchronization schemes discussed above, the received spread-spectrum signal was correlated with a receiver generated replica of the spreading waveform. The output of this correlator was processed to determine whether in fact despreading had occurred, indicating that the phase of the replica was correct. Recall that the output $y(t)$ of any filter is the convolution of its input $x(t)$ and its impulse response $h(t)$, that is,

$$y(t) = \int_0^t x(\lambda)h(t - \lambda)\, d\lambda \qquad (10\text{-}165)$$

where the filter is assumed to be causal and the input is assumed to begin at $t = 0$. If the filter input is a direct-sequence modulated spread-spectrum signal and the filter impulse response is a time-reversed segment of the receiver despreading waveform, the MF continuously correlates the received signal with this segment of the spreading waveform and generates a maximum output when it receives the corresponding segment of the received spreading waveform. This maximum output can be sensed and used to start the receiver code generators at the appropriate phase and synchronization will be accomplished. This acquisition strategy is called *matched filter synchronization*.

For simplicity suppose that the received signal is BPSK direct-sequence modulated and that the data modulation is also BPSK. In the noiseless case, the received signal is

$$r(t) = \sqrt{2P}\, c(t - T_d)d(t - T_d) \cos(\omega_0' t + \varphi') \qquad (10\text{-}166)$$

where the received signal power is P, and the carrier frequency and phase are ω_0' and θ', respectively. The synchronization matched filter has an impulse response

$$h(t) = \begin{cases} 2c_R(t) \cos(\omega_0 t) & \text{for } 0 \le t \le T_M \\ 0 & \text{elsewhere} \end{cases} \qquad (10\text{-}167)$$

where $c_R(t)$ is a time-reversed segment T_M seconds long of the spreading waveform $c(t)$. Figure 10-27 illustrates one possible spreading waveform $c(t)$ and baseband MF impulse response $c_R(t)$. The MF impulse response is matched to the segment of $c(t)$ between T_A and T_B which has duration T_M seconds. The output of the MF is

$$\begin{aligned} y(t) &= \int_{t-T_M}^t \sqrt{2P}\, c(\lambda - T_d)d(\lambda - T_d) \cos(\omega_0'\lambda + \varphi') \\ &\quad \times 2c_R(t - \lambda) \cos(\omega_0 t - \omega_0\lambda)\, d\lambda \\ &= \sqrt{2P} \int_{t-T_M}^t d(\lambda - T_d)c(\lambda - T_d)c_R(t - \lambda) \\ &\quad \times \cos(\omega_0 t + \Delta\omega\,\lambda)\, d\lambda \end{aligned} \qquad (10\text{-}168)$$

where $\Delta\omega = \omega_0' - \omega_0$ accounts for doppler and other frequency offsets of the received signal. At time $t = T_B + T_d$ the function $c_R(t - \lambda)$, which is a function of λ, will be phase synchronized with the segment of $c(\lambda)$ to which it is matched.

At this time, the product $c(\lambda - T_d)c_R(T_B + T_d - \lambda) = 1.0$ over the range of integration. Assuming that the data sequence is slowly varying (i.e., high processing gain), the integral above becomes

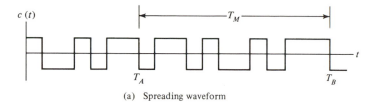

(a) Spreading waveform

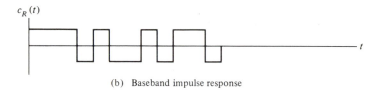

(b) Baseband impulse response

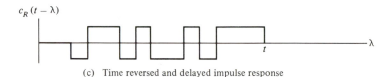

(c) Time reversed and delayed impulse response

FIGURE 10-27. Typical baseband spreading waveform and matched filter impulse response.

$$y(t) = \sqrt{2P}\, d \int_{t-T_M}^{t} \cos(\omega_0 t + \Delta\omega\lambda)\, d\lambda$$

$$= \sqrt{2P}\, T_M d\, \frac{\sin[\Delta\omega\,(T_M/2)]}{\Delta\omega(T_M/2)} \cos\left[(\omega_0 + \Delta\omega)t - \Delta\omega\,\frac{T_M}{2}\right] \quad (10\text{-}169)$$

where $d = \pm 1$ is the value of the data modulation over the period of integration. This result is strictly valid only for $t = T_d + T_B$; however, for carrier frequencies that are large relative to the spreading code fundamental frequency, the result is approximately correct for $T_d + T_B - \Delta t \le t \le T_d + T_B + \Delta t$. At some point in this range, the maximum value of $y(t)$ is

$$y_{\max} = \sqrt{2P}\, T_M\, \frac{\sin[\Delta\omega\,(T_M/2)]}{\Delta\omega\, T_M/2} \quad (10\text{-}170)$$

This result indicates that as the impulse response of the MF becomes long, the received frequency must be carefully controlled in order to have a maximum response from the matched filter.

A practical MF synchronization system is illustrated in Figure 10-28. This circuit employs two matched filters. The second MF is used to provide a reference level for the synchronization decision. The impulse response of the second MF is matched to a sequence of symbols which are hopefully not contained in the received code sequence. When the signal channel MF output exceeds the reference channel MF output by k times, a start signal is sent to the receiver code generator. The receiver code generator is preset to begin at a state corresponding to the MF impulse response.

The matched filter necessary for synchronization can be implemented in many ways. One implementation that has received much attention recently uses surface

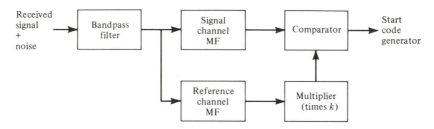

FIGURE 10-28. Practical matched filter synchronization circuit. (From Ref. 35.)

acoustic wave (SAW) technology. The student is referred to Reible [36], Toplicar [37], Hickernell et al. [38], and Milstein and Das [39] for additional discussion of the application of SAW filters to spread spectrum communications. Another implementation makes use of charged-coupled device (CCD) delay lines and is capable of being used for BPSK or QPSK spreading modulation. A particular CCD implementation used for global positioning system (GPS) receivers is described in Grieco [40].

If the required MF can be implemented, the synchronization time of a MF synchronizer will usually be orders of magnitude less than any of the previously discussed systems. The reduced synchronization time is explained by noting that the MF continuously correlates over a time interval T_M. That is, the MF merely waits for the preselected waveform to be received. Synchronization will usually be achieved within several repetitions of the spreading waveform except in very noisy or highly jammed situations. When the spreading waveform has essentially an infinite period, the MF must be programmable as discussed in Hickernell [38]. The figure of merit normally used to compare various MF synchronizers is their time–bandwidth product. This is the product of the received spread-spectrum signal bandwidth and the duration of the MF impulse response. The processing gain of the MF synchronizer is approximately equal to this time–bandwidth product. At this writing, the application of matched filter synchronizers is limited by the achievable time–bandwidth products of the filters.

10-5

SYNCHRONIZATION BY ESTIMATING THE RECEIVED SPREADING CODE

Recall from Chapter 8 that the shift register for one form (Figure 8-6) of a maximal-length sequence generator always contains r symbols of the code sequence. Thus, if r symbols of the spreading code can be estimated from the received waveform, these symbols can be loaded into the shift register generator to synchronize the system. At medium to high received SNR, these r symbols can be estimated with sufficient accuracy that this method of initial synchronization outperforms all the serial search techniques described earlier. This technique was first described in Ward [1] and was later expanded in Kilgus [41] and Ward and Yiu [2]. Although these references and the discussion to follow are limited to baseband spread-spectrum systems, the technique can in principle be used for any spreading modulation.

The system first described in Ward [1] is called recursion aided sequential estimation (RASE) and is described for a baseband signaling scheme. Figure 10-29 illustrates the RASE synchronization system along with a full-time early-late code

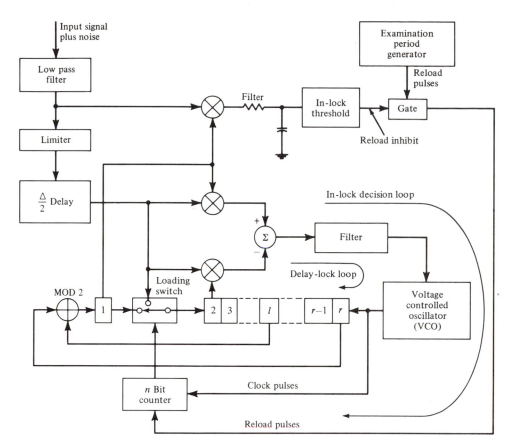

FIGURE 10-29.   RASE synchronization system with full-time early-late code tracking loop. (From Ref. 1.)

tracking loop and an on-time lock detector correlator. The input signal is assumed to be a sequence symbol from the alphabet ± 1 plus AWGN. The input is lowpass filtered to eliminate the majority of the noise and then limited. The limiter output is the estimate of the received sequence, which is loaded into the second stage of the linear feedback shift register generator. After r symbol estimates are loaded into the shift register, the contents of all stages, including the first, are a function of the received signal. At this time the shift register loading switch is repositioned so that the code generator operates normally. If the load was correct, the reference code generator outputs will be properly phased to begin code tracking using stages 1 and 2 for the early and late reference signals. The 1/2 chip delay in the received signal path is required to minimize the tracking loop pull-in transient which occurs when the tracking loop is closed. Correct code tracking is sensed by the on-time correlator, lowpass filter, and threshold comparator. The lowpass filter integrates over many code symbols and thus provides an accurate estimate of whether or not the tracking loop is functioning. If the shift register load was incorrect, the phase of the reference code will be incorrect and tracking loop pull-in will be impossible. The RASE system reloads the shift register each time an incorrect load is sensed. This process continues until a correct load and code tracking is achieved.

Suppose that the probability of correctly estimating a particular received symbol is denoted p. This probability is a function of received SNR. Then the probability of correctly loading the shift register of length r in a single trail is p^r, and the probability of an incorrect shift register load is $1.0 - p^r$. The probability of obtaining a correct load on exactly the kth trial is [1]

$$\text{Pr}(k) = p^r(1 - p^r)^{k-1} \tag{10-171}$$

The average number of trials required to achieve a correct load is

$$\bar{k} = \sum_{k=1}^{\infty} k \, \text{Pr}(k)$$

$$= \sum_{k=1}^{\infty} kp^r(1 - p^r)^{k-1} \tag{10-172}$$

Using the substitutions $q = 1 - p^r$, and $m = k - 1$ along with the identity [5]

$$\sum_{m=0}^{\infty} (a + m)q^m = \frac{a}{1 - q} + \frac{q}{(1 - q)^2} \tag{10-173}$$

Equation (10-172) can be evaluated. The result is

$$\bar{k} = \frac{1}{p^r} \tag{10-174}$$

Each shift register load requires rT_c seconds plus T_e seconds to evaluate the correctness of the load. Ignoring the possibility that a correct load could be misidentified as an incorrect load, the mean synchronization time for this system is

$$\bar{T}_s = \bar{k}(rT_c + T_e) = \frac{rT_c + T_e}{p^r} \tag{10-175}$$

For the example system, the symbol error probability is calculated from the statistics of the signal at the output of the received signal lowpass filter. Prior to synchronization, no symbol synchronization signal is available, so that an average error probability over all sample times must be calculated. The reference contains a plot of p versus received SNR. For very low and very high SNR the probability p is 0.5 and 1.0, respectively. Thus for very low SNR the mean synchronization time is

$$\bar{T}_s = 2^r(rT_c + T_e) \tag{10-176a}$$

and for very high SNR the mean synchronization time is

$$\bar{T}_s = rT_c + T_e \tag{10-176b}$$

A crude comparison can be made between a serial search synchronization system and RASE by assuming that both use identical evaluation times. Ignoring false-alarm penalty time and assuming that $1 - P_d << 1$, the average synchronization time for the serial search system is approximately

$$\bar{T}_s \approx 2 \frac{2^r - 1}{2} T_e \approx 2^r T_e \tag{10-177}$$

No a priori information has been assumed about the received code phase, so that the phase uncertainty is the full period of the code. The serial search step size is $\frac{1}{2}$ chip. For very low SNR, comparison of (10-177) and (10-176a) show that average synchronization time is approximately the same for RASE and serial search. If a priori information about the received code phase is available, serial search performs better than RASE. At high SNR, however, the RASE system will achieve synchronization in much less time than serial search. Detailed numerical comparisons are given in Ward [1].

Extensions of the RASE system are possible. The most straightforward of these [2] compares the sequence generator output with received symbols for several chip times prior to initiating a comprehensive evaluation requiring T_e seconds. The goal of this rapid check is to discard the majority of the incorrect loads rapidly, thus speeding up the search. This concept is exactly the same as that used in multiple-dwell synchronization. Another extension [41] employs forward error correction techniques to correct the shift register load.

10-6

TRACKING LOOP PULL-IN

The synchronization schemes described above are designed to position the phase of the receiver-generated spreading waveform within a fraction of a chip of the received spreading waveform phase. It is assumed that this positioning is accurate enough that the tracking loop can take over the synchronization process at this point and pull the receiver spreading waveform to precisely the correct phase. This pull-in process is identical to the pull-in process of a conventional phase-locked loop, and it is analyzed using the same nonlinear analysis techniques [42]. Since the tracking loop error may be large during the pull-in transient, the linear analysis described in Chapter 9 is not applicable. The nonlinear analytical technique used in this case is to generate a family of curves relating the rate of change of the tracking error and the tracking error itself. Each member of this family is associated with a different set of initial conditions and/or different dynamics of the received waveform. Stable tracking will occur at points on these curves where the rate of change of the error signal is zero and where small perturbations away from the point will result in the proper direction of change of the error signal to return to the original point. The plots generated are called *phase-plane plots* and are described in detail in Viterbi [42] and Cunningham [43]. The general idea is to generate a family of phase plane trajectories related to a particular synchronization scenario and to see what trajectories lead to stable lock points. This will in turn lead to a set of initial conditions and input dynamics for which tracking loop pull-in is possible.

The analysis begins with the nonlinear equivalent circuit for the tracking loop illustrated in Figure 10-30. This model is very general in that it can be used to represent any of the tracking loops discussed in Chapter 9 with proper selection of the discriminator characteristic $D_\Delta(\delta)$ and loop gain K. This model was derived in Chapter 9. In this model, $\delta(t) = [T_d(t) - \hat{T}_d(t)]/T_c$ is the normalized tracking error, which is of primary interest in this analysis. The discriminator characteristic $D_\Delta(\delta)$ and gain K relate the voltage at the input to the loop filter to the normalized tracking error. The loop filter is described by (9-20), repeated here for convenience:

$$a_m \frac{d^m \epsilon}{dt^m} + a_{m-1} \frac{d^{m-1} \epsilon}{dt^{m-1}} + \cdots + a_0 \epsilon$$

$$= b_n \frac{d^n v}{dt^n} + b_{n-1} \frac{d^{n-1} v}{dt^{n-1}} + \cdots + b_0 v \qquad (9\text{-}20)$$

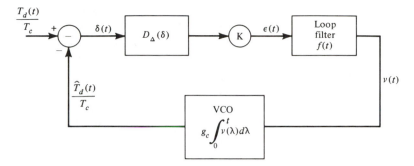

FIGURE 10-30. Noiseless nonlinear-equivalent circuit of spread-spectrum code tracking loop.

where $\epsilon = \epsilon(t)$ is the filter input and $v = v(t)$ is the filter output. By inspection of Figure 10-30 it is seen that

$$g_c \int_0^t v(\lambda)\, d\lambda = \frac{\hat{T}_d(t)}{T_c} = \frac{T_d(t)}{T_c} - \delta(t) \qquad (10\text{-}178)$$

and taking one derivative with respect to time yields

$$g_c v(t) = \frac{d}{dt}\left[\frac{T_d(t)}{T_c}\right] - \frac{d}{dt}[\delta(t)] \qquad (10\text{-}179)$$

The loop filter input is $\epsilon(t) = KD_\Delta(\delta)$, so that the complete nonlinear differential equation describing the tracking loop is

$$\left(a_m \frac{d^m}{dt^m} + a_{m-1} \frac{d^{m-1}}{dt^{m-1}} + \cdots + a_0\right) KD_\Delta(\delta)$$

$$= \left(b_n \frac{d^n}{dt^n} + b_{n-1}\frac{d^{n-1}}{dt^{n-1}} + \cdots + b_0\right)\frac{1}{g_c}\left\{\frac{d}{dt}\left[\frac{T_d(t)}{T_c}\right] - \frac{d}{dt}[\delta(t)]\right\} \qquad (10\text{-}180)$$

To simplify the following discussion, only second-order tracking loops will be considered which have a loop filter transfer function

$$F(s) = \frac{s\tau_2 + 1}{s(\tau_1 + \tau_2) + 1} = \frac{sa_1 + a_0}{sb_1 + b_0} \qquad (10\text{-}181)$$

It is convenient to use normalized variables in the desired phase plane plots. The customary [44–46] normalized variables are

$$\tau = \omega_n t \qquad (10\text{-}182)$$

$$y(t) = \frac{T_d(t)}{T_c} \qquad (10\text{-}183)$$

and

$$x(t) = \delta(t) \qquad (10\text{-}184)$$

where ω_n is the natural radian frequency of the linearized loop. Using this normalization and the specific filter of (10-181), equation (10-180) becomes

$$\left(\tau_2 \omega_n \frac{d}{d\tau} + 1 \right) KD_\Delta[x(\tau)]$$

$$= \left[(\tau_2 + \tau_1)\omega_n \frac{d}{d\tau} + 1 \right] \frac{1}{g_c} \left\{ \omega_n \frac{d}{d\tau} [y(\tau)] - \omega_n \frac{d}{d\tau} [x(\tau)] \right\} \quad (10\text{-}185)$$

or

$$\tau_2 \omega_n KD'_\Delta[x(\tau)]\dot{x}(\tau) + KD_\Delta[x(\tau)]$$

$$= (\tau_2 + \tau_1) \frac{\omega_n^2}{g_c} \ddot{y}(\tau) + \frac{\omega_n}{g_c} \dot{y}(\tau) - (\tau_2 + \tau_1) \frac{\omega_n^2}{g_c} \ddot{x}(\tau) - \frac{\omega_n}{g_c} \dot{x}(\tau) \quad (10\text{-}186)$$

where

$$\dot{x}(\tau) = \frac{dx(\tau)}{d\tau} \qquad \ddot{x}(\tau) = \frac{d^2 x(\tau)}{d\tau^2}$$

$$\dot{y}(\tau) = \frac{dy(\tau)}{d\tau} \qquad \ddot{y}(\tau) = \frac{d^2 y(\tau)}{d\tau^2}$$

$$D'_\Delta[x(\tau)] = \frac{d}{dx} D_\Delta[x(\tau)]$$

Algebraic manipulation of (10-186) yields

$$\frac{\ddot{x}}{\dot{x}} = -\frac{Kg_c D_\Delta(x)}{(\tau_2 + \tau_1)\omega_n^2 \dot{x}} - \left[\frac{1}{(\tau_2 + \tau_1)\omega_n} + \left(\frac{\tau_2}{\tau_2 + \tau_1} \right) \frac{g_c}{\omega_n} KD'_\Delta(x) \right]$$

$$+ \frac{\ddot{y}}{\dot{x}} + \frac{1}{(\tau_2 + \tau_1)\omega_n} \frac{\dot{y}}{\dot{x}} \quad (10\text{-}187)$$

Simplification of (10-187) is accomplished by using the relationships between the second-order loop damping, ζ, natural frequency, ω_n, and the gains and time constants of the linearized loop. The linearized loop model is illustrated in Figure 10-31. In this figure, K_n is the slope of the discriminator characteristic $D_\Delta(x)$ in volts per chip at $x = \delta = 0$. This linear servo loop model has been analyzed in many texts. With $F(s)$ given by (10-181) the loop damping and natural frequency are [47]

$$\zeta = \frac{1}{2} \left(\frac{K_n Kg_c}{\tau_2 + \tau_1} \right)^{1/2} \left(\tau_2 + \frac{1}{K_n Kg_c} \right) \quad (10\text{-}188a)$$

$$\omega_n = \left(\frac{K_n Kg_c}{\tau_2 + \tau_1} \right)^{1/2} \quad (10\text{-}188b)$$

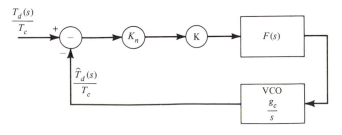

FIGURE 10-31. Noiseless linear-equivalent circuit of spread-spectrum code tracking loop.

For high loop gain, $\zeta \approx \frac{1}{2}\omega_n\tau_2$. Equation (10-187) can be written

$$\frac{\ddot{x}}{\dot{x}} = -\frac{K_nKg_c}{(\tau_2 + \tau_1)\omega_n^2}\left(\frac{D_\Delta(x)}{K_n}\right)\frac{1}{\dot{x}}$$

$$-\left[\frac{K_nKg_c}{(\tau_2 + \tau_1)\omega_n}\left(\frac{1}{K_nKg_c}\right) + \left(\frac{K_nKg_c}{(\tau_2 + \tau_1)\omega_n}\right)\tau_2\left(\frac{D_\Delta'(x)}{K_n}\right)\right]$$

$$+\frac{\ddot{y}}{\dot{x}} + \frac{K_nKg_c}{(\tau_2 + \tau_1)\omega_n}\frac{1}{K_nKg_c}\frac{\dot{y}}{\dot{x}} \qquad (10\text{-}189)$$

Defining a normalized loop gain $g = K_nKg_c/\omega_n$ and using (10-188), this equation simplifies to

$$\frac{\ddot{x}}{\dot{x}} = -\frac{D_\Delta(x)}{K_n}\frac{1}{\dot{x}} - \left[\frac{1}{g} + 2\zeta\frac{D_\Delta'(x)}{K_n}\right] + \frac{\ddot{y}}{\dot{x}} + \frac{1}{g}\frac{\dot{y}}{\dot{x}}$$

$$= \frac{D_n(x) - [(1/g) + 2\zeta D_n'(x)]\dot{x} + \ddot{y} + (1/g)\dot{y}}{\dot{x}} \qquad (10\text{-}190)$$

where $D_n(x) = D_\Delta(x)/K_n$ is the normalized discriminator characteristic which has a slope of 1.0 at $x = 0$. Noting that $\ddot{x}/\dot{x} = d\dot{x}/dx$, this equation is identical to the result derived in [44, 46, 48] and commented on in Nielson [45]. Similar relations can be derived for other loop filters.

Equation (10-190) is used to plot paths through the phase plane. This is done by choosing an arbitrary initial point $(x, \dot{x})$ and calculating the slope $d\dot{x}/dx$ of the trajectory passing through this point using (10-190). An appropriately small value of dx is chosen and the next point,

$$\left(x + dx, \dot{x} + \frac{d\dot{x}}{dx}dx\right)$$

is calculated. A new slope is calculated for the new point on the trajectory and the process repeats. At the end of many iterations, a single curve in the desired family will have been generated. Other curves are generated in an identical manner using other initial conditions and/or different normalized input signal dynamics $\dot{y}$ and $\ddot{y}$. Example phase plane plots have been given in [Nielsen, 1975]; these plots have been reproduced as Figures 10-32 through 10-34 for the convenience of the student. In all these plots, the loop is a critically damped second-order loop, and normalized input is $y(\tau) = T_d(\tau)/T_c = K_1\tau + K_0$, so that $\dot{y} = K_1$ and $\ddot{y} = 0$. Recall from Chapter 9 that for large phase errors, the discriminator output is zero, so that no correction signal exists in the tracking loop and the rate of change of the error signal $\dot{x}$ is equal to the rate of change of the input $\dot{y}$. Therefore, $\dot{y}$ determines the value of $\dot{x}$ for large x. Figures 10-32 and 10-33 illustrate the phase plane trajectories for baseband loops having early-late delay differences of 1.0 chip and 2.0 chips, respectively. These curves illustrate that the loop is unable to pull-in at all if the input dynamics are too large. Those trajectories that do not converge to $\dot{x} = 0$ never achieve code tracking. Thus, even if the synchronization hardware placed the receiver code phase in exactly the correct position, large signal dynamics due to doppler could cause the loop to never achieve lock. Figure 10-34 illustrates phase trajectories for the full-time early-late noncoherent tracking loop using a delay difference of 1.0 chip. This same technique can be used to calculate the transient response to input dynamics after the loop is locked. In this case, the initial condition is $(x, \dot{x}) = (0, 0)$.

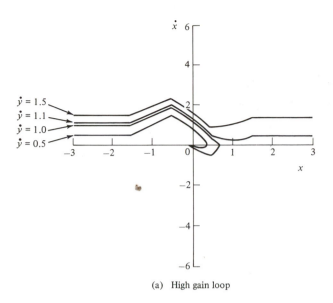

(a) High gain loop

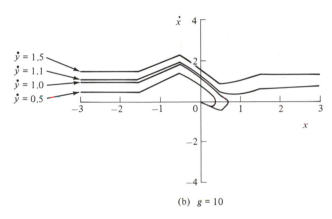

(b) $g = 10$

FIGURE 10-32. Phase-plane trajectories for baseband full-time early-late tracking loops using a delay difference of 1.0 chip: (a) high-gain loop; (b) $g = 10$. (From Ref. 45.)

Finally, the phase plane plots give no indication of the time required for tracking loop pull-in. If pull-in time is important, plots of x versus τ can be easily generated from the same data used to calculate the phase trajectories. Example results are given in Nielsen [45].

10-7

SUMMARY

Initial spreading waveform synchronization is an extremely important problem in spread-spectrum system design. In fact, overall system performance is often limited by the performance of the synchronizer. The simplest synchronization scheme is to sweep the receiver spreading waveform phase until the proper phase is sensed. Stepped serial search over all potential waveform phases will usually achieve lower

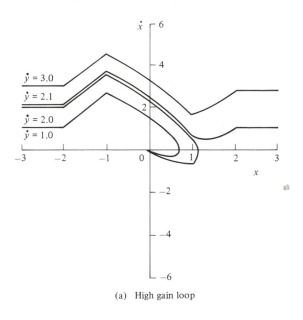

(a) High gain loop

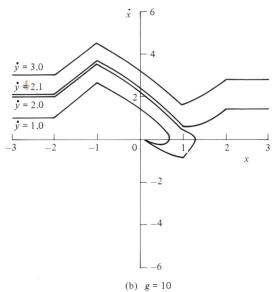

(b) $g = 10$

FIGURE 10-33. Phase-plane trajectories for baseband full-time early-late tracking loops using a delay difference of 2.0 chips: (a) high-gain loop; (b) $g = 10$. (From Ref. 45.)

synchronization times than the swept search. After each phase step in a stepped serial search system, the correctness of the phase must be evaluated. This is accomplished by attempting to despread the received waveform using the trial spreading waveform phase. When the phase is correct, the input signal spectrum is collapsed and energy appears at the output of a narrowband filter. This energy is sensed using one of a number of techniques. Fixed integration-time energy detection is the simplest technique. Improved performance is achieved using multiple-dwell techniques, since incorrect phases can be quickly discarded without serious risk of paying a large false-alarm penalty. The most sophisticated energy detection

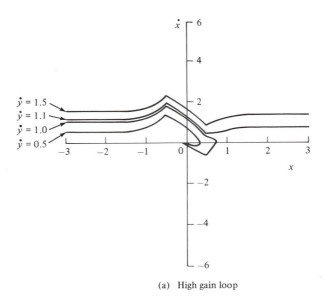

(a) High gain loop

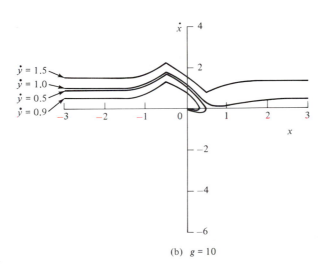

(b) $g = 10$

FIGURE 10-34. Phase-plane trajectories for full-time early-late noncoherent tracking loop using a delay difference of 1.0 chip: (a) high-gain loop; (b) $g = 10$. (From Ref. 45.)

technique is the Wald sequential probability ratio test. The performance of all these energy detection techniques can be calculated using the methods described in this chapter.

Matched filter synchronization promises to yield the lowest average synchronization time of all the schemes studied; however, at this time, technology limits the application of matched filters to systems having moderate processing gains. At high SNR, sequential estimation of the code generator state from the wideband received signal is a viable synchronization technique. Tracking loop pull-in has been discussed since all the synchronization techniques finish with a phase error which is potentially a large fraction of a chip. Pull-in must occur before the linear tracking results of Chapter 9 apply.

REFERENCES

[1] R. B. WARD, "Acquisition of Pseudonoise Signals by Sequential Estimation," *IEEE Trans. Commun. Technol.*, December 1965.

[2] R. B. WARD and K. P. YIU, "Acquisition of Pseudonoise Signals by Recursion-Aided Sequential Estimation," *IEEE Trans. Commun.*, August 1977.

[3] A. PAPOULIS, *Probability, Random Variables, and Stochastic Processes* (New York: McGraw-Hill, 1965).

[4] J. K. HOLMES and C. C. CHEN, "Acquisition Time Performance of PN Spread Spectrum Systems," *IEEE Trans. Commun.*, August 1977.

[5] I. S. GRADSHTEYN and I. W. RYZHIK, *Tables of Integrals Series and Products* (New York: Academic Press, 1965).

[6] W. R. BRAUN, "Performance Analysis for the Expanding Search PN Acquisition Algorithm," *IEEE Trans. Commun.*, March 1982.

[7] G. F. SAGE, "Serial Synchronization of Pseudonoise Systems," *IEEE Trans. Commun. Technol.*, December 1964.

[8] R. E. ZIEMER and W. H. TRANTER, *Principles of Communications* (Boston: Houghton Mifflin, 1976).

[9] S. O. RICE, "Mathematical Analysis of Random Noise," *Bell Syst. Tech. J.*, Vol. 23, 1944; Vol. 24, 1945.

[10] M. J. SKOLNIK, *Introduction to Radar Systems* (New York: McGraw-Hill, 1962).

[11] D. A. SHNIDMAN, "Evaluation of the Q Function," *IEEE Trans. Commun.*, March 1974.

[12] H. URKOWITZ, "Energy Detection of Unknown Deterministic Signals," *Proc. IEEE*, April 1967.

[13] J. M. WOZENCRAFT and I. M. JACOBS, *Principles of Communication Engineering* (New York: Wiley, 1965).

[14] U. GRENANDER, H. POLLAK, and D. SLEPIAN, "The Distribution of Quadratic Forms in Normal Variates: A Small Sample Theory With Applications to Spectral Analysis," *J. Soc. Industr. Appl. Math.*, Vol. 7, No. 4, Dec. 1959.

[15] P. B. PATNAIK, "The Non-central Chi-Square and F-Distribution and Their Applications," *Biometrika*, Vol. 36, p. 202, 1949.

[16] M. ZELEN and N. C. SEVERO, "Probability Functions," in *Handbook of Mathematical Functions,* M. Abramowitz and I. Stegun, eds., NBS Applied Mathematics, Series 55 (Washington, D.C.: U.S. Government Printing Office, 1964).

[17] E. FIX, "Tables of Noncentral Chi-Square," *Publications in Statistics*, Vol. 1, No. 2, U. of California Press, Berkeley, 1949.

[18] S. V. SMIRNOV and M. K. POTAPOV, "A Nomogram for the Chi-Square Probability Function," *J. SIAM*, Vol. 6, No. 1, p. 124, 1961.

[19] G. H. ROBERTSON, "Computation of the Noncentral Chi-Square Distribution," *Bell Syst. Tech. J.*, January 1969.

[20] H. P. BAER, "The Calculation of the Statistical Properties of the Synchronization Times in a PN Spread Spectrum System for Minimized Acquisition Time Designs," *IEEE Trans. Commun.*, May 1980.

[21] W. B. DAVENPORT and W. L. ROOT, *An Introduction to the Theory of Random Signals and Noise* (New York: McGraw-Hill, 1958).

[22] W. FELLER, *An Introduction to Probability Theory and Its Applications* (New York: Wiley, 1950).

[23] P. M. HOPKINS, "A Unified Analysis of Pseudonoise Synchronization by Envelope Correlation," *IEEE Trans. Commun.*, August 1977.

[24] A. WALD, *Sequential Analysis* (New York: Wiley, 1947). (Also reprinted in 1973 by Dover Publications.)

[25] C. W. HELSTROM, "Sequential Detection," in *Communication Theory*, A. V. Balakrishnan, ed. (New York: McGraw-Hill, 1968).

[26] J. C. HANCOCK and P. A. WINTZ, *Signal Detection Theory* (New York: McGraw-Hill, 1966).

[27] A. WALD and J. WOLFOWITZ, "Optimum Character of the Sequential Probability Ratio Test," *Ann. Math. Statist.*, Vol. 19, p. 326, 1948.

[28] R. F. COBB and A. D. DARBY, "Acquisition Performance of Simplified Implementations of the Sequential Detection Algorithm," *Conf. Rec.,* IEEE Natl. Telecommun. Conf., 1978.

[29] M. G. KENDALL and A. STUART, *The Advanced Theory of Statistics*, Vol. 2 (New York: Hafner Press, 1961).

[30] F. W. J. OLVER, "Bessel Functions of Integer Order," in *Handbook of Mathematical Functions*, M. Abramowitz and J. Stegun, eds., NBS Applied Mathematics Series 55 (Washington, D.C.: U.S. Government Printing Office, 1964).

[31] L. CARSON, "A Microprocessor-Based Spread Spectrum Processor for Low Signal-to-Noise Ratios," *Conf. Rec.,* Phoenix Conf. Comput. Commun., 1982.

[32] J. J. BUSSGANG and W. L. MUDGETT, "A Note of Caution on the Square-Law Approximation to an Optimum Detector," *IEEE Trans. Inf. Theory*, September 1960.

[33] W. B. KENDALL, "Performance of the Biased Square Law Sequential Detector in the Absence of Signal," *IEEE Trans. Inf. Theory*, January 1965.

[34] J. K. HOLMES, *Coherent Spread Spectrum Systems* (New York: Wiley, 1982).

[35] C. R. CAHN, "Spread Spectrum Applications and State-of-the-Art Equipments," in *Spread Spectrum Communications*, NATO AGARD Lecture Series 58, June 6, 1973 (AD 766 914).

[36] S. A. REIBLE, "Acoustoelectric Convolver Technology for Spread-Spectrum Communications," *IEEE Trans. Microwave Theory Tech.*, May 1981.

[37] J. R. TOPLICAR, "Wide Band Monolithic SAW Convolver for Asynchronous Communications," *Conf. Rec.*, NAECON, 1981.

[38] F. S. HICKERNELL et al., "SAW Programmable Matched Filter Signal Processor," *Conf. Rec.,* Ultrasonics Symp., 1980.

[39] L. B. MILSTEIN and P. K. DAS, "Spread Spectrum Receiver Using Surface Acoustic Wave Technology," *IEEE Trans. Commun.*, August 1977.

[40] D. M. GRIECO, "Application of Charge-Coupled Devices to GPS Acquisition and Data Modulation," *Conf. Rec.,* IEEE Natl. Telecommun. Conf., 1978.

[41] C. C. KILGUS, "Pseudonoise Code Acquisition Using Majority Logic Decoding," *IEEE Trans. Commun.*, June 1973.

[42] A. J. VITERBI, *Principles of Coherent Communication* (New York: McGraw-Hill, 1966).

[43] W. J. CUNNINGHAM, *Introduction to Nonlinear Analysis* (New York: McGraw-Hill, 1958).

[44] J. J. SPILKER, "Delay-Lock Tracking of Binary Signals," *IEEE Trans. Space Electron. Telem.*, March 1963.

[45] P. T. NIELSEN, "On the Acquisition Behavior of Binary Delay-Lock Loops," *IEEE Trans. Aerosp. Electron. Syst.*, May 1975.

[46] J. K. HOLMES, *Coherent Spread Spectrum Systems* (New York: Wiley-Interscience, 1982).

[47] F. M. GARDNER, *Phaselock Techniques* (New York: Wiley, 1966).

[48] J. J. SPILKER, *Digital Communications by Satellite* (Englewood Cliffs, N.J.: Prentice-Hall, 1977).

PROBLEMS

10-1. Consider a BPSK spread spectrum system using a spreading code clock $f_c = 100$ MHz. Suppose that the spreading code is a maximal length sequence generated using a shift register of length 11, and that no a priori information is available about the transmitter to receiver propagation delay. The system data rate is 100 BPS and the maximum uncertainty in the received carrier frequency is ± 5 KHz. Assume that the post despreading IF filter bandwidth is large enough to accommodate the frequency uncertainty. Calculate the average synchronization time for a single-dwell serial search synchronizer which uses $\frac{1}{2}$ chip steps and which is designed for a single sweep probability of detection of $P_d = 0.8$, and for a probability of false alarm $P_{fa} = 10^{-3}$. Assume that the false alarm penalty is $100T_i$ where T_i is the fixed integration time and that the received SNR in the data bandwidth is $+10$ dB. Plot the cumulative probability of being synchronized as a function of time.

10-2. Repeat Problem 10-2 assuming that the received carrier frequency is precisely known.

10-3. A synchronizer uses a stepped serial search with step size of $\frac{1}{2}$ chip and a fixed integration time energy detector. It is designed to achieve $P_d = .9$ with a time bandwidth product of 25 when the reference waveform phase error is zero and the SNR is 0 dB in the post despreading bandwidth. Assume BPSK spreading modulation and calculate the worst case total probability of detection on a single sweep. Note that the synchronizer does not necessarily step the reference waveform phase to exactly zero phase error.

10-4. Repeat Problem 10-3 assuming MSK spreading modulation.

10-5. A BPSK direct sequence spread spectrum system is used to provide line-of-sight communications between two aircraft whose maximum relative velocity is ± 2-mach. The nominal carrier frequency is 950 MHz and the spreading rate is 5 M chips/sec. The receiver synchronization system uses a correlator, bandpass filter, square law envelope detector, and integrator. Calculate the output of the integrator as a function of integration time and transmitter-to-receiver relative velocity. Ignore noise. Explain your result and predict how it might affect system noise performance.

10-6. Compare the times required to synchronize a pure frequency hop spread spectrum system and a pure BPSK direct sequence spread spectrum system. The data modulation is differential binary PSK for both systems and the data rate is 4800 BPS. The frequency hop rate is 100×10^3 hops/sec and the direct chip rate is 100×10^6 chips/sec. Assume that the initial system timing uncertainty is ± 0.5 ms and that both spreading code periods are long.

10-7. Calculate the average number of sweeps through the phase uncertainty region for a serial search synchronizer as a function of single sweep probability of detection P_d.

10-8. Plot the probability density function of the output of a linear envelope detector with a sinewave plus additive white Gaussian noise input for signal-to-noise ratios of -20 dB, 0 dB, 3 dB, and 6 dB.

10-9. A direct sequence spread spectrum system uses BPSK spreading modulation and a chip rate of 50×10^6 chips/sec. The received carrier power to noise power spectral density ratio is $+35$ dB-Hz. The receiver synchronizer uses a swept serial search with a matched filter detector. What is the maximum sweep rate which may be used to obtain $P_d = 0.99$ and $P_{fa} = 10^{-6}$? Does this result apply if the spreading modulation was changed to MSK? If not, what modifications to the analysis of Section 10-2 must be made?

10-10. The input to a fixed integration time energy detector identical to that illustrated in Figure 10-8 is taken from the RF receiver front end illustrated below. The energy detector filter is a 3rd order Butterworth filter with a -3 dB bandwidth of 5 KHz. The received signal power is -125 dBm and the receiver input stage is at room temperature. What is the signal-to-noise ratio at the energy detector filter output? What is the variance of the noise waveform at the filter output?

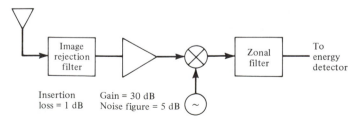

PROBLEM 10-10. Receiver front end.

10-11. Reconsider the system of Problem 10-10. (a) Determine the absolute energy detector threshold required to obtain $P_{fa} = 10^{-3}$ assuming an integrator time bandwidth product of 40. Do not assume the integrator output pdf is Gaussian. What is the resultant P_d? (b) Repeat (a) assuming that the integrator output pdf is Gaussian.

10-12. Develop the state transition diagram and matrix representation for the multiple-dwell energy detector defined by the logic diagram below.

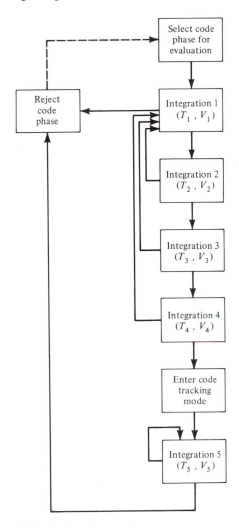

PROBLEM 10-12. Multiple-dwell detector.

10-13. Evaluate the matrix expressions for T_{da} and P_d for the state transition diagrams of the figure on the next page in general. That is, develop algebraic expressions for T_{da} and P_d in terms of the integration times and transition probabilities.

10-14. Consider a direct sequence spread spectrum system which uses BPSK spreading modulation and a chip rate of 3×10^6 chips/sec. Suppose that no timing information is available to the receiver and that the spreading code is an m-sequence generated using a 13 stage shift register. The synchronizer uses a double-dwell detector configured as in figure b of Problem 10-13. The energy detector bandpass filter noise bandwidth is 5 KHz and the received $C/N_0 = 32$ dB-Hz. The detector thresholds are chosen to yield $p'_{12} = 0.1$ and $p'_{23} = 10^{-2}$ in the noise alone case, and $p_{33} = 0.999$ with signal present. Assume that the integrator

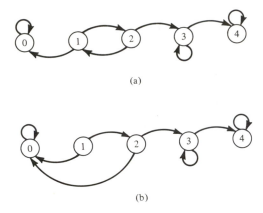

(a)

(b)

PROBLEM 10-13.   State transition diagrams.

time bandwidth products for states 1, 2, 3 are 100, 100, 200 respectively and calculate the average synchronization time.

10-15. Repeat Problem 10-14 for a synchronizer using sequential detection designed for $P_d = .90$ and $P_{fa} = 10^{-3}$.

10-16. The sequential detector discussed in Section 10-2 uses a linear envelope detector at the IF filter output. What modifications must be made to the log-likelihood ratio calculation when the linear detector is replaced by a square law envelope detector?

10-17. Determine the magnitude of the error made in calculating average synchronization time for a serial search synchronization system using sequential detection when the fourth order term of Equation (10-154) is ignored.

10-18. Design a sequential detector used to distinguish between two Gaussian probability density functions. Both pdfs have variance σ^2. One pdf has zero mean and the other has mean μ. Derive the operating characteristic function for this test.

10-19. Develop a detailed procedure for calculating P_d and P_{fa} as a function of received SNR, frequency offset, and filter integration time for a matched filter synchronizer. Assume BPSK direct sequence spread spectrum modulation.

10-20. Consider a BPSK spread spectrum system. Assume that the receiver has carrier synchronization and must obtain code synchronization using recursion aided sequential estimation. The spreading code generator is an 11-stage maximal length feedback shift register. Each shift register load is evaluated using a fixed integration time detector employing an IF filter with a 10 KHz noise bandwidth. Integration time and thresholds are chosen to yield $P_d = .95$ and $P_{fa} = 10^{-3}$ at a received SNR of 0 dB in the IF bandwidth. The code clock rate is 10 MHz. Determine average synchronization time as a function of received SNR.

CHAPTER 11

Performance of Spread-Spectrum Systems in a Jamming Environment

11-1

INTRODUCTION

The purpose of most spread-spectrum systems is to transfer information from one place to another. One figure of merit for these systems is the probability of correctly communicating a message in a particular noise and/or jamming environment. The techniques described in this chapter will enable the student to evaluate the transmission error probabilities for many common spectrum spreading and data modulation combinations in a number of different friendly and unfriendly signal environments. It will be shown in this chapter and the next that even the most sophisticated jammer can be almost completely countered by a combination of spectrum spreading, interleaving, and forward error correction (FEC).

In this chapter the error probabilities for systems without FEC will be derived. Spectrum spreading by itself will be shown to provide large performance improvements; however, the smart jammer is still able to produce significant performance degradations. In particular, the familiar exponential relationship between error probability and signal-to-noise ratio will be degraded to an inverse linear relationship between error probability and signal-to-jammer power ratio by the smart jammer. Further performance improvement requires FEC and interleaving and is the subject of Chapter 12.

SPREAD-SPECTRUM COMMUNICATION SYSTEM MODEL

Figure 11-1 illustrates the components of a spread-spectrum (SS) communication system. The discrete memoryless source (DMS) was discussed in Chapter 1 and is assumed to output a sequence of independent equally likely symbols $\mathbf{u} = \ldots,$ $u_{-1}, u_0, u_1, \ldots$ from the alphabet $\{0, 1\}$. The channel coder receives a sequence of input symbols $\mathbf{u}$ from the DMS and outputs a sequence of channel symbols $\mathbf{x} = \ldots, x_{-1}, x_0, x_1, \ldots$ from the alphabet $\{0, 1, \ldots, X - 1\}$. The sequence of encoder output symbols has a very carefully controlled structure which enables the detection and correction of transmission errors by the channel decoder. The decoder output sequence is an estimate $\hat{\mathbf{u}}$ of the coder input sequence $\mathbf{u}$. As discussed in Chapter 1, the channel coding may be either block coding or convolutional coding; both will be discussed in Chapter 12. Note that the channel coder input and output alphabets are not always the same. Also, the decoder input alphabet is often not the same as the coder output alphabet. The reason for choosing different alphabets for the coder output and decoder input is to assist the decoder by making additional information available to it. Additional reliability information can be obtained for the decoder by measuring the channel state as shown by the dashed line of Figure 11-1. This is especially important in a hostile environment where information on whether a jammer is "on" or "off" can be input to the decoder.

In some interference environments, channel transmission errors will occur in bursts. Most FEC schemes function better when channel errors are independent,

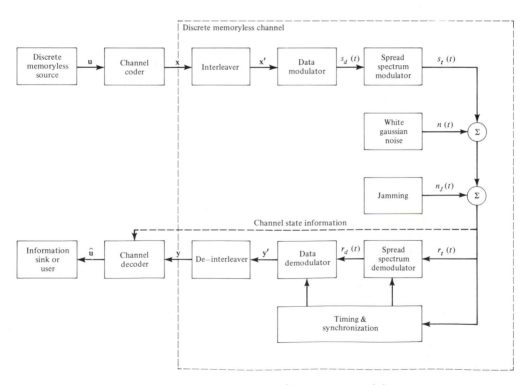

FIGURE 11-1. Spread spectrum communication system model.

and the function of the interleaver is to distribute channel errors randomly throughout the decoder input sequence y. This is accomplished by changing the order in which the coder output symbols are transmitted over the channel. With the transmission order changed, it is unlikely that a burst of contiguous channel errors will affect contiguous coder output symbols. Prior to decoding, the reordering performed by the interleaver is inverted by the de-interleaver. In the de-interleaving process, the burst of channel errors is distributed throughout the decoder input sequence. In all cases, the interleaver input and output alphabets are identical. The de-interleaver input and output alphabets are also identical but may be different from the interleaver alphabet.

The data modulator generates a signal waveform $s_d(t)$ in response to an input symbol sequence $\mathbf{x}'$. The modulation scheme may be any one of the digital modulation schemes discussed in the first half of this text. However, because of the large number of potential combinations of data modulation, spreading modulation, and jamming strategies, attention will be limited in this chapter to the following data modulations: (1) coherent binary phase-shift keying (BPSK), (2) coherent quaternary phase-shift keying (QPSK), (3) differentially coherent binary phase-shift keying (DPSK), and (4) noncoherent M-ary frequency-shift keying (MFSK). The modulator output $s_d(t)$ is transmitted to the data demodulator input over a channel which includes the spread-spectrum spreading and despreading function. The demodulator observes the received waveform $r_d(t)$ and produces either an estimate of the data modulator input or a symbol that includes the data modulator input estimate and reliability information. To perform the demodulation function, timing information must be available. This information is simply the carrier phase and symbol timing for coherent modulations, and the symbol timing for differentially coherent and noncoherent modulations. This information is obtained from the timing and synchronization hardware shown. Note that, for spread-spectrum systems, symbol timing can usually be obtained from the spreading waveform phase, which must be known for the system to function.

The spread-spectrum modulator is one of the modulators discussed in Chapter 7. Attention in this chapter is limited to the following types of spreading modulations: (1) direct-sequence (DS) coherent binary phase-shift keying, (2) direct-sequence coherent quaternary phase-shift keying, (3) direct-sequence coherent minimum-shift keying, (4) noncoherent slow frequency hop (FH), (5) noncoherent fast frequency hop, and (6) hybrid binary phase-shift keying direct sequence and noncoherent slow frequency hop. These spreading modulations combined with the data modulations listed above are typical of the modulation schemes used in modern spread-spectrum communication systems.

The spreading modulator output, denoted $s_t(t)$, is transmitted via the waveform channel which includes the transmitter up-conversion and power amplifier, the antennas, the receiver front end, and the receiver down-conversion. In the waveform channel the signal may be distorted, and Gaussian noise $n(t)$ and perhaps jamming $s_j(t)$ are added to the signal. In all that follows the additive noise process is assumed to have a two-sided power spectral density of $N_0/2$ watts/hertz.

A number of jamming signals have been postulated for examination in this chapter. Some of the postulated threats are clearly real, whereas others are considered only because of their optimal nature. The most benign jammer is the *barrage noise jammer*. This jammer transmits bandlimited white Gaussian noise with one-sided power spectral density of N_J watts/hertz, as shown in Figure 11-2a. It is usually assumed that the jammer power spectrum covers exactly the same frequency range as the SS signal. The effect of the barrage noise jammer on the system is

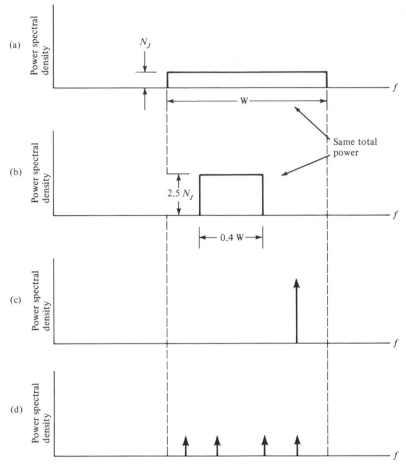

FIGURE 11-2. Typical jammer one-sided power spectral densities: (a) barrage noise jammer; (b) partial band noise jammer; (c) tone jammer; (d) multiple-tone jammer.

simply to increase the Gaussian noise level at the output of the receiver down-converter.

When the SS modulation has a frequency-hop component, jamming power can be more efficiently used by transmitting all the available power in a limited band-width which is smaller than the SS signal bandwidth. The jammer which uses this strategy is called a *partial band jammer* and the fraction of the SS signal bandwidth which is jammed is denoted by ρ. If the total jammer power is J and the SS signal bandwidth is W, the barrage noise jammer one-sided psd is $N_J = J/W$ over the entire band, while the partial band jammer one-sided psd is $N_J' = N_J/\rho = J/\rho W$ over a bandwidth ρW as illustrated in Figure 11-2b for the special case $\rho = 0.4$. The partial band jammer is particularly effective against a FH SS system because the signal will hop in and out of the jamming band and can be seriously degraded when in the jamming band. It will be shown later that there is an optimum (for the jammer) ρ which is a function of the ratio of signal power to total jammer power (P/J).

A third type of jammer is the single-tone jammer. The *single-tone jammer* trans-mits an unmodulated carrier with power J somewhere in the SS signal bandwidth. The one-sided power spectrum of this jamming signal is shown in Figure 11-2c.

The single-tone jammer is important because the jamming signal is easy to generate and is rather effective against DS SS systems. The analysis of this jammer with coherent SS systems will show that the jammer should place the tone at the center of the SS signal bandwidth to achieve maximum effectiveness. The single-tone jammer is somewhat less effective against a FH signal since the FH instantaneous bandwidth is small and for large processing gains the probability of being jammed on any one hop is small. For FH systems, a better tone jamming strategy is to use several tones which share the power of the single-tone jammer; a jammer using this technique is called a *multiple-tone jammer* and the one-sided power spectrum of a typical jammer is shown in Figure 11-2d for a four-tone jammer. The jammer selects the number of tones so that the optimum degradation occurs when the SS signal hops to a tone frequency. There is little to be gained by annihilating the use of a single hop frequency when the same total power can be used to degrade a number of frequencies by a smaller but still significant amount. The optimum number of tones is a function of the received ratio of signal power to jammer power (P/J). The multiple-tone jammer is also effective against hybrid direct-sequence frequency-hop systems.

Another technique for concentrating the jamming power is to pulse the jammer "on" and "off" as discussed briefly in Chapter 7. The jamming philosophy is the same as for partial-band and multiple-tone jamming. Specifically, the jammer turns "on" with just sufficient power to degrade SS system performance significantly, but does not totally annihilate system performance when "on." The *pulsed noise jammer* transmits a pulsed bandlimited white Gaussian noise signal whose psd just covers the SS system bandwidth W. The duty factor for the jammer is the fraction of time during which the jammer is "on" and is denoted by ρ. A subscript will be used on ρ whenever necessary to distinguish between the pulse duty factor and the fractional bandwidth of a partial band jammer. When the jammer is "on," the received jammer power spectral density is $N'_J = J/\rho W$. Figure 11-3a and b compare the transmitted waveforms of the full-time barrage noise jammer and the pulsed noise jammer. The pulse duty factor used in Figure 11-3b is $\rho = 0.5$ and the rms amplitude of the signal is $\sqrt{2}$ larger than the rms amplitude of the full-time jammer. It is possible to use partial band techniques simultaneously with pulse techniques, although in this text they are considered independently. Note that pulsed jamming assumes a jammer final power amplifier which is average-power limited rather than peak-power limited. In the analysis to follow, peak-power limitations are ignored to simplify the calculations. These limits must be considered by the spread-spectrum system designer. Pulse techniques can also be used for tone or multiple-tone jammers.

The last type of intentional jamming that is considered in this text is repeater jamming. As the name implies, a *repeater jammer* receives the SS signal, distorts it in some well-defined manner, and retransmits the signal at high power. The SS receiver then receives the distorted signal at high power may track and demodulate this distorted signal. Note that the jammer *must* distort the signal; otherwise, the jammer acts simply as a power amplifier for the desired signal. Also, receiving and transmitting simultaneously on the same band of frequencies presents formidable practical problems for the jammer.

For the jammer to be most effective, the jamming signal must be tailored to the SS communication system and to the actual received signal power P. A jammer which has knowledge of the type of signaling being used, which can accurately predict the received signal power, and which can adapt to transmit the optimum jamming signal is called a *smart jammer*. A smart jammer is usually assumed in

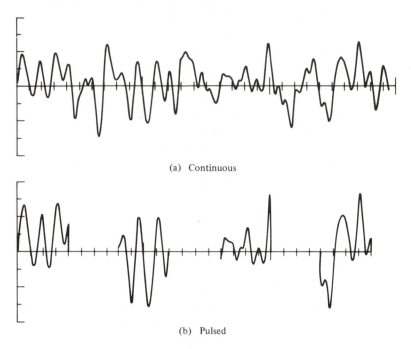

(a) Continuous

(b) Pulsed

FIGURE 11-3. Typical waveforms for continuous and pulsed noise jammer.

all worst-case system designs. The field of study that includes the design and analysis of jammers and jamming strategies is called *electronic counter measures* (ECM). Since one of the purposes of spread-spectrum communication systems is to counter specific jamming threats, some SS design and analysis is included within the general area of *electronic counter counter measures* (ECCM).

There are two other important types of interference to SS communication systems. The first of these is multipath interference. Multipath exists when there is more than one transmission path between the transmitter and receiver. If the secondary transmission paths can be characterized by the summation of several delayed and attenuated replicas of the desired signal, the multipath is called *specular multipath*. This type of multipath can occur when the receiving antenna sees reflections from such obstacles as buildings, the earth, the ionosphere, and so on. When the minimum delay between the desired and any delayed path is larger than the chip duration of a DS SS system, the receiver despreading operation will not despread the delayed signals and they can be rejected. However, a problem may occur during the code synchronization process since there is no way of distinguishing between the desired and reflected signals except by their relative amplitudes. A second type of nonhostile interference occurs when the spectrum spreading is used to provide a multiple-access capability for the system. In this case, different SS spreading codes are assigned to different users so that the despreading operation in any one receiver may be used to distinguish between users. All users use the same band of frequencies at the same time, and, to a first approximation, all received signals except the desired signal may be considered additional receiver noise. Multiple-access spread-spectrum systems are not considered in detail in this text because their analysis is complex and because the number of topics covered had to be limited. With the tools presented in this text the interested student will be able to study multiple-access applications with little difficulty. The student is referred to [1–8] for further information on this interesting subject.

Returning now to the discussion of Figure 11-1, the waveform channel output is input to the spread-spectrum demodulator and the timing and synchronization hardware. These functions have been discussed in great detail in Chapters 5, 6, 9, and 10. The purpose of the spread-spectrum demodulator is to remodulate the received waveform with the SS spreading waveform to wipe the spreading modulation off the received signal. This despreading operation is the key function of any SS system. Despreading can be accomplished only if accurate synchronization information is available. This information is derived in the timing and synchronization hardware. The operations of spreading waveform synchronization, carrier recovery, symbol synchronization, and frame synchronization are all included in the synchronization block of Figure 11-1.

The remaining functions in the path to the information sink of Figure 11-1 are the inverse operations of functions already discussed. When forward error correction is discussed later, it will be convenient to group all the functions from the interleaver input to the de-interleaver output into a single block called a discrete memoryless channel (DMC). The concept of the DMC, introduced in Chapter 1, enables the analyst to consider the function of error correction separately from the modulation/demodulation functions. That is, the analyst first considers the data modulation, spreading modulation, and the waveform channel (including jamming) and generates a transition matrix or diagram which models the entire transmission path. This model is the DMC. Several examples of DMC models are given in Chapter 1. A discrete memoryless channel is completely defined by its transition probabilities $p(i|j)$ which are the probabilities of the output i occurring when the channel input is the letter j. The output of a discrete input digital channel may have any number symbols and may, in fact, be continuous. The continuous output channel is characterized by probability density functions rather than discrete probabilities. With the DMC defined, the analyst is able to calculate the system transmission error probability using the known error correction code structure and decoding algorithm. This calculation varies significantly for different coding strategies.

The principal goal of the remainder of this chapter and the next is to describe the tools necessary to enable the student to predict the transmission error probability for most spread-spectrum communication systems in most interference environments. This will be accomplished using the system model described in this section. In most instances perfect synchronization will be presumed and the student is cautioned that some systems will be synchronization limited rather than transmission error probability limited. When system decisions are made, failure to consider the synchronization problem as well as transmission errors may lead to serious difficulty.

11-3

PERFORMANCE OF SPREAD-SPECTRUM SYSTEMS WITHOUT CODING

The results of this section predict the message error probability for many common spread-spectrum systems. These results are grouped by interference type rather than by system type since a number of system types respond in the same way to a particular interference.

11-3.1 Performance in AWGN or Barrage Noise Jamming

Because the barrage noise jammer transmits bandlimited white Gaussian noise at high power, the performance of any SS system is the same in either AWGN or barrage noise. When the noise is unintentional (i.e., thermal), the AWGN one-sided psd is denoted by N_0, and when the noise is intentional (i.e., jamming), the AWGN psd is denoted by N_J.

Coherent DS Systems. Consider an arbitrary coherent DS system transmitting a signal described by

$$s_t(t) = \sqrt{2P} \cos [2\pi f_0 + \theta_d(t) + \theta_{SS}(t)] \qquad (11\text{-}1)$$

where $\theta_d(t)$ is an arbitrary coherent data modulation and $\theta_{SS}(t)$ is an arbitrary coherent spreading modulation. The noise interference is bandlimited AWGN and the received signal is, assuming zero transmission delay,

$$r(t) = s_t(t) + \sqrt{2}n_I(t) \cos 2\pi f_0 t - \sqrt{2}n_Q(t) \sin 2\pi f_0 t \qquad (11\text{-}2)$$

The received noise power spectrum is assumed to overlap the received signal spectrum completely so that $n_I(t)$ and $n_Q(t)$ are independent and each has a two-sided psd of $N_0/2$ or $N_J/2$ watts/hertz.

A simplified model of the receiver for this signal is illustrated in Figure 11-4. In this figure, the first mixer performs the despreading operation and its output near the IF is

$$x(t) = \sqrt{2P} \cos [2\pi f_{IF} t + \theta_d(t)]$$
$$+ [\sqrt{2}n_I(t) \cos 2\pi f_0 t - \sqrt{2}n_Q(t) \sin 2\pi f_0 t]$$
$$\times \{2 \cos [2\pi(f_0 + f_{IF})t + \theta_{SS}(t)]\} \qquad (11\text{-}3)$$

Perfect synchronization has been assumed. The second term of this equation is the product of two independent noise-like waveforms and has a psd equal to the convolution of the psd's of each term. The noise psd $N(f)$ is $\frac{1}{2}N_0$ over frequency limits defined by $|f \pm f_0| \le \frac{1}{2}W$, where W is the null-to-null spread-spectrum bandwidth. The psd of the despreading reference signal is a function of the specific type of spreading used. For BPSK or QPSK spreading the reference psd has the form

$$S_R(f) = \frac{2}{W} \text{sinc}^2 \left[(f - f_0 - f_{IF}) \frac{2}{W} \right] + \frac{2}{W} \text{sinc}^2 \left[(f + f_0 + f_{IF}) \frac{2}{W} \right]$$

$$(11\text{-}4)$$

The bandpass filter following the despreader eliminates all frequency components not close to f_{IF}, and these terms may be ignored when evaluating the convolution.

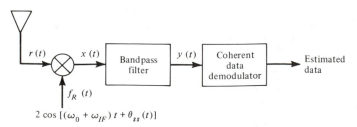

FIGURE 11-4. Simplified model of a coherent spread-spectrum receiver.

The terms of interest of $N(f) * S(f)$ are

$$N(f) * S_R(f) \approx \frac{N_0}{W} \int_{f_0-(1/2)W}^{f_0+(1/2)W} \text{sinc}^2 \left[(f - \lambda + f_0 + f_{IF}) \frac{2}{W} \right] d\lambda$$

$$+ \frac{N_0}{W} \int_{-f_0-(1/2)W}^{-f_0+(1/2)W} \text{sinc}^2 \left[(f - \lambda - f_0 - f_{IF}) \frac{2}{W} \right] d\lambda \quad (11\text{-}5)$$

Making the change of variables $\gamma' = f - \lambda + f_0 + f_{IF}$ in the first integral and $\gamma = f - \lambda - f_0 - f_{IF}$ in the second integral results in

$$N(f) * S_R(f) \simeq \frac{N_0}{W} \int_{f+f_{IF}-(1/2)W}^{f+f_{IF}+(1/2)W} \text{sinc}^2 \left(\gamma' \frac{2}{W} \right) d\gamma'$$

$$+ \frac{N_0}{W} \int_{f-f_{IF}-(1/2)W}^{f-f_{IF}+(1/2)W} \text{sinc}^2 \left(\gamma \frac{2}{W} \right) d\gamma \quad (11\text{-}6)$$

This convolution is an even function of f, so evaluation for either positive or negative frequencies is sufficient. Assuming that f_{IF} is large relative to W, the positive frequency component of (11-6) is

$$N(f) * S_R(f) \approx \frac{N_0}{W} \int_{f-f_{IF}-(1/2)W}^{f-f_{IF}+(1/2)W} \text{sinc}^2 \left(\gamma \frac{2}{W} \right) d\gamma \qquad \text{for } f > 0 \quad (11\text{-}7)$$

Assuming now that the data modulation bandwidth and hence the IF bandpass filter bandwidth is small relative to W, the noise two-sided psd near $f = \pm f_{IF}$ may be approximated by a constant $\frac{1}{2}N_n$ given by

$$\frac{1}{2}N_n = N(f_{IF}) * S_R(f_{IF})$$

$$= \frac{N_0}{W} \int_{-(1/2)W}^{+(1/2)W} \text{sinc}^2 \left(\gamma \frac{2}{W} \right) d\gamma = KN_0 \quad (11\text{-}8)$$

where K is a constant. The integral is the area under the main lobe of the $\text{sinc}^2[\gamma(2/W)]$ function. This area is $(0.903)W/2$. Observe that, if the noise bandwidth is much larger than the spread-spectrum bandwidth, the limits of integration of (11-8) include many lobes of the $\text{sinc}^2[\gamma(2/W)]$ function and $N_n/2$ approaches the limit $N_0/2$ or $N_J/2$. A similar analysis for MSK spreading modulation yields $N_n/2 \approx (0.995)W/2$ for a noise bandwidth equal to the null-to-null spreading bandwidth. Even when jamming is present, the psd $N_n/2$ contains a thermal noise component as well as a jamming component, and

$$\frac{1}{2}N_n = K(\frac{1}{2}N_0 + \frac{1}{2}N_J) \quad (11\text{-}9)$$

where K is a function of the spreading modulation and is usually close to unity.

The noise process at the filter output is Gaussian. Thus the input to the coherent data demodulator is

$$y(t) = \sqrt{2P} \cos \left[2\pi f_{IF} t + \theta_d(t) \right] + n_n(t) \quad (11\text{-}10)$$

where $n_n(t)$ is a bandlimited white Gaussian noise process with two-sided psd of $N_n/2$. From this point on, the analysis is identical to the analysis of any phase coherent digital modulation scheme. For the particular case where BPSK or QPSK data modulation is assumed, the bit error probability is

$$P_b = Q\left(\sqrt{\frac{2E_b}{N_n}} \right) = Q\left(\sqrt{\frac{2}{K[(N_0 R/P) + (J/P)(R/W)]}} \right) \quad (11\text{-}11)$$

where (11-9) is used with $N_J = J/W$, $E_b = P/R$, and R the bit rate.

In AWGN, the spectrum spreading has no effect at all since the noise bandwidth is much larger than W. However, when the noise is due to a jammer, the total jammer power required to reduce system performance to a given level is greatly increased by spreading the spectrum. Suppose, for example, that the desired system performance is achieved with a demodulator input signal-to-noise ratio of $\mathrm{SNR}_a = E_b/N_{na}$. For the normal non-spread-spectrum system the transmission bandwidth is W_d, which is equal to one or two times the data symbol rate, and the jammer power required to achieve SNR_a is $J_1 = N_{na}W_d$. For a spread-spectrum system the transmission bandwidth is $W >> W_d$ and the jammer power required to achieve SNR_a is approximately $J_2 = KN_{na}W$ where $K \approx 1$ and is a function of the spreading modulation. The processing gain of the system is $J_2/J_1 \approx W/W_d$ and is due to the fact that the jammer must fill a much larger bandwidth with noise for the SS system than for the non-SS system. Figure 11-5 is a plot of (11-11) as a function of $(P/J)(W/R)$ for various signal-to-thermal noise ratios $P/N_0R = E_b/N_0$ using the approximation $K = 1.0$. Observe that processing gain is included in this figure through the use of the factor W/R in the independent variable.

FH/MFSK. With slow FH spreading modulation, the transmitted signal is

$$s_t(t) = \sqrt{2P} \sum_{n=-\infty}^{\infty} p_{T_c}(t - nT_c) \cos\left[2\pi f_n t + \varphi_n + \theta_d(t)\right] \quad (11\text{-}12)$$

where f_n is the hop frequency during the nth hop interval, $p_{T_c}(t)$ is a unit pulse of duration T_c, φ_n is a random phase during the nth hop interval, and $\theta_d(t)$ is the M-ary FSK data modulation. The received signal is given by (11-2) with f_0 equal to the center frequency of the transmission bandwidth and the despreading reference signal is

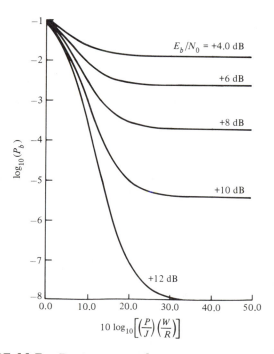

FIGURE 11-5. Performance of a coherent direct-sequence spread-spectrum system in barrage noise jamming.

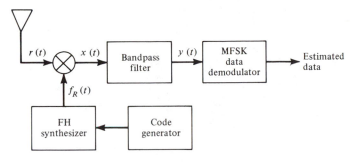

FIGURE 11-6. Simplified model of a FH/MFSK spread-spectrum receiver.

$$f_R(t) = 2 \sum_{n=-\infty}^{\infty} p_{T_c}(t - nT_c) \cos [2\pi(f_n + f_{IF})t] \qquad (11\text{-}13)$$

A simplified block diagram of the receiver is shown in Figure 11-6.

The components of the output of the despreading mixer near the IF are

$$x(t) = \sqrt{2P} \sum_{n=-\infty}^{\infty} p_{T_c}(t - nT_c) \cos [2\pi f_{IF} t + \varphi_n + \theta_d(t)]$$

$$+ [\sqrt{2}\, n_I(t) \cos (2\pi f_c t) - \sqrt{2}\, n_Q(t) \sin (2\pi f_c t)]$$

$$\times \left\{ 2 \sum_{n=-\infty}^{\infty} p_{T_c}(t - nT_c) \cos [2\pi(f_n + f_{IF})t] \right\} \qquad (11\text{-}14)$$

The first term above is a M-ary FSK signal with random phase on each data symbol. The second term is a frequency-hopped noise signal with instantaneous one-sided bandwidth W.

The despreader output noise psd is the convolution of the psd's of the input noise and the despreading signal. Since slow FH has been assumed, it is quite accurate to calculate the despreader output psd near the IF using quasi-static analysis. That is, consider one hop interval at a time to calculate the desired result. During any one hop interval, the despreader output-noise two-sided psd near the IF is $\frac{1}{2}N_n = \frac{1}{2}N_0 + \frac{1}{2}N_J$. Since the received noise spectrum covers the entire FH band, the despreader output noise psd near the IF is constant for all hop frequencies, so that the bandpass filter output may be written

$$y(t) = \sqrt{2P} \cos [2\pi f_{IF} t + \theta_d(t)] + n_n(t) \qquad (11\text{-}15)$$

where the FH random phase has been included with the MFSK random phase and $n_n(t)$ is bandlimited AWGN with two-sided psd of $\frac{1}{2}N_n$. Thus the data demodulator input noise psd is the same as the received noise psd.

The symbol error probability for M-ary FSK in an AWGN environment is calculated in Arthurs and Dym [9]. The result is

$$P_s = \frac{1}{M} \exp \left(-\frac{E_s}{2N_n} \right) \sum_{q=2}^{M} \binom{M}{q} (-1)^q \exp \left[\frac{E_s(2 - q)}{2N_n q} \right] \qquad (11\text{-}16)$$

where E_s is the symbol energy and orthogonal signaling is assumed. Orthogonal signaling means that the symbol tone spacing for adjacent tones is at least $1/T_s$, where T_s is the symbol duration. The demodulator assumed in deriving this result is illustrated in Figure 11.7. Both implementations shown in Figure 11.7 perform identically and both are optimal noncoherent detectors. When orthogonal signaling

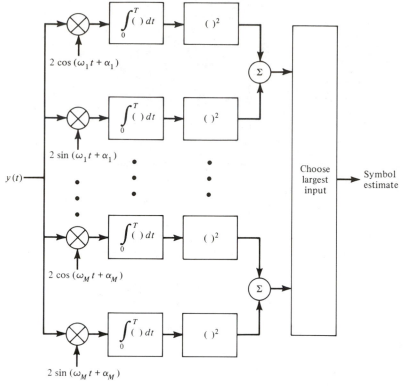

(a) Correlate and integrate implementation

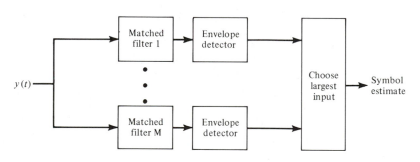

(b) Bandpass filter and envelope detect implementation

FIGURE 11-7. *M*-ary FSK noncoherent demodulators.

is used, the distance in signal space between any signal and all others is the same. Thus, when an error is made, it is equally likely that the error symbol is any of the $M - 1$ other symbols.

The bit error probability is calculated by noting that there are $l = \log_2 M$ bits associated with each data symbol. Since all possible symbol errors are equally likely, the average number of bit errors made when a symbol error occurs is given by

$$\frac{1}{M-1} \sum_{i=1}^{l} \binom{l}{i} i = \frac{l 2^{l-1}}{M-1} = \frac{M}{2(M-1)} l \qquad (11\text{-}17)$$

This expression is the sum over all possible symbol error events of the probability of that error event times the number of bit errors in that error event. Since l bits are demodulated with each symbol, the bit error probability, or average number of bit errors for each bit which is demodulated, is

$$
\begin{aligned}
P_b &= \frac{1}{l}\left[\frac{M}{2(M-1)}l\right]P_s \\
&= \frac{1}{2(M-1)}\exp\left(-\frac{lE_b}{2N_n}\right)\sum_{q=2}^{M}\binom{M}{q}(-1)^q\exp\left[\frac{lE_b(2-q)}{2N_nq}\right] \quad (11\text{-}18)
\end{aligned}
$$

Note that $E_s = lE_b$. Using the relations $\frac{1}{2}N_n = \frac{1}{2}N_0 + \frac{1}{2}N_J$, $E_b = P/R$, and $N_J = J/W$, it can be shown that

$$
\frac{E_b}{N_n} = \frac{1}{(N_0R/P) + (J/P)(R/W)} \quad (11\text{-}19)
$$

Figure 11-8 is a plot of (11-18) using $(P/J)(W/R)$ as the independent variable for various signal-to-thermal noise ratios and $l = 1$.

Once again the spectrum spreading and despreading operation has not affected the form of the bit error probability expression. The system processing gain may be calculated using the same steps used above for coherent systems. Specifically, suppose that $\text{SNR}_a = E_b/N_{na}$ is required to achieve the desired performance level. Without spread spectrum, the signal bandwidth is $W_d \approx MR/\log_2 M$ and the jammer power required is $J_1 = N_{na}W_d$. With spread spectrum, the jammer must fill the entire transmission bandwidth with noise having the same psd, and the jammer power is $J_2 = N_{na}W$. The FH/MSK processing gain is $J_2/J_1 = W/W_d$.

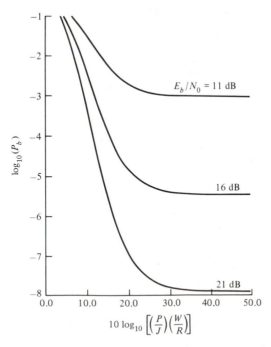

FIGURE 11-8. Performance of a FH/MFSK spread-spectrum system in barrage noise jamming.

FH/DPSK. The receiver for a frequency-hop spread-spectrum system using differential PSK data modulation is identical to the receiver of Figure 11-6 with the MFSK demodulator replaced by a DPSK demodulator. The analysis of the effect of the despreading operation on AWGN or the barrage noise jamming is identical to the FH/MFSK analysis. The conclusion is that the noise psd at the data demodulator input is the same as the receiver input noise psd over the transmission bandwidth.

In this chapter, consideration is limited to binary DPSK modulation. The optimum demodulator for binary DPSK, illustrated in Figure 11-9a, was discussed in Chapter 4. This demodulator is also analyzed in Arthurs and Dym [9], Park [10], Miller [11], and Lindsey and Simon [12], where the bit error probability is shown to be

$$P_b = \frac{1}{2} \exp\left(-\frac{E_b}{N_n}\right) = \frac{1}{2} \exp\left[\frac{-1}{(N_0 R/P) + (J/P)(R/W)}\right] \quad (11\text{-}20)$$

Note that Arthurs and Dym and Lindsey and Simon also consider M-ary DPSK systems, and Miller provides some interesting practical results on degradations due to timing and frequency offsets.

An alternative demodulator for binary DPSK is shown in Figure 11-9b. This demodulator is discussed in many modern texts, where it is usually concluded that its noise performance is also given by (11-20). Park [11] has shown that (11-20) applies to Figure 11-9b only if intersymbol interference is ignored. When ISI is considered, the corrected expression is

$$P_b = \frac{1}{2} \exp\left(-k\frac{E_b}{N_n}\right) \quad (11\text{-}21)$$

where k depends on the specific characteristics of the input bandpass filter. In all cases, the analysis of the demodulator of Figure 11-9b presumes that the length of the delay line is simultaneously equal to a symbol time T_s and chosen so that

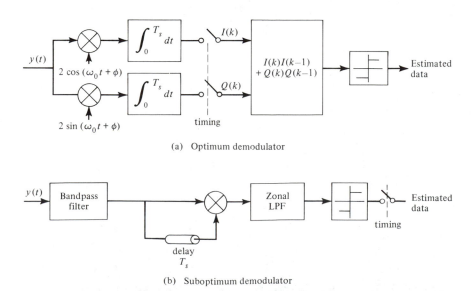

(a) Optimum demodulator

(b) Suboptimum demodulator

FIGURE 11-9. Binary differential phase-shift-keeping demodulators.

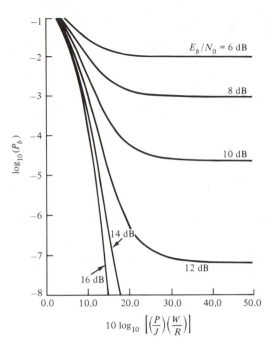

FIGURE 11-10. Performance of FH/DPSK spread-spectrum system in barrage noise jamming.

$\omega_0 T_s = 2\pi l$, where l is any integer. For high carrier frequencies this implies extremely precise control of the length of the delay line. If this factor is a problem in system design, the equivalent baseband implementations can be used. Figure 11-10 is a plot of (11-20) using E_b/N_n from (11-19).

The effect of spectrum spreading on this system is identical to its effect on FH/MFSK. Frequency hopping forces the barrage jammer to fill the full spread transmission band W with noise rather than just the data bandwidth W_d. Thus the processing gain is approximately W/W_d. Finally, recall that slow-frequency-hop systems are being considered. The demodulators discussed each compare the phases of pairs of transmitted waveforms. If the transmission frequency changes between a transmitted symbol waveform and its reference waveform, phase comparison information is lost and a transmission error may result. Thus it is highly desirable to transmit a large number of data symbols on each hop frequency or to provide a means for transmitting "dummy bits" at frequency-hop transition times.

DS-FH/MFSK or DS-FH/DPSK. With the barrage noise jammer or in AWGN, there is little performance improvement which can be realized using hybrid DS-FH spreading modulation over FH spreading modulation. However, it may be more convenient for the hardware designer to obtain very wide transmission bandwidths using hybrid modulations. The presence of the frequency-hop component forces the data modulation choice to be either noncoherent or differentially coherent. The demodulator for DS-FH/MFSK is illustrated in Figure 11-11. The received signal is first mixed with a hopped reference signal to remove the frequency-hop component of the spreading modulation. The output of the first mixer is centered at the first IF and is still DS spread and has a phase modulation due to the noncoherence of the transmitter and receiver frequency synthesizers. The bandwidth of the first bandpass filter must be large enough to pass the DS modulation without significant

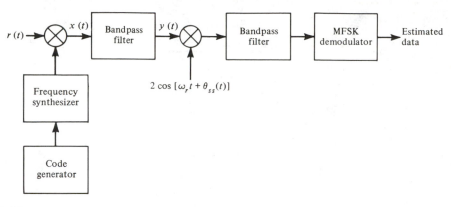

FIGURE 11-11. Simplified model of a DS-FH/MFSK spread-spectrum receiver.

distortion. The second mixer removes the DS phase modulation and downconverts the signal to the second IF where filtering to the data bandwidth and data demodulation occurs.

The same arguments used for FH to show that the noise psd at the first bandpass filter output is the same as the noise psd over the transmission band apply here. Thus the signal $y(t)$ may be written

$$y(t) = \sqrt{2P} \cos [2\pi f_{\mathrm{IF}}t + \theta_{\mathrm{SS}}(t) + \theta_d(t)] + n_n(t) \qquad (11\text{-}22)$$

where $\theta_{\mathrm{SS}}(t)$ is the DS modulation, $\theta_d(t)$ is the data modulation including a random phase component due to the dehopping, and $n_n(t)$ is a bandlimited white Gaussian noise process with two-sided psd of $\frac{1}{2}N_n = \frac{1}{2}N_0 + \frac{1}{2}N_J$. The bandwidth of this noise process is obviously set by the bandpass filter. Since the filter bandwidth is roughly the same as the DS spreading bandwidth, the analysis of the effect of the DS despreading on the noise process is identical to the analysis for coherent systems given above. The result is that the despreader output noise psd at the second IF is given by a convolution of the input noise psd and the spreading waveform psd. The resultant density is dependent on the details of the spreading waveform and is given by $KN_n/2$, where K is a constant near unity.

The performance of the system is given by (11-18) with N_n replaced by KN_n. Once again the spectrum spreading has little effect in an AWGN environment, but does force the barrage jammer to use a very wide bandwidth. The processing gain of the system is approximately W/W_d. An identical result can be derived for DS-FH/DPSK except that the noise performance is given by (11-20) or (11-21), depending on which DPSK demodulator is used.

11-3.2 Performance in Partial Band Jamming

Except for the transmission bandwidth, the partial band jammer is the same as the wideband barrage noise jammer. Because of the smaller bandwidth, the partial band jamming signal is easier to generate than the barrage jamming signal, and with some types of spectrum spreading, the partial band jammer is considerably more effective than the barrage noise jammer. For these reasons, the partial band jamming is a commonly used ECM technique. In the analysis that follows, the communication system bandwidth is denoted by W, the nonspread data transmission bandwidth is denoted by W_d, and the jammer transmission bandwidth is denoted by W_J. The fraction of the communication bandwidth that is jammed is denoted by $\rho = W_J/W$. The jammer one-sided psd over the transmission band W_J is $N'_J = J/W_J = J/(\rho W)$ and the full-band jammer one-sided psd is $N_J = J/W$. Thermal noise with one-sided psd of N_0 also degrades system performance. Therefore, over one part of the transmission band the total noise psd will be $N_0 + N'_J$, and over the remaining part, the total noise psd will be N_0.

Coherent DS Systems. Consider an arbitrary DS system transmitting a signal described by $s_t(t)$ of (11-1), and using a receiver which can be modeled by Figure 11-4. The received signal is

$$r(t) = s_t(t) + n(t) + n_J(t) \tag{11-23}$$

where $n(t)$ represents the thermal noise and $n_J(t)$ represents the partial band noise jammer. Both noise processes are narrowband relative to the carrier frequency. The receiver despreader output signal is

$$x(t) = \sqrt{2P} \cos [2\pi f_{IF} t + \theta_d(t)]$$

$$+ n(t)\{2 \cos [2\pi(f_0 + f_{IF})t + \theta_{SS}(t)]\}$$

$$+ n_J(t)\{2 \cos [2\pi(f_0 + f_{IF})t + \theta_{SS}(t)]\} \tag{11-24}$$

The middle term of this expression is identical to the noise term of (11-3) and its two-sided psd near f_{IF} is $N_0/2$. This psd is obtained from (11-8) by noting that the AWGN bandwidth is much wider than the spreading waveform bandwidth, so that the limits of integration of (11-8) include many lobes of the $\text{sinc}^2(\cdot)$ function. Equation (11-8) applies only to BPSK and QPSK spreading modulation; however, a similar equation for MSK spreading leads to the same conclusion for the AWGN psd at the despreader output.

The last term of (11-24) is due to the jammer. The psd due to the jammer at the despreader output is the convolution of the jammer psd and the despreading waveform psd. Usually, the jammer will not know the precise center frequency of the communication signal. The jammer center frequency will be denoted by f_J. In general, the despreader output psd is given by

$$S_R(f) * S_J(f) = \int_{-\infty}^{+\infty} S_R(f - \lambda) S_J(\lambda)\, d\lambda \tag{11-25}$$

where $S_J(f)$ is the jammer input psd and $S_R(f)$ is the psd of the reference spreading waveform. For BPSK or QPSK spreading, $S_R(f)$ is given by (11-4). Observe that the total power in the reference waveform is 2.0. The jammer two-sided psd $S_J(f)$ is $\frac{1}{2}N'_J$ over the frequency limits $|f \pm f_J| < \frac{1}{2}W_J$ and, for QPSK or BPSK spreading,

$$S_R(f) * S_J(f) = \frac{N'_J}{W} \int_{f_J-(1/2)W_J}^{f_J+(1/2)W_J} \left\{ \text{sinc}^2 \left[(f - \lambda - f_0 - f_{IF})\frac{2}{W} \right] \right.$$

$$+ \text{sinc}^2 \left[(f - \lambda + f_0 + f_{IF})\frac{2}{W} \right] \right\} d\lambda$$

$$+ \frac{N'_J}{W} \int_{-f_J-(1/2)W_J}^{-f_J+(1/2)W_J} \left\{ \text{sinc}^2 \left[(f - \lambda - f_0 - f_{IF})\frac{2}{W} \right] \right.$$

$$+ \text{sinc}^2 \left[(f - \lambda + f_0 + f_{IF})\frac{2}{W} \right] \right\} d\lambda$$

$$\triangleq S(f) \tag{11-26}$$

These integrals are depicted graphically in Figure 11-12. A typical spreading waveform psd shown in Figure 11-12a and a jammer psd in Figure 11-12b. The jammer center frequency does not equal the signal center frequency. The reference (i.e., despreading) waveform psd is shown in Figure 11-12c; the same psd is shifted and reversed for the convolution of (11-25) in Figure 11-12d and e. The frequency shifts shown in Figure 11-12d and e are $-f_{IF}$ and $+f_{IF}$, respectively. These shifts are used to evaluate $S(f)$ at $f = \pm f_{IF}$, the primary frequencies of interest. The shaded areas indicate the range of integration for (11-26). Observe that half of the terms of (11-26) are approximately zero for $f = \pm f_{IF}$, and that $S(f)$ is an even function. Thus

$$S(f_{IF}) = S(-f_{IF}) \approx \frac{N_J'}{W} \int_{f_J-(1/2)W_J}^{f_J+(1/2)W_J} \text{sinc}^2\left[(f_0 - \lambda)\frac{2}{W}\right] d\lambda \qquad (11\text{-}27)$$

A similar result is easily derived for MSK spreading.

The integral of (11-27) may be evaluated numerically for a specific jammer. For extremely narrowband jammers (i.e., $\rho \ll 1$), the $\text{sinc}^2(\cdot)$ function may be assumed constant over the range of integration and

$$S(\pm f_{IF}) \approx \frac{N_J'}{W} W_J \, \text{sinc}^2\left[(f_0 - f_J)\frac{2}{W}\right]$$

$$= N_J \, \text{sinc}^2\left[(f_0 - f_J)\frac{2}{W}\right] \qquad (11\text{-}28)$$

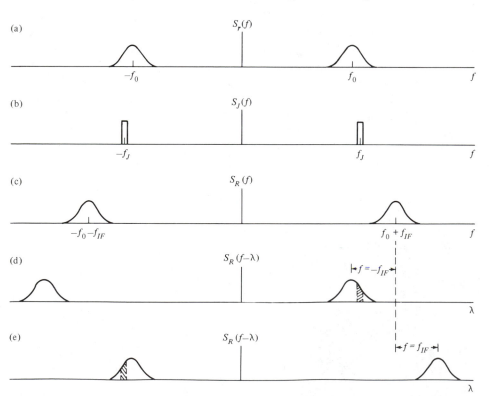

FIGURE 11-12. Typical power spectral densities used to calculate despreader output density for partial band jamming: (a) received signal psd; (b) received jammer psd; (c) reference waveform psd; (d) reference psd reversed and shifted by $-f_{IF}$; (e) reference psd reversed and shifted by $+F_{IF}$.

For maximum effectiveness, the very narrowband jammer center frequency must equal the carrier frequency f_0. It may also be deduced from (11-27) for any partial band jammer that the jammer center frequency needs to equal f_0 for maximum effectiveness. When $f_J = f_0$, evaluation of (11-27) as a function of W_J will show that $S(f_{IF})$ is maximized for small W_J; in particular, for $W_J << W$, $S(f_{IF}) = N_J$ and for $W_J = W$, $S(f_{IF}) \approx N'_J/2 = N_J/2$. Thus the jammer gains about 3 dB by using a very narrowband signal, but this 3 dB is quickly lost if the jammer center frequency is not properly selected.

The system bit error probability is calculated by assuming that the IF filter output noise may be modeled as bandlimited AWGN with two-sided psd equal to the sum of the thermal noise component and a jammer noise component. For BPSK or QPSK data modulation, the error probability is

$$P_b = Q\left(\sqrt{\frac{E_b}{(N_0/2) + S(f_{IF})}}\right) \tag{11-29}$$

and for very narrowband jamming,

$$P_b = Q\left(\sqrt{\frac{2E_b}{N_0 + 2N_J}}\right) = Q\left(\sqrt{\frac{2}{(RN_0/P) + 2(J/P)(R/W)}}\right) \tag{11-30}$$

This relationship is plotted in Figure 11-13 using $(P/J)(W/R)$ as the independent variable for various signal-to-thermal noise ratios.

Recall that N_J is the one-sided psd that would be generated by the jammer if the total jammer power J were spread uniformly across the transmission bandwidth W; thus $J = N_J W$. Ignoring thermal noise, the system processing gain is calculated by comparing total jammer power required to increase system error probability to

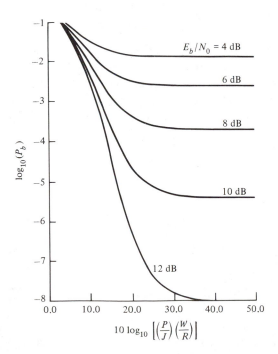

FIGURE 11-13. Performance of coherent direct-sequence spread-spectrum systems in partial band jamming.

a specified level with and without spectrum spreading. Suppose that the specified error probability is obtained at $\text{SNR}_a = E_b/N_{Ja}$ without spectrum spreading. In this case, the jammer power is uniformly spread over a bandwidth W_d and the total jammer power is $J_1 = N_{Ja}W_d$. With spectrum spreading and narrowband jamming, the same error probability is achieved with $\text{SNR}_b = E_b/N_{Jb}$, and the total jammer power is $J_2 = N_{Jb}W$. SNR_a and SNR_b are related by equating the arguments of the Q-functions for the spread and nonspread cases. Without spreading, the error probability is given by $P_b = Q(\sqrt{2E_b/N_J})$. With spreading the error probability is given by (11-30). Thus equal error probabilities imply that $2\text{SNR}_a = \text{SNR}_b$ and that $2N_{Jb} = N_{Ja}$. Using these relationships, the processing gain is $J_2/J_1 = W/W_d$, which is half as large as the comparable result for barrage noise jamming. Thus the jammer obtains a 3-dB advantage by using narrowband jamming rather than barrage jamming. Similar results can be derived for MSK spreading modulation.

FH/MFSK. The received signal for the FH/MFSK system is defined by (11-12) and the despreading reference is defined by (11-13). A very simplified block diagram of the receiver was illustrated in Figure 11-6. The system analysis for partial band jamming is the same as the analysis for barrage jamming through (11-14). The received FH signal hops over the entire transmission bandwidth W. On some hops the signal will be at a frequency which is jammed, while on other hops, the signal will be at a frequency where the only interference is thermal noise. When the system is properly synchronized, the despreader output may be represented by

$$x(t) = \sqrt{2P} \sum_{k=-\infty}^{\infty} p_{T_c}(t - kT_c) \cos\left[2\pi f_{\text{IF}}t + \varphi_k + \theta_d(t)\right] + n_n(t) \quad (11\text{-}31)$$

Because the signal is hopping in and out of the jamming noise, $n_n(t)$ is not stationary.

For slow-frequency-hop systems, the system message error probability is calculated using quasi-static analysis. That is, the error probability is calculated separately for thermal noise interference and thermal plus jamming noise interference, and the two results are averaged. Both error probabilities are calculated using (11-18). For the thermal noise calculation use $N_n = N_0$, and for the thermal plus jamming noise calculation use $N_n = N_0 + N_J'$. Since the fraction of the band that is jammed is ρ, the average bit error probability is

$$\overline{P}_b = (1 - \rho)P_b\left(\frac{E_b}{N_0}\right) + \rho P_b\left(\frac{E_b}{N_0 + N_J'}\right)$$

$$= (1 - \rho)P_b\left(\frac{P}{RN_0}\right) + \rho P_b\left(\frac{1}{(RN_0/P) + (J/P)(R/W)(1/\rho)}\right) \quad (11\text{-}32)$$

where $P_b(\cdot)$ represents (11-18). This equation is plotted in Figure 11-14 for large signal-to-thermal noise ratio and $M = 2$ using the familiar $(P/J)(W/R)$ as the independent variable and using ρ as a parameter.

Observe in Figure 11-14 that for any value of $(P/J)(W/R)$ there is an optimum value of ρ which maximizes the system bit error probability. The jammer has control of ρ and the smart jammer will be able to adjust ρ dynamically to always maximize system bit error probability. The system performance with this worst-case smart jammer has been calculated in Houston [13] for the special case where thermal noise is negligible. In this case, (11-32) simplifies to

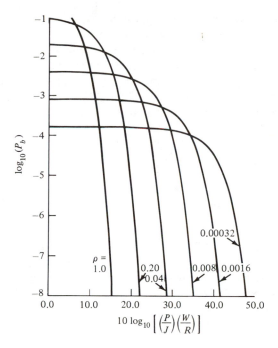

FIGURE 11-14. Performance of FH/MFSK spread spectrum in partial band jamming.

$$\overline{P}_b = \rho P_b\left(\frac{E_b}{N'_J}\right) = \rho P_b\left(\rho\,\frac{E_b}{N_J}\right)$$

$$= \frac{\rho}{2(M-1)}\exp\left(-\frac{l\rho E_b}{2N_J}\right)\sum_{q=2}^{M}\binom{M}{q}(-1)^q\exp\left[\frac{l\rho E_b(2-q)}{2N_J q}\right]$$

$$= \frac{\rho}{2(M-1)}\sum_{q=2}^{M}\binom{M}{q}(-1)^q\exp\left[\frac{l\rho E_b(1-q)}{N_J q}\right] \tag{11-33}$$

The jammer wishes to maximize $\overline{P}_b$ by choosing the fractional bandwidth appropriately. In order to find the maximizing ρ, (11-33) is differentiated with respect to ρ and the result is set equal to zero. The result is

$$\frac{d\overline{P}_b}{d\rho} = \frac{\rho}{2(M-1)}\sum_{q=2}^{M}\binom{M}{q}(-1)^q\exp\left[\frac{l\rho E_b(1-q)}{N_J q}\right]\frac{lE_b(1-q)}{N_J q}$$

$$+ \frac{1}{2(M-1)}\sum_{q=2}^{M}\binom{M}{q}(-1)^q\exp\left[\frac{l\rho E_b(1-q)}{N_J q}\right] = 0 \tag{11-34}$$

which can be written [13] as

$$\frac{\displaystyle\sum_{q=1}^{M}\binom{M}{q}\frac{(-1)^q}{q}\exp\left(\frac{y}{q}\right)}{\displaystyle\sum_{q=1}^{M}\binom{M}{q}(-1)^q\exp\left(\frac{y}{q}\right)} = \frac{y-1}{y} \tag{11-35}$$

where

$$y = l\rho\,\frac{E_b}{N_J} \tag{11-36}$$

This equation can be solved numerically for y as a function of M. The results of this calculation are given in Table 11-1 for several M.

The value of y_0 from Table 11-1 is used to calculate the optimum ρ as a function of E_b/N_J using (11-36). Of course, the fractional bandwidth can be no larger than 1.0 so $\rho = 1.0$ is used when $\rho > 1.0$ is found from the calculation. The values of E_b/N_J below which $\rho > 1.0$ are calculated from (11-36) are also given in Table 11-1. For $E_b/N_J > (E_b/N_J)_0$, the worst case $\bar{P}_b$ is found by substituting $\rho = y_0/(lE_b/N_J)$ into (11-33). The result is

$$(\bar{P}_b)_{max} = \frac{y_0}{2l(M-1)} \left(\frac{1}{E_b/N_J}\right) \sum_{q=2}^{M} \binom{M}{q}(-1)^q \exp\left(y_0 \frac{1-q}{q}\right)$$

$$\triangleq \frac{k'}{E_b/N_J} \tag{11-37}$$

where k' is a constant which is also given in Table 11-1 for several M. For $E_b/N_J \leq (E_b/N_J)_0$, the average error probability is calculated directly from Equation (11-33) with $\rho = 1.0$. In summary, the worst case average bit error probability for partial band jamming of FH/MFSK is given by

$$(\bar{P}_b)_{max} = \begin{cases} \dfrac{k'}{E_b/N_J} & \dfrac{E_b}{N_J} > \left(\dfrac{E_b}{N_J}\right)_0 \\[2em] \dfrac{1}{2(M-1)} \sum_{q=2}^{M} \binom{M}{q}(-1)^q \exp\left[\dfrac{lE_b}{N_J}\left(\dfrac{1-q}{q}\right)\right] & \dfrac{E_b}{N_J} \leq \left(\dfrac{E_b}{N_J}\right)_0 \end{cases} \tag{11-38}$$

where k' and $(E_b/N_J)_0$ are given in Table 11-1. Equation (11-38) is plotted in Figure 11-15 as a function of $E_b/N_J = (P/J)(W/R)$ for $M = 2, 4, 8,$ and 16. Comparing Figure 11-8 with Figure 11-15 shows that the jammer can be considerably more effective using partial band jamming than using barrage jamming. This increased effectiveness is obtained by causing a large degradation to a fraction of the transmitted symbols rather than a little degradation to all symbols. The inverse linear relationship between $(\bar{P}_b)_{max}$ and E_b/N_J is typical of the performance of uncoded spread-spectrum communication systems in optimized jamming. The in-

TABLE 11-1. Solution of (11-35) as a Function of M

M	y_0	k'	$\left(\dfrac{E_b}{N_J}\right)_0$
2	2.00	0.3679	2.000
3	2.19		
4	2.34	0.2329	1.170
5	2.48		
6	2.59		
7	2.70		
8	2.79	0.1954	0.930
16	3.49	0.1803	0.872
32	3.99	0.1746	0.798

Source: Ref. 13.

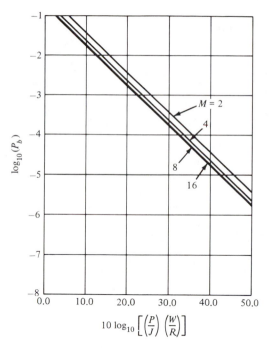

$$10 \log_{10}\left[\left(\frac{P}{J}\right)\left(\frac{W}{R}\right)\right]$$

FIGURE 11-15. Performance of FH/MFSK spread spectrum in worst-case partial band jamming.

verse linear relationship is also evident in Figure 11-14 as the envelope of the family of curves shown.

The calculation of the spread-spectrum system processing gain for FH/MFSK in partial band jamming is somewhat different than the processing gain calculation performed earlier. The reason for this difference is that the mathematical expression for bit error probability has a different form in the jamming environment than it has in the nonjamming environment. Thus the arguments of the error probability expressions can no longer be equated to obtain the same performance with and without spectrum spreading. Processing gain is most easily calculated by plotting the bit error probability expressions with and without spectrum spreading as a function of $(E_b/N_J)_{\mathrm{dB}}$ on the same grid as shown in Figure 11-16 for $M = 2$. At any given $\overline{P}_b$, the difference Δ_{dB} in $(E_b/N_J)_{\mathrm{dB}}$ between the two curves of Figure 11-16 can be found, and is attributable to a change in N_J due to spectrum spreading. The N_J required without spectrum spreading is denoted by N_{J1} and the total jammer power required is $J_1 = N_{J1}W_d$. The N_J required with spectrum spreading is denoted by N_{J2} and the total jammer power required is $J_2 = N_{J2}W$. The relationship between N_{J1} and N_{J2} is

$$\frac{N_{J1}}{N_{J2}} = 10^{0.1\Delta_{\mathrm{dB}}} \tag{11-39}$$

and the system processing gain is

$$\frac{J_2}{J_1} = \frac{N_{J2}W}{N_{J1}W_d} = \left(\frac{W}{W_d}\right)10^{-0.1\Delta_{\mathrm{dB}}} \tag{11-40}$$

This equation indicates that no matter how large W/W_d is, processing gain can be less than unity for large enough Δ_{dB}. Large Δ_{dB} is seen at low $\overline{P}_b$ on Figure 11-16.

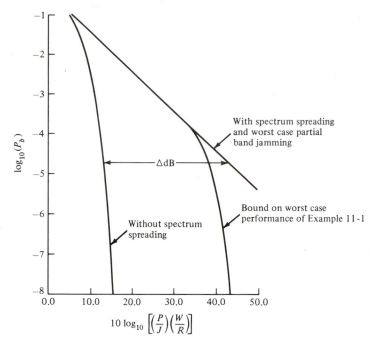

$$10 \log_{10} \left[\left(\frac{P}{J} \right) \left(\frac{W}{R} \right) \right]$$

FIGURE 11-16. Calculation of processing gain for FH/MFSK spread spectrum in worst-case partial band jamming.

Fortunately, however, Δ_{dB} is limited by the fact that the derivation of (11-38) assumes that ρ can be arbitrarily small. The range of useful ρ is bounded below by the fact that no additional performance improvement is obtained by the jammer by using $W_J < W_d$. Thus $W_d/W \leq \rho \leq 1.0$ is the useful range of ρ, and Δ_{dB} cannot be arbitrarily large.

EXAMPLE 11-1

Consider a FH/MFSK system with $W/W_d = 1000$ and using binary FSK data modulation. The minimum useful $\rho = 0.001$ for this system. The E_b/N_J at which this ρ is required is calculated from (11-36) with $y = y_0$ from Table 11-1; the result is

$$\frac{E_b}{N_J} = \frac{2}{0.001} = 2000 = 33 \text{ dB}$$

Above this value of E_b/N_J, the optimum $\rho = 0.001$ and the complete average error probability expression is

$$(\overline{P}_b)_{\max} = \begin{cases} \dfrac{0.001}{2} \exp\left(-\dfrac{0.001}{2} \dfrac{E_b}{N_J} \right) & 2000 < \dfrac{E_b}{N_J} \\[3mm] \dfrac{0.3679}{E_b/N_J} & 2.0 < \dfrac{E_b}{N_J} < 2000 \\[3mm] \dfrac{1}{2} \exp\left(-\dfrac{1}{2} \dfrac{E_b}{N_J} \right) & \dfrac{E_b}{N_J} < 2.0 \end{cases}$$

This result is also plotted in Figure 11-16. The maximum Δ_{dB} can be read from the figure or calculated from the equation above. The result is about 26 dB. ∎

In all cases, the inverse linear relationship of (11-38) returns to the exponential relationship for very large E_b/N_J. The specific E_b/N_J at which this happens is a function of the bandwidth ratio W/W_d so that totally general results cannot be derived. The conclusion that the partial band jammer is considerably more effective than the wideband barrage jammer remains valid in spite of the bounding of this improvement.

FH/DPSK. Once again consideration will be limited to binary DPSK data modulation. The analysis of the performance of FH/DPSK is identical to that of FH/MFSK except that the data modulation of (11-31) is DPSK rather than MFSK and the data demodulator is modified appropriately. Slow frequency hop is assumed and the noise term $n_n(t)$ of 11-31 is the same as for MFSK; that is, the noise is bandlimited AWGN with psd dependent on whether or not the transmitted carrier frequency is in or out of a jamming band. The average bit error probability is given by

$$\overline{P}_b = (1 - \rho)P_b\left(\frac{E_b}{N_J}\right) + \rho P_b\left(\frac{E_b}{N_0 + N_J'}\right) \tag{11-41}$$

with $P_b(\cdot)$ given by (11-20) when the optimum binary DPSK demodulator is used. Recall that, in partial band jamming, $N_n = N_0$ when the transmitted frequency is not in a jamming band and $N_n = N_0 + N_J' = N_0 + N_J/\rho$ when the transmitted frequency is within a jamming band. Substituting (11-20) into equation (11-41) yields

$$\overline{P}_b = \tfrac{1}{2}(1 - \rho)\exp\left[-\left(\frac{P}{N_0 R}\right)\right] + \tfrac{1}{2}\rho\exp\left[\frac{-1}{(N_0 R/P) + (J/P)(R/W)(1/\rho)}\right] \tag{11-42}$$

This result is plotted in Figure 11-17 for high signal-to-thermal noise ratio using ρ as a parameter and $(P/J)(W/R)$ as the independent variable. Observe that there is an optimum ρ for any specific value of $(P/J)(W/R)$. The smart jammer will choose ρ to maximize system average bit error probability.

Suppose now that the signal-to-thermal noise ratio is very high so that thermal noise induced errors can be neglected. Equation (11-42) becomes

$$\overline{P}_b \approx \frac{\rho}{2}\exp\left[-\rho\left(\frac{P}{J}\right)\left(\frac{W}{R}\right)\right] \tag{11-43}$$

and the worst case ρ is found by differentiating with respect to ρ and setting the result equal to zero.

$$\frac{d\overline{P}_b}{d\rho} = \left[-\frac{\rho}{2}\left(\frac{P}{J}\right)\left(\frac{W}{R}\right) + \frac{1}{2}\right]\exp\left[-\rho\left(\frac{P}{J}\right)\left(\frac{W}{R}\right)\right] = 0 \tag{11-44}$$

Solving for the optimum ρ yields

$$\rho_0 = \frac{1}{(P/J)(W/R)} \tag{11-45}$$

and substituting this optimum ρ into (11-43) yields

$$(\overline{P}_b)_{\max} = \frac{1}{2(P/J)(W/R)}\exp(-1) = \frac{0.1839}{(P/J)(W/R)} \tag{11-46}$$

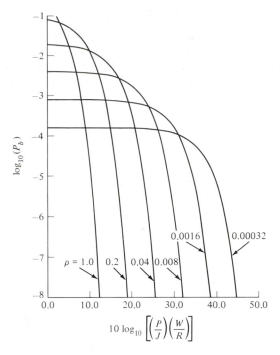

$$10 \log_{10}\left[\left(\frac{P}{J}\right)\left(\frac{W}{R}\right)\right]$$

FIGURE 11-17. Performance of FH/DPSK spread spectrum in partial band jamming.

This relationship is only valid for $\rho_0 \leq 1.0$. For cases where (11-45) indicates that $\rho_0 > 1.0$, use $\rho_0 = 1.0$. The worst-case average bit error probability is therefore given by

$$(\overline{P}_b)_{\max} = \begin{cases} \dfrac{e^{-1}}{2(P/J)(W/R)} & \dfrac{P}{J}\left(\dfrac{W}{R}\right) \geq 1.0 \\[4mm] \dfrac{1}{2}\exp\left[-\dfrac{P}{J}\left(\dfrac{W}{R}\right)\right] & \dfrac{P}{J}\left(\dfrac{W}{R}\right) < 1.0 \end{cases} \qquad (11\text{-}47)$$

This result is plotted in Figure 11-18. The now familiar inverse linear relationship between $(P_b)_{\max}$ and $(P/J)(W/R)$ is seen again. Comparing Figures 11-10 and 11-18 shows that the partial band jammer can cause significantly worse system performance than the barrage noise jammer. Similar results can be derived for nonbinary DPSK systems.

The system processing gain is calculated using the same procedure used for FH/MFSK. That is, plot $\overline{P}_b$ versus E_b/N_J for the non-spread-spectrum and spread spectrum cases on the same grid. For a particular $\overline{P}_b$, the difference Δ_{dB} in E_b/N_J between the two curves is found. This difference is due to a change in N_J required to degrade system performance to $\overline{P}_b$. Denote the N_J required without SS by N_{J1} and the associated jammer total power by $J_1 = N_{J1}W_d$. The N_J required with SS is denoted N_{J2} and the associated jammer power by $J_2 = N_{J2}W$. The densities N_{J1} and N_{J2} are related by (11-39) and the processing gain is given by (11-40). Again it appears that the processing gain might be reduced to values less than unity for very low $\overline{P}_b$. This is not true, however, since the jammer fractional bandwidth is practically limited to $W_d/W \leq \rho$. The particular E_b/N_J above which $\rho \leq W_d/W$ is $E_b/N_J = W/W_d$ calculated from (11-45). Using this lower bound on ρ, the error

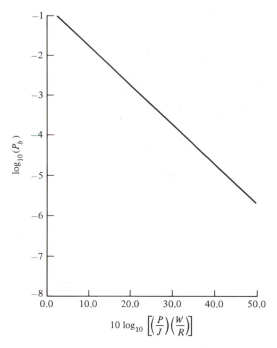

$$10 \log_{10}\left[\left(\frac{P}{J}\right)\left(\frac{W}{R}\right)\right]$$

FIGURE 11-18. **Performance of FH/DPSK spread spectrum in worst-case partial band jamming.**

probability expression can be rewritten

$$\overline{(P_b)}_{max} = \begin{cases} \dfrac{1}{2}\dfrac{W_d}{W}\exp\left[-\dfrac{P}{J}\left(\dfrac{W}{R}\right)\left(\dfrac{W_d}{W}\right)\right] & \dfrac{W}{W_d} < \dfrac{P}{J}\left(\dfrac{W}{R}\right) \\[3ex] \dfrac{e^{-1}}{2(P/J)(W/R)} & 1.0 \le \dfrac{P}{J}\left(\dfrac{W}{R}\right) \le \dfrac{W}{W_d} \\[3ex] \dfrac{1}{2}\exp\left[-\dfrac{P}{J}\left(\dfrac{W}{R}\right)\right] & \dfrac{P}{J}\left(\dfrac{W}{R}\right) < 1.0 \end{cases} \qquad (11\text{-}48)$$

System processing gain should be calculated from this result. These calculations will show that processing gain is significantly reduced by a partial band jammer.

DS-FH/MFSK and DS-FH/DPSK. The receiver for the hybrid DS-FH systems is illustrated in Figure 11-11 and was described earlier. In partial band jamming, the noise component at the output of the frequency dehopper may be thermal noise alone or may be the frequency-translated partial band jammer. When the instantaneous bandwidth of the transmitted DS-FH signal overlaps the spectrum of the partial band jammer, a portion of the jammer spectrum will appear at the output of the first IF bandpass filter. For simplicity, assume that the partial band jammer spectrum fills the entire DS bandwidth whenever the two spectra overlap at all. This assumption is equivalent to the assumption that there are a large number of frequency hops across the transmission bandwidth.

In this special case, the DS despreader input noise component is either thermal noise with two-sided psd of $\frac{1}{2}N_0$ or thermal noise plus jamming noise which fills the entire DS bandwidth with two-sided psd of $\frac{1}{2}N_0 + \frac{1}{2}N_J'$. With this input noise, the DS despreader output noise has been calculated in the discussion of barrage noise jamming. The result is that the output noise has a two-sided psd of $KN_0/2$ or $K(N_0 + N_J')/2$ depending on whether or not the transmission is in a jammed frequency band. The constant K is a function of the details of the DS modulation and is usually near unity. Let $P_b(\cdot)$ denote the bit error probability for the data modulation of interest. Then the average bit error probability is

$$\overline{P}_b = (1 - \rho)P_b\left(\frac{E_b}{KN_0}\right) + \rho P_b\left(\frac{E_b}{KN_0 + KN_J'}\right) \tag{11-49}$$

Except for the constant, K, this equation is the same as the result for FH/MFSK or FH/DPSK. The analysis from this point on is identical to that already performed and will not be repeated. The final result is that (11-38) and (11-47) apply with signal-to-noise ratio increased by a factor $1/K$.

11-3.3 Performance in Pulsed Noise Jamming

The pulsed noise jammer uses essentially the same strategy as the partial band jammer to degrade system performance. This strategy is to degrade performance significantly for a small fraction of the time in order to produce the maximum possible increase in average bit error probability. The pulsed noise jammer is on for a fraction ρ of the time and produces a wideband noise signal with two-sided psd of $N_J'/2 = N_J/2\rho$ when "on." The two-sided psd $N_J/2$ is the density the jammer would produce if "on" continuously. The average jammer power is denoted by J.

It will be shown that the pulsed jammer can produce the same type of degradation for coherent DS systems that the partial band jammer produced for pure FH systems. A disadvantage of the pulsed noise jammer is that very high peak power may be required from the jammer. Peak power constraints will limit the minimum value of ρ and thereby limit jamming effectiveness.

It is assumed that the jammer pulse repetition frequency is slow enough that a large number of information bits are transmitted between each change of state of the jammer. With this assumption, the situation where the jammer turns "on" or "off" during the transmission of a bit may be ignored, and a quasi-static analysis of the average bit error probability is appropriate. For any of the modulations to be considered, let $P_b(E_b/N_n)$ denote the data demodulator bit error probability, where N_n is the one-sided noise psd at the input to the data demodulator. The magnitude of N_n depends on the particular spreading modulation used and whether or not the jammer is "on". Let N_{n1} and N_{n2} denote the value of N_n when the jammer is "off" and "on," respectively. N_{n1} is not zero because, in general, thermal noise is not neglected. The average bit error probability $\overline{P}_b$ is given by

$$\overline{P}_b = (1 - \rho)P_b\left(\frac{E_b}{N_{n1}}\right) + \rho P_b\left(\frac{E_b}{N_{n2}}\right) \tag{11-50}$$

The jammer selects ρ to maximize $\overline{P}_b$ subject to peak power constraints. The density N_{n2} is also a function of ρ.

Coherent DS Systems. For all the coherent data modulations of interest in this chapter (BPSK, QPSK, OQPSK, and MSK), the bit error probability is given by $P_b(E_b/N_n) = Q(\sqrt{2E_b/N_n})$. The value of N_n is calculated using the same steps used in the barrage noise analysis with the end result that $N_{n1} = KN_0$ and

$N_{n2} = K(N_0 + N_J')$, where K is a function of the spreading modulation. For BPSK and QPSK spreading modulations K is given in (11-8). Substituting into (11-50) yields

$$\overline{P}_b = (1 - \rho)Q\left(\sqrt{\frac{2E_b}{KN_0}}\right) + \rho Q\left[\sqrt{\frac{2E_b}{K(N_0 + N_J/\rho)}}\right] \quad (11\text{-}51)$$

For the special case where thermal noise is negligible relative to jamming noise, Equation (11-51) may be approximated by

$$\overline{P}_b \approx \rho Q\left(\sqrt{\frac{2\rho E_b}{KN_J}}\right) = \rho Q\left[\sqrt{\frac{2\rho}{K}\left(\frac{P}{J}\right)\left(\frac{W}{R}\right)}\right] \quad (11\text{-}52)$$

This expression is plotted in Figure 11-19 as a function of $(P/J)(W/R)$ using $K = 1.0$ and various ρ. The remainder of this discussion considers only this special case where thermal noise is negligible.

Calculation of the worst-case ρ could be accomplished by differentiating (11-52) with respect to ρ and setting the result equal to zero. Unfortunately, the derivative of the Q-function is not easily calculated so that this method fails. The worst case $\overline{P}_b$ can, however, be easily determined graphically from the plot of Figure 11-19. The envelope of the family of curves shown represents the worst case $\overline{P}_b$. This envelope exhibits the usual inverse linear relation between $(\overline{P}_b)_{max}$ and $(P/J)(W/R)$. An alternative means of calculating the worst case ρ is to use an upper bound for the Q-function. This bound was given in (7-4), repeated here for convenience:

$$\overline{P}_b \le \frac{\rho}{\sqrt{4\pi\rho y}} \exp\left(-\rho y\right) \quad (7\text{-}4)$$

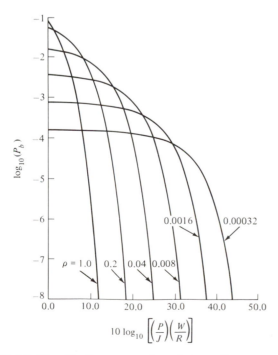

FIGURE 11-19. Performance of a coherent direct-sequence spread spectrum in pulsed noise jamming.

where $y = E_b/N_J = (P/J)(W/R)$. Taking the first derivative and setting it equal to zero and solving for ρ yields $\rho = 1/2y$. The duty factor can be no larger than unity so that $\rho = 1.0$ is used whenever $y < 0.5$. Substituting the optimum ρ into (11-52) yields

$$(\bar{P}_b)_{max} = \begin{cases} \dfrac{1}{2(P/J)(W/R)} Q\left(\sqrt{\dfrac{1}{K}}\right) & 0.5 < \dfrac{P}{J}\left(\dfrac{W}{R}\right) \\[4mm] Q\left\{\sqrt{\dfrac{2}{K}\left(\dfrac{P}{J}\right)\left(\dfrac{W}{R}\right)}\right\} & \dfrac{P}{J}\left(\dfrac{W}{R}\right) < 0.5 \end{cases} \tag{11-53}$$

This equation is plotted in Figure 11-20. A similar result is discussed in Viterbi [14] using the bound of (7-4) in the final $(\bar{P}_b)_{max}$ result.

The calculation of the processing gain for coherent direct-sequence systems must take into account the fact that pulsed jamming is also effective against the nonspread system. In fact, the analysis of the pulse-jammed nonspread system is identical to the analysis above except that there is no factor K in the equations comparable to (11-51) and (11-52). The final result is identical to (11-53) with $K = 1.0$ and total bandwidth W_d. The processing gain is then calculated by equating E_b/N_J for the spread and nonspread systems. For both systems the jammer is pulsed with duty factor ρ. The average jammer power without spectrum spreading is $J_1 = \rho N_J' W_d = N_J W_d$ and the average jammer power with spectrum spreading is $J_2 = \rho N_J' W = N_J W$. The processing gain is therefore $J_2/J_1 = W/W_d$.

FH/MFSK. The bit error probability for MFSK is given by (11-18). The data demodulator input noise psd is calculated using the same steps used for barrage

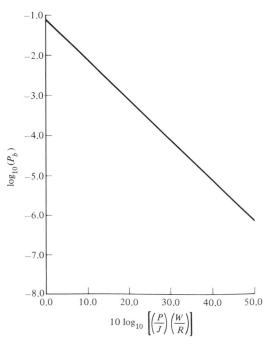

FIGURE 11-20. Performance of coherent direct-sequence spread spectrum in worst-case pulsed noise jamming.

noise jamming. The results are $N_{n1} = N_0$ and $N_{n2} = N_0 + N'_J = N_0 + N_J/\rho$ and the average bit error probability is

$$\overline{P}_b = (1 - \rho)P_b\left(\frac{E_b}{N_0}\right) + \rho P_b\left(\frac{E_b}{N_0 + N_J/\rho}\right) \qquad (11\text{-}54)$$

where $P_b(\cdot)$ is defined in (11-18). This equation is identical to (11-32), which was derived for FH/MFSK in partial band jamming. Thus the analysis to find the worst case ρ is the same as was done for partial band jamming and the final result is given by (11-38) which is plotted in Figure (11-15).

The calculation of processing gain for the pulsed jammer is, however, different since the pulsed jammer can also force the nonspread system to an inverse linear relationship. The bit error probability for the pulse jammed nonspread system is also specified by (11-38), and the processing gain is calculated by equating E_b/N_J for the spread and nonspread systems. Once again, jamming resistance is achieved by forcing the jammer to spread its power over a bandwidth W using spread spectrum rather than W_d without spread spectrum. The system processing gain is $J_2/J_1 = W/W_d$.

FH/DPSK. The bit error probability for DPSK is given by (11-20), and the data demodulator input noise psd N_n is calculated using the same steps used for barrage jamming. When the pulse jammer is "on," $N_{n2} = N_0 + N'_J$, and when the pulse jammer is "off," $N_{n1} = N_0$. The average bit error probability is

$$\overline{P}_b = \frac{1 - \rho}{2} \exp\left(-\frac{E_b}{N_0}\right) + \frac{\rho}{2} \exp\left(-\frac{E_b}{N_0 + N_J/\rho}\right) \qquad (11\text{-}55)$$

which is identical to (11-42). The calculation of the worst case ρ was described above and the final averge bit error probability is given by (11-47) and is plotted in Figure 11-18.

The system processing gain is not the same as calculated for the partial band jamming case. The pulsed jammer can degrade the nonspread system in the same way as the spread system. Thus the equations describing the bit error probability of the spread and nonspread systems are the same and the processing gain is equal to the bandwidth ratio W/W_d.

DS-FH/MFSK and DS-FH/DPSK. The pulsed jamming performances of DS-FH/MFSK and of DS-FH/DPSK are calculated using (11-50) with N_{n1} and N_{n2} calculated as in the barrage jamming analysis of these hybrid systems. Therefore, when the jammer is "off," $N_n = KN_0$, and when the jammer is "on," $N_n = K(N_0 + N'_J)$, where K is dependent on the details of the DS modulation. The average bit error probability is

$$\overline{P}_b = (1 - \rho)P_b\left(\frac{E_b}{KN_0}\right) + \rho P_b\left(\frac{E_b}{KN_0 + KN'_J}\right) \qquad (11\text{-}56)$$

where $P_b(\cdot)$ is given by (11-18) for DS-FH/MFSK and by (11-20) for DS-FH/MFSK. From this point on the analysis is the same as the analysis for the pure FH systems pulsed jamming except that the factor K is included. When signal-to-thermal noise ratio is high, thermal noise can be neglected and the worst case performance is given by (11-38) for DS-FH/MFSK and by (11-47) for DS-FH/MFSK with E_b/N_J in each case increased by $1/K$.

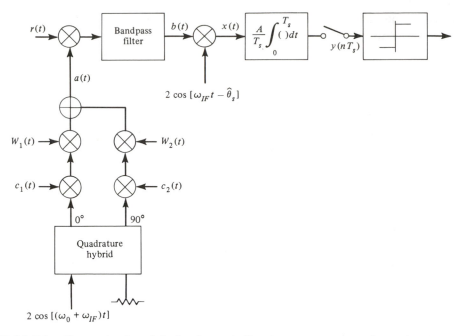

FIGURE 11-21. Conceptual model of coherent direct-sequence spread-spectrum receiver.

11-3.4 Performance in Single-Tone Jamming

The single-tone jammer is perhaps the easiest of all jamming signals to generate. In early spread-spectrum literature, the jammer was analyzed by assuming that after despreading, the jamming power was equivalent to Gaussian noise. The analysis was simply to compute the jamming power spectrum at the despreader output and to use conventional Gaussian noise techniques thereafter. More recently, a number of authors [15–18] have analyzed the effect of the tone jammer in greater detail and their example is followed here. A single-tone jammer affects only a single hop frequency in a FH system. For this season, this jammer is not a very effective countermeasure against FH systems and FH systems will be ignored in this section.

Receiver Description and General Analysis. Only coherent direct-sequence SS systems are considered in this section. A generalized receiver model that applies to all the DS modulations of interest, including BPSK/BPSK, MSK/BPSK, QPSK/BPSK, is illustrated in Figure 11-21. In this figure, $c_1(t)$ and $c_2(t)$ are independent inphase and quadrature channel spreading waveforms, respectively, and $W_1(t)$ and $W_2(t)$ are inphase and quadrature weighting functions. For BPSK/BPSK modulation, set $c_2(t) = 0$ to obtain a single channel. QPSK/QPSK is considered as two independent BPSK/BPSK signals and is therefore a slight modification of the figure. For all modulations except MSK/BPSK, the weighting functions may be set equal to unity while for MSK/BPSK use half-sine functions*

---

*Observe that only the positive half-cycle of the sine-wave weighting function is used here, in contrast to the alternating positive and negative half cycles of Figure 7-14. In addition, a $\sqrt{2}$ normalization factor is included here which does not appear in Figure 7-14. Either format is valid for MSK.

$$W_1(t) = \sqrt{2} \sum_i p_{T_c}(t - iT_c) \sin \left[\frac{\pi}{T_c} (t - iT_c) \right]$$

$$W_2(t) = \sqrt{2} \sum_i p_{T_c}\left(t - iT_c - \frac{T_c}{2} \right) \sin \left[\frac{\pi}{T_c} \left(t - iT_c - \frac{T_c}{2} \right) \right] \quad (11\text{-}57)$$

where T_c is the spreading waveform chip duration and $p_{T_c}(t)$ is the unit pulse. The received spread-spectrum signal is

$$s_t(t) = \sqrt{P} d(t)\{c_1(t)W_1(t) \cos (\omega_0 t + \theta_s) + c_2(t)W_2(t) \sin (\omega_0 t + \theta_s)\} \quad (11\text{-}58)$$

where $d(t)$ is the data modulation, and the received tone jammer is

$$n_J(t) = \sqrt{2J} \cos (\omega_J t + \varphi_J) \quad (11\text{-}59)$$

The received signal power is P and the received jammer power is J. The signal frequency ω_0 and the jammer frequency ω_J are not assumed to be equal. The thermal noise component of the received signal is denoted $n(t)$ and the total received signal is $r(t) = s_t(t) + n_J(t) + n(t)$. The purpose of the analysis to follow is to develop an approximate expression for the statistics of the integrator output so that bit error probability can be calculated.

The integrator input will be calculated separately for the three received signal components. Consider the signal component first. The DS despreader input reference $a(t)$ is

$$a(t) = \sqrt{2}\, c_1(t)W_1(t) \cos [(\omega_0 + \omega_{IF})t]$$

$$+ \sqrt{2}\, c_2(t)W_2(t) \sin [(\omega_0 + \omega_{IF})t] \quad (11\text{-}60)$$

$$b_1(t) = [s_t(t)a(t)]_{\substack{\text{diff.} \\ \text{freq.}}}$$

$$= \frac{\sqrt{2P}}{2} d(t)[W_1^2(t) + W_2^2(t)] \cos (\omega_{IF} t - \theta_s) \quad (11\text{-}61)$$

Perfect synchronization has been assumed. The system is phase coherent so that $\hat{\theta}_s \approx \theta_s$ and the second mixer output due to signal is

$$x_1(t) = \frac{\sqrt{2P}}{2} d(t)\{W_1^2(t) + W_2^2(t)\} \quad (11\text{-}62)$$

The integrator input due to jamming is calculated next. The despreading mixer output component which passes through the bandpass filter is

$$b_2(t) = [n_J(t)a(t)]_{\substack{\text{diff.} \\ \text{freq.}}}$$

$$= \sqrt{J}\, c_1(t)W_1(t) \cos [(\omega_{IF} + \Delta\omega)t - \varphi_J]$$

$$+ \sqrt{J}\, c_2(t)W_2(t) \sin [(\omega_{IF} + \Delta\omega)t - \varphi_J] \quad (11\text{-}63)$$

where $\Delta\omega = \omega_0 - \omega_J$. The output of the second mixer due to jamming is

$$x_2(t) = \sqrt{J}\, c_1(t)W_1(t) \cos (\Delta\omega\, t - \varphi_J + \theta_s)$$

$$+ \sqrt{J}\, c_2(t)W_2(t) \sin (\Delta\omega\, t - \varphi_J + \theta_s) \quad (11\text{-}64)$$

Finally, the integrator input due to noise is calculated. The received noise is assumed to be bandlimited to approximately the signal bandwidth so that the noise analysis for the barrage noise jammer can be applied directly. The result is that the bandpass filter output is bandlimited white Gaussian noise whose psd is the convolution of the input noise spectrum and the spreading waveform power spectrum evaluated at the IF. For all cases of interest, the two-sided psd is approximately $\frac{1}{2}N_0$, which is identical to the received noise two-sided pdf. Thus

$$b_3(t) = \sqrt{2}\, n_I(t)\, \cos \omega_{IF}t - \sqrt{2}\, n_Q(t)\, \sin \omega_{IF}t \qquad (11\text{-}65)$$

The second mixer output is

$$x_3(t) = \sqrt{2}\, n_I(t)\, \cos \hat{\theta}_s + \sqrt{2}\, n_Q(t)\, \sin \hat{\theta}_s \qquad (11\text{-}66)$$

Since $n_I(t)$ and $n_Q(t)$ are independent, the right-hand side of this equation may be considered a single Gaussian noise process whose two-sided power spectrum is

$$\frac{N_n}{2} = \frac{N_0}{2}\, [(\sqrt{2}\, \cos \hat{\theta}_s)^2 + (\sqrt{2}\, \sin \hat{\theta}_s)^2]$$

$$= N_0 \qquad (11\text{-}67)$$

over the range of frequencies where it is nonzero.

The integrator output due to noise at the sampling time is

$$y_3(nT_s) = \frac{A}{T_s} \int_0^{T_s} x_3(t)\, dt \qquad (11\text{-}68)$$

Since $x_3(t)$ is Gaussian and integration is a linear operation, $y_3(nT_s)$ is also Gaussian. The mean of $y_3(T_s)$ is zero and the variance is calculated using (10-74), which is repeated here in slightly modified form for convenience [19]

$$\sigma_{y_3}^2 = A^2 \frac{2}{T_s} \int_0^{T_s} \left(1 - \frac{\lambda}{T_s}\right)\{R_{x_3}(\lambda) - E^2[x_3(t)]\}\, d\lambda \qquad (10\text{-}74)$$

where $x_3(t)$ is the integrator input noise process, T_s is the integration time, and $R_{x_3}(\lambda)$ is the autocorrelation function of $x_3(t)$. Denote the bandpass filter bandwidth by B, then

$$R_{x_3}(\lambda) = BN_0\, \text{sinc}\,(B\lambda) \qquad (11\text{-}69)$$

It is reasonable to assume that B is large relative to the inverse of the integration time since the matched filtering operation is accomplished by the integrator and not the bandpass filter. With this assumption the integral may be approximated by

$$\sigma_{y_3}^2 \approx A^2 \frac{2}{T_s} \int_0^{\infty} BN_0\, \text{sinc}\,(B\lambda)\, d\lambda = \frac{N_0}{T_s} A^2 \qquad (11\text{-}70)$$

The total integrator input is $x(t) = x_1(t) + x_2(t) + x_3(t)$. The integrator output due to signal and jamming will be calculated for each modulation of interest.

BPSK/BPSK. In order to model BPSK/BPSK using the equations above, set $W_2(t) = 0$ and $W_1(t) = \sqrt{2}$. With this substitution, all the signal power is placed in the I channel of both the received signal and the despreading reference. Thus

$$x_1(t) + x_2(t) = \sqrt{2P}\, d(t) + \sqrt{2J}\, c_1(t)\, \cos (\Delta\omega\, t - \varphi_J + \theta_s) \qquad (11\text{-}71)$$

The integrator output is normalized using $A = 1/\sqrt{2P}$, so that the output data component is ± 1; the result is (signal and jammer only)

$$y(nT_s) = \pm 1 + y_2(nT_s) \tag{11-72}$$

where

$$y_2(nT_s) = \frac{1}{T_s} \sqrt{\frac{J}{P}} \int_0^{T_s} c_1(t) \cos{(\Delta\omega\, t - \varphi_J + \theta_s)}\, dt \tag{11-73}$$

Calculation of the bit error probability requires knowledge of the statistics of y_2. As a first step in evaluating the statistics of y_2, the mean and variance will be calculated. The mean is the ensemble average over all possible spreading code sequences during the integration time,

$$\bar{y}_2 = E\left[\frac{1}{T_s} \sqrt{\frac{J}{P}} \int_0^{T_s} c_1(t) \cos{(\Delta\omega\, t + \theta_s - \varphi_J)}\, dt\right] \tag{11-74}$$

Since $c_1(t)$ takes on values of ± 1 with equal probability, this average is zero for each code chip, and $\bar{y}_2 = 0$.

The evaluation of the variance is considerably more complicated. The first step in this calculation is to evaluate the integral y_2 over each chip interval to express y_2 as a summation. Thus

$$y_2 = \frac{1}{T_s} \sqrt{\frac{J}{P}} \sum_{i=0}^{N-1} c_i \int_{iT_c}^{(i+1)T_c} \cos{(\Delta\omega\, t + \theta_s - \varphi_J)}\, dt$$

$$= \frac{1}{T_s} \sqrt{\frac{J}{P}} \sum_{i=0}^{N-1} c_i \frac{1}{\Delta\omega} \{\sin{[\Delta\omega\,(i+1)T_c + \theta_s - \varphi_J]}$$

$$- \sin{(\Delta\omega i T_c + \theta_s - \varphi_J)}\} \tag{11-75}$$

where $c_i = \pm 1$ represents the spreading code sequence and N is the number of spreading waveform chips in a data symbol. This expression can be simplified by expressing the sine function in exponential form, factoring common phase angles from all terms, and then recombining. The final result is

$$y_2 = \frac{T_c}{T_s} \sqrt{\frac{J}{P}} \frac{\sin{(\Delta\omega\, T_c/2)}}{\Delta\omega\, T_c/2} \sum_{i=0}^{N-1} c_i \cos{(\Delta\omega\, T_c i + \varphi)} \tag{11-76}$$

In this equation, $\varphi = -\varphi_J + \Delta\omega T_c/2 + \theta_s$.

Since the mean of y_2 is zero, the variance of y_2 is the ensemble average over all spreading waveforms of y_2^2. Thus

$$\sigma_{y_2}^2 = E\left[\left(\frac{T_c}{T_s}\right)^2 \left(\frac{J}{P}\right) \frac{\sin^2{(\Delta\omega\, T_c/2)}}{(\Delta\omega\, T_c/2)^2} \sum_{i=0}^{N-1} c_i \cos{(\Delta\omega\, T_c i + \varphi)}\right.$$

$$\left. \times \sum_{k=0}^{N-1} c_k \cos{(\Delta\omega\, T_c k + \varphi)}\right]$$

$$= \frac{1}{N^2} \frac{J}{P} \frac{\sin^2{(\Delta\omega\, T_c/2)}}{(\Delta\omega\, T_c/2)^2} E\left[\sum_{i=0}^{N-1} \sum_{k=0}^{N-1} c_i c_k \cos{(\Delta\omega\, T_c i + \varphi)}\right.$$

$$\left. \times \cos{(\Delta\omega\, T_c k + \varphi)}\right]$$

$$= \frac{1}{N^2} \frac{J}{P} \frac{\sin^2{(\Delta\omega\, T_c/2)}}{(\Delta\omega\, T_c/2)^2} \sum_{i=0}^{N-1} \sum_{k=0}^{N-1} E(c_i c_k) \cos{(\Delta\omega\, T_c i + \varphi)}$$

$$\times \cos{(\Delta\omega\, T_c k + \varphi)} \tag{11-77}$$

Assume now that the spreading code is an infinite sequence of independent equally likely random binary digits so that $E(c_i c_k) = 1$ for $i = k$ and $E(c_i c_k) = 0$ for $i \neq k$. Thus,

$$\sigma_{y_2}^2 = \frac{1}{N^2} \frac{J}{P} \frac{\sin^2(\Delta\omega \, T_c/2)}{(\Delta\omega \, T_c/2)^2} \sum_{i=0}^{N-1} \cos^2(\Delta\omega \, T_c i + \varphi)$$

$$= \frac{1}{N^2} \frac{J}{P} \frac{\sin^2(\Delta\omega \, T_c/2)}{(\Delta\omega \, T_c/2)^2} \sum_{i=0}^{N-1} [\tfrac{1}{2} + \tfrac{1}{2} \cos (2\Delta\omega \, T_c i + 2\varphi)] \quad (11\text{-}78)$$

This expression is simplified using the identity [20]

$$\sum_{i=0}^{N-1} \cos (iy + x) = \frac{\cos \{x + [(N-1)/2] \, y\} \sin (Ny/2)}{\sin (y/2)} \quad (11\text{-}79)$$

the final result is [16]

$$\sigma_{y_2}^2 = \frac{J}{2PN} \frac{\sin^2(\Delta\omega \, T_c/2)}{(\Delta\omega \, T_c/2)^2} \left\{ 1 + \frac{\cos [2\varphi + (N-1) \, \Delta\omega \, T_c] \sin (N \, \Delta\omega \, T_c)}{N \sin (\Delta\omega \, T_c)} \right\}$$

$$(11\text{-}80)$$

The mean and variance of a probability density do not completely specify the pdf. The exact pdf for the case where the jammer frequency is offset from the received carrier frequency has not been found. However, when $\omega_J = \omega_0$ the pdf can be found exactly, as shown by Levitt [15]. In this special case, (11-73) becomes

$$y_2 = \frac{1}{T_s} \sqrt{\frac{J}{P}} \int_0^{T_s} c_1(t) \cos (-\varphi_J + \theta_s) \, dt$$

$$= \frac{T_c}{T_s} \sqrt{\frac{J}{P}} \cos (\theta_s - \varphi_J) \sum_{i=0}^{N-1} c_i \quad (11\text{-}81)$$

The integral y_2 is a discrete random variable whose value depends on the particular sequence of N spreading code symbols associated with the information symbol being demodulated. Let

$$K = \frac{1}{N} \sum_{i=0}^{N-1} c_i \quad (11\text{-}82)$$

and assume that the spreading code is a sequence of independent equally likely binary symbols. Let N_1 be the number of $+1$'s and N_2 be the number of -1's in the spreading code sequence. Then

$$K = \frac{1}{N} (N_1 - N_2) = \frac{1}{N} (2N_1 - N) \quad (11\text{-}83)$$

All spreading code sequences of length N chips are equally likely. The probability of a particular sequence is $(1/2)^N$ and the probability of a sequence having N_1 ones and N_2 minus ones is $\binom{N}{N_1}(1/2)^N$. For any specific value of K, say k, (11-83) can be solved for N_1. The result is $N_1 = (N/2)(k + 1)$, and the probability that $K = k$ is

$$\Pr (K = k) = \binom{N}{(N/2)(k + 1)} \frac{1}{2^N} \quad (11\text{-}84)$$

The possible values for K are $k = -1 + 2n/N$, $n = 0, 1, 2, \ldots, N$, and the pdf of K is recognized as the binomial distribution. The mean and variance of K are easily calculated [19]; the results are

$$E(K) = 0 \tag{11-85a}$$

$$\sigma_K^2 = \frac{1}{N} \tag{11-85b}$$

Since y_2 and K are linearly related, the exact pdf of y_2 in this special case is a discrete binomial pdf with zero mean, and variance $[J\cos^2(\varphi_J - \theta_s)/PN]$.

A convenient approximation for the discrete binomial probability density is the continuous Gaussian pdf having the same mean and variance. Figure 11-22 compares the Gaussian pdf and the binomial pdf for the special case where $J\cos^2(\varphi_J - \theta_s)/P = 1.0$ and $N = 16$. In this figure, the Gaussian pdf is normalized by the spacing between the discrete lines of the binomial distribution, so that both functions have comparable magnitudes.

Return now to the case where $\omega_J \neq \omega_0$. The discussion of the coherent jammer was intended to make plausible the assumption that the integrator output pdf is Gaussian. The mean of this Gaussian pdf is data dependent and equals ± 1. The variance of the output is the sum of the variance due to thermal noise from (11-70) and the variance due to the spread jammer from (11-80). Assuming that a -1 was transmitted, an error is made whenever the integrator output is positive and

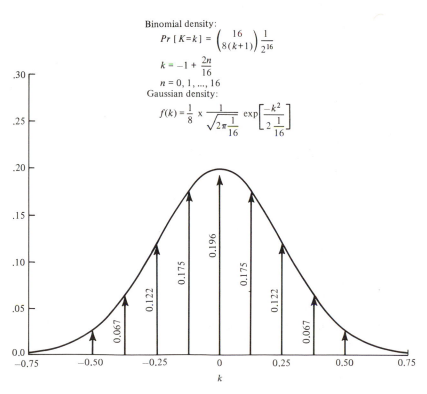

FIGURE 11-22. Comparison of Gaussian and binomial distribution functions having the same mean and variance.

$$P_b = \Pr\left[y_2(nT_s) + y_3(nT_s) > 1.0\right]$$

$$= \int_{1.0}^{\infty} \frac{1}{\sqrt{2\pi}\sigma}\, e^{-\alpha^2/2\sigma^2}\, d\alpha$$

$$= Q\left(\frac{1}{\sigma}\right) \tag{11-86}$$

where $\sigma^2 = \sigma_{y_2}^2 + \sigma_{y_3}^2$, and $\sigma_{y_3}^2$ is evaluated using the normalization $A^2 = 1/2P$. An identical result is obtained for a transmitted $+1$, so that (11-86) represents the total transmission error probability when equally likely symbols are transmitted. For the special case where $\omega_J = \omega_0$,

$$\sigma_{y_2}^2 = \frac{J}{PN}\cos^2(\varphi_J - \theta_s) \tag{11-87a}$$

$$\sigma_{y_3}^2 = \frac{N_0}{2PT_s} \tag{11-87b}$$

and

$$P_b = Q\left[\sqrt{\frac{1}{(J/PN)\cos^2(\varphi_J - \theta_s) + (N_0/2PT_s)}}\right]$$

$$= Q\left[\sqrt{\frac{2E_b}{2JT_c\cos^2(\varphi_J - \theta_s) + N_0}}\right] \tag{11-88}$$

which agrees with Singh [16] and Levitt [15]. Observe that for certain $\varphi_J - \theta_s$ the cosine in the denominator equals zero. This occurs when the jammer is phased such that it affects only the quadrature carrier channel. The worst-case jammer phasing results in $\cos^2(\varphi_J - \theta_s) = 1.0$. Equation (11-88) is valid only when $\Delta\omega = 0$.

Whenever $\Delta\omega \neq 0$, (11-80) must be used for the variance of y_2 when substituting into (11-86). For high signal-to-thermal noise ratio, the resulting error probability is

$$P_b = Q\left(\sqrt{\frac{2PN}{J\dfrac{\sin^2(\Delta\omega\, T_c/2)}{(\Delta\omega\, T_c/2)^2}\left\{1 + \dfrac{\cos\left[2\varphi + (N-1)\,\Delta\omega\, T_c\right]\sin(N\,\Delta\omega\, T_c)}{N\sin(\Delta\omega\, T_c)}\right\}}}\right) \tag{11-89}$$

BPSK spreading is being used so that the transmission bandwidth $W = 2/T_c$. Thus $N = T_s/T_c = W/2R$ and $2PN/J = (P/J)(W/R)$. Using $(P/J)(W/R)$ as the independent variable and $\Delta\omega$ as a parameter, (11-89) is plotted in Figure 11-23. Each curve of the figure represents an average over all possible jammer phases. For all curves, $N = 1024$.

QPSK/BPSK. For QPSK spreading modulation and BPSK data modulation, set $W_1(t) = W_2(t) = 1.0$. In this case, the signal component at the integrator input is

$$x_1(t) = \sqrt{2P}\, d(t) \tag{11-90}$$

and the jamming component is

$$x_2(t) = \sqrt{J}\, c_1(t)\cos(\Delta\omega\, t - \varphi_J + \theta_s) + \sqrt{J}\, c_2(t)\sin(\Delta\omega\, t - \varphi_J + \theta_s) \tag{11-91}$$

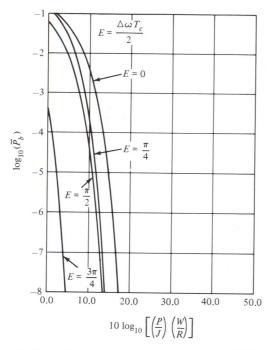

$$10 \log_{10}\left[\left(\frac{P}{J}\right)\left(\frac{W}{R}\right)\right]$$

FIGURE 11-23. **Performance of coherent BPSK/BPSK spread spectrum in continuous tone jamming.**

The integrator output is again normalized to obtain a data component at the output of ± 1. The normalizing factor is $A^2 = \frac{1}{2}P$ and the normalized integrator output signal and jammer components are

$$y_1(nT_s) + y_2(nT_s) = \pm 1 + \frac{1}{T_s}\sqrt{\frac{J}{2P}}\int_0^{T_s} c_1(t)\cos(\Delta\omega\, t - \varphi_J + \theta_s)\, dt$$

$$+ \frac{1}{T_s}\sqrt{\frac{J}{2P}}\int_0^{T_s} c_2(t)\sin(\Delta\omega\, t - \varphi_J + \theta_s)\, dt \tag{11-92}$$

The first integral above is identical to the integral of (11-73) and may be approximated by a zero mean Gaussian random variable with a variance of one-half the magnitude given in (11-80). The one-half factor is due to the jamming power being split between the inphase and quadrature channels. The mean and variance of the second integral of (11-92) are calculated using steps identical to those used to arrive at (11-80). The final result is that the mean is zero and the variance is

$$\sigma_{y_{2b}}^2 = \frac{J}{4PN}\frac{\sin^2(\Delta\omega\, T_c/2)}{(\Delta\omega\, T_c/2)^2}\left\{1 - \frac{\cos[2\varphi + (N-1)\,\Delta\omega\, T_c]\sin(N\,\Delta\omega\, T_c)}{N\sin(\Delta\omega\, T_c)}\right\} \tag{11-93}$$

Because the spreading waveforms $c_1(t)$ and $c_2(t)$ are independent, the two integrals of (11-92) are independent and their sum may be considered a single Gaussian random variable with zero mean and variance equal to the sum of (11-93) and one-half of (11-80). The variance of the sum is

$$\sigma_{y_2}^2 = \frac{J}{2PN}\frac{\sin^2(\Delta\omega\, T_c/2)}{(\Delta\omega\, T_c/2)^2} \tag{11-94}$$

The integrator output due to noise was calculated earlier. The total integrator output at the sampling time is the sum of a ± 1 signal component and Gaussian noise with zero mean and variance $\sigma^2 = \sigma_{y_2}^2 + \sigma_{y_3}^2$ with $\sigma_{y_3}^2 = N_0/2PT_s$. A transmission error is made when a -1 is transmitted and $y_2 + y_3 > 1.0$. The probability of this event is

$$P_b = \Pr\,(y_2 + y_3 > 1.0)$$

$$= \int_{1.0}^{\infty} \frac{1}{\sqrt{2\pi}\sigma}\, e^{-\alpha^2/2\sigma^2}\, d\alpha = Q\!\left(\sqrt{\frac{1}{\sigma^2}}\right)$$

$$= Q\left\{\sqrt{\frac{1}{\dfrac{J}{2PN}\dfrac{\sin^2(\frac{1}{2}\Delta\omega\, T_c)}{(\frac{1}{2}\Delta\omega\, T_c)^2} + \dfrac{N_0}{2PT_s}}}\right\} \qquad (11\text{-}95)$$

Since transmitted information symbols are equally likely and the same error probability expression applies to both, Equation (11-95) also represents the total bit error probability. Observe that the jammer phase does not appear in this equation.

MSK/BPSK. Equations (11-57) are used for MSK spreading modulation and BPSK data modulation. The receiver is defined in Figure 11-21. For MSK spreading the two spreading waveforms are offset from one another by one-half chip just as $W_1(t)$ and $W_2(t)$ are offset. Over each spreading waveform chip interval, the weighting function is a half-cycle of a sinusoid whose magnitude is chosen to equal $\sqrt{2}$ so that the total received power is P and the total power in the despreading reference is 2.0. The signal component of the integrator input is calculated using (11-62) and (11-57).

$$x_1(t) = \frac{\sqrt{2P}}{2}\, d(t)\left\{ 2 \sum_i \sum_j p_{T_c}(t - iT_c) p_{T_c}(t - jT_c) \sin\left[\frac{\pi}{T_c}(t - iT_c)\right]\right.$$

$$\times \sin\left[\frac{\pi}{T_c}(t - jT_c)\right]$$

$$+ 2 \sum_k \sum_l p_{T_c}\!\left(t - kT_c - \frac{T_c}{2}\right) p_{T_c}\!\left(t - lT_c - \frac{T_c}{2}\right)$$

$$\left. \times \sin\left[\frac{\pi}{T_c}\!\left(t - kT_c - \frac{T_c}{2}\right)\right] \sin\left[\frac{\pi}{T_c}\!\left(t - lT_c - \frac{T_c}{2}\right)\right]\right\}$$

$$= \sqrt{2P}\, d(t)\left\{ \sum_i p_{T_c}(t - iT_c)\sin^2\!\left[\frac{\pi}{T_c}(t - iT_c)\right]\right.$$

$$\left. + \sum_k p_{T_c}\!\left(t - kT_c - \frac{T_c}{2}\right)\sin^2\!\left[\frac{\pi}{T_c}\!\left(t - kT_c - \frac{T_c}{2}\right)\right]\right\}$$

$$= \sqrt{2P}\, d(t)\left\{\frac{1}{2} - \frac{1}{2}\sum_i p_{T_c}(t - iT_c)\cos\left[\frac{2\pi}{T_c}(t - iT_c)\right]\right.$$

$$\left. + \frac{1}{2} - \frac{1}{2}\sum_k p_{T_c}\!\left(t - kT_c - \frac{T_c}{2}\right)\cos\left[\frac{2\pi}{T_c}\!\left(t - kT_c - \frac{T_c}{2}\right)\right]\right\}$$

$$= \sqrt{2P}\, d(t)\left\{1.0 - \frac{1}{2}\cos\left(\frac{2\pi}{T_c}t\right) - \frac{1}{2}\cos\left[\frac{2\pi}{T_c}\!\left(t - \frac{T_c}{2}\right)\right]\right\}$$

$$= \sqrt{2P}\, d(t) \qquad (11\text{-}96)$$

As expected, the despreading operation has completely eliminated the MSK modulation from the data signal. The normalized integrator output due to signal is the usual ± 1.

The integrator input due to jamming is calculated from (11-64) and (11-57). Taking into account that the two spreading waveforms are offset in phase,

$$x_2(t) = \sqrt{2J} \cos (\Delta \omega \, t - \varphi_J + \theta_s) \sum_i p_{T_c}(t - iT_c)c_{1i} \sin \left[\frac{\pi}{T_c} (t - iT_c) \right]$$

$$+ \sqrt{2J} \sin (\Delta \omega \, t - \varphi_J + \theta_s) \sum_j p_{T_c}\left(t - jT_c - \frac{T_c}{2} \right)c_{2j}$$

$$\sin \left[\frac{\pi}{T_c} \left(t - jT_c - \frac{T_c}{2} \right) \right] \qquad (11\text{-}97)$$

The normalized integrator output is

$$y_2(nT_s) = \sqrt{\frac{J}{P}} \frac{1}{T_s} \sum_{i=0}^{N-1} \int_{iT_c}^{(i+1)T_c} c_{1i} \sin \left[\frac{\pi}{T_c} (t - iT_c) \right] \cos (\Delta \omega \, t - \varphi_J + \theta_s) \, dt$$

$$+ \sqrt{\frac{J}{P}} \frac{1}{T_s} \sum_{j=0}^{N-1} \int_{jT_c+T_c/2}^{(j+1)T_c+T_c/2} c_{2j} \sin \left[\frac{\pi}{T_c} \left(t - jT_c - \frac{T_c}{2} \right) \right]$$

$$\times \sin (\Delta \omega \, t - \varphi_J + \theta_s) \, dt \qquad (11\text{-}98)$$

Consider the first sum of integrals above. For convenience let $\varphi = -\varphi_J + \theta_s$, and expand the trigonometric product to obtain

$$y_{2a}(nT_s) = \sqrt{\frac{J}{P}} \frac{1}{2T_s} \sum_{i=0}^{N-1} c_{1i} \int_{iT_c}^{(i+1)T_c} \left\{ \sin \left[\left(\frac{\pi}{T_c} - \Delta \omega \right) t - i\pi - \varphi \right] \right\}$$

$$+ \left\{ \sin \left[\left(\frac{\pi}{T_c} + \Delta \omega \right) t - i\pi + \varphi \right] \right\} dt \qquad (11\text{-}99)$$

Straightforward evaluation of the integrals and trigonometric manipulations result in

$$y_{2a}(nT_s) = \sqrt{\frac{J}{P}} \frac{1}{T_s} \frac{\pi/T_c}{(\pi/T_c)^2 - (\Delta \omega)^2}$$

$$\times \sum_{i=0}^{N-1} c_{1i}\{\cos [(i + 1)\Delta \omega \, T_c + \varphi] + \cos [i\Delta \omega \, T_c + \varphi]\}$$

$$= \sqrt{\frac{J}{P}} \frac{1}{T_s} \frac{\pi/T_c}{(\pi/T_c)^2 - (\Delta \omega)^2} \sum_{i=0}^{N-1} c_{1i} \cos \left(\frac{\Delta \omega \, T_c}{2} \right)$$

$$\times 2 \cos \left(\Delta \omega \, T_c i + \varphi + \frac{\Delta \omega \, T_c}{2} \right) \qquad (11\text{-}100)$$

The value of $y_{2a}(nT_s)$ is a random variable which depends on the particular sequence of spreading code symbols associated with the data being demodulated. For $\Delta \omega = 0$ this random variable has a binomial distribution which may be accurately approximated by a continuous Gaussian distribution for large N. With this as justification, assume that the random variable $y_{2a}(nT_s)$ may also be approximated by a Gaussian random variable. The mean of $y_{2a}(nT_s)$ is zero and the variance will now be calculated. By definition

$$\sigma_{y_{2a}}^2 = E[y_{2a}^2(nT_s)]$$

$$= \frac{J}{PT_s^2}\left[\frac{\pi/T_c}{(\pi/T_c)^2 - (\Delta\omega)^2}\right]^2 \sum_{i=0}^{N-1}\sum_{k=0}^{N-1} E[c_{1i}c_{1k}]$$

$$\times\ 4\cos^2\left(\frac{\Delta\omega\, T_c}{2}\right)\cos\left[\Delta\omega\, T_c i + \varphi + \frac{\Delta\omega\, T_c}{2}\right]$$

$$\cos\left(\Delta\omega\, T_c k + \varphi + \frac{\Delta\omega\, T_c}{2}\right) \qquad (11\text{-}101)$$

If the spreading code symbols are independent and equally likely, this expression simplifies to

$$\sigma_{y_{2a}}^2 = \frac{4J}{PT_s^2}\left[\frac{\pi/T_c}{(\pi/T_c)^2 - (\Delta\omega)^2}\right]^2 \sum_{i=0}^{N-1}\cos^2\left(\frac{\Delta\omega\, T_c}{2}\right)$$

$$\cos^2\left(\Delta\omega\, T_c i + \varphi + \frac{\Delta\omega\, T_c}{2}\right)$$

$$= \frac{4J}{PT_s^2}\left[\frac{\pi/T_c}{(\pi/T_c)^2 - (\Delta\omega)^2}\right]^2 \cos^2\left(\frac{\Delta\omega\, T_c}{2}\right)$$

$$\times \sum_{i=0}^{N-1}[\tfrac{1}{2} + \tfrac{1}{2}\cos(2\Delta\omega\, T_c i + 2\varphi + \Delta\omega T_c)] \qquad (11\text{-}102)$$

Using (11-79), this equation becomes

$$\sigma_{y_{2a}}^2 = \frac{4J}{PT_s^2}\left[\frac{\pi/T_c}{(\pi/T_c)^2 - (\Delta\omega)^2}\right]^2 \cos^2\left(\frac{\Delta\omega\, T_c}{2}\right)$$

$$\times\left[\frac{N}{2} + \frac{\cos(2\varphi + N\,\Delta\omega\, T_c)\sin(N\,\Delta\omega\, T_c)}{2\sin(\Delta\omega\, T_c)}\right]$$

$$= \left(\frac{J}{4PN}\right)\left(\frac{8}{\pi^2}\right)\left[\frac{\cos(\Delta\omega\, T_c/2)}{1 - (\Delta\omega\, T_c/\pi)^2}\right]^2\left[1.0 + \frac{\cos(2\varphi + N\,\Delta\omega\, T_c)\sin(N\,\Delta\omega\, T_c)}{N\sin(\Delta\omega\, T_c)}\right]$$

$$(11\text{-}103)$$

An identical derivation yields the variance $\sigma_{y_{2b}}^2$ of the second integral of Equation (11-98). The result is

$$\sigma_{y_{2b}}^2 = \left(\frac{J}{4PN}\right)\left(\frac{8}{\pi^2}\right)\left[\frac{\cos(\Delta\omega\, T_c/2)}{1 - (\Delta\omega\, T_c/\pi)^2}\right]^2$$

$$\left[1.0 - \frac{\cos(2\varphi + N\,\Delta\omega\, T_c)\sin(N\,\Delta\omega\, T_c)}{N\sin(\Delta\omega\, T_c)}\right] \qquad (11\text{-}104)$$

Because the two spreading codes are independent, the sum of the two integrals may be assumed to be another Gaussian random variable with variance $\sigma_{y_2}^2 = \sigma_{y_{2a}}^2 + \sigma_{y_{2b}}^2$. Using the equations above,

$$\sigma_{y_2}^2 = \left(\frac{J}{2PN}\right)\left(\frac{8}{\pi^2}\right)\left[\frac{\cos(\Delta\omega\, T_c/2)}{1 - (\Delta\omega\, T_c/\pi)^2}\right]^2 \qquad (11\text{-}105)$$

The thermal noise component of the integrator output is also a Gaussian random variable with variance $\sigma_{y_3}^2 = N_0/2PT_s$. Bit errors are made when a -1 is transmitted and $y_2 + y_3 > 1.0$ or when a $+1$ is transmitted and $y_2 + y_3 < -1.0$. These error events have the same probabilities of occurrence and the total bit error probability for this system is

$$
\begin{aligned}
P_b &= \Pr\left[y_2(nT_s) + y_3(nT_s) > 1.0\right] \\[2mm]
&= Q\left(\sqrt{\frac{1}{\sigma_{y_2}^2 + \sigma_{y_3}^2}}\right) \\[2mm]
&= Q\left\{\sqrt{\dfrac{1.0}{\left(\dfrac{J}{2PN}\right)\left(\dfrac{8}{\pi^2}\right)\left[\dfrac{\cos\left(\frac{1}{2}\Delta\omega\, T_c\right)}{1 - (\Delta\omega\, T_c/\pi)^2}\right]^2 + \dfrac{N_0}{2PT_s}}}\right\}
\end{aligned}
\qquad (11\text{-}106)
$$

This expression should be compared to (11-95) for QPSK/BPSK. For $\Delta\omega = 0$ it appears that the MSK/BPSK jammer power is multiplied by $8/\pi^2$. This translates into a 0.91-dB advantage for MSK spreading. Note, however, that the null-to-null transmission bandwidth for MSK spreading is $3.0/T_c$, whereas the same bandwidth for QPSK spreading is $2.0/T_c$. If the transmission bandwidths of the two systems are made equal, the MSK system performs slightly worse than the QPSK system with this jammer.

11-3.5 Performance in Multiple-Tone Jamming

The multiple-tone jammer is the tone equivalent of the partial band noise jammer and is most effective against FH systems. An estimate of the performance of DS-FH systems may be obtained by considering the output of the DS despreader due to jamming to be Gaussian noise [13]. With this assumption, the DS-FH performance in multiple-tone jamming is the same as the performance in partial band noise jamming. Only FH systems will be considered in the remainder of this section. The total jamming power is denoted by J, and this power is equally divided between q jamming tones. The jammer has complete knowledge of the transmitted signal structure as well as the exact received signal and jammer power. The jammer does not, however, know the FH pattern. The task of the jammer is to choose q and the tone spacing such that the received bit error probability is maximized.

FH/MFSK. A number of analyses of multiple-tone jamming for FH/MFSK have appeared in the literature [13, 21–25]. The most complicated of these analyses considers MFSK with nonorthogonal frequency spacing and jammer tones which are frequency locked to the communication system MFSK tones. Another analysis presumes that more than one tone can be placed in a single MFSK bandwidth, and the simplest permits only a single tone in each MFSK bandwidth.

First, consider a system that employs orthogonal MFSK with the receiver of Figures 11-6 and 11-7. The jammer has complete knowledge of the signal structure and can therefore select jamming frequencies so that no more than one tone appears in each FH bandwidth. Slow frequency hopping is assumed so that a quasi-static analysis can be used. Orthogonal signaling is achieved using a MFSK tone spacing which is equal to the MFSK symbol rate $R_s = R/\log_2 M$ where R is the system bit rate. There are M tones spaced by R_s hertz, so that the bandwidth for each frequency hop is $W_d = MR/\log_2 M$. The jammer transmits q tones each having power $J_q = J/q$ and spaced in frequency by W_d hertz.

For simplicity, suppose that thermal noise is negligible. Then no symbol errors will be made if $J_q < P$ since even when a jamming tone is within the FH band, the energy detector will still identify the correct signaling tone. When $J_q > P$, a symbol error will be made if the jamming tone is within the FH band but not in the same energy detector filter bandwidth as the desired signaling tone. When $J_1 = P$, a symbol error occurs with probability 0.5 under the conditions described for $J_q > P$. The jammer can therefore jam the most FH bands if $J_q = P + \epsilon$, where $\epsilon \ll P$, and the number of jammer tones is approximately given by

$$q \approx \left\lfloor \frac{J}{P} \right\rfloor \tag{11-107}$$

where $\lfloor \cdot \rfloor$ denotes the largest integer less than or equal to J/P. Of course, there must be at least one jammer tone so that $q_{\min} = 1.0$, and $q_{\max}$ is limited by the total number of FH bands or $q_{\max} = W/W_d$. When $\lfloor J/P \rfloor \leq 1.0$, no errors will be made since there is insufficient jammer power to jam even a single FH band. Equation (11-107) is approximate only because ϵ has been set to zero. This value of q is optimum since a larger q results in $J_q < P$ and no errors, and since a smaller q results in fewer FH bands being jammed but no larger error probability for any one band.

The total transmission band W contains W/W_d FH bands, so that the probability that any one FH band is jammed is

$$p = \frac{q}{W/W_d} \tag{11-108}$$

When a FH band is jammed and q is chosen using (11-107), the symbol error probability is the probability that the jammer and the signal are not in the same band. This probability is $(M - 1)/M$ and the total symbol error probability P_s is

$$P_s = \begin{cases} \dfrac{M-1}{M} & \dfrac{W}{W_d} < \left\lfloor \dfrac{J}{P} \right\rfloor \\[2em] \dfrac{M-1}{M}\dfrac{qW_d}{W} & 1.0 \leq \left\lfloor \dfrac{J}{P} \right\rfloor \leq \dfrac{W}{W_d} \\[2em] 0.0 & \left\lfloor \dfrac{J}{P} \right\rfloor < 1.0 \end{cases} \tag{11-109}$$

The symbol error probability is related to the bit error probability by $P_b = MP_s/2(M - 1)$ for orthogonal signals [see (11-18)]. Substituting $q \approx J/P'$ and $W_d = MR/\log_2 M$ into (11-109) results in the following expression for bit error probability for worst-case (optimum q) tone jamming:

$$P_b = \begin{cases} 0.5 & \dfrac{P}{J}\left(\dfrac{W}{R}\right) < \dfrac{M}{\log_2 M} \\[2em] \dfrac{M}{2\log_2 M}\dfrac{1}{(P/J)(W/R)} & \dfrac{M}{\log_2 M} \leq \dfrac{P}{J}\left(\dfrac{W}{R}\right) \leq \dfrac{W}{R} \\[2em] 0.0 & \dfrac{W}{R} < \dfrac{P}{J}\left(\dfrac{W}{R}\right) \end{cases} \tag{11-110}$$

This relationship is plotted in Figure 11-24 for $M = 2$, 4, and 8 and $W/R = 1000$. Observe that the usual inverse linear relationship is evident in the mid-portion of these curves. Equation (11-110) should be compared to (11-38) for partial band noise jamming. In the inverse linear region, it is seen that the same bit error probability is achieved for values of $(P/J)(W/R)$ having a ratio

$$\frac{[(P/J)(W/R)]_{noise}}{[(P/J)(W/R)]_{tone}} = \frac{k'}{M/(2 \log_2 M)} \tag{11-111}$$

with k' given in Table 11-1. Thus tone jamming is 4.3, 6.3, 8.3, and 10.5 dB more efficient than noise jamming for $M = 2$, 4, 8, and 16, respectively.

A second type of tone jamming permits the jammer to place more than one tone in each FH band. The jammer will not use this strategy if the precise FH bands are known. Knowledge of the precise FH bands can be denied to the jammer by hopping each symbol frequency independently so that the best that the jammer can do is to jam symbol frequencies randomly across the entire band. The performance of the FH/MFSK system for this case was calculated in Houston [13] and corrected by Levitt [26]. In these references, it is shown that (11-107) is not necessarily optimum because additional error mechanisms are possible. For example, both the signaling tone and one other MFSK tone may be jammed simultaneously. In this case the relative phase of the jammer and signaling tone is a factor and errors may occur even with $J_q < P$. It is concluded in Levitt [26] that the worst-case system bit error probability for this jammer is identical to that specified by (11-110) for large $(P/J)(W/R)$.

FH/DPSK. The analysis of binary and quaternary DPSK with frequency hopping and multitone jamming was first published in Houston [13]. This early work

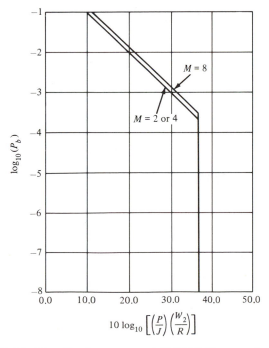

FIGURE 11-24. Performance of FH/MFSK spread spectrum with worst-case multiple-tone jamming.

was generalized in Simon [27] to M-ary DPSK with FH. Only the binary case is considered here. Slow frequency hopping is assumed so that a quasi-static analysis may be used. The total jammer power, J, is partitioned into q tones each with power $J_q = J/q$. The signal power is P and each FH band has bandwidth R where R is the transmitted symbol rate. The transmission bandwidth is W and the total number of FH bands is W/R. The probability that a particular FH band is jammed is

$$\rho = \frac{q}{W/R} \qquad (11\text{-}112)$$

Assume now that thermal noise is negligible, and consider a single FH band which is jammed. The DPSK demodulator functions by comparing the phase of two successive received symbols. Denote this phase difference by α. When $-\pi/2 < \alpha \leq \pi/2$, the receiver estimates that a one has been transmitted, and when $\pi/2 < \alpha \leq 3\pi/2$, the receiver estimates that a zero has been transmitted. The received signal is the phasor sum of the desired tone and the jamming tone as illustrated in Figure 11-25. In Figure 11-25a a one has been transmitted and the signal does not change phase between the two symbol times being compared. Thus the phasor sum $R_1 \underline{/\theta_1}$ during the first interval is identical to the phasor sum $R_2 \underline{/\theta_2}$ during the second interval. The receiver calculates $\alpha = \theta_2 - \theta_1 = 0$ and correctly estimates that a one has been transmitted regardless of the magnitude of the jammer.

In Figure 11-25b a zero has been transmitted and the desired signal changes phase by π radians between the two signaling intervals being compared. In this case, $R_1 \underline{/\theta_1}$ is significantly different from $R_2 \underline{/\theta_2}$ as shown. Using the law of cosines on the triangle OAB of the figure yields

$$\cos \alpha = \frac{R_1^2 + R_2^2 - (2\sqrt{2P})^2}{2R_1 R_2} \qquad (11\text{-}113a)$$

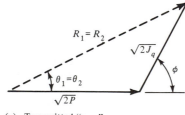

(a) Transmitted "one"

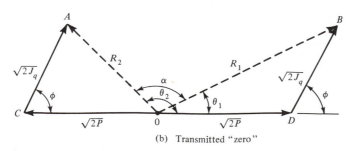

(b) Transmitted "zero"

FIGURE 11-25. Phasor diagrams showing effect of tone jammer on binary DPSK receiver. [13]

Similarly, for triangles OCA and OBD, respectively,

$$\cos \varphi = \frac{2P + 2J_q - R_2^2}{2\sqrt{4J_q P}} \qquad (11\text{-}113\text{b})$$

$$\cos (\pi - \varphi) = -\cos \varphi = \frac{2P + 2J_q - R_1^2}{2\sqrt{4J_q P}} \qquad (11\text{-}113\text{c})$$

Solving (11-113b) and (11-113c) for $R_1^2 + R_2^2$ and substituting into (11-113a) results in

$$\cos \alpha = \frac{2(J_q - P)}{R_1 R_2} \qquad (11\text{-}114)$$

When $\pi/2 < \alpha \leq 3\pi/2$, the demodulator produces the correct output, but when $-\pi/2 < \alpha \leq \pi/2$, an error is made. Equivalently, an error is made whenever $\cos \alpha \geq 0$, which occurs whenever $J_q \geq P$. No error occurs when $J_q < P$.

The optimum jamming strategy is to place just sufficient power in each tone so that an error will always be made when a zero is transmitted using a hop frequency which is jammed. The optimum jamming power is $J_q = P$. The jammer cannot produce any errors when ones are transmitted. Using this strategy, the optimum number of tones is

$$q = \left\lfloor \frac{J}{P} \right\rfloor \qquad (11\text{-}115)$$

The total bit error probability is

$$P_b = \tfrac{1}{2} \Pr (\text{error}|\text{zero transmitted})$$

$$+ \tfrac{1}{2} \Pr (\text{error}|\text{one transmitted})$$

$$= \tfrac{1}{2} \Pr (\text{error}|\text{zero transmitted}) \qquad (11\text{-}116)$$

The conditional error probability given that a zero was transmitted is simply the probability ρ that the hop frequency is jammed. Combining (11-112), (11-115), and (11-116) yields

$$P_b \simeq \frac{1}{2} \frac{J/P}{W/R} = \frac{1}{2(P/J)(W/R)} \qquad (11\text{-}117)$$

The usual limits apply to q, that is, $1 \leq q \leq W/R$. For large jammer power, all FH bands can be jammed and $P_b = \tfrac{1}{2}$. When $J < P$, not even a single FH band can be adequately jammed to create errors and $P_b = 0$. Thus

$$P_b = \begin{cases} 0.5 & \dfrac{P}{J}\left(\dfrac{W}{R}\right) < 1 \\[2ex] \dfrac{1}{2(P/J)(W/R)} & 1 \leq \dfrac{P}{J}\left(\dfrac{W}{R}\right) < \dfrac{W}{R} \\[2ex] 0 & \dfrac{W}{R} < \dfrac{P}{J}\left(\dfrac{W}{R}\right) \end{cases} \qquad (11\text{-}118)$$

Once again, the inverse linear relationship is evident in the midrange. Equation (11-118) is plotted in Figure 11-26. Observe that binary FH/DPSK outperforms binary FH/MFSK by exactly 3 dB in the inverse linear region.

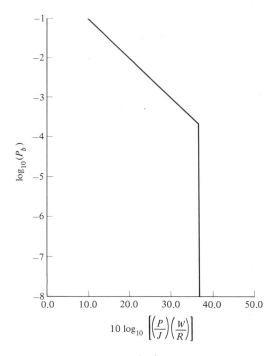

FIGURE 11-26. Performance of FH/DPSK spread spectrum in worst-case multiple-tone jamming.

11-3.6 Conclusions

It has been demonstrated above that a proper choice of jamming strategy can produce large performance degradations in all the modulation combinations considered. In all instances, the worst-case relationship between bit error probability and $(P/J)(W/R)$ is inverse linear. Values of $(P/J)(W/R)$ in the range $+30$ to $+50$ dB are required to achieve reasonable performance. These large values of $(P/J)(W/R)$ in no way imply that the benefits of spectrum spreading have been nullified by the smart jammer. The same inverse linear relationships are obtained for the nonspread system without the benefit of the W/R factor. Thus, when comparing the spread and nonspread systems with both optimally jammed, the spread system has a significant power advantage. Meaningful calculations of system processing gains must always include worst-case jamming for both the spread and nonspread systems.

REFERENCES

[1] G. LONGO, ed., *Multi-user Communication Systems* (New York: Springer-Verlag, 1981).

[2] J. K. SKWIRZYNSKI, ed., *New Concepts in Multi-user Communications*, (Alphen aan den Rijn, The Netherlands: Sijtoff en Noordhoff, 1981).

[3] M. B. PURSLEY, D. SARWATE, and W. STARK, ''Error Probability for Direct-Sequence Spread-Spectrum Multiple-Access Communications, Part I,'' *IEEE Trans. Commun.*, May 1982.

[4] E. A. GERANIOTIS and M. B. PURSLEY, "Error Probability for Direct Sequence Spread-Spectrum Multiple-Access Communications, Part II," *IEEE Trans. Commun.*, May 1982.

[5] P. HENRY, "Spectrum Efficiency of a Frequency-Hopped-DPSK Mobile Radio System," 29th Vehicular Technol. Conf., March 1979.

[6] A. J. VITERBI, "A Processing Satellite Transponder for Multiple Access by Low Rate Mobile Users," *Conf. Rec.*, Int. Conf. Digit. Satellite Commun., Montreal, October 1978.

[7] D. J. GOODMAN, P. S. HENRY, and V. K. PRABHU, "Frequency Hopped Multilevel FSK for Mobile Radio," *Bell Sys. Tech. J.*, September 1980.

[8] G. R. COOPER and R. W. NETTLETON, "A Spread-Spectrum Technique for High-Capacity Mobile Communications," *IEEE Trans. Veh. Technol.*, November 1978.

[9] E. ARTHURS and H. DYM, "On the Optimum Detection of Digital Signals in the Presence of White Gaussian Noise—A Geometric Interpretation and a Study of Three Basic Data Transmission Systems," *IRE Trans. Commun. Syst.*, December 1962.

[10] J. H. PARK, "On Binary DPSK Detection," *IEEE Trans. Commun.*, April 1978.

[11] T. W. MILLER, "Imperfect Differential Detection of a Biphase Modulated Signal—An Experimental and Analytical Study," Ohio State University Research Foundation, February 1972. (Available through DTIC: AD 740617.)

[12] W. C. LINDSEY and M. K. SIMON, *Telecommunication Systems Engineering* (Englewood Cliffs, N.J.: Prentice-Hall, 1973).

[13] S. W. HOUSTON, "Modulation Techniques for Communication, Part 1: Tone and Noise Jamming Performance of Spread Spectrum M-ary FSK and 2, 4-ary DPSK Waveforms," *Conf. Rec.*, NAECON, 1975.

[14] A. J. VITERBI, "Spread Spectrum Communications—Myths and Realities," *IEEE Commun. Mag.*, May 1979.

[15] B. K. LEVITT, "Effect of Modulation Format and Jamming Spectrum on Performance of Direct Sequence Spread Spectrum Systems," *Conf. Rec.*, IEEE Natl. Telecommun. Conf., 1980.

[16] R. SINGH, "Performance of a Direct Sequence Spread Spectrum System with Long Period and Short Period Code Sequences," *Conf. Rec.*, Int. Conf. Commun., 1981.

[17] D. L. SCHILLING, et al., "Optimization of the Processing Gain of an M-ary Direct Sequence Spread Spectrum Communication System," *IEEE Trans. Commun.*, August 1980.

[18] L. COUCH, "Performance of DS Spread Spectrum Systems," *Proc. IEEE*, February 1980.

[19] A. PAPOULIS, *Probability, Random Variables and Stochastic Processes* (New York: McGraw-Hill, 1965).

[20] I. S. GRADSHTEYN and I. W. RYZHIK, *Tables of Integrals Series and Products* (New York: Academic Press, 1965).

[21] D. J. TORRIERI, "Frequency Hopping, Multiple-Frequency Shift Keying, Coding, and Optimal Partial-Band Jamming," Naval Air Systems Command, CM/CCM Center, Tech. Rep. CM/CCM-82-1, August 1982 (AD A118585).

[22] M. P. RISTENBATT and J. L. DAVIS, JR., "Performance Criteria for Spread Spectrum Communications," *IEEE Trans. Commun.*, August 1977.

[23] L. B. MILSTEIN, R. L. PICKHOLTZ, and D. S. SCHILLING, "Optimization of the Processing Gain of an FSK-FH System," *IEEE Trans. Commun.*, July 1980.

[24] H. R. PETTIT, *ECM and ECCM Techniques for Digital Commmunication Systems*, (Belmont, Calif.: Lifetime Learning Publications, 1982).

[25] A. J. VITERBI, "A Robust Ratio-Threshold Technique to Mitigate Tone and Partial Band Jamming in Coded Frequency Hopped Communication Links," M/A-COM Linkabit, Inc., Final Report on Contract N00019-81-C-0451, September 1982 (AD A 123559).

[26] B. K. LEVITT, "Use of Diversity to Improve FH/MFSK Performance in Worst Case Partial Band Noise and Multitone Jamming," *Conf. Rec.*, IEEE Milit. Commun. Conf., 1982.

[27] M. K. SIMON, "The Performance of M-ary FH-DPSK in the Presence of Partial Band Multitone Jamming," *IEEE Trans. Commun.*, May 1982.

PROBLEMS

11-1. A communicator uses BPSK direct-sequence spreading modulation and BPSK data modulation. The DS chip rate is 10 Mchips/s, and the data rate is 75 bits per second. Calculate the ratio of barrage noise average jamming power to pulse noise average jamming power assuming that both jammers increase the communicator's average bit error probability to $P_b = 10^{-2}$. Calculate the same ratio for $P_b = 10^{-5}$.

11-2. A jammer is to be designed as an ECM against the communicator of Problem 11-1. The jammer will estimate the communicator's carrier frequency with an accuracy of ± 5 MHz. The jammer must increase the communicator's error probability to $P_b = 10^{-3}$ to be effective. What is a good jamming strategy?

11-3. A communicator uses FH/DPSK modulation to transmit a digital voice signal using a data rate of 32 kbps. The maximum acceptable transmission bit error probability is $P_b = 10^{-3}$. A smart partial band noise jammer has a power advantage of 30 dB over the communicator. What spread-spectrum bandwidth must be used to obtain the required bit error probability? How many different frequencies must the FH synthesizer generate?

11-4. A communicator uses FH/DPSK modulation. The rate of information transmission is 1 Mbps and the total spread transmission bandwidth is 20 MHz. Is a pulse noise jammer more effective or less effective than a partial band noise jammer against this system? Why?

11-5. Repeat Problem 11-4 for BPSK/BPSK modulation.

11-6. Consider the direct sequence spread spectrum receiver illustrated below. The receiver input is a BPSK/BPSK signal using chip rate $R_c = 100$ Mchips/s and data rate $R = 1$ Mbps, plus a tone jammer at frequency $f_J = f_0 + \Delta f$. What are reasonable choices for the image reject and IF bandpass filter bandwidths and center frequencies? Assume a signal-to-jammer power ratio of $10 \log (P/J) = -10$ dB. What is the received bit error probability for $\Delta f = 1.5, 5, 10,$ and 50 MHz?

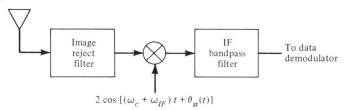

PROBLEM 11-6. Direct sequence spread spectrum receiver.

11-7. Consider the frequency-hop spread-spectrum receiver shown below. Binary FSK data modulation is used with tone spacing of 100 kHz. The data rate is 100 kbps. The FH tone

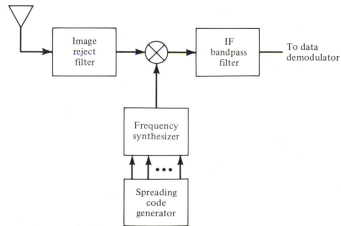

PROBLEM 11.7 Frequency-hop spread-spectrum receiver.

spacing is 200 kHz and there are a total of 1024 tones. Select reasonable image reject and IF bandpass filter bandwidths. The received signal is jammed by a multiple-tone jammer, and the total received signal-to-jammer power ratio is $10 \log (P/J) = -10$ dB. What is the optimal number of jamming tones? What is the optimal jammer tone spacing and tone power level? Select a frequency-hop rate. What is the received bit error probability?

11-8. Consider the FH communication system described in Problem 11-7.
(a) Calculate the bit error probability assuming the signal is jammed by a single-tone jammer with $10 \log (P/J) = -10$ db.
(b) Calculate the bit error probability assuming that the signal is jammed by a barrage noise jammer with $10 \log (P/J) = -10$ dB.

11-9. Consider a BPSK/BPSK spread-spectrum system which is jammed by a single-tone jammer whose frequency is identical to the BPSK/BPSK carrier frequency but which is not phase coherent. Suppose that the spreading code is a maximal-length sequence whose period is exactly equal to the information bit period. Calculate the bit error probability as a function of (P/J) and the jammer phase error θ_J [16].

11-10. Consider a BPSK/BPSK spread-spectrum system jammed by a single-tone jammer. Assume there is no thermal noise interference and calculate a lower bound on $10 \log (P/J)$ above which the demodulator makes no errors. *Hint:* Do not use the Gaussian approximation for the integrator output statistics.

11-11. A communicator uses FH/MFSK as an ECCM against a pulsed noise jammer. The MFSK modulator transmits 4-ary symbols. Assume that the received signal-to-thermal noise ratio is $10 \log (E_b/N_0) = +15$ dB and the received signal-to-jammer noise ratio is $10 \log (E_b/N_J) = +3$ dB. Calculate the optimum jammer duty factor and plot the received bit error probability as a function of time. Draw and label the discrete memoryless channel transition diagram representing this channel.

11-12. A communicator has a transmitter power amplifier whose power output is 10 W. A jammer has a power amplifier whose average power output is 100 W and which can be pulsed using duty factors in the range $0.01 \le \rho \le 1.0$. The bandwidth of these two power amplifiers is the same. Propose a spectrum spreading strategy which does not use error correction coding and which will enable the communicator to achieve a bit error probability of 10^{-5}.

11-13. Consider a BPSK/BPSK coherent spread-spectrum system operating in a pulsed noise jamming environment. Assume that the information bit rate is 1 kbps and that the direct-sequence chip rate is 10 Mchips/s, and plot the worst-case system bit error probability performance as a function of $10 \log (P/J)$. Calculate and plot the error probability performance for non-spread-spectrum BPSK communications system operating in the same environment. What is the spread-spectrum processing gain?

CHAPTER 12

Performance of Spread-Spectrum Systems with Forward Error Correction

12-1

INTRODUCTION

Spectrum spreading by itself produces large communication system performance improvements by effectively spreading the jammer power over the full spread communications bandwidth. The resultant performance was analyzed in Chapter 11. The effect of worst-case jamming can be further mitigated using one or more of the powerful forward error correction (FEC) techniques which have been developed following Shannon's pioneering work. Error correction coding is an extremely complex topic to which entire texts are dedicated. In this text a small number of the most basic concepts of FEC are discussed in order to give the student a preliminary idea of the power of these techniques.

The most effective jamming strategies are those which concentrate jamming resources on some fraction of the transmitted symbols using either pulsed or partial band techniques. A result of these jamming strategies is that demodulator output errors occur in bursts. Because the FEC techniques to be discussed perform best when channel errors are independent from one signaling interval to the next, interleaving is assumed for all FEC schemes. The purpose of the interleaver is to rearrange the order in which coder output symbols are transmitted so that bursts of transmission errors will not appear as a burst at the decoder input. The system

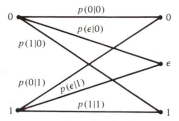

FIGURE 12-1. Transition diagram for soft decision binary erasure channel.

model for the coded spread-spectrum system is the same model illustrated in Figure 11-1 and discussed in Section 11-2. This model includes the interleaver and deinterleaver within the discrete memoryless channel (DMC). It is the interleaver that makes the channel memoryless.

The performance of the FEC schemes in this chapter is calculated using the DMC transition probabilities $p(i|j)$. These transition probabilities are calculated using the basic techniques discussed in Chapter 11. These techniques must be slightly extended, however, to include instances where the DMC input and output alphabets are not the same. Additional information can be given to the decoder by permitting the DMC input and output alphabets to be different. For example, in addition to outputting a binary 1 or 0, the DMC may also output information indicating the reliability of the 1/0 decision. When additional information is output, the channel is called a soft decision channel. An example of the transition diagram for a soft decision channel is shown in Figure 12-1, where a binary input channel has three outputs: (1) one, (2) zero, or (3) erasure. An erasure means that the demodulator is not sure what was transmitted and chooses not to guess. Decoders using reliability information typically require 1 to 2 dB less transmitter power than decoders not using this information. The limiting case of a soft decision channel results from a demodulator whose output is a real number; in this case, the channel is specified by a conditional (on the input) probability density function.

In the remainder of this chapter, the fundamental concepts of both block and convolutional codes are discussed. The student is referred to Lin and Costello [1] or Clark and Cain [2] for detailed introductory discussions of FEC, and to Viterbi and Omura [3], Peterson and Weldon [4], Gallager [5], and Berlekamp [6] for advanced discussions of both information-theoretic and algebraic foundations of FEC.

12-2

ELEMENTARY BLOCK CODING CONCEPTS

Block codes can be either linear or nonlinear; discussion here will be limited to linear codes. Consider only encoders whose input and output is binary. A block encoder groups k input binary symbols into a word and outputs a n-bit binary codeword. The code rate is $R = k/n$. There are 2^k possible input words and each has a unique output codeword associated with it. Since there are 2^n possible output codewords and $k < n$, not all possible output codewords are used.

The error correction capability of any error correction code is due to the fact that not all possible encoder output n-tuples are used. Because of this, it is possible to generate codes with codewords selected so that a number of transmission errors must occur before one codeword will be confused with another. Codewords are represented by binary n-tuples $\mathbf{x}_m = (x_{m1}, x_{m2}, \ldots, x_{mn})$ where $m = 0, 1,$ $2, \ldots, 2^k - 1$ is the message associated with the codeword. Any two codewords $\mathbf{x}_m$ and $\mathbf{x}_{m'}$ which differ from one another in d_H places and agree in $n - d_H$ places are said to be separated by *Hamming distance d_H*. A total of d_H transmission errors must occur before $\mathbf{x}_m$ is changed into $\mathbf{x}_{m'}$. The minimum Hamming distance between any two of the 2^k codewords in a code is the *minimum distance $d_{\min}$* of the code. An (n, k) binary block code uses a fraction $2^k/2^n = 2^{k-n}$ of all possible output codewords. A low-rate code uses a smaller fraction of the possible output words than a high rate code and codewords can therefore be separated further from one another. Thus low rate codes typically have more error correction capability than high-rate codes.

An important property of linear block codes is that the symbol-by-symbol mod-ulo-2 sum of any two codewords is another codeword. This implies that the all zeros codeword is a codeword in all linear codes. Consider any two codewords $\mathbf{x}_a$ and $\mathbf{x}_b$ separated by Hamming distance d_H, and a third arbitrary codeword $\mathbf{x}_c$. Let $\mathbf{x}_{a'} = \mathbf{x}_a \oplus \mathbf{x}_c$ and $\mathbf{x}_{b'} = \mathbf{x}_b \oplus \mathbf{x}_c$, and notice that the Hamming distance between $\mathbf{x}_{a'}$ $\mathbf{x}_{b'}$ is also d_H. Using these two properties, it can be demonstrated that the set of Hamming distances between a codeword $\mathbf{x}_m$ and all other codewords $\mathbf{x}_{m'}$, $m' \neq m$, in the code is the same for all m. Denote the number of codewords which are Hamming distance k from the all zeros codeword by W_k. The *weight distribution* of the code is the set of all W_k for $k = d_{\min}, \ldots, n$.

Consider the binary block code defined in Table 12-1. This code uses $k = 4$, $n = 7$, and has rate $R = 4/7$ and minimum distance 3. There are a total of $2^k = 16$ codewords. The weight distribution of this code is $W_3 = 7$, $W_4 = 7$, and

TABLE 12-1. Typical Block Code

	Message	Codeword
0	0 0 0 0	0 0 0 0 0 0 0
1	1 0 0 0	1 1 0 1 0 0 0
2	0 1 0 0	0 1 1 0 1 0 0
3	1 1 0 0	1 0 1 1 1 0 0
4	0 0 1 0	1 1 1 0 0 1 0
5	1 0 1 0	0 0 1 1 0 1 0
6	0 1 1 0	1 0 0 0 1 1 0
7	1 1 1 0	0 1 0 1 1 1 0
8	0 0 0 1	1 0 1 0 0 0 1
9	1 0 0 1	0 1 1 1 0 0 1
10	0 1 0 1	1 1 0 0 1 0 1
11	1 1 0 1	0 0 0 1 1 0 1
12	0 0 1 1	0 1 0 0 0 1 1
13	1 0 1 1	1 0 0 1 0 1 1
14	0 1 1 1	0 0 1 0 1 1 1
15	1 1 1 1	1 1 1 1 1 1 1

$W_7 = 1$. Thus there are a total of seven codewords $\mathbf{x}_{m'}$ with distance 3, seven codewords $\mathbf{x}_{m'}$ with distance 4, and one codeword $\mathbf{x}_{m'}$ with distance 7 from any particular codeword $\mathbf{x}_m$. A codeword is transmitted one symbol at a time over the digital channel. Each symbol is corrupted by noise and/or jamming on the channel. In all cases, the channel can be described by the probability $p(\mathbf{y}|\mathbf{x}_m)$ of receiving a particular output n-tuple $\mathbf{y} = (y_1, y_2, \ldots, y_n)$ given that the input was a particular codeword n-tuple $\mathbf{x}_m$. The channel is assumed to be memoryless so that

$$p(\mathbf{y}|\mathbf{x}_m) = \prod_{i=1}^{n} p(y_i|x_{mi}) \tag{12-1}$$

12-2.1 Optimum Decoding Rule

Given a particular known block encoding procedure and a particular received n-tuple $\mathbf{y}$, the decoder must estimate the actual encoder input k-tuple. Denote the encoder input message (k-tuple) by m and the decoder estimate by $\hat{m}$. Since each message is uniquely associated with a codeword $\mathbf{x}_m$, the decoder may be viewed as estimating $\mathbf{x}_m$ by a codeword $\hat{\mathbf{x}}$. The decoding procedure is chosen to result in the minimum possible decoding error probability. For well-designed codes, minimization of the block decoding error probability P_B also minimizes the information bit decoding error probability P_b. It is conjectured that,* when a block decoding error occurs, on the average about one-half of the decoded information bits are incorrect so that $P_b \approx \frac{1}{2}P_B$. It can be shown that $P_b \leq P_B \leq kP_b$. The decoder will be designed to minimize P_B.

The decoder input is an n-tuple $\mathbf{y}$. The optimum decoder must use its knowledge of the codeword set, the message probabilities $\Pr(m)$, and the channel transition probabilities $p(\mathbf{y}|\mathbf{x}_m)$ to select which one of the channel input codewords was most likely to have been sent. That is, the decoder will select as its estimate $\hat{\mathbf{x}}$ of the transmitted codeword the codeword with the largest a posteriori probability $\Pr(\mathbf{x}_m|\mathbf{y})$. The conditional probability of a correct decision given that $\mathbf{y}$ has been received and the decoder estimate is $\hat{\mathbf{x}} = \mathbf{x}_m$ is

$$\Pr(\text{correct decision}|\mathbf{y}) = \Pr(\mathbf{x}_m|\mathbf{y}) \tag{12-2a}$$

The unconditional probability of a correct decision is the sum of (12-2a) over all possible received $\mathbf{y}$ or

$$\Pr(\text{correct decision}) = \sum_{\mathbf{y}} \Pr(\mathbf{x}_m|\mathbf{y}) \Pr(\mathbf{y}) \tag{12-2b}$$

Since $\Pr[\mathbf{y}] \geq 0$ for all $\mathbf{y}$, this summation is maximized by choosing $\hat{\mathbf{x}}$ so that $\Pr(\mathbf{x}_m|\mathbf{y})$ is maximized for each $\mathbf{y}$. Thus the minimum error probability decoding rule is: Choose $\hat{m} = m'$ if

$$\Pr(\mathbf{x}_{m'}|\mathbf{y}) \geq \Pr(\mathbf{x}_{m''}|\mathbf{y}) \tag{12-3}$$

for all $m'' \neq m'$.

The a posteriori probabilities are related to the a priori probabilities using Bayes' rule,

$$\Pr(\mathbf{x}_m|\mathbf{y}) \Pr(\mathbf{y}) = p(\mathbf{y}|\mathbf{x}_m) \Pr(m) \tag{12-4}$$

---

*It is assumed that the code is ''good'' in this sense. Counterexamples exist for which this conjecture is not true.

and the decoding rule may also be stated: Choose $\hat{m} = m'$ if

$$\frac{\text{Pr}(m')p(\mathbf{y}|\mathbf{x}_{m'})}{\text{Pr}(\mathbf{y})} \geq \frac{\text{Pr}(m'')p(\mathbf{y}|\mathbf{x}_{m''})}{\text{Pr}(\mathbf{y})} \tag{12-5}$$

for all $m'' \neq m'$. The denominator of both sides of (12-5) is independent of the message m' or m'' and may be ignored in the decoder decision. In all cases of practical interest, all messages are equally probable so that $\text{Pr}(m') = \text{Pr}(m'')$ and the decoding rule simplifies to: Choose $\hat{m} = m'$ if

$$p(\mathbf{y}|\mathbf{x}_{m'}) \geq p(\mathbf{y}|\mathbf{x}_{m''}) \tag{12-6}$$

for all $m'' \neq m'$.

Since the decision rule just derived does not take into account the possibility of different message probabilities, the rule is a maximum-likelihood rule. Consider now the special case where the channel is a hard decision binary symmetric channel (BSC) with crossover probability denoted by p. Both $\mathbf{y}$ and $\mathbf{x}_m$ are binary n-tuples. Suppose that $\mathbf{y}$ and $\mathbf{x}_m$ differ in d_H positions. When $\mathbf{x}_m$ is transmitted, $\mathbf{y}$ is received only if d_H specific transmission errors occur. Therefore,

$$p(\mathbf{y}|\mathbf{x}_m) = p^{d_H}(1-p)^{n-d_H} \tag{12-7}$$

For $p < 0.5$, $p(\mathbf{y}|\mathbf{x}_m)$ increases with decreasing d_H and is maximum for $d_H = 0$. Thus, for the BSC the maximum-likelihood decoding rule is simply to choose as the decoder estimate $\hat{\mathbf{x}}$ the codeword which is closest to the received n-tuple $\mathbf{y}$ in Hamming distance.

EXAMPLE 12-1
Consider the $n = 7$, $k = 4$ binary code of Table 12-1. Suppose that the demodulator output is $\mathbf{y} = (1, 0, 1, 1, 0, 1, 0)$. Using the decoding rule just described, the decoder calculates the Hamming distance between $\mathbf{y}$ and all possible codewords. These distances are compared and $\hat{\mathbf{x}}$ is the codeword with the minimum Hamming distance from $\mathbf{y}$. This minimum distance is 1 and $\hat{\mathbf{x}} = (0, 0, 1, 1, 0, 1, 0)$. ■

Consider next a soft decision memoryless AWGN channel where the channel input is ± 1 and the channel output is a real number with Gaussian statistics. Specifically, the channel is defined by

$$p(y_i|x_i) = \frac{1}{\sqrt{\pi N_0}} \exp\left[-\frac{(y_i - x_i)^2}{N_0}\right] \tag{12-8}$$

This model is based on the assumption that an ideal matched filter receiver is employed and that the channel noise is ideal AWGN. Since the channel is memoryless,

$$p(\mathbf{y}|\mathbf{x}_m) = \prod_{i=1}^{n} p(y_i|x_{mi}) \tag{12-9}$$

The decoder could implement the decoding rule of (12-6) directly using (12-9); however, it is more convenient to take the logarithm of both sides of (12-6), so that the product of (12-9) becomes a summation. Since the logarithm is a monotonic increasing function of its argument, decisions based on $\ln[p(y|x)]$ will be identical to decisions based on $p(y|x)$. Using (12-8) and (12-9), and taking the logarithm, (12-6) becomes

$$-\frac{n}{2} \ln (\pi N_0) - \frac{1}{N_0} \sum_{i=1}^{n} (y_i - x_{m'i})^2 \geq -\frac{n}{2} \ln (\pi N_0) - \frac{1}{N_0} \sum_{i=1}^{n} (y_i - x_{m''i})^2$$

$$(12\text{-}10a)$$

Canceling common terms and reversing the inequality to eliminate the minus signs, the decoding rule is: Choose $\hat{m} = m'$ if

$$\sum_{i=1}^{n} (y_i - x_{m'i})^2 \leq \sum_{i=1}^{n} (y_i - x_{m''i})^2 \qquad (12\text{-}10b)$$

for all $m'' \neq m'$. The summations above equal the square of the Euclidean distance between the received real n-tuple and the codeword real n-tuples, and the decoder chooses the output estimate $\hat{\mathbf{x}}$ to be the codeword closest to $\mathbf{y}$ in Euclidean distance.

EXAMPLE 12-2

Consider the code of Table 12-1 and suppose that message 0 is transmitted. For the hard decision BSC, the corresponding codeword is $\mathbf{x}_0 = (0, 0, 0, 0, 0, 0, 0)$ and, for the real vector AWGN channel, the corresponding codeword is represented by $\mathbf{x}_0 = (+1, +1, +1, +1, +1, +1, +1)$. Suppose further that the received real n-tuple (i.e., the sampled output of the matched filter) is $y = (-0.01, -0.01, 1.0, 1.0, 1.0, 1.0, 1.0)$. In the hard decision demodulator, a decision threshold is placed at zero volts and the hard decision output is $\mathbf{y} = (1, 1, 0, 0, 0, 0)$. This n-tuple is closest to code n-tuple $(1, 1, 0, 1, 0, 0, 0)$ in Hamming distance and the hard decision decoder outputs message 1 and makes an error. The soft decision decoder calculates the Euclidean distances shown in Table 12-2. The real vector $\mathbf{y}$

TABLE 12-2. Soft Decision Decoding Calculation for Example 12-2

Message	Channel Input Vector	Euclidean Distance Between Codeword and y = (-0.01, -0.01, 1, 1, 1, 1, 1)
0	(1, 1, 1, 1, 1, 1, 1)	2.04
1	(-1, -1, 1, -1, 1, 1, 1)	5.96
2	(1, -1, -1, 1, -1, 1, 1)	10.00
3	(-1, 1, -1, -1, -1, 1, 1)	14.00
4	(-1, -1, -1, 1, 1, -1, 1)	9.96
5	(1, 1, -1, -1, 1, -1, 1)	14.04
6	(-1, 1, 1, 1, -1, -1, 1)	10.00
7	(1, -1, 1, -1, -1, -1, 1)	14.00
8	(-1, 1, -1, 1, 1, 1, -1)	10.00
9	(1, -1, -1, -1, 1, 1, -1)	14.00
10	(-1, -1, 1, 1, -1, 1, -1)	9.96
11	(1, 1, 1, -1, -1, 1, 1)	10.04
12	(1, -1, 1, 1, 1, -1, -1)	10.00
13	(-1, 1, 1, -1, 1, -1, -1)	14.00
14	(1, 1, -1, 1, -1, -1, -1)	18.04
15	(-1, -1, -1, -1, -1, -1, -1)	21.96

is closest to real vector $(+1, +1, +1, +1, +1, +1, +1)$ in Euclidean distances and the soft decision decoder outputs message 0 and does not make an error. ∎

The purpose of Example 12-2 is to illustrate a case where a hard decision decoder makes an error but a soft decision decoder does not. Because the soft decision decoder is capable of making correct decisions in cases like this, the soft decision decoder always performs better than the hard decision decoder.

12-2.2 Calculation of Error Probability

The word error probability for any specific code is dependent on the details of the code as well as the decoding rule used. In the following, the maximum-likelihood or minimum distance decoding rule is assumed. It is convenient to describe the decoder as a partitioning of the space of all possible received n-tuples $\mathbf{y}$. This space has n dimensions and is partitioned by using the decoding rule on all points in the space, thereby associating a decoder output message with each point. The partitioning results in M subspaces Λ_m, where $m = 0, 1, \ldots, 2^k - 1$. When a decoder input $\mathbf{y}$ is within Λ_m, the decoder output is $\hat{m} = m$. A decoding error occurs whenever message m is transmitted and $\mathbf{y} \in \overline{\Lambda}_m$.

Given that message m was transmitted, the block error probability is given by Viterbi and Omura [3]

$$P_{Bm} = \text{Pr } (\mathbf{y} \in \overline{\Lambda}_m | \mathbf{x}_m)$$

$$= \sum_{\mathbf{y} \in \overline{\Lambda}_m} p(\mathbf{y}|\mathbf{x}_m) \tag{12-11}$$

where the sum is over all n-tuples $\mathbf{y}$ which are not within the decoding region for message m. The condition on message m is removed by averaging over all messages, yielding

$$P_B = \sum_{m=0}^{M-1} \text{Pr } (m)P_{Bm} \tag{12-12}$$

The error probability could be calculated directly using (12-11) and (12-12); however, this calculation is tedious except in trivial cases. The calculation is simplified by examining the region $\overline{\Lambda}_m$ in more detail. Assuming that all messages are equally likely, write [3]

$$\overline{\Lambda}_m = \{\mathbf{y} | \ln [p(\mathbf{y}|\mathbf{x}_{m'})] \geq \ln [p(\mathbf{y}|\mathbf{x}_m)] \text{ for some } m' \neq m\}$$

$$= \bigcup_{m' \neq m} \{\mathbf{y} | \ln [p(\mathbf{y}|\mathbf{x}_{m'})] \geq \ln [p(\mathbf{y}|\mathbf{x}_m)]\}$$

$$= \bigcup_{m' \neq m} \Lambda_{mm'} \tag{12-13}$$

where

$$\Lambda_{mm'} = \left\{ \mathbf{y} \middle| \ln \left[\frac{p(\mathbf{y}|\mathbf{x}_{m'})}{p(\mathbf{y}|\mathbf{x}_m)} \right] \geq 0 \right\} \tag{12-14}$$

With this definition, the error probability of (12-11) can be overbounded using the union bound. Recall that the union bound is a statement of the fact that the probability of a union of events is less than or equal to the sum of the probabilities of the component events. Thus

$$P_{Bm} = \text{Pr} \, (\mathbf{y} \in \overline{\Lambda}_m | \mathbf{x}_m)$$

$$= \text{Pr} \, (\mathbf{y} \in \bigcup_{m' \neq m} \Lambda_{mm'} | \mathbf{x}_m)$$

$$\leq \sum_{m' \neq m} \text{Pr} \, (\mathbf{y} \in \Lambda_{mm'} | \mathbf{x}_m)$$

$$= \sum_{m' \neq m} P_B(m, m') \tag{12-15}$$

where $P_B(m, m')$ is the block error probability under the assumption that there are only two codewords in the code. The last equality follows from the observation that (12-14) partitions the demodulator output space into two regions which are the decoding regions for a code with two codewords $\mathbf{x}_m$ and $\mathbf{x}_{m'}$. From this point two different methods may be used to proceed. Both methods calculate $P_B(m, m')$ for all codeword pairs of the code. One method calculates $P_b(m, m')$ exactly and the other uses a convenient overbound. Both methods are considered below.

Suppose that the channel is the BSC with crossover probability p. The BSC decoding rule is to choose as the decoder output estimate the codeword which is closest to the received n-tuple $\mathbf{y}$ in Hamming distance. Consider a code having only two codewords, $\mathbf{x}_1$ and $\mathbf{x}_2$, which differ in d positions. Suppose that $\mathbf{x}_1$ is transmitted. A decoding error occurs whenever enough transmission errors are made to cause $\mathbf{y}$ to be closer to $\mathbf{x}_2$ than to $\mathbf{x}_1$. Note that transmission errors in positions where the two codewords are identical increase the distance between $\mathbf{x}_1$ and $\mathbf{y}$ and the distance between $\mathbf{x}_2$ and $\mathbf{y}$ equally and therefore have no effect on the decoding result. Thus, for d even, a decoding error is made whenever $(d/2) + 1$ or more transmission errors occur in the positions where $\mathbf{x}_1$ and $\mathbf{x}_2$ differ. When only $(d/2)$ errors occur, $\mathbf{y}$ is equidistant from $\mathbf{x}_1$ and $\mathbf{x}_2$ and a decoding error is assumed to be made one-half of the time. Thus the two-codeword error probability for d even is

$$P_B(1, 2) = \sum_{e = (d/2)+1}^{d} \binom{d}{e} p^e (1 - p)^{d-e} \qquad d \text{ even}$$

$$+ \frac{1}{2} \binom{d}{d/2} p^{d/2} (1 - p)^{d/2} \tag{12-16a}$$

For d odd, a decoding error is made whenever $(d + 1)/2$ or more transmission errors occur in the k positions where $\mathbf{x}_1$ and $\mathbf{x}_2$ differ, and the two-codeword error probability for d odd is

$$P_B(1, 2) = \sum_{e = (d+1)/2}^{d} \binom{d}{e} p^e (1 - p)^{d-e} \qquad d \text{ odd} \tag{12-16b}$$

Using (12-11), (12-12), and (12-15), the total block decoding error probability is bounded by

$$P_B \leq \sum_{m=0}^{M-1} \text{Pr} \, (m) \sum_{\substack{m'=0 \\ m' \neq m}}^{M-1} P_B(m, m') \tag{12-17}$$

Since linear codes are being considered, the set of Hamming distances between x_m and $x_{m'}$ is the same for all codewords m, and the second summation of (12-17) is the same for all m. Thus, using $m = 0$ to represent any of the 2^k possible second summations,

$$P_B < \sum_{\substack{m'=0 \\ m' \neq 0}}^{M-1} P_B(0, m') \sum_{m=0}^{M-1} \Pr(m)$$

$$= \sum_{m'=1}^{M-1} P_B(0, m') \qquad (12\text{-}18)$$

where $P_B(0, m')$ is given by (12-16). This says that the total average block error probability for linear codes is equal to the block error probability given that the all zeros codeword is transmitted.

In principle, the codewords for any specific code can be enumerated. In addition, when codeword 0 is transmitted, the number of bit errors $B(m')$ which are made when codeword $m' \neq 0$ is decoded incorrectly can be determined together with the Hamming distance between each codeword and the all zeros codeword. The probability of information bit error P_b is calculated by weighting each two-code-word block error probability in (12-18) by the number of information bit errors which occur with that block error event and dividing by the total number of bits decoded. Thus

$$P_b < \sum_{m'=1}^{M-1} \frac{1}{k} B(m') P_B(0, m') \qquad (12\text{-}19)$$

Let B_d denote the total number of bit errors that occur in *all* block error events involving codewords which are distance d from the all-zeros codeword. $P_B(0, m')$ is the same for all m' which are distance d from the all-zeros codeword. Let P_d denote this error probability. Then (12-19) can be written

$$P_b < \frac{1}{k} \sum_{d=d_{\min}}^{n} B_d P_d \qquad (12\text{-}20)$$

where $d_{\min}$ is the minimum distance of the code. Equation (12-20) provides a good bound on bit error probability for any particular code. The values of B_d may be given or may be calculated from the code definition. For high signal-to-noise ratio, the bit error probability is often bounded using only the first term of this equation.

EXAMPLE 12-3

Calculate the bit error probability bound as a function of crossover probability p for the code of Table 12-1.

Solution: From the list of all codewords, B_3 is the total number of ones in all messages whose codewords are distance 3 from the all-zeros codeword. Thus $B_3 = 12$. Similarly, $B_4 = 16$, and $B_7 = 4$. From (12-16)

$$P_3 = \sum_{e=2}^{3} \binom{3}{e} p^e (1 - p)^{3-e}$$

$$P_4 = \frac{1}{2}\binom{4}{2} p^2 (1 - p)^2 + \sum_{e=3}^{4} \binom{4}{e} p^e (1 - p)^{4-e}$$

$$P_7 = \sum_{e=4}^{7} \binom{7}{e} p^e (1 - p)^{7-e}$$

The total average error probability is bounded by

$$P_b < \tfrac{1}{4}(12P_3 + 16P_4 + 4P_7)$$ ∎

EXAMPLE 12-4

Calculate a bound on bit error probability for a FH/MFSK-coded spread-spectrum system which is operating in a worst-case partial band jamming environment. Assume that the error correction code used is the code given in Table 12-1.

Solution: In all error probability calculations for coded systems, care must be exercised to attribute the proper amount of energy to information bits and to encoder output symbols. For binary (n, k) block codes, n binary symbols are transmitted for each k binary information bits so that the energy per encoder output symbol is $E_s = kE_b/n$ and E_s is always less than E_b. Symbols with energy E_s are transmitted over the discrete binary channel. The error probability for this channel is given by (11-38) with E_b replaced by kE_b/n. Using Table 11-1, the symbol error probability, which is the BSC crossover probability, is

$$P_s = p = \begin{cases} \dfrac{0.3679}{(4/7)(E_b/N_J)} & \dfrac{4}{7}\dfrac{E_b}{N_J} > 2.00 \\[3mm] \dfrac{1}{2}\exp\left(-\dfrac{4}{7}\dfrac{E_b}{N_J}\dfrac{1}{2}\right) & \dfrac{4}{7}\dfrac{E_b}{N_J} \le 2.00 \end{cases}$$

The error probability bound is calculated using this equation for p in the expression for P_b found in Example 12.3. The result of this calculation is illustrated in Figure 12-2 together with the error probability for the same system without coding. ∎

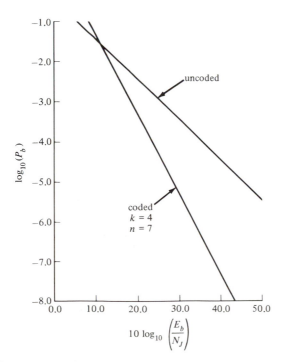

FIGURE 12-2. Comparison of coded and uncoded performance of FH/MFSK in worst-case partial band jamming.

Although the two-codeword error probability calculation of (12-16) is not difficult, it is sometimes desirable to have an even simpler expression available. It can be shown [3] that the two-codeword error probability P_d for binary codewords which differ in d positions is overbounded by

$$P_d \leq D_0^d \tag{12-21}$$

where

$$D_0 = \sum_{y_i} \sqrt{p(y_i|1)p(y_i|0)} \tag{12-22}$$

The sum in (12-22) is over all possible received symbols. For the BSC, $y_i = 0$ or 1, $p(1|1) = p(0|0) = 1 - p$, $p(0|1) = p(1|0) = p$, and

$$D_0 = 2\sqrt{p(1 - p)} \tag{12-23}$$

Using these results, the bit error probability is overbounded by

$$P_b < \frac{1}{k} \sum_{d=d_{\min}}^{n} B_d D_0^d \tag{12-24}$$

Equations (12-21) and (12-22) apply to any binary input memoryless channel and can therefore be used to bound the error probability for soft decision decoders as well.

In summary, a linear binary block code may be thought of as a mapping of encoder input messages into encoder output codewords. The codewords are carefully selected to be separated as far as possible in Hamming distance. The decoder input is a distorted version of the encoder output. The decoder operates by choosing as its estimate of the transmitted codeword the codeword which is "closest" to the received n-tuple **y.** The distance measure used by the decoder is, in general, the a posteriori probability. For the BSC an equivalent distance measure is Hamming distance and for the AWGN channel an equivalent distance measure is Euclidean distance. The design of hardware efficient coders and decoders is the subject of coding theory and is not discussed here. The primary problem of error correction coding is the development of good codes which are, at the same time, reasonably easy to decode. A decoder that operates directly on the principles defined above will be inefficient from a hardware standpoint in almost all cases. The student is referred to Lin and Costello [1] for further discussion of coding and decoding techniques.

12-3

ELEMENTARY CONVOLUTIONAL CODING CONCEPTS

A convolutional code is similar to a block code in that each time k binary input symbols are collected, the encoder outputs n binary symbols. In contrast to the block encoder, the mapping of input k-tuples to output n-tuples are not independent from one mapping to the next, that is, the convolutional codes possesses memory. The amount of memory varies for different codes and is specified by the constraint length of the code. The *constraint length* of a convolutional code is equal to the number of output n-tuples which are influenced by a particular input k-tuple [3].*

One convenient representation of a convolutional encoder is as a finite-state

---

*A number of different definitions of constraint length are used in the literature.

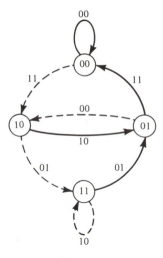

FIGURE 12-3. Finite-state machine representation of convolutional encoder.

machine. Consider, for example, the $k = 1$, $n = 2$ convolutional code represented in Figure 12-3. The finite-state machine here has four states 00, 01, 10, 11 which are connected by branches labeled with two binary symbols. Suppose that the encoder is in state 00 and that the encoder input message is a binary 1. The encoder changes from state 00 to state 10 moving along the dashed path and outputs the codeword 11. If the input message had been a 0, the encoder would have moved along the solid path leaving state 00, outputting 00, and returning to state 00. In general, the encoder moves from one state to another, following dashed paths when the input is a 1 and solid paths when the input is a 0, and outputting the pairs of symbols encounted on the paths between states. Observe that an encoder input 1 results in different encoder outputs depending on the state of the encoder. Other convolutional codes may have many more states as well as more paths leaving each state and more labels on each path.

An alternative representation for a convolutional code makes use of trellis diagrams [7]. The trellis representing the code of Figure 12-3 is shown in Figure 12-4. The encoding procedure may be described by moving from left to right through the trellis, traversing a single branch for each encoder input. The dashed branch is followed out of a state if the input is a 1 and the solid branch if the input is a 0. The encoder outputs correspond to the branch labels. Suppose, for example, that the encoder begins in state 00 and that the input sequence is 1011000. The

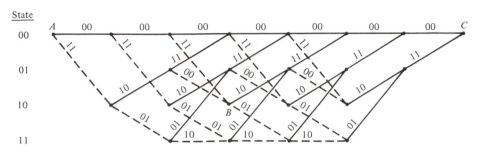

FIGURE 12-4. Trellis diagram representing a convolutional code.

encoder goes through the sequence of states (00, 10, 01, 10, 11, 01, 00, 00) and the encoder output sequence is (11, 10, 00, 01, 01, 11, 00). In this special case, the encoder begins and ends in state 00 and there are only 14 output symbols encountered in any path through the trellis. In typical convolutionally coded systems, the central part of the trellis is much longer and may be considered effectively infinite in length. Each path through the trellis corresponds to a unique encoder input message, and the labels on each path correspond to the codeword sequence associated with the input message. The code illustrated in Figure 12-4 is a rate $\frac{1}{2}$ code since two symbols are output for each input symbol.

The code of Figure 12-4 can be implemented using the digital circuit of Figure 12-5. During each input symbol time the output commutator is cycled through its two positions. In the first (upper) position, the output is the modulo-2 sum of the current input and two previous inputs. In the second (lower) position, the output is the modulo-2 sum of the current input and the input two input symbol times earlier. The state of the encoder is the contents of the two-stage shift register. Observe that each input symbol affects three pairs of output symbols, so that the code has constraint length 3. Figure 12-6 is the conceptual circuit diagram for a constraint length 7 convolutional encoder which is used in some modern communications systems. The trellis diagram for this code has 64 states with two branches leaving and two branches entering each state.

12-3.1 Decoding of Convolutional Codes

The decoding of convolutional codes is most easily explained using the trellis diagram. The encoder input is a sequence of binary symbols which the encoder uses to select a particular path m through the trellis. The length of the trellis may be very long and the number of possible paths from one end to the other may be extremely large. The encoder output sequence, denoted $\mathbf{x}_m = (x_{m1}, x_{m2}, \ldots, x_{mn})$, is the sequence of path labels encountered along this path. The output sequence is transmitted over a discrete memoryless channel which is characterized by a set of transition probabilities $p(y_i|x_i)$. The channel distorts the transmitted sequence and the received sequence is $\mathbf{y} = (y_1, y_2, \ldots, y_n)$. The decoder uses its knowledge of the channel characteristics, the code structure, and $\mathbf{y}$ to estimate which path the encoder followed through the trellis. Using arguments similar to those used for developing the block decoding rule, the optimum decoding rule can be shown to

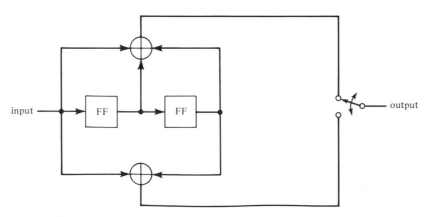

FIGURE 12-5. Typical rate-½ convolutional encoder.

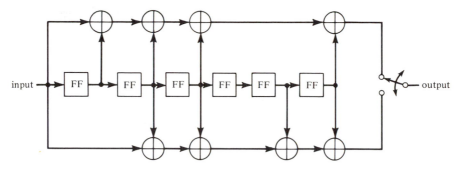

FIGURE 12-6. Rate-½ constraint length 7 convolutional encoder.

be: Choose $\hat{m} = m'$ if

$$p(\mathbf{y}|\mathbf{x}_{m'}) \geq p(\mathbf{y}|\mathbf{x}_{m''}) \qquad (12\text{-}25)$$

for any $m'' \neq m'$, where $\hat{m}$ represents the decoder estimate of the path through the encoder trellis. Thus the decoder chooses as its estimate $\hat{m}$ the path which was most likely to have been followed through the encoder trellis given that $\mathbf{y}$ was received. For the BSC, an equivalent decoding rule is to choose $\hat{m} = m$ if $\mathbf{x}_m$ is the closest codeword to $\mathbf{y}$ in Hamming distance. For the AWGN channel, the equivalent decoding rule is to choose $\hat{m}$ to be that m for which the Euclidean distance between $\mathbf{x}_m$ and $\mathbf{y}$ is the smallest.

It was noted earlier that a convolutional encoder may be thought of as a finite-state machine. As this machine goes from state to state it outputs channel symbols which are observed by the receiver after being disturbed by additive memoryless noise. The end state of a state transition is dependent only on the state immediately prior to the transition and the encoder input symbol. Thus the encoder output sequence is a finite-state Markov process. It has been shown that the Viterbi algorithm [3] is an optimum solution to the problem of estimating the state sequence of a finite-state Markov process observed in memoryless noise. It is therefore concluded that the Viterbi algorithm can also be used to decode a received sequence which is the result of transmitting a convolutionally coded signal over a DMC. The Viterbi algorithm is discussed in detail in Appendix D.

Given the convolutional code trellis, the received sequence, and the channel model, the Viterbi algorithm is able to determine the most likely path to have been followed through the trellis by the encoder. This most likely path is the path whose associated output sequence is closest to the received sequence in either Hamming distance, Euclidean distance, or some other distance measure. The algorithm functions exactly as described in Appendix D, with the generalized metric replaced by the appropriate distance measure.

Viterbi decoding is not the only method of decoding convolutional codes. In fact, for trellises with a large number of states, Viterbi decoding becomes impractical and other methods are used. Sequential decoding of convolutional codes makes use of the tree structure described in Chapter 1. A description of sequential decoding can be found in Lin and Costello [1] together with a list of additional references. Certain convolutional codes can be decoded using threshold decoding [8]. Threshold decoding is useful whenever extremely high data rates are required since the decoder is very simple. The coding gain obtainable using threshold decoding is relatively small for the AWGN channel. However, large coding gains are possible for a pulse jamming channel, and threshold decoding should be considered for any spread-spectrum system.

12-3.2 Error Probability for Convolutional Codes

The calculation of a bound on bit error probability for convolutional codes is accomplished using the same concepts used for block codes. The calculation is complicated by the structure imposed by the trellis, but the final result is not difficult. The derivation of this bound can be found in Viterbi [9]. The reference cited is a tutorial on convolutional codes which is recommended for any reader wishing to apply these codes to modern communication systems. For rate-$1/n$ codes the bound on bit error probability is

$$P_b \le \sum_{d=d_{\text{free}}}^{\infty} B_d P_d \qquad (12\text{-}26)$$

In this bound, P_d is the probability of making an error when comparing two paths through the trellis whose codeword symbol sequences differ in d positions.

The encoder constants B_d are similar to the comparable constants of (12-20) and account for all bit errors made when a particular type of decoding error event occurs. The constants B_d can be determined by considering the decoding trellis in detail. Assume that the central part of the decoding trellis is infinite in length and consider the last decoder decision on the right of the trellis. Assume also that the transmitted message is the all-zeros message which results in an encoder path that never leaves state 00. B_d is found by considering all incorrect paths through the trellis which first remerge with the correct path in the last state of the trellis. Specifically, B_d is the total number of bit errors in all incorrect paths through the trellis which are Hamming distance d from the all-zeros path and which first remerge with the all-zeros path in the last state of the trellis. The constants B_d are given in the specification of some convolutional codes and can be determined using a straightforward computer program for any convolutional code.

The limits of summation in (12-26) are d_{free} and infinity. The free distance, d_{free}, of a convolutional code is comparable to the minimum distance of a block code. The *free distance* is the minimum Hamming distance between the encoder output sequences for any two distinct paths through the trellis. The upper limit of infinity is due to the assumption that the central portion of the trellis was infinite in length. Thus there are paths through the trellis which are infinitely distant from the all-zeros path and hence the upper limit of infinity. In most cases the error probability calculation may be truncated after a small number of terms of (12-26). The number of terms required depends on the relative rates of increase of B_d and decrease of P_d as d increases. For high signal-to-noise ratios, a single term may be adequate. The two-codeword error probability P_d used in (12-26) may be calculated using (12-16) or may be overbounded using (12-21). More accurate results are obtained using (12-16) but computations are simplified using (12-21).

12-4

RESULTS FOR SPECIFIC ERROR CORRECTION CODES

The discussions of Sections 12-2 and 12-3 were general and were intended to introduce the fundamental concepts of error correction coding. In this section specific error correction codes are considered. The codes selected for this discussion are a small sample of a very large range of codes which are currently available.

12-4.1 BCH Codes

The family of BCH codes are powerful linear block codes for which excellent decoding algorithms exist. This family of codes contains codes of many rates and a wide range of error correction capability. Note that the Hamming codes are single error-correcting binary BCH codes. All texts on error correction coding contain detailed discussions of this family of codes. The student is referred to Blahut [10] or Lin and Costello [1] for detailed encoder and decoder designs. The spread-spectrum system designer must understand two aspects of any error correction scheme. The first aspect is the complexity of the encoder and decoder. Extremely complex decoders can be implemented only at low speeds and therefore are not usable in many systems. The second aspect is the performance of the code in the anticipated communications environment. Only the latter aspect is considered here. Only binary BCH codes are considered.

The error probability bounds derived in Section 12-2 are based on the assumption that maximum-likelihood decoding is used. Unfortunately, the most common decoding techniques for binary BCH codes are not true maximum-likelihood procedures, so that the performance bounds are optimistic. The standard BCH decoding algorithms are bounded distance algorithms and error patterns with more than t errors can never be corrected. The bit error probability for these decoders is [2]

$$P_b \leq \sum_{i=t+1}^{n} \left(\frac{i+t}{n}\right)\binom{n}{i}p^i(1-p)^{n-i} \tag{12-27}$$

where t is the number of channel errors that can be corrected by the code. The values of n, k, and t for all known BCH codes up to length $n = 1023$ are given in Lin and Costello [1]. The code minimum distance d_{min} and t are related by

$$t = \begin{cases} \frac{1}{2}d_{min} - 1 & \text{for } d_{min} \text{ even} \\ \dfrac{d_{min} - 1}{2} & \text{for } d_{min} \text{ odd} \end{cases} \tag{12-28}$$

The error probability bound just given is convenient in that knowledge of the code weight distribution is not required.

EXAMPLE 12-5

A BCH code with $n = 63$, $k = 30$, and $t = 6$ exists. This code has a rate of $30/63 = 0.4762$. This code is used in a FH/DPSK spread-spectrum system. Calculate the upper bound on bit error probability when the system is operating in worst-case tone jamming.

Solution: The channel symbol error probability in worst-case tone jamming is calculated using (11-118). When using this equation for coded systems, it gives the symbol (not bit) error probability and R is the symbol rate. That is,

$$p = \begin{cases} 0.5 & \dfrac{P}{J}\left(\dfrac{W}{R_s}\right) < 1 \\[2ex] \dfrac{1}{2(P/J)(W/R_s)} & 1 \leq \dfrac{P}{J}\left(\dfrac{W}{R_s}\right) < \dfrac{W}{R_s} \\[2ex] 0 & \dfrac{W}{R_s} < \dfrac{P}{J}\left(\dfrac{W}{R_s}\right) \end{cases}$$

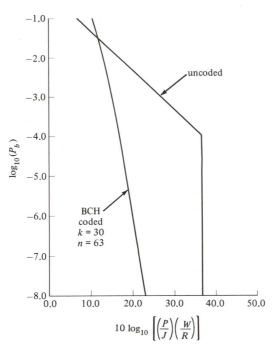

FIGURE 12-7. Comparison of coded and uncoded performance of FH/DPSK spread-spectrum system in worst-case tone jamming.

For the code being considered, the bit rate R and the symbol rate are related by $R_s = 63R/30$. Substituting for R_s in the equation above yields the BSC crossover probability as a function of $(P/J)(W/R)$. Using p in (12-27) yields a bound on information bit error probability. This result is plotted in Figure 12-7 together with the previously calculated result without error correction coding. ∎

12-4.2 Reed–Solomon Codes

The Reed–Solomon codes are nonbinary BCH codes which have found numerous applications in spread-spectrum systems. The encoder input is a message k-tuple made of k symbols from an alphabet of $q = 2^m$ symbols. The encoder output is a codeword n-tuple with symbols from the same q-ary alphabet. Observe that since the input–output alphabet size is a power of 2, input and output symbols may be represented by m-ary binary words. That is, the input message may be considered a km-bit binary vector and the output codeword a nm-bit binary vector. The Reed–Solomon codes are capable of correcting t channel symbol errors where $n - k = 2t$. The block length of standard Reed–Solomon codes is $n = q - 1$. Reed–Solomon codes exist for $1 \le k \le n - 2$. The Reed–Solomon codes can also be extended to have block length $n = q$ and $n = q + 1$.

The bit error probability for Reed–Solomon codes is given in Clark and Cain [2]. The result is an overbound

$$P_b < \frac{2^{k-1}}{2^k - 1} \sum_{j=t+1}^{n} \left(\frac{j + t}{n} \right) \binom{n}{j} p_s^j (1 - p_s)^{n-j} \tag{12-29}$$

where p_s is the channel symbol error probability. The symbol error probability may be the error probability for an actual nonbinary channel such as in FH/MFSK or may be the probability of one or more binary errors in an m-bit word on a binary channel. When the channel is binary, then

$$p_s = \sum_{e=1}^{m} \binom{m}{e} p^e (1 - p)^{m-e} \qquad (12\text{-}30)$$

is used for the symbol error probability.

EXAMPLE 12-6

Consider a Reed–Solomon code having the parameters $n = 2^m - 1 = 63$, $k = 31$, $t = 16$, and $m = 6$. This code is used in a FH/DPSK spread-spectrum system. Calculate the error probability bound for worst-case tone jamming.

Solution: The worst-case jamming binary symbol error probability for this channel was given in Example 12-5. Since the encoder input and output alphabets are the same, the energy per transmitted information bit E_b is related to the energy per transmitted binary symbol E_s by $E_s = 31E_b/63$. Therefore, substitute $R_s = 63R/31$ into the equation for p in the last example to obtain the BSC crossover probability as a function of $(P/J)(W/R)$. This p is used in (12-30) to obtain the symbol error probability p_s. The symbol error probability is then used in (12-29) to obtain the desired bound on bit error probability. The result of this calculation is illustrated in Figure 12-8 together with the comparable result without coding. ∎

Reed–Solomon codes are often used in *concatenated coding* schemes [11], wherein two or more codes are used to improve the system performance. In these systems, the output of the first encoder, rather than being transmitted directly over the

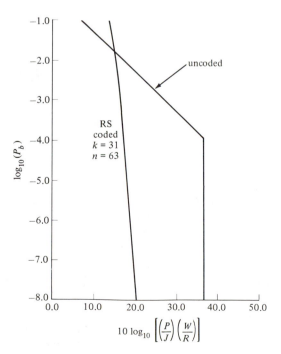

FIGURE 12-8. Comparison of coded and uncoded performance of FH/DPSK spread spectrum in worst-case tone jamming.

channel, is further encoded. The code for the second coding is called the *inner code* and the code for the first coding is called the *outer code*. The student is referred to Clark and Cain [2] or Lin and Costello [1] for additional discussion of concatenated coding schemes.

12-4.3 Maximum Free-Distance Convolutional Codes

Many researchers have used a variety of techniques to search for and find "good" convolutional codes. The quality factor used in these efforts is always a measure of the weight structure of the code. When maximum-likelihood decoding is used, the optimum weight structure is that which has the minimum number of bit errors in the paths through the code trellis which are closest to one another in Hamming distance. The weight structure of the best rate-$\frac{1}{2}$ and rate-$\frac{1}{3}$ codes in this sense were published in Odenwalder [11] and later in Clark and Cain [2]. For convenient reference, these results are presented in Table 12-3 for rate-$\frac{1}{2}$ codes and in Table 12-4 for rate-$\frac{1}{3}$ codes. The constraint length of the code is denoted by ν and the number of states in the code trellis is $2^{\nu-1}$. The free distance is the minimum Hamming distance between the codewords on any two paths through the trellis. The code generators are given in octal notation. This notation gives the connections between the encoder shift register stages and the modulo-2 adders. Consider, for example, the constraint length 7 rate-$\frac{1}{2}$ code. The generators are (171, 133) in octal or (1111001, 1011011) in binary. This means that connections to the first modulo-2 adder are from shift register stages 0, 1, 2, 3, and 6 and to the second modulo-2 adder are from shift register stages 0, 2, 3, 5, and 6. This encoder is illustrated in Figure 12-6.

The performance of these convolutional codes in spread-spectrum systems is found using (12-26), with B_d given in Table 12-3 or 12-4 and P_d given by (12-16) or bounded by (12-21). The result of these calculations for a FH/DPSK spread-spectrum system in worst-case tone jamming and using a rate-$\frac{1}{2}$ constraint length 5 and constraint length 7 codes are given in Figure 12-9. Note that a Viterbi decoder for the constraint length 5 rate-$\frac{1}{2}$ code is available in a single integrated circuit [12].

12-4.4 Repeat Coding for the Hard Decision FH/MFSK Channel

Although repeat coding is not useful for the AWGN channel, significant gains can be achieved with its use in a FH/MFSK system which operates in a partial band or pulsed noise jamming environment. Consider a system using orthogonal MFSK and assume that the modulator divides each symbol interval into L chips and transmits the symbol L times using a different FH frequency for each chip. Denote the MFSK uncoded symbol rate by R_s and the chip rate by $R_c = LR_s$. The jammer is smart and knows exactly how much power to transmit over a particular bandwidth or over a particular time interval to cause the maximum possible average chip error probability. Assume that the modem employs ideal interleaving so that all chip errors may be considered independent.

The transmitted signal is illustrated conceptually in Figure 12-10. For clarity, the FH pattern is not shown in this figure. The message to be transmitted is one of $M = 2^b$ frequencies, and is repeated L times as shown. The jammer optimizes its bandwidth or duty factor just as it would if normal uncoded FH/MFSK with

TABLE 12-3. Best Rate-$\frac{1}{2}$ Convolutional Codes and Their Partial Weight Structure

Constraint Length, v	Code Generators	Free Distance, d_f	B_d for $d =$							
			d_f	$d_f + 1$	$d_f + 2$	$d_f + 3$	$d_f + 4$	$d_f + 5$	$d_f + 6$	$d_f + 7$
3	(7, 5)	5	1	4	12	32	80	192	448	1024
4	(17, 15)	6	2	7	18	49	130	333	836	2069
5	(35, 23)	7	4	12	20	72	225	500	1324	3680
6	(75, 53)	8	2	36	32	62	332	701	2342	5503
7	(171, 133)	10	36	0	211	0	1404	0	11633	0
8	(371, 247)	10	2	22	60	148	340	1008	2642	6748
9	(753, 561)	12	33	0	281	0	2179	0	15035	0

Source: Ref. 11.

TABLE 12-4. Best Rate-$\frac{1}{3}$ Convolutional Codes and Their Partial Weight Structure

Constraint Length, ν	Code Generators	Free Distance, d_f	B_d for $d =$							
			d_f	$d_f + 1$	$d_f + 2$	$d_f + 3$	$d_f + 4$	$d_f + 5$	$d_f + 6$	$d_f + 7$
3	(7, 7, 5)	8	3	0	15	0	58	0	201	0
4	(17, 15, 13)	10	6	0	6	0	58	0	118	0
5	(37, 33, 25)	12	12	0	12	0	56	0	320	0
6	(75, 53, 47)	13	1	8	26	20	19	62	86	204
7	(171, 145, 133)	14	1	0	20	0	53	0	184	0
8	(367, 331, 225)	16	1	0	24	0	113	0	287	0

Source: Ref. 11.

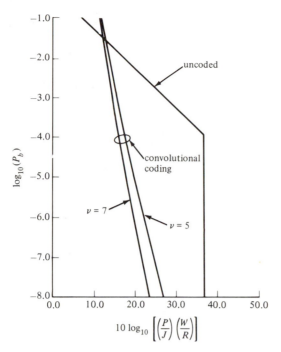

FIGURE 12-9. Comparison of uncoded and coded performance of FH/DPSK spread spectrum in worst-case tone jamming.

symbol rate LR_s were being used. The chip error probability for worst-case partial band or pulsed noise jamming of FH/MFSK is calculated using identical steps to those used to calculate the bit error probability of (11-38). With appropriate adjustments to obtain MFSK symbol (as opposed to bit) error probability and to account for the reduced energy per chip due to the repeat coding, (11-38) becomes

$$P_c = \begin{cases} \dfrac{2(M-1)}{M} \dfrac{Lk'}{E_b/N_J} & \dfrac{E_b}{N_J} > L\left(\dfrac{E_b}{N_J}\right)_0 \\[2em] \dfrac{1}{M} \displaystyle\sum_{m=2}^{M} (-1)^m \binom{M}{m} \exp\left[\dfrac{bE_b}{LN_J}\left(\dfrac{1-m}{m}\right)\right] & \dfrac{E_b}{N_J} \le L\left(\dfrac{E_b}{N_J}\right)_0 \end{cases} \qquad (12\text{-}31)$$

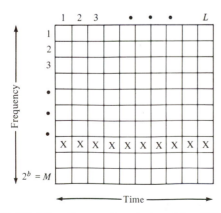

FIGURE 12-10. Transmitted signal for FH/MFSK with repeat coding.

where k' and $(E_b/J_J)_0$ are given in Table 11-1 and P_c denotes the channel chip error probability.

The receiver demodulator output is a sequence of L chip estimates for each information symbol. The demodulator makes a hard decision for each chip interval. One possible received symbol sequence is shown in Figure 12-11. The decoding rule used by the receiver is to choose as the decoder output symbol the symbol with the largest number of votes out of L transmissions. For the example of Figure 12-11, the decoded output symbol is a six. Given that P_c is the received chip error probability, the probability p_e of an entry in the correct row and column of the decoding matrix is

$$p_e = 1 - P_c$$

and the probability of l entries in the correct row of the decoding matrix is

$$\Pr (l \text{ entries in correct row}) = \binom{L}{l} p_e^l (1 - p_e)^{L-l}$$

$$= P_l(l) \qquad (12\text{-}32)$$

A symbol error is made whenever the number of entires in *some* incorrect row of the decoding matrix is greater than the number of entries in the correct row. A symbol error *may* be made when the number of entries in an incorrect row equals the number of entries in the correct row.

There are at most L entries in all rows of the decoding matrix. Let $P_{s|l}$ denote the symbol error probability given that there are l entries in the correct row of the decoding matrix. Then the total symbol error probability is

$$P_s = \sum_{l=0}^{L} P_{s|l} P_l(l) \qquad (12\text{-}33)$$

With l entries in the correct row, there are $L - l$ entries in all incorrect rows. There are $(M - 1)L$ incorrect positions in the decoding matrix of which $(M - 1)l$ cannot contain entries since there are l entries in the correct row. Thus there are a total of $(M - 1)(L - l)$ incorrect positions in the decoding matrix which may contain entries. It can be shown that the probability of j entries in a block which has $(L - l)$ elements, given a total of $(L - l)$ entries in $(M - 1)(L - l)$ elements, is given by

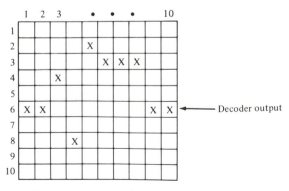

FIGURE 12-11. Possible FH/MFSK decoder input and decoder estimate.

$$q(j|l) = \frac{\binom{L-l}{j}\binom{(M-2)(L-l)}{L-l-j}}{\binom{(M-1)(L-l)}{L-l}}$$ (12-34)

Two cases must be considered when calculating total symbol error probability. First consider binary FSK, where whenever a chip error is made an entry appears in the one incorrect row of the decoding matrix. For L even,

$$P_s = \sum_{l=0}^{(L/2)-1} P_I(l) + \frac{1}{2} P_I\left(\frac{L}{2}\right)$$ (12-35a)

and for L odd,

$$P_s = \sum_{l=0}^{(L-1)/2} P_I(l)$$ (12-35b)

The first expression is based on the fact that a decoding error always occurs when there is from zero to $(L/2) - 1$ entires in the correct row and occurs with probability $\frac{1}{2}$ when there are $L/2$ entries in the correct row. The second expression is similar but accounts for the fact that there cannot be $L/2$ entries in any row since L is odd. For $M \geq 4$ the error probability expression is, for L even:

$$P_s = P_I(0) + \sum_{l=1}^{L/2-1} (M-1)P_I(l)\left[\frac{1}{2} q(l|l) + \sum_{j=l+1}^{L-l} q(j|l)\right]$$

$$+ (M-1)P_I\left(\frac{L}{2}\right)\frac{1}{2} q\left(\frac{L}{2}\bigg|\frac{L}{2}\right)$$ (12-36a)

and for L odd,

$$P_s = P_I(0) + \sum_{l=1}^{(L-1)/2} (M-1)P_I(l)\left[\frac{1}{2} q(l|l) + \sum_{j=l+1}^{L-l} q(j|l)\right]$$ (12-36b)

The first term in the first equation accounts for the case where there are no entries in the correct row, so some other row (incorrect) must have more entries than the correct row. The last term in the first equation accounts for the case where there are $L/2$ entries in the correct row. An error then occurs with probability $\frac{1}{2}$ only when some incorrect row also has $L/2$ entries. The probability of a particular incorrect row having $L/2$ entries is $q(L/2|L/2)$ and the probability of any one of the $(M-1)$ incorrect rows having $L/2$ entries is $(M-1)q(L/2|L/2)$. The first summation in the first equation is over all remaining values of l for which errors can occur. The term in brackets is the probability of a particular incorrect row being decoded given l entries in the correct row. The $(M-1)$ factor with the middle term accounts for the $(M-1)$ possible incorrect rows.

Using the equations derived here, the symbol error probability for diversity transmission and worst-case jamming can be calculated. Symbol error probability for orthogonal symbols is related to information bit error probability by

$$P_b = \frac{M}{2(M-1)} P_s$$ (12-37)

so that bit error probability can also be calculated. The results of this calculation are illustrated in Figure 12-12. Observe that significant performance improvements

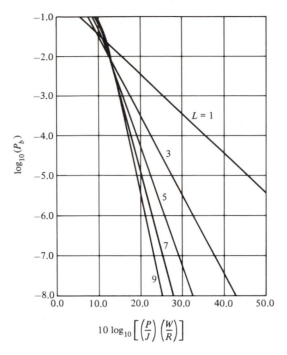

$$10 \log_{10}\left[\left(\frac{P}{J}\right)\left(\frac{W}{R}\right)\right]$$

FIGURE 12-12. Performance of FH/MFSK in worst-case partial band jamming with *L*-times repeat coding.

are obtained with this simple repeat coding procedure, and that a different L is optimum for different input signal-to-jammer power ratio. Similar conclusions are derived in Viterbi [13] for a case where soft decision decoding is used.

12-5

INTERLEAVING

Most forward error correction codes perform well only when the channel errors are completely independent from one signaling interval to the next. All the bounds on bit error probability discussed in Sections 12-2 through 12-4 were based on the assumption that the channel is memoryless. The worst-case jammers analyzed in Chapter 11 are the pulsed noise or pulsed tone jammers for direct-sequence systems or the partial band noise or tone jammers for frequency-hop systems. These jammers produce bursts of errors and therefore the channel is not memoryless. To counter this difficulty, an interleaver is placed between the encoder and the modulator and a de-interleaver is placed between the demodulator and the decoder.

One of two types of interleaving commonly used is called *block interleaving*. A block interleaver uses four N row by B column random access memories to randomize errors. Two of these memories are in the transmitter and the others are in the receiver. The transmitter reads encoder output symbols into a memory by columns until it is full. Then the memory is read out to the modulator by rows. While one memory is filling the other is being emptied, so two memories are needed. In the receiver, the inverse operation is effected by reading the demodulator output into a memory by rows and reading the decoder input from the memory by columns.

B = 10; N = 10

1	11	21	31	**41**	**51**	**61**	**71**	**81**	91
2	12	22	32	42	52	62	72	82	92
3	13	23	33	43	53	63	73	83	93
4	14	24	34	44	54	64	74	84	94
5	15	25	35	45	55	65	75	85	95
6	16	26	36	46	56	66	76	86	96
7	17	27	37	47	57	67	77	87	97
8	18	28	38	48	58	68	78	88	98
9	19	**29**	**39**	**49**	**59**	**69**	79	89	99
10	20	30	40	50	60	70	80	90	100

FIGURE 12-13. Operation of a block interleave.

This operation is illustrated in Figure 12-13 for a $N = 10$, $B = 10$ interleaver. The encoder output symbols are numbered consecutively 1 through 100. These symbols are transmitted over the channel by rows. Thus the order of transmission for the first 20 symbols is 1, 11, 21, 31, 41, 51, 61, 71, 81, 91, 2, 12, 22, 32, 42, 52, 62, 72, 82, 92, . . . A burst of channel errors is assumed to hit channel symbols 41, 51, 61, 71, 81 and symbols 29, 39, 49, 59, 69, as shown by the shaded blocks. The demodulator output is read into the memory by rows resulting in the same array of symbols shown in Figure 12-13. This memory is then read into the decoder by columns. The end result of this operation is that adjacent channel errors are spaced by N symbols at the decoder input. N is chosen to be large enough that errors spaced by N will affect different codewords and may be considered independent.

Another type of interleaving is convolutional interleaving. A convolutional interleaver is depicted in Figure 12-14. All multiplexers in the transmitter and receiver are assumed to operate synchronously. The multiplexer switches change position after each symbol time so that successive encoder outputs enter different rows of the interleaver memory. Each interleaver and de-interleaver row is a shift register

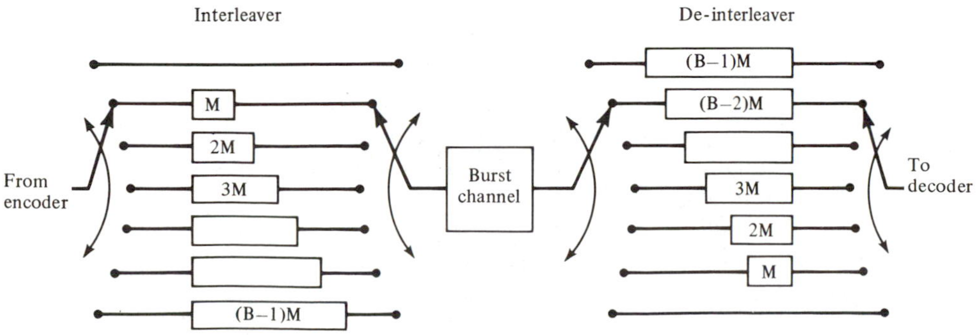

FIGURE 12-14. Operation of a convolutional interleaver.

memory with the number of units of delay shown within the rectangle. The first encoder output enters the top interleaver row, is transmitted over the channel immediately, and enters the de-interleaver memory of $(B - 1)M$ symbols. The second encoder output enters the second row of the interleaver and is delayed M symbol times before transmission. Thus adjacent encoder outputs are transmitted M symbol times apart and are not affected by the same channel error burst. Upon reception, the second encoder symbol is delayed by an additional $(B - 2)M$ symbol time for a total delay of $(B - 1)M$. Observe that all symbols have the same delay after passing through both the interleaver and de-interleaver, so that the decoder input symbols are in the same order as the encoder output symbols.

Interleaving is an essential part of all spread-spectrum systems which are expected to operate in worst-case jamming environments. Additional discussion on the subject of interleaver design can be found in Clark and Cain [2], Forney [14], and Cain and Geist [15].

12-6

RANDOM CODING BOUNDS

All the discussion about forward error correction codes up to this point has been directed to the performance evaluation of specific codes. In the initial stages of spread spectrum system design, it is useful to know whether any error correction code at all will have adequate power to satisfy system specifications. It is also helpful to have a preliminary idea of the complexity of the code required to satisfy system specifications. Both of these issues can be addressed using the concept of random coding error probability bounds first derived by Shannon [16]. In his classic paper, Shannon proved that error correction coding schemes exist which permit digital communications at arbitrarily low bit error probability provided the transmission rate is below channel capacity as discussed in Chapter 1.

A key element in this work is the coding theorem, which gives an upper bound on the achievable error probability using forward error correction with maximum-likelihood decoding. For discrete memoryless channels defined by a transition probability matrix $p(y|x)$ and block coding, the probability of a bit error $\overline{P}_{Be}$ averaged over all possible block codes is bounded by

$$\overline{P}_{Be} < 2^{-n(R_0 - R)} \tag{12-38}$$

where n is the code block length, R is the code rate in bits per channel use, and R_0 is called the *computational cutoff rate*. For a symmetric discrete memoryless channel the cutoff rate is given by

$$R_0 = -\log_2\left\{ \frac{1}{M^2} \sum_y \left[\sum_x \sqrt{p(y|x)} \right]^2 \right\} \tag{12-39}$$

where M is the number of elements in the input alphabet. It can be shown that R_0 is a positive number, so that $(R_0 - R)$ is also positive for $R < R_0$. Thus, when $R < R_0$, the bit error probability can be made arbitrarily small by choosing the block length n to be sufficiently large.

A similar result is valid for time-varying convolutional codes. In particular, for binary rate $R = 1/n$ time-varying convolutional codes, the bit error probability averaged over all possible codes is bounded by [9]

$$\overline{P}_{Be} < \frac{2^{-\nu R_0/R}}{\{1 - 2^{-[(R_0/R) - 1]}\}^2} \qquad R < R_0 \tag{12-40}$$

where ν is the constraint length of the code. Thus, for a fixed code rate R and channel cutoff rate R_0, the bit error probability can be made arbitrarily small by making the code constraint length sufficiently large. Observe that convolutional code constraint length plays the same role in (12-40) that block length played in (12-38). The student is reminded that (12-40) applies only to time-varying convolutional codes, that is, convolutional codes which change after each trellis branch. The result is, however, commonly used to bound the performance of fixed convolutional codes with little error.

One method of using these results [3] is to select an acceptable block length or constraint length and an acceptable code rate R, and to use (12-38) or (12-40) to find the required R_0 which yields the desired error probability. The required signal-to-jammer power ratio can then be calculated from the relationships between R_0 and the DMC transition probabilities and between the transition probability and P/J as calculated in Chapter 11. For example, suppose that a block code with block length $n = 63$ and rate $R = 30/63$ have been selected. Then the error probability bound yields

$$P_b < 2^{-63(R_0 - 0.476)} \tag{12-41}$$

If the system requires a bit error probability of 10^{-5}, the solving for R_0 results in $R_0 = 0.740$. For the binary symmetric channel with crossover probability p, (12-39) becomes

$$R_0 = 1 - \log_2[1 + 2\sqrt{p(1 - p)}] \tag{12-42}$$

which can be solved for $p \approx 0.01$. This crossover probability is related to the required P/J using the applicable relationship to Chapter 11.

REFERENCES

[1] S. Lin and D. J. Costello, Jr., *Error Control Coding: Fundamentals and Applications* (Englewood Cliffs, N.J.: Prentice-Hall, 1983).

[2] G. C. Clark and J. B. Cain, *Error-Correction Coding for Digital Communications* (New York: Plenum, 1981).

[3] A. J. Viterbi and J. K. Omura, *Principles of Digital Communication and Coding* (New York: McGraw-Hill, 1979).

[4] W. W. Peterson and E. J. Weldon, *Error-Correcting Codes* (Cambridge, Mass.: MIT Press, 1972).

[5] R. G. Gallager, *Information Theory and Reliable Communication* (New York: Wiley, 1968).

[6] E. R. Berlekamp, *Algebraic Coding Theory* (New York: McGraw-Hill, 1968).

[7] G. D. Forney, Jr., "The Viterbi Algorithm," *Proc. IEEE*, March 1973.

[8] J. L. Massey, *Threshold Decoding* (Cambridge, Mass.: MIT Press, 1963).

[9] A. J. Viterbi, "Convolutional Codes and Their Performance in Communication Systems," *IEEE Trans. Commun.*, October 1971.

[10] R. Blahut, *Theory and Practice of Error Control Codes* (Reading, Mass.: Addison-Wesley, 1983).

[11] J. P. Odenwalder, "Optimal Decoding of Convolutional Codes," Ph.D. dissertation, University of California, Los Angeles, 1970.

[12] R. D. McCallister and J. J. Crawford, "A Low-Power, High Throughput Maximum Likelihood Convolutional Decoder Chip for NASA's 30/20 GHz Program," *Conf. Rec.*, IEEE Natl. Telecommun. Conf., 1981.

[13] A. J. Viterbi and I. M. Jacobs, "Advances in Coding and Modulation for Noncoherent Channels Affected by Fading, Partial Band, and Multiple-Access Interference," in *Adv. Commun. Syst.*, McGraw-Hill, 1975.

[14] G. D. Forney, "Burst-Correcting Codes for the Classic Bursty Channel," *IEEE Trans. Commun. Technol.,* October 1971.

[15] J. B. Cain and J. M. Geist, "Modulation, Coding, and Interleaving Tradeoffs for Spread Spectrum Systems," *Conf. Rec.*, IEEE Natl. Telecommun. Conf., 1981.

[16] C. E. Shannon, "A Mathematical Theory of Communication," *Bell Syst. Tech. J.,* July 1948.

PROBLEMS

12-1. Consider an error correction code whose codewords are generated using a four-stage maximal-length shift register whose generator is $g(D) = 1 + D + D^4$ and which is illustrated in Figure 8-16. A codeword is generated by loading the shift register with the information word and clocking the register 15 times. The codeword appears serially in the right stage of the shift register. What is the output message estimate of a maximum-likelihood hard decision decoder whose input 15-tuple is $\mathbf{y}$ = (101100101011000)?

12-2. Repeat Problem 12-1 for a Gaussian channel with a continuous output using a soft decision decoder with input y = (+0.8, −0.1, +1.1, +0.25, +1.0, −1.0, −0.35, +0.95, +0.87, −0.01, +0.20, +1.0, +1.7, −0.82, +0.13).

12-3. Consider the convolutional code defined by the trellis diagram of Figure 12-4. What is the output message estimate for a hard decision maximum-likelihood decoder whose input is y = (00101011000100)?

12-4. Repeat Problem 12-3 for a Gaussian channel with a continuous output using a soft decision maximum likelihood decoder whose input is $\mathbf{y}$ = (+1.1, +0.95, −0.78, +1.0, −1.0, +0.95, +0.98, −0.82, −0.90, −1.0, +0.01, −0.90, −1.0, −1.0).

12-5. Consider a BPSK/BPSK direct-sequence spread-spectrum system designed for a data rate of 1 MBPS and a direct sequence chip rate of 100 Mchips/s. An n = 15, k = 5, t = 3 BCH code together with a block interleaver are used to improve system performance. The system operates in a pulsed noise environment and the maximum acceptable average bit error probability is $P_b = 10^{-5}$.
(a) What is the maximum BSC crossover probability that will yield the required system P_b?
(b) What is the minimum required signal-to-jammer power ratio?
(c) What is the processing gain of the system relative to a nonspread, noncoded BPSK system operating in the same environment?

12-6. Repeat Problem 12-5 using the best constraint length 4, rate-$\frac{1}{3}$ convolutional code of Table 12-4.

12-7. Draw a single stage of the convolutional code trellis for the best constraint length 4, rate-$\frac{1}{3}$ code of Table 12-4. Draw a block diagram of this convolutional encoder.

12-8. Consider a FH/DPSKK spread-spectrum system using a data rate of 1 Mbps, a frequency-hop tone spacing of 2 MHz, and using 64 tones. This system is jammed by a smart partial band jammer.
(a) Calculate the computational cutoff rate R_0 as a function of 10 log (P/J) and plot the result.
(b) Calculate and plot the upper bound on bit error probability of (12-38) assuming a block code having block length 7 and rate R = 0.571.

12-9. Derive an expression for decoded bit error probability for a FH/DPSK spread-spectrum system which uses repeat coding. Assume the system operates in a worst-case jamming environment.

12-10. Calculate the error probability upper bound for the best constraint length 7 rate-$\frac{1}{2}$ convolutional code as a function of E_b/N_J using (12-26) and using (12-40) and compare the results. Assume FH/DPSK modulation and barrage noise jamming.

CHAPTER 13

Example Spread-Spectrum Systems

INTRODUCTION

A goal of this text is to develop the tools needed to evaluate and to design modern spread-spectrum systems. These tools should now be in place so that complete spread-spectrum systems can be considered. The systems described in this chapter are either operational spread-spectrum systems or are systems which are in the final stages of engineering development. Many of the concepts discussed in Chapters 7 through 12 are demonstrated within these example systems. The first two examples describe coherent direct-sequence systems used with the Tracking and Data Relay Satellite System (TDRSS). These two examples illustrate specific spreading codes, both fixed integration time and sequential detection search, and tau-dither non-coherent tracking. The third system discussed is the Global Positioning System (GPS), which is a direct-sequence spread-spectrum satellite ranging system used by mobile users to determine accurately their position on the earth's surface. The military name for GPS is NAVSTAR. The last example is the Joint Tactical Information Distribution System (JTIDS). Because JTIDS is still in development and because it is a military system, the discussion of JTIDS is more limited in scope than the discussion of the TDRSS-related systems.

SPACE SHUTTLE SPECTRUM DESPREADER

Communications between the NASA Space Shuttle and ground stations is accomplished via a number of different paths. Some of these paths are direct between the Shuttle and ground and some are indirect and include the Tracking and Data Relay Satellites (TDRS) as relay stations. The TDRSS is a system that includes two relay satellites in geostationary orbits about 22,300 miles above the earth's equator. The orbit of these satellites is selected so that their period of revolution is exactly one sidereal day. Thus the satellites appear to be stationary to an observer on the ground. The two TDRS are always able to communicate with a large ground station at White Sands, New Mexico, and one of the two TDRS is able to communicate with low orbiting satellites or the Space Shuttle most of the time. The TDRSS enables nearly continuous communication with low orbiting satellites without an expensive network of ground stations. The Space Shuttle can make use of the TDRSS using either S-band or Ku-band direct sequence spread-spectrum links. The particular link to be discussed here is the S-band link from the TDRS to the Space Shuttle. The Space Shuttle receiver was designed and built by TRW Systems Group and detailed information is available from several sources, including Alem [1] and TRW [2].

The Space Shuttle S-band forward link (TDRS to shuttle) operates at one of two carrier frequencies (2106.406/2287.5 MHz). The transmitted signal is a BPSK/BPSK spread-spectrum signal with convolutionally coded data modulo-2 assynchronously added to the DS spreading code. The particular convolutional code used is the constraint length 7 rate-$\frac{1}{3}$ code which was defined in Chapter 12. The critical parameters of the S-band signal are given in Table 13-1. Observe that two data rates must be accommodated. For operation at low signal-to-noise ratios, this implies that two IF filter bandwidths must be used. A conceptual block diagram of a portion of the receiver designed by TRW for this application is illustrated in Figure 13-1.

Only the spread-spectrum code acquisition and tracking functions are performed by the portion of the receiver shown in Figure 13-1. The received spread signal is

TABLE 13-1.   TDRS to Space Shuttle S-Band Signal Parameters

Carrier frequency	2106.406 MHz or 2287.500 MHz
Data rates	32 kbps or 72 kbps
Forward error correction	Rate-$\frac{1}{3}$; constraint length 7; convolutional code
Data modulation	BPSK–Manchester
Data modulator symbol rate	96 kbps or 216 kbps
Spreading code rate	11.232 Mchips/s
Spreading code length	2047 chips
Received signal-to-noise density ratio (P/N_0)	48 dB-Hz for low data rate; 51 dB-Hz for high data rate
Code doppler offset	± 100 Hz

Source: Ref. 1.

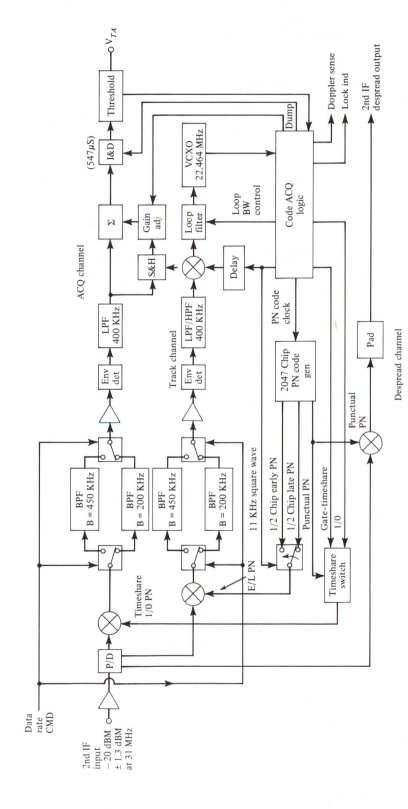

FIGURE 13-1. Space Shuttle S-band spread spectrum receiver. (From Ref. 1.)

down-converted from S-band to 31 MHz by the receiver front end (not shown) and the coherent carrier tracking and demodulation is performed by hardware (not shown) whose input is the "second IF despread output." The received IF signal is power divided and routed to three different channels. The upper channel is the acquisition channel, the middle channel is the code tracking channel, and the lower channel is the data demodulator channel.

Consider the acquisition channel of Figure 13-1. The code acquisition strategy selected is a stepped serial search using $\frac{1}{2}$-chip steps and using fixed-integration-time energy detection. The mixer labeled "timeshare" in Figure 13-1 is the acquisition despreading mixer. If the spreading code phase selected for evaluation by the code acquisition logic is incorrect, the output of this mixer is a wideband signal most of whose energy will be rejected by the bandpass filters. If the phase is correct, however, the 31-MHz IF signal will be collapsed to the Manchester-coded data bandwidth and will pass through the bandpass filter. The one-sided IF bandwidths necessary to accommodate the Manchester-coded data are approximately four times the symbol rate or 384 and 864 kHz for 96 and 216 kbps, respectively. The bandwidths shown in Figure 13-1 are half-bandwidths or, in other words, the one-sided bandwidths of the equivalent lowpass filters. The bandwidths shown are slightly larger than one-half the values just calculated so that the despread signal will pass through the filter when doppler offsets are present. The correctness of the spreading code phase is indicated by the presence or absence of energy at the IF filter output. This energy is sensed by the envelope detector, lowpass filter, integrate-and-dump circuitry, and the threshold detector. The integration time is 547 μs.

The predetection signal-to-noise ratio is calculated at the output of the IF filter. For the wider filter (i.e., high data rate), the output signal-to-noise ratio is 51 dB-Hz - 10 log (900 kHz) = −8.5 dB, and for the narrow filter, the output signal-to-noise ratio is 48 dB-Hz - 10 log (400 kHz) = −8.0 dB. Because these SNRs are relatively low, the energy detection time–bandwidth product will have to be large to achieve large detection probabilities simultaneously with low false-alarm rates. For the high and low data rates the time–bandwidth products are 492 and 219, respectively. The probability density of the integrator output is accurately modeled as Gaussian at either of these time–bandwidth products.

An important feature of the acquisition strategy is the automatic gain control hardware consisting of a sample/hold, a gain-adjust circuit, and a summing junction. This circuitry senses and compensates for variations in energy detector input noise levels. These variations may be due to day-to-day hardware changes or other factors. A gain control feature is essential in all spread-spectrum systems. Another method of accomplishing this task will be described later.

The acquisition integrator output after 547 μs is compared with a fixed threshold. If the output is above the threshold, the code phase is declared correct. Otherwise, the acquisition logic selects a new spreading code phase and the process repeats. After a correct detection, code tracking is attempted and the code phase is re-evaluated using an integration time of 12.6 ms. The long integration is used as an initial lock detector as well as a drop lock indicator. The entire acquisition strategy can be represented by the signal flow graph of Figure 13-2. The tracking loop is closed when the system enters state 2. In state 2, the integration increases to 12.6 ms and a single missed detection will cause the system to reject the trial code phase. A single detection in state 2 will cause the system to enter state 3, where it can remain indefinitely. Once in state 3, five consecutive missed detections are

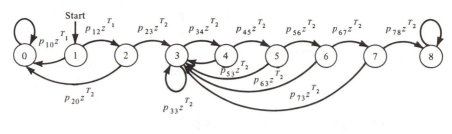

$T_1 = 547\ \mu s$
$T_2 = 12.6$ ms

FIGURE 13-2.   Space Shuttle spectrum despreader acquisition signal flow diagram.

required before the code phase is rejected. The acquisition strategy is clearly a multiple-dwell strategy using two different integration times.

The middle channel of Figure 13-1 functions as the spreading code tracking discriminator. The tracking loop uses a noncoherent tau-dither early-late phase discriminator with a dither magnitude of $\pm\frac{1}{2}$ chip and a dither rate of 22 kHz. Note that the 11-kHz square wave shown in Figure 13-1 switches the correlator between its early and late functions at a 22-kHz rate. The reference code is dithered by switching between early and late versions of the code. The dither discriminator energy detector is identical to the acquisition energy detector. The dither is removed and the error signal is generated by the mixer preceding the loop filter in the middle channel. The loop filter drives the code clock voltage-controlled oscillator, which is operating at twice the clock frequency. Observe that the dither rate is much lower than the IF filter bandwidth. The parameters of the code tracking loop are summarized in Table 13-2 for convenience.

When the code tracking loop is functioning properly, the PN code generator outputs a spreading code replica which is precisely phase synchronous with the received spreading waveform. This punctual spreading code is used in the bottom channel of Figure 13-1 to despread the received waveform for later coherent data demodulation. This is the only function of the bottom channel of this figure.

Several other factors in the receiver design merit discussion. First, observe that the reference input to all of the despreading mixers is the spreading code directly and not a modulated carrier as used in all the examples of Chapter 9. The reason for using the modulated carrier reference is to avoid a potential problem with RF to IF feedthrough in nonideal mixers. Second, the code tracking loop of Figure 13-1 actually has four different loop bandwidths. The largest bandwidth is used

TABLE 13-2.   TDRS to Space Shuttle S-band Spread-Spectrum Code Tracking Loop Parameters

Tracking loop type	Second order
Phase discriminator type	Noncoherent tau-dither early-late discriminator
Dither magnitude	$\pm\frac{1}{2}$ chip
Dither rate	22 kHz
Closed-loop bandwidth (one-sided)	500 Hz for acquisition; 10 Hz for tracking

Source: Ref. 1.

for acquisition and the smallest for final code tracking. The step from a 500- to a 10-Hz bandwidth is too large to maintain lock reliably during the switching transient so that two intermediate bandwidths of 135 and 37 Hz are included. Finally, the references cited give performance data on this system for the interested reader.

13-3

TDRSS USER TRANSPONDER

In addition to providing continuous communication between the Space Shuttle and White Sands, New Mexico, the Tracking and Data Relay Satellite System also provides continuous communication between low-orbiting satellites and the ground. The TDRSS User Transponder is placed on the low-orbiting satellite to enable that satellite to communicate through the TDRS. The User Transponder was designed and built by Motorola Inc., Government Electronics Group.*

The communication link discussed in this section is the S-band forward link from the TDRS to the user satellite. The modulation used on this link is QPSK direct-sequence spread spectrum. Both the in-phase and quadrature carrier are direct sequence modulated, however, only one of these carriers is modulated with data. The forward link signal is represented mathematically by

$$s(t) = \sqrt{2P_c} \, d(t) \, c_c(t) \cos \omega_0 t - \sqrt{2P_r} \, c_r(t) \sin \omega_0 t \qquad (13\text{-}1)$$

where the subscripts c and r refer to the command and ranging channels, respectively. The spreading waveform $c_c(t)$ is a code from the family of Gold codes with period 1023 chips, and the spreading waveform $c_r(t)$ is a truncated 18-stage maximal-length code. The code chip rate is approximately 3 M chips/s. The ranging channel m-sequence is truncated so that the code epochs of the short Gold code and the m-sequence occur simultaneously. The resultant period of the range channel code is $1023 \times 256 = 261,888$ chips, where the period of the complete m-sequence is $2^{18} - 1 = 262,143$ chips. The command channel data $d(t)$ is asynchronously modulo-2 added to $c_c(t)$. There are four possible data rates which are 125, 250, 500, or 1000 bps. Finally, the command channel and ranging channel power levels are not equal. Specifically, the command channel power is 10 dB larger than the ranging channel power. The important signal parameters are summarized in Table 13-3.

The User Transponder includes both a receiver and a transmitter; only the receiver is considered here. Also, the receiver is designed to demodulate both the TDRS spread-spectrum signal as well as a nonspread ground-to-satellite uplink signal. The features of the design to accommodate this nonspread signal are not discussed. The system specification which was the primary motivation for many of the design choices was the command channel synchronization time specification. The User Transponder is required to achieve initial Gold code synchronization in less than 20 s at threshold signal-to-noise ratios. To meet this specification, a multiple-channel synchronizer using sequential detection on all channels was implemented.

A simplified block diagram of the spread-spectrum receiver of the TDRSS User Transponder is illustrated in Figure 13-3. This diagram does not show a number of functions that have been implemented, such as sidelobe search and multipath

*Other manufacturers besides TRW and Motorola have made major contributions to the TDRSS. In particular, Harris Corp. is responsible for the ground station at White Sands, N.M.

TABLE 13-3. TDRS-to-User Satellite Signal Parameters

Carrier frequency	2025–2118 MHz
Spread-spectrum modulation	Unbalanced QPSK (direct sequence)
Spreading code rates	2.95–3.09 Mchips/sec (synchronized with carrier frequency)
Spreading codes	
Command channel	1023-bit Gold codes
Ranging channel	Truncated 18-stage m-sequence (period 1023×256)
Data modulation	BPSK on command channel only
Data rates	125, 250, 500, or 1000 bps
Data format	NRZ

rejection, and therefore is not complete. The primary initial code synchronization, code tracking, and carrier tracking functions are shown. The S-band spread-spectrum signal enters the receiver at the upper left corner. After image reject filtering the signal is down-converted to an IF of about 121.9 MHz. The reference for this initial downconversion is phase locked to the carrier VCO via the coherent frequency synthesizer. The bandpass filter following the first mixer rejects the mixer sum frequency output as well as any out of band interference.

The IF spread-spectrum signal is power divided for input to five independent direct-sequence despreaders. Four of these despreaders are used to search simultaneously and independently for the correct command channel Gold code phase. Consider the upper output of the power divider. This output is amplified and then command channel despreading is attempted. The reference spreading waveform is the output of one of the four Gold code generators shown on the right. The Gold code phase being evaluated in the top despreader is selected by the code synchronization logic, entered into one of the four Gold code generators via the initial condition, and routed through the matrix switch to $c_1(t)$. The waveform $c_1(t)$ phase modulates (BPSK) a reference carrier at 109.65 MHz, which is then used to despread the IF signal. The command channel despreader output is a narrowband data-modulated carrier at 12.25 MHz if the spreading code phase is correct and is wideband otherwise. The one-sided bandwidth of IF filters 1 to 4 is approximately 5 kHz. This bandwidth is necessary to accommodate doppler shifts on the received signal prior to carrier tracking.

The presence or absence of energy at the IF bandpass filter output is sensed by the synchronization detector. This detector is a square-law envelope detector followed by circuitry which approximates a sequential detector. A bias is added to the envelope detector output and the result is integrated. If noise alone were present at the IF filter output, the integrator output will ramp downward to the lower threshold V_{T1}. When V_{T1} is reached, a dismiss pulse is transmitted to the code synchronization logic, which then instructs the appropriate Gold code generator to step $\frac{1}{2}$ chip to a new phase. If the IF filter output contains a signal component, the sync detector integrator ramps upward. In this implementation there is no upper sequential detection threshold. Rather, a hit is declared if the lower threshold has not been crossed within a fixed time interval. This time interval is set in the block labeled "timeout." The detector bias and threshold are set for a particular envelope detector output due to noise alone. The noise alone output will vary due to changes

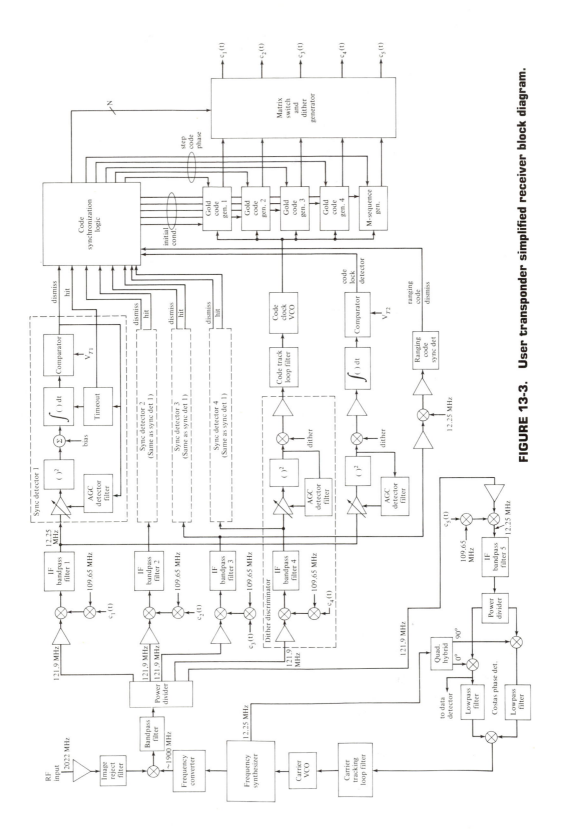

FIGURE 13-3. User transponder simplified receiver block diagram.

in system input noise, changes in amplifier gains, and so on. An AGC is provided to compensate for these changes. In this design, the AGC senses the average dismiss rate and sets the AGC amplifier gain appropriately. This is in contrast to the TRW approach using a noise reference channel to compensate for input noise changes.

At the beginning of the synchronization process, the four Gold code generators are set to four phases of the code which are 90° apart. The four code generator outputs are routed to the top four despreading correlators so that four phases of the spreading code are evaluated simultaneously. An important feature of the synchronization hardware is that the four code generators and the four synchronization detectors operate independently. Thus, when a particular detector dismisses a phase, its associated code generator can immediately step to a new phase. It does not have to wait for all four synch detectors to dismiss. A synchronization system with dependent code generators would not realize a 4:1 improvement in synchronization time.

The four Gold code generators step through the code phase cells in $\frac{1}{2}$-chip increments until one of them finds a correct phase. When this event occurs, the matrix switch immmediately re-routes the output of the generator with the correct phase to a dither generator and then to the $c_4(t)$ output. The $c_4(t)$ output is the reference input to the dither discriminator of the tracking loop. A phase delayed version of $c_4(t)$ is also routed to $c_1(t)$ to be used in the on-time correlator for code lock detection. The code tracking loop for this system is a noncoherent tau-dither early-late tracking loop using a dither magnitude of $\pm\frac{1}{2}$ chip and a dither rate of 200 Hz. Once again, the dither rate is significantly below the IF bandwidth as required to avoid degradation due to transient effects. The dither discriminator output is input to the code-tracking loop filter. The filter output controls the code clock VCO.

With Gold code synchronization established, carrier recovery can begin. Carrier recovery is accomplished by routing an ''on-time'' version of the command channel spreading code through the matrix switch to $c_5(t)$. The command channel IF signal is despread and the despreading mixer output passes through bandpass filter 5. The filter output is a BPSK data-modulated carrier at 12.25 MHz. This signal is input to a Costas-type phase discriminator. The Costas phase discriminator generates a dc output proportional to the carrier phase error which is input to the carrier tracking loop filter. The filter output controls the carrier VCO and all carriers within the receiver are synchronized to this VCO.

With carrier tracking established, the code synchronization logic begins searching for the correct range channel code phase. The range code is epoch synchronous with the Gold code epoch so that there are only 256 potential range code phase cells which need to be evaluated. The m-sequence code generator output is routed through the matrix switch to $c_3(t)$. Range code despreading is accomplished in the third despreading mixer. The output of IF bandpass filter 3 contains the range channel carrier if the range code phase is correct. This carrier is synchronously detected using the mixer preceding the ''ranging code lock detector.'' Observe that the ranging code search does not affect the command channel code tracking or the carrier tracking.

The receiver just described makes use of a large number of the concepts described throughout this text. The principal problem addressed by this design is that of achieving rapid synchronization at low signal-to-noise ratios. New designs with improved performance are currently being investigated.

GLOBAL POSITIONING SYSTEM

One of the many applications of direct-sequence spread spectrum is to ranging systems. Recall that the spread-spectrum code tracking circuitry is actually tracking the propagation delay between the transmitter and the receiver. If the transmission delay is known, the range from the transmitter to the receiver can be determined. The Global Positioning System (GPS) enables users to determine their position on or above the earth's surface by measuring their range to four GPS satellites whose position is accurately known. In its mature configuration, there will be 18 GPS satellites in circular orbits 12,211 miles above the earth. The orbital period of these satellites is exactly 12 hours. There will be six satellites in each of three orbits. These orbits are inclined 55 degrees with respect to the equator and are separated in longitude by 120 degrees. The mature orbital configuration is illustrated in Figure 13-4. This configuration has been selected so that a GPS user anywhere on the earth is able to receive signals from at least four satellites at any time.

Each satellite maintains a highly accurate clock as well as data on the orbits of all 18 satellites. The clocks of all satellites are precisely synchronized through a ground station in California. This ground station also updates the orbital position information for all satellites using information supplied to it by a network of satellite tracking stations. For the purposes of this discussion, all satellite clocks may be considered synchronous and all satellite orbital data may be considered completely accurate. Thus each satellite knows the position of all GPS satellites at any instant in time.

Two direct-sequence spread-spectrum signals, L1 and L2, are transmitted from each satellite to the ground. The L1 and L2 carrier frequencies are 1575.42 and 1227.60 MHz, respectively. Consider the L1 signal first. The spread-spectrum modulation is unbalanced QPSK and I-channel spreading code is a Gold code of length 1023 chips. The chip rate is 1.023 MHz and the code period is 1.0 ms. A

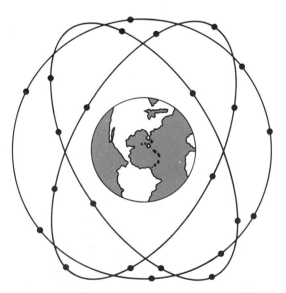

FIGURE 13-4. Global Positioning system operational satellite configuration. (From Ref. 3.)

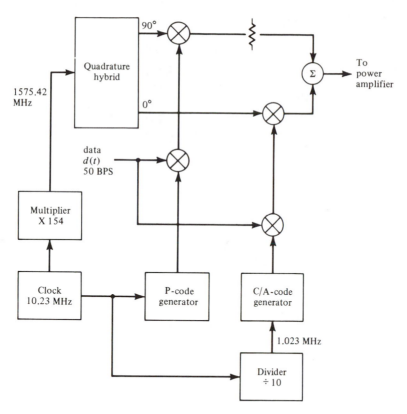

FIGURE 13-5.   Generation of the GPS L1 spread-spectrum signal.

different Gold code is used for each satellite so that a ground receiver can distinguish satellites. This short spreading code is called the Clear/Acquisition or C/A
code. The Q-channel spreading code is a very long nonlinear code whose clock
rate is 10.230 MHz. This code is called the Precise code or P code, and its period
is measured in days. Each satellite uses a different phase of the long code. The
P-channel power is 3 dB less than the C/A channel power. Both the P-channel
and C/A-channel signals are modulated by a 1500-bit data message which is transmitted at 50 bps and repeated cyclically. This 1500-bit message contains position
information as well as data used to correct for errors in propagation time, satellite
clock bias, and so on. One method of generating L1 is shown in Figure 13-5.
Consider now the L2 signal. The spread-spectrum modulation for L2 is BPSK
direct sequence using the P code with a chip rate of 10.23 MHz. The data modulation is also BPSK and the data is identical to the L1 channel data. The L2 signal
is intended for military users of GPS.

Suppose that the GPS user receives the L1 signal. The most sophisticated user
receivers will have four separate channels for simultaneously demodulating the L1
signals from four different GPS satellites. The code generators for each channel
will be programmed to the codes for the satellite of interest. A less complex receiver
will have a single channel and will demodulate the signals from each of four
satellites one after the other. In either case, the receiver first synchronizes to and
tracks the C/A code. With C/A code synchronization established, carrier recovery
circuits can phase lock to the C/A carrier and data demodulation can begin. Using
either information obtained from the data message or previously known data,
P-code synchronization can now be accomplished.

A receiver may determine the range to a satellite using either the C/A code or the P code. Of course, the P code range information is more accurate because its chip rate is higher. Suppose that the C/A code is being used, however, to simplify this discussion. Then, after synchronization to the C/A code, the receiver records the time at which the code epoch* occurs. The time reference used for this measurement is the receiver's own clock, which is assumed to be offset from the system clock by Δ. The time of arrival (TOA) for the C/A code epoch from all four satellites is recorded.

The data transmitted from each satellite gives the exact position of the satellite and the system time when the code epoch event occurred. Figure 13-6 illustrates the time relationship between the system time and the receiver time. The code epoch occurred at the satellite transmitter at system time t_1. The receiver records the time of the received epoch event as t_2 (in receiver time) and calculates the propagation time $T_i' = t_2 - t_1$. Because of the error between the system and receiver clocks, T_1' is incorrect and the actual propagation time is $T_i = T_i' + \Delta$. The value of T_i' is recorded for all four satellites, and is used to calculate a pseudorange $\rho_i = T_i'c$ for each. Because the satellite clocks are all synchronous, the difference between the actual range $R_i = T_ic$ and the pseudorange ρ_i is the same for all satellites. This range error is denoted by $B = c\Delta$.

The coordinate system used to calculate the user position is a rectangular system with origin at the earth's center. Let x, y, and z denote the user position and U_i, V_i, and W_i denote the position of the ith satellite in this coordinate system. Then the following set of four equations relate the pseudoranges calculated by the receiver to the user position, the satellite positions, and the unknown range error:

$$R_1 = \rho_1 + B = [(U_1 - x)^2 + (V_1 - y)^2 + (W_1 - z)^2]^{1/2}$$

$$R_2 = \rho_2 + B = [(U_2 - x)^2 + (V_2 - y)^2 + (W_2 - z)^2]^{1/2}$$

$$R_3 = \rho_3 + B = [(U_3 - x)^2 + (V_3 - y)^2 + (W_3 - z)^2]^{1/2}$$

$$R_4 = \rho_4 + B = [(U_4 - x)^2 + (V_4 - y)^2 + (W_4 - z)^2]^{1/2} \qquad (13\text{-}2)$$

This is a set of four equations in four unknowns which can be solved for the user position and range error. Iterative techniques are normally used to solve these equations. The translation from the (x, y, z) coordinates to latitude, longitude, and altitude is straightforward and is discussed in [4].

*A code epoch is marked by any known code generator state. It is usually the all-ones state.

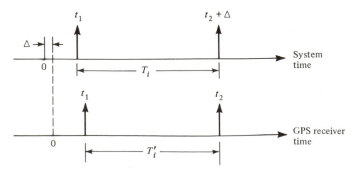

FIGURE 13-6. **Time lines illustrating the relationship between system time and receiver time.**

This discussion of GPS is oversimplified in that many error sources are not discussed. The simplest receivers perform a calculation just as described; however, the more sophisticated receivers use a more complicated algorithm. Position accuracy on the order of tens of meters is possible and has been demonstrated. The student is referred to [5–8] for additional information about the Global Positioning System.

JOINT TACTICAL INFORMATION DISTRIBUTION SYSTEM

The Joint Tactical Information Distribution System (JTIDS) is a tactical military spread-spectrum radio network which is currently being developed in the United States. This system will provide jam-resistant communications and location for troops in combat situations, and is being developed with support from all the military services and NATO. Jamming resistance is achieved using a hybrid direct-sequence/frequency-hop/time-hop transmission strategy. The description that follows is extracted entirely from the open literature cited.

The motivation for the development of JTIDS was a realization that the battlefield of the future will contain a large number of electronic systems all of which must be coordinated to produce maximum effectiveness. Air and land units must communicate with one another as well as with distant command centers and with large radar systems such as the Airborne Warning and Control System (AWACS). In addition, it is important for all field units to know their relative position and the positions of hostile forces. This communications and information gathering must be accomplished in an electromagnetic environment which contains many signals, some of which may be hostile. Even without hostile signal interference, the large number of friendly communications must be highly coordinated to avoid significant interference with one another. The rather complex JTIDS signal structure provides an efficient means of signaling in the anticipated battlefield environment of the future.

All JTIDS carrier frequencies are within the frequency bands 969 to 1008 MHz and 1113 to 1206 MHz [9]. The choice of carrier frequencies in these bands limits transmitted energy to roughly the band 960 to 1215 MHz and avoids interference with the IFF frequencies of 1030 and 1090 MHz. Within this band a number of JTIDS networks may be established. These networks are distinguished from one another by using different spreading codes for each net. Thus JTIDS is a spread-spectrum multiple-access system. Consider first a single JTIDS network. All members of this network use the same spreading codes and can therefore communicate with one another. The members of this network coordinate their transmissions using time-division multiple-access techniques. The time-division structure for a single JTIDS network is illustrated in Figure 13-7. The first level of time division is the EPOCH, which is illustrated on the top line. A JTIDS EPOCH is 12.8 minutes long. All network events occur at least once every 12.8 minutes but some events occur much more often than this. Each EPOCH is subdivided into 64 FRAMES of duration 12 seconds as illustrated in the second line of Figure 13-7. A FRAME is further subdivided into TIME SLOTS. There are 1536 TIME SLOTS in a FRAME and each TIME SLOT lasts 7.8125 milliseconds.

JTIDS has developed in two phases. The phase I program (called JTIDS I or JTIDS/TDMA) signal structure assigns an entire time slot to a single JTIDS network member. In JTIDS I, the smallest signaling unit that could be assigned to a

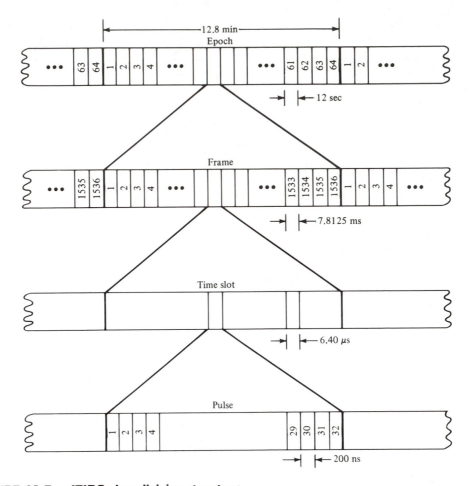

FIGURE 13-7. JTIDS time-division structure.

single member was the TIME SLOT and each active member had to be assigned at least one TIME SLOT per EPOCH. Most users will, however, require more channel capacity than one TIME SLOT per EPOCH and will therefore be assigned many TIME SLOTS. Within a TIME SLOT a network member transmits a single JTIDS I message which has a precisely defined format and meaning. Standardized messages are used so that often used communications consume the minimum possible system resources. Nonstandard messages can also be transmitted when required. Within a TIME SLOT there are message PULSES. Each of these PULSES is a minimum-shift-keyed direct-sequence spread-spectrum burst consisting of 32 chips each 200 nanoseconds long. Thus the direct-sequence spreading chip rate is 5×10^6 chips/second. The total pulse length is 6.4 microseconds. A 5-bit message is transmitted with each pulse by associating each message with a different phase of the 32-bit direct-sequence spreading code burst. Frequency hopping is implemented by hopping the carrier to a new frequency for each transmitted pulse. There are 51 possible hop frequencies and all network members hop synchronously. Within each time slot there are a total of 129 pulses or 129 double pulses. Double pulses, each transmitting the same information, are used to provide higher jam resistance when necessary.

In the JTIDS phase II signal structure an entire TIME SLOT is no longer assigned to a particular network member. Rather, time-division multiple-access occurs on a pulse-by-pulse basis. Each user message consists of pulses which are pseudorandomly spaced in time. This modified time-division strategy is called distributed TDMA and the system is called JTIDS II or JTIDS/DTDMA. For both JTIDS I and II the basic channel is a 32-ary channel since one of 32 phases of the MSK spreading code burst is transmitted. Error correction coding is used by all JTIDS members. The particular code used is a $n = 31$, $k = 15$ Reed–Solomon block code. This code provides both error correction and error detection. Both JTIDS I and JTIDS II are advanced spread-spectrum systems which use many of the concepts discussed throughout this text. Further information about this important system can be found in [10–12].

REFERENCES

[1] W. K. ALEM et al., "Spread Spectrum Acquisition and Tracking Performance for Shuttle Communication Links," *IEEE Trans. Commun.*, November 1978.

[2] TRW Systems Group, "Shuttle Spectrum Despreader," Final Report, May 21, 1976, Contract NAS-9-14690.

[3] J. J. SPILKER, *Digital Communications by Satellite* (Englewood Cliffs, N.J.: Prentice-Hall, 1977).

[4] K. P. YIU et al., "Land Navigation with a Low Cost GPS Receiver," *Conf. Rec.*, IEEE Natl. Telecommun. Conf., 1980.

[5] K. M. JOSEPH et al., "NAVSTAR GPS Receiver for Satellite Applications," AGARD Guidance and Control Panel Symp., London, October 1980.

[6] J. D. CARDALL and RICHARD S. CNOSSEN, "Civil Application of Differential GPS," *Conf. Rec.*, Int. Telem. Conf., October 1981.

[7] R. P. DENARO, "NAVSTAR: The All-Purpose Satellite," *IEEE Spectrum*, May 1981.

[8] A. J. VAN DIERENDONCK et al., "Time Transfer Using NAVSTAR GPS," *Conf. Rec.*, IEEE Natl. Telecommun. Conf., 1981.

[9] INTERAVIA, "JTIDS: A Major Effort Towards Jam Resistant Secure Communications," November 1978.

[10] COL. W. S. JONES, "Army Firms Up JTIDS Planning," *Defense Electron.*, August 1982.

[11] A. HAUPTSCHEIN, "Practical, High Performance Concatenated Coded Spread Spectrum Channel for JTIDS," *Conf. Rec.*, Natl. Telecommun. Conf., 1977.

[12] D. B. BRICK and F. W. ELLERSICK, "Future Air Force Tactical Communications," *IEEE Trans. Commun.*, September 1980.

Probability and
Random Variables

In this appendix, basic theory regarding probability and random variables is reviewed. Also developed are random signal and system analysis concepts used in the analysis of digital communication systems.

A-1

PROBABILITY THEORY

A-1.1 Definitions

In considering the theory of probability, as with any theory, it is convenient to introduce several definitions. Definitions that will be useful in the consideration of the theory of probability are listed below:

1. *Outcome*: The end result of an experiment.
2. *Random or chance experiment*: An experiment whose outcome is not known in advance.
3. *Event*: An event A is an outcome or collection of outcomes of a random experiment.
4. *Sample space*: The sample space S of a chance experiment is the set of all possible outcomes of the chance experiment.

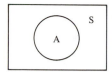

Sample space, S, and
Event, A.

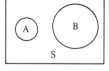

Disjoint events, or
sets, A and B.

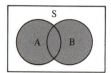

Union of two events,
or sets, A and B
(shaded).

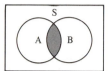

Intersection of two
events or sets, A and
B (shaded).

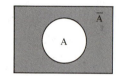

Complement of set A (shaded)

FIGURE A-1.   Various definitions involving events in probability.

5. *Mutually exclusive events*: Events A and B are said to be mutually exclusive (disjoint) if they cannot occur simultaneously (contain no common outcomes).
6. *Union of events*: The union of two events A and B, denoted by $A \cup B$ (sometimes also denoted by ''A or B'' or ''$A + B$''), is the set of all outcomes that belong to A or B, or both.
7. *Null event*: The null event, ϕ, represents an impossible outcome or impossible collection of outcomes of a chance experiment.
8. *Intersection of events*: The intersection of two events A and B, denoted by $A \cap B$ (also denoted by ''A and B'' or AB), is the set of all outcomes which belong to both A and B.
9. *Complement of an event*: B is the complement of A if it consists of all outcomes not included in A. (Notations used for the complement are A' and A^c.)
10. *Occurrence*: Event A of a chance experiment is said to have occurred if the experiment terminates in an outcome that belongs to event A.

In probability theory, it is useful to represent events as sets in a sample space. The relationships between several events can be shown by a *Venn diagram*, which represents each event or set as an area inside a box representing S. Venn diagrams illustrating the definitions given above are shown in Figure A-1.

A-1.2   Axioms

The whole of probability is based on three axioms. Let A be an event, or outcome of a chance experiment, and $P(A)$ a real number called the probability of A. $P(A)$ satisfies the following axioms:

(a) $P(A) \geq 0$.

(b) $P(S) = 1$, where S is the sample space.

(c) If $A \cap B = 0$ (i.e., A and B are mutually exclusive), then

$$P(A \cup B) = P(A) + P(B) \tag{A-1}$$

The reasonableness of these axioms can be seen by considering the relative frequency definition of the probability of A, which is

$$P(A) \triangleq \lim_{n \to \infty} \frac{n_A}{n} \tag{A-2}$$

where n is the number of times a chance experiment is performed, and n_A the number of times event A occurs during the repeated performance of this experiment. (*Note:* There are difficulties with this definition. What are they?) Axiom (a) simply states that n_A and n are nonnegative; axiom (b) states that if A always occurs (i.e., is the same as S), its probability must be unity; axiom (c) says that the number of times that either of two mutually exclusive events occur in n repetitions of a chance experiment is equal to the sum of their separate occurrences.

Using only the axioms above, it is possible to show that

$$P(A + B) = P(A) + P(B) - P(AB) \tag{A-3}$$

where A and B are not necessarily mutually exclusive. Furthermore, if $\overline{A}$ is the *complement* of A (i.e., $A + \overline{A} = S$ and $A\overline{A} = 0$), then

$$P(\overline{A}) = 1 - P(A) \tag{A-4}$$

A-1.3   Joint, Marginal, and Conditional Probabilities

Let A, B, C, B_1, B_2, and so on, be events. Let $P(A|B)$ denote the probability of event A *given* the occurrence of event B. Then the following relationships hold:

(a) $P(AB) = P(A|B)P(B) = P(B|A)P(A)$ $\tag{A-5}$

(b) $P(B|A) = \dfrac{P(A|B)P(B)}{P(A)} \qquad P(A) \neq 0$ $\tag{A-6}$

   [called Bayes' rule; obtained by solving (A-5) for $P(B|A)$]

(c) $P(ABC) = P(C|AB)P(AB)$
$$= P(C|AB)P(B|A)P(A) \tag{A-7}$$

   (the chain rule)

(d) $P(A) = \displaystyle\sum_{j=1}^{m} P(A|B_j)P(B_j)$ $\tag{A-8}$

   where $B_1 + B_2 + \cdots B_m = S$ and $A(B_1 + B_2 + \cdots + B_m) = A$ (i.e., mutually exclusive and exhaustive) $P(A)$ is often called a *marginal probability*.

Intuitively, if A and B are events associated with the outcomes of an experiment such that the occurrence of A does not influence the occurrence of B, they are said to be *statistically independent*. For such events, the statements

$$P(AB) = P(A)P(B) \tag{A-9a}$$

$$P(B|A) = P(B) \tag{A-9b}$$

$$P(A|B) = P(A) \tag{A-9c}$$

are equivalent and, in fact, any one of them defines statistical independence.

The following example illustrates some of the foregoing relationships in a communications system context.

EXAMPLE A-1

Consider the transmission of binary digits through a communications channel. As is customary, the two possible symbols are denoted as 0 and 1. Let the probability of receiving a 0, given that a 0 was sent, $P(0r|0s)$, and the probability of receiving a 1, given that a 1 was sent, $P(1r|1s)$, be

$$P(0r|0s) = P(1r|1s) = 0.99$$

From (A-4), it follows that

$$P(1r|0s) = 1 - P(0r|0s) = 0.01$$

and

$$P(0r|1s) = 1 - P(1r|1s) = 0.01$$

respectively. These probabilities characterize the channel and would be obtained through experimental measurement or analysis. Techniques for calculating them for particular situations are discussed in Chapters 3 and 4.

In addition to these probabilities, suppose that we have determined through measurement that the probability of sending a 0 is

$$P(0s) = 0.6$$

and, therefore, from (A-4) the probability of sending a 1 is

$$P(1s) = 1 - P(0s) = 0.4$$

The probability of a 1 having been sent given a 1 was received, from (A-6), is

$$P(1s|1r) = \frac{P(1r|1s)P(1s)}{P(1r)}$$

To find $P(1r)$, note that

$$P(1r) = P(1r, 1s) + P(1r, 0s)$$

where

$$P(1r, 1s) = P(1r|1s)P(1s) = (0.99)(0.4) = 0.396$$

$$P(1r, 0s) = P(1r|0s)P(0s) = (0.01)(0.6) = 0.006$$

where a comma is used to separate the joint event 1 received and 0 sent. Thus

$$P(1r) = P(1r, 1s) + P(1r, 0s)$$

$$= 0.396 + 0.006 = 0.402$$

and $P(1s|1r)$ is found to be

$$P(1s|1r) = \frac{(0.99)(0.4)}{0.402} = 0.985$$

Similarly, one could calculate $P(0s|1r)$, $P(0s|0r)$, and $P(1s|0r)$. The necessary calculations are left to the student as an exercise. ∎

RANDOM VARIABLES, PROBABILITY DENSITY FUNCTIONS, AND AVERAGES

A-2.1 Random Variables

A random variable is a rule, or functional relationship, which assigns real numbers to each possible outcome of a chance experiment. An example of a random variable is the assignment of 1 to the occurrence of a head up and a 0 to the occurrence of a tail up when a coin is tossed. The assignment of random variables to the outcomes of chance experiments is convenient from the standpoint of analysis.

A standard notation is to denote random variables by capital letters (X, Y, etc.) and the values that they take on are denoted by the corresponding lowercase letters (y, x, etc.).

Random variables may be discrete, continuous, or mixed, depending on whether they take on a countable (discrete) or uncountable (continuous) number of values, or both.

A-2.2 Probability Density Functions

A *probability density function*, $f_X(x)$, characterizes a continuous random variable, and is defined by the following properties:

(1) $f_X(x) \geq 0, \quad -\infty < x < \infty$ $\qquad\qquad$ (A-10)

(2) $\displaystyle\int_{-\infty}^{\infty} f_X(x)\, dx = 1$ $\qquad\qquad$ (A-11)

(3) $\displaystyle\Pr(X \leq x) = \int_{-\infty}^{x} f_X(\lambda)\, d\lambda \triangleq F_X(x)$ $\qquad\qquad$ (A-12)

The function $F_X(x)$ is called the *distribution function* of X.

From (A-12), and noting that the events $-\infty < X \leq a$ and $a < X \leq b$ are disjoint and that the union of these two events is the event $-\infty < X \leq b$, it is concluded that

$$\Pr(a < X \leq b) = \int_{a}^{b} f_X(\lambda)\, d\lambda$$

$$= F_X(b) - F_X(a) \qquad\qquad \text{(A-13)}$$

Letting $a = x$ and $b = x + \Delta x$ with Δx small, it follows that if $f_X(x)$ is continuous in $[x, x + \Delta x]$, then

$$f_X(x)\, \Delta x \simeq \Pr(x < X \leq x + \Delta x)$$

$$= F_X(x + \Delta x) - F_X(x) \qquad\qquad \text{(A-14)}$$

or, dividing by Δx and taking the limit $x \to 0$,

$$f_X(x) = \frac{dF_X(x)}{dx} \qquad\qquad \text{(A-15)}$$

The distribution function, $F_X(x)$, of a continuous random variable X is a *continuous, nondecreasing* function of x with the end-point values $F_X(-\infty) = 0$ and

$F_X(\infty) = 1$. The distribution function of a discrete random variable is a *nondecreasing stair step* function of x.

In a strict mathematical sense, the probability density function of a discrete random variable is undefined. If the use of delta functions is accepted, it is possible to define probability density functions for discrete random variables as well as for continuous random variables. Figure A-2 illustrates density and distribution functions for continuous and discrete random variables.

So far, probability density and distribution functions have been discussed only for single random variables. They may be defined for more than one random variable as well. For example, if X and Y are random variables, their joint probability density function, $f_{XY}(x, y)$, is a function of two variables x and y with the following properties:

$$f_{XY}(x, y) \geq 0, \qquad -\infty < x, y < \infty \tag{A-16a}$$

$$\iint\limits_{-\infty}^{\infty} f_{XY}(x, y) \, dx \, dy = 1 \tag{A-16b}$$

$$\mathrm{Pr}\,(X \leq x, Y \leq y) \triangleq F_{XY}(x, y)$$

$$= \int_{-\infty}^{x} \int_{-\infty}^{y} f_{XY}(u, v) \, du \, dv \tag{A-16c}$$

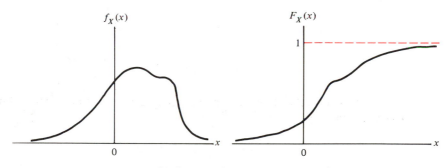

(a) Continuous random variable functions

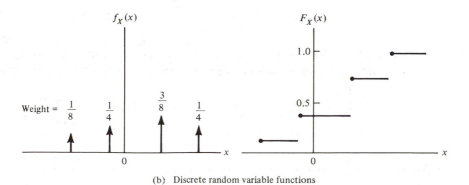

(b) Discrete random variable functions

FIGURE A-2. Probability density and distribution functions for continuous and discrete random variables.

$$\int_{-\infty}^{\infty} f_{XY}(x, y) \, dy = f_X(x) \tag{A-16d}$$

$$\int_{-\infty}^{\infty} f_{XY}(x, y) \, dx = f_Y(y) \tag{A-16e}$$

where $f_X(x)$ and $f_Y(y)$ are sometimes referred to as "marginal" density functions.

Common examples of marginal and joint probability density functions are those for *Gaussian random variables*. X is Gaussian if it is described by the probability density function

$$f_X(x) = \frac{\exp\left[-(x - m)^2/2\sigma^2\right]}{\sqrt{2\pi\sigma^2}} \tag{A-17}$$

where m and σ are parameters of the density function. The error function, erf (x), is related to the distribution function of a Gaussian random variable, and is defined as

$$\text{erf}\,(x) = \frac{2}{\sqrt{\pi}} \int_0^x e^{-u^2} \, du \tag{A-18}$$

The *complementary error function* is defined as erfc $(x) = 1 - $ erf (x). For a Gaussian random variable, it can be shown that

$$\Pr\,(m - a \le X \le m + a) = \text{erf}\left(\frac{a}{\sqrt{2}\sigma}\right) \tag{A-19}$$

where

$$m = \int_{-\infty}^{\infty} x f_X(x) \, dx \tag{A-20a}$$

$$\sigma^2 + m^2 = \int_{-\infty}^{\infty} x^2 f_X(x) \, dx \tag{A-20b}$$

Another function related to the complementary error function is the Q-function, which is defined as

$$Q(x) = \frac{1}{\sqrt{2\pi}} \int_x^{\infty} e^{-v^2/2} \, dv \tag{A-21}$$

By the change of $u = v/\sqrt{2}$ it is readily shown that

$$\tfrac{1}{2}\,\text{erfc}\,(x) = \frac{1}{\sqrt{2\pi}} \int_{\sqrt{2}x}^{\infty} e^{-v^2/2} \, dv = Q(\sqrt{2}x) \tag{A-22}$$

Appendix E provides several useful relations for $Q(x)$ as well as a table of values.

If X and Y are jointly Gaussian, they have the joint probability density function

$$f_{XY}(x, y) = \frac{\exp\left[-\dfrac{\left(\dfrac{x - m_X}{\sigma_X}\right)^2 - 2\rho\left(\dfrac{x - m_X}{\sigma_X}\right)\left(\dfrac{y - m_Y}{\sigma_Y}\right) + \left(\dfrac{y - m_Y}{\sigma_Y}\right)^2}{2(1 - \rho^2)}\right]}{2\pi\sigma_x\sigma_y\sqrt{1 - \rho^2}} \tag{A-23}$$

It can be shown that (A-16d) and (A-16e) hold between (A-23) and (A-17). In (A-23) m_X, m_Y, σ_X, σ_Y, and ρ are parameters.

The *conditional probability density function* of Y given X is defined as

$$f_{Y|X}(y|x) = \frac{f_{XY}(x, y)}{f_X(x)}$$ (A-24)

with a similar definition for the probability density function of X given Y.

A-2.3 Averages of Random Variables

The average of some function $g(\cdot)$ of a random variable, X, is obtained by evaluating the integral

$$E[g(X)] = \overline{g(X)} = \int_{-\infty}^{\infty} g(x)f_X(x)\, dx$$ (A-25)

The average of a function of two random variables, say $h(X, Y)$, is obtained by evaluating the double integral

$$E[h(X, Y)] = \overline{h(X, Y)} = \int_{-\infty}^{\infty}\int_{-\infty}^{\infty} h(x, y)f_{XY}(x, y)\, dx\, dy$$ (A-26)

If a random variable is discrete (A-25) and (A-26) are used with the probability density functions being sums of unit impulses $\delta(x - x_j)$ or $\delta(x - x_j)\delta(y - y_k)$ with weights equal to the probability of the particular value of X or X jointly with Y being taken on. For a discrete random variable, or set of discrete random variables, (A-25) or (A-26), as the case may be, revert to summations. The notation $E\{\cdot\}$ means "expectation of" and either it or the overbar denote a statistical average. Several of the most commonly occurring averages are listed below:

(1) mean $= m_1 = m_X = E(X) = \int_{-\infty}^{\infty} xf_X(x)\, dx$ (A-27a)

(2) mean square $= m_2 = E(X^2) = \int_{-\infty}^{\infty} x^2 f_X(x)\, dx$ (A-27b)

(3) variance $= \sigma_X^2 = \overline{(X - \overline{X})^2} = E[\{X - E(X)\}^2]$ (A-27c)

(4) covariance $= \mu_{XY} = E[(X - \overline{X})(Y - \overline{Y})]$ (A-27d)

(5) correlation coefficient $= \rho_{XY} = \dfrac{\mu_{XY}}{\sigma_X \sigma_Y}$ (A-27e)

EXERCISE

(a) Show that the variance can be written as

$$\sigma_X^2 = m_2 - m_X^2$$

(b) Show that the covariance can be expressed as

$$\mu_{XY} = E(XY) - m_X m_Y$$ ∎

A random variable X for which $m_X = 0$ is called *zero mean*. Two random variables for which $\mu_{XY} = 0$ are said to be *uncorrelated*. If two random variables have a joint probability density function that factors, that is,

$$f_{XY}(x, y) = f_X(x)f_Y(y)$$ (A-28)

they are said to be *statistically independent*. Statistically independent random variables are uncorrelated; the reverse is not necessarily true [it is for Gaussian random variables, however, since $\rho = 0$ in (A-23) means that $f_{XY}(x, y)$ factors].

Convenient theorems pertaining to averages of random variables are listed below.

1. The average of a sum of random variables is equal to the sum of their respective averages. That is,

$$\overline{X_1 + X_2 + \cdots + X_N} = \overline{X}_1 + \overline{X}_2 + \cdots + \overline{X}_N \qquad \text{(A-29)}$$

2. The average of a constant is the constant itself; that is, $E(A) = A$, where A is a constant.
3. The average of a constant times a random variable is the constant times the average of the random variable; that is, $E(AX) = AE(X)$ where A is a constant.
4. The average of the product of two random variables is equal to the product of their separate averages *if they are statistically independent* (sufficient condition); that is,

$$\overline{XY} = \overline{X}\,\overline{Y} \qquad (X \text{ and } Y \text{ statistically independent}) \qquad \text{(A-30)}$$

A-3

CHARACTERISTIC FUNCTION AND PROBABILITY GENERATING FUNCTIONS

A-3.1 Characteristic Function

An extremely useful average is the *characteristic function*. For a single random variable, it is obtained by letting $g(X) = \exp(jvX)$ in (A-25), which results in the definition

$$M_X(v) = \int_{-\infty}^{\infty} f_X(x)e^{jvx}\, dx \qquad \text{(A-31)}$$

For several random variables, say $X_1, X_2, \ldots, X_N$, the *N-fold characteristic function* is given by

$$M_{X_1 \cdots X_N}(v_1, v_2, \ldots, v_N) = E\left[\exp\left(j \sum_{i=1}^{N} v_i X_i\right)\right] \qquad \text{(A-32)}$$

Note that setting $v_N = v_{N-1} = \cdots = v_2 = 0$ results in (A-31). Equation (A-32) can be written compactly by defining the column matrices or vectors

$$\mathbf{v} = \begin{bmatrix} v_1 \\ v_2 \\ \cdot \\ \cdot \\ \cdot \\ v_N \end{bmatrix} \qquad \text{and} \qquad \mathbf{X} = \begin{bmatrix} X_1 \\ X_2 \\ \cdot \\ \cdot \\ \cdot \\ X_N \end{bmatrix} \qquad \text{(A-33)}$$

which results in

$$M_{\mathbf{X}}(\mathbf{v}) = E[\exp(\mathbf{v}^t \mathbf{X})] \qquad \text{(A-34)}$$

where the superscript t denotes transpose. The usefulness of the characteristic function is due primarily to three properties:

1. By comparing (A-31) with the definition of the Fourier transform, it is seen that the characteristic function is the Fourier transform of the probability density function of random variable with the exception that a minus sign appears in the exponent of the normal definition of the Fourier transform. From this observation, it follows that the probability density function of a random variable can be obtained as the inverse Fourier transform of the characteristic function (accounting, of course, for the difference of sign in the exponent in the table of Fourier transforms used). Similarly, if dealing with N random variables, the Nth order joint probability density function may be obtained as the N-fold inverse Fourier transform of (A-32). Obtaining the probability density function in this fashion may be a useful approach in cases where the characteristic function is easy to obtain.* The inverse Fourier transform operation is not necessarily easy, but perhaps series approximations for it are possible.

2. Differentiation of (A-31) underneath the integral sign n times with respect to v and setting $v = 0$ shows that

$$E(X^n) = (-j)^n \left. \frac{d^n M_X(v)}{dv^n} \right|_{v=0} \tag{A-35}$$

A similar procedure using partial differentiation on the N-fold characteristic function (A-32) shows that

$$E(X_1^{n_1} X_2^{n_2} \cdots X_N^{n_N}) = (-j)^N \left. \frac{\partial^N M_{\mathbf{X}}(\mathbf{v})}{\partial^{n_1} v_1 \, \partial^{n_2} v_2 \cdots \partial^{n_N} v_N} \right|_{\mathbf{v}=\mathbf{0}} \tag{A-36}$$

where $n_1 + n_2 + \cdots + n_N = N$. Thus the moments of a random variable or set of random variables can be obtained through differentiation of the characteristic functions.

3. Expansion of the exponentials in (A-31) or (A-32) in a power series shows that the moments of a random variable or set of random variables are the coefficients of a Taylor series expansion for the characteristic function.

The observations above are useful in cases where the characteristic function of a set of random variables is easier to obtain or work with than the probability density function. The following examples develop a useful relation for Gaussian random variables.

EXAMPLE A-2

The characteristic function of a Gaussian random variable, by using (A-17) in (A-31), is

$$M_X(v) = \int_{-\infty}^{\infty} \exp\left[\frac{-(x - m)^2}{2\sigma^2} + jvx \right] dx$$

$$= (2\pi\sigma^2)^{-1/2} \exp\left(jmv - \tfrac{1}{2}\sigma^2 v^2 \right) \int_{-\infty}^{\infty} \exp\left[-\frac{(u - j\sigma^2 v)^2}{2\sigma^2} \right] du$$

which follows by the change of variables $x = m + u$ and completing the square in the exponent. Making the change of variables $y = u - j\sigma^2 v$ and noting that $\int_{-\infty}^{\infty} (2\pi\sigma^2)^{-1/2} \exp(y^2/2\sigma^2) \, dy = 1$ results in

$$M_X(v) = \exp\left(jmv - \tfrac{1}{2}\sigma^2 v^2 \right) \tag{A-37}$$

---

*By using the fact that the area (volume) under the probability density function is unity, one can show that (A-31) or (A-32) converge absolutely.

The characteristic function of the new zero-mean Gaussian random variable $Y = X - m$ is

$$M_Y(v) = \exp\left(-\tfrac{1}{2}\sigma^2 v^2\right) = \exp\left[\tfrac{1}{2}\sigma^2(jv)^2\right]$$

which results simply by setting $m = 0$ in (A-37). Its moments can be obtained by repeated differentiation of $M_Y(v)$ or expansion in a power series in jv. Use of the latter approach results in

$$M_Y(v) = \sum_{n=0}^{\infty} \frac{1}{n!}\left[\frac{1}{2}\sigma^2(jv)^2\right]^n$$

The nth-order moment of $Y = X - m$ is the coefficient of the term $(jv)^n/n!$. These *central moments*, as they are called, are given by

$$E\{Y^{2k}\} = E\{(X - m)^{2k}\} = \frac{(2k)!}{2^k k!}\,\sigma^{2k}, \ k = 0, 1, 2, \ldots \tag{A-38}$$

for the even-order moments, with the odd-order moments being zero. ∎

EXAMPLE A-3

The joint probability density function of N jointly Gaussian random variables is [1, p. 208]

$$f_X(x) = (2\pi)^{-n/2}|\det\,\mathbf{C}|^{-1/2}\exp\left[-\tfrac{1}{2}(\mathbf{x} - \mathbf{m})^t\mathbf{C}^{-1}(\mathbf{x} - \mathbf{m})\right] \tag{A-39}$$

where $\mathbf{x}$ and $\mathbf{m}$ are column matrices defined as

$$\mathbf{x} = \begin{bmatrix} x_1 \\ x_2 \\ \cdot \\ \cdot \\ \cdot \\ x_N \end{bmatrix} \quad \text{and} \quad \mathbf{m} = \begin{bmatrix} m_1 \\ m_2 \\ \cdot \\ \cdot \\ \cdot \\ m_N \end{bmatrix} \tag{A-40}$$

respectively, and $\mathbf{C}$ is the positive definite matrix of covariances with elements

$$C_{ij} = \frac{E[(X_i - m_i)(X_j - m_j)]}{\sigma_{X_i}\sigma_{X_j}} \tag{A-41}$$

The joint characteristic function of the Gaussian random variables $X_1, X_2, \ldots, X_N$ is

$$M_X(\mathbf{v}) = \exp\left(j\mathbf{m}^t\mathbf{v} - \tfrac{1}{2}\mathbf{v}^t\mathbf{C}\mathbf{v}\right) \tag{A-42}$$

From the power series expansion of (A-42) it follows that for any four zero-mean Gaussian random variables

$$E(X_1 X_2 X_3 X_4) = E(X_1 X_2)E(X_3 X_4) + E(X_1 X_3)E(X_2 X_4) + E(X_1 X_4)E(X_2 X_3) \tag{A-43}$$

which is a rule that is sufficiently useful to commit to memory. ∎

A-3.2 Probability Generating Function

For a discrete random variable, X, taking on uniformly spaced values, a convenient tool for analyzing probability distributions is the *probability generating function*.

Clearly, if a random variable takes on uniformly spaced values, it can be represented as an integer-valued random variable. Let $\Pr (X = k) = P_k$ be the probabilities that X takes on the integer value k, and assume that $k \geq 0$.* The probability generating function, defined by

$$h(z) = \sum_{k=0}^{\infty} P_k z^k \qquad \text{(A-44)}$$

is recognized as the z-transform of the sequence of probabilities $\{p_k\}$ with z^{-1} replaced by z. Differentiation of (A-44) n times with respect to z and setting z in the result equal to zero shows that

$$P_n = \frac{1}{n!} \frac{d^n}{dz^n} h(z) \bigg|_{z=0} \qquad \text{(A-45)}$$

Furthermore, the derivatives of $h(z)$ at $z = 1$ result in the factorial moments so that

$$E[X(X - 1)(X - 2) \cdots (X - n + 1)] \triangleq c_n = \frac{d^n}{dz^n} h(z) \bigg|_{z=1} \qquad \text{(A-46)}$$

From the definition of c_n, it is seen that

$$c_1 = E(X) = m_1 \qquad \text{(A-47)}$$

$$c_2 = E[X(X - 1)] = m_2 - m_1 \qquad \text{(A-48)}$$

$$\sigma_X^2 = m_2 - m_1^2 = c_2 + c_1 - c_1^2 \qquad \text{(A-49)}$$

The Maclaurin series expansion of $h(z)$ about $z = 1$ is

$$h(z) = \frac{\displaystyle\sum_{m=0}^{\infty} c_m(z - 1)^m}{m!} \qquad \text{(A-50)}$$

EXAMPLE A-4

An integer-valued random variable with probabilities

$$P_k = \binom{N}{k} p^k q^{N-k} \qquad 0 \leq k \leq N$$

where k and N are integers and $p + q = 1$ with $p > 0$ is called a *binomial random variable*. Its probability generating function is

$$\begin{aligned} h(z) &= \sum_{k=0}^{N} \binom{N}{k} p^k q^{N-k} z^k \\ &= \sum_{k=0}^{N} \binom{N}{k} (pz)^k q^{N-k} \\ &= (pz + q)^N \\ &= [1 + p(z - 1)]^N \end{aligned}$$

---

*In what follows, negative values of k could also be accommodated, but would be slightly more complicated mathematically. See [2] for a discussion of the two-sided z-transform.

Using the binomial theorem again, $h(z)$ can be expanded as

$$h(z) = \sum_{m=0}^{N} \binom{N}{m} p^m (z - 1)^m$$

$$= \frac{\displaystyle\sum_{m=0}^{N} \frac{N!}{(N - m)!} p^m (z - 1)^m}{m!}$$

Comparing this series with (A-50), it is seen that

$$c_m = \frac{N!}{(N - m)!} p^m$$

Therefore, the mean and variance of a binomial random variable are, from (A-47) and (A-49), given by

$$m_1 = c_1 = Np$$

$$\sigma^2 = Np(1 - p) = Npq \qquad \blacksquare$$

EXAMPLE A-5

A Poisson random variable is an integer-valued random variable taking on values in the range $(0, \infty)$ with probabilities

$$P_k = \frac{\lambda^k e^{-\lambda}}{k!} \qquad 0 \le k < \infty; \quad \lambda > 0$$

The probability generating function is

$$h(z) = \frac{\displaystyle\sum_{k=0}^{\infty} \lambda^k z^k e^{-\lambda}}{k!}$$

$$= e^{-\lambda} e^{\lambda z}$$

$$= e^{\lambda(z - 1)}$$

$$= \frac{\displaystyle\sum_{m=0}^{\infty} \lambda^m (z - 1)^m}{m!}$$

Comparing the second expansion with (A-50) shows that

$$c_m = \lambda^m$$

Therefore, the mean and variance of a Poisson random variable are given, respectively, by

$$m_1 = \lambda$$

$$\sigma^2 = \lambda$$

That is, the mean and variance of a Poisson random variable are equal. $\qquad \blacksquare$

TRANSFORMATIONS OF RANDOM VARIABLES

A-4.1 General Results

The transformation of a random variable or set of random variables to another random variable or set of random variables is a frequently occurring problem. Examples are the following:

1. The input at time $t = t_1$ to a nonlinear device, say a diode, is a random variable with Gaussian probability density function. What is the probability density function of the output?
2. In shooting at a target, it is known that the errors along orthogonal axes between the bullet's impact point and the center of the target are statistically independent Gaussian random variables. What is the probability that the distance away from the target center in any direction will be greater than some value, say 10 cm?
3. Given the same situation as in (2), what is the joint probability density of the impact point of the bullet being between 10 and 11 cm away from the target center and between the angles of 90 and 95 degrees?

All of these examples are cases where it is necessary to transform the distribution of one random variable, or set of random variables, to the distribution of another random variable, or set of random variables. The reasoning used to obtain the transformation rule makes use of (A-14) for single random variables, or its generalization for multiple random variables, which is

$$\Pr(x_1 < X_1 \le x_1 + \Delta x_1, x_2 < X_2 \le x_2 + \Delta x_2, \ldots, x_N < X_N \le x_N + \Delta x_N)$$

$$= f_{X_1 X_2 \cdots X_N}(x_1, x_2, \ldots, x_N)\, \Delta x_1 \Delta x_2 \cdots \Delta x_N \quad \text{(A-51)}$$

If a transformation of a single random variable, say

$$Y = g(X) \quad \text{(A-52)}$$

is monotonic, the probability mass associated with the event $x < X \le x + \Delta x$ is the same as that associated with the event $y < Y \le y + \Delta y$. Using (A-14), this observation can be expressed as

$$f_X(x)\, \Delta x = f_Y(y)\, \Delta y$$

or in the limit as $\Delta x \Delta y \to dx dy$,

$$f_Y(y) = f_X(x)\left|\frac{dx}{dy}\right|\Bigg|_{x=g^{-1}(y)} \quad \text{(A-53)}$$

where $g^{-1}(\cdot)$ is the inverse of $g(\cdot)$ and the absolute-value signs are necessary to ensure that $f_Y(y) > 0$. If $g(\cdot)$ is not monotonic, (A-53) generalizes to

$$f_Y(y) = \sum_{i=1}^{n} f_X(x)\left|\frac{dx}{dy}\right|\Bigg|_{x=g_i^{-1}(y)} \quad \text{(A-54)}$$

where $x = g_i^{-1}(y)$ is the ith solution to the inverse transformation. An example will make the procedure clearer.

EXAMPLE A-6

Consider a square-law device with transfer characteristic

$$Y = X^2$$

The input, X, is a Gaussian random variable with mean zero and variance σ^2. Use (A-54) with the two inverse functions

$$x = \sqrt{y} \quad \text{and} \quad x = -\sqrt{y}$$

Because $Y = X^2$ is nonnegative, it follows that $f_y(y) = 0, y < 0$. For both inverses, the absolute value of the derivative is

$$\left| \frac{dx}{dy} \right| = \frac{1}{2\sqrt{y}}$$

Therefore, (A-54) is

$$f_Y(y) = \frac{1}{\sqrt{2\pi\sigma^2}} \left[\exp\left(\frac{-x^2}{2\sigma^2} \right) \Bigg|_{x=\sqrt{y}} + \exp\left(\frac{-x^2}{2\sigma^2} \right) \Bigg|_{x=-\sqrt{y}} \right] \frac{1}{2\sqrt{y}} \quad y > 0$$

$$= \frac{1}{\sqrt{2\pi\sigma^2 y}} \exp\left(\frac{-y}{2\sigma^2} \right) \quad y > 0$$ ■

With multiple random variables, it is easiest to assume the same number of output variables as input variables and that the transformation is one-to-one. Let such a transformation be represented by the set of equations

$$y_1 = g_1(x_1, x_2, \ldots, x_N)$$

$$y_2 = g_2(x_1, x_2, \ldots, x_N)$$

$$\vdots$$ (A-55)

$$y_N = g_N(x_1, x_2, \ldots, x_N)$$

and let the joint density functions of the input and output random variables be denoted by $f_{\mathbf{X}}(x_1, x_2, \ldots, x_N)$ and $f_{\mathbf{Y}}(y_1, y_2, \ldots, y_N)$, respectively. It can be shown [1] that the joint density of the output random variables is

$$f_{\mathbf{Y}}(y_1, y_2, \ldots, y_N) = f_{\mathbf{X}}(x_1, x_2, \ldots, x_N) \left| J\left(\begin{matrix} x_1, x_2, \ldots, x_N \\ y_1, y_2, \ldots, y_N \end{matrix} \right) \right| \quad \text{(A-56)}$$

where

$$J\left(\begin{matrix} x_1, x_2, \ldots, x_N \\ y_1, y_2, \ldots, y_N \end{matrix} \right) = \begin{vmatrix} \dfrac{\partial x_1}{\partial y_1} & \dfrac{\partial x_2}{\partial y_1} & \cdots & \dfrac{\partial x_N}{\partial y_1} \\[2mm] \dfrac{\partial x_1}{\partial y_2} & \dfrac{\partial x_2}{\partial y_2} & \cdots & \dfrac{\partial x_N}{\partial y_2} \\[2mm] & & \cdots & \\[2mm] \dfrac{\partial x_1}{\partial y_N} & \dfrac{\partial x_2}{\partial y_N} & \cdots & \dfrac{\partial x_N}{\partial y_N} \end{vmatrix} \quad \text{(A-57)}$$

is the Jacobian. In (A-56) the variables $x_1, x_2, \ldots, x_N$ are replaced by the inverse transformation equations

$$x_1 = g_1^{-1}(y_1, y_2, \ldots, y_N)$$

$$x_2 = g_2^{-1}(y_1, y_2, \ldots, y_N)$$

$$\vdots$$

$$x_N = g_N^{-1}(y_1, y_2, \ldots, y_N) \qquad \text{(A-58)}$$

which exist by virtue of the transformation being one-to-one.

EXAMPLE A-7

Consider the transformation of two statistically independent, Gaussian random variables with zero means to polar coordinates. Thus

$$f_{\mathbf{X}}(x_1, x_2) = \frac{\exp\left[-(x_1^2 + x_2^2)/2\sigma^2\right]}{2\pi\sigma^2}$$

$$y_1 = r = \sqrt{x_1^2 + x_2^2} \qquad 0 \le r < \infty$$

$$y_2 = \theta = \tan^{-1}\frac{x_2}{x_1} \qquad -\pi < \theta \le \pi$$

With the restrictions placed on the ranges of r and θ, the transformation is one-to-one. It follows that

$$x_1 = r\cos\theta$$

$$x_2 = r\sin\theta$$

From (A-57), the Jacobian is r. The transformed density function is

$$f_{R\Theta}(r, \theta) = \frac{r}{2\pi\sigma^2}\exp\left(\frac{-r^2}{2\sigma^2}\right) \qquad r \ge 0; \quad -\pi < \theta \le \pi$$

The density over r alone is obtained by integration over θ. The result is

$$f_R(r) = \frac{r}{\sigma^2}\exp\left(\frac{-r^2}{2\sigma^2}\right) \qquad r \ge 0$$

which is known as a *Rayleigh* density. ∎

Example A-7 illustrates a procedure that can be used to go from many-to-one random variables. Auxiliary transformations can be defined such that

$$y_1 = g_1(x_1, x_2, \ldots, x_N)$$

$$y_2 = x_2$$

$$y_3 = x_3$$

$$\vdots$$

$$y_N = x_N$$

The transformed density for $y_1, y_2, \ldots, y_N$ is obtained and then the unwanted variables are "integrated out."

EXAMPLE A-8

Given two independent random variables, X and Y, with density functions $f_X(x)$ and $f_Y(y)$, respectively. What is the density function of their product?

Solution: Let

$$W = X$$

$$Z = XY$$

Then the inverse transformation is

$$X = W$$

$$Y = \frac{Z}{W}$$

which is one-to-one. The Jacobian is

$$J\left(\begin{matrix} x, y \\ w, z \end{matrix}\right) = \begin{vmatrix} \dfrac{\partial x}{\partial w} & \dfrac{\partial y}{\partial w} \\[2mm] \dfrac{\partial x}{\partial z} & \dfrac{\partial y}{\partial z} \end{vmatrix} = \begin{vmatrix} 1 & \dfrac{-z}{w^2} \\[2mm] 0 & \dfrac{1}{w} \end{vmatrix} = \frac{1}{w}$$

The joint density function of W and Z is therefore

$$f_{WZ}(w, z) = f_X(x)f_Y(y) \left. \frac{1}{|w|} \right|_{\substack{x=w \\ y=z/w}}$$

$$= f_X(w)f_Y\left(\frac{z}{w}\right) \frac{1}{|w|}$$

To obtain the density function of Z, alone, this joint density is integrated over all values of w:

$$f_Z(z) = \int_{-\infty}^{\infty} f_X(w)f_Y\left(\frac{z}{w}\right) \frac{dw}{|w|} \tag{A-59}$$

■

A-4.2   Linear Transformations of Gaussian Random Variables

If a set of jointly Gaussian random variables is transformed to a new set of random variables by a linear transformation, the resulting random variables are jointly Gaussian. To show this, consider the linear transformation

$$\mathbf{y} = \mathbf{A}\mathbf{x} \tag{A-60}$$

where $\mathbf{y}$ and $\mathbf{x}$ are column matrices of dimension N and $\mathbf{A}$ is a nonsingular $N \times N$ square matrix with elements $[a_{ij}]$. From (A-57), the Jacobian is

$$J\left(\begin{matrix} x_1, x_2, \ldots, x_N \\ y_1, y_2, \ldots, y_N \end{matrix}\right) = \det \mathbf{A}^{-1} \tag{A-61}$$

where $\mathbf{A}^{-1}$ is the inverse matrix of $\mathbf{A}$. But $\det \mathbf{A}^{-1} = 1/\det \mathbf{A}$. Using this in (A-39) together with

$$\mathbf{x} = \mathbf{A}^{-1}\mathbf{y} \tag{A-62}$$

gives

$$f_{\mathbf{Y}}(\mathbf{y}) = (2\pi)^{-N/2}|\det \mathbf{C}|^{-1/2}|\det \mathbf{A}|^{-1}$$

$$\times \exp\left[-\tfrac{1}{2}(\mathbf{A}^{-1}\mathbf{y} - \mathbf{m})^t \mathbf{C}^{-1}(\mathbf{A}^{-1}\mathbf{y} - \mathbf{m})\right] \qquad \text{(A-63)}$$

Now $\det \mathbf{A} = \det \mathbf{A}^t$ and $\mathbf{A}\mathbf{A}^{-1} = \mathbf{I}$, the identity matrix, so that (A-63) can be written as

$$f_{\mathbf{Y}}(\mathbf{y}) = (2\pi)^{-N/2}|\det \mathbf{A}\mathbf{C}\mathbf{A}^t|^{-1/2}$$

$$\times \exp\left\{-\tfrac{1}{2}[\mathbf{A}^{-1}(\mathbf{y} - \mathbf{A}\mathbf{m})]^t \mathbf{C}^{-1}[\mathbf{A}^{-1}(\mathbf{y} - \mathbf{A}\mathbf{m})]\right\} \qquad \text{(A-64)}$$

But the rules $(\mathbf{A}\mathbf{B})^t = \mathbf{B}^t\mathbf{A}^t$ and $(\mathbf{A}^{-1})^t = (\mathbf{A}^t)^{-1}$ allow the exponent to be written as

$$-\tfrac{1}{2}[(\mathbf{y} - \mathbf{A}\mathbf{m})^t(\mathbf{A}^t)^{-1}\mathbf{C}^{-1}\mathbf{A}^{-1}(\mathbf{y} - \mathbf{A}\mathbf{m})]$$

Finally, the rule $(\mathbf{A}\mathbf{B})^{-1} = \mathbf{B}^{-1}\mathbf{A}^{-1}$ allows the exponent to be rearranged as

$$-\tfrac{1}{2}[(\mathbf{y} - \mathbf{A}\mathbf{m})^t(\mathbf{A}\mathbf{C}\mathbf{A}^t)^{-1}(\mathbf{y} - \mathbf{A}\mathbf{m})]$$

so that (A-64) becomes

$$f_{\mathbf{Y}}(\mathbf{y}) = (2\pi)^{-N/2}|\det \mathbf{A}\mathbf{C}\mathbf{A}^t|^{-1/2}$$

$$\times \exp\left\{-\tfrac{1}{2}(\mathbf{y} - \mathbf{A}\mathbf{m})^t(\mathbf{A}\mathbf{C}\mathbf{A}^t)^{-1}(\mathbf{y} - \mathbf{A}\mathbf{m})\right\} \qquad \text{(A-65)}$$

which is recognized as a joint Gaussian density for a random vector $\mathbf{Y}$ with mean vector

$$E(\mathbf{Y}) = \mathbf{A}\mathbf{m}$$

and covariance matrix $\mathbf{A}\mathbf{C}\mathbf{A}^t$.

A-5

CENTRAL LIMIT THEOREM

The widespread use of Gaussian random variables as representations for signal and noise in communication systems is due in part to the relative ease with which analyses can be carried out. For example, the derivation in Section A-4 showed that the mean and covariance matrices of a linearly transformed Gaussian vector are easily obtained, and that knowledge of them allows the joint density function of the transformed variables to be written down immediately using (A-39).

A second reason for the widespread use of Gaussian random variables as signal or noise models is due to the central limit theorem, which states that if X_1, $X_2, \ldots, X_N$ are N identically distributed, independent random variables, all with mean m and finite variance σ^2, the random variable

$$Y = N^{-1/2} \sum_{i=1}^{N} (X_i - m) \qquad \text{(A-66)}$$

approaches a zero-mean Gaussian random variable with variance σ^2 as $N \to \infty$. The conditions of independence and identically distributed may be relaxed with suitable other restrictions imposed [3].

RANDOM PROCESS

In Chapter 2, techniques for describing and analyzing deterministic signals and the effects of systems on such signals were considered. In this Section, signal and system analysis procedures for *random signals* or random processes will be described. These analysis techniques require the use of probabilistic concepts.

As discussed in Chapter 1, the requirement for representing waveforms in communication systems in terms of random processes arises due to two considerations. First, noise is, by its very nature, unpredictable and must be represented probabilistically. Second, by the nature of information itself, the signals in a communication system are random.

In addition to signals and noise in communication systems, examples of random processes are the pressure fluctuations at some point on the surface of an airplane wing, the height of ocean waves in the ocean, and the power delivered to its customers by an electric power plant. These examples serve to illustrate that a random process is normally a function of time, but may also be a function of other independent variables such as Cartesian spatial coordinates. The mathematical description of such processes is given in the next section.

A-6.1 Mathematical Description of Random Processes

It is convenient to assign to the outcomes of a chance experiment numerical values. As discussed in Section A-1, this assignment together with the underlying probability assignment is referred to as a random variable.

Imagine now a chance experiment where functions of time are assigned to each possible outcome of the experiment. An example is where a pivoted pointer is spun and the angle with respect to some reference position measured when it stops. Let this angle be represented by the random variable Θ. Suppose that the probability density function of Θ, either through measurement or physical considerations, is hypothesized to be uniform over the interval $[0, 2\pi]$. For a given realization of Θ, say θ, the time function

$$X(t) = A \cos (\omega_0 t + \theta) \qquad -\infty < t < \infty \qquad \text{(A-67)}$$

is assigned, where A and ω_0 are constants and t is time. The collection of all possible such waveforms together with the underlying probability assignment is called a *random process*. Each realization of a particular waveform is referred to as a *sample function*. Several possible sample functions are sketched in Figure A-3. Several other examples of random processes will be given shortly.

The complete statistical description of a random process is specified when the joint probability density function of its values at an arbitrary number, N, of time instants is available. Let these time instants be $t_1 < t_2 < \cdots < t_N$ and let the possible values of the random process at these time instants be $X_1 = X(t_1)$, $X_2 = X(t_2)$, . . ., $X_N = X(t_N)$. Then the joint density

$$f_X(x_1, t_1; x_2, t_2; \ldots; x_N, t_N) \qquad \text{(A-68)}$$

conveys all the information about the random process that can practically be known. Note that if the joint density function at $N - 1$ instants is desired, it can be obtained

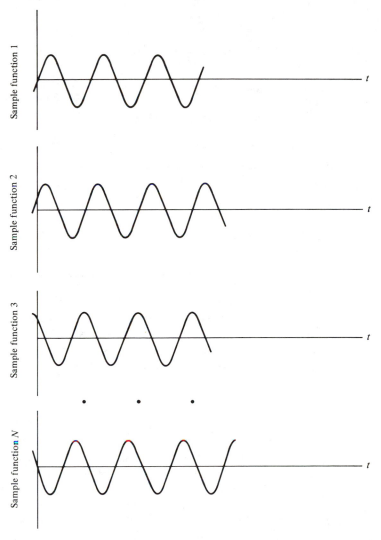

FIGURE A-3.   Several sample functions of the random process [$A \cos (\omega_0 + \theta)$] where θ is a random variable in (0, 2π).

by integrating over the undesired variable. In some cases, this description would be enormously difficult to give. In other cases, such as the example given by (A-67), it is relatively simple, for knowledge of the waveform at two instants in time (not separated by a period of the sinusoidal waveform) means that it is known at all other instants, provided ω_0 and A are known.

Although the N-fold joint density (or distribution) function constitutes a complete statistical description of a random process, it may not be possible to obtain in some cases. Or, one may not be interested in a complete statistical description of the random process. Often, the availability of only the first-order density function or perhaps only certain moments of the random process is required. Moments of interest are the mean and the variance. These, in general, are functions of time. The mean at time $t = t_1$ is

$$E[X(t_1)] = \overline{X(t_1)} = \int_{-\infty}^{\infty} x f_X(x, t_1) \, dx = m(t_1) \qquad \text{(A-69)}$$

and the second moment at time $t = t_1$ is

$$E[X^2(t_1)] = \overline{X^2(t_1)} = \int_{-\infty}^{\infty} x^2 f_X(x, t_1) \, dx \qquad (A\text{-}70)$$

In terms of these moments, the variance is

$$\sigma_X^2(t_1) = E[X^2(t_1)] - \{E[X(t_1)]\}^2 \qquad (A\text{-}71)$$

These averages provide a partial description of the process at a single time instant. Another average, which gives a partial description of how the process at one time instant depends statistically on the process at a second time instant, is provided by the *autocorrelation function*, $R_X(t_1, t_2)$, defined by

$$R_X(t_1, t_2) \triangleq E[X(t_1)X(t_2)]$$

$$= \int_{-\infty}^{\infty} \int_{-\infty}^{\infty} x_1 x_2 f_X(x_1, t_1; x_2, t_2) \, dx_1 \, dx_2 \qquad (A\text{-}72)$$

It is, in general, a function of the two time instants t_1 and t_2. The *autocovariance function* is obtained by subtracting the mean at time t_1 from $X(t_1)$ and the mean at time t_2 from $X(t_2)$ before finding the average of their product. The autocovariance function, $C_X(t_1, t_2)$, and autocorrelation function are related by

$$C_X(t_1, t_2) = R_X(t_1, t_2) - m(t_1)m(t_2) \qquad (A\text{-}73)$$

Note that the variance at time t_1 is equal to $C_X(t_1, t_1)$.

While the mean and variance of an arbitrary random process are, in general, functions of the time instant chosen to compute the average, and the autocorrelation function is a function of two time instants, there is a class of random processes, referred to as *stationary*, for which the mean and variance are time independent and $R_X(t_1, t_2)$ is a function only of the time difference $t_2 - t_1$.

A *strictly stationary* process is one for which the N-fold joint probability density function is not dependent on the time origin chosen for the sampling instants. In particular, it is a function only of the *time differences* $t_2 - t_1, t_3 - t_1, \ldots,$ $t_N - t_1$. Strict-sense stationarity implies that the mean and variance are independent of time and that the autocorrelation function is a function only of the time difference $t_2 - t_1$. For an even larger class of random processes, referred to as *wide-sense stationary*, the mean and variance are time independent and the autocorrelation function is dependent only on the time difference $t_2 - t_1$, but the N-fold probability density is not necessarily independent of the choice of time origin. Quite often, wide-sense stationarity, or stationarity of second order, is all that is required in a given situation.

EXAMPLE A-9
Consider the random-phase sinusoidal process given by (A-67). Using the definition of the average of a function of a random variable, the mean and mean square are

$$E[X(t)] = \int_0^{2\pi} A \cos(\omega_0 t + \theta) \frac{d\theta}{2\pi} = 0$$

and

$$E[X^2(t)] = \int_0^{2\pi} A^2 \cos^2(\omega_0 t + \theta) \frac{d\theta}{2\pi}$$

$$= \int_0^{2\pi} \frac{A^2}{2} \frac{d\theta}{2\pi} + \int_0^{2\pi} \frac{A^2}{2} \cos 2(\omega_0 t + \theta) \frac{d\theta}{2\pi}$$

$$= \frac{A^2}{2} = \sigma_X^2$$

respectively. (Recall that the density function of θ is $1/2\pi$ in the interval 0 to 2π.) Because the mean is zero, the variance and the mean-square value are equal. The autocorrelation function is

$$R_X(t, t + \tau) = E[X(t)X(t + \tau)]$$

$$= \int_0^{2\pi} A^2 \cos(\omega_0 t + \theta) \cos[\omega_0(t + \tau) + \theta] \frac{d\theta}{2\pi}$$

$$= \frac{A^2}{2} \cos \omega_0 \tau$$

where $t_1 = t$ and $t_2 = t + \tau$. The autocorrelation function is seen to be function only of the time difference $t_2 - t_1 = \tau$, and the mean and variance are independent of time. This process is therefore wide-sense stationary. ∎

Several properties of the autocorrelation function of a stationary random process are illustrated by Example A-9. First, it is evident that

$$E[X^2(t)] = R_X(0) \tag{A-74}$$

which follows by setting $\tau = 0$ in the definition of $R_X(\tau) = E[X(t)X(t + \tau)]$. Second, by considering

$$E\{[X(t + \tau) \pm X(t)]^2\} \geq 0 \tag{A-75}$$

which, being the expectation of a nonnegative quantity is itself nonnegative, it follows that

$$|R_X(\tau)| \leq R_X(0) \tag{A-76}$$

Third, since $X(t)$ is assumed stationary, it follows that

$$R_X(\tau) = E[X(t)X(t + \tau)] = E[X(t' - \tau)X(t')] = R_X(-\tau) \tag{A-77}$$

where the substitution $t' = t + \tau$ has been used. Finally, it will be shown in the next section that the Fourier transform of $R_X(t)$ is nonnegative; that is,

$$S_X(f) \triangleq \int_{-\infty}^{\infty} R_X(\tau)e^{-j2\pi f\tau} d\tau \geq 0 \tag{A-78}$$

The function $S_X(f)$ is called the *power spectral density* of the random process $X(t)$.

It is useful to talk about the joint statistics of two random processes, say $X(t)$ and $Y(t)$. One may represent the input to a system and the other the system's output. Complete statistical characterization of the joint chance experiment giving rise to the processes then requires the specification of not only the density functions of $X(t)$ and $Y(t)$ at N time instants, but also the joint density function of both processes at an arbitrarily selected set of time instants. If, for all values of N and $t_1, t_2, \ldots, t_N$, this joint density function factors into the product of the density functions of the separate processes, $X(t)$ and $Y(t)$, they are said to be *statistically independent random processes*. If the joint density function of $X(t)$ and $Y(t)$ is independent of time origin, the processes are said to be *jointly stationary*.

One final property of many random processes is that of *ergodicity*. For ergodic random processes, statistical averages may be found as a corresponding time average. The mean, mean square, and autocorrelation function as time averages, for example, are given by

$$\langle X(t) \rangle = \lim_{T \to \infty} \frac{1}{2T} \int_{-T}^{T} X(t) \, dt \tag{A-79}$$

$$\langle X^2(t) \rangle = \lim_{T \to \infty} \frac{1}{2T} \int_{-T}^{T} X^2(t) \, dt \tag{A-80}$$

and

$$\langle X(t)X(t + \tau) \rangle = \lim_{T \to \infty} \frac{1}{2T} \int_{-T}^{T} X(t)X(t + \tau) \, dt \tag{A-81}$$

respectively, where the angular brackets signify a time average. For an ergodic process, it is true that

$$\overline{X(t)} = \langle X(t) \rangle \tag{A-82}$$

$$\overline{X^2(t)} = \langle X^2(t) \rangle \tag{A-83}$$

and

$$\overline{X(t)X(t + \tau)} = \langle X(t)X(t + \tau) \rangle \tag{A-84}$$

with similar statements holding for all other possible averages. From (A-82) through (A-84) it is seen that an ergodic process must be at least wide-sense stationary and, in fact, strict-sense stationary because *all* time averages are interchangeable with the corresponding statistical averages. For a process to be stationary, it is apparent that each sample function must be statistically representative of the entire collection of all possible sample functions. The assumption of ergodicity is a necessity in almost any experiment involving the measurement of the statistics of a random process, such as the measurement of average power, for the measurement is by necessity made on a single sample function.

EXAMPLE A-10

The measurement of the autocorrelation function of an ergodic (and therefore stationary) random process can be approximated by the system shown in Figure A-4. A sample function (realization) of the random process is delayed by an amount τ and multiplied by an undelayed version, and the product averaged by a lowpass filter with impulse response $h(t)$. Assume that this filter is an ideal integrator with impulse response

$$h(t) = \frac{1}{T}[u(t) - u(t - T)]$$

where T is the integration time. The output of the integrator, by the superposition integral, is

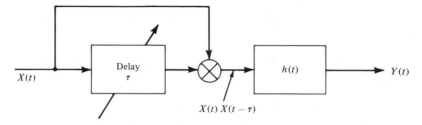

FIGURE A-4. System for the measurement of the autocorrelation function of a random process.

$$Y(t) = \int_{-\infty}^{\infty} h(\lambda)X(t - \lambda)X(t - \lambda - \tau)\, d\lambda$$

$$= \frac{1}{T} \int_{0}^{T} X(t - \lambda)X(t - \lambda - \tau)\, d\lambda$$

Measures of goodness of the estimate $Y(t)$ are its mean and variance. The mean is

$$E[Y(t)] = \frac{1}{T} \int_{0}^{T} E[X(t - \lambda)X(t - \lambda - \tau)]\, d\lambda$$

where the expectation can be taken inside the integral because both integrals exist and form an iterated double integral. The expectation is recognized as the auto-correlation function of $X(t)$, $R_X(\tau)$. Therefore,

$$E[Y(t)] = \frac{1}{T} \int_{0}^{T} R_X(\tau)\, d\lambda = R_X(\tau)$$

Any estimate whose expectation is equal to the quantity being estimated is called *unbiased*. Therefore, the system of Figure A-4 produces an unbiased estimate of the autocorrelation function.

The variance of $Y(t)$ can be found by first computing its mean-square value, which is

$$E[Y^2(t)] = \frac{1}{T^2} E\left\{ \left[\int_{0}^{T} X(t - \lambda)X(t - \lambda - \tau)\, d\lambda \right]^2 \right\}$$

$$= \frac{1}{T^2} \int_{0}^{T} \int_{0}^{T} E[X(t - \lambda)X(t - \alpha)X(t - \lambda - \tau)X(t - \alpha - \tau)\, d\lambda\, d\alpha]$$

where the double integral follows by writing the squared integral as an iterated integral. Without further assumptions this result for $E[Y^2(t)]$ cannot be simplified further. If $X(t)$ is Gaussian and zero mean, however, the expectation in the integrand can be simplified with the help of (A-43). It can be shown that $E[Y^2(t)]$ simplifies to

$$E[Y^2(t)] = R_X^2(\tau) + \frac{1}{T} \int_{-T}^{T} \left(1 - \frac{|u|}{T} \right) R_X^2(u)\, du$$

$$+ \frac{1}{T} \int_{-T}^{T} \left(1 - \frac{|u|}{T} \right) R_X(u - \tau)R_X(u + \tau)\, du$$

where the evenness property of $R_X(\tau)$ has been used. Subtracting the mean of $Y(t)$ squared, which is $R_X^2(\tau)$, results in the variance. The result is

$$\sigma_Y^2 = \frac{1}{T} \int_{-T}^{T} \left(1 - \frac{|u|}{T}\right) [R_X^2(u) + R_X(u - \tau)R_X(u + \tau)] \, du$$

In the limit as $T \to \infty$, $\sigma_Y^2 \to 0$ if the integral

$$\int_{-\infty}^{\infty} [R_X^2(u) + R_X(u - \tau)R_X(u + \tau)] \, du$$

is bounded. This is indeed the case for many autocorrelation functions of interest, as will be shown in the problems. An unbiased estimate whose variance approaches zero as the interval of data from which the estimate is made approaches infinity is said to be *consistent*. This estimate for the autocorrelation function is therefore consistent. ∎

A-7

INPUT/OUTPUT RELATIONSHIPS FOR FIXED LINEAR SYSTEMS WITH RANDOM INPUTS; POWER SPECTRAL DENSITY

A-7.1 Partial Descriptions

Consider a fixed linear system with an input, $X(t)$, assumed present since $t = -\infty$, which is a sample function of a stationary random process as shown in Figure A-5. The output, $Y(t)$, is given in terms of $X(t)$ by the superposition integral:

$$Y(t) = \int_{-\infty}^{\infty} h(\alpha)X(t - \alpha) \, d\alpha \qquad (A-85)$$

Because of the presence of the input since $t = -\infty$, any system transients have long since died out and the output is stationary also. A partial description of the output is provided by its mean and autocorrelation function. From (A-85) the mean is

$$m_Y = E[Y(t)] = E\left[\int_{-\infty}^{\infty} h(\alpha)X(t - \alpha) \, d\alpha\right]$$

$$= \int_{-\infty}^{\infty} h(\alpha)E[X(t - \alpha)] \, d\alpha = m_X \int_{-\infty}^{\infty} h(\alpha) \, d\alpha \qquad (A-86)$$

which follows because m_X is time independent for a stationary process.

The autocorrelation function of the output is

$$R_Y(\tau) = E[Y(t)Y(t + \tau)]$$

$$= E\left[\int_{-\infty}^{\infty} h(\alpha)X(t - \alpha) \, d\alpha \int_{-\infty}^{\infty} h(\beta)X(t + \tau - \beta) \, d\beta\right]$$

$$= \int_{-\infty}^{\infty}\int_{-\infty}^{\infty} h(\alpha)h(\beta)E[X(t - \alpha)X(t + \tau - \beta)] \, d\alpha \, d\beta$$

$$= \int_{-\infty}^{\infty}\int_{-\infty}^{\infty} h(\alpha)h(\beta)R_X(\tau - \beta + \alpha) \, d\alpha \, d\beta \qquad (A-87)$$

The double integral in this equation appears difficult to evaluate, even for simple autocorrelation functions. Because of its similarity to a convolution, it appears that

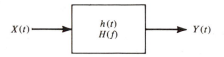

$X(t) \longrightarrow \boxed{\begin{array}{c} h(t) \\ H(f) \end{array}} \longrightarrow Y(t)$

FIGURE A-5. Fixed, linear system with a sample function of a random process as an input.

it might be advantageous to express it in terms of frequency domain quantities. Accordingly, let

$$H(f) = \mathcal{F}[h(t)] \tag{A-88}$$

be the Fourier transform of the impulse response, or transfer function, of the system and let

$$S_X(f) = \mathcal{F}[R_X(\tau)] \tag{A-89}$$

From the evenness of $R_X(\tau)$ it follows that $S_X(f)$ is real and even.

From the inverse Fourier transform integral, it follows that

$$R_X(0) = E[X^2(t)] = \int_{-\infty}^{\infty} S_X(f) \, df \tag{A-90}$$

so that the integral of $S_X(f)$ over all frequency is the mean-square value, or average power, of the input. The function $S_X(f)$ is called the power spectral density of the process $X(t)$. It is a function that describes the distribution of power with frequency of a random process. This interpretation will be given more justification shortly.

If $R_X(\tau - \beta + \alpha)$ in (A-87) is represented in terms of the inverse Fourier transform of $S_X(f)$, the resulting triple integral can be reduced to

$$R_Y(\tau) = \int_{-\infty}^{\infty} H(f)H^*(f)S_X(f)e^{j2\pi f\tau} \, d\tau \tag{A-91}$$

which is recognized as the inverse Fourier transform of $|H(f)|^2 S_X(f)$. That is,

$$\mathcal{F}[R_Y(\tau)] = S_Y(f) = |H(f)|^2 S_X(f) \tag{A-92}$$

or the power spectral density of the output is the magnitude squared of the transfer function times the power spectral density of the input. This extremely useful result, which is illustrated in Figure A-6, can be used in the analysis of linear systems with random inputs that have been present so long that the transients have died out.

That $S_X(f)$ physically represents the density of power of $X(t)$ versus frequency can be seen by considering the output of an ideal bandpass filter of bandwidth $\Delta W \ll 1$ centered at frequency f_0 with passband gain of unity. According to (A-91) with $\tau = 0$, the power in $X(t)$ in bandwidth ΔW is

$$R_Y(0) = E[Y^2(t)] = 2 \int_{f_0 - \Delta W/2}^{f_0 + \Delta W/2} S_X(f) \, df$$

$$\simeq 2S_X(f_0) \, \Delta W \tag{A-93}$$

which follows by the evenness of $S_X(f)$ and the definition of an integral. Since both $R_Y(0)$ and ΔW are positive, it follows that

$$S_X(f_0) = \frac{R_Y(0)}{2 \, \Delta W} > 0 \tag{A-94}$$

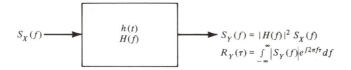

$$S_Y(f) = |H(f)|^2 S_X(f)$$

$$R_Y(\tau) = \int_{-\infty}^{\infty} |S_Y(f)| e^{j2\pi f\tau} df$$

(a) Input/output relations for a linear system with
     random input

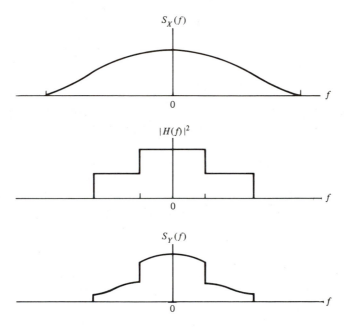

(b) Illustration of the relationship between input and
     output spectra

**FIGURE A-6.   Obtaining the output spectrum of a fixed linear system with stationary
random process at its input.**

which represents the power density in a very small bandwidth ΔW centered at
frequency f_0. Since f_0 is arbitrary and power is nonnegative, it follows that the
power spectral density function is a nonnegative function of frequency.

The fact that $S_X(f)$, which represents power density with frequency of the process
$X(t)$, is the Fourier transform of the autocorrelation function is referred to as the
Weiner–Khintchine theorem. Its proof proceeds by considering the time average
of the energy density of a time-truncated version of the random process. The reader
may consult other references for the proof of the Wiener–Khintchine theorem
[1, 3].

EXAMPLE A-11

A useful mathematical artifice is white noise, which has a constant power spectral
density for all frequencies. Let the spectral density of a white-noise process be

$$S_W(f) = \frac{N_0}{2} \qquad -\infty < f < \infty$$

where N_0 is the single-sided (positive frequencies only) spectral density. By the Wiener–Khintchine theorem,

$$R_W(\tau) = \mathscr{F}^{-1}\left(\frac{N_0}{2}\right)$$

$$= \frac{N_0}{2}\,\delta(\tau) \qquad\qquad (A\text{-}95)$$

Now consider a filter with transfer function $H(f)$ with white noise at its input. What is the bandwidth of an equivalent ideal filter with the same maximum gain, say H_0, that passes the same power as the original filter? From (A-92) the power out of the original filter is

$$\sigma_Y^2 = N_0 \int_0^\infty |H(f)|^2\,df$$

and from the equivalent filter, it is

$$\sigma_{Y'}^2 = N_0 H_0^2 B_N$$

where B_N is its bandwidth (to be found). Setting these two powers equal results in

$$B_N = \frac{1}{H_0^2} \int_0^\infty |H(f)|^2\,df \qquad\qquad (A\text{-}96)$$

This is referred to as the *noise equivalent bandwidth* of the filter. Figure A-7 illustrates this concept. The ratios of noise equivalent bandwidth to a 3-dB cutoff frequency of several filters are given in Table A-1. ■

EXAMPLE A-12

A useful relationship involving the correlation functions and spectral densities of statistically independent random processes is developed in this example. Let $\{X(t)\}$ and $\{Y(t)\}$ be independent and stationary. Let $Z(t) = X(t)Y(t)$. Then

$$R_Z(\tau) = E[X(t)X(t + \tau)Y(t)Y(t + \tau)]$$

$$= \overline{X(t)X(t + \tau)}\;\overline{Y(t)Y(t + \tau)} = R_X(\tau)R_Y(\tau) \qquad (A\text{-}97)$$

where $R_X(\tau)$ and $R_Y(\tau)$ are the respective autocorrelation functions of $\{X(t)\}$ and $\{Y(t)\}$. It follows from the multiplication theorem of Fourier transforms that the

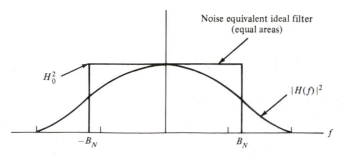

FIGURE A-7. Illustration of the concept of noise equivalent bandwidth.

TABLE A-1. Ratio of Noise Bandwidths to 3-dB Bandwidth

Filter Type	Order	B_N/B_3
Butterworth	1	$\pi/2 \approx 1.57$
Exact: $\dfrac{B_n}{B_3} = \dfrac{\pi/2n}{\sin(\pi/2n)}$	2	1.11
	3	1.04
	4	1.03
	5	1.02
	6	1.01
Bessel	1	1.57
	2	1.16
	3	1.08
	4	1.04
	5	1.04
	6	1.04
Chebyshev ($\frac{1}{2}$ dB)	1	1.57
	2	1.15
	3	1.00
	4	1.08
	5	0.96
	6	1.07
Chebyshev (1 dB)	1	1.57
	2	1.21
	3	0.96
	4	1.15
	5	0.92
	6	1.13
Chebyshev (2 dB)	1	1.57
	2	1.33
	3	0.86
	4	1.28
	5	0.82
	6	1.27
Chebyshev (3 dB)	1	1.57
	2	1.48
	3	0.78
	4	1.43
	5	0.73
	6	1.41

Source: After Ref. 4.

power spectral density of $\{Z(t)\}$ is given by

$$S_Z(f) = S_X(f) * S_Y(f) = \int_{-\infty}^{\infty} S_X(\lambda) S_Y(f - \lambda) \, d\lambda \qquad \text{(A-98)}$$

Figure A-8 illustrates this idea. ∎

A-7.2 Output Statistics of Linear Systems

Given a linear system with transfer function $H(f)$ and impulse response $h(t)$ with random input $X(t)$, what can one say about the probability density function of the output at a single time instant t if one knows the probability density function of

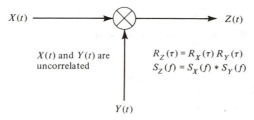

$X(t)$ and $Y(t)$ are uncorrelated

$R_Z(\tau) = R_X(\tau) R_Y(\tau)$
$S_Z(f) = S_X(f) * S_Y(f)$

$Y(t)$

FIGURE A-8. Correlation function and power spectral density relationships for a multiplier with statistically independent inputs.

the input? *This problem has no general solution.* However, *if the input is Gaussian,* (i.e., has a density function of the form (A-17) at a single time instant, and of the form (A-39) at N arbitrarily chosen time instants), *the output is Gaussian also.* If the input has zero mean, so does the output. The variance of the output is then given by

$$\sigma_Y^2 = R_Y(0) = \int_{-\infty}^{\infty} |H(f)|^2 S_X(f)\, df \qquad (A\text{-}99)$$

and, using (A-17), one may immediately write down the first-order density function of the output. This procedure will be used extensively in the analysis of digital communication systems.

EXAMPLE A-13

Consider the system shown in Figure A-9. The inputs to the first-order lowpass filters are assumed to be independent white Gaussian noise processes with two-sided power spectral densities $N_1/2$ and $N_2/2$, respectively. The filter transfer functions are

$$H_1(f) = \frac{1}{1 + j(f/f_1)}$$

and

$$H_2(f) = \frac{1}{1 + j(f/f_2)}$$

respectively, where f_1 and f_2 are their 3-dB frequencies.

The following will be found: (1) the autocorrelation functions, $R_X(\tau)$ and $R_Y(\tau)$, of the signals at the outputs of the two filters; (2) the power spectral densities, $S_X(f)$ and $S_Y(f)$, of the two signals at the outputs of the two filters; and (3) the autocorrelation function, $R_Z(\tau)$, and the power spectral density, $S_Z(f)$, at the output of the multiplier.

Solution: Using (A-92), the power spectral densities at the filter outputs are found to be

$$S_X(f) = \frac{N_1/2}{1 + (f/f_1)^2}$$

$$S_Y(f) = \frac{N_2/2}{1 + (f/f_2)^2}$$

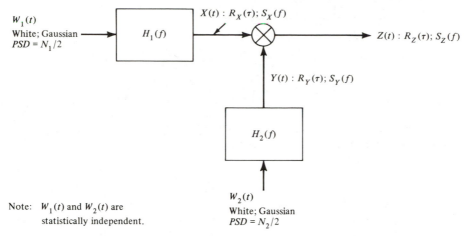

Note: $W_1(t)$ and $W_2(t)$ are
statistically independent.

$W_1(t)$
White; Gaussian
$PSD = N_1/2$

$X(t) : R_X(\tau); S_X(f)$

$Z(t) : R_Z(\tau); S_Z(f)$

$Y(t) : R_Y(\tau); S_Y(f)$

$W_2(t)$
White; Gaussian
$PSD = N_2/2$

FIGURE A-9.   System analyzed in Example A-13.

Using the Fourier transform pair

$$Ae^{-\alpha|\tau|} \leftrightarrow \frac{2\alpha A}{\alpha^2 + (2\pi f)^2}$$

the inverse Fourier transforms $S_X(f)$ and $S_Y(f)$ are readily found to give $R_X(\tau)$ and $R_Y(\tau)$, respectively. They are

$$R_X(\tau) = N_1\left(\frac{\pi f_1}{2}\right)e^{-2\pi f_1|\tau|}$$

$$R_Y(\tau) = N_2\left(\frac{\pi f_2}{2}\right)e^{-2\pi f_2|\tau|}$$

The variances of $X(t)$ and $Y(t)$ are

$$\sigma_X^2 = R_X(0) = N_1\left(\frac{\pi f_1}{2}\right)$$

and

$$\sigma_Y^2 = R_Y(0) = N_2\left(\frac{\pi f_2}{2}\right)$$

respectively. Note that $\pi f_1/2$ and $\pi f_2/2$ are the equivalent noise bandwidths of the two filters.

From (A-97) the autocorrelation function of the multiplier output is

$$R_Z(\tau) = R_X(\tau)R_Y(\tau)$$

$$= \tfrac{1}{4}\ \pi^2 f_1 f_2 N_1 N_2 e^{-2\pi(f_1 + f_2)|\tau|}$$

The Fourier transform of $R_Z(\tau)$ gives $S_Z(f)$, which is

$$S_Z(f) = \frac{\pi N_1 N_2[f_1 f_2/4(f_1 + f_2)]}{1 + [f/(f_1 + f_2)]^2}$$

The fact that the inputs to the filters were given as Gaussian have not yet been used. Using the fact that a linear transformation of a Gaussian process results in a

Gaussian process, one could use the variances calculated here, and means (they are zero; why?) of $X(t)$ and $Y(t)$ to write down their marginal probability density functions according to (A-17). The result for the probability density function of $Z(t)$ does not follow immediately, however, but requires a transformation of random variables as outlined in Example A-8. ∎

EXAMPLES OF RANDOM PROCESSES

In this section, several examples of random processes are investigated in terms of their probability density functions, means, and autocorrelation functions. It is not the intent to exhaustively catalog all possible random processes, but to provide intuitive insight into the properties of various random waveforms through the consideration of several examples. Also introduced will be the concept of a cyclostationary process, which is a process exhibiting periodicity in its mean and autocorrelation functions.

EXAMPLE A-14
Consider a random process whose sample functions are constant functions of time, but random variables, for $-\infty < t < \infty$. That is, the sample functions are of the form

$$X(t) = A \qquad -\infty < t < \infty$$

where A is a random variable with density function $f_A(a)$. The mean of this random process is

$$E[X(t)] = \int_{-\infty}^{\infty} a f_A(a) \, da = E(A)$$

while its autocorrelation function is

$$R_X(t_1, t_2) = E[X(t_1)X(t_2)] = \int_{-\infty}^{\infty} a^2 f_A(a) \, da = E(A^2)$$

Both are independent of the particular time instants chosen for evaluation. Because the sample functions are time independent (constants, but random variables), the marginal density function at any arbitrarily selected time $t = t_1$ is simply $f_A(a)$. The joint density function at two arbitrarily selected times, t_1 and t_2, can be written as

$$f_X(x_1, t_1; x_2, t_2) = f(x_2, t_2 | x_1, t_1) f_A(x_1)$$

where $f(x_2, t_2 | x_1, t_1)$ is the conditional density function at time t_2 given the value of x_1 at time t_1. Because the sample functions are time independent, knowledge of the value of the random process at time t_1 completely specifies its value at time t_2. In particular, since the sample functions are time independent, it follows that

$$f(x_2, t_2 | x_1, t_1) = \delta(x_2 - x_1)$$

which is a mathematical statement that the value $X(t)$ at time t_2 is equal to the value at time t_1. Consequently, the joint density function is

$$f_X(x_1, t_1; x_2, t_2) = f_A(x_1)\delta(x_2 - x_1)$$

and the autocorrelation function can be computed according to

$$R_X(t_1, t_2) = \int\!\!\int_{-\infty}^{\infty} x_1 x_2 f_A(x_1) \delta(x_2 - x_1) \, dx_1 \, dx_2$$

$$= \int_{-\infty}^{\infty} x_1^2 f_A(x_1) \, dx_1 = E(A^2)$$

as before.

This random process, although relatively simple, is useful for demonstrating the computation of various statistical characterizing functions. It is also an example of a random process which is stationary in the strict sense but *not* ergodic. ■

EXAMPLE A-15

The next random process to be considered is characterized by sample functions of the form

$$X(t) = A \cos(\omega_0 t + \theta)$$

Initially, ω_0 and θ will be taken as constants and A is a random variable with density function $f_A(a)$. The mean is

$$E[X(t)] = \int_{-\infty}^{\infty} a \cos(\omega_0 t + \theta) f_A(a) \, da$$

$$= E(A) \cos(\omega_0 t + \theta)$$

which is a periodic function of time. The autocorrelation function is calculated from

$$R_X(t_1, t_2) = E[X(t_1)X)(t_2)]$$

$$= \int_{-\infty}^{\infty} a^2 \cos(\omega_0 t_1 + \theta) \cos(\omega_0 t_2 + \theta) f_A(a) \, da$$

$$= \tfrac{1}{2}E(A^2)\{\cos[\omega_0(t_2 - t_1)] + \cos[\omega_0(t_1 + t_2) + 2\theta]\}$$

which has the property that

$$R_X(t_1 + T, t_2 + T) = R_X(t_1, t_2) \tag{A-100}$$

where $T = 2\pi/\omega_0$. This periodicity property, together with the mean being a periodic function of period $2\pi/\omega_0$, defines a nonstationary class of random processes referred to as *cyclostationary* in the wide sense.

A way to remove the nonstationary character of the random process is through *phase randomizing*. For example, if θ is a uniformly distributed random variable in the interval $(0, 2\pi)$, then

$$E[X(t)] = 0$$

$$R_X(t_1, t_2) = \tfrac{1}{2}E(A^2) \cos \omega_0(t_2 - t_1)$$

In other words, the process is now wide-sense stationary. While phase randomizing removes the nonstationary character, it is important to consider whether the model is a good description of the physical process. If, for example, the process is being observed with total lack of knowledge of the time origin of the observed process, phase randomizing is probably a good model of the physical process. If, on the other hand, a model is desired for the acquisition of a phase reference for the observed sinusoid, phase randomization should not be performed since it implies total lack of knowledge of the phase of the observed process. ∎

As another example, a pulse-amplitude-modulated (PAM) signal will be considered.

EXAMPLE A-16
Consider a random process with sample functions of the form

$$X(t) = \sum_{k=-\infty}^{\infty} A_k p(t - kT - \Delta) \tag{A-101}$$

where $p(t)$ is a pulse of arbitrary shape and the A_k are independent random variables with correlation $E(A_k A_{k+m}) = \alpha_{|m|}$ and means $E(A_k) = a$ for all k. The parameters T and Δ are both constants. Later Δ will be taken as a random variable uniformly distributed in the interval $(-T/2, T/2)$. In the first case, $X(t)$ will be found to be cyclostationary, and in the second case, wide-sense stationary.

Case 1: Δ is a constant. The mean and autocorrelation functions are, respectively,

$$E[X(t)] = a \sum_{k=-\infty}^{\infty} p(t - kT - \Delta) \tag{A-102}$$

$$R_X(t + \tau, t) = E\left[\sum_{j,k=-\infty}^{\infty} A_j A_k p(t + \tau - jT - \Delta) p(t - kT - \Delta) \right]$$

$$= \sum_{m=-\infty}^{\infty} \alpha_{|m|} \sum_{k'=-\infty}^{\infty} p(t + \tau - k'T - \Delta) p[t - (k' + m)T - \Delta] \tag{A-103}$$

Replacing t by $t + T$ in (A-102) and (A-103) shows that this process is cyclostationary.

Case 2: Δ is a uniformly distributed random variable in $(-T/2, T/2)$. In this case, an average of (A-102) and (A-103) over Δ must be carried out. The mean becomes

$$E[X(t)] = a \sum_{k=-\infty}^{\infty} \int_{-T/2}^{T/2} p(t - kT - \delta) \frac{d\delta}{T}$$

$$= \frac{a}{T} \sum_{k=-\infty}^{\infty} \int_{t-[k-(1/2)]T}^{t-[k+(1/2)]T} p(u) \, du$$

$$= \frac{a}{T} \int_{-\infty}^{\infty} p(t) \, dt \tag{A-104}$$

where the change of variables $u = t - kT - \delta$ has been made. The autocorrelation function can be written as

$$R_X(t + \tau, t) = \sum_{k,m=-\infty}^{\infty} \alpha_{|m|} \int_{-T/2}^{T/2} p(t + \tau - kT - \delta)p[t - (k + m)T - \delta]\frac{d\delta}{T}$$

$$= \frac{1}{T} \sum_{m=-\infty}^{\infty} \alpha_{|m|} \sum_{k=-\infty}^{\infty} \int_{t-[k-(1/2)]T}^{t-[k+(1/2)]T} p(u + \tau)p(u - mT)\, du$$

$$= \frac{1}{T} \sum_{m=-\infty}^{\infty} \alpha_{|m|} \int_{-\infty}^{\infty} p(t' + \tau + mT)p(t')\, dt'$$

$$= \frac{1}{T} \sum_{m=-\infty}^{\infty} \alpha_{|m|} r(\tau + mT) \qquad \text{(A-105)}$$

where

$$r(\tau) = \int_{-\infty}^{\infty} p(t + \tau)p(t)\, dt \qquad \text{(A-106)}$$

Since the mean is time independent and the autocorrelation function is dependent on τ only, the process is wide-sense stationary. Using the delay theorem of Fourier transforms and noting that the Fourier transform of (A-106) is $|P(f)|^2$, where $P(f) = \mathcal{F}\{p(t)\}$, the Fourier transform of (A-105) or power spectral density becomes

$$S_X(f) = \frac{1}{T}|P(f)|^2 \sum_{m=-\infty}^{\infty} \alpha_{|m|}e^{j2\pi mTf} \qquad \text{(A-107)}$$

As a further specialization, suppose that the random variables $\{A_k\}$ constitute an independent, identically distributed (iid) sequence of random variables. Then

$$\alpha_{|m|} = \begin{cases} a^2 & m \neq 0 \\ a^2 + \sigma_a^2 & m = 0 \end{cases} \qquad \text{(A-108)}$$

where $a = E(A_k)$ and $\sigma_a^2 = \text{var}(A_k)$. In this case, (A-107) becomes

$$S_X(f) = \frac{\sigma_a^2}{T}|P(f)|^2 + \frac{a^2}{T}|P(f)|^2 \sum_{m=-\infty}^{\infty} e^{j2\pi mTf} \qquad \text{(A-109)}$$

Now the Poisson sum formula states that

$$\sum_{n=-\infty}^{\infty} g(nT)e^{-j2\pi nTf} = \frac{1}{T} \sum_{m=-\infty}^{\infty} G\left(\frac{f - m}{T}\right) \qquad \text{(A-110)}$$

where $\mathcal{F}[g(t)] = G(f)$. Using this in the sum of (A-109) with $g(t) = 1$ and $G(f) = \delta(f)$ results in the power spectrum expression

$$S_X(f) = \frac{\sigma_a^2}{T}|P(f)|^2 + \left(\frac{a}{T}\right)^2 \sum_{m=-\infty}^{\infty} \left|P\left(\frac{m}{T}\right)\right|^2 \delta\left(\frac{f - m}{T}\right) \qquad \text{(A-111)}$$

This shows that the power spectrum of a PAM signal is composed of a continuous part [the first term of (A-111)] plus a discrete spectral part with power concentrated at harmonics of the pulse repetition frequency $1/T$. The power concentrated at these frequencies depends both on the mean of the pulse amplitudes and on the harmonically sampled values of the Fourier transform of the pulse shape function. Correlation between the pulse amplitudes changes the distribution of power. ∎

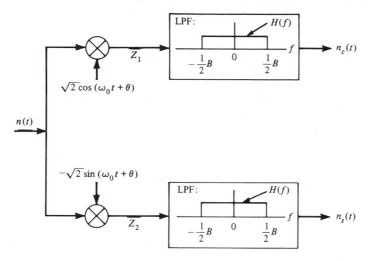

FIGURE A-10. Block diagram of a system for determining narrowband noise spectra.

NARROWBAND NOISE REPRESENTATION

Given a stationary random process $\{n(t)\}$ with power spectrum $S_n(f)$, it is often convenient to represent its sample functions in the form

$$n(t) = n_c(t)\sqrt{2} \cos (2\pi f_0 t) + n_s(t)\sqrt{2} \sin (2\pi f_0 t) \qquad \text{(A-112)}$$

where $\{n_c(t)\}$ and $\{n_s(t)\}$ are lowpass random processes called the *quadrature components* of $n(t)$. If $\{n(t)\}$ is narrowband in the sense that the bandwidth of $S_n(f)$ is less than $f_0/2$, then $n_c(t)$ and $n_s(t)$ can be found by the operations depicted in Figure A-10. Since these operations are linear, the quadrature components are Gaussian if $\{n(t)\}$ is Gaussian.

The spectra of $\{n_c(t)\}$ and $\{n_s(t)\}$ are identical and can be found from the operation*

$$S_{nc}(f) = S_{ns}(f) = \tfrac{1}{2}\text{LP}[S_n(f - f_0) + S_n(f + f_0)] \qquad \text{(A-113)}$$

where $\text{LP}(\cdot)$ denotes the lowpass part of the spectrum in the argument. The Fourier transform of the cross-correlation function between $\{n_c(t)\}$ and $\{n_s(t)\}$, referred to as the cross spectrum, is given by

$$S_{ncns}(f) = \tfrac{1}{2} j \, \text{LP}[S_n(f - f_0) - S_n(f + f_0)] \qquad \text{(A-114)}$$

From (A-114) it follows that if the power spectral density of $n(t)$ is symmetrical about $f = f_0$ for $f > 0$ (or $f < 0$), the cross spectrum is zero and the processes $\{n_c(t)\}$ and $\{n_s(t)\}$ are uncorrelated. An example will illustrate the procedure for obtaining the spectra pertaining to $\{n_c(t)\}$ and $\{n_s(t)\}$.

---

*For proofs of these properties, see [3, pp. 239–244].

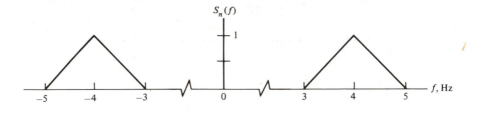

(a) Narrowband noise spectrum

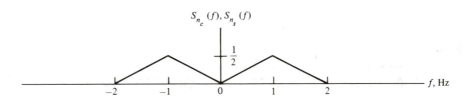

(b) Quadrature component spectra for $f_0 = 4$ Hz

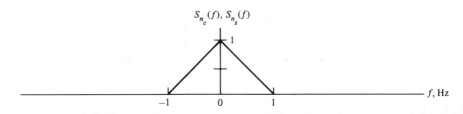

(c) Quadrature component spectra for $f_0 = 3$ Hz

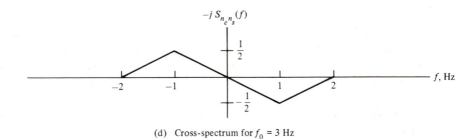

(d) Cross-spectrum for $f_0 = 3$ Hz

FIGURE A-11. Narrowband noise spectra.

EXAMPLE A-17

Consider the narrowband noise spectrum shown in Figure A-11a. Two choices will be made for f_0, the first of which gives uncorrelated quadrature components and the second of which does not.

Case 1: $f_0 = 4$ Hz. In this case, $S_n(f)$ is symmetrical about $f = f_0$ for $f > 0$ and the quadrature components are uncorrelated. Construction of the spectrum for these components is illustrated in Figure A-11b.

Case 2: $f_0 = 3$ Hz. The spectra for this case are shown in Figures A-11c and d. In this case, the quadrature components are correlated. In fact, the cross-correlation function of $n_c(t)$ and $n_s(t)$ is

$$R_{n_c n_s}(\tau) = \mathscr{F}^{-1}\left\{\left(\frac{j}{2}\right)[\Lambda(f+1) + \Lambda(f-1)]\right\}$$

$$= \frac{j}{2}\,[\text{sinc}^2\,(\tau)e^{j2\pi\tau} + \text{sinc}^2\,(\tau)e^{-j2\pi\tau}]$$

$$= -\text{sinc}^2\,(\tau)\,\sin 2\pi\tau$$

Although $R_{n_c n_s}(\tau) = 0$ for $\tau = 0$ it is not zero for all τ and $n_c(t)$ and $n_s(t)$ are therefore correlated. ∎

REFERENCES

[1] C. W. HELSTROM, *Probability and Stochastic Processes for Engineers* (New York: Macmillan, 1984).

[2] A. V. OPPENHEIM and R. W. SCHAFER, *Digital Signal Processing* (Englewood Cliffs, N.J.: Prentice-Hall, 1975), Chap. 2.

[3] R. E. ZIEMER and W. H. TRANTER, *Principles of Communications* (Boston: Houghtor Mifflin, 1976), Chap. 4.

[4] R. D. SHELTON and A. F. ADKINS, *IEEE Trans. Commun.*, December 1970, pp. 828–830.

ADDITIONAL READING

L. E. FRANKS, *Signal Theory* (Englewood Cliffs, N.J.: Prentice-Hall, 1969), Chap. 8.

PROBLEMS

A-1. Given two chance experiments, A and B: Experiment A can result in either of the exhaustive, mutually exclusive outcomes A_1 or A_2. Experiment B can result in any one of the exhaustive, mutually exclusive outcomes B_1, B_2, or B_3. A table listing the joint probabilities of all possible combinations of events, one from A and one from B, is given below.
(a) Compute $P(A_1)$ and $P(A_2)$.
(b) Compute $P(B_1)$, $P(B_2)$, and $P(B_3)$.
(c) Compute the conditional probabilities $P(A_i|B_j)$, $i = 1, 2$ and $j = 1, 2, 3$.
(d) Compute the conditional probabilities $P(B_j|A_i)$, $j = 1, 2, 3$, and $i = 1, 2$.
(e) Are A and B statistically independent?

Table of $P(A_i,\ B_j)$

	B_1	B_2	B_3
A_1	0.1	0.2	0.3
A_2	0.3	0.05	0.05

A-2. The conditional probabilities $P(0r|0s) = 0.8$ and $P(1r|1s) = 0.9$ are obtained for a certain communications channel (see Example A-1 for an explanation of the notation). If $P(0s) = 0.4$ and $P(1s) = 0.6$, find the conditional probabilities of a one sent, given that a one was received, and a zero sent, given that a zero was received.

A-3. The *Rayleigh probability density function* is given by

$$f_R(r) = \begin{cases} \dfrac{r}{\sigma^2}\, e^{-r^2/2\sigma^2} & 0 \le r < \infty \\ 0 & r < 0 \end{cases}$$

(a) Find the distribution function of a Rayleigh random variable.
(b) Find Pr $(0.1\sigma < r \le 0.2\sigma)$ for a Rayleigh random variable.
(c) Show that the mean and mean square of a Rayleigh random variable are given by $\sqrt{\pi/2}\,\sigma$ and $2\sigma^2$, respectively.
(d) Find the variance of a Rayleigh random variable.

A-4. Two random variables, R and Φ, have the joint probability density function

$$f_{R\Phi}(r, \varphi) = \begin{cases} \dfrac{r}{2\pi\sigma^2}\exp\left[\dfrac{-(r^2 - 2Ar\cos\varphi + A^2)}{2\sigma^2}\right] & 0 \le r < \infty;\ -\pi < \varphi \le \pi \\ 0 & \text{otherwise} \end{cases}$$

Show that the marginal probability density function of R is

$$f_R(r) = \dfrac{r}{\sigma^2}\exp\left[\dfrac{-(A^2 + r^2)}{2\sigma^2}\right]I_0\left(\dfrac{Ar}{\sigma}\right) \qquad r \ge 0$$

where $I_0(u)$, the modified Bessel function of the first kind and order zero, is given by

$$I_0(u) = \dfrac{1}{2\pi}\int_0^{2\pi} e^{u\,\cos\varphi}\,d\varphi$$

This probability density function is known as the Rice–Nakagami, or simply Rician, density function.

A-5. (a) Show that m_X, m_Y, σ_X, and σ_Y in (A-23) are the means and variance of X and Y, respectively.
(b) Show that the covariance of X and Y is $\mu_{XY} = \rho\sigma_X\sigma_Y$.

A-6. Prove (A-29) and (A-30).

A-7. A random variable has probability density function

$$f_X(x) = A\exp(-b|x|) \qquad b > 0$$

(a) Obtain the relationship between A and b.
(b) Obtain an expression for the moments $E(X^m)$ of X.
(c) Obtain an expression for the characteristic function of X.
(d) What is the probability that $|X| \ge 1/b$?
(e) Find the characteristic function of X. Use it to check your result for the moments found in part (b).

A-8. Show that the probability density function of the sum of two independent random variables, $Z = X + Y$, is

$$f_Z(z) = \int_{-\infty}^{\infty} f_X(z - \lambda)f_Y(\lambda)\,d\lambda$$

where f_X and f_Y are the density functions of X and Y.

A-9. (a) Using the transformation of random variables technique discussed in Section A-4, show that the probability density function for the random process described in connection with (A-67) at any time t is given by

$$f_X(x, t) = \begin{cases} \dfrac{1}{\pi\sqrt{1 - (x/A)^2}} & |x| \le A \\ 0 & |x| > A \end{cases}$$

(b) Given the value of $X(t)$ at time $t = t_1$, obtain an expression for the conditional density function at time $t_2 > t_1$. *Hint:* Note that if $X(t_1)$ is known, so is $X(t_2)$.

A-10. A random process is defined as in (A-67) but with Θ having the density function

$$f_\Theta(\theta) = \begin{cases} \dfrac{2}{\pi} & 0 \le \theta \le \dfrac{\pi}{2} \\ 0 & \text{otherwise} \end{cases}$$

(a) Find the mean and variance of this random process.
(b) Obtain its autocorrelation function.
(c) Is the process stationary? Cyclostationary?

A-11. Which of the following functions are suitable for autocorrelation functions? If a function is not suitable, tell why it isn't.
(a) $A \exp(-\alpha|\tau|)$, $\alpha > 0$ and $A > 0$
(b) $A \sin \omega_0 \tau$
(c) $A \exp(-\beta t)u(t)$, $\beta > 0$
(d) $A \Pi(\tau/\tau_0)$
(e) $A \operatorname{sinc}(2W\tau)$
(f) $A \exp[-(\tau/2\tau_0)^2]$
(g) $A \tau^2 \exp(-\alpha|\tau|)$

A-12. Show that the time average mean and variance of the process defined by (A-67) with Θ uniform in $(0, 2\pi)$ are equal to the statistical average mean and variance.

A-13. Derive the expression for the variance of the output of the system shown in Figure A-4 assuming that the input is a stationary Gaussian random process.

A-14. Obtain the autocorrelation functions and power spectral densities of the outputs of the following systems and input autocorrelation functions or power spectral densities.
(a) $H(f) = \Pi(f/2B)$
    $R_X(\tau) = (N_0/2)\,\delta(\tau)$
    B and N_0 positive constants
(b) $h(t) = A \exp(-\alpha t)u(t)$
    $S_X(f) = B/[1 + (2\pi\beta f)^2]$
    A, B, β, and α positive constants

A-15. **(a)** Derive the general result for the ratio of equivalent noise to 3-dB bandwidth of an nth order Butterworth filter given in Table A-1.
(b) Numerically verify the results given for Butterworth filters of order 1 through 6.

A-16. The two statistically independent inputs to a multiplier have autocorrelation functions

$$R_X(\tau) = A \cos \omega_0 \tau$$

$$R_Y(\tau) = B \operatorname{sinc}(2W\tau)$$

Obtain and plot the power spectrum of the multiplier output.

A-17. Suppose that the inputs to a multiplier, as illustrated in Figure A-8, are independent random processes with power spectral densities $S_X(f) = 5\Pi(f/10)$ and $S_Y(f) = 2\Pi(f/6)$. Compute and sketch the power spectral density of the output of the multiplier.

A-18. The input to a lowpass filter with impulse response

$$h(t) = \exp(-10t)u(t)$$

is white, Gaussian noise with single-sided power spectral density 1 W/Hz. Obtain the following:
(a) The mean of the output.
(b) The autocovariance function of the output.

(c) The probability density function of the output at a single time instant, t_1.

(d) The joint probability of the output at time instants t_1 and $t_1 + 0.1$ s.

A-19. Referring to Example A-13, obtain the probability density function of $Z(t)$ at an arbitrary time t if $N_1 = N_2 = N_0$ and $f_1 = f_2 = f_0$. Assume that both inputs are stationary Gaussian inputs with zero means.

A-20. Derive Equation (A-103).

A-21. **(a)** Derive (A-105).

(b) If $\alpha_{|m|} = a^{|m|}$ and $p(t) = A \exp(-\beta t)u(t)$, where a, A, and β are positive constants, sum (A-105) to obtain a closed-form result for the autocorrelation function. Assume a is less than 1.

Characterization of Internally Generated Noise

Noise in a communication system can be attributed to external sources or noise generated by components and subsystems making up the communication system itself. It is the purpose of this appendix to discuss briefly the characterization of internally generated noise. The internally generated noise originates from the random (Brownian) motion of charged carriers within the system components and can be categorized as flicker, shot, or thermal noise. The latter is due to the random motion of charged carriers within resistive materials and will be characterized briefly in this appendix in terms of *noise figure* or *noise temperature*. Shot noise is due to the motion of charge carriers through a junction, such as the anode of a vacuum tube or the junction between *p*- and *n*-type material of a semiconductor diode. Its name is derived from the analogy with shot falling on a metal plate. The mechanism for the generation of flicker noise is not well understood, but it is characterized by increasing intensity with decreasing frequency and is also referred to as $1/f$ noise.

The noise figure of a device or subsystem may be defined in either one of two ways. The first is that it is the ratio of signal-to-noise *power* ratio at the *input* of the device to the signal-to-noise ratio at its *output*. The second is that it is the ratio of *available noise power* at the *output* terminals of a device to that at its *input*. In equation form, these definitions may be expressed as

$$F = \frac{(S/N)_{\text{in}}}{(S/N)_{\text{out}}} = \frac{N_{\text{out}}}{N_{\text{in}}} \qquad \text{(B-1)}$$

691

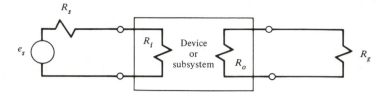

FIGURE B-1. Device or subsystem model pertaining to noise characterization.

where $(S/N)_{\text{in}}$ = input signal-to-noise power ratio
$(S/N)_{\text{out}}$ = output signal-to-noise power ratio
N_{in} = available noise power at the input
N_{out} = available noise power at the output

Matched conditions do not need to be specified for the case of signal-to-noise ratios since the measurement of signal and noise power is performed at the same point.

The circumstances regarding these definitions are illustrated in Figure B-1, which illustrates a device with input resistance R_i and output resistance R_0, driven by a source with internal resistance R_s, and with a load resistance R_g across its output terminals.

It will be recalled that maximum power transfer takes place under *matched conditions*; i.e., $R_s = R_i$ and $R_o = R_g$. By Nyquist's theorem,* the mean-square thermal noise voltage produced by R_s is

$$v_{\text{rms}}^2 = 4kTR_sB \ \ V^2 \tag{B-2}$$

where k = Boltzmann's constant = 1.38×10^{-23} J/K, T is the temperature in kelvin, and B is the bandwidth in hertz in which the noise voltage is measured. Under matched conditions with $R_s = R_i$, half this rms voltage appears across R_s and half across R_i. The *available noise power* of the source resistance is therefore

$$P_a = \frac{(\frac{1}{2} v_{\text{rms}})^2}{R_s} = \frac{v_{\text{rms}}^2}{4R_s} = \frac{4kTR_sB}{4R_s} = kTB \tag{B-3}$$

which is *independent of the source resistance*. If $T = T_0 = 290$ K, which will be referred to as *standard temperature*, the available noise power per hertz of bandwidth is very nearly

$$kT_0 = 4 \times 10^{-21} \ \text{W/Hz}$$

or, when expressed in decibels referenced to 1 W, the result is

$$10 \log_{10} kT_0 = -204 \ \text{dBW/Hz} \tag{B-4}$$

Returning to (B-1), it is seen that the noise figure of a device can be viewed as a measure of the degradation in signal-to-noise ratio imposed by the device or subsystem under consideration, or as a measure of the increase in output noise power over input noise power due to the internal noise added by the device. According to the latter viewpoint, the noise figure can be rewritten as

$$F = \frac{kT_0BG + \Delta N}{kT_0BG} = 1 + \frac{\Delta N}{kT_0BG} \tag{B-5}$$

*For a proof of Nyquist's theorem, see [1, pp. 7–21].

where G is the *available power gain* of the subsystem and ΔN is its internally generated noise power. It is important to note that T_0 is used in (B-5) to give a standardized definition for noise figure. To see how this expression generalizes to two subsystems in cascade, let G_1 and G_2 be their respective available power gains, and let ΔN_1 and ΔN_2 be their internally generated noise powers. Then the following contributions to the output noise power the cascade may be identified:

$kT_0BG_1G_2$ = available noise power at the input amplified by the cascade and appearing at the output

ΔN_1G_2 = internally generated noise of the first subsystem of the cascade, amplified by the second subsystem and appearing at the output of the cascade

ΔN_2 = internally generated noise of the second subsystem of the cascade and appearing at its output

The overall noise figure of the cascade, according to (B-5), is the total noise power appearing at its output divided by the available noise power at its input. Using the definitions above, these can be written as

$$F = \frac{kT_0BG_1G_2 + \Delta N_1G_2 + \Delta N_2}{kT_0BG_1G_2}$$

$$= 1 + \frac{\Delta N_1}{kT_0BG_1} + \frac{1}{G_1}\left(1 + \frac{\Delta N_2}{kT_0BG_2} - 1\right)$$

$$= F_1 + \frac{F_2 - 1}{G_1} \tag{B-6}$$

where $F_1 = 1 + \Delta N_1/kT_0BG_1$ is the noise figure of subsystem 1 and $F_2 = 1 + \Delta N_2/kT_0BG_2$ is the noise figure of subsystem 2. The noise figure of N subsystems in cascade with noise figures $F_1, F_2, \ldots, F_N$ and available power gains $G_1, G_2, \ldots, G_N$ can be shown to be

$$F = F_1 + \frac{F_2 - 1}{G_1} + \frac{F_3 - 1}{G_1G_2} + \cdots + \frac{F_N - 1}{G_1G_2 \cdots G_{N-1}} \tag{B-7}$$

This result, known as *Friis' formula* [2], shows that it is important that the first stage in the cascade have a low noise figure and a large gain for a low overall noise figure.

Since F is dimensionless, it follows from (B-5) that the quantity $\Delta N/kBG$ has the dimensions of temperature. Accordingly, the *effective noise temperature* of a device is defined as

$$T_e = \frac{\Delta N}{kBG} \tag{B-8}$$

where ΔN is its internally generated noise power appearing at the output terminals, G is its power gain, and B is its bandwidth. This definition allows its noise figure to be written as

$$F = 1 + \frac{T_e}{T_0} \tag{B-9}$$

which, when solved for T_e, gives the effective noise temperature in terms of noise figure as

$$T_e = (F - 1)T_0 \tag{B-10}$$

For a cascade of subsystems with noise temperatures $T_1, T_2, T_3, \ldots$ and available power gains $G_1, G_2, \ldots$, the effective noise temperature of the cascade is

$$T_e = T_1 + \frac{T_2}{G_1} + \frac{T_3}{G_1 G_2} + \cdots \qquad \text{(B-11)}$$

which follows by using the definition of T_e and Friis' formula.

For a receiver, it is important to include the effect of noise generated by the antenna or noise intercepted by the antenna generated by hot bodies within its field of view, such as the sun, stars, or the earth. If T_a is the antenna temperature and T_e is the effective noise temperature of the receiver excluding the antenna, the *system noise* temperature, T_s, is

$$T_s = T_a + T_e \qquad \text{(B-12)}$$

which includes the effect of noise power due to the antenna and the internally generated noise of the receiver.

It is important that the first stage of any cascade of subsystems not be an attenuator, for the noise figure of an attenuator of loss factor $L = 1/G$ is simply

$$F_{\text{atten}} = L \qquad \text{(B-13)}$$

if the attenuator is at standard temperature, T_0.*

This discussion of the characterization of internally generated noise will be closed with an example to illustrate the use of the developed relationships in characterizing the noise properties of communications systems.

EXAMPLE B-1

Crystal mixers are sometimes characterized by a *noise temperature ratio*, t_r, defined as

$$t_r = \frac{\text{actual available IF noise power}}{\text{available noise power from an equivalent resistance}}$$

$$= \frac{F_c k T_0 B G_c}{k T_0 B} = F_c G_c = \frac{F_c}{L_c} \qquad \text{(B-14)}$$

where F_c = crystal mixer noise figure

$L_c = 1/G_c$ = mixer conversion loss

Due to flicker noise, the noise temperature ratio of a crystal mixer varies inversely with frequency from about 100 kHz down to less than 1 Hz. Above about 500 kHz, the noise temperature ratio approaches a constant value which is typically in the range 1.3 to 2.0.

Consider, first, the case where a crystal mixer with $t_r = 1.4$ and $L_c = 6$ dB is cascaded with an IF amplifier of noise figure $F_{\text{IF}} = 3$ dB. The noise figure of the cascade, using Friis' formula and the definition of t_r, is

$$F = F_1 + \frac{F_2 - 1}{G} = L_c(t_r + F_{\text{IF}} - 1)$$

$$= 4(1.4 + 2 - 1) = 9.6 = 9.8 \text{ dB}$$

As a second part of the example, suppose that the mixer is preceded by a GaAs field-effect transistor (FET) amplifier with a noise figure of $F_0 = 2$ dB and available

*A derivation may be found in [3, pp. 483–484].

power gain of $G_0 = 10$ dB. The overall noise figure now becomes

$$F = F_0 + \frac{F_1 - 1}{G_0} + \frac{F_2 - 1}{G_0 G_1}$$

$$= F_0 + \frac{L_c t_r - 1}{G_0} + \frac{L_c(F_{IF} - 1)}{G_0}$$

$$= F_0 + \frac{L_c}{G_0}(t_r + F_{IF} - 1) - \frac{1}{G_0}$$

$$= 1.58 + \left(\frac{4}{10}\right)(1.4 + 1.99 - 1) - \frac{1}{10}$$

$$= 2.44$$

$$= 3.87 \text{ dB} \qquad \blacksquare$$

The importance of having a first stage with a moderately high gain and low noise figure is clearly evident from Example B-1.

REFERENCES

[1] A. VAN DER ZIEL, *Noise* (Englewood Cliffs, N.J.: Prentice-Hall, 1954).
[2] H. F. FRIIS, *Proc. Inst. Radio Engs.,* Vol. 32, p. 419, 1944.
[3] R. E. ZIEMER and W. H. TRANTER, *Principles of Communications* (Boston: Houghton Mifflin, 1976).

PROBLEMS

B-1. Obtain the available power at standard temperature in a 1-MHz bandwidth for the following resistances. Express in dBW.
(a) 1 k Ω
(b) 10 k Ω
(c) 1 M Ω
(d) Two 20-k Ω resistors in parallel

B-2. (*Nyquist's Formula*)   As one might suspect, the available noise power for an arbitrary network of resistors, capacitors, and inductors can be found by finding its equivalent impedance looking back into the port (terminal pair) of interest. Let this equivalent impedance be

$$Z(f) = R(f) + jX(f)$$

Then the mean-square thermal noise voltage at the port of interest is

$$v_{rms}^2 = 4kT \int_0^\infty R(f)\, df \qquad \text{V}^2$$

For a purely resistive network, this simplifies to

$$v_{rms}^2 = 4kTR_{eq}B \qquad \text{V}^2$$

where R_{eq} is the equivalent resistance of the network looking into the port of interest.
    Using this theorem, find the mean-square noise voltage at standard temperature in a 1-MHz bandwidth of a 200-Ω resistor in series with a 1000-Ω resistor, with the series combination paralleled by 4800 Ω.

B-3. A receiver with effective noise temperature of 600 K is connected to an antenna of temperature 70 K. Show that the available noise power per hertz for this combination is -170.3 dBm.

B-4. **(a)** A crystal mixer with noise temperature ratio $t_r = 1.3$ and a conversion loss of $L_c = 6.5$ dB is connected to an antenna of temperature 80 K by a transmission line with a loss of $L_t = 1$ dB. The mixer output is connected to an IF amplifier with a noise figure of 5.3 dB. Find the overall effective noise temperature and the noise figure of the cascade.

(b) The mixer is now preceded by a low-noise amplifier with a gain of 12 dB and a noise figure of 1.9 dB. Everything else remains the same. Find the noise temperature and noise figure of the new arrangement.

Communication Link Performance Calculations

In this appendix, the equations used to compute signal-to-noise power ratio in a radio-frequency propagation communications link are derived and applied to a typical satellite relay communications system. The equations and calculational procedures are also applicable to other types of links.

Consider a transmitter that produces P_T watts at the final power amplifier output. If radiated isotropically, the power density at a distance d from the transmitter antenna would be

$$p_t = \frac{P_T}{4\pi d^2} \qquad \text{W/m}^2 \tag{C-1}$$

Every antenna, however, has directivity associated with it. Let this directivity, in the direction of desired propagation, be described by an antenna power gain G_T over the isotropic radiation level. Thus the power density at distance d from the transmitter in the desired direction is

$$p_t' = \frac{P_T G_T}{4\pi d^2} \qquad \text{W/m}^2 \tag{C-2}$$

The power P_R intercepted by the receiving antenna of aperture area A_R is the product of p_t' and A_R:

$$P_R = p_t A_R = \frac{P_T G_T A_R}{4\pi d^2} \tag{C-3}$$

For antennas with aperture area A_R that is large compared with the square of the wavelength, λ, of the transmitted signal, it can be shown that the maximum gain is*

$$G_R = \frac{4\pi A_R}{\lambda^2} \tag{C-4}$$

This allows the received signal power to be written as

$$P_R = \left(\frac{\lambda}{4\pi d}\right)^2 \frac{P_T G_T G_R}{L_t} \tag{C-5}$$

where L_t represents the total system losses. The factor $(4\pi d/\lambda)^2$ is referred to as the *free-space loss* and the product $P_T G_T$ is referred to as the *effective isotropic radiated power* (EIRP).

If the receiver is characterized by an effective noise temperature, T_R, the internally generated noise power, referred to the receiver terminals (receiver input), is

$$\Delta N = kT_R B \tag{C-6}$$

where B is the bandwidth of interest and k is Boltzmann's constant.† For binary signaling, the bandwidth of interest is a *bit-rate bandwidth*, defined as

$$B = \frac{1}{T_b} = R_b \tag{C-7}$$

where R_b is the bit rate.

The ratio of (C-5) and (C-6) with (C-7) substituted gives the signal-to-noise ratio (SNR):

$$\frac{P_R}{\Delta N} = \left(\frac{\lambda}{4\pi d}\right)^2 \frac{P_T G_T G_R}{L_t k T_R R_b} \tag{C-8}$$

A quantity of interest in specifying the performance of digital communication systems is the ratio of energy per bit to noise power-spectral density, E_b/N_0. For a received power of P_R, the energy per bit is $E_b = P_R T_b$ while $\Delta N = N_0 R_b$, where N_0 is the noise power spectral density in W/Hz. Thus

$$\frac{E_b}{N_0} = \frac{P_R T_b}{\Delta N/R_b} = \frac{P_R}{\Delta N} \tag{C-9}$$

which can be related to the various link parameters through (C-8). It is most convenient, however, to express all quantities in decibels. This results in

$$(E_b/N_0)_{\text{dB}} = 20 \log_{10}\left(\frac{\lambda}{4\pi d}\right) + 10 \log_{10} P_T + 10 \log_{10} G_T$$

$$+ 10 \log_{10} G_R - 10 \log_{10} L_t$$

$$- 10 \log_{10} kT_R - 10 \log_{10} R_b \tag{C-10}$$

Often, (C-10) is written without the $10 \log_{10}(\cdot)$ explicitly and decibels are understood. It is also usual to have E_b/N_0 specified and P_T to be found. The following example illustrates the use of (C-10) for this purpose.

---

*Equation (C-4) is often multiplied by an *antenna aperture efficiency*, $\rho_a \leq 1$, to reflect the fact that no antenna is 100% efficient.
†$k = 1.38 \times 10^{-23}$ J/K

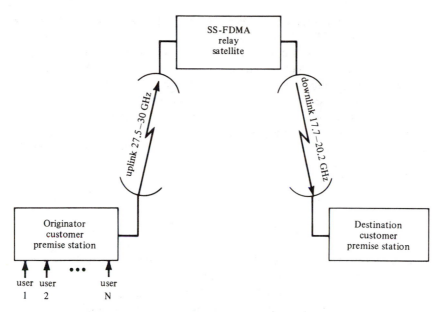

FIGURE C-1. Model for a single path of a satellite-switched frequency-division multiple-access satellite communication link.

EXAMPLE C-1

Shown in Figure C-1 is one RF path* for a satellite-switched frequency-division multiple-access (SS-FDMA) satellite relay link. The path consists of an uplink, assumed to be in the band 27.5 to 30 GHz, a frequency-translation receiver-transmitter pair on board the FDMA satellite, and a downlink in the band 17.7 to 20.2 GHz. "Satellite switched" refers to the fact that N_1 receive antenna beams and N_2 transmit antenna beams are used on board the satellite which cover certain "spots" on the earth's surface.† Any path can be routed through any receive-transmit antenna beam by means of a switch matrix. The transmitted signals originate from N users which are combined in a suitable fashion for transmission through the relay satellite at a Customer Premise Station (CPS). This combining of user transmissions is done in frequency as illustrated in Figure C-2.

---

*An RF path is a band of frequencies routed through the satellite through two antenna beams. Each path may include many users or channels.

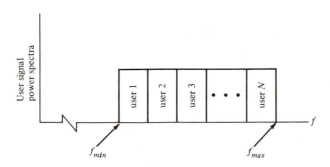

FIGURE C-2. Combining several users in frequency for transmission through a SS-FDMA satellite path.

Now it is desired to attain a certain level of bit-error-rate (BER) performance at the destination CPS. This can be related directly to a required E_b/N_0 at the destination CPS. For illustrative purposes, suppose that the required E_b/N_0 is 14 dB, including hardware and channel impairments. Noise enters the system due to two major sources: the satellite relay and destination CPS receiver front ends. Also, the transmitted power can be apportioned between the originator CPS and the relay satellite transmitter. Thus two E_b/N_0 ratios are available for trade-off purposes; that associated with the uplink, $(E_b/N_0)_u$, and that associated with the downlink, $(E_b/N_0)_d$. Equation (C-10) applies to both the uplink and the downlink. For the uplink, it is convenient to write it as

$$(E_b/N_0)_u = \text{EIRP} + G_s - L_{pu} - L_{tu} - kT_s - R_b \qquad \text{(C-11)}$$

where dB quantities are used throughout and

EIRP $= 10 \log_{10} P_T G_T$ is the effective radiated power of the CPS station
$\quad G_s =$ gain of the satellite receive antenna
$\quad L_{pu} = -20 \log_{10}(\lambda_u/4\pi d)$ is the path loss on the uplink
$\quad L_{tu} =$ total system losses or the uplink (path loss excluded)
$\quad kT_s =$ satellite receive noise power density
$\quad R_b =$ data rate

Similarly, (C-10) for the downlink becomes

$$(E_b/N_0)_d = P_s + G_T + G_R - L_{pd} - L_{td} - kT_R + R_b \qquad \text{(C-12)}$$

where all quantities are in dB and

$\quad P_s =$ satellite transmit power
$\quad G_T =$ satellite transmit antenna gain
$\quad G_R =$ gain of the CPS user receive antenna
$\quad L_{pd} = 20 \log_{10}(\lambda_d/4\pi d)$ is the path loss on the downlink
$\quad L_{td} =$ total system losses on the downlink
$\quad kT_R =$ CPS user receive noise power density

To demonstrate the use of (C-11) and (C-12), assume that $(E_b/N_0)_u = (E_b/N_0)_d = 17$ dB, so that $E_b/N_0 = 14$ dB. Also assume that the other parameters are as follows:

$\quad\quad G_R = 53$ dB (3-m parobolic antenna; aperture efficiency $= 54\%$)
$G_T = G_s = 45$ dBI
$\quad\quad L_{pu} = 214$ dB (41,000-km path length with $\lambda_u = 1$ cm)
$\quad\quad L_{pd} = 210.5$ dB ($\lambda_d = 1.5$ cm)
$\quad\quad L_{tu} = 2.5$ dB (2.5-dB antenna pointing loss; modem and channel impairments included in E_b/N_0)
$\quad\quad L_{td} = 5$ dB (pointing loss plus 2.5-dB line loss)
$\quad\quad kT_s = -169$ dBm/Hz (5-dB noise figure plus 2-dB line losses)
$\quad\quad kT_R = -170.3$ dBm/Hz ($T_R = 600$ K plus 70 K antenna temperature)
$\quad\quad R_b = 60$ dB-Hz (1 Mbps data rate assumed)

Using these parameters and the assumed apportionment between $(E_b/N_0)_u$ and $(E_b/N_0)_d$ to give the desired $E_b/N_0 = 14$ dB, the following values are obtained for the ground station EIRP and satellite transmit power:

$$\text{EIRP} = (E_b/N_0)_u - G_s + kT_s + L_{\text{pu}} + L_{\text{tu}} + R_b$$

$$= 17 - 45 - 169 + 214 + 2.5 + 60$$

$$= 79.5 \text{ dBm}$$

$$P_s = (E_b/N_0)_d - G_T + kT_R + L_{\text{pd}} + L_{\text{td}} - G_R + R_d$$

$$= 17 - 45 - 170.3 + 210.5 + 5 - 53 + 60$$

$$= 24.2 \text{ dBm} \qquad \blacksquare$$

EXAMPLE C-2

In Example C-1 the required transmitter EIRP and satellite transmit power was calculated to give an E_b/N_0 at the CPS user station of 14 dB with the noise contribution of the relay satellite and CPS user station equal. This example generalizes that example in that the noise contributions of satellite and user station are not necessarily equal, but the noise division is left as a parameter.

Let the available power gain of the satellite be G_a and let the signal power received by its antenna be P_R. Assume that the noise power density of the user satellite at its antenna terminals is N_{0a}, so that the noise power in a bit-rate bandwidth is $N_{0a}R_b$. At the output of the satellite relay, these powers are G_aP_R and $G_aN_{0a}R_b$, respectively, and at the user station antenna terminals, they are both scaled by the path attenuation of the downlink, α. The user station adds noise of its own, which is $N_{0b}R_b$ referred to the receiver input. The SNR, E_b/N_0, accounting for both noise sources at the ground station, is

$$\frac{E_b}{N_0} = \frac{\alpha G_a P_R}{\underbrace{\alpha G_a N_{0a} R_b}_{\substack{\text{uplink} \\ \text{noise}}} + \underbrace{N_{0b} R_b}_{\substack{\text{downlink} \\ \text{noise}}}}$$

$$= \frac{E_b}{\alpha G_a N_{0a} + N_{0b}}$$

$$= \left[\left(\frac{E_b}{N_0} \right)_u^{-1} + \left(\frac{E_b}{N_0} \right)_d^{-1} \right]^{-1} \qquad \text{(C-13)}$$

where $E_b = \alpha G_a P_R / R_b$ is the energy per bit at the ground station receiver antenna terminals and $\alpha G_a N_{0a}$ is the noise due to the satellite relay. By varying G_a, one can make the uplink (relay satellite input) noise dominant or the user station noise dominant. Setting E_b/N_0 equal to an appropriate value in order to give a predetermined BER performance then allows a parametric curve to be obtained, showing

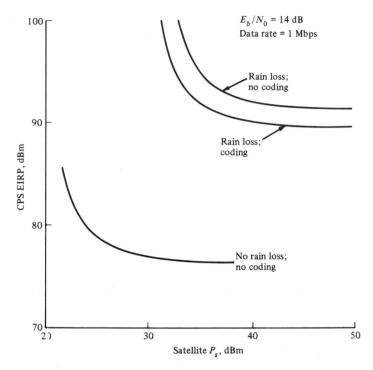

FIGURE C-3. Trade-off between CPS and satellite power transmission requirements for an SS-FDMA link.

the trade-off between EIRP and P_s. Such a curve is shown in Figure C-3 for $E_b/N_0 = 14$ dB and the parameter values used previously. The expressions used to plot the graph of Figure C-3 are

$$\text{EIRP} = (E_b/N_0)_u + 62.5 \text{ dBm}$$

$$P_S = (E_b/N_0)_d + 7.2 \text{ dBm}$$

together with (C-13). Several numerical values are given in the following table.

$(E_b/N_0)_u$ (dB)	EIRP (dBm)	$(E_b/N_0)_d$ (dB)	P_s (dB)
15	77.5	20.9	28.1
17	79.5	17	24.2
20	82.5	15.3	25.5
23	85.5	14.6	21.8

EXAMPLE C-3

In Examples C-1 and C-2 transmission in the band 30/20 GHz was assumed. In these bands, rain can impose severe degradation in both the uplink and the downlink. In this example, a rain loss of 15 dB is assumed in the uplink, which uses a frequency of approximately 30 GHz. Using scaling proportional to the square of the frequency gives rain loss for the downlink, which is at a frequency of approximately 20 GHz, of 11.5 dB.

There are various ways that this loss can be combatted. One way is simply by increasing the CPS EIRP and the satellite transmit power, P_s. Using the same parameters as before with the additional rain losses included gives

$$\text{EIRP} = (E_b/N_0)_u = 62.5 + 15$$

$$= (E_b/N_0)_u + 77.5 \text{ dBm}$$

$$P_s = (E_b/N_0)_d + 7.2 + 11.5$$

$$= (E_b/N_0)_d + 18.7 \text{ dBm}$$

Often, this is considered too expensive an approach. Another way to combat the rain loss is to introduce time diversity by FEC coding. Assume a coding gain for a rate-$\frac{1}{2}$ convolutional code of 4.6 dB. To keep the data rate constant, R_b must be doubled because two code bits are sent for each data bit. Thus the required E_b/N_0 is decreased by 4.6 dB, to 9.4 dB, but the equations for EIRP and P_s have an additional 3 dB added to them due to R_b doubling, giving

$$\text{EIRP} = (E_b/N_0)_u + 80.5 \text{ dBm}$$

$$P_s = (E_b/N_0)_d + 21.7 \text{ dBm}$$

(rain loss and bandwidth expansion due to coding are included). These curves are also shown in Figure C-3. Although constructed for a data rate of 1 Mbps, they are easily adjusted for other data rates by simple scaling procedures. A set of such curves therefore becomes a useful tool in making comparisons between data-rate requirements and determining the trade-off between power requirements for the CPS and relay satellite. ■

PROBLEMS

C-1. Verify that the path loss for a 41,000-km path for a signal with a wavelength of 1 cm (30 GHz) is 214 dB.

C-2. Verify the curves of Figure C-3.

Overview of the Viterbi Algorithm*

The Viterbi algorithm (VA) is an algorithm that produces a MAP estimate of the state sequence of a finite-state discrete-time Markov process observed in memoryless noise.

A discrete-time Markov process has the property that the conditional probability of a particular amplitude, x_{k+1}, at time $k + 1$ given the sequence of amplitudes $x_0, x_1, \ldots, x_k$ up to time k is dependent only on the immediately preceding amplitude, x_k:

$$P(x_{k+1}|x_k, x_{k-1}, \ldots, x_1, x_0) = P(x_{k+1}|x_k) \tag{D-1}$$

In discussing the applications of the VA it is useful to view the amplitude sequence $x_0, x_1, \ldots, x_k$ as being generated by a finite-state machine. The state of any system at time k_0 can be defined as the minimum information necessary to specify completely the condition of the system at time k_0 and to allow its output to be determined for times $k > k_0$ when the inputs up to time k_0 are specified. An example will illustrate this idea.

EXAMPLE D-1

Consider the shift register arrangement shown in Figure D-1. Such a circuit is used to generate a rate-$\frac{1}{2}$ convolutional code. Its operation is as follows. Consider

---

*The summary of the Viterbi algorithm given here follows [1].

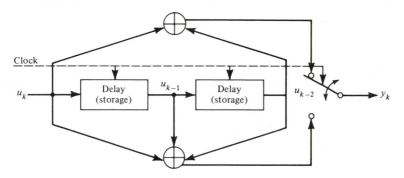

FIGURE D-1.   Feedback shift register circuit illustrating the concept of a state.

the storage registers to be initially loaded with zeros and assume that the sequence 1100101 is the input (the most recent symbol is assumed on the right). The output, y_k, is taken alternately with the switch in the upper position and then in the lower. A new input digit is stored in the left-hand register for every two outputs. Thus the code generated is referred to as *rate-$\frac{1}{2}$*. The contents of the storage register and the outputs are given in the following table.

| Time Index, k | Input, u_k | Storage Registers | | Output |
		u_{k-1}	u_{k-2}	
0	1	0	0	1
		0	0	1
1	1	1	0	1
		1	0	0
2	0	1	1	1
		1	1	0
3	0	0	1	1
		0	1	1
4	1	0	0	1
		0	0	1
5	0	1	0	0
		1	0	1
6	1	0	1	0
		0	1	0

The contents of the storage registers correspond to the states. In particular, the states assume the possible values (0, 0), (1, 0), (0, 1), and (1, 1). Only certain states may follow from a given state. This can be summarized by a state diagram as illustrated in Figure D-2. Each node represents a possible state and branches represent transitions between states. The encoder follows the solid line branches out of a state if the input is a zero and the dashed line branches if the input is a one. The labels on the branches are the encoder outputs. Over the course of time the process traces a path from state to state through the state diagram. Note that

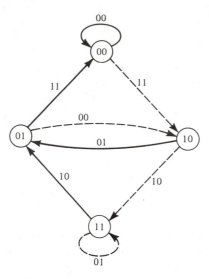

FIGURE D-2. State diagram for a four-state process.

the time index is suppressed. Note also that the present state and the future input are sufficient to trace out the progression of future states. The number of states is denoted by Q, the number of storage registers by D, and the number of values for each stored quantity by M; it follows that $Q = M^D$. ∎

Instead of a state diagram, wherein the time variable is suppressed, it is often convenient to utilize a trellis diagram as used in Chapter 4 for the excess phase of a multi-h CPM signal and for convolutional codes in Chapter 12. The trellis diagram corresponding to the state diagram of Figure D-2 is illustrated in Figure D-3 assuming that the initial state was 00. An important property of the trellis diagram is that *each sequence of states,* $\mathbf{s} = (s_1, s_2, s_3, \ldots, s_k)$, *corresponds to a unique path through the trellis*. Each unique path through the trellis is associated with a sequence of elements given by the labels on the branches that connect the states. For a convolutional code these branch labels represent the encoder output and for multi-h CPM these labels represent the sequence of receiver matched filter output samples for the noiseless case. In all cases, the sequence of branch labels represents a sample function of the finite-state Markov process $\mathbf{x} = (x_0, x_1, x_2, \ldots, x_{k-1})$. The problem addressed by the VA is this. Given a noise-corrupted Markov process, $\mathbf{z} = \mathbf{x} + \mathbf{n}$, where the noise samples are independent from one instant to the next, find the unique path $\mathbf{x}$ through the trellis giving rise to the best approximation to $\mathbf{z}$ in the sense of maximizing $P(\mathbf{x}|\mathbf{z})$. Using Bayes' rule and assuming that all paths through the trellis are equally likely, it can be shown* that maximizing $P(\mathbf{x}|\mathbf{z})$ is equivalent to maximizing $P(\mathbf{z}|\mathbf{x})$. The VA finds the path that maximizes $P(\mathbf{z}|\mathbf{x})$.

The solution to this problem is to assign a length $-\ln P(\mathbf{z}|\mathbf{x})$ to each path in the trellis, and choose the path through the trellis with the minimum length. Since $-\ln(\cdot)$ is a monotonically decreasing function of its argument, minimization of $-\ln P(\mathbf{z}|\mathbf{x})$ results in the maximization of $P(\mathbf{z}|\mathbf{x})$. When the Markov sequence is a sequence of real numbers and the noise is AWGN, an equivalent path length is the Euclidian distance between the received sequence and the Markov sequence. When

*See Chapter 12 for a detailed example of this calculation.

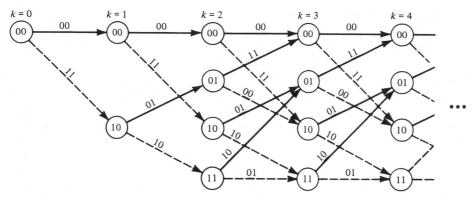

FIGURE D-3. Trellis diagram for the four-state Markov process representing the convolutional code of Figure D-1.

the VA is applied to hard decision convolutional decoding, an equivalent path length is the Hamming distance between the received sequence and the trellis path sequence.

Since the noise is memoryless, $P(\mathbf{z}|\mathbf{x})$ factors as follows:

$$P(\mathbf{z}|\mathbf{x}) = \prod_{k=0}^{K-1} P(z_k|x_k) \tag{D-2}$$

Taking the negative of the logarithm results in

$$-\ln P(\mathbf{z}|\mathbf{x}) = -\sum_{k=0}^{K-1} \ln P(z_k|x_k)$$

$$\triangleq \sum_{k=0}^{K-1} \lambda(z_k, x_k) \tag{D-3}$$

Observe that because the logarithm is being used, the total path length is the sum of the incremental path length accumulated over each branch of the path. The incremental path length is called the *branch metric*. The branch metric may be the Euclidean distance, Hamming distance, or any other appropriate [i.e., monotone-increasing function of $P(z_k|x_k)$] distance measure. The VA, given the received sequence $\mathbf{z}$, searches through the entire trellis to find the particular path with the minimum length or, equivalently, the minimum total metric.

Clearly, calculation of the lengths of all possible paths through the trellis can be an enormous task for several states and K large. This computational burden is decreased considerably through a clever observation. Consider the possible paths through a trellis illustrated in Figure D-3 starting at time $k = 0$ and ending at state j at time $k = K$. Each of these paths will have a different length (metric). The shortest such path segment is called the *survivor*, corresponding to the state j at time K and is denoted $\hat{\mathbf{x}}_j^k$. For any $k > 0$, there are Q such survivors, one for each state. The important observation is this: *The shortest complete path* $\hat{\mathbf{x}}$ *from the beginning node at* $k = 0$ *to the ending state at* $k = K_m$ *must begin with one of these survivors.* If it did not, we could replace its initial segment by a shorter segment and get a shorter path, which is a contradiction. Thus, to proceed from the kth set of states to the $(k + 1)$st set of states and find the new survivors, we need only compute the new distances using the lengths for the old survivors and adding to each the length of the appropriate transitions. An example will illustrate the procedure.

EXAMPLE D-2

The numbers beside each transition of the trellis in Figure D-4a represent the branch metrics. The beginning state at $k = 0$ is $s_0 = 00$ and the final state at $k = 5$ is also $s_5 = 00$. In going from time $k = 0$ to $k = 1$, two transitions are possible, namely (00, 00) and (00, 10). The metrics are both unity. From each of the nodes 00 and 10 at time $k = 1$, two transitions are also possible. The total path metric for the four possible paths after two time units is the sum of the branch metrics for the first two branches on each path. Figure D-4b shows the cumulative path metric for the survivor path into each state for time units 1 through 5. After the third time unit, there is more than a single path into each state. Consider state 00 at $k = 3$. Paths enter this state from state 00 at time $k = 2$ with a metric $2 + 0 = 2$ and from state 01 at time $k = 2$ with metric $3 + 2 = 5$. The first term in the sums is the cumulative metric at $k = 2$ and the second term is the branch metric for the most recent branch. The path with metric 2 is chosen by the VA as the survivor and is the only path shown entering state 00 at $k = 3$. This procedure continues until there is a single path remaining at $k = 5$. ■

The Viterbi algorithm consists of the following steps:

1. At time k, find the path metric for each path entering each state by adding the path metric of the survivor at time $k - 1$ to the most recent branch metric.
2. For each state of time k, choose the path with the minimum metric as the survivor.
3. Store the state sequence and the path metric for the survivor at each state.
4. Increment the index k and repeat steps 1 through 3.

If the state sequence is not finite, it is necessary to truncate survivors to some reasonable length, referred to as the *decision depth*, δ. At time k, a decision is forced at time $k - \delta$ using some reasonable criterion in case the choice is not unique. As time progresses it may also be necessary to renormalize the metrics. The last consideration is how to start the algorithm if the initial state is unknown. Again, any reasonable strategy may be used; a random initial choice will result in all survivors merging with the correct path with high probability after an initial transient.

A final example illustrates the use of the VA for decoding convolutional codes. An example of the use of the VA for multi-h CPM is given in Chapter 4.

EXAMPLE D-3

Suppose that the Viterbi decoder input sequence for the trellis of Figure 12-4 is $\mathbf{y} = (11, 10, 10, 00, 01, 11, 00)$. What is the most likely codeword sequence $\mathbf{x}_m$ to have been transmitted?

Solution: The decoding begins by considering the paths through the trellis up to the first set of states where paths merge, as shown in Figure D-5a. The received signal is binary, indicating that a hard decision demodulator is being used and it is correct to use Hamming distance to compare paths. Consider state 00. The Hamming distance between the upper path into state 00 and the received sequence is 4, the number of positions in which the two sequences differ. The Hamming distance between the lower path into state 00 and the received sequence is 1. Therefore, the lower path into state 00 is more likely to have been transmitted and is retained. The discarded path is crossed out with an X in Figure D-5a, and the distance between the retained path and the received sequence is written in paren-

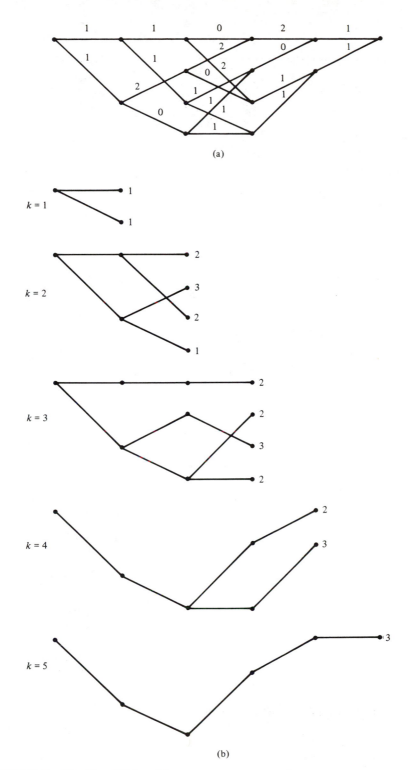

(a)

$k = 1$

$k = 2$

$k = 3$

$k = 4$

$k = 5$

(b)

FIGURE D-4. Trellis with lengths.

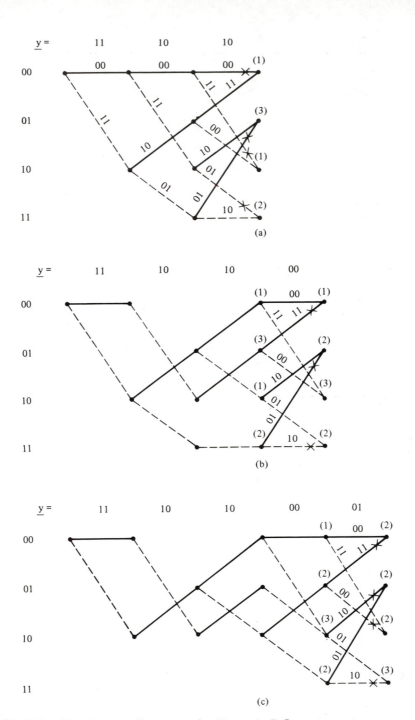

FIGURE D-5. Viterbi decoding steps for Example D-3.

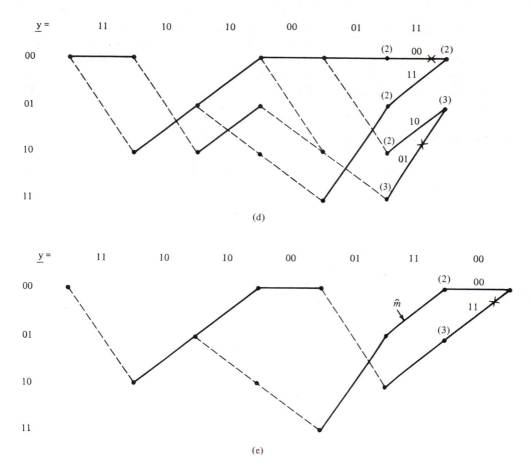

FIGURE D-5. continued.

theses above the state. A similar comparison of Hamming distance to the received sequence for the two paths entering state 01 yields distance 3 for the upper path and distance 4 for the lower path. Thus the upper path and its distance are retained as shown. Similar calculations for nodes 10 and 11 yield the results shown.

The decoder now moves one branch to the right and performs another set of distance calculations which will be described using Figure D-5b. In this figure, the paths that were discarded in the preceding decoding step are no longer shown. Consider the two paths entering state 00. This distance between the path coming from state 00 and the received sequence is found by adding the stored distance 1 to the distance accumulated over the new branch which is 0. Thus the new distance is 1. Adding the stored distance, 3, to the distance accumulated on the new branch, 2, for the path entering state 00 from state 01 yields a new distance 5. Thus the path entering 00 from 00 is more likely to have been transmitted and is retained for future comparison. Similar comparisons for the other states yield the distances and retained paths shown. The distance comparison for node 10 resulted in the same distance for both paths so both paths were retained. In actual hardware implementation a selection would have to be made even in the case of ties. This selection is arbitrary and does not affect the error probability.

The remaining decoding steps are illustrated in Figures D-5c, d, and e. In the final two decoding steps, the encoder trellis is merging toward state 00 so that fewer states have to be considered. When the final state is reached, only two paths through the trellis remain to be compared. The path entering node 00 from node 00 is closer to the received sequence and is the decoder output estimate $\hat{m}$. The actual encoder output was $x_m = (11, 10, 00, 01, 01, 11, 00)$ and two transmission errors occurred in positions 5 and 8. ■

REFERENCE

[1] G. D. FORNEY, "The Viterbi Algorithm," *Proc. IEEE*, Vol. 61, pp. 268–278, March 1973.

Gaussian Probability Function*

The Gaussian probability function of unit variance and zero mean is

$$Z(x) = \frac{e^{-x^2/2}}{\sqrt{2\pi}} \tag{E-1}$$

and the corresponding cumulative distribution function is

$$P(x) = \int_{-\infty}^{x} Z(t)\, dt \tag{E-2}$$

The Q-function first defined in Chapter 3 is

$$Q(x) = 1 - P(x) = \int_{x}^{\infty} Z(t)\, dt \tag{E-3}$$

An asymptotic expansion for $Q(x)$ valid for large x is

$$Q(x) = \frac{Z(x)}{x}\left[1 - \frac{1}{x^2} + \frac{1\cdot 3}{x^4} + \cdots + \frac{(-1)^n 1\cdot 3\ \ldots\ \cdot(2n-1)}{x^{2n}}\right] + R_n \tag{E-4}$$

where

$$R_n = (-1)^{n+1}1\cdot 3\ \ldots\ \cdot(2n+1)\int_{x}^{\infty}\frac{Z(t)}{t^{2n+2}}\, dt \tag{E-5}$$

*The notation used here is that of Abramowitz and Stegun [1, pp. 931 ff.].

which is less in absolute value than the first neglected term. For moderate values of x, several rational approximations are available. One such approximation is

$$1 - Q(x) = P(x) = 1 - Z(x)(b_1 t + b_2 t^2 + b_3 t^3 + b_4 t^4 + b_5 t^5) + \epsilon(x)$$

$$t = \frac{1}{1 + px} \qquad \text{(E-6)}$$

$$|\epsilon(x)| < 7.5 \times 10^{-8}$$

$$p = 0.2316419$$

$$b_1 = 0.319381530 \qquad b_4 = -1.821255978$$

$$b_2 = -0.356563782 \qquad b_5 = 1.330274429$$

$$b_3 = 1.781477937$$

The error function can be related to the Q-function by

$$\operatorname{erf}(x) \triangleq \frac{2}{\sqrt{\pi}} \int_0^x e^{-t^2} \, dt = 1 - 2Q(\sqrt{2}x) \qquad \text{(E-7)}$$

The complementary error function can be approximated similarly to the Q-function.

TABLE E-1. Abbreviated Table of Values for $Q(x)$ and $Z(x)$

x	$Q(x)$	$Z(x)$
0.0	0.50000	0.39894
0.1	0.46017	0.39695
0.2	0.42074	0.39104
0.3	0.38209	0.38138
0.4	0.34458	0.36827
0.5	0.30854	0.35206
0.6	0.27425	0.33322
0.7	0.24196	0.31225
0.8	0.21186	0.28969
0.9	0.18406	0.26608
1.0	0.15866	0.24197
1.1	0.13567	0.21785
1.2	0.11507	0.19419
1.3	0.09680	0.17137
1.4	0.08076	0.14973
1.5	0.06681	0.12952
1.6	0.05480	0.11092
1.7	0.04457	0.09405
1.8	0.03593	0.07895
1.9	0.02872	0.06562

TABLE E-1. (continued)
**Abbreviated Table of Values
for $Q(x)$ and $Z(x)$**

x	$Q(x)$	$Z(x)$
2.0	0.02275	0.05399
2.1	0.01786	0.04398
2.2	0.01390	0.03547
2.3	0.01072	0.02833
2.4	0.00820	0.02239
2.5	0.00621	0.01753
2.6	0.00466	0.01358
2.7	0.00347	0.01042
2.8	0.00256	0.00792
2.9	0.00187	0.00595
3.0	0.00135	0.00443
3.1	0.00097	0.00327
3.2	0.00069	0.00238
3.3	0.00042	0.00723
3.4	0.00034	0.00123
3.5	0.00023	0.00087
3.6	0.00016	0.00061
3.7	0.00011	0.00042
3.8	7.24×10^{-5}	0.00029
3.9	4.81×10^{-5}	0.00020
4.0	3.17×10^{-5}	0.00013

A short table of values for $Q(x)$ and $Z(x)$ is given in Table E-1. Extensive tables of $P(x)$, $Z(x)$ and its derivatives can be found in Abramowitz and Stegun [1].

REFERENCE

[1] M. ABRAMOWITZ and I. STEGUN, *Handbook of Mathematical Functions* (New York: Dover). Originally published in 1964 as NBS Applied Mathematics Series 55.

Power Spectral Densities for Sequences of Random Binary Digits and Random Tones

Calculation of the transmitted spread-spectrum power spectral density requires knowledge of the power spectral density of the spreading function. The psd of a random binary sequence is usually calculated from the time autocorrelation function as was done in the text. The psd of a random sequence of tones is normally approximated by a sum of delta functions at the tone frequencies, with each weighted by the probability of that tone being transmitted. The exact psd for the special case where the tones are phase coherent from one transmission to the next was given previously (Lindsey and Simon [1]). In this appendix, the ensemble autocorrelation function for a sequence of random binary symbols is calculated and is used in the calculation of the psd of a sequence of noncoherent tones.

Consider an infinite random sequence of binary symbols $\ldots, a_{-1}, a_0, a_1, \ldots$, where $a_n \in \{+1, -1\}$ and the random process generated using these symbols:

$$s(t, \mathbf{a}, T) = \sum_{n=-\infty}^{\infty} a_n p(t + T - nT_c) \qquad \text{(F-1)}$$

In this equation, $\mathbf{a}$ represents a random vector, T is a random phase required to make $s(t)$ stationary, T_c is the sequence chip duration, and $p(t)$ is the unit pulse of duration T_c. One possible sample function of this random process is illustrated in Figure F-1. The a_n's above are independent and it is equally likely that any a_n is $+1$ or -1. The phase T is uniformly distributed over the interval $(0, T_c)$.

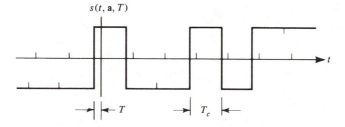

FIGURE F-1. Sample function of the random process $s(t, a, T)$.

The ensemble autocorrelation function is defined by

$$R_s(t_1, t_2) = E[s(t_1)s(t_2)] \tag{F-2}$$

where the expected value is over all **a** and all T. Substituting (F-1) into (F-2) yields

$$R_s(t_1, t_2) = E\left[\sum_{n=-\infty}^{\infty} \sum_{n'=-\infty}^{\infty} a_n a_{n'} p(t_1 + T - nT_c)p(t_2 + T - n'T_c)\right]$$

$$= \sum_{n} \sum_{n'} E_1[a_n a_{n'}] E_2[p(t_1 + T - nT_c)p(t_2 + T - n'T_c)] \tag{F-3}$$

since the expectation is linear and the random variables a_n and T are independent. In (F-3), expectation E_1 is over a_n and E_2 is over T. It is easily shown that $E_1(a_n a_n') = 0$ for $n \neq n'$ and $E_1(a_n a_{n'}) = 1$ for $n = n'$, so that

$$R_s(t_1, t_2) = \sum_{n} E_2[p(t_1 + T - nT_c)p(t_2 + T - nT_c)] \tag{F-4}$$

At this time it is convenient to return the summation to within the expected value and to write the expected value explicitly. This yields

$$R_s(t_1, t_2) = \frac{1}{T_c} \int_0^{T_c} \sum_{n} p(t_1 + T - nT_c)p(t_2 + T - nT_c) \, dT \tag{F-5}$$

The product within the summation is illustrated in Figure F-2 for a particular $t_1 - t_2 < T_c$ and n. For $t_1 - t_2 > T_c$, the product is zero since the pulses are nonoverlapping. The area under the product pulse is $T_c - |t_1 - t_2|$. The complete time function under the integral of (F-5) is a sum of time translations of the function shown in Figure F-2d as illustrated in Figure F-2e. The integral of (F-5) can now be calculated by inspection. The result is the shaded area of Figure F-2e and

$$R_s(t_1, t_2) = \begin{cases} \dfrac{1}{T_c}(T_c - |t_1 - t_2|) & \text{for } |t_1 - t_2| < T_c \\ 0 & \text{for } |t_1 - t_2| > T_c \end{cases} \tag{F-6}$$

This function is a function only of $|t_1 - t_2| = \tau$ and not the absolute value of t_1 and t_2 as expected. The result is identical to the result obtained previously using the time autocorrelation, so that the power spectral density, the Fourier transform of $R_s(\tau)$, is given by

$$S_s(f) = T_c \, \text{sinc}^2 (fT_c) \tag{F-7}$$

as in Example (7-1).

Consider next a random sequence of tones having frequencies chosen from a set of 2^k possible frequencies and having a random phase each time a new frequency is selected. Assume that the random phase is uniformly distributed over $(0, 2\pi)$

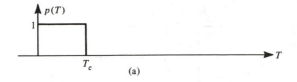

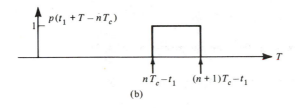

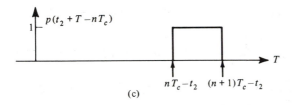

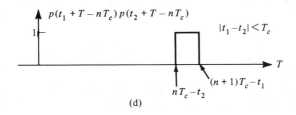

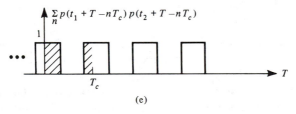

FIGURE F-2. Functions used to calculate $R_s(t_1, t_2)$.

and that the phase of the time of frequency change is uniformly distributed over $(0, T_c)$. Thus the random process being considered is

$$s(t) = 2 \sum_{n=-\infty}^{\infty} p(t + T - nT_c) \cos(\omega_n t + \varphi_n) \qquad \text{(F-8)}$$

The ensemble autocorrelation function is given by

$$R_s(t_1, t_2) = E[s(t_1)s(t_2)]$$

$$= 4E\left[\sum_n \sum_{n'} p(t_1 + T - nT_c)p(t_2 + T - n'T_c)\right.$$

$$\left. \times \cos(\omega_n t_1 + \varphi_n) \cos(\omega_{n'} t_2 + \varphi_{n'})\right] \qquad \text{(F-9)}$$

when the expectation is over all ω_n, φ_n, and T. The linearity of the ensemble average, and the independence of ω_n, φ_n, and T, are used to obtain

$$R_s(t_1, t_2) = 2 \sum_n \sum_{n'} E_1[p(t_1 + T - nT_c)p(t_2 + T - n'T_c)]$$

$$\times \{E_2[\cos(\omega_n t_1 + \omega_{n'} t_2 + \varphi_n + \varphi_{n'})]$$

$$+ E_2[\cos(\omega_n t_1 - \omega_{n'} t_2 + \varphi_n - \varphi_{n'})]\} \tag{F-10}$$

where E_1 is over all T and E_2 is over all ω_n and φ_n. For $n \neq n'$, the average over all φ_n and φ_n' in E_2 implies that these terms are zero. This is also true for the sum frequency term even when $n = n'$. Thus

$$R_s(t_1, t_2) = 2 \sum_n E_1[p(t_1 + T - nT_c)p(t_2 + T - nT_c)]$$

$$\times E_2[\cos(\omega_n [t_1 - t_2])] \tag{F-11}$$

All frequencies are assumed to be discrete and equally probable, so that the second expected value for any n is simply a summation over all 2^k frequencies, yielding

$$R_s(t_1, t_2) = \frac{1}{2^{k-1}} \sum_{m=1}^{2^k} \cos(\omega_m[t_1 - t_2]) \sum_n E_1[p(t_1 + T - nT_c)$$

$$\times p(t_2 + T - nT_c)] \tag{F-12}$$

The remaining expected value was evaluated above so that the final result can be written directly and is

$$R_s(t_1, t_2) = R_s(\tau) = \begin{cases} \dfrac{1}{2^{k-1}} \displaystyle\sum_{m=1}^{2^k} \cos(\omega_m \tau)\left(1 - \dfrac{|\tau|}{T_c}\right) & \text{for } |\tau| \leq T_c \\ 0 & \text{for } |\tau| > T_c \end{cases} \tag{F-13}$$

The power spectral density of $s(t)$ is the Fourier transform of (F-13). The Fourier transform is a linear operation, so that each term of the sum can be independently evaluated. Each term of (F-14) is a product of terms each of whose Fourier transform is known. The frequency convolution theorem is invoked, yielding

$$S_s(f) = \frac{T_c}{2^k} \sum_{m=1}^{2^k} \{\text{sinc}^2 [(f - f_m)T_c] + \text{sinc}^2 [(f + f_m)T_c]\} \tag{F-14}$$

This expression is equivalent to the continuous frequency term of (7-41). In this case none of the signal power is in discrete components as was anticipated.

REFERENCE

[1] W. C. LINDSEY and M. K. SIMON, *Telecommunication Systems Engineering* (Englewood Cliffs, N.J.: Prentice-Hall, 1973).

Calculation of the Power Spectrum of the Product of Two *M*-Sequences

The following development is a minor modification of the development by [1]. The power spectrum of $c(t)c(t + \epsilon)$ is calculated by first determining the autocorrelation function and then using the Wiener–Khintchine theorem. Denote the product $c(t)c(t + \epsilon)$ by $b(t, \epsilon)$. The autocorrelation function of this periodic function is

$$R_b(\tau, \epsilon) = \frac{1}{T} \int_0^T b(t, \epsilon)b(t + \tau, \epsilon)\, dt \qquad \text{(G-1)}$$

The functions $c(t)$, $c(t + \epsilon)$, and $b(t, \epsilon)$ are illustrated in Figure G-1 for a 7-bit *m*-sequence. For any particular *m*-sequence and any ϵ, $R_b(\tau, \epsilon)$ can be calculated directly. This calculation is facilitated, however, by recognizing [1] that $b(t, \epsilon)$ can be represented by the sum of two functions $p(t, \epsilon) + q(t, \epsilon)$, which are also illustrated in Figure G-1. The function $p(t, \epsilon)$ is a binary-valued function with period T_c, the code chip duration. The function $q(t, \epsilon)$ is a three-valued function whose nonzero values are the same as the original *m*-sequence shifted in phase. Using this decomposition, the autocorrelation function can be written as the sum

$$R_b(\tau, \epsilon) = R_p(\tau, \epsilon) + R_{pq}(\tau, \epsilon) + R_{qp}(\tau, \epsilon) + R_q(\tau, \epsilon) \qquad \text{(G-2)}$$

Each of these autocorrelations can be separately evaluated and separately Fourier transformed to obtain the desired power spectrum. The following discussion is limited to values of $|\epsilon| \leq T_c$.

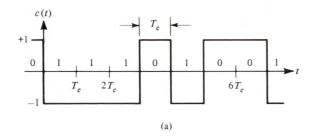

(a)

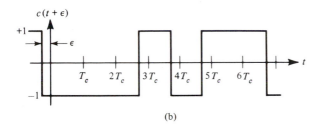

(b)

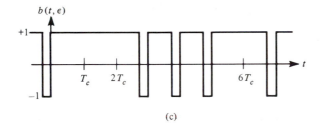

(c)

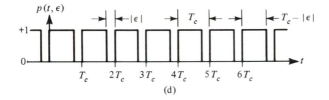

(d)

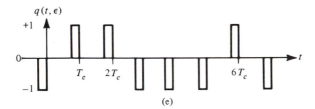

(e)

FIGURE G-1. Waveform used in the calculation of the power spectrum of $c(T)$ $c(t + \epsilon)$. (From Ref. 1.)

Consider first the autocorrelation of $p(t, \epsilon)$. This autocorrelation

$$R_p(\tau, \epsilon) = \frac{1}{T_c} \int_0^{T_c} p(t, \epsilon)p(t + \tau, \epsilon) \, dt \qquad \text{(G-3)}$$

is illustrated in Figure G-2a for $0 \leq |\epsilon| \leq T_c/2$ and in Figure G-2b for $T_c/2 < |\epsilon| \leq T_c$. These functions are calculated by inspection from Figure G-1d. Each of the functions is a periodic sequence of triangular pulses. For $0 \leq |\epsilon| \leq T_c/2$,

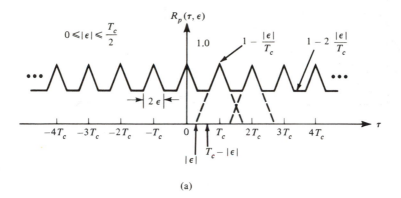

(a)

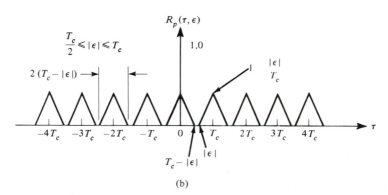

(b)

FIGURE G-2. Autocorrelation function of $p(t, \epsilon)$: (a) $0 \leq |\epsilon| < T_c/2$; (b) $T_c/2 \leq |\epsilon| \leq T_c$. (From Ref. 1.)

$$R_p(\tau, \epsilon) = \left(1 - 2\frac{|\epsilon|}{T_c}\right) + \frac{|\epsilon|}{T_c}\sum_{n=-\infty}^{\infty}\Lambda(\tau - nT_c, |\epsilon|) \qquad \text{(G-4a)}$$

and for $T_c/2 < |\epsilon| \leq T_c$,

$$R_p(\tau, \epsilon) = \left(1 - \frac{|\epsilon|}{T_c}\right)\sum_{n=-\infty}^{\infty}\Lambda(\tau - nT_c, T_c - |\epsilon|) \qquad \text{(G-4b)}$$

In (G-4) the function $\Lambda(\tau, B)$ is a triangular pulse of height 1.0 and width $2B$, as illustrated in Figure G-3. When $\epsilon \leq T_c/2$ the triangular pulses rest on a constant pedestal of height $(1 - 2|\epsilon/T_c|)$.

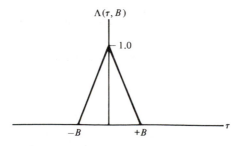

FIGURE G-3. Triangular pulse waveform.

It was shown earlier that the Fourier transform of a periodic waveform $x(t)$ with period T, for example,

$$x(t) = \sum_{m=-\infty}^{\infty} y(t - mT) \tag{G-5a}$$

is

$$X(f) = \frac{1}{T} \sum_{n=-\infty}^{\infty} Y(nf_0)\delta(f - nf_0) \tag{G-5b}$$

where $Y(f)$ is the Fourier transform of a single pulse $y(t)$ and $f_0 = 1/T$. The Fourier transform of the triangular pulse $\Lambda(\tau, B)$ is $S_\Lambda(f, B) = B\ \text{sinc}^2\ (fB)$. Therefore, substituting the triangular pulse for $y(t)$ in (G-5) yields the power spectrum of $p(t, \epsilon)$. The result for $0 \leq |\epsilon| \leq T_c/2$ is

$$S_p(f, \epsilon) = \left(1 - 2\frac{|\epsilon|}{T_c}\right)\delta(f) + \left(\frac{|\epsilon|}{T_c}\right)^2 \sum_{n=-\infty}^{\infty} \text{sinc}^2\ (nf_c|\epsilon|)\delta(f - nf_c) \tag{G-6a}$$

and for $T_c/2 < |\epsilon| \leq T_c$ is

$$S_p(f, \epsilon) = \left(1 - \frac{|\epsilon|}{T_c}\right)^2 \sum_{n=-\infty}^{\infty} \text{sinc}^2\ [nf_c(T_c - |\epsilon|)]\delta(f - nf_c) \tag{G-6b}$$

where $f_c = 1/T_c$.

The autocorrelation function of $q(t, \epsilon)$ is illustrated in Figure G-4a for $0 \leq |\epsilon| < T_c/2$ and Figure G-4b for $T_c/2 \leq |\epsilon| \leq T_c$. These autocorrelation functions are calculated by inspection of Figure G-1e and from the fact that the values of $q(t, \epsilon)$, where it is nonzero, are the same as the original m-sequence values shifted in phase. The functions of Figure G-4a or G-4b can be decomposed into two periodic sequences of triangular functions, that is, $R_q(\tau, \epsilon) = R_{qa}(\tau, \epsilon) + R_{qb}(\tau, \epsilon)$, where, for $0 \leq |\epsilon| \leq T_c/2$,

$$R_{qa}(\tau, \epsilon) = -\frac{|\epsilon|}{NT_c} \sum_{n=-\infty}^{\infty} \Lambda(\tau - nT_c, |\epsilon|) \tag{G-7a}$$

and for $T_c/2 < |\epsilon| \leq T_c$,

$$R_{qa}(\tau, \epsilon) = -\frac{1}{N}\left(2\frac{|\epsilon|}{T_c} - 1\right) - \frac{1}{N}\left(1 - \frac{|\epsilon|}{T_c}\right) \sum_{n=-\infty}^{\infty} \Lambda(\tau - nT_c, T_c - |\epsilon|) \tag{G-7b}$$

and for all $0 \leq |\epsilon| \leq T_c$,

$$R_{qb}(\tau, \epsilon) = \left(1 + \frac{1}{N}\right)\frac{|\epsilon|}{T_c} \sum_{m=-\infty}^{\infty} \Lambda(\tau - mNT_c, |\epsilon|) \tag{G-7c}$$

Now consider the cross-correlation functions $R_{pq}(\tau, \epsilon)$ and $R_{qp}(\tau, \epsilon)$. By definition

$$R_{qp}(\tau, \epsilon) = \frac{1}{NT_c} \int_0^{NT_c} q(t, \epsilon)p(t + \tau, \epsilon)\ dt \tag{G-8}$$

Using the substitution $\lambda = t + \tau$, this equation becomes

$$R_{qp}(\tau, \epsilon) = \frac{1}{NT_c} \int_\tau^{\tau + NT_c} q(\lambda - \tau, \epsilon)p(\lambda, \epsilon)\ d\lambda$$

$$= R_{pq}(-\tau, \epsilon) \tag{G-9}$$

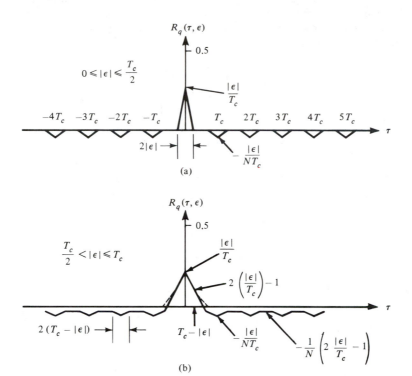

FIGURE G-4. Autocorrelation function of $q(t, \epsilon)$: (a) $0 < |\epsilon| \leq T_c/2$; (b) $T_c/2 < |\epsilon| \leq T_c$. (From Ref. 1.)

since both $p(t, \epsilon)$ and $q(t, \epsilon)$ are periodic with period NT_c. The function $R_{pq}(\tau, \epsilon)$ is found by inspection of Figure G-1 and knowledge that values of $q(t, \epsilon)$ follow the original m-sequence. The function is illustrated in Figure G-5a for $0 \leq |\epsilon| \leq T_c/2$ and in Figure G-5b for $T_c/2 < |\epsilon| < T_c$. These functions can be written as a

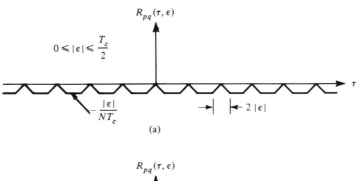

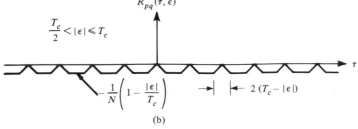

FIGURE G-5. Crosscorrelation function of $p(t, \epsilon)$ and $q(t, \epsilon)$: (a) $0 < |\epsilon| \leq T_c/2$; (b) $T_c/2 < |\epsilon| \leq T_c$. (From Ref. 1.)

sum of triangle functions on a constant pedestal yielding for $0 \le |\epsilon| \le T_c/2$,

$$R_{pq}(\tau, \epsilon) = -\frac{|\epsilon|}{NT_c} + \frac{|\epsilon|}{NT_c} \sum_{n=-\infty}^{\infty} \Lambda(\tau - nT_c, |\epsilon|) \qquad \text{(G-10a)}$$

and for $T_c/2 \le |\epsilon| \le T_c$,

$$R_{pq}(\tau, \epsilon) = -\frac{1}{N}\left(1 - \frac{|\epsilon|}{T_c}\right) + \frac{1}{N}\left(1 - \frac{|\epsilon|}{T_c}\right) \sum_{n=-\infty}^{\infty} \Lambda(\tau - nT_c, T_c - |\epsilon|)$$

$$\text{(G-10b)}$$

The sum of the final three terms $R_\Sigma(\tau, \epsilon)$ of (G-2) can now be written for $0 \le |\epsilon| \le T_c/2$,

$$R_\Sigma(\tau, \epsilon) = R_{pq}(\tau, \epsilon) + R_{qp}(\tau, \epsilon) + R_q(\tau, \epsilon)$$

$$= 2R_{pq}(\tau, \epsilon) + R_{qa}(\tau, \epsilon) + R_{qb}(\tau, \epsilon)$$

$$= -\frac{2|\epsilon|}{NT_c} + \frac{2|\epsilon|}{NT_c} \sum_{n=-\infty}^{\infty} \Lambda(\tau - nT_c, |\epsilon|)$$

$$- \frac{|\epsilon|}{NT_c} \sum_{n=-\infty}^{\infty} \Lambda(\tau - nT_c, |\epsilon|) + \left(1 + \frac{1}{N}\right)\frac{|\epsilon|}{T_c} \sum_{m=-\infty}^{\infty} \Lambda(\tau - mNT_c, |\epsilon|)$$

$$= -\frac{2|\epsilon|}{NT_c} + \frac{|\epsilon|}{NT_c} \sum_{n=-\infty}^{\infty} \Lambda(\tau - nT_c, |\epsilon|)$$

$$+ \left(1 + \frac{1}{N}\right)\frac{|\epsilon|}{T_c} \sum_{m=-\infty}^{\infty} \Lambda(\tau - mNT_c, |\epsilon|) \qquad \text{(G-11)}$$

Observe that the triangle functions have been combined in the last line. For $T_c/2 < |\epsilon| \le T_c$ this expression becomes

$$R_\Sigma(\tau, \epsilon) = R_{pq}(\tau, \epsilon) + R_{qp}(\tau, \epsilon) + R_q(\tau, \epsilon)$$

$$= 2R_{pq}(\tau, \epsilon) + R_{qa}(\tau, \epsilon) + R_{qb}(\tau, \epsilon)$$

$$= -\frac{2}{N}\left(1 - \frac{|\epsilon|}{T_c}\right) + \frac{2}{N}\left(1 - \frac{|\epsilon|}{T_c}\right) \sum_{n=-\infty}^{\infty} \Lambda(\tau - nT_c, T_c - |\epsilon|)$$

$$- \frac{1}{N}\left(2\frac{|\epsilon|}{T_c} - 1\right) - \frac{1}{N}\left(1 - \frac{|\epsilon|}{T_c}\right) \sum_{n=-\infty}^{\infty} \Lambda(\tau - nT_c, T_c - |\epsilon|)$$

$$+ \left(1 + \frac{1}{N}\right)\frac{|\epsilon|}{T_c} \sum_{m=-\infty}^{\infty} \Lambda(\tau - mNT_c, |\epsilon|)$$

$$= -\frac{1}{N} + \frac{1}{N}\left(1 - \frac{|\epsilon|}{T_c}\right) \sum_{n=-\infty}^{\infty} \Lambda(\tau - nT_c, T_c - |\epsilon|)$$

$$+ \left(1 + \frac{1}{N}\right)\frac{|\epsilon|}{T_c} \sum_{m=-\infty}^{\infty} \Lambda(\tau - mNT_c, |\epsilon|) \qquad \text{(G-12)}$$

Equation (G-5) is employed to find the Fourier transforms and complete the calculation of the power spectrum of $c(t)c(t + \epsilon)$. Using (G-5) and $S_\Lambda(f, B) = B \,\text{sinc}^2(fB)$, for $0 < |\epsilon| \le T_c/2$,

$$S_{\Sigma}(f, \epsilon) = -\frac{2|\epsilon|}{NT_c}\,\delta(f) + \frac{1}{N}\left(\frac{|\epsilon|}{T_c}\right)^2 \sum_{n=-\infty}^{\infty} \text{sinc}^2\,(nf_c|\epsilon|)\delta(f - nf_c)$$

$$+ \frac{1}{N}\left(1 + \frac{1}{N}\right)\left(\frac{|\epsilon|}{T_c}\right)^2 \sum_{m=-\infty}^{\infty} \text{sinc}^2\left(m\frac{f_c}{N}|\epsilon|\right)\delta\left(f - m\frac{f_c}{N}\right)$$

$$\text{(G-13a)}$$

and for $T_c/2 < |\epsilon| \le T_c$,

$$S_{\Sigma}(f, \epsilon) = -\frac{1}{N}\delta(f) + \frac{1}{N}\left(1 - \frac{|\epsilon|}{T_c}\right)^2 \sum_{n=-\infty}^{\infty} \text{sinc}^2\,[nf_c(T_c - |\epsilon|)]\delta(f - nf_c)$$

$$+ \frac{1}{N}\left(1 + \frac{1}{N}\right)\left(\frac{|\epsilon|}{T_c}\right)^2 \sum_{m=-\infty}^{\infty} \text{sinc}^2\left(m\frac{f_c}{N}|\epsilon|\right)\delta\left(f - m\frac{f_c}{N}\right)$$

$$\text{(G-13b)}$$

Finally, to simplify the final expression for the power spectral density, examine the function $\text{sinc}^2\,[nf_c(T_c - |\epsilon|)]$ which appears in both (G-6) and (G-13). For $n \ne 0$,

$$\text{sinc}^2\,[nf_c(T_c - |\epsilon|)] \triangleq \frac{\sin^2\,[\pi nf_c(T_c - |\epsilon|)]}{[\pi nf_c(T_c - |\epsilon|)]^2}$$

$$= \frac{\sin^2\,[\pi n - \pi nf_c|\epsilon|]}{[\pi nf_c|\epsilon|]^2}\,\frac{[\pi nf_c|\epsilon|]^2}{[\pi nf_c(T_c - |\epsilon|)]^2}$$

$$= \frac{\cos^2\,(\pi n)\sin^2\,(\pi nf_c|\epsilon|)}{[\pi nf_c|\epsilon|]^2}\left(\frac{|\epsilon|}{T_c - |\epsilon|}\right)^2$$

$$= \text{sinc}^2\,(nf_c|\epsilon|)\,\frac{(|\epsilon|/T_c)^2}{(1 - |\epsilon|/T_c)^2} \qquad \text{(G-14)}$$

and for $n = 0$,

$$\text{sinc}^2\,[nf_c(T_c - |\epsilon|)] = 1 = \text{sinc}^2\,(nf_c|\epsilon|)$$

At this point the final expression for the power spectrum can be written by adding (G-6) and (G-13) and using (G-14). This result is, for $0 \le |\epsilon| < T_c/2$,

$$S_b(f, \epsilon) = \left(1 - 2\frac{|\epsilon|}{T_c}\right)\delta(f) + \left(\frac{|\epsilon|}{T_c}\right)^2 \sum_{n=-\infty}^{\infty} \text{sinc}^2\,(nf_c|\epsilon|)\delta(f - nf_c)$$

$$- 2\frac{|\epsilon|}{NT_c}\,\delta(f) + \frac{1}{N}\left(\frac{|\epsilon|}{T_c}\right)^2 \sum_{n=-\infty}^{\infty} \text{sinc}^2\,(nf_c|\epsilon|)\delta(f - nf_c)$$

$$+ \frac{1}{N}\left(1 + \frac{1}{N}\right)\left(\frac{|\epsilon|}{T_c}\right)^2 \sum_{m=-\infty}^{\infty} \text{sinc}^2\left(\frac{mf_c}{N}|\epsilon|\right)\delta\left(f - \frac{mf_c}{N}\right)$$

$$= \left[1 - \left(1 + \frac{1}{N}\right)\left(\frac{|\epsilon|}{T_c}\right)\right]^2 \delta(f)$$

$$+ \left(1 + \frac{1}{N}\right)\left(\frac{|\epsilon|}{T_c}\right)^2 \sum_{\substack{n=-\infty\\ n\ne 0}}^{\infty} \text{sinc}^2\,(nf_c|\epsilon|)\delta(f - nf_c)$$

$$+ \frac{N + 1}{N^2}\left(\frac{|\epsilon|}{T_c}\right)^2 \sum_{\substack{m=-\infty\\ m\ne 0}}^{\infty} \text{sinc}^2\left(\frac{mf_c|\epsilon|}{N}\right)\delta\left(f - \frac{mf_c}{N}\right)$$

$$\text{(G-15)}$$

and for $T_c/2 \le |\epsilon| \le T_c$,

$$S_b(f, \epsilon) = \left(\frac{|\epsilon|}{T_c}\right)^2 \sum_{\substack{n=-\infty \\ n\neq 0}}^{\infty} \text{sinc}^2\left(nf_c|\epsilon|\right)\delta(f - nf_c) + \left(1 - \frac{|\epsilon|}{T_c}\right)^2 \delta(f)$$

$$- \frac{1}{N}\delta(f) + \frac{1}{N}\left(\frac{|\epsilon|}{T_c}\right)^2 \sum_{\substack{n=-\infty \\ n\neq 0}}^{\infty} \text{sinc}^2\left(nf_c|\epsilon|\right)\delta(f - nf_c)$$

$$+ \frac{1}{N}\left(1 - \frac{|\epsilon|}{T_c}\right)^2 \delta(f)$$

$$+ \frac{1}{N}\left(1 + \frac{1}{N}\right)\left(\frac{|\epsilon|}{T_c}\right)^2 \sum_{\substack{m=-\infty \\ m\neq 0}}^{\infty} \text{sinc}^2\left(\frac{mf_c}{N}|\epsilon|\right)\delta\left(f - \frac{mf_c}{N}\right)$$

$$+ \frac{1}{N}\left(1 + \frac{1}{N}\right)\left(\frac{|\epsilon|}{T_c}\right)^2 \delta(f)$$

$$= \left[1 - \left(1 + \frac{1}{N}\right)\left(\frac{|\epsilon|}{T_c}\right)\right]^2 \delta(f)$$

$$+ \left(1 + \frac{1}{N}\right)\left(\frac{|\epsilon|}{T_c}\right)^2 \sum_{\substack{n=-\infty \\ n\neq 0}}^{\infty} \text{sinc}^2\left(nf_c|\epsilon|\right)\delta(f - nf_c)$$

$$+ \left(\frac{N+1}{N^2}\right)\left(\frac{|\epsilon|}{T_c}\right)^2 \sum_{\substack{m=-\infty \\ m\neq 0}}^{\infty} \text{sinc}^2\left(\frac{mf_c}{N}|\epsilon|\right)\delta\left(f - \frac{mf_c}{N}\right)$$

$$\text{(G-16)}$$

Before using (G-14) to simplify (G-13b) and (G-6b), the zero-frequency term must first be removed from the summation. Observe that (G-15) and (G-16) are identical, so that the desired power spectrum can be calculated from either for any $0 \le |\epsilon| \le T_c$.

REFERENCE

[1] W. J. Gill, "Effect of Synchronization Error in Pseudo-random Carrier Communications," *Conf. Rec.*, First Annual IEEE Commun. Conf., June 1965.

Evaluation of Phase Discriminator Output Autocorrelation Functions and Power Spectra

FULL-TIME EARLY-LATE NONCOHERENT TRACKING LOOP

The discriminator output autocorrelation function is defined in Equation (9-65) with $\epsilon(t, \delta)$ given in Equation (9-64). In order to shorten the following equations, define

$$C_1 = \frac{1}{2} K_1 P \left\{ R_c^2 \left[\left(\delta - \frac{\Delta}{2} \right) T_c \right] - R_c^2 \left[\left(\delta + \frac{\Delta}{2} \right) T_c \right] \right\}$$

$$C_2 = \sqrt{2 K_1 P}$$

$$R_- = R_c \left[\left(\delta - \frac{\Delta}{2} \right) T_c \right]$$

$$R_+ = R_c \left[\left(\delta + \frac{\Delta}{2} \right) T_c \right]. \tag{H-1}$$

Then

$$E[\epsilon(t, \delta)\epsilon(t + \tau, \delta)] = E\{[C_1 + C_2\{R_- n_{2I}(t) - R_+ n_{1I}(t)\}\cos[\varphi - \varphi' + \theta_d(t)]$$

$$+ C_2\{R_- n_{2Q}(t) - R_+ n_{1Q}(t)\} \sin [\varphi - \varphi' + \theta_d(t)]$$

$$+ [n_{2I}(t)]^2 + [n_{2Q}(t)]^2 - [n_{1I}(t)]^2 - [n_{1Q}(t)]^2]$$

$$\cdot [C_1 + C_2\{R_- n_{2I}(t + \tau) - R_+ n_{1I}(t + \tau)\}$$

$$\cos [\varphi - \varphi' + \theta_d(t + \tau)]$$

$$+ C_2\{R_- n_{2Q}(t + \tau) - R_+ n_{1Q}(t + \tau)\} \sin [\varphi - \varphi' + \theta_d(t + \tau)]$$

$$+ [n_{2I}(t + \tau)]^2 + [n_{2Q}(t + \tau)]^2$$

$$- [n_{1I}(t + \tau)]^2 - [n_{1Q}(t + \tau)]^2]\}. \tag{H-2}$$

Expansion of the products within this expected value yields the sum of a large number of terms. The expected value of each term can be separately evaluated. The random noise process is independent of the local oscillator phase so that the expectations can be factored. For example

$$E[C_1 C_2\{R_- n_{2I}(t + \tau) - R_+ n_{1I}(t + \tau)\} \cos [\varphi - \varphi' + \theta_d(t + \tau)]]$$

$$= C_1 C_2 \, E[R_- n_{2I}(t + \tau) - R_+ n_{1I}(t + \tau)]E[\cos [\varphi - \varphi' + \theta_d(t + \tau)]]. \tag{H-3}$$

The cosine term can be further expanded yielding

$$E[\cos [\varphi - \varphi' + \theta_d(t + \tau)]] = E[\cos (\varphi - \varphi')]E[\cos (\theta_d(t + \tau))]$$

$$- E[\sin(\varphi - \varphi')]E[\sin(\theta_d(t + \tau))]. \tag{H-4}$$

The received carrier phase and the local oscillator phase are uniformly distributed over $(0, 2\pi)$ so that

$$E[\cos (\varphi - \varphi')] = E[\sin (\varphi - \varphi')] = 0. \tag{H-5}$$

Therefore, all expressions having the form of Equation (H-3) may be set to zero in the expression. Expanding Equation (H-2) yields

$$E[\epsilon(t, \delta)\epsilon(t + \tau, \delta)] = C_1^2 + C_1 E[\{n_{2I}(t + \tau)\}^2 + \{n_{2Q}(t + \tau)\}^2$$

$$- \{n_{1I}(t + \tau)\}^2 - \{n_{1Q}(t + \tau)\}^2]$$

$$+ C_2^2 \, E[\{R_- n_{2I}(t) - R_+ n_{1I}(t)\}$$

$$\{R_- n_{2I}(t + \tau) - R_+ n_{1I}(t + \tau)\}]$$

$$\cdot E[\cos \{\varphi - \varphi' + \theta_d(t)\}$$

$$\cos \{\varphi - \varphi' + \theta_d(t + \tau)\}]$$

$$+ C_2^2 \, E[\{R_- n_{2I}(t) - R_+ n_{1I}(t)\}$$

$$\{R_- n_{2Q}(t + \tau) - R_+ n_{1Q}(t + \tau)\}]$$

$$\cdot E[\cos \{\varphi - \varphi' + \theta_d(t)\}$$

$$\sin \{\varphi - \varphi' + \theta_d(t + \tau)\}]$$

$$+ C_2^2 \, E[\{R_- n_{2Q}(t) - R_+ n_{1Q}(t)\}$$

$$\{R_- n_{2I}(t + \tau) - R_+ n_{1I}(t + \tau)\}]$$

$$\cdot E[\sin \{\varphi - \varphi' + \theta_d(t)\}$$

$$\cos \{\varphi - \varphi' + \theta_d(t + \tau)\}]$$

$$+ C_2^2 \, E[\{R_- n_{2Q}(t) - R_+ n_{1Q}(t)\}$$

$$\{R_- n_{2Q}(t + \tau) - R_+ n_{1Q}(t + \tau)\}]$$

$$\cdot E[\sin \{\varphi - \varphi' + \theta_d(t)\}$$

$$\sin \{\varphi - \varphi' + \theta_d(t + \tau)\}]$$

$$+ C_1 \, E[\{n_{2I}(t)\}^2 + \{n_{2Q}(t)\}^2 - \{n_{1I}(t)\}^2 - \{n_{1Q}(t)\}^2]$$

$$+ E[\{[n_{2I}(t)]^2 + [n_{2Q}(t)]^2 - [n_{1I}(t)]^2 - [n_{1Q}(t)]^2\}$$

$$\{[n_{2I}(t + \tau)]^2 + [n_{2Q}(t + \tau)]^2 \cdot$$

$$- [n_{1I}(t + \tau)]^2 - [n_{1Q}(t + \tau)]^2\}]. \qquad \text{(H-6)}$$

The baseband noise processes $n_{1I}(t)$, $n_{2I}(t)$, $n_{1Q}(t)$, and $n_{2Q}(t)$ have been assumed to be white Gaussian noise processes but have not been assumed to be independent. It was demonstrated earlier that, under the assumptions that $\Delta \geq 1.0$ and that the input noise have a significantly wider bandwidth than the received signal, these noise processes are uncorrelated and therefore independent. In order to complete this analysis, assume that the processes are independent so that the products of expected value factor. With this assumption and because

$$E[n_{1I}(t)] = E[n_{1Q}(t)] = E[n_{2I}(t)] = E[n_{2Q}(t)] = 0,$$

Equation (H-6) simplifies to

$$E[\epsilon(t, \delta) \, \epsilon \, (t + \tau, \delta)] = C_1^2 + C_1 \, E[\{n_{2I}(t)\}^2 + \{n_{2Q}(t)\}^2$$

$$- \{n_{1I}(t)\}^2 - \{n_{1Q}(t)\}^2]$$

$$+ \{C_2^2 \, R_-^2 \, E[n_{2I}(t)n_{2I}(t + \tau)]$$

$$+ C_2^2 \, R_+^2 \, E[n_{1I}(t)n_{1I}(t + \tau)]\}$$

$$\cdot E[\cos \{\varphi - \varphi' + \theta_d(t)\} \cos \{\varphi - \varphi' + \theta_d(t + \tau)\}]$$

$$+ \{C_2^2 R_-^2 \, E[n_{2Q}(t)n_{2Q}(t + \tau)]$$

$$+ C_2^2 R_+^2 \, E[n_{1Q}(t)n_{1Q}(t + \tau)]\}$$

$$\cdot E[\sin \{\varphi - \varphi' + \theta_d(t)\} \sin \{\varphi - \varphi' + \theta_d(t + \tau)\}]$$

$$+ C_1 \, E[\{n_{2I}(t + \tau)\}^2 + \{n_{2Q}(t + \tau)\}^2$$

$$- \{n_{1I}(t + \tau)\}^2 - \{n_{1Q}(t + \tau)\}^2]$$

$$+ E[\{[n_{2I}(t)]^2 + [n_{2Q}(t)]^2 - [n_{1I}(t)]^2 - [n_{1Q}(t)]^2\}$$

$$\cdot \{[n_{2I}(t + \tau)]^2 + [n_{2Q}(t + \tau)]^2$$

$$- [n_{1I}(t + \tau)]^2 - [n_{1Q}(t + \tau)]^2\}]. \qquad \text{(H-7)}$$

Further simplification of this result is achieved if it is noticed that all of the noise processes have identical statistics. Combining similar terms yields

$$E[\epsilon(t, \delta) \, \epsilon \, (t + \tau, \delta)] = C_1^2 + C_2^2(R_-^2 + R_+^2)E[n_b(t)n_b(t + \tau)]$$

$$\cdot E[\cos \{\varphi - \varphi' + \theta_d(t)\} \cos \{\varphi - \varphi' + \theta_d(t + \tau)\}]$$

$$+ C_2^2(R_-^2 + R_+^2)E[n_b(t)n_b(t + \tau)]$$

$$\cdot E[\sin \{\varphi - \varphi' + \theta_d(t)\} \sin \{\varphi - \varphi' + \theta_d(t + \tau)\}]$$

$$+ E[4\{n_b(t)n_b(t + \tau)\}^2 - 4\{n_b(t)\}^2\{n_b(t + \tau)\}^2]$$

$$= C_1^2 + C_2^2(R_-^2 + R_+^2)E[n_b(t)n_b(t + \tau)]$$

$$E[\cos \{\theta_d(t) - \theta_d(t + \tau)\}]$$

$$+ 4 E[\{n_b(t)n_b(t + \tau)\}^2] - 4 E^2[\{n_b(t)\}^2] \qquad \text{(H-8)}$$

where $n_b(t)$ is any one of the four baseband noise processes. Observe that this function is not independent of the data modulation because of the term $E[\cos \{\theta_d(t) - \theta_d(t + \tau)\}]$. Using Equation (H-1), Equation (H-8) becomes

$$R_\epsilon(\tau) = \tfrac{1}{4}K_1^2P^2\left\{R_c^2\left[\left(\delta - \frac{\Delta}{2}\right)T_c\right] - R_c^2\left[\left(\delta + \frac{\Delta}{2}\right)T_c\right]\right\}^2$$

$$+ 2K_1P\left\{R_c^2\left[\left(\delta - \frac{\Delta}{2}\right)T_c\right] + R_c^2\left[\left(\delta + \frac{\Delta}{2}\right)T_c\right]\right\}E[n_b(t)n_b(t + \tau)]$$

$$\cdot E[\cos \{\theta_d(t) - \theta_d(t + \tau)\}]$$

$$+ 4E[\{n_b(t)n_b(t + \tau)\}^2] - 4 E^2[\{n_b(t)\}^2]. \qquad \text{(H-9)}$$

The power spectrum at the discriminator output is found by taking the Fourier transform of Equation (H-9). This Fourier transform will be considered one term at a time. The first term is not a function of τ so that its transform results in a delta function at DC. This DC component is the desired phase correction term.

The second term of Equation (H-9) is a (signal × noise) term which is the result of the squaring operation. It is the product of two functions of τ so that the Fourier multiplication theorem can be used to calculate the Fourier transform. By definition

$$E[n_b(t)n_b(t + \tau)] = R_{n_b}(\tau) \qquad \text{(H-10)}$$

and by the Wiener-Khintchine theorem

$$S_{n_b}(f) = \int_{-\infty}^{\infty} R_{n_b}(\tau)e^{-j2\pi f\tau} d\tau \qquad \text{(H-11)}$$

The expected value of $\cos [\theta_d(t) - \theta_d(t - \tau)]$ is evaluated by considering the data modulated carrier

$$a(t) = \cos [\omega_{IF}t + \theta_d(t) + \beta] \qquad \text{(H-12)}$$

with complex envelope

$$A(t) = \exp [j\theta_d(t) + j\beta] \qquad \text{(H-13)}$$

The complex autocorrelation function is defined [Stein, 1967] by

$$R_A(\tau) = \tfrac{1}{2} E[A^*(t)A(t + \tau)] \qquad \text{(H-14)}$$

where the complex conjugate arises from the fact that the difference frequency term of the product of $a(t)a(t + \tau)$ is the desired term in the real autocorrelation $R_a(\tau)$. Substituting Equation (H-13) into (H-14) yields

$$R_A(\tau) = \tfrac{1}{2} E\{\exp [-j\theta_d(t) + j\theta_d(t + \tau)]\} \qquad \text{(H-15)}$$

so that

$$E[\cos \{\theta_d(t) - \theta_d(t + \tau)\}] = 2 \text{ Re } [R_A(\tau)]. \qquad \text{(H-16)}$$

The Fourier transform of Equation (H-16) is

$$S_{\theta_d}(f) = \int_{-\infty}^{\infty} 2 \text{ Re } [R_A(\tau)]e^{-j2\pi f\tau}d\tau$$

$$= \int_{-\infty}^{\infty} [R_A(\tau) + R_A^*(\tau)]e^{-j2\pi f\tau}d\tau \qquad \text{(H-17)}$$

It is easily shown that $R_A^*(\tau) = R_A(-\tau)$ using a change of variable in Equation (H-16) so that

$$S_{\theta_d}(f) = \int_{-\infty}^{\infty} R_A(\tau)e^{-j2\pi f\tau}d\tau + \int_{-\infty}^{\infty} R_A(-\tau)e^{-j2\pi f\tau}d\tau$$

$$= S_A(f) + S_A(-f) \qquad \text{(H-18)}$$

Finally, the total contribution of the (signal × noise) term of Equation (H-9) to the discriminator output power spectrum is

$$S_2(f) = 2K_1P\left\{R_c^2\left[\left(\delta - \frac{\Delta}{2}\right)T_c\right] + R_c^2\left[\left(\delta + \frac{\Delta}{2}\right)T_c\right]\right\}S_{n_b}(f)*S_{\theta_d}(f). \qquad \text{(H-19)}$$

The third term of Equation (H-9) is four times the autocorrelation function of the output of a square law device with $n_b(t)$ as its input. The Fourier transform yields the power spectrum of the output of this same device. This output spectrum has been calculated in detail in [Davenport, 1958]. The square law device considered in this analysis is characterized by

$$y = ax^2 \qquad \text{(H-20)}$$

where the input x is a bandlimited Gaussian noise process which has power spectrum $S_x(f)$ and total power σ_x^2. The power spectrum of y is shown to be

$$S_y(f) = a^2\sigma_x^4 \delta(f) + 2a^2 S_x(f)*S_x(f). \qquad \text{(H-21)}$$

Using this result, the contribution of the third term in Equation (H-9) is

$$S_3(f) = 4\sigma_{n_b}^4 \delta(f) + 8 S_{n_b}(f)*S_{n_b}(f) \qquad \text{(H-22)}$$

where

$$\sigma_{n_b}^2 = \int_{-\infty}^{\infty} S_{n_b}(f)df. \qquad \text{(H-23)}$$

The fourth and last term of Equation (H-9) is the square of the expected value of the square of $n_b(t)$. The expected value of the square of a random process is the power in that process. This term is not a function of τ so that its Fourier transform yields a delta function whose magnitude is

$$S_4(f) = 4\{\sigma_{n_b}^2\}^2 \delta(f) \qquad \text{(H-24)}$$

Observe that this DC component exactly cancels the DC component of $S_3(f)$. Combining all four terms of the discriminator output power spectrum yields

$$S_\epsilon(f) = \tfrac{1}{4} K_1^2 P^2 \left\{ R_c^2 \left[\left(\delta - \frac{\Delta}{2} \right) T_c \right] - R_c^2 \left[\left(\delta + \frac{\Delta}{2} \right) T_c \right] \right\}^2$$

$$+ 2K_1 P \left\{ R_c^2 \left[\left(\delta - \frac{\Delta}{2} \right) T_c \right] + R_c^2 \left[\left(\delta + \frac{\Delta}{2} \right) T_c \right] \right\} S_{n_b}(f) * S_{\theta_d}(f)$$

$$+ 8 S_{n_b}(f) * S_{n_b}(f) \tag{H-25}$$

TAU-DITHER EARLY-LATE NONCOHERENT TRACKING LOOP

The discriminator output autocorrelation function is defined in Equation (9-101) with $\epsilon(t, \delta)$ defined in Equation (9-100). In order to shorten the following equations, define

$$C_1 = K_1 P$$

$$C_2 = 2\sqrt{K_1 P}$$

$$R_+ = R_c \left[\left(\delta + \frac{\Delta}{2} \right) T_c \right]$$

$$R_- = R_c \left[\left(\delta - \frac{\Delta}{2} \right) T_c \right] \tag{H-26}$$

Then

$$E[\epsilon(t, \delta)\epsilon(t + \tau, \delta)] = E\{[C_1\{R_-^2 q_2(t) - R_+^2 q_1(t)\}$$

$$+ C_2 \cos [\varphi - \varphi' + \theta_d(t - T_d)]$$

$$\cdot \{n_{2I}(t)R_- q_2(t) - n_{1I}(t)R_+ q_1(t)\}$$

$$+ C_2 \sin [\varphi - \varphi' + \theta_d(t - T_d)]\{n_{2Q}(t)R_- q_2(t)$$

$$- n_{1Q}(t)R_+ q_1(t)\} + [n_{2I}(t)]^2 q_2(t) + [n_{2Q}(t)]^2 q_2(t)$$

$$- [n_{1I}(t)]^2 q_1(t) - [n_{1Q}(t)]^2 q_1(t)]$$

$$\cdot [C_1\{R_-^2 q_2(t + \tau) - R_+^2 q_1(t + \tau)\}$$

$$+ C_2 \cos [\varphi - \varphi' + \theta_d(t + \tau - T_d)]$$

$$\cdot \{n_{2I}(t + \tau)R_- q_2(t + \tau) - n_{1I}(t + \tau)R_+ q_1(t + \tau)\}$$

$$+ C_2 \sin [\varphi - \varphi' + \theta_d(t + \tau - T_d)]$$

$$\{n_{2Q}(t + \tau)R_- q_2(t + \tau)$$

$$- n_{1Q}(t + \tau)R_+ q_1(t + \tau)\}$$

$$+ [n_{2I}(t + \tau)]^2 q_2(t + \tau)$$

$$+ [n_{2Q}(t + \tau)]^2 q_2(t + \tau)$$

$$- [n_{1I}(t + \tau)]^2 q_1(t + \tau) - [n_{1Q}(t + \tau)]q_1(t + \tau)]\} \tag{H-27}$$

Observe that the function $q(t)$ and therefore $q_1(t)$ and $q_2(t)$ are stationary random processes only if a random delay or phase shift is included in their argument. All expected values involving $q_1(t)$ or $q_2(t)$ will be assumed to be over this implied random phase. When Equation (H-27) is expanded, many terms can be immediately set equal to zero because of the independence of all random processes and because all of the noise processes have zero mean. In particular, any term including a factor of the form $E\{\cos[\varphi - \varphi' + \theta_d(t - T_d)]\}$ may be set to zero as shown earlier. After considerable but straightforward manipulation,

$$E[\epsilon(t, \delta)\epsilon(t + \tau, \delta)] = E\{C_1^2[R_-^2 q_2(t) - R_+^2 q_1(t)]$$

$$[R_-^2 q_2(t + \tau) - R_+^2 q_1(t + \tau)]\}$$

$$+ E\{C_1[R_-^2 q_2(t) - R_+^2 q_1(t)][n_{2I}(t + \tau)]^2 q_2(t + \tau)\}$$

$$+ E\{C_1[R_-^2 q_2(t) - R_+^2 q_1(t)][n_{2Q}(t + \tau)]^2 q_2(t + \tau)\}$$

$$- E\{C_1[R_-^2 q_2(t) - R_+^2 q_1(t)][n_{1I}(t + \tau)]^2 q_1(t + \tau)\}$$

$$- E\{C_1[R_-^2 q_2(t) - R_+^2 q_1(t)][n_{1Q}(t + \tau)]^2 q_1(t + \tau)\}$$

$$+ E\{C_2^2 \cos[\varphi - \varphi' + \theta_d(t - T_d)]$$

$$\cos[\varphi - \varphi' + \theta_d(t + \tau - T_d)]\}$$

$$\cdot E\{[n_{2I}(t)R_- q_2(t) - n_{1I}(t)R_+ q_1(t)]$$

$$[n_{2I}(t + \tau)R_- q_2(t + \tau) - n_{1I}(t + \tau)R_+ q_1(t + \tau)]\}$$

$$+ E\{C_2^2 \sin[\varphi - \varphi' + \theta_d(t - T_d)]$$

$$\sin[\varphi - \varphi' + \theta_d(t + \tau - T_d)]\}$$

$$\cdot E\{[n_{2Q}(t)R_- q_2(t) - n_{1Q}(t)R_+ q_1(t)]$$

$$[n_{2Q}(t + \tau)R_- q_2(t + \tau) - n_{1Q}(t + \tau)R_+ q_1(t + \tau)]\}$$

$$+ E\{C_1[n_{2I}(t)]^2 q_2(t)[R_-^2 q_2(t + \tau) - R_+^2 q_1(t + \tau)]\}$$

$$+ E\{C_1[n_{2Q}(t)]^2 q_2(t)[R_-^2 q_2(t + \tau) - R_+^2 q_1(t + \tau)]\}$$

$$- E\{C_1[n_{1I}(t)]^2 q_1(t)[R_-^2 q_2(t + \tau) - R_+^2 q_1(t + \tau)]\}$$

$$- E\{C_1[n_{1Q}(t)]^2 q_1(t)[R_-^2 q_2(t + \tau) - R_+^2 q_1(t + \tau)]\}$$

$$+ E\{[[n_{2I}(t)]^2 q_2(t) + [n_{2Q}(t)]^2 q_2(t) - [n_{1I}(t)]^2 q_1(t)$$

$$- [n_{1Q}(t)]^2 q_1(t)]$$

$$\cdot [[n_{2I}(t + \tau)]^2 q_2(t + \tau) + [n_{2Q}(t + \tau)]^2 q_2(t + \tau)$$

$$- [n_{1I}(t + \tau)]^2 q_1(t + \tau) - [n_{1Q}(t + \tau)]^2 q_1(t + \tau)]\}.$$

$$(H-28)$$

All of the baseband noise processes above have identical statistics so that a number of the terms of this equation can be combined to obtain

$$E[\epsilon(t, \delta)\epsilon(t + \tau, \delta)] = C_1^2 R_-^4 E[q_2(t)q_2(t + \tau)] - C_1^2 R_-^2 R_+^2 E[q_1(t)q_2(t + \tau)]$$

$$- C_1^2 R_-^2 R_+^2 E[q_2(t)q_1(t + \tau)] + C_1^2 R_+^4 E[q_1(t)q_1(t + \tau)]$$

$$+ 4C_1 \sigma_n^2 R_-^2 E[q_2(t)q_2(t + \tau)]$$

$$- 2C_1 \sigma_n^2 R_+^2 E[q_1(t)q_2(t + \tau)]$$

$$- 2C_1 \sigma_n^2 R_-^2 E[q_2(t)q_1(t + \tau)]$$

$$+ 4C_1 \sigma_n^2 R_+^2 E[q_1(t)q_1(t + \tau)]$$

$$- 2C_1 \sigma_n^2 R_+^2 E[q_2(t)q_1(t + \tau)]$$

$$- 2C_1 \sigma_n^2 R_-^2 E[q_1(t)q_2(t + \tau)]$$

$$+ C_2^2 E[\cos \{\theta_d(t - T_d) - \theta_d(t + \tau - T_d)\}]$$

$$\cdot E[n_b(t)n_b(t + \tau)]$$

$$\cdot \{R_-^2 E[q_2(t)q_2(t + \tau)] + R_+^2 E[q_1(t)q_1(t + \tau)]\}$$

$$+ 2E[\{n_b(t)n_b(t + \tau)\}^2]E[q_2(t)q_2(t + \tau)]$$

$$+ 2\sigma_n^4 E[q_2(t)q_2(t + \tau)] - 4\sigma_n^4 E[q_2(t)q_1(t + \tau)]$$

$$+ 2\sigma_n^4 E[q_1(t)q_1(t + \tau)] - 4\sigma_n^4 E[q_1(t)q_2(t + \tau)]$$

$$+ 2E[\{n_b(t)n_b(t + \tau)\}^2]E[q_1(t)q_1(t + \tau)] \qquad \text{(H-29)}$$

where

$$\sigma_n^2 = E[\{n_{1I}(t)\}^2] = E[\{n_{1Q}(t)\}^2] = E[\{n_{2I}(t)\}^2] = E[\{n_{2Q}(t)\}^2] \quad \text{(H-30)}$$

for any time t since these processes are stationary.

At this point it is convenient to define the crosscorrelation and autocorrelation functions involving $q_1(t)$ and $q_2(t)$ in terms of correlation functions of $q(t)$. These correlation functions can be calculated using the expected value over a random phase α, that is,

$$E[q(t)q(t + \tau)] = \int_{-\infty}^{\infty} q(t + \alpha)q(t + \tau + \alpha)p_T(\alpha)d\alpha \qquad \text{(H-31)}$$

where $p_T(\alpha)$ is the probability density function of the random phase. The phase is assumed to be uniformly distributed over the period $2T_q$ of $q(t)$. Therefore

$$E[q(t)q(t + \tau)] = \frac{1}{2T_q} \int_{-T_q}^{T_q} q(t + \alpha)q(t + \tau + \alpha)d\alpha = R_q(\tau) \quad \text{(H-32)}$$

The result of this integration is independent of t and is a periodic triangle function as illustrated in Figure H-1a. Using Equation (9-91) it is easy to show that

$$E[q_1(t)q_1(t + \tau)] = E[q_2(t)q_2(t + \tau)] = \tfrac{1}{4} + \tfrac{1}{4}R_q(\tau) = R_{q_1}(\tau), \quad \text{(H-33)}$$

that

$$E[q_1(t)q_2(t + \tau)] = E[q_2(t)q_1(t + \tau)] = \tfrac{1}{4} - \tfrac{1}{4}R_q(\tau), \qquad \text{(H-34)}$$

and that

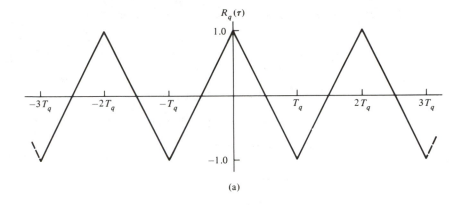

(a)

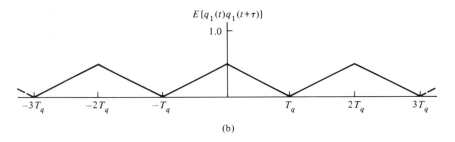

(b)

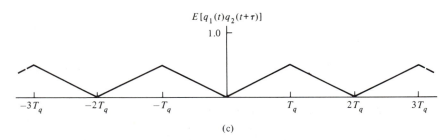

(c)

FIGURE H-1. Autocorrelation and crosscorrelation function for tau-dither tracking-loop switching functions: (a) $R_q(\tau)$; (b) $R_{q1}(\tau)$ or $R_{q2}(\tau)$; (c) $R_{q1q2}(\tau)$.

$$E[q_1(t)q_2(t + \tau)] = \tfrac{1}{2} - E[q_1(t)q_1(t + \tau)] = \tfrac{1}{2} - R_{q_1}(\tau). \quad \text{(H-35)}$$

These functions are plotted in Figure H-1b and H-1c.

Substituting the expressions of Equations (H-33) through (H-35) into Equation (H-29) and simplifying yields

$$E[\epsilon(t, \delta)\epsilon(t + \tau, \delta)] = C_1^2(R_-^2 + R_+^2)^2 R_{q1}(\tau) - C_1^2 R_-^2 R_+^2$$

$$+ 8C_1\sigma_n^2(R_-^2 + R_+^2)R_{q1}(\tau) - 2C_1\sigma_n^2(R_-^2 + R_+^2)$$

$$+ C_2^2 E[\cos\{\theta_d(t - T_d) - \theta_d(t + \tau - T_d)\}]$$

$$(R_-^2 + R_+^2)R_{n_b}(\tau)R_{q1}(\tau)$$

$$+ 4E[\{n_b(t)n_b(t + \tau)\}^2]R_{q1}(\tau)$$

$$+ 12\sigma_n^4 R_{q1}(\tau) - 4\sigma_n^4. \quad \text{(H-36)}$$

Using Equation (H-26) the desired autocorrelation becomes

$$R_\epsilon(\tau) = K_1^2 P^2 \left\{ R_c^2\left[\left(\delta - \frac{\Delta}{2}\right)T_c\right] + R_c^2\left[\left(\delta + \frac{\Delta}{2}\right)T_c\right]\right\}^2 R_{q1}(\tau)$$

$$- K_1^2 P^2 R_c^2\left[\left(\delta - \frac{\Delta}{2}\right)T_c\right]R_c^2\left[\left(\delta + \frac{\Delta}{2}\right)T_c\right]$$

$$+ 8K_1 P\sigma_n^2\left\{ R_c^2\left[\left(\delta - \frac{\Delta}{2}\right)T_c\right] + R_c^2\left[\left(\delta + \frac{\Delta}{2}\right)T_c\right]\right\}R_{q1}(\tau)$$

$$- 2K_1 P\sigma_n^2\left\{ R_c^2\left[\left(\delta - \frac{\Delta}{2}\right)T_c\right] + R_c^2\left[\left(\delta + \frac{\Delta}{2}\right)T_c\right]\right\}$$

$$+ 4K_1 P\left\{ R_c^2\left[\left(\delta - \frac{\Delta}{2}\right)T_c\right] + R_c^2\left[\left(\delta + \frac{\Delta}{2}\right)T_c\right]\right\}R_{n_b}(\tau)R_{q1}(\tau)$$

$$\cdot E[\cos\{\theta_d(t - T_d) - \theta_d(t + \tau - T_d)\}]$$

$$+ 4E[\{n_b(t)n_b(t + \tau)\}^2]R_{q1}(\tau)$$

$$+ 12\sigma_n^4 R_{q1}(\tau) - 4\sigma_n^4. \tag{H-37}$$

The Fourier transform of $R_\epsilon(\tau)$ is the power spectrum needed to complete the analysis of the tau-dither tracking loop. The Fourier transform of the first term is a function of δ times the power spectrum $S_{q1}(f)$ of $q_1(t)$ which is given by

$$S_{q1}(f) = \sum_{n=-\infty}^{\infty} D_n \,\delta(f - nf_q) \tag{H-38}$$

where

$$D_n = \begin{cases} \dfrac{1}{4} & \text{for } n = 0 \\ 0 & \text{for } n \text{ even} \\ \left(\dfrac{1}{n\pi}\right)^2 & \text{for } n \text{ odd.} \end{cases}$$

Thus

$$S_1(f) = K_1^2 P^2\left\{ R_c^2\left[\left(\delta - \frac{\Delta}{2}\right)T_c\right] + R_c^2\left[\left(\delta + \frac{\Delta}{2}\right)T_c\right]\right\}^2 S_{q1}(f). \tag{H-39}$$

The second term is not a function of τ so that its Fourier transform is a delta function at DC,

$$S_2(f) = -K_1^2 P^2 R_c^2\left[\left(\delta - \frac{\Delta}{2}\right)T_c\right]R_c^2\left[\left(\delta + \frac{\Delta}{2}\right)T_c\right]\delta(f). \tag{H-40}$$

The third term is similar to the first in form so that

$$S_3(f) = 8K_1 P\sigma_n^2\left\{ R_c^2\left[\left(\delta - \frac{\Delta}{2}\right)T_c\right] + R_c^2\left[\left(\delta + \frac{\Delta}{2}\right)T_c\right]\right\}S_{q1}(f), \tag{H-41}$$

and the fourth term is again a delta function

$$S_4(f) = -2K_1 P\sigma_n^2\left\{ R_c^2\left[\left(\delta - \frac{\Delta}{2}\right)T_c\right] + R_c^2\left[\left(\delta + \frac{\Delta}{2}\right)T_c\right]\right\}\delta(f). \tag{H-42}$$

The Fourier transform of the fifth term results in a convolution of three power spectra. The Fourier transform of $E[\cos\{\theta_d(t - T_d) - \theta_d(t + \tau - T_d)\}]$ is denoted by $S_{\theta_d}(f)$ and is defined in Equations (H-13) through (H-16). Then

$$S_5(f) = 4K_1P\left\{R_c^2\left[\left(\delta - \frac{\Delta}{2}\right)T_c\right] + R_c^2\left[\left(\delta + \frac{\Delta}{2}\right)T_c\right]\right\}S_{n_b}(f)*S_{q1}(f)*S_{\theta_d}(f).$$

(H-43)

The transform of the sixth term is the convolution of the power spectrum discussed in Section 9-3 for the output of a square law device and the power spectrum of $q_1(t)$. From Equation (H-21),

$$S_6(f) = \{4\sigma_n^4\,\delta(f) + 8S_{n_b}(f)*S_{n_b}(f)\}*S_{q1}(f)$$

$$= 4\sigma_{n_b}^4 S_{q1}(f) + 8S_{n_b}(f)*S_{n_b}(f)*S_{q1}(f).$$ (H-44)

The transform of the seventh and eighth terms are

$$S_7(f) = 12\sigma_n^4 S_{q1}(f)$$ (H-45)

and

$$S_8(f) = -4\sigma_n^4\delta(f).$$ (H-46)

The desired discriminator output power spectrum is therefore

$$S_\epsilon(f) = \sum_{j=1}^{8} S_j(f)$$ (H-47)

The complete discriminator output power spectrum can be calculated from Equations (H-39) through (H-47).

In order to obtain results which are reasonably simple, the exact analysis of $S_\epsilon(f)$ must be abandoned at this point. Assume now that the dithering frequency is chosen such that the only components of $S_{q1}(f)$ which influence tracking performance are those at $f = 0$, f_q, and $3f_q$. Assume also that there is no data modulation so that $S_{\theta_d}(f) = \delta(f)$. With these assumptions

$$S_{q1}(f) = \frac{1}{9\pi^2}\,[\delta(f + 3f_q) + \delta(f - 3f_q)]$$

$$+ \frac{1}{\pi^2}\,[\delta(f + f_q) + \delta(f - f_q)] + \tfrac{1}{4}\delta(f)$$ (H-48)

and after some simplification

$$S_\epsilon(f) = \tfrac{1}{4}K_1^2P^2\left\{R_c^2\left[\left(\delta - \frac{\Delta}{2}\right)T_c\right] - R_c^2\left[\left(\delta + \frac{\Delta}{2}\right)T_c\right]\right\}^2\delta(f)$$

$$+ \left[K_1P\left\{R_c^2\left[\left(\delta - \frac{\Delta}{2}\right)T_c\right] + R_c^2\left[\left(\delta + \frac{\Delta}{2}\right)T_c\right]\right\} + 4\sigma_n^2\right]^2$$

$$\cdot\left[\frac{1}{\pi^2}\,\delta(f - f_q) + \frac{1}{\pi^2}\,\delta(f + f_q) + \frac{1}{9\pi^2}\,\delta(f + 3f_q)\right.$$

$$\left. + \frac{1}{9\pi^2}\,\delta(f - 3f_q)\right]$$

$$+ \left[K_1 P \left\{ R_c^2 \left[\left(\delta - \frac{\Delta}{2} \right) T_c \right] + R_c^2 \left[\left(\delta + \frac{\Delta}{2} \right) T_c \right] \right\} \right]$$

$$\cdot \left[S_{n_b}(f) + \frac{4}{\pi^2} S_{n_b}(f - f_q) + \frac{4}{\pi^2} S_{n_b}(f + f_q) + \frac{4}{9\pi^2} S_{n_b}(f - 3f_q) \right.$$

$$\left. + \frac{4}{9\pi^2} S_{n_b}(f + 3f_q) \right]$$

$$+ 2S_T(f) + \frac{8}{\pi^2} S_T(f - f_q) + \frac{8}{\pi^2} S_T(f + f_q)$$

$$+ \frac{8}{9\pi^2} S_T(f - 3f_q) + \frac{8}{9\pi^2} S_T(f + 3f_q) \tag{H-49}$$

where

$$S_T(f) = S_{n_b}(f) * S_{n_b}(f).$$

REFERENCES

STEIN, S. and JONES, J., *Modern Communications Principles*, McGraw-Hill Book Co., New York, 1967.

DAVENPORT, W. B. and ROOT, W. L., *An Introduction to the Theory of Random Signals and Noise*, McGraw-Hill Book Co., New York, 1958.

AUTHOR INDEX

Aaron, M. R., 2
Abramowitz, M., 198, 713
Adkins, A. F., 678
Alem, W. K., 636–640
Amoroso, F., 9, 130
Anderson, J. B., 231, 243
Aprille, T. J., 141
Arthurs, E., 565, 568
Aulin, T., 233, 236

Baer, H. P., 507, 509
Batson, B. H., 384
Bello, P. A., 35
Bennett, W. R., 241
Berelekamp, E. R., 607
Bhargava, V. K., 23, 31, 174, 295
Birkhoff, G., 368, 412
Blahut, R., 621
Blanchard, A., 254, 272
Braun, W. R., 492–494
Brick, D. B., 649
Bussgang, J. J., 537

Cain, J. B., 607, 621, 622, 624, 632
Cardall, J. D., 647

Carter, C. R., 324
Chang, H., 280
Chen, C. C., 490, 492
Chu, T. S., 34
Clark, G. C., 607, 621, 622, 624, 632
Cobb, R. F., 528
Cnossen, R. S., 647
Collin, R., 23
Cooper, G. R., 561
Costello, D. J., 17, 368, 369, 372, 412, 607, 616, 619, 621, 624
Couch, L., 586
Crane, R. K., 34
Crawford, J. J., 624
Cunningham, W. J., 543

Darby, A. D., 528
Das, P. K., 540
Davenport, W. B., 509
Davey, J. R., 241
Daws, J. L., 597
Denaro, R. P., 647
Didday, R. L., 280
Diffie, W., 15, 411
Dillard, R. A., 331
Dixon, R. C., 420
Dym, H., 565, 568

Egan, W. F., 254, 257, 260, 281, 283
Ellersick, F. W., 649

Feller, W., 516, 520
Fix, E., 505
Forney, G. D., 617, 632, 704–712
Franks, L. E., 315
Frazier and Page, 277
Friis, H. F., 693

Gallager, R. G., 607
Gardner, F. M., 254, 256, 264, 266, 267, 275, 277, 545
Gaus, R. C., 233, 236
Geist, J. M., 632
Geraniotis, E. A., 561
Gill, W. J., 414, 420, 720–727
Golomb, S. W., 386, 412, 413, 414
Goodman, D. J., 561
Gorski-Popiel, J., 283
Gradshteyn, I.S., 491, 498, 509, 603
Grenander, U., 504
Grieco, D. M., 540
Groth, E. J., 411–413, 414

Haccoun, D., 23, 174, 295, 314
Hancock, J. C., 528, 529, 530
Harris, R. L., 364
Hartman, H. P., 420, 447
Hauptschein, A., 649
Hedin, G., 280
Heller, J. A., 19
Hellman, M. E., 15, 411
Helstrom, C. W., 528, 529, 533, 536, 660, 664, 676
Henry, P. S., 561
Hickernell, F. S., 540
Hogg, D. C., 34
Holmes, J. K., 124, 280, 294–295, 390, 407, 415, 420, 490, 492, 537, 544, 546
Hopkins, P. M., 456, 459, 517
Houston, S. W., 574–576, 597, 599

Ippolito, L. J., 34, 35

Jacobs, I. M., 19, 196, 223, 224, 329, 362, 502, 630
Jacobson, N., 412
Jones, J. J. 168–172, 357, 461
Jones, W. S., 649
Joseph, K. M., 647

Kendall, M. G., 533
Kendall, W. B., 537
Key, E. L., 414
Kilgus, C. C., 540, 543
Kivett, J. A., 130
Klein, S. A., 280
Kleinberg, S. T., 280

Lender, A., 143
Lereim, A. T., 233
Levitt, B. K., 342, 586, 591, 592, 599

Lin, S., 17, 368, 369, 372, 412, 607, 616, 619, 621, 624
Lindholm, J. H., 395, 414
Lindsey, W. C., 202, 207, 222, 254, 280, 295, 303, 313, 349, 568, 716
Longo, G., 561
Lucky, R. W., 207
Luecke, E. J., 306, 307, 308

MacLane, S., 368, 412
MacWilliams, F. J., 413
Magill, D. T., 420, 421, 422
Manassewitsch, V., 254, 281
Massey, J. L., 403, 619
Matthaei, G., 86
Matyas, R., 23, 174, 295, 314
Mazo, J. E., 236
Mazur, B. A., 243
McCallister, R., 323, 624
Meyer, H., 278, 281
Miller, T. W., 568
Milstein, L. B., 540, 597
Monsen, P., 36
Mudgett, W. L., 537

Neilsen, P. T., 544, 546–549
Nettleton, R. W., 561
Nuspl, P., 23, 174, 295, 314

Odenwalder, J. P., 623, 625, 626
Oliver, B. M., 2
Olsen, R. L., 34
Olson, M. L., 280
Olver, F. W. J., 534
Omura, J. K., 607, 612, 616, 619, 633
Oppenheim, A. V., 661

Palmer, L. C., 280
Papoulis, A., 340, 489, 490, 503, 508, 510, 588, 591
Park, J. H., 217, 568
Pasupathy, S., 121, 143
Patnaik, P. B., 505
Peterson, W. W., 377–379, 383, 389–391, 412, 607
Pettit, R., 597
Pickholtz, R. L., 384, 597
Pierce, J. R., 2
Pollak, H., 504
Potapov, M. K., 506
Prabhu, V. K., 222, 561
Pursley, M. B., 367, 393, 404, 405, 407, 413, 561

Quereshi, S., 147, 157

Raghavan, S. H. R., 158
Reible, S. A., 540
Rice, S. O., 498, 535
Ristenbatt, M. P., 2, 597
Robertson, G. H., 507
Root, W. L., 509
Ryan, C. R., 129–131, 151, 165
Ryzhik, I. W., 491, 498, 509, 603

Sage, G. F., 494
Salz, J., 207, 236
Sarwate, D. V., 367, 393, 404, 405, 407, 413, 561
Schafer, R. W., 661
Schetzen, S. M., 81
Schilling, D. L., 384, 586, 597
Scholtz, R. A., 324, 358
Shannon, C. E., 2, 7, 632
Shelton, R. D., 678
Shnidman, D. A., 501
Simon, M. K., 202, 207, 222, 280, 295, 303, 313,
    349, 415, 420, 447, 568, 600, 716
Singh, R., 586, 592
Skolnik, M. I., 500
Skwirzynski, J. K., 561
Slepian, D., 504
Sloan, N. J. A., 413
Smirnov, S. V., 506
Spilker, J. J., 178, 254, 261, 264, 277, 279, 407,
    414, 420, 421, 422, 544, 546, 644
Stark, W., 561
Stegun, I., 198, 713
Stein, S., 357, 461
Stiffler, J. J., 295, 407, 414
Stuart, A., 533
Sundberg, C. E., 233, 236

Taylor, D. P., 231, 243
Titsworth, R. C., 349, 414
Toplicar, J. R., 540
Torrieri, D. J., 597
Tranter, W. H., 215, 296, 298, 366, 495, 667, 676,
    685, 693

Urkowitz, H., 501, 505, 506

Van der Ziel, A., 692
VanDierendonck, A. J., 647
VanTrees, H. L., 165, 298
Viterbi, A. J., 254, 269, 543, 561, 584, 597, 607,
    612, 616, 619, 620, 630, 632, 633

Wald, A., 528, 529
Ward, R. B., 420, 485, 540–543
Welch, L. R., 349
Weldon, E. J., 207, 377–379, 383, 389–391, 412,
    607
Williams, A. B., 86
Wilson, S. G., 233, 236
Wintz, P. A., 306, 307, 308, 528, 529, 530
Woo, K. T., 280
Wozencraft, J. M., 196, 223, 224, 329, 362, 502

Yiu, K. P., 485, 540, 543
Young, L., 86

Zelen, M., 505
Ziemer, R. E., 129–131, 151, 158, 165, 215, 296,
    298, 366, 495, 667, 676, 685, 693
Zvereu, A. I., 88–99

SUBJECT INDEX

Acquisition for spread spectrum (*see* Code synchronization)
Allan variance, 259–260
 defined, 260
 relation to frequency noise power spectrum, 260
AM-to-PM conversion, 80
Analytic signal, 66
Antenna
 aperture efficiency, 698
 gain degradations, 33
 power gain, 697
Antipodal signaling, 113
ASCII, 11
ASK, 114
Attenuation
 due to absorption, 33
 due to scattering, 33
 of Bessel filter, 99
 of Butterworth filter, 88
 of Chebyshev filter, 91–92
Available noise power, 692
Available power gain, 693
Averages
 statistical, 657
  central moments, 660
  covariance, 657
  mean, 657
  variance, 657

Bandlimited channels
 digital signaling, 133
 intersymbol interference, 134
Bandwidth
 bit rate, 698
 definition, 3
 efficiency, 211
 limitations, 33
 limited transmission, 8
Baud, 3
Binary data formats, 294
Binary messages, 3
Binary symmetric channel (*see* Channel)
Bit
 definition, 3, 5
 error rate, 15
 least significant, 21
 synchronizer (*see also* Synchronizer, symbol
  (bit)), 36
 most significant, 21
Block coding (*see* Code, block)
BPSK, 113, 199, 328

Capacity, 4
 channel, definition of, 26, 228
 error free, 7
 exponential bound parameter, 228
 theorem, 228

Carrier phase estimator
  BPSK, 301
  QPSK, 305
  sinusoid in AWGN, 299
Carrier recovery (*see also* Phase locked loops), 299, 301, 305, 315
Central limit theorem, 667
Channel, 2
  bandlimited, 133
  binary symmetric, 26
  capacity theorem, 228
  coding theorem, 224
  discrete, 23
  discrete memoryless, 24, 556, 561
  fading, 33
  frequency selective, 33
  hard decision, 23, 561
  linear, 29
  single channel per carrier, 29
  soft decision, 23, 561, 607
  transition diagram, 24
  transition probabilities, 24
  transition probability matrix, 24
Code (encoding)
  ASCII, 11
  BCH, 17
  block, 16, 556, 607
    BCH, 17, 621
    elementary concepts, 607
    error probability calculation, 612
    Euclidean distance, 611
    Hamming distance, 608
    hard decision decoding, 610
    minimum distance, 608
    optimum decoding rule, 609
    Reed-Solomon, 622
    soft decision decoding, 610
    weight distribution, 608
  channel, 15, 556
  combined modulation and encoding, 217
  computational cutoff rate, 632
  concatenated, 623
  convolutional 16, 17, 556, 704
    as finite state machine output, 616
    constraint length, 616
    elementary concepts, 616
    error probability, 620
    free distance, 620
    minimum distance, 624
    optimum decoding rule, 618
    performance of, 19
    sequential decoding, 619
    state diagram, 705
    table of optimum codes, 625, 626
    threshold decoding, 619
    trellis diagram, 617, 706
    Viterbi decoding, 619, 704
  differential, 144, 217
  forward error correction, 2, 4, 15, 555, 606
  gain (coding), 15
  Gold, 366, 404
  Gray, 20, 21, 124, 209
  Hamming, 17
  Huffman, 12
  interleaving, 557
    block, 630
    convolutional, 631
  linear, 16, 608

maximal length sequences (*see also* Spreading codes), 365ff
Morse, 11
nonlinear, 411
prefix, 12
pseudorandom, 366
random coding bounds, 632
rate, 16, 607
redundancy, 10
repeat, 16, 624
source, 2, 4, 12
spreading (*see* Spreading codes)
systematic, 16
tracking (*see* Code tracking loops)
tree, 18
variable-length source, 12
use in proof of Shannon's theorem, 225
Code synchronization (spread spectrum)
  detection strategies
    fixed integration time, 501ff
    multiple dwell, 513
    sequential detection (*see* Sequential detection)
  loop pull-in, 486, 543
  matched filter, 538
  optimum, 486
  recursion aided sequential estimation, 485, 540
  serial search, 485, 488
    continuous sweep, 494
    mean synchronization time, 488
    modified sweep strategies, 492
    variance of synchronization time, 488
Code tracking loops (spread spectrum)
  baseband full-time early-late, 423
    linear equivalent circuit, 430
    noise performance, 432
    nonlinear equivalent circuit, 430
    s-curve, 425
    self noise in, 424
  double-dither early-late noncoherent, 456
  frequency hop systems, 467
    full-time loop, 467
    noise performance, 477
    s-curve, 476
    time shared loop, 476
  noncoherent full-time early-late, 433
    arbitrary spreading modulation, 459
    complex envelope model, 460–467
    linear equivalent circuit, 445
    using MSK spreading, 466
    noise performance, 447
    nonlinear equivalent circuit, 445
    s-curve, 439
    self noise in, 437
  optimum, 420
  performance summary, 479
  tau-dither early-late noncoherent, 447
    dithering, 451
    noise performance, 454
    s-curve, 451
  types, 420
Codeword, average length, 13
Coding theorem, 224
Complex ennvelope, 66–72
Confluent hypergeometric function, 535
Constraint length, 231, 616
Continuous phase modulation (*see* Multi-*h*)

Conversion
   AM-to-PM, 80
   PM-to-AM, 81
Convolution, 46
Convolutional coding (see Code, convolutional)
Correlation receiver, 116, 192
Correlator, realization of matched filter, 115
Costas loop (see Phase locked loops)
CPFSK, 230
CPM, 229
Crosstalk
   due to depolarization, 35
   in QAM, 142
Cyclostationary process
   defined, 682
   use in carrier recovery, 315
   use in synchronization, 295, 314

Data rate
   symbol, 211
   bit, 211
Decoding (decoder)
   channel, 556, 606ff
   differential, 217
Degradation
   due to imperfect phase reference, 162
   due to nonideal detection filter, 165, 169
   due to phase and amplitude imbalance, 159
   due to power loss, 162
   due to predetection filtering, 169
   due to timing error, 171
   due to transmitter/channel filtering, 170
   from ideal performance, 106
   in duobinary signaling, 145
Demodulation, 5
Demodulator
   coherent, 116
   hard decision, 23
   soft decision, 23
Depolarization, 33
Detection (detector)
   binary, 36, 106, 112
   data, 36, 106, 112
   definition, 5
   integrate-and-dump, 116
   matched filter, 106, 112
   maximum a posteriori (MAP), 186, 192, 195
Digital communication system, 6
Digital transmission
   M-ary, 184
   memoryless, 184
   phase noncoherent, 184
Dimensionality theorem, 223
Discrete memoryless channel (see Channel)
Distortion, 72–86
   amplitude, 73
   harmonic, 80, 83
   intermodulation, 83
   nonlinear, 73, 81
   phase, 73, 74
Distortionless transmission, 73
Diversity
   frequency, 33
   polarization, 33
   space, 33
Doppler
   frequency shift, 36
   spread, 35

DPSK, 213, 215, 569ff
Duobinary signaling, 143

ECM and ECCM, 560
Eigenfunction, 47
Electronic countermeasures, 560
Encryption (encryptor), 2, 15
Energy
   out-of-band, 62, 128
   spectrum, 60
Entropy, 24
   conditional, 25
   function, 27
   source, 9
Envelope functions, 177
Equalization, 147–158
   adaptive weight adjustment, 155
   decision directed, 157
   decision feedback, 157
   equalizer, 33
      least mean squares, 151
      preset weights, 156
      zero forcing, 147
Error function (see also Q-function), 656
   complementary, 109, 656, 714
Estimation (estimate, estimator)
   consistent, 674
   continuous waveform, 298
   Cramer-Rao inequality, 297
   efficient, 297
   of joint parameters, 309
   maximum a posteriori (MAP), 295
      condition for estimate, 296
   maximum likelihood (ML), 295
      condition for estimate, 296
      performance of symbol synchronizer, 308
   multiple symbol intervals, 302
   of phase of sinusoid, 299
   timing epoch, 307
   unbiased, 297, 673
Exclusive OR truth table, 21
Eye diagrams, use in computer simulation, 146

Fading, 30
FEC (forward error correction), 15, 606–633
FFSK, 230
Filter
   bandpass, 87
   band edge frequencies, 87
   band reject, 87, 88
   Bessel (maximally flat delay), 86, 97
   Butterworth (maximally flat), 86, 88
   Chebyshev, 86, 90
   group delay, 72
   matched, 109, 111, 113, 194
   notch, 87
   optimum, 136
   phase delay, 72
   pole locations, 98
Final power amplifier, 22
Finite field
   arithmetic defined, 368
   extension, 369, 371–373
   Galois, 369
   integer, 374
   prime, 369

Finite field (cont.)
    primitive polynomial, 371
    reciprocal polynomial, 383
Fixed integration time detection, 501ff
Fourier series
    complex exponential, 50
    cosine, 51
    examples, 53
    generalized, 48
    Parseval's theorem, 50
    properties, 52
    trigonometric sine-cosine, 51
Fourier transform
    conditions for existence, 53
    definition, 53
    inverse, 53
    Parseval's theorem, 58
    of periodic signals, 56
    transform pairs, 55
    transform theorems, 54
Free distance, 231, 620
Free space loss, 698
Frequency
    deviation, 261
    division multiple access, 699
    multiplier/divider effect on phase, 260
    synthesis
        digital, 281ff
        direct, 283ff
        output phase noise, 288
        phase locked (indirect), 287ff
        spurs on direct, 285
        spurs on indirect, 288
Friis's formula, 693
FSK, 114, 213, 199

Gaussian integral, 109
    Q-function, 112, 656, 713
Generalized vector space, 48
Global Positioning System (GPS), 635
Gram-Schmidt procedure, 189
Group delay
    Bessel filter, 99
    Butterworth filter, 89
    Chebyshev filter, 93

Hartley, 5
Hilbert transform, 65

Impairments, 106
Impulse response, 46, 63
    of Bessel filter, 100
    of Butterworth filter, 89
    of Chebyshev filter, 95
    complex envelope, 69
Information
    content, 5, 7
    mutual, 26
    source, 9
Integral-square error, 48, 49
    minimization by Fourier series, 51
Interleaving, 630
Intersymbol interference, 30, 106, 134

Jamming (jammer)
    barrage noise, 557, 562
    by pulse noise in BPSK, 328
    in spread spectrum, 327, 555ff
    multiple tone, 559, 621
    partial band, 558, 571
    performance of communication system in (see
        Spread spectrum performance)
    pulsed noise, 328, 559, 582
    repeater, 559
    resistance, 7
    single tone, 337, 558, 588
    smart, 559
JTIDS, 647

Link, line-of-sight microwave, 32
Low probability of intercept (detection), 7, 330

MAP hypothesis test, 186
    minimizes average probability of error, 187
    multiple hypothesis, 187
    multiple observations, 187
    performance calculations, 195
    symbol (word) error, 196
Markov chain (finite state), 516
Markov process (discrete time), 704
Matched filter
    for binary signal detection, 113
    definition, 109, 194
    impulse response, 111
    in MAP receiver, 194
    peak output signal, 111
Maximal length sequence (see also Spreading codes),
        365ff
Message, 5
Mixer (mixing), 83
    compression, 85
    conversion loss, 84
    harmonic distortion, 85
    high-side frequencies, 83
    intermodulation distortion, 85
    intermodulation products, 84
    isolation, 85
    low-side frequencies, 83
    noise figure, 86
    third-order intercept, 85
    VSWR, 86
Modified Bessel function, 268
Modulation
    amplitude, 22, 65
        double sideband (DSB), 65
    analog, 65
    apparent carrier, 130
    CPFSK, 230
    definition, 5
    digital
        antipodal baseband, 113
        ASK, 114
        biphase (BPSK), 113, 199
        differential (DPSK), 213, 215
        FSK (coherent), 114, 199
        FSK (noncoherent), 213
        M-ary orthogonal, 200
        M-ary PSK, 192, 204
        MSK, 117, 120
        multiple amplitude/phase shift, 141, 207

OQPSK, 118
QAM, 141
QASK, 207
QPSK, 118
  quadrature bandpass, 141
  quadrature multiplexed, 117
  performance, 120ff
  serial MSK, 128
frequency, 22
linear, 22, 105
multi-*h* CPM, 228
noncoherent, defined, 36
nonlinear, 22
phase, 22
terminology, 22
Modulator
  data, 15, 19
  spread spectrum, 15, 22
  structures, 174
MSK, 117, 120
Multi-*h* continuous phase modulation (CPM)
  constraint length, 231
  CPFSK, 230
  detection using Viterbi algorithm, 246
  difference sequence, 234
  envelope decay of spectrum, 230
  excess phase, 229
  FFSK, 230
  free distance, 231
  frequency pulse shape function, 229
  full response, 229
  MLSE, 233
  partial response, 229
  performance bounds, 233
  phase pulse shape function, 229
  phase trellis, 231
  power spectrum calculation, 236
  superbaud timing, 243
  synchronization, 243
  trellis period, 233
Multipath, 7, 30
  atmospheric, 35
  delay, 35, 328
  diffuse, 33, 560
  specular, 33
Multiple access, 7, 560
Multiple dwell detection, 513

Nat, 10
Noise, 5
  AWGN, 7, 105
  emission, 33
  equivalent bandwidth, 677
    of filters in terms of 3-db bandwidth, 678
  externally generated, 29
  figure, 35, 86, 691
  internally generated, 29
  orthogonal component, 189
  relevant component, 189
  temperature, 692
    effective, 693
    ratio, 694
    system, 694
Nyquist, 106
  pulse shaping criterion, 135
  raised cosine spectra, 137
  theorem (noise), 692

Optimization criteria, minimum probability of decision error, 107
OQPSK, 117, 118
Orthogonal functions
  complete sets, 49
  definition, 48
  generalized Fourier series, 49
  normalized, 48
Oscillator fractional frequency stability, 259
Out-of-band energy, 60, 62
Out-of-band power, 127

Parseval's theorem, 50, 191
Phase delay, 73
Phase noise (jitter), 255
Phase jitter power spectral density
  cutoff, 257
  flat phase, 256
  frequency flicker, 256
  measured, 257
  phase flicker, 256
  white frequency, 256
Phase locked loops, 261ff
  acquisition, 275
  acquisition aids, 277
  average time to cycle slip, 269
  coherent amplitude detector, 278
  coherent reference generation, 264
  Costas loop, 280
  detector characteristics, 262
  dithering signal, 278
  double Costas loop, 280
  error response, 266
  false lock, 280
  frequency lock, 275
  frequency response, 264, 266
  FM demodulation, 264
  Fokker Plank analysis, 268
  hang-up, 277
  hold-in range, 273
  lock-in range, 276
  loop filters, 263
  maximum sweep rate, 275, 277
  noise equivalent bandwidth, 265, 267
  noise equivalent model, 267
  optimum bandwidth, 272
  pdf of phase error, 268
  phase error variance, 270
  pull-in time, 277
  pull-in voltage, 276
  quadrupling loop, 280
  signal-to-noise ratio, 268
  spread spectrum (*see* Code tracking loops)
  squaring loop, 280
  synchronized linear model, 261
  time to settle, 277
  tracking error, 272
  transfer functions, 263, 265
  transient response, 272
  transport delay, 281
  VCO phase variance, 268
PM-to-AM conversion, 81
Power
  effective isotropic radiated (EIRP), 698
  efficiency, 5
  limited, 8

Power (cont.)
  spectra
    BPSK, 125
    dicode, 143
    duobinary, 143
    frequency hop signals, 351, 719
    MSK, 126
    OQPSK, 126
    partial response, 143
    product of M-sequences, 720
    QPSK, 126
    random binary sequences, 716
Probability(ies)
  axioms of, 651
  Bayes rule, 652
  central limit theorem, 667
  chance experiment, 650
  characteristic function, 658
  conditional, 652
  density function, 196, 654
    Chi-square, 503
    conditional, 657
    noncentral chi-square, 505
    Rayleigh, 665
    Rician, 498
  distribution function, 654
  of error
    for coherent binary signaling, 178
    equivalent word, 203
    M-ary FSK, 204
    M-ary PSK, 204
    orthogonal signaling, 198
    16-QASK, 211
    union bound, 196
  event(s)
    complement of, 651
    intersection of, 651
    marginal, 652
    mutually exclusive, 651
    null, 651
    statistically independent, 652
  generating function, 660
  outcome, 650
  random variables, 654
    average of, 657
    binomial, 661
    Gaussian, 655
    Poisson, 662
    statistically independent, 658
    transformation of, 663
    uncorrelated, 657
  relative frequency definition, 652
  sample space, 650
PSK, 117–119, 204
Pulse shaping, 2

QAM, 141
QASK, 207
Q-function, 656, 713
  asymptotic expansion, 713
  rational approximation, 714
  tables, 714
QPSK, 117, 118
Quadrature multiplexing, 117
Quantization, 9

Radiometer detection of large TW signal, 331

Raised cosine, 136, 230
Random coding bounds, 632
Random processes, 668
  autocorrelation function, 669
  cyclostationary, 682
  ergodic, 672
  examples, 681
  jointly stationary, 672
  through linear systems, 678
  narrowband representation, 685
  power spectral density, 670
  sample function, 668
  statistically independent, 672
  strictly stationary, 669
  widesense stationary, 669
Receiver, 2
  Bayes, 186
  coherent, 36
  maximum a posteriori (MAP), 186
  noncoherent, 36
  optimization, 107
  performance calculation of MAP, 195
Redundant (parity) symbols, 16
Regenerative repeaters, 2

Sampling, 9
Schwarz's inequality
  proof, 190
    uses in receiver optimization bandlimited channels, 138
  vector analogy, 110
Scintillation, 33, 35
Sequential detection, 524–537
  average sample number, 526, 531
  correct detection, probability of, 527
  excess overboundary problem, 528
  false alarm, 525, 527
  likelihood ratio, 524
  operating characteristic function, 528
  practical considerations, 536
  thresholds, 526
  using linear envelope detector, 530
  using log likelihood ratio, 530
Sequential probability ratio test (SPRT), 526
Serial MSK, 128ff
  conversion filter, 129
  excess phase, 130
  spectra for, 131
  trellis diagram, 132
Shannon's second (coding) theorem, 8, 224
Shift register sequences (see Spreading codes)
  maximal length, 365ff
  as spreading codes, 365
Signal
  analog, 43
  analytic, 66
  aperiodic, 44
  bandpass, 64
  complex envelope, 68
  complex exponential, 47
  continuous time, 43
  deterministic, 44
  discrete-time, 44
  finite energy, 44
  finite power, 44
  highpass, 64
  Kronecker delta, 191

lowpass, 64
narrowband, 67
periodic, 44
pulse, 45
quadrature components, 68
random, 44
signum function, 45
sinc function, 45
symmetrical rectangular pulse, 45
unit impulse, 45
unit step, 45
Signal-to-noise ratio, 8
at matched filter output, 109
Signal space, 188
applied to data transmission, 186
basis function sets, 189
Gram-Schmidt procedure, 189, 245
M-ary PSK signal set, 192
norm, 189
Parseval's theorem, 191
representation of signals, 188
scalar product, 189
Schwarz's inequality, 190
signal point constellation, 208
triangle inequality, 189
Signaling
block orthogonal, 223
duobinary, 143
exponential bound parameter, 228
vertices of a hypercube, 223
M-ary orthogonal, 200, 213
Source
extension, 13
memoryless, 9, 184
reduced, 12
Spectrum
amplitude, 57
energy, 57, 60
phase, 57
Spread spectrum
BPSK direct sequence, 332
despreading in, 333
interference rejection, 337
model, 333
processing gain, 338
transmitted power spectrum, 334
chip defined, 335
code acquisition (*see* Code synchronization)
code sequence generator (*see also* Spreading codes), 15, 365ff
code tracking (*see* Code tracking loops)
complex envelope representation, 357–359
BPSK/MSK, 358
slow FH/DPSK, 360
various modulation types, 359
defined, 5, 328
direct sequence, 332
frequency hopped, 348–355
fast, noncoherent, 354
slow, coherent, 348
slow, noncoherent, 352
transmitted power spectrum, 350
hybrid, 355
modulator, 15
MSK direct sequence, 344–348
excess phase trellis, 345, 347
parallel implementation, 345
serial implementation, 346

nonlinear, 411
performance in barrage noise jamming
coherent DS, 562
DS-FH/DPSK, 569
DS-FH/MFSK, 569
FH/DPSK, 568
FH/MFSK, 564
performance in multi-tone jamming
FH/DPSK (with FEC coding), 621, 623
FH/DPSK (without coding), 599
FH/MFSK, 597
performance in partial band jamming
coherent DS, 571
DS-FH/DPSK, 581
DS-FH/MFSK, 581
FH/DPSK, 579
FH/MFSK (with FEC coding), 615, 627
FH/MFSK (without coding), 574
performance in pulsed noise jamming
coherent DS, 582
DS-FH/DPSK, 585
DS-FH/MFSK, 585
FH/DPSK, 585
FH/MFSK, 584
performance in single-tone jamming
BPSK/BPSK, 588
general analysis, 586
MSK/BPSK, 594
QPSK/BPSK, 592
PN code defined, 366
processing gain defined, 328
QPSK direct sequence, 340–343
balanced, 340–342
dual channel, 342
used in TDRSS, 343
spreading code defined, 366
spreading waveform defined, 366
system model, 556
time bandwidth product, 331
types, 328
uses, 7
Spreading codes
autocorrelation function
defined, 366
discrete periodic, 367
autocorrelation spectrum, 404
crosscorrelation function
defined, 366
discrete periodic, 367
crosscorrelation spectrum, 404
delay operator, 366
finite fields, use in (*see* Finite field)
Gold, 404–407
maximal length sequences, 385–404
autocorrelation, 386, 387
partial autocorrelation properties, 392
decimation of, 405
generation of specific delays, 396
polynomial tables for, 388
power spectra of, 387
power spectrum of products, 396,720
preferred pairs, 405
properties, 385
security of, 403
polynomial representation (*see* Finite field)
rapid acquisition, 407
sequence generators, 375–385
calculation of output, 380

Spreading codes, sequence generators (cont.)
    maximum period, 383
    recursion relationships, 380
  used in TDRSS, 406
Squaring loss, 319
Staggered QPSK, 119
Statistical decision theory
  a posteriori probability, 194
  average cost, 186
  Bayes, 186
  likelihood ratio test, 186
  MAP detector, 187
  probability of a false alarm, 186
  probability of a miss, 186
Step response of filters
  Bessel, 100
  Butterworth, 90
  Chebyshev, 96
Superposition integral, 46, 107
Synchronizer (synchronization)
  carrier, 117, 293
  carrier phase, 299, 301, 305
  cross spectrum, 295
    performance, 322
  data aided, 309
  data transition tracking loop (DTTL), 314
  delay and multiply, 319
  early-late gate, 313
  feedback structures, 310
  frame, 117, 293
  inphase/midphase, 314
  non-data aided, 309
  parallel structures, 311
  sequential structures, 312
  serial structures, 311
  spreading code, 293
  symbol (bit), 36, 117, 293, 306, 319
System
  amplitude response, 47
  causal, 46
  distortionless, 72
  gain, 73
  group delay, 73
  impulse response, 47
  linear, 45
  lumped, 46
  narrowband, 69
  nonlinear, 181
  phase delay, 73
  phase response, 47
  realizable, 46
  time invariant, 46
  transfer function, 61
  wideband, 64
  zero memory, 81

TDRSS, 636
TDRSS User Transponder, 640–643
Telegraph, 2
Telephone, 2
Threshold comparator, 106
Transfer function, 47, 61–62
  properties, 64
  lowpass equivalent, 71
Transmitted signal, 3
Transmitter, 2
Traveling wave tube amplifier, 23

Unique word, 23

Variance, Allan (*see* Allan variance)
VCO, 261
Vector space (*see* Signal space)
Viterbi algorithm, 18, 229, 704–712
  applied to convolutional coding, 618
  applied to detection of multi-h, 246ff
  branch metric, 707
  cumulative metric, 249
  decision depth, 708
  estimation of discrete-time Markov process, 704
  survivor phase state sequence, 248
  survivor paths, 249, 707
  trellis diagram, 249, 707
Volterra series, 81